The Comprehensive Guide to Nature's Sunshine Products

Produced by Tree of Light Publishing

The information in this book was compiled by
Steven Horne, Registered Herbalist (AHG)
with contributions from
Kimberly Balas, ND

Editing and data entry for this 2009 edition done by Carolyn Hughes & Hughes, David Horne, Elena Horne, Katie Horne, Leslie Lechner and Sharon Grimes

Cover Design by David Horne
Layout by Steven Horne and David Horne

Important Notice: This material is for educational purposes only. It is not intended to replace the services of licensed health care providers. Always obtain competent medical advice for all serious or persistent illness. If you use any of the procedures in this material to treat any disease in yourself or others without the assistance of licensed health care providers, you are doing so at your own risk.

This book is the work of Tree of Light Publishing and was not authorized by, edited by or endorsed by Nature's Sunshine Products, Inc. Tree of Light Publishing is solely responsible for the contents, and has no financial connections to Nature's Sunshine Products, Inc. Product names are trademarks of Nature's Sunshine Products, Inc.

Although we have made an effort to make the information contained herein as up-to-date and accurate as possible, there are constant adjustments in the NSP product line. New products are added, old products are discontinued and ingredients may change. Please check with NSP for the most current list of products available.

Tree of Light Publishing

P.O. Box 911239

St. George, UT 84791

800-416-2887

Table of Contents

This book is dedicated to Gene and Christine Hughes and the other people at NSP Home Office who have been supportive of our efforts to educate people about herbs and naturally healing over the past 23 years.

Introduction

Why Nature's Sunshine?

My introduction to Nature's Sunshine Products came in 1975 when a lady in Tustin, California sold me a bottle of Special Formula #1 (now known as All Cell Detox). She told me I needed to clean out my colon to get rid of my sinus problems. I took the product and started having three bowel movements a day. I thought the product was giving me diarrhea so I quit taking it.

In 1978, I used the product again while recovering from a Moped accident, and found the herbs really helped speed my recovery. In 1981, I signed up as a Distributor and in 1984, I went to work at the Home Office of NSP in Spanish Fork, Utah as the editor of their corporate magazine, Sunshine Horizons. I started publishing a newsletter for NSP Managers in 1986 called Nature's Field, and in 1990 I struck out on my own and started producing educational materials for NSP Managers and Distributors full time. Tree of Light Publishing is the result of those efforts.

As a professional herbalist, I'm fully capable of making my own herbal formulas (which I have done and continue to do now and then). I'm also familiar with a number of excellent herbal companies, whose products I have tried and use. However, for some reason I continue to stick with NSP as the primary product line I use and recommend. There are several reason why I like Nature's Sunshine.

First, as one of the first companies to encapsulate herbs on a commercial scale, NSP was also one of the pioneers in quality control in the herbal industry. They continue to be a leader in this area, constantly testing their products to ensure the highest quality product possible in commercial herbal manufacturing. Because of this, I get consistently good results in using their product and recommending it to my clients.

Second, Nature's Sunshine has always been committed to education. While so many companies rely on hype to market their herbs and supplements, NSP continually hires top notch herbalists and health professionals to train their people. They are committed to giving their sales force a solid understanding of natural health care.

Third, while you may find the occasional NSP distributor who cares more about selling product than about helping people, most NSP Distributors I have met are interested in learning how to genuinely help their customers and clients improve their health and well-being. Many are involved in "cutting edge" technologies in natural healing and willingly share their expertise with others.

I also find NSP's salespeople (Managers and Distributors) to be an extremely ethical group of people. In almost 20 years of doing tens of thousands of dollars of business with NSP distributors we've only had one or two checks each year that have bounced and haven't been made good, and we've seen countless examples of honesty and genuine concern for others. NSP's motto of "caring and sharing" isn't just a marketing slogan. Their Managers and Distributors really do tend to care, and share of themselves and their knowledge with others.

Fourth, NSP provides a solid opportunity for people who would like to help others improve their health with herbs, nutrition and natural healing methods to start a business with very little capital and literally "earn while they learn." While I know that many professional healers have a negative attitude about network marketing, it is still the easiest way for someone to get started in building a successful herbal business. I've known many people who have taken classes in herbs and nutrition and have no idea how to earn a living as an herbalist or natural health consultant.

There are some valid reasons for professional healers to be suspicious of network marketing companies. Many have a limited number of products, which are sold primarily through hype. The major product isn't "health," it's greed. Promises of "get rich quick" on a product where "one-size-fits all" are a turn off to me, too. But NSP is different. I've already pointed out some of these differences in their commitment to education and the genuine desire to help people embodied by the majority of their Managers and Distributors. Another way NSP is different is their incredibly broad product line. With over 500 products ranging from herbal formulas and nutritional supplements to essential oils, homeopathics and home health products (such as waterless cookware and the best reverse-osmosis water treatment appliance in the marketplace) Nature's Sunshine has enough products to provide people with solid choices for most of their natural health care needs.

Furthermore, making a living just as a practitioner, means that you have no cash flow for retirement, vacations or other times when you can't or don't want to work. NSP's network marketing system allows you to not only consult with people and help them one on one, but also to build a business you can retire from someday. This is because network marketing allows one to develop a stream of passive income through training other people to duplicate your successes.

It is this wide range of products that makes this book valuable. The range of products available through NSP can be somewhat intimidating to a beginner. It was to me when I signed up as a Distributor in 1980, which is why I've devoted over 20 years to finding ways to train people to become herbalists using NSP as a base. This guide provides an overview of our ABC+D system, which makes it easy to begin matching people to products. So, this guide can help NSP Managers, Distributors, Members and even customers to discover which products will help them get the results they desire. In fact, even if you don't use NSP products at all, this guide can help you learn how to help yourself to better health the natural way.

One final note. Before you start looking up your health problems in the Conditions section of this guide, please read the introductory material. It provides you with the information you need to put together a supplement program that maximizes your chances for success.

Steven Horne
President of Tree of Light Publishing

Section One

Start Here...
How to Get the Most Out of This Book

This section explains the basic principles of putting together a health and supplement program that will provide you with the results you are looking for.

Modern medical care isn't really health care. It's disease care. It focuses on disease treatment and symptomatic relief, not on building health. As a result, people have been trained to think in terms of treating specific diseases or seeking "instant" relief from annoying disease symptoms. People carry this attitude with them when they start using herbs and nutritional supplements. They want to use natural products as alternatives to drugs and expect supplements to achieve the same results—symptomatic relief.

Unfortunately, that isn't what natural healing is really all about. So, before you look up one of your ailments in the Condition's section of this book and try to figure out what you need to do to work on that ailment naturally, please read this introductory material. It will not only help you get the most out of this book, it will also maximize your chances for success in improving and maintaining your health.

Treating the Effect Versus Removing the Cause

The following fictitious story illustrates the basic problem people encounter with modern medical care.

> There was once a carpenter who was a little clumsy. He regularly hit his thumb with his hammer. Soon, his thumb became very swollen and inflamed. He went to a doctor who said, "That finger is badly inflamed, let me write you a prescription for an anti-inflammatory."
>
> The man took the medication and noticed that it helped the thumb a little, but because he kept striking it with the hammer, it continued to get worse. The pain was becoming difficult to bear. So the man went to another doctor. This one prescribed a painkiller.
>
> The painkiller really helped take the pain away, but it also made the man's fingers a little numb so that he wound up hitting his thumb more than ever. Soon the thumb was very raw and badly damaged. So, the man sought out a third doctor, a surgeon, who said, "That thumb is badly diseased, I think we should cut it off before it damages the rest of the body.
>
> Finally, the man went to an herbalist, who took a case history and suggested the man find a different job, since he was obviously too clumsy to be a good carpenter. The carpenter became a salesman and his thumb cleared up in just a few days."

Although the causes of most people's health problems are much more subtle than this fictitious character's, the story clearly illustrates the problem of treating the effect without removing the cause. Unfortunately, much of what we encounter in modern medicine is exactly that—treating the effect without removing the cause. We call it symptomatic relief. The medical mind set is often oriented towards symptomatic relief without regard to the cause.

So, the first thing one needs to understand if one wants to get consistent, effective results with natural health care is that natural health care isn't about easing symptoms. It's about dealing with root causes. As the 18th century herbalist Samuel Thomson put it so succinctly, "Remove the cause and the effect will cease."

A New Model of Disease

To help clients understand what we are doing and how to get to the root causes of disease, the ABC+D Approach utilizes a model of disease we call The Disease Tree™. The idea for this model came from the writings of Samuel Thomson, an American herbalist who lived in the early part of the 19th century. Thomson used a systematic approach to treating disease, which he summarized in a short poem in his book, *New Guide to Health*. Part of that poem reads:

> Let names of all disorders be,
> Like to the limbs, joined to the tree,
> Work on the root, and that subdue
> And all the limbs will bow to you.
> The limbs are colic, pleurisy,
> Worms and gravel, gout and stone,
> Remove the cause and they are gone.

Thomson's poetic metaphor helps us realize that focusing on specific disease symptoms is only attacking the "branches" of disease. Unless we work on the "roots" or underlying causes, the disease will simply manifest in a new form. Whenever a person takes a drug to relieve a symptom, or has

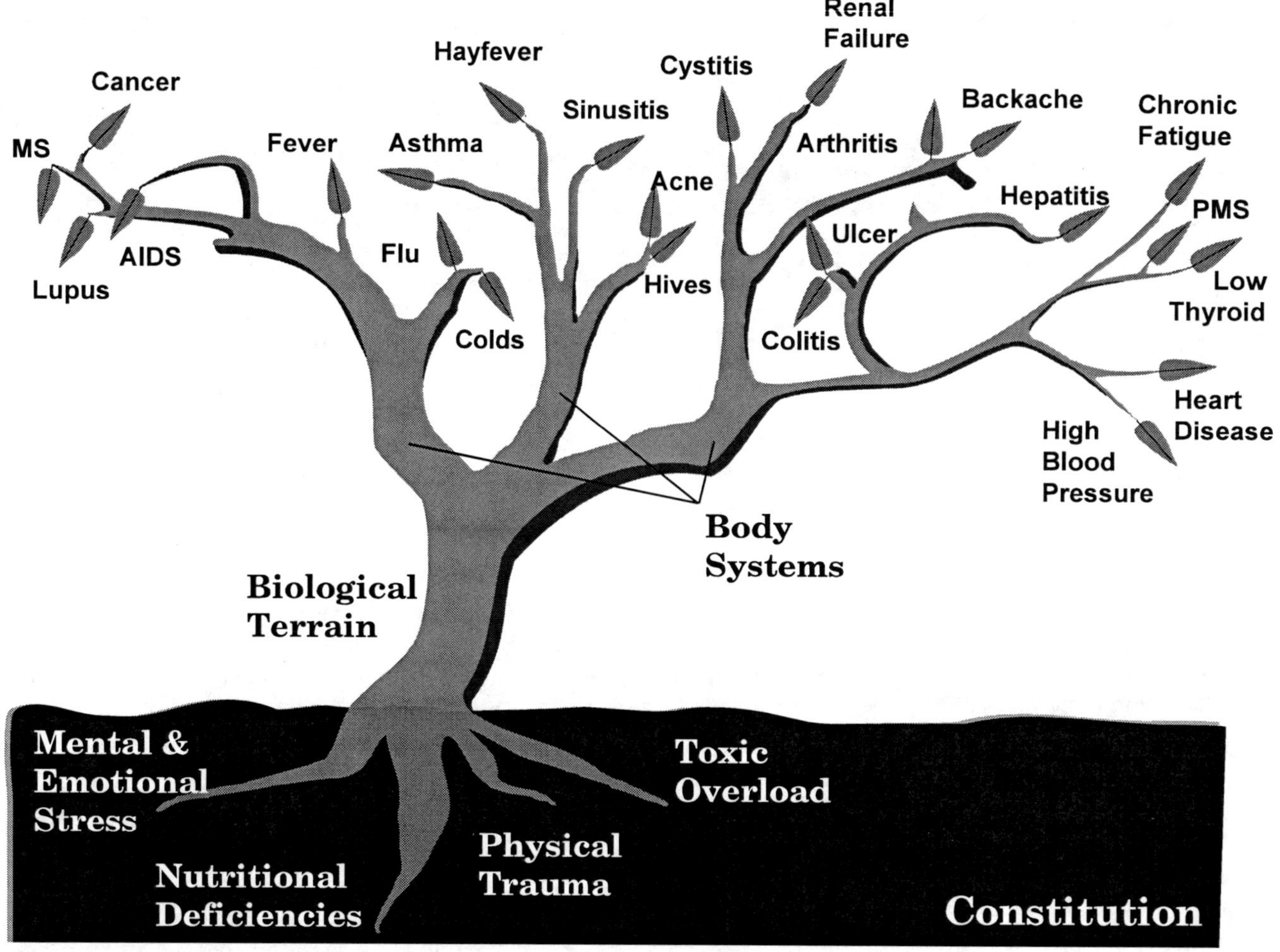

Figure 1—The Disease Tree™

some part of the body cut out, they are not being healed. To be healed is to be made whole, that is, to be restored to wholeness. Drugs and surgery only "chop off" branches. The root causes of the problem remain and will grow "suckers"—new diseases that will spring up in place of the old.

Most newcomers to natural health care are still thinking in terms of the disease/treatment model. They are looking for symptomatic relief. They want the branches of their disease tree to be pruned. Many inexperienced herbalists and natural health consultants try to accommodate this desire by recommending herbs and supplements to replace their medications. However, if the root causes are ignored, the results will be disappointing even if natural substances are used. In order to really correct the problem, the consultant has to help a person shift his or her focus away from disease symptoms and begin looking at root causes.

Shown in Figure 1, The Disease Tree™ is a model you can use to help clients understand what good health is really all about. Many successful consultants are now using this model to explain what they are doing with their clients. Here is a brief breakdown of the elements in this model.

Soil—Constitution

The first element in the model is the soil, which represents our constitution. This is a person's inherent physical and emotional makeup. A person who has a strong constitution can handle more physical stress than a person with a weaker constitution. In other words, under the same physical conditions, one person may thrive, while another will be taken ill. Constitution is our basic ability to cope with our environment. Part of this is our genetic inheritance, but it also includes early childhood conditioning and our personality.

Roots—Environmental Stress

The roots of disease are the environmental stresses that overwhelm the ability of the physical body to maintain balance. There are four basic root causes of disease, as follows.

1. Trauma. The body can be mechanically damaged. All physical damage, from burns and frostbite to cuts, lacerations and broken bones, causes damage to organs and tissues that results in disease. In fact, physical trauma, if not addressed appropriately, can lay the foundation for more serious diseases later in life. For example, a damaged joint is more susceptible to developing arthritis than a joint that was never subjected to trauma.

2. Toxicity or poisoning. Mechanical damage isn't the only way the body can be injured. One can be bitten by a snake, stung by a bee, or encounter a bed of poison ivy, all of which introduce chemical poisons that damage tissues. This toxicity can also occur from food additives, pesticides, household cleaning products, heavy metals, drugs and other toxic substances present in modern society. In fact, the toxic waste can even come from the body. The body produces waste in the process of metabolism, which must be eliminated from the system or it will damage tissues.

3. Nutritional deficiencies. The body needs nutrients in order to function correctly. When these nutrients are not present in the person's diet, then the body does not have what it needs to stay healthy. If levels of these nutrients are low, then the body may not be able to resist other environmental stresses as efficiently, either.

4. Mental and Emotional Stress. Everyone encounters mental and emotional stress in relationships and life situations. Financial problems, marital problems and other emotional stresses also lay the foundation for disease. This critical root cause is very often overlooked in both orthodox and alternative healing circles. People who continue to experience health problems, no matter what they do, often have unresolved mental and emotional stress that must be dealt with before healing can occur.

As these four environmental stressors overwhelm the capacity of a person's constitution to cope with them, the process of disease begins. This concept is expanded on in Figure 2. Disease arises from the disharmony that results when our constitution cannot adapt or cope with our environment. Dis-ease, that is "lack of ease," develops as these stresses overwhelm the adaptive mechanisms in the body.

This even applies to trauma. A young person slipping and falling on the ice might experience only a bruise, while an elderly person might fracture their hip. So, disease is always a combination of constitutional weakness interacting with environmental stresses.

Trunk—Biological Terrain

All of these root causes feed into the same trunk. The trunk of the disease tree is biological terrain. All of the cells of the body live in an internal ocean of lymphatic fluid. It is the composition of this fluid that determines the health of our cells.

Just as the health of a plant depends on the health of the soil in which it is grown, so the health of our tissues depends on the composition of our bodily fluids, lymph and blood. When these are imbalanced, the whole organism is affected. There are six basic imbalances in biological terrain. Since natural remedies help to balance biological terrain, determining which remedies will balance this terrain is a critical part of the natural healing process.

Branches—Body Systems

The cells of our body need five things to survive. They need nutrients, water, oxygen, waste removal and a regulated temperature. In the body, these needs are met by groups of specialized tissues that perform the various functions required to maintain the health of the whole body. These specialized tissues form organs and body systems.

As the imbalances in biological terrain affect the structure and function of these body systems, the major limbs and branches of disease begin to form. So, another part of natural healing is to determine which major body organs and systems are in need of nutritional support.

Twigs and Leaves—Specific Disease Symptoms

Finally, we come to the twigs and leaves of our metaphorical disease tree. These represent the specific disease symptoms the body manifests. As we have already suggested, most people are only concerned with the treatment of symptoms. Treating symptoms is like pruning the tree. You may be able to eliminate certain branches by suppressing specific symptoms with drugs or by removing diseased tissues with surgery, but these actions do little or nothing to eliminate root causes, bring our biological terrain back into balance, or restore the normal function of body systems. They certainly do nothing to strengthen constitution.

When we focus on the leaves and twigs, our efforts are misplaced. When a person is obese, diabetic, suffering from

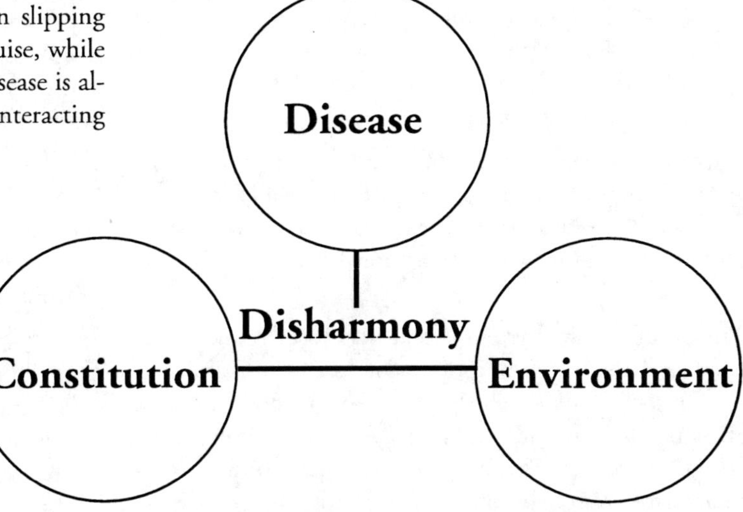

Figure 2—Disease as Disharmony

depression, high blood pressure, low thyroid and arthritis—things seem very complicated. But this complication only exists when we are looking at the situation from the perspective of disease symptoms. All of these conditions are arising from the same root causes, the same biological terrain and the same imbalanced body systems. The body does not exist in pieces; it exists as a whole, and all of these disease symptoms are simply part of the pattern of the whole.

There is no point in focusing our efforts on symptoms. First of all, we can't heal anything to begin with. The body is self-healing when provided with the right tools and environment. So, if we simply remove the environmental stressors that are overcoming the person's natural constitution and support weak body systems while balancing the biological terrain, the body will heal itself of whatever ails it.

Working on the Roots

To help people learn to remove the roots of disease from their lives, we use a modification of the ABC+D healing system created by a Master Herbalist named Edward Milo Millet, who was one of Steven's first teachers.

The ABC+D Approach™ is part of a comprehensive system of natural health care. Based on The Disease Tree™ model, this system addresses the root causes of disease with a four-step approach: Activate, Build, Cleanse and give Direct Aid. The program is designed to work on the underlying causes of disease, instead of treating disease symptoms.

Everyone needs to Activate, Build and Cleanse, no matter what health problems they may be facing. The first chart, found on the next page, shows the basic ABC+D steps. There are certain basic supplements and health practices listed which you should try first, regardless of the nature of the specific disease symptoms you may be experiencing. These products and natural therapies address the underlying causes of all the illnesses you are experiencing.

The +D of the system involves strengthening weakened body systems while balancing biological terrain. A chart is also provided that gives an overview of the whole system of biological terrain imbalances.

There are six imbalances that take place in biological terrain and another chart is provided that provides general symptoms of each imbalance. The six tissue states represent simple excesses and deficiencies in three basic physiological factors.

The beauty of this system is that instead of thinking in terms of diseases, we think in terms of tissue states and organ systems. We look at which systems of the body are not functioning properly and then assess what terrain imbalance exists in those tissues. Then, we select remedies from the appropriate list to restore balance to the different body systems by restoring balance to their biological terrain.

Basic Therapies: The ABC's of Healing

Natural healing does not rely on the disease diagnosis/treatment model. That is, we do not diagnose diseases or prescribe treatments for them. We are helping to balance the body so it will heal itself. We rely on certain basic therapies that remove underlying causes of disease and use them to help the body heal itself. These basic therapies fall under the ABCs of Healing: Activate, Build and Cleanse.

Here is a brief overview of the basic principles of natural healing and some of these basic therapies. You will find mention of these basic therapies listed under the various ailments in the *Conditions Section* of this book.

A = Activating Therapies

Activation is one of the most critical steps in the healing process. Start by working on the belief that you can be well. Effective natural healing acknowledges the simple truth that the body has a natural capacity to heal itself. In fact, the body's ability to heal is so remarkable that sometimes miracles occur that baffle the scientific community. Often, if a person simply believes they are going to get well, they will, regardless of whatever else they do. What we want to do is to enhance this innate self-healing ability.

This remarkable self-healing ability has been called placebo effect. While many healers look at placebo effect with skepticism, it is not a negative thing. It actually shows the power of a person's beliefs. When it comes right down to it, faith is the greatest healer of all. Faith is laying hold of what we do not already have as if we already had it. Hence, to exercise genuine faith in healing is to see oneself as whole, even when one is injured or sick.

There are many ways to exercise faith or belief to help activate the healing process, but they all involve helping people acquire positive attitudes and emotions. Thoughts and emotions are vibrations or energies which have a ripple effect that runs through the whole body, creating a positive (or negative) response.

Positive thinking provides an electrical stimulus through the nervous system and activates the healing energy of the tissues. In contrast, negative thoughts send a different kind of stimulation to tissues, which can actually result in the tissues becoming dysfunctional. Our language is loaded with examples of these thought/illness connections:

That child is such a headache.
She makes me sick to my stomach.
You're a pain in the neck.
I feel stabbed in the back.
That makes me want to throw up.

The ABC+D Body Systems Approach

GENERAL THERAPY FOR PREVENTION & RECOVERY

A - ACTIVATE

- Stimulate the healing response with positive attitudes, affirmations, visualization, faith, prayer, etc.
- Learn to manage stress
- Identify and address unresolved emotional conflicts
- Use energetic remedies that balance the mind and body

• Flower Remedies
• Homeopathic Remedies
• Essential Oils
• Affirmations and Visualizations
• Faith and Prayer

B - BUILD

- Hydrate the body with an adequate intake of pure water
- Eat a balanced, low-glycemic diet
- Correct hiatal hernia
- Select foods appropriate for your blood type
- Eat natural, locally-grown and/or organic foods
- Use appropriate basic nutritional supplements

• Multiple Vitamin and Mineral or SuperFood
• Trace Minerals
• Digestive Enzymes
• Probiotics
• Fiber
• Essential Fatty Acids
• Antioxidants

C - CLEANSE

- Do a general cleanse at least twice each year
- Avoid chemicals such as food additives, pesticides, toxic cleaning and personal care products, electromagnetic pollution, etc.
- Avoid xenoestrogens
- Use other cleansing procedures as needed

• Colon and liver cleansing
• Oral Chelation
• Heavy Metal Detoxification
• Gall Bladder Flush
• Avoid Xenoestrogens
• Sweat Baths, Drawing Baths, Enemas, Colonics, Foot Baths and Other Hydrotherapies

SPECIFIC BODY SYSTEM AND TERRAIN SUPPORT

D - DIRECT AID

- Use the results from the Body Systems Questionnaire to identify one or two body systems that require nutritional support
- Determine how each body system is energetically out of balance using the six tissue terrains and the appropriate chart(s)
- Use the chart(s) to select appropriate direct aids to support the specific body system(s)

Six Tissue Terrains

Irritation, Depression, Stagnation, Atrophy, Constriction, Relaxation

• Digestive System
• Hepatic System
• Respiratory System
• Intestinal System
• Immune System
• Urinary System
• Circulatory System
• Nervous System
• Structural System
• Glandular System
• Reproductive System

Continually repeated in our thought processes, our body can literally respond to these subtle messages by causing the very pains and illnesses our thoughts are telling us to create. That's why it is absolutely essential to healing that we don't identify directly with a disease. When we see disease for what it is, a process demonstrating that one's system is out of balance, it is easier to heal.

When we see the disease as something that is part of us, something that has taken control of us, then a feeling of helplessness and hopelessness, takes control. For instance, it has been documented that tumors typically start growing more rapidly after a person has been told they have cancer. This is because the person's fear and worry depress the immune system, making the disease worse. Here are two basic activating therapies you can use to promote healing in the body.

Activating Therapy #1: *Affirmation and Visualization*

There are two basic ways that we can create positive thoughts and emotions that will enhance our healing process. The first is *affirmation* and the second is *visualization.*

Affirmation is a present tense statement that affirms what I want as if I actually had it. Thus, if I had a broken bone and I wanted to speed the healing of that bone I would affirm "my bone is whole and strong." Notice the present tense statement, "My bone IS." This is a vital key to making affirmations work because it is laying hold of what you desire in the present tense. If you say, "My bone will be whole and strong," this places the fulfillment in the future, not the present and does not convey the same power. If a direct statement like that is difficult to use, a less direct, but equally effective statement would be, "My bone is healing as it should."

Examples of healing affirmations for yourself or others would include statements such as:

My body is healthy and strong.
My body is healing as it should.
You will feel better starting now.
Your body is recovering nicely.

The second method for helping to create positive thoughts and emotional responses is *visualization.* Visualization involves getting into a relaxed state and breathing deeply while you picture the final result you desire in your mind. Again, it is important to see yourself having what you want right now, not in the future. For instance, cancer patients have practiced visualizing their white blood cells gobbling up the cancer cells and destroying them. It has been proven in studies that such visualization actually enhances immunity.

Both of these techniques are enhanced by practicing deep breathing and relaxation, as is done in meditation. As one breathes deeply, tissues are oxygenated, which fans the spark of life and increases the flame of life throughout the body. Breath is called the "breath of life" because it is intimately connected with the vital force. To breathe is to connect with feelings, to be alive. Shallow breathing causes a person to stifle his or her feelings by deadening the body.

So, start the healing process by breathing deeply and allowing the body to relax. Then you can pick one of the foregoing methods for "laying hold" on the health you want before you actually have it.

Activating Therapy #2: *Stress Management*

Worry. Tension. Stress. In modern society, it's difficult to avoid stress. From debts and unpaid bills to traffic jams and deadlines at work, everyday we're faced with situations that can cause us to "tense up" or start worrying. It has been estimated that 75-90% of all visits to primary care physicians are for stress-related health problems. So, learning how to manage the stress in our lives is a major key to maintaining good health.

To understand how to manage stress, we first need to understand what stress is. When the brain perceives stress, it sends a chemical message to the pituitary via the hypothalamus which triggers the release of the adrenocorticotropic hormone (ACTH). ACTH causes the adrenals to start producing hormones like epinephrine (adrenaline) and cortisol. Epinephrine is both a hormone and neurotransmitter. It tenses our muscles, increases our heart rate and blood pressure, dilates the bronchials and speeds up our breathing, shuts down digestion and other functions not essential to immediate survival, and otherwise prepares the body for action.

Cortisol reduces inflammation, enabling us to deal better with injury and pain. Although it's role in reducing inflammation is important, too much cortisol causes premature aging, depresses immune function and causes us to lose muscle and gain weight. These stress hormones also cause a rise in blood sugar levels and an increase in blood clotting factors.

With this understanding, it's easy to see how chronic, long term stress can become a factor in numerous health problems, including poor digestive function, constipation, tension headaches, neck and shoulder pain, low back pain, ulcers, high blood pressure, blood clotting, increased risk of infections, asthma, diabetes, excess weight and even cancer and autoimmune disorders. In fact, it is probable that a large percentage of all the illness we experience has a stress component. So, reducing stress is one of the basic things we can do to improve our overall health. In fact, relaxation is necessary for the healing process.

If stress can cause so many health problems, it's obvious that we need to learn how to reduce stress in our lives. We may not be able to eliminate the stressful situations in our life, but we can reduce the stressful effects these problems cause in our body.

Here are seven keys to reducing the effects of stress on the body.

1. Breathe Deeply

One of the simplest things you can do to reduce your stress level, calm your anxiety and relieve the tension in your body is to just breathe. If you stop and notice what happens when you are feeling stressed, you will probably notice that you are either holding your breath or breathing very rapidly and shallowly. By concentrating on breathing slowly and deeply, you will activate the parasympathetic nervous system and help reduce your stress levels. You can also try breathing in while thinking, "I am," and out while thinking, "relaxed."

2. Practice the Relaxation Response

You can take the breathing a step further by utilizing what Dr. Herbert Benson dubbed "The Relaxation Response." In 1975, Dr. Benson published his book of that title showing how a simple, non-religious meditation technique could help patients with insomnia, heart problems, high blood pressure and chronic pain. Dr. Benson demystified the subject, showing that all one needed to do was consciously relax the muscles, breathe slowly and deeply and find a repetitive phrase to keep the brain occupied (such as repeating the word "one" over and over in one's mind).

Taking just 20 minutes a day for this process will dramatically reduce one's stress level. Start by finding a comfortable place to sit or lie down. Consciously allow all the muscles of your body to relax. Start breathing slowly and deeply while counting your breath (in, one, two, three, four and out, one, two, three, four). Then pick a single focus for your mind, such as the word "one" or "peace" and simply repeat this word over and over again in your mind. This causes the "monkey chatter" in the brain to stop and quiets the mind.

3. Avoid Caffeine and Sugar

Have you ever noticed how attracted you are to junk food when you are under stress? Sugar and caffeine may give you a quick "pick up," but they'll let you down just as fast. Even worse, they tend to further stress the adrenal glands, which eventually will tire and give you that "burned-out" feeling. To reduce stress, avoid sugar-sweetened, high carbohydrate snacks in favor of snacks high in protein and good quality fats (like nuts). If you feel tired without caffeine, consider taking Adrenal Support to rebuild your adrenals and increase your energy.

4. Hydrate

This may seem strange, but drinking more water can actually make your nerves feel calmer and help you sleep more soundly. Dehydration increases anxiety levels, so drink plenty of purified water when you are under stress.

5. Exercise

What are those stress hormones for? They're gearing your body up to take physical action, and that's what makes modern stress such a big problem. The stress hormones gear our body to run, fight or physically work to combat the problem, but our sedentary lifestyle doesn't allow us to burn off these stress hormones in physical activity. Exercise gives us the opportunity to "work off" those stressful feelings.

6. Feed Your Nerves and Take Adaptagens

Nerves, like any other part of the body need nutrition. For starters, nerves need good quality fats like butter, coconut oil, nuts, olive oil, flax seed oil and Omega-3 essential fatty acids.

Vitamins are also important for nerve functions. Many people have found that B-complex vitamins help them cope with stress more easily. Vitamin C and pantothenic acid are also helpful because they support the adrenal glands. Nutri-Calm is a great supplement for nutritionally supporting the nerves. Silica, found in HSN-W helps the nerves become more resilient because it strengthens the myelin sheath.

There is a specific class of herbs that has been shown to greatly reduce the impact of stress on our health. These herbs are called adaptagens. Adaptagenic herbs modulate the signals that are sent from the hypothalamus and pituitary glands causing a reduction of adrenal output of adrenaline and cortisol, thus lowering overall stress levels. They help to break the damaging fight-or-flight chain reaction patterns in which the body gets stuck due to chronic stress. By reducing cortisol levels, these herbs also help boost the immune system.

Eleuthero root was the first to be identified as an adaptagen. Russian studies proved it helps increase stamina, endurance and energy, improve concentration and stimulate male hormone production. It also helps the immune system response.

Other single herbs that have been identified as possessing adaptagenic properties include gotu kola, American and Korean ginseng, suma and schizandra berries. Some of the adaptagenic formulas available from NSP include AdaptaMax, Nervous Fatigue Formula and SUMA Combination.

7. Make Time for Rest and Relaxation

Telling someone to reduce stress is like telling them to avoid death and taxes. It just isn't going to happen. The good news is one doesn't have to try to avoid stress to reduce its effects. It turns out that a pleasurable experience causes the release of hormones and neurotransmitters that counteract the effects of stress. And, a pleasurable experience creates more positive benefits than a stressful experience causes harm. So, instead of reducing stress, we should be deliberately creating pleasure and enjoyment in our lives.

It's likely that a major part of the reason anxiety-related disorders are epidemic in our society is because we are just too busy. We are constantly on the go, and take very little time for pleasure and recreation. Making sure we plan time to do enjoyable things is very important to our emotional and physical health.

Many people feel they are too busy for this. Well, the truth is, that the busier you are, the more important it is for you to make time for rest and relaxation. If a woodcutter doesn't take time to sharpen his saw or his axe, he will find himself working harder and harder while becoming less and less productive. Rest and relaxation is "saw-sharpening" time, it makes you more productive with the rest of your day. If you are busy, you can't afford to not take time for rest and relaxation.

Watching TV doesn't count. Generally speaking, TV is designed to be stimulating, not relaxing. Instead, look for activities that feel pleasurable to the body, such as a warm bath, a soak in a hot tub, a massage, listening to relaxing music or taking a walk in the park. Find things that make you laugh and awaken a child-like delight in life. Slow down when you eat and really enjoy the flavor of the food. Remember that anything that brings a sensation of bodily pleasure counteracts the effects of stress and reduces anxiety.

Part of this is also making sure you are getting a good night's sleep. If you aren't, look up *Insomnia* in the *Conditions Section* and follow some of the suggestions for getting a good night's sleep.

B = Building Therapies

Healing requires nutrition. This is a simple fact that seems to get overlooked in our society. Many people want to eat their junk food diet and then expect some "magic pill" to cure the ailments they get from their high sugar, nutrient depleted diet. It just doesn't work that way.

So, don't by-pass the general building process and skip directly to using direct aids for specific body systems or imbalances. Build a foundational program first. Here are five basic building therapies. Therapy number five includes seven basic supplements most people can benefit from taking. These basic building therapies and supplements are referred to throughout the *Conditions Section* as basic aids to overcoming many health problems.

Building Therapy #1: *Hydration*

We don't often think of water as a nutrient, but it is the most important thing the body's needs besides oxygen. We can survive for weeks without food, but without water we would last a few days at best. A loss of just 15-20% of our body's water can be fatal. Only a lack of oxygen could kill us faster. Adequate intake of pure water is one of the simplest and cheapest health insurance policies you can buy. Without water, you don't have the ability to properly utilize either the food you eat or the supplements you take.

It has been estimated that 75% of all Americans are chronically dehydrated. So, of all underlying causes of ailments, dehydration is probably the most common and frequently overlooked. In about one-third of all Americans, the thirst mechanism is so weak that it is often mistaken for hunger. In one University of Washington study, a glass of water eliminated hunger in almost 100% of all the dieters studied.

Dehydration contributes to a wide variety of ailments, including indigestion, colitis, appendicitis, heart burn, rheumatoid arthritis, back and neck pain, headaches, stress, depression, high blood pressure, asthma, fatigue, memory loss and allergies. Many people have found that increasing their water intake reduces pain of all kinds, but especially headache, back and neck pain. Preliminary research indicates that 8-10 glasses of water a day could significantly ease back and joint pain for up to 80% of sufferers. Lack of water is the number one trigger of daytime fatigue.

Drinking water can also help to prevent disease. Water is necessary to flush waste products, particularly acid waste products, from the system. There is research to suggest that drinking five glasses of water daily could decrease the risk of colon cancer by 45%. Increased water intake could also slash the risk of breast cancer by 79% and reduce bladder cancer risk by 50%.

The brain is 80% water, so proper hydration is essential to its function. A mere 2% drop in body water can trigger fuzzy short-term memory, trouble with basic math, and difficulty focusing on the computer screen or on a printed page.

How much water do you need? A good rule of thumb is to divide your body weight in half and drink that many ounces of water per day. So, if you weigh 160 pounds, you need to drink about 80 ounces of water each day. There are 32 ounces in a quart, so this would equate to a little less than three quarts of water per day.

If you aren't drinking 1/2 ounce of water per pound of body weight per day and don't feel thirsty, you really need to drink more water anyway. When the human body is sufficiently dehydrated, its thirst mechanism shuts off. This means that senior citizens are at greater risk for dehydration than younger people because their bodies are less effective at letting them know when they need water.

However, it isn't just the amount of water that is important. The kind of water we drink is critical, too. Increasingly, our water supplies are being polluted and poisoned, with disastrous consequences to our health and well-being. So, drink the purest water you can find. At the least, use a carbon filter and change it regularly. Better yet, invest in a reverse osmosis or water ionizing unit.

Building Therapy #2: *Low Glycemic Diet*

The single biggest problem with our modern diet is the huge amount of refined carbohydrates we consume. There are many problems with this obsession we have with refined sugars, white flour and processed grains. Here's why:

The pancreas, adrenals and liver work together to maintain a stable blood sugar level for healthy body and brain function. When you get up in the morning, your body has been fasting all night and your blood sugar is low. You "break your fast" by eating breakfast. And how you breakfast in the morning will set your metabolism for the day.

If you start off with coffee and donuts or other pastries, sugar sweetened breakfast cereal or other simple carbohydrates, you raise your blood sugar level quickly, but it comes at a cost. Your blood sugar goes too high and your pancreas has to secrete high levels of insulin to get this sugar out of your bloodstream. Insulin causes the body to store carbohydrates in the liver and fat cells (causing weight gain). Insulin also increases inflammation, which is the root cause of heart disease, cancer and numerous other degenerative ailments.

Once that sugar is gone, your blood sugar is low again and you crave your next high carbohydrate and/or caffeine fix. This situation of low blood sugar is called hypoglycemia. 'Hypo' meaning low, 'gly' for sugar and 'cemia' referring to the bloodstream.

When blood sugar levels start to dip below normal, the body gives certain subtle clues that it needs help. These may include suddenly feeling cold or getting a cold nose, strong craving for sweets or caffeine, sudden fatigue or mental confusion, the inability to concentrate, a mild headache or sense of pain around the eyes. If not dealt with soon, the symptoms may worsen into irritability, severe fatigue, dizziness or shakiness.

As the above symptoms suggest, hypoglycemia affects far more than our physical bodies. It also affects our mind and emotions. As the sugar in candy and chocolate rushes into the blood stream, it produces a sugar "high." The pancreas hypersecretes insulin to remove the excess sugar from the blood.

Since insulin depresses glucagon production, this results in a corresponding "downer" as the blood sugar level falls below normal. The person craves sugar and the cycle begins again, putting the body, and especially the brain, on a blood sugar roller coaster ride. All day long your blood sugar goes up and down like a roller coaster, and your energy and mood goes up and down with it.

These refined carbohydrates also rob your body of vitamins and minerals, since it requires these nutrients to process the carbohydrates into energy. This steadily reduces nutrient stores and depletes the health not only of bones and teeth, but the brain, heart, liver and other vital organs.

In fact, half of most people's health problems would go away if they just stopped eating refined carbohydrates and processed fats. It's actually easier to change than most people think. It starts with breakfast.

Start the day by eating some high quality fat and protein. Eggs, avocados, organic meats or unsweetened yoghurt are all good choices. If you're in a hurry, take a spoonful of coconut oil and make a protein shake with SynerProtein, Nutri-Burn, Love and Peas or some other protein powder.

Protein and good fat cause your pancreas to secrete a different hormone called glucagon. Glucagon mobilizes sugar stored in your liver to enter the blood stream. This raises your blood sugar, but it also sets your metabolism to start burning fats instead of storing them. The result, your blood sugar stays more stable throughout the day and so does your energy and mood.

You can also reduce sugar cravings by taking Super Algae and licorice root. Two capsules of each at breakfast, two again at lunch and two in the middle of the afternoon.

If you also start selecting complex carbohydrates (such as fruits, vegetables and whole grains) instead of simple carbohydrates (foods with refined sugars and grains) your metabolism will adjust in about two weeks. Your cravings for sugar and caffeine will cease and your mood, energy and overall health will be much better. You'll be amazed at how much better you feel and how much more clearly you'll be able to think. Try it!

Building Therapy #3: *Hiatal Hernia Correction*

It doesn't do much good to eat good food or take supplements if you aren't digesting them properly. Many people suffer from a hiatal hernia, which inhibits proper digestion. In fact, just about all chronically ill people have this problem, and correcting it goes a long way towards improving their health.

A hiatal hernia occurs when the stomach moves up into the opening in the diaphragm for the esophagus. This may be due to stress and repeated bouts of bloating and gas. It can also be caused by obesity and pregnancy.

The hiatal hernia stresses the stomach by inhibiting the vagus nerve and blood flow to the stomach. Protein digestion is impaired and the resulting lack of essential amino acids causes glandular malfunction, immune system deficiency, poor muscle tone, excessive weight loss or gain, cold limbs and general physical weakness.

Symptoms of a hiatal hernia include the inability to breath from the diaphragm, tension in the solar plexus, difficulty swallowing capsules, the sensation of a "lump" in the throat and an over-stimulated thyroid gland (high metabolism). Chronic intestinal gas may occur as the ileocecal valve becomes permanently swollen and irritated and unable to close properly. Most people suffering from general poor health have this condition.

This problem can be overcome using a variety of self-help techniques. Here are some of them.

Check your breathing. Follow this simple test to access your pattern of breathing as a first step in treating the hiatal hernia. Put your hand on your abdomen as you breathe. If your abdomen moves in and out more than your chest, you are probably handling your stress well, or at least, you aren't letting stress control you.

If you are breathing from the top of your lungs, just sit back and relax to allow your breathing apparatus to revert to normal abdominal breathing. If it doesn't, then you need to relax the diaphragm. To do this, take lobelia or blue vervain in liquid form. Then, practice breathing from the abdomen again. You can also practice abdominal breathing while relaxing in a bath with lavender essential oil and Epsom salt.

Find healthy ways to vent your repressed anger and frustration. This releases tension from the diaphragm and will help defuse much of the tension maintaining the hiatal hernia problem. For example, try taking a long, slow deep breath and feel the tension build up in your diaphragm (like you are starting to get angry). Make your hands into fists and raise them up in front of you as if you want to punch somebody. Exhale forcefully with an angry "huh!" sound while shaking your fists downward like you are hitting something. Do this several times, safely discharging your inner tension and frustrations.

Other methods of dealing with stress include changing your environment, finding new ways to resolve problems and communicating your thoughts and feelings honestly with others.

You can also find a chiropractor or a massage therapist who knows how to manually manipulate a hiatal hernia. It usually takes 4-6 treatments, combined with self-help techniques to bring down a hiatal hernia.

As an alternative to having someone work on the problem for you, you can use the following technique.

Drink a pint of warm water first thing in the morning. Next, stand on your toes and drop suddenly to your heels several times. The force of this little jump and the weight of the water help pull the stomach down in place while the warm temperature of the water relaxes the stomach area. Taking a dropperful of lobelia essence with the water will relax the stomach and make the treatment more effective.

If you're adventurous, jump off a chair or down a short flight of stairs to get the same effect. The idea behind this technique is to get your stomach to "drop" as if you were in an elevator that suddenly started going down.

If this doesn't solve the problem, place both hands under your breastbone in the center of your rib cage. Take a deep breath, press your fingers firmly into the solar plexus area (just under the breastbone). Then, as you forcefully exhale, push your fingers downward and bend forward slightly. Be careful not to push your fingers up under the rib cage. Repeat this action several times. Do this before meals on an empty stomach.

You can find more information on dealing with a hiatal hernia, including a video demonstrating how to work on the problem in the article database at www.healityourself.com.

Building Therapy #4: *Blood Type Diet*

One of the puzzling problems in working with nutrition is that individuals can react differently to the exact same diets and supplement plans. A nutritional program that produces positive benefits for one person may elicit a negative reaction from another. Fortunately, there are tools available that can help us understand why this is the case, and help someone find the appropriate program for their type.

Research done by Dr. Peter D'Adamo and his father, Dr. James D'Adamo, has demonstrated there is a strong correlation between a person's blood type (O, A, B or AB) and the foods and supplements they need to consume for optimal health. Dr. D'Adamo has widely promoted this concept in several popular books, including *Eat Right for Your Type* and *Live Right for Your Type.*

To understand the blood type diet, we need to realize that all organisms are equipped with a system that enables their bodies to identify structures that are part of their organism from structures belonging to other organisms. Our blood type is one of the most fundamental methods the body utilizes to separate "self from not self." This ability to determine what is self and what is not self is at the core of our immune function.

All foods contain chemicals known as lectins. Lectins that are incompatible with one's blood type create negative reactions. For starters, they cause agglutination of the blood, meaning they make red blood cells "sticky" so they clump together. This results in a reduction in oxygen supply in the body and lowered immunity. Lectins can also interfere with protein digestion, block hormones, trigger immune reactions and impair absorption.

Understanding which foods contain lectins compatible with the body and which foods contain lectins that are not is the basis of the blood type diet. A food is classified as an Avoid because it produces negative lectin reactions in that blood type, which act as toxins. On the other hand, foods that are Beneficial for a particular blood type are highly compatible with that type. These foods actually serve as a type of medicine, strengthening health and preventing disease. Foods that are Neutral merely supply nourishment. They do not have negative lectin reactions, but they also do not have positive, healing properties either.

In addition to diet, the four blood types have other unique differences that require specialized supplementation. Each blood type is prone to a unique set of health challenges which can be mediated or eliminated through appropriate herbs and nutrients. So, with that basic understanding, here are some basic things people of each blood type can do nutritionally to enhance their health and wellbeing.

Blood Type O

Blood Type O generally does well on a diet of meat and other high protein foods. They can eat most types of animal protein, but, like all blood types, they must prepare these protein-rich foods properly to be able to derive positive benefits from them. Overcooking protein denatures it and taxes the digestive system. Meat should be organic and eaten rare or slow cooked to obtain maximum benefits.

Large portions of non-starchy vegetables that are Neutral or Beneficial should also be consumed with protein foods. Examples of Beneficial non-starchy vegetables for Blood Type O include beet greens, broccoli, romaine lettuce, okra and Swiss chard. Eating generous portions of these vegetables also helps maintain proper acid/alkaline balance.

Grains, especially corn and wheat, as well as dairy products, are major Avoids for Blood Type O. Other Avoids include pinto beans, lentils, cashews, black olives and avocados.

Blood Type O people need protein and have a special need for the amino acid l-tyrosine. L-tyrosine helps maintain levels of dopamine and epinephrine, two neurotransmitters involved in mood. The best sources of this amino acid are red meat and wheat. If Blood Type O people do not eat red meat, they may crave wheat, which is an Avoid for their type. Super Algae and Free Amino Acids can supply l-tyrosine and help keep protein levels stable in Blood Type O.

Because of their tendency to an overactive immune system, Blood Type O should generally avoid immune stimulants. ADD and ADHD are most prevalent in Blood Type O.

Blood Type A

Being the more agrarian type, Blood Type A doesn't do well with animal protein. Beef, lamb and other red meats are Avoids for them. Dairy foods are also Avoids for this type. So Blood Type A people must make certain they get adequate protein from other sources. Fortunately, poultry and eggs are Neutral for them. Many types of fish are Beneficial, making fish a good protein food for this type. Fish is also an important source of Omega 3 essential fatty acids which reduce the risk of heart disease for Blood Type A.

There are also vegetable sources of protein which are highly compatible with the metabolism of Blood Type A. In particular, soy products are Beneficial. (Avoid GMO soy products, however.) Given that the A antigen is more susceptible to the xenoestrogens in the environment, the phytoestrogens in soy can reduce cancer risk and improve glandular function, too. Also Beneficial to Blood Type A are lentils, pinto beans and black beans.

Blood Type A is generally more tolerant of grains than other blood types. Grains and beans can be combined to form a complete vegetarian-source protein. Their systems are also compatible with a wide variety of fruits and vegetables, which should be used in large quantities to balance protein, beans and grains so that proper acid/alkaline balance is maintained.

Blood Type A has a difficult time with protein digestion. They generally need to supplement with digestive enzymes. One of the biggest enemies of Blood Type A is stress. People with this blood type are very prone to high levels of cortisol, which results in anxiety, lowered immune response, compulsive behavior and premature aging. Under stress, Blood Type A loses muscle mass and gains fat. Adaptagens and other remedies for stress can be very helpful for people with Blood Type A.

Blood Type B

More balanced than the other types, Blood Type B people will find that wild game provides their best protein. Chicken is a major Avoid for this blood type. Eggs do not possess the lectin found in chicken, so eggs are a Neutral for this blood type.

Blood Type B people are unique in their ability to handle dairy products. The B antigen and the dairy sugar, d-galactosamine, possess a similar structure, which makes milk, yogurt and even some types of cheese Beneficials for this type. (Make sure these are organic.) However, those of Asian or African descent may have difficulty digesting lactose, the sugar in milk, due to a lack of the enzyme lactase, because dairy products were not a part of the dietary pattern in those early cultures.

Corn, wheat, and certain legumes like lentils, pinto beans, peanuts and soy are Avoids for Blood Type B. Beneficial vegetables include beets, broccoli, cabbage, eggplant, parsnips, peppers and yams .

Blood Type B people often have problems with their adrenal glands and stress. Blood Type B people assimilate calcium very easily, which often results in a magnesium deficiency. They are more prone to certain types of infections. For starters, they are more susceptible to the influenza (flu) virus. Blood Type B people, especially those of Asian descent, are also at higher risk for tuberculosis. Another bacteria they are susceptible to is the streptococcal type. Finally, urinary tract infections (UTIs) are common in this type.

Blood Type AB

Possessing the rarest blood type, people with Blood Type AB have the antigen structure of both the A and B types. Because of their A antigen structure, red meat is an Avoid for them, and because of their B antigen structure, chicken is also an Avoid.

Due to the B antigen, Blood Type AB can utilize milk and eggs, as these foods are Neutral (when organic). Yogurt and egg whites are actually Beneficials. Too much dairy, however, can contribute to increased mucus production.

A number of fish are Beneficial to Blood Type AB. These include cod, mahi mahi, salmon, red snapper and tuna, among others. Some legumes are also Beneficial, including pinto beans, soy beans, peanuts and green lentils. Blood Type AB people have some of Blood Type A's tolerance for grain, with rice, oats and rye being Beneficial and sprouted wheat being Neutral.

Again, Beneficial and Neutral fruits and vegetables must be consumed in larger quantities than protein foods in order to keep pH balanced in the body. Beneficial vegetables for Blood Type AB include beets, broccoli, celery, cucumber, eggplant, parsnip and yams.

Blood Type AB people have better protein digestion than Blood Type A, but not as good as Blood Types O and B, so they may need some assistance with digestion. Blood Type AB people have a tendency to high cholesterol levels and environmental sensitivity, resulting in reduced immunity. High blood pressure may also be a problem with Blood Type AB, especially from overly emotional reactions to stress. Stress can also be a serious issue for people with Blood Type AB.

Following the Program

Space does not allow for a complete list of foods that are Beneficial, Neutral or Avoid for each blood type in this book, but Tree of Light Publishing has blood type charts you can get that tell which foods are Beneficial, Neutral or Avoid for each type. These charts also provide information for pH balance and "zoning" your diet (balancing fats, proteins and carbohydrates).

You don't have to follow the blood type diet perfectly to obtain results. If you focus on eating mostly Beneficial and Neutral foods seventy-five percent of the time, this will allow you to eat the Avoid foods once in a while. The more serious your illness, the more you need to adhere to the program by following it ninety to one hundred percent of the time.

Our clinical experience suggests that following the blood type diet is not only an excellent way to lose weight, it is also an important key to recovering from many types of chronic and degenerative illness.

Building Therapy #5: *Create a Basic Supplement Program*

Even when you eat right, many of our modern foods just aren't as nutritionally dense as foods eaten by our ancestors. That's why supplements can make up for weaknesses in our diets. The following are seven basic supplements to consider. These are the most commonly needed supplements. In fact, as you look up many of these supplements you may find that they have historically been helpful for many of the problems you are experiencing.

Basic Supplement #1: SuperFood Supplement

When most people think of supplements, they immediately think of vitamins, or vitamin and mineral supplements. You can take a multiple vitamin and mineral supplement like Super Supplemental, but whole foods and herbs as the better choice. Plant and animal foods are highly complex substances containing thousands of chemicals. New compounds are constantly discovered in these plants that play critical roles in human health, so a "one-a-day" vitamin and mineral pill is not an adequately substitute for the rich diversity of nutrients in whole foods.

A good SuperFood Supplement from Nature's Sunshine is GreenZone, which can be taken as basic whole food nutrition to support the body. Your other option is, of course, to take Super Supplemental or Super Trio (which contains Super Supplemental and two more of the seven basic supplements listed here.)

Basic Supplement #2: Minerals

Since our bodies are literally composed of minerals (i.e., we are made of the "dust of the earth"), healthy bodies are connected to healthy soil. If any element is missing from the soil, then it will be missing from the foods we eat and as a result, we will not be properly nourished.

Unfortunately, our commercial methods of agriculture are not only depleting the soil of precious trace minerals, they are also destroying the ability of plants to be able to utilize those elements. Hence, our food is nutritionally deficient right from the start. To make matters worse, as our food gets refined, more of its nutritional content is removed.

Modern food is lacking in minerals because modern farming methods reduce mineral content in food. This is partly because chemical fertilizers focus on three nutrients nitrogen, phosphorus and potassium and neglect other elements needed in the soil for healthy plants. Agricultural chemicals and a lack of organic matter in the soil also reduce mineral availability. This is because minerals are made available to plants through microorganisms in the soil.

Bacteria in the soil break down organic material from dead animals and plants to recycle it for use by other plants. Mycorrhiza fungi, which grow on the roots of plants, protect growing plants against these bacteria (fungus and bacteria are natural antagonists). Both these fungi and the bacteria in the soil help the plants assimilate the minerals they need. Chemical agriculture destroys these microbes.

This is why most people need something to supplement their mineral intake. When seeking to obtain minerals, the first source people should be encouraged to use is mineral-rich plants. This is because the minerals in plants are more bioavailable to the body. Mineral rich herbal formulas include HSN-W, Herbal CA and Herbal Trace Minerals.

Another way human beings get minerals is through water. Mineral springs have long been sought out for their healing benefits. Colloidal mineral products are generally a form of concentrated mineral water and can also be used to increase mineral intake. NSP has two good products here, Colloidal Minerals and Mineral-Chi Tonic. Mineral-Chi Tonic also contains adaptagenic herbs for reducing stress and balancing the body.

Basic Supplement #3: Enzymes

Enzymes are the "spark plugs" of the life process. They regulate numerous body functions. Minerals act as catalysts for enzymes, so the two work hand in hand to promote health. Enzymes are natural components of living things, so they are found in raw foods. However, heat deactivates or destroys enzymes.

Cultured foods like live-culture yogurt and sauerkraut are also rich in enzymes. These foods are regularly consumed in many cultures and appear to contribute greatly to gastrointestinal health, but few Americans consume them regularly. And, if these foods have been heated past 120 degrees, the enzymes are deactivated. To compound the problem, most processed foods contain enzyme inhibitors, which are added to processed foods to prevent spoilage and increase shelf life.

Ideally, a large percentage of a person's food should be raw, but few people are able to do this in modern society. This fact, compounded with the other problems we've discussed, means most people need to take an enzyme supplement.

There are two basic types of enzyme supplements—those that contain the actual digestive secretions produced by the body (pancreatic enzymes, hydrochloric acid and bile salts), such as Food Enzymes, and those that contain plant-based enzymes (amylases, lipases, proteases, etc.), such as Proactazyme Plus.

Generally speaking, we recommend plant-based enzymes as a basic supplement as these replace the enzymes lost through cooking and processing. We suggest you reserve the use of Food Enzymes for people who are very elderly or weak and have lost the ability to digest food.

It is also possible to stimulate the body's own digestive secretions. Generally speaking, bitter-tasting herbs and pungent or aromatic herbs will stimulate digestive secretions. This includes bitter greens eaten raw in a salad, as long as they are not coated with sweet or creamy salad dressings, just an oil and vinegar dressing. Digestive Bitters Tonic can be helpful, but it has too much sweetener to be really effective. Placing the powder from about 1/2 capsule of goldenseal in the month is more effective. Spleen Activator, Anti-Gas Formula with Lobelia and the Chinese Anti-Gas Formula can also be used to improve digestive function herbally.

Basic Supplement #4: Probiotics

Most of us associate bacteria with disease. We think of bacteria as something to be eliminated and destroyed. This has created an almost obsessive use of disinfectants in our culture. But, not all bacteria are bad. It is the action of bacteria, for example, that allows milk to be fermented to create cheese, yogurt and kiefer. Bacteria also create other fermented foods such as sauerkraut and tofu. Another benefit of bacteria is that they breakdown minerals in the soil and make them available to the roots of plants. So, plants need bacteria to be healthy.

Our "roots," i.e., the place where we absorb water and nutrients, is our intestinal tract, and bacteria play an important role in our "root" system, too. In fact, there are about three to four pounds of friendly microorganisms living in the intestinal tract, most of them bacteria. A proper balance of these microbes is essential to one's health, because we live in a symbiotic relationship with microorganisms. Many strains of bacteria are actually part of our body's natural ecosystem. They serve to help protect the body against unfriendly microbes.

There are many different species of beneficial bacteria inhabiting our intestines. Many belong to the genus *Lactobacillis*. These include *L. acidophilus*, one of the first strains sold as a supplement. Another genus containing species of friendly bacteria is *Bifidobacterium*, sometimes referred to as bifidophilus. A third major group belong to the *Streptococcus* genus. There are many others.

The good bacteria inhabiting the intestines are called *friendly flora* or *probiotics*. *Biotic* is from a Greek word that refers to life. So *pro*-biotic means favorable to life. This is in contrast to the word *anti*-biotic, which literally means against life.

Antibiotics weaken the immune system because they destroy the friendly flora. These friendly flora are actually part of the immune system. Friendly bacteria enhance the immune system in several ways. First of all, they form a sort of living "blanket" that coats the intestinal tract and inhibits other species of microorganisms from "gaining a foothold" on the intestinal mucosa. They compete with other microbes for food, which also holds down the growth of infectious organisms.

Friendly bacteria even produce chemicals that are deadly to harmful forms of bacteria, so they act as natural antibiotic agents against harmful bacteria. Another benefit of friendly bacteria is that they have a stimulating effect on the body's immune system. For instance, animal studies showed that *S. thermophilus* and *L. bulgaricus* increased proliferation of lymphocytes, stimulated B lymphocytes and activated macrophages.

A well-known benefit of friendly flora is their ability to prevent yeast such as *Candida albicans* in check. When antibiotics, chemotherapy, chlorine or other chemicals or drugs destroy the friendly flora, yeasts multiple out of control. The yeast secrete a toxin that weakens the intestinal membranes and reduces the immune response. Probiotics are the antidote to this problem, helping to restore a healthy intestinal microflora.

Probiotics also help overall colon health. They reduce the risk of inflammatory bowel disorders such as colitis, Crohn's disease, and irritable bowel syndrome. They also reduce the risk of colon cancer. They should be used as part of a natural treatment plan for these diseases.

Healthy intestinal microflora improve the body's ability to digest fats and proteins. Probiotics synthesize certain vitamins the body needs, including B1, B2, B6, B12, folic acid and biotin. The synthesis of B12 by probiotics is particularly important for vegetarians who are not getting this vitamin in their diets.

The friendly flora also help detoxify certain poisons in the digestive tract. For instance, they help break down ammonia, cholesterol and excess hormones.

The anthraquinone glycosides in stimulant laxative herbs like cascara sagrada and senna are activated by the intestinal microflora. In fact, these herbs are much less effective if the intestinal microflora is out of balance. Probiotics by themselves help to overcome constipation, too.

Finally, about 70% of the energy requirements of the intestinal mucosa come from fatty acids produced as a by-product of bacterial fermentation. This means that the intestinal microflora actually helps feed the intestinal lining demonstrating how vital this synergistic relationship is to health.

In fact, a healthy intestinal microflora is such an important part of total health, that some health researchers feel it should be considered as an independent body system.

The intestinal microflora is a highly adaptable system, as it changes constantly, adapting itself to one's diet and environment. It is easy to see why a balanced intestinal microflora is such an important factor in a healthy body.

When looking for a probiotic product look for one that contains several strains of bacteria such as acidophilus, bifidophilus, etc. Make certain that the product is stored in the refrigerator, as the bacteria are living organisms and will die quickly when stored a room temperatures. For more information, look under Probiotics in the Product Section.

Basic Supplement #5: Fiber

This brings us to the next essential supplement just about everyone needs—fiber. Of course, if people were eating fruits and vegetables every day, including edible skins and seeds, and if they were eating only whole grains, then they'd be getting enough fiber. However, very few people do this.

Fiber has numerous benefits. It absorbs bile from the gallbladder to help reduce cholesterol levels, slows the release of sugar into the blood to regulate hypoglycemia and diabetes, absorbs toxins in the intestinal tract to help detoxify the body, reduces inflammation in the gut and provides food for friendly bacteria. Fiber also reduces the risk of colon cancer and prevents diverticulitis and hemorrhoids. And, of course, it helps assure regular elimination.

Many people think that cleansing the colon means taking a stimulant laxative. These products just stimulate peristalsis, which is something that is rarely needed. Most people are constipated from lack of fiber, lack of water, and magnesium deficiency. Besides, fiber is what really cleanses the colon, because fiber is what binds the toxins so they can't be absorbed into the bloodstream. So, fiber is the one cleansing product that can be taken regularly by most Americans.

I find that taking just one heaping teaspoonful of a fiber blend, first thing in the morning, along with a large glass of water, can make a dramatic difference.

The only problem you might run into with fiber is if you don't drink enough water with it. Without water, fiber can actually bind you, so make certain you drink plenty of water when you take fiber. Stay hydrated. If you are taking fiber and stop having regular bowel movements, discontinue the fiber and drink plenty of water. You many need to take some magnesium or a stimulant laxative to get the colon working properly again.

NSP has four fiber blends to pick from, Psyllium Hulls Combination, Nature's Three, LOCLO and Everybody's Fiber. The gentlest of these is EveryBody's Fiber, which is the most suitable fiber for people with irritable bowel or inflammatory bowel disorders.

Basic Supplement #6: Essential Fatty Acids

Contrary to all the propaganda that suggests otherwise, we need fats in our diet to stay healthy. Fats play critical roles in our health. Brain and nerve tissue, for instance, requires the proper kind of fats, and low fat diets can lower the intelligence of children. The heart burns fat as its primary source of fuel. Fats are burned to keep the body warm in cold weather and are necessary for the production of many hormones.

Extremely low fat diets aren't good for us and can actually raise cholesterol, since about half of the cholesterol in our body is used to make bile to digest fats. However, we need to get the right kinds of fats in our diet. Unfortunately, most Americans are eating the wrong kind of fats, which include margarine, shortening, processed vegetable oils and animal fat from non-organically raised animals.

Americans tend to get too many Omega-6 fatty acids and not enough Omega-3 fatty acids in their diets. Omega-3 fatty acids protect us against heart disease. They also benefit the immune system because they help control the chronic inflammation that underlies the development of hardening of the arteries, arthritis, memory loss in aging and other degenerative disease. The best sources of Omega-3 are wild game, grass fed beef, eggs and poultry and deep ocean fish (not farm raised). Avocados and nuts, especially walnuts, also contain good fats.

Because most Americans get too many bad fats, and not enough good fats, most Americans can benefit from supplementing their diet with Omega-3 essential fatty acids. Super Omega 3 EPA and flax seed oil are two good options.

Basic Supplement #7: Antioxidants

If you've ever cut an apple and left it sitting on the counter, you've seen oxidation at work. Oxidation is what causes a cut apple to turn brown. It's also what causes a fire to burn, oils to go rancid, iron to rust and copper to develop a green patina. Most researchers now believe that oxidation is also what causes the body to deteriorate and develop degenerative diseases as we age. The free radicals that cause oxidative stress can be thought of as tiny "arsonists" waiting to start little inflammatory "fires" in the body.

Experts suggest that about 50-80% of all chronic and degenerative diseases, including heart disease, cancer, diabetes, arthritis, macular degeneration, Alzheimer's and dementia are caused by oxidative stress, also known as free radical damage. Oxidative stress also causes the cosmetic problems we associate with aging, dry skin, wrinkles, age spots and so forth.

Oxidation is not bad per se. The body uses oxidation to break down the food we eat and convert it to energy. The immune system also uses oxidation to destroy microbes and fight infection. Free radicals are also produced during the process the body uses to break down toxins for elimination.

Just like an automobile needs a cooling system (radiator) to keep the heat generated by the engine from destroying the engine parts, the body needs a cooling system to keep the oxidative processes it generates under control. Antioxidants can be thought of as the body's radiator cooling system.

Antioxidant nutrients are abundant in fresh plant foods. For example, fresh wheat contains an antioxidant called vitamin E, but this vitamin deteriorates within a few days after grinding the grain. Similar processes take place in all the foods we eat. So, because so little of the food we eat is fresh, most of us don't get enough antioxidant nutrients to keep our body's cooling system working properly. This allows oxidative stress and the resulting inflammation to "burn up" our health.

To compound this problem, we actually eat foods that increase oxidation and inflammation. Refined sugar, white flour, hydrogenated oils, trans fats and food additives put additional stress on our already overtaxed cooling system and further overheat the body's metabolic engine.

Fortunately, just like the car has a temperature gauge that warns you when the engine is getting too hot, the body also has warning indicators that tell you the metabolic engine is overheating. These warnings include: chronic, low grade aches and pains, fatigue (especially after exercise), constipation, headaches, difficulty concentrating ("brain fog"), excess weight, carbohydrate cravings, gum disease and loss of energy.

If these warning lights are going off in your body, don't ignore them. It's time to add some antioxidant "coolants" to protect your metabolic engine from continuing to "overheat" because of oxidative stress and chronic inflammation. The best way to do this is to eat 5-7 servings of fresh fruits and vegetables every day. And no, the ketchup and lettuce on that hamburger and those greasy french fries don't count! We're talking about eating generous portions (about 1/2 cup) of 5-7 different fresh fruits and vegetables every day.

The fact is, most people primarily eat processed (canned and frozen) produce (if they eat produce at all!). Studies suggest that the average American consumes only 1-1/2 servings of vegetables and no servings of fruit on a typical day.

While eating processed produce is better than not eating any produce at all, much of the antioxidant potential of these foods is lost in processing. And as for that word "fresh," how much of the produce at your local mega-mart is actually fresh? Because it typically takes 7-14 days to ship produce from the farm to the store, many foods, such as tomatoes or peaches, are picked green, which means they've never been allowed to develop their full nutritional value (or flavor).

That's part of the reason most of us don't eat more fresh fruits and vegetables; the mega-mart produce doesn't really taste that good. If you've ever eaten home-grown tomatoes or tree-ripened peaches, you'll know what we're talking about. That fresh, fully-ripe flavor that makes these foods so delicious is actually a sign that the food is loaded with nutritional (including antioxidant) value.

It's a good idea for most of us to supplement our diet with antioxidant nutrients. And, no, taking supplements isn't an excuse to eat whatever junk food we want, thinking that supplements alone are going to keep us healthy. We should still eat those 5-7 servings of the best produce (fresh, frozen or even canned) we can find. Supplements are meant to supplement, not replace, a healthy diet.

C = Cleansing Therapies

Besides good nutrition, the body needs to remove metabolic wastes and toxins we may be exposed to through our food, air and water. If you are not pregnant, nursing, feeble or emaciated, or extremely fatigued, you should probably find that doing a cleanse will improve your energy and your mental clarity. You will also be amazed at the number of health problems that disappear after doing a good cleanse.

Here are some basic cleansing therapies you can use to detoxify your body. Like our nutritional building therapies, these therapies are referred to under numerous conditions as basic aids to overcoming various ailments.

Cleansing Therapy #1: *Colon Cleansing*

Every day, our body manufactures waste in the process of metabolism. Every day, we also ingest substances through food, water and air that are potentially harmful for our system. Fortunately, the body has the capacity to deal with this problem. The body rids itself of metabolic waste and chemical irritants through various eliminative systems.

Although medical science tends to discredit the idea of cleansing, natural healers have long stressed the importance of maintaining good elimination for better health. After all, it makes sense that the body will be healthier if waste and toxins are eliminated quickly. In fact, just about any system or machine needs some kind of regular cleaning to run properly.

Plumbers know that pipes can get clogged and drains need to be cleaned. Auto mechanics realize that oil and other fluids need to be regularly changed to keep engines running smoothly. Even electronic equipment needs to be cleaned periodically to keep dust from damaging circuits. It makes sense that this is also true for our bodies.

Most of us strive to keep the outside of the body clean, but few pay much attention to keeping clean on the inside. Most people who have done some internal cleansing, however, have noted numerous improvements in their general health.

Cleansing is about two things. One is minimizing your exposure to toxins in the first place and the other is using herbs, supplements, hydrotherapy, fasting or other natural means to improve the function of eliminative organs.

Since, cleansing is the process of getting rid of what is no longer useful, doing a cleanse simply involves supporting the body's natural detoxification systems to eliminate metabolic waste and environmental toxins more efficiently. Generally, this means using herbs that have been found historically or scientifically to improve liver and kidney function, bind toxins, increase lymphatic flow, open the sweat glands and encourage elimination from the bowels. It may also involve destroying harmful organisms (yeast, bacteria or parasites).

Here are the basic elements of a good cleanse.

Water

The most important tool for cleansing is water. Most people do not drink enough water. Experts suggest we should have about 1/2 ounce of water per pound of body weight each day. On a cleanse, one might need a little more. Since the quality of water is also important, drink the purest water you can find.

Fiber

The second most important tool for cleansing is fiber. Dietary fiber binds toxins in the intestinal tract and bulks the stool to promote normal and healthy elimination. Most Americans do not get enough fiber in their diet, so a good fiber supplement like Psyllium Hulls Combination or Everybody's Fiber is important for maintaining normal elimination.

Detoxifying Herbs

The third tool needed for a good cleanse is a blend of herbs that support the liver, kidneys, colon and lymphatics. The liver utilizes enzyme systems that neutralize toxins and prepare them to be flushed through the kidneys or colon (via the gallbladder). Water and fiber then carry these toxins away.

Many good herbal formulas for supporting this process are available. A few good examples are All Cell Detox, Chinese Liver Balance and Enviro-Detox.

For people with extremely sluggish elimination, an herbal laxative may also be helpful. LBS II is a popular choice. However, herbal laxatives should not be used long term because people tend to become dependent on them. For long term problems with sluggish elimination, consider using Gentle Move.

Depending on your specific needs, herbal and nutritional formulas that help destroy yeast, parasites or bacteria may also be used as part of a cleanse. Formulas are available to assist the body in ridding itself of more specific toxins such as heavy metals.

Diet

Traditionally, a cleanse has involved fasting or at least partially fasting. In fact, fasting for 24 hours consuming only water one day per month is a good practice for general health.

A modified form of fasting is called the juice fast, where a person consumes nothing but fresh, raw vegetable and/or fruit juices for three days. A popular version of this is the master cleanse, promoted by Stanley Burroughs, where a person drinks natural lemonade made with fresh lemons and sweetened with real maple syrup.

However, since most cleansing programs last two weeks or more, it is better to adopt a semi-fasting state during the cleanse. Simply avoid all refined and processed foods during the cleanse and eat lots of fresh fruits and vegetables. This aids the cleansing process by not burdening the body with more cooked, processed foods and chemical additives.

The nice thing about making this a time to clean up your diet is that cleaning out the body tends to reduce your craving for junk food, anyway. So, by the time you have finished the cleansing program it will be easier to maintain a healthy eating program.

Other Helps

Cleansing can also be aided by various forms of hydrotherapy. For instance, enemas and colonics can be helpful as long as they aren't overdone. Generally speaking, enemas and colonics shouldn't be done more than once or twice per week and never for a period longer than a few months without taking a break. Sweat baths, steam baths or saunas are also useful as they encourage detoxification through the skin. These may be done more frequently. Foot soaks or foot spa baths are also helpful to the detoxification process. Again, once or twice a week for a limited period of time is more than enough.

Remember that cleansing takes the good out with the bad, so you should alternate cleansing with a program of rebuilding and good nutrition. When it comes to cleansing more is not better.

Prepackaged Cleansing Programs

Doing a cleanse is easy with a pre-packaged cleansing program. Here are four great cleansing programs. Choose the one that's right for you.

The Tiao He Cleanse

The Tiao He Cleanse is one of the best basic cleansing programs in the marketplace. It is not a harsh cleanse, but is still very effective.

The Tiao He Cleanse gets its name from the Chinese herbal formula Tiao He, sold under the trade name Chinese Liver Balance. Tiao He means to "mediate harmony," referring to the formula's ability to harmonize the function of internal organs by helping to ease congestion in the liver.

Besides the Chinese Liver Balance (Tiao He) formula, the cleanse contains All Cell Detox, a general cleansing formula. LBS II, a stimulant laxative herbal blend, and Psyllium Hulls, a dietary fiber.

Two single herbs, Burdock Root and Black Walnut Hulls ATC Concentrate, complete the program. Burdock root is a traditional blood purifier that has been used to clear up skin conditions and toxic conditions in the blood and lymph, including cancer. Black Walnut has antiparasitic and antimicrobial action.

The Tiao He Cleanse not only helps improve colon transit time and clean out the colon, it also aids liver detoxification and improves kidney and lymphatic drainage. It stimulates digestion and has a mild antiparasitic action.

This 14-day cleanse contains convenient packets you can carry with you in your pocket or purse. It is even more effective if you add more fiber. Take 1-2 heaping teaspoons of a fiber blend like Psyllium Hulls Combination or Everybody's Fiber first thing in the morning along with a large glass of water.

Taking an enzyme formula like Proactazyme Plus between meals will also enhance the action of the cleanse. And, remember to drink plenty of water during the cleanse, at least 1/2 ounce per pound of body weight per day.

Clean Start

A simpler and more basic cleanse is CleanStart. This program contains a fiber packet with psyllium hulls and hydrated bentonite. The packet can be mixed with water or juice and taken before both breakfast and dinner. It comes in two flavors, Apple/Cinnamon and Wild Berry.

CleanStart contains Enviro-Detox, a liver detoxifying formula that helps the body get rid of environmental pollutants, and LBS II, the stimulant laxative formula. Both of these formulas are also in a convenient packet that can be taken at the same time as the fiber.

CleanStart is a two-week cleansing program. It is great for people who are just getting started on the path to improving their health with herbs and supplements. It is not a good cleanse, however, for people who suffer from inflammatory bowel disorders or autoimmune conditions. Again it is important to drink plenty of water.

Dieter's Cleanse

A good cleanse can be a very effective aid in weight loss, Dieter's Cleanse is the perfect one to help you get started on your weight loss program. Besides the LBS II and Enviro-Detox found in CleanStart, Dieter's Cleanse contains Bowel Detox, LIV-A, Chromium and MasterGland.

Bowel Detox is a fiber and enzyme formula that helps cleanse the colon, while LIV-A is a formula for supporting liver function. Chromium helps to balance blood sugar levels and MasterGland helps balance glandular function to support increased metabolism and weight loss.

Dieter's Cleanse is also a conveniently packaged two-week cleansing program. It is best used to kick-start a good weight loss program. Like the Tiao He Cleanse, it also works best when taken with extra fiber, and, of course, plenty of water.

ParaCleanse

A final cleanse to consider is the ParaCleanse. This parasite cleanse contains two antiparasitic herbal formulas, Herbal Pumpkin and Artemesia Combination, along with Paw Paw Cell-Reg and ATC Concentrated Black Walnut Hulls. It is a great cleanse to knock down intestinal bacteria and parasites.

A good way to do this cleanse is to start with one package of the Tiao He Cleanse to remove toxins and improve bowel function. Then, do one package of the Para-Cleanse program. Take a one-week break and then do a second package of Para-Cleanse. This program has helped many people regain their health after exposure to various parasites.

Daily Internal "Housekeeping"

Doing a complete cleansing program is like doing a major clean-up job in your home, washing walls and carpets while throwing away things you no longer need. However, we also do little housekeeping chores every day to keep our homes clean, such as washing dishes, vacuuming or putting away clothes.

Our body also needs to do its daily cleaning for us to stay in good health. We can help our body do its "housekeeping" by taking some supplements that keep the colon and eliminative organs working properly. This prevents the build-up of toxins in the first place. Drinking plenty of water and making sure we get enough fiber are the most important aspects of daily cleansing.

Most people also need a fiber supplement like Psyllium Hulls Combination or Everybody's Fiber. This is best taken in the morning before breakfast with water or juice. Fiber will help to prevent toxins from being absorbed into your body and will help you maintain healthy cholesterol and blood sugar levels. It will even protect you against colon cancer and other chronic diseases.

In addition to the fiber, you may also want to consider taking digestive enzymes such as Proactazyme Plus or Protease Plus with your fiber. Enzymes break down toxins and undigested food to keep your intestinal tract healthy. They also boost your immune system.

If you have problems with regularity, take 2-3 capsules of Gentle Move in the morning with your enzymes and fiber. Gentle Move will help improve your bowel tone when you take it regularly.

Cleansing Therapy #2: *Oral Chelation*

Hardening of the arteries is a precursor to cardiovascular disease, the leading cause of death in the United States. Intravenous chelation is a controversial, but effective therapy, many people have used to help reverse hardening of the arteries. Oral chelation is an alternative to intravenous chelation.

The idea that one can reverse arterial plaque is very controversial. Most medical people think it can't be done. However, there are many people who have experienced dramatic improvement in their circulation (verified by doctors) and major improvements in their health using this procedure.

Oral chelation isn't just for improving circulation. It can help a wide variety of health problems, including helping to detoxify the body from heavy metals.

It is very important to start slowly with this program and work up as instructed. Otherwise, symptoms, such as nausea, dizziness, headaches and skin eruptions, may occur. It is also important to taper off as instructed, or fatigue and temporary nutritional deficiencies may result.

For the first week, take the following with breakfast and dinner:

1 tablet of Mega-Chel

1/2 ounce of Mineral Chi Tonic or Colloidal Minerals

Each week increase the dosage of Mega-Chel by one tablet. Hence, the second week, take 2 tablets of Mega-Chel, twice daily and on the third week take three tablets of Megal-Chel twice daily. Gradually increase the amount of minerals as well, until you are taking 1 ounce in the morning and 1 ounce at night.

A full dose of Mega-Chel is 4-6 tablets twice daily, depending on body weight. Persons over 200-225 pounds should probably take the full 6 tablets twice per day. People who weigh less than 125-150 pounds will probably need only 4 tablets twice daily. Individuals of average height and weight should find 5 tablets two times per day (for a total of 10 per day) sufficient. When you reach full dose, you will be taking the following with breakfast and dinner:

4-6 Mega-Chel Tablets

1 ounce of minerals

You will need to stay on this full dose for a minimum of one month for each ten years of your age. Thus, if you are 40 you need to stay on the full dose for at least four months, six months if you are 60, etc. If you have serious problems, you should consider staying on the program for one and half months for every ten years of your life. That is, six months if you are 40 and nine months if you are 60 and so forth.

It is important to taper off in a similar manner to building up. On the full program you are taking very large doses of certain vitamins and minerals, and the body gets lazy about extracting them from food. So, if you quit all at once, your body may experience a sudden drop in nutrient levels until it readjusts to absorbing these vitamins and minerals from food.

Taper off by reducing the amount you take by two tablets each week. So, if you were taking 5 tablets twice daily, then take 4 tablets twice daily for a week. The next week drop it to 3 tablets twice daily and so forth until you reach 1 tablet twice daily. After that, you can either discontinue the program entirely, or stay on a maintenance dose of 1 tablet twice daily.

When using Mega-Chel, a multivitamin and mineral is not necessary as the Mega-Chel takes the place of the multivitamin. Many elderly people find that they do better using Mega-Chel as their multi.

As the body removes the plaque from the walls of the arteries, the cholesterol level in the blood will temporarily rise. This is normal. The kidneys and liver will remove the calcium, cholesterol and other impurities from the body. If there are indications that these organs are weak it may be necessary to give them extra support as follows:

For persons with kidney weakness (history of symptoms like arthritis, chronic back pain, urinary infections, etc.) take 2 KB-C with each meal, or put one-half teaspoon each of Lymphatic Drainage and Kidney Drainage into a quart of water and sip this throughout the day.

For persons with liver weakness (history of high cholesterol, skin problems, digestive upset, etc.) take one heaping teaspoon of LOCLO or Nature's Three in a large glass of water or juice upon arising and before retiring. Also take 2 Chinese Liver Balance with each meal.

You may also wish to add some of the following supplements for special problems. These are suggested full doses. You can work up gradually on taking these supplements, as well.

For heart problems take 2 HS II or 2 Hawthorn Berries with each meal. Also take 2-6 capsules of CoQ10 75 daily.

For senility, Alzheimer's or other problems with memory, take 2 Ginkgo/Hawthorn or 2 Ginkgo/Gotu Kola with each meal.

For varicose veins and high risk of stroke take 2 Butcher's Broom with each meal or 1 Vari-Gone twice daily.

For heavy metal detoxification take 1 Heavy Metal Detox and 2-3 Algin twice daily.

For dissolving calcium deposits or calcifications, take 2 hydrangea and 2 Magnesium Complex twice daily.

For high blood pressure consider using one scoop of RG-Max daily or taking 1 Blood Pressurex three times daily.

Cleansing Therapy #3: *Heavy Metal Detoxification*

Because the body is continually exposed to small amounts of heavy metals and other toxins, even in natural foods, it has defensive mechanisms eliminate these substances from our body. By nutritionally supporting these mechanisms, while keeping the body's channels of elimination open, one can help the body remove excess heavy metals from the system.

One of the principal detoxifying agents in the body is a substance called glutathione. It is an antioxidant, produced in the liver from three amino acids: cysteine, glutamic acid and glycine. Glutathione helps cells eliminate drugs and heavy metals and protects the body from damage from smoking, radiation and alcohol.

N-Acetyl Cysteine (NAC) contains the amino acid cysteine, which is a building block for glutathione, a powerful antioxidant that protects tissues including the liver, respiratory and immune systems and the eyes. By enhancing glutathione, NAC protects healthy cells from damage by heavy metals and other toxic chemicals.

Another nutrient that helps the body detox from heavy metals is alpha lipoic acid, a powerful antioxidant. Because it is soluble in both water and fat, it has an especially wide range of protective actions. Even more, it enhances the function of other antioxidants like vitamin C, vitamin E and glutathione. It also helps increase energy production in the cells.

Sodium alginate or algin is a mucilage derived from kelp. Kelp is a purifier of the oceans because the alginate in it bonds to heavy metals and other toxins to neutralize them. Sodium alginate binds to heavy metals such as lead and mercury in the intestinal tract and carries them out of body with regular bowel movements. Other fiber products such as apple pectin can also help bind heavy metals.

The Heavy Metal Detox formula contains all these substances—N-Aceytl Cysteine, alpha lipoic acid and sodium alginate. It also contains l-methionine and cilantro, an herb reported to help the body detoxify from mercury. This formula is a potent heavy metal cleanser. Do not exceed 1 capsule with a meal 2 times a day, at least in the beginning. If you develop a strong cleansing reaction (rash, diarrhea, nausea, dizziness, weakness, etc.), stop taking it for a couple of days. Take some fiber and water to clear the system and then restart with only 1 capsule per day.

Many heavy metals (and other environmental toxins) are not water-soluble. This means the body must remove them by binding them to fats. Thus, essential fatty acids are very helpful in detoxifying from heavy metals. Daily supplementation with 1-2 Tablespoons of flax seed oil or Super Omega-3 EPA will help bind these toxins in the system for removal.

Once the heavy metals, pesticides or other chemicals have been bound to fats, the toxic fats are taken to the liver where they are eliminated by dumping them into the small intestines through the bile. Fiber is needed to bind these fats and heavy metals so they will be carried out of the body. While any kind of fiber helps, a particular form of mucilage known as sodium alginate is especially good at binding heavy metals.

If you know or suspect you have heavy metal poisoning, it's probably a good idea to work with an experienced doctor, naturopath or herbalist to custom design a program for your individual needs. However, as a starting point, here's a mercury and heavy metal detox program I've used on myself and others:

- 1 Tablespoon of flax seed oil twice daily
- 1 Heavy Metal Detox twice daily
- 2-4 Algin three times daily or 1 Tablespoon Nature's Three in a glass of water or juice twice daily
- Once or twice each week take a drawing bath with Hydrated Bentonite or another fine clay or take a foot spa bath

Make certain the bowels are moving at least two to three times per day. If not, you may wish to take some LBS II at bedtime or 2-3 Gentle Move twice daily

- Optional: For a stronger effect add 1-2 MegaChel twice daily

Heavy metal detoxification is important for anyone who has worked around a lot of chemicals in their job (including painters beauticians, lab technicians, dry cleaners, carpet cleaners, farmers and factory workers in many industries). It's also a good thing for people suffering from any kind of chronic inflammatory disorder or problem that involves nerve damage.

Cleansing Therapy #4: *Gall Bladder Flush*

The gall bladder flush is a natural procedure that has been used to ease gall bladder attacks and potentially pass gall stones. It has been around for years and many people have had good success in using it. How it works isn't totally clear, but one explanation is that the large amount of olive oil ingested on the gall bladder flush sends the gall bladder into spasms which eject small stones and may also clear bile ducts.

The procedure is controversial and there is a slight risk, which is that a large stone may get stuck in the bile ducts resulting in the need for surgery. However, in nearly 30 years of experience I have only had one report of this happening. Besides, surgery is the standard treatment for this condition and surgery carries a much higher risk than this cleanse, which makes me think that it is worth trying first. If it fails, then go for the surgery.

Here's the standard way to do a gall bladder flush. Start by fasting for 24 to 48 hours on fresh, raw apple juice or fresh squeezed grapefruit juice to clear the colon. Malic acid, an ingredient in the apple juice, is reported to soften the stones, but persons with hypoglycemia or yeast infections will do better on grapefruit juice. If using grapefruit juice take Fibralgia, which contains malic acid and magnesium, for a similar effect.

Just before going to bed at the close of the fast, drink 1/2 cup of olive oil and 1/2 cup of lemon (or grapefruit) juice. Mix these together thoroughly like you would shake up a salad dressing. The lemon or grapefruit juice cuts the olive oil and makes it more palatable. It sounds and smells worse than it tastes. Next, lie on your right side for a half hour before going to sleep. In the morning, if you don't have a bowel movement, take an enema. This procedure may need to be repeated 2 days in a row.

Generally, you will pass some dark black or green objects that look like shriveled peas the day after drinking the olive oil and lemon juice. These objects are not gallstones. Gallstones that can be passed are much smaller than this, generally less than 2 millimeters in diameter. Chemical analysis of these objects shows they are composed of soap, and are created by the bile interacting with the oil.

The controversy of this procedure is whether the stones actually pass, or it just eases the pain and discomfort of the gall bladder attack and allows the problem to become asymptomatic again. Most people with gallstones don't know they have them, because they cause no symptoms. Whether stones are passing or not, this procedure typically eases gall bladder pain and allows the person to resume a normal life without surgery.

There are a number of versions of this procedure, but they all rely on olive oil. This may be because olive oil acts as a solvent of cholesterol, the chief constituent of most gall stones. Some gall stones, however, are calcium based.

One variation that seems to work particularly well is to take a dose of Epsom salt about two or three hours prior to taking the olive oil and lemon juice. Follow the directions on the box of Epsom salts as per the dosage.

Certain herbs may also enhance the procedure. Herbs called cholagogues increase the flow of bile and help to dissolve stones slowly over a period of weeks and months. Herbs that have this property include dandelion root, artichoke, barberry bark, yellow dock root, fringetree bark, turmeric and celandine. Any of these can be taken before attempting the gallbladder flush to increase its effectiveness, or afterwards to continue improving gallbladder function.

In fact, taking these herbs regularly for a year or more may even help to dissolve larger stones. A combination of artichoke, barberry, fringetree and turmeric is especially good at helping the gallbladder to heal and potentially removing stones. Unfortunately, no such combination exists in the NSP product line at the time this book was written.

If gall stones are calcium-based, then hydrangea and magnesium will be helpful in dissolving them. Take these herbs during the fast and for several months after the gall bladder flush.

If the procedure does not bring relief, medical help should be sought.

Cleansing Therapy #5: *Avoid Xenoestrogens*

Over 50 years ago it was discovered that chemicals in our environment were having a negative impact on the reproductive capability of wild animals. In spite of this, our society has continued to accept and use these chemicals because they offer "quick fixes" in modern agriculture. Some of these chemicals have now been dubbed as xenoestrogens.

The term estrogen does not refer to a specific hormone. An estrogen is any natural or artificial substance that induces estrus (female fertility and desire to mate). The human body makes three different estrogens—estriol, estrone, and estradiol. Xenoestrogens are chemical compounds from environmental pollutants that bond to estrogen receptor sites. Xeno is a Greek word meaning foreigner, stranger or alien. So xenoestrogens are foreign or alien estrogens.

Xenoestrogens bond to receptor sites within cells to make specific changes to cellular activity. These chemical estrogens can disrupt the function of the endocrine system in two ways. First, they can mimic natural hormones and turn on cellular processes at the wrong time or simply overstimulate them. A second way they can disrupt the body's hormonal processes is to bond to receptor sites without stimulating them, blocking normal hormonal processes.

The results of this bonding can be cellular damage, the inappropriate activation of genes, or the disruption of normal hormonal processes.

Although these chemicals have been "tested" for safety, they have all been tested individually, not collectively. One experiment showed that when 10 commonly encountered chemicals were mixed at a tenth of their individually active dose, the potency (measured as cell proliferation) was 10 times higher than expected. So, the synergistic effect of these chemicals is dangerous.

Furthermore, they do not readily degrade nor break down in the environment. In fact, they tend to accumulate in the fatty tissues of animals and concentrate the higher up the food chain you go. This may be one of the reasons that vegetarians have a lower incidence of breast cancer than meat eaters.

Xenoestrogens have been documented as causes in reproductive dysfunction and mutations in wild birds, frogs, reptiles and even mammals. However, the first species of animals to be affected were birds of prey, because they sit at the top of the food chain. The problems these chemicals have caused in wild animals should have clued us into the harm they are causing human beings, but commercial interests have continued to push for their use. Some of the possible effects these xenoestrogens are having on human beings include:

1. Earlier onset of puberty in young girls.
2. Increases in breast and prostate cancer. These tissues contain estrogen receptor sites and are extremely prone to genetic damage and the stimulation of excess growth by xenoestrogens. Other cancers of the reproductive organs may also be caused by xenoestrogens.
3. Uterine fibroids and other reproductive disorders in women. By overstimulating uterine tissue, excessive tissue growth is encouraged.
4. A world-wide decrease in male fertility.

Some of the chemicals that appear to have serious reproductive and endocrine disruptive effects include:

Pesticides (such as 2,4-D, DDT and many others)

Organochlorides (dioxin, PPBs, PCBs and others)

Heavy metals (cadmium, lead and mercury)

Plastic ingredients (particularly soft plastics)

Hormones fed to chickens and cows to increase egg and milk production

Both men and women need to become keenly aware of xenoestrogens, avoiding them as much as they possibly can. Organic fruits and vegetables should be purchased whenever available and commercial produce should be washed in Sunshine Concentrate or another natural soap to remove pesticide residues. Use only organic meat, dairy and eggs. Use glass or paper cartons instead of plastic containers where possible. Do not microwave food in plastic containers or put hot food in plastic containers. Avoid chemicals in general wherever possible.

Another strategy to minimize exposure to xenoestrogens is to use natural plant-based estrogens to tie up estrogen receptor sites. Phytoestrogens are chemicals in plants that also bond to estrogen receptor sites. However, phytoestrogens have a much weaker estrogenic effect than natural estrogens or xenoestrogens. The theory is that by consuming foods rich in phytoestrogens, receptor sites will be tied up, resulting in less estrogen stimulation.

Soy products and other legumes (beans and peas) are rich in phytoestrogens. Other good sources include dark green vegetables and whole grains. Herbal sources include Phyto-Soy, red clover, licorice, black cohosh and hops. The formula Breast Assured is a concentrated source of these phytoestrogens.

Because our exposure to chemicals is so high in modern society, it is also wise to cleanse the body periodically. All Cell Detox, Lymphatic Drainage, and Enviro-Detox are also good formulas for assisting the body in eliminating these chemicals.

The liver detoxifies excess estrogens and a variety of supplements can be used to enhance liver detoxification of estrogens. Indole-3 Carbinol, Liver Balance and Milk Thistle Combination are useful supplements to help detoxify xenoestrogens.

D = Direct Aids

In addition to doing a cleanse and taking some of the basic supplements listed under Building, you may also wish to add one or two Direct Aids for a specific body system or organ you want to support, but keep your starting program fairly simple. Generally, start with three to five carefully chosen products. (This includes some of your basic building supplements.) Read the description of each product you plan to take and be sure to note any warnings or contraindications that are listed.

To determine what direct aids you need, start by filling out a copy of the Body Systems Questionnaire, found on pages 26 and 27. You are free to make as many copies of this form as you like provided you do not alter it in anyway. A copy of this form is also available in spreadsheet form on our website—www.treelite.com.

After totaling the Body Systems Questionnaire determine which system has the highest score. Select the Body System's Chart for that system and see which set of indications most closely matches the symptoms you or the person you are working with is experiencing. Some of the indications involve iridology and tongue and pulse analysis. If you are not familiar with these modalities simply ignore them. If you would like to learn more about them, Tree of Life Publishing offer instructional programs that will help you acquire these skills. For more information, visit our website at www.treelite.com.

Most of these charts are based on the six imbalances in biological terrain. There are a few systems that didn't fit this model and are organized slightly differently. The nervous and glandular charts, for instance, are organized around only two of the six imbalances—overactive or irritated, and underactive or depressed.

The Body Systems Questionnaire helps you determine which system you want to give direct aid to, while the Body System's charts help you determine what tissue condition you are dealing with. Once you have found the body system and tissue condition, you will find a list of remedies for that system and that biological terrain imbalance on the back of the chart.

If you are familiar with muscle testing, you can muscle test some or all of the products listed to determine which is best for you or the person you are trying to help. If not, you can select whichever product you feel might be the best for them. If you are interested in learning about muscle testing, Tree of Light Publishing also provide educational materials on this topic to help you. Many NSP Managers can also teach you this skill.

As you find a basic program that works, you can improve it by adding more direct aids for problems that still need help. You can look up specific health problems and read about some of the possible root causes and view lists of supplements people have used for those health problems in the *Conditions* Section. You can also look up various therapeutic actions of plants and supplements in the *Properties* Section. Finally, you can discover supplements that have affinity for specific organs or body parts in the *Systems* Section.

Always read the complete description of each product before you take it. Where there are multiple dosage forms for a particular herb, formula or supplement, they are all listed in the Products Section, along with stock numbers and the country they are available in, for convenience in ordering them.

Also, keep in mind that, generally speaking, it is a good idea to keep your program to less than ten products. Taking more things isn't always better and in many cases is less effective than a smaller number of carefully chosen supplements.

Summary

Remember that the body does not heal selectively. As we are able to create the correct environment for healing, there is a cascade effect—as overall health improves all of a person's symptoms start to improve. So, it is not necessary to understand every disease that can possibly afflict a person to understand how to promote healing. It is only necessary to understand how to promote health, or the normal and natural balance that exists in the body. This means we need to study health, not disease, so we can recognize the healthy state of the body and see how the body has deviated from that healthy state.

So, don't be tempted to skip the basic ABC+D principles taught in this introductory section and skip directly to the *Conditions* section. And, don't focus on figuring out what to do for a specific disease. It simply isn't necessary. The ABC+D principles are the foundation of health and are universally applicable to everyone, no matter what disease(s) they may have had diagnosed.

Every problem cannot be helped by herbs or supplements. Sometimes medical intervention is necessary. We have noted throughout the text situations where medical attention is warranted. Please heed these suggestions and seek appropriate professional help when it is necessary. And always seek professional help for serious or persistent health problems.

May your future be filled with health and happiness.

Body Systems Questionnaire

Client Name ____ **Date** __________________

Directions: If you have problems with any of the symptoms listed on the left hand side of the table, then circle all the numbers on the row to the right of that symptom. When you have finished reviewing the list of symptoms, total all the numbers you have circled in each column. There are spaces provided at the bottom of each page to total the columns on that page and a space on page two for the Grand Totals. The higher the number, the more likely it is that the body system associated with that column needs nutritional support.

Abdominal Pain or Discomfort	1	1	1								
Absent-mindedness or forgetfulness						2	2	1			
Acid indigestion or heartburn	2										
Anxiety, nervousness or tension	1						2	1			
Asthma	1	1		2							
Bad breath or body odor			1		1						
Brittle fingernails									2		1
Burning or painful urination					2						
Cold hands and feet						1		1			
Colitis or other bowel irritations		1	2							1	
Congested air passages			1	2						1	
Constipation or dry stools			2				1				
Cravings for fat or high fat diet		1				2					
Cravings for sugar	1							2			
Dark circles or puffiness under eyes		1			1			1			
Difficulty getting to sleep		1					2	1			
Dizziness or light headedness.						1		1			
Dry skin								1	1		
Excess mucus production			1	2							
Family history of heart disease						2					
Fatigue in the afternoons								1			
Fatigue or low energy levels		1	1			1	1	1		2	
Food allergies	1	1								2	
Food sits heavy on stomach after eating	2										
Frequent backache					2				2		
Frequent cough				2						1	
Frequent infections				1						2	
Frequent urinary tract infections					2						
General weakness or chronic illness	2									2	1
Hayfever		1	1	2						2	
Heart problems						2					
High blood pressure					1	2	2				
High cholesterol		1				2					
Impotency (males only)											2
TOTALS FOR SIDE ONE											

Body Systems	Digestive	Hepatic	Intestinal	Respiratory	Urinary	Circulation	Nerves	Glandular	Structural	Immune	Reproductive
Infertility											2
Intestinal gas or bloating	2	1	1								
Itchy nose and ears				1						1	
Joint pain, arthritis or gout					1				2		
Leg cramps or pains					1				2		
Less than 1 bowel elimination per day		1	2								
Loose stool or diarrhea		1	2								
Loss of appetite or poor appetite	2						1	1			
Loss of sexual desire											2
Menopause Problems (Females only)											3
Menstrual problems (females only)							1				3
Mental/emotional stress							2	2			1
Migraine headaches		2				1	2				
Muddled thinking, confusion or mental sluggishness			1				1	2			1
Osteoporosis					1				2		2
Pale complexion and/or anemia	1					1				1	
Prostate problems (males only)											3
Restless dreams or nightmares			1				1	1			1
Scant or excessive urination					2						
Sinus congestion			1	2						1	
Sinus headaches			1	2							
Skin problems (acne, rashes, etc.)		2			1				2	1	2
Stiff, aching or painful muscles		1	1		1				2	1	
Swollen lymph glands		1		2						2	
Ulcers	2										
Underweight or unable to gain weight	2							1			
Urinating at night					1		1	1			
Varicose veins		1				2			1		
Waking up frequently at night							1	1			
Water retention or edema					2						
Weak legs, knees or ankles					1				2		1
Wheezing or shortness of breath				2							
Wounds won't heal in extremities						1			2		
TOTALS FOR SIDE TWO											
TOTALS FOR SIDE ONE											
Grand Totals											

Prepared by The Tree of Light Publishing, P.O. Box 911239, St. George, UT 84791, 1-800-416-2887. (www.treelite.com)

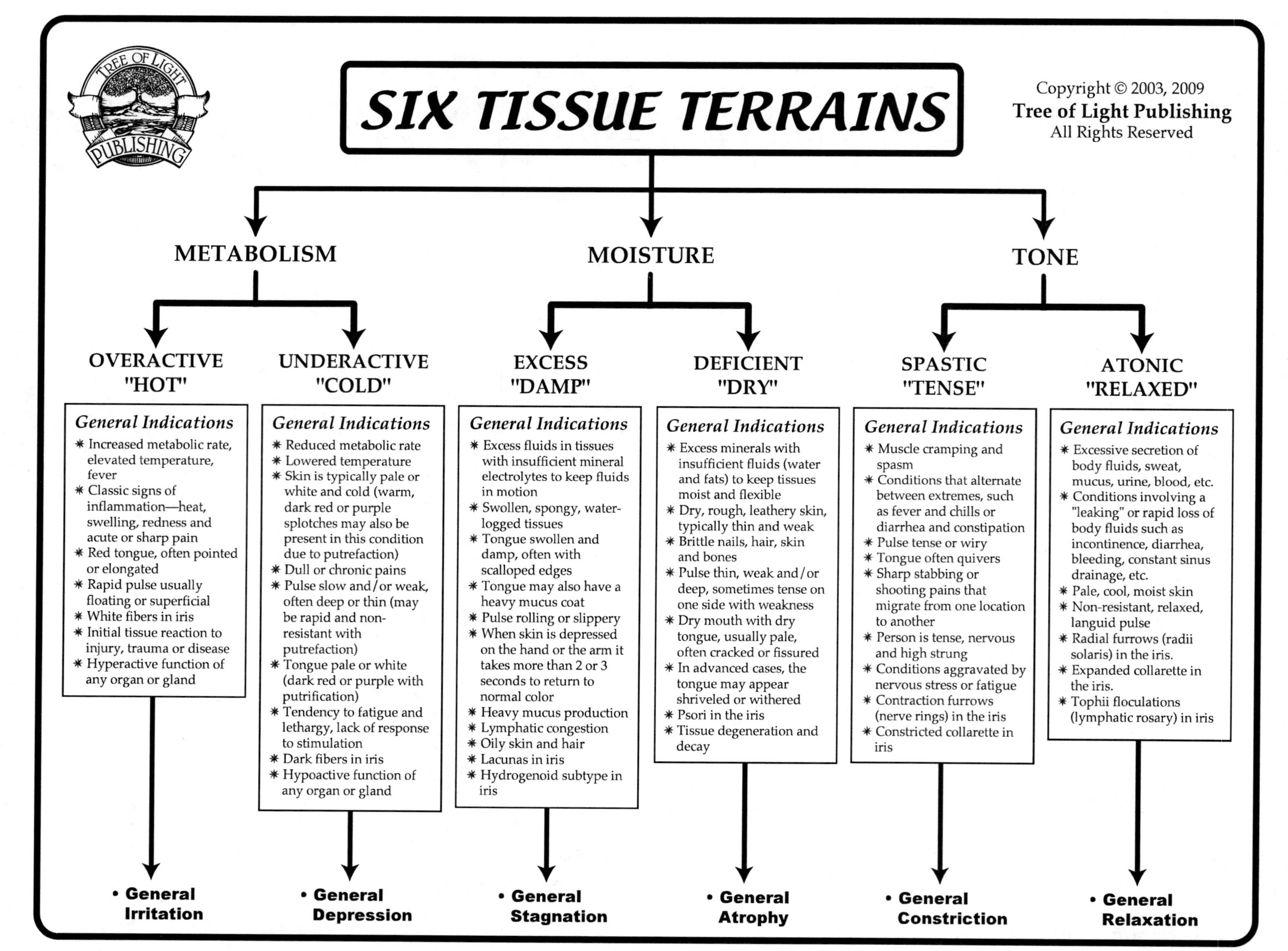

SIX TISSUE TERRAINS
TREE OF LIGHT PUBLISHING
Copyright © 2003, 2009
Tree of Light Publishing
All Rights Reserved
METABOLISM
MOISTURE
TONE
OVERACTIVE "HOT"
General Indications
* Increased metabolic rate, elevated temperature, fever
* Classic signs of inflammation—heat, swelling, redness and acute or sharp pain
* Red tongue, often pointed or elongated
* Rapid pulse usually floating or superficial
* White fibers in iris
* Initial tissue reaction to injury, trauma or disease
* Hyperactive function of any organ or gland
• General Irritation
UNDERACTIVE "COLD"
General Indications
* Reduced metabolic rate
* Lowered temperature
* Skin is typically pale or white and cold (warm, dark red or purple splotches may also be present in this condition due to putrefaction)
* Dull or chronic pains
* Pulse slow and/or weak, often deep or thin (may be rapid and non-resistant with putrefaction)
* Tongue pale or white (dark red or purple with putrification)
* Tendency to fatigue and lethargy, lack of response to stimulation
* Dark fibers in iris
* Hypoactive function of any organ or gland
• General Depression
EXCESS "DAMP"
General Indications
* Excess fluids in tissues with insufficient mineral electrolytes to keep fluids in motion
* Swollen, spongy, water-logged tissues
* Tongue swollen and damp, often with scalloped edges
* Tongue may also have a heavy mucus coat
* Pulse rolling or slippery
* When skin is depressed on the hand or the arm it takes more than 2 or 3 seconds to return to normal color
* Heavy mucus production
* Lymphatic congestion
* Oily skin and hair
* Lacunas in iris
* Hydrogenoid subtype in iris
• General Stagnation
DEFICIENT "DRY"
General Indications
* Excess minerals with insufficient fluids (water and fats) to keep tissues moist and flexible
* Dry, rough, leathery skin, typically thin and weak
* Brittle nails, hair, skin and bones
* Pulse thin, weak and/or deep, sometimes tense on one side with weakness
* Dry mouth with dry tongue, usually pale, often cracked or fissured
* In advanced cases, the tongue may appear shriveled or withered
* Psori in the iris
* Tissue degeneration and decay
• General Atrophy
SPASTIC "TENSE"
General Indications
* Muscle cramping and spasm
* Conditions that alternate between extremes, such as fever and chills or diarrhea and constipation
* Pulse tense or wiry
* Tongue often quivers
* Sharp stabbing or shooting pains that migrate from one location to another
* Person is tense, nervous and high strung
* Conditions aggravated by nervous stress or fatigue
* Contraction furrows (nerve rings) in the iris
* Constricted collarette in iris
• General Constriction
ATONIC "RELAXED"
General Indications
* Excessive secretion of body fluids, sweat, mucus, urine, blood, etc.
* Conditions involving a "leaking" or rapid loss of body fluids such as incontinence, diarrhea, bleeding, constant sinus drainage, etc.
* Pale, cool, moist skin
* Non-resistant, relaxed, languid pulse
* Radial furrows (radii solaris) in the iris.
* Expanded collarette in the iris.
* Tophii floculations (lymphatic rosary) in iris
• General Relaxation

General Irritation

Herbs

- ✗ Acai berry
- Bilberry
- ✗ Cranberry
- ✓ ✗ Elderberry
- ✗ Elderflower
- Hawthorn
- ✓ ✗ Lemon
- ✗ Lycium (wolfberry)
- ✓ ✗ Mangosteen
- Rose Hips
- ✗ Willow Bark
- Yarrow
- Yellow Dock

Supplements

- Grapine
- Co-Q 10
- Vitamin C

Combinations

- APS II with White Willow
- CC-A with Yerba Santa
- Bone/Skin Poultice
- Elderberry Defense
- EW
- I-X
- ✓ IF Relief
- ✓ IF-C
- ✓ Super ORAC
- ✓ Thai-Go

Topical Remedies

- ✓ Healing AC Cream
- ✓ Nature's Fresh
- Tei Fu Oil

General Depression

Herbs

- ✓ Capsicum
- Chamomile
- ✓ Garlic
- ✓ Ginger
- ✗ Horseradish
- Juniper Berries
- Oregano EO
- Peppermint herb and EO
- Rosemary EO
- Safflowers
- Sage herb and EO
- Thyme herb and EO

Supplements

- B-Complex
- Magnesium
- Niacin

Combinations

- Artemesia Combination
- ✓ Capsicum & Garlic with Parsley
- ✓ Cellular Energy
- Fenugreek and Thyme
- GastroHealth
- GC-X
- HCP-X
- HS II
- Papaya Mint
- ✓ Target Endurance

Topical Remedies

- ✓ Deep Relief
- Herbal Trim Skin Treatment
- ✓ Tei Fu Oil

General Stagnation

Herbs

- Alfalfa
- Black Walnut
- ✓ Burdock
- Chickweed
- Dandelion
- ✓ Echinacea
- Goldenseal
- ✗ Myrrh Gum
- ✓ Red Clover
- ✓ Yellow Dock
- Yucca

Supplements

- Chlorophyll

Combinations

- ALJ
- ✓ All Cell Detox
- ✓ Enviro-Detox
- IN-X
- ✓ Kidney Activator (Chinese)
- ✓ Kidney Drainage
- LB-X
- LBS II
- LIV-J
- ✓ Liver Balance
- Lymph Gland Cleanse
- Lymph Gland Cleanse - HY
- ✓ Lymphatic Drainage Formula
- Sinus Support
- Skin Detox

Topical Remedies

- Black Ointment

General Atrophy

Herbs

- Aloe Vera
- ✓ Cordyceps
- Dulse
- ✓ Ginseng, Korean
- ✓ Ginseng, Wild American
- Ho Shou Wu
- Licorice Root
- ✓ Marshmallow
- Mullein
- Saw Palmetto
- ✓ Slippery Elm
- Wild Yam

Supplements

- Collatrim
- Evening Primrose Oil
- Flax Seed Oil
- ✓ Omega-3
- Super GLA
- ✓ Vitamin B12
- Vitamin E

Combinations

- Blood Build
- Bone/Skin Poultice
- CA, Herbal
- HSN-W
- ✓ Intestinal Soothe and Build
- ✓ KB-C
- LOCLO
- ✓ Lung Support
- ✓ Mineral Chi Tonic
- PLS II
- ✓ Spleen Activator
- ✓ Trigger Immune
- V-X

Topical Remedies

- Herbal Trim Skin Treatment

General Constriction

Herbs

- ✓ ✗ Agrimony
- Black Cohosh
- Blue Vervain
- Chamomile
- ✓ Kava Kava
- ✓ Lavender
- ✓ Lobelia
- Valerian
- ✓ Wild Yam

Supplements

- B-Complex
- ✓ Magnesium

Combinations

- Adrenal Support
- Breathe EZ
- Cellular Energy
- CLT-X
- ✓ Cramp Relief
- Fibralgia
- ✓ Gall Bladder Formula
- Gentle Move
- Nutri-Calm
- Stress Relief
- Stress-J
- Target Endurance

General Relaxation

Herbs

- ✓ Bayberry
- Black Walnut
- Eyebright
- Green Tea
- Horsetail
- Red Raspberry
- Rose Hips
- Sage
- Uva Ursi
- ✓ White Oak Bark
- ✓ Yarrow

Supplements

- Calcium
- Charcoal, Activated
- Collatrim
- ✓ Co-Q 10
- Grapine
- ✓ Vitamin C

Combinations

- CA, Herbal
- Carotenoid Blend
- ✓ EW
- ✓ HCP-X
- Hista-Block
- ✓ HSN-W
- Kudzu/St. John's wort
- ✓ Menstrual Reg
- Skeletal Strength
- Super Antioxidant
- Vari-Gone

Topical Remedies

- Vari-Gone Cream

Important Notice: This chart is a reference tool for the ABC+D approach to nutritional support of each body system. It does not replace the services of qualified practitioners. For serious issues and emergencies, please seek the advice of a certified or licensed health professional.

Key: ✓-Top choices ✗-Not available as a single through NSP
EO-Essential Oil (topical use) ❊-Toxic botanical for professional use only

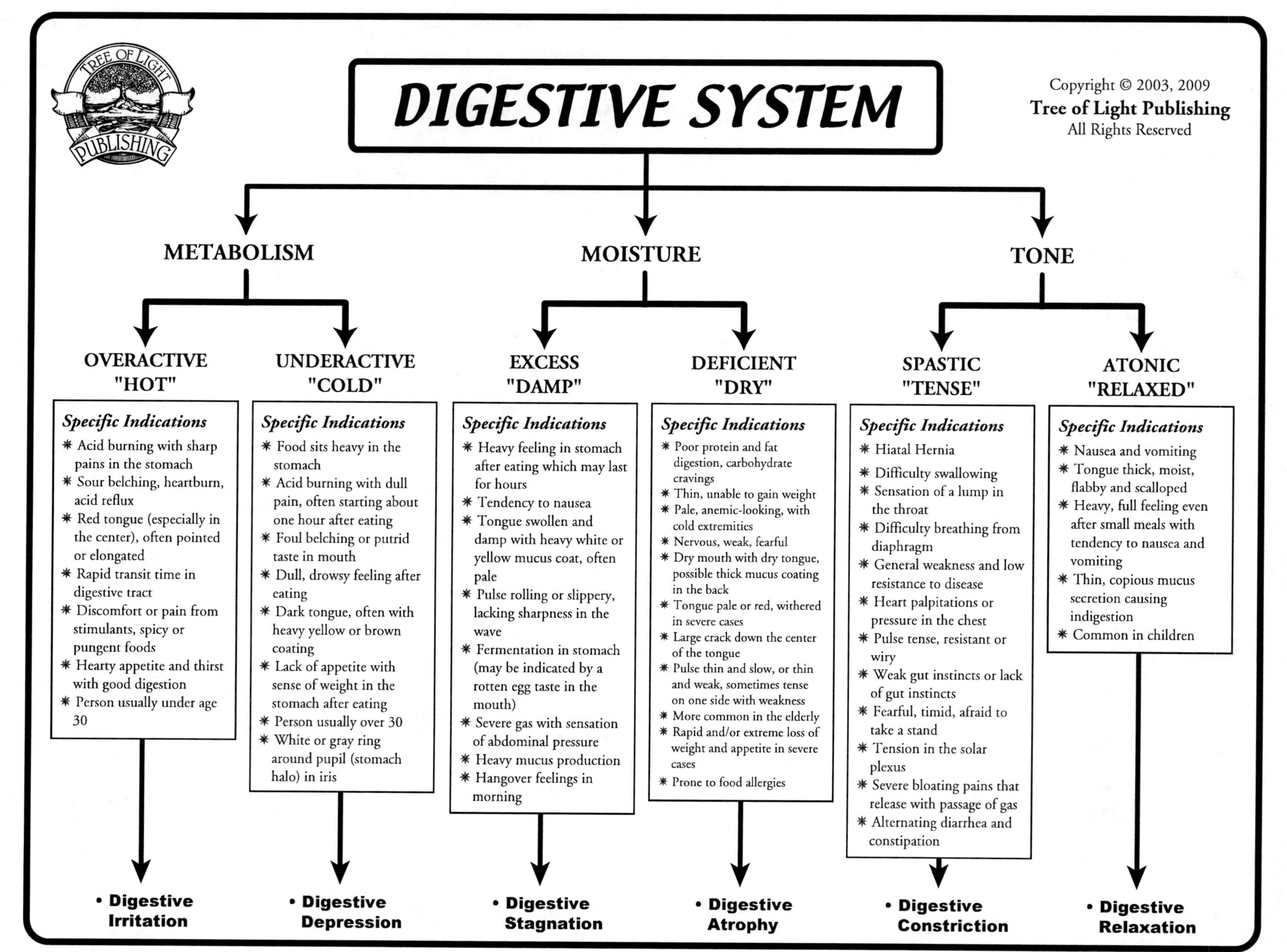
TREE OF LIGHT PUBLISHING
DIGESTIVE SYSTEM
Copyright © 2003, 2009
Tree of Light Publishing
All Rights Reserved
METABOLISM
MOISTURE
TONE
OVERACTIVE "HOT"
Specific Indications
✱ Acid burning with sharp pains in the stomach
✱ Sour belching, heartburn, acid reflux
✱ Red tongue (especially in the center), often pointed or elongated
✱ Rapid transit time in digestive tract
✱ Discomfort or pain from stimulants, spicy or pungent foods
✱ Hearty appetite and thirst with good digestion
✱ Person usually under age 30
• Digestive Irritation
UNDERACTIVE "COLD"
Specific Indications
✱ Food sits heavy in the stomach
✱ Acid burning with dull pain, often starting about one hour after eating
✱ Foul belching or putrid taste in mouth
✱ Dull, drowsy feeling after eating
✱ Dark tongue, often with heavy yellow or brown coating
✱ Lack of appetite with sense of weight in the stomach after eating
✱ Person usually over 30
✱ White or gray ring around pupil (stomach halo) in iris
• Digestive Depression
EXCESS "DAMP"
Specific Indications
✱ Heavy feeling in stomach after eating which may last for hours
✱ Tendency to nausea
✱ Tongue swollen and damp with heavy white or yellow mucus coat, often pale
✱ Pulse rolling or slippery, lacking sharpness in the wave
✱ Fermentation in stomach (may be indicated by a rotten egg taste in the mouth)
✱ Severe gas with sensation of abdominal pressure
✱ Heavy mucus production
✱ Hangover feelings in morning
• Digestive Stagnation
DEFICIENT "DRY"
Specific Indications
✱ Poor protein and fat digestion, carbohydrate cravings
✱ Thin, unable to gain weight
✱ Pale, anemic-looking, with cold extremities
✱ Nervous, weak, fearful
✱ Dry mouth with dry tongue, possible thick mucus coating in the back
✱ Tongue pale or red, withered in severe cases
✱ Large crack down the center of the tongue
✱ Pulse thin and slow, or thin and weak, sometimes tense on one side with weakness
✱ More common in the elderly
✱ Rapid and/or extreme loss of weight and appetite in severe cases
✱ Prone to food allergies
• Digestive Atrophy
SPASTIC "TENSE"
Specific Indications
✱ Hiatal Hernia
✱ Difficulty swallowing
✱ Sensation of a lump in the throat
✱ Difficulty breathing from diaphragm
✱ General weakness and low resistance to disease
✱ Heart palpitations or pressure in the chest
✱ Pulse tense, resistant or wiry
✱ Weak gut instincts or lack of gut instincts
✱ Fearful, timid, afraid to take a stand
✱ Tension in the solar plexus
✱ Severe bloating pains that release with passage of gas
✱ Alternating diarrhea and constipation
• Digestive Constriction
ATONIC "RELAXED"
Specific Indications
✱ Nausea and vomiting
✱ Tongue thick, moist, flabby and scalloped
✱ Heavy, full feeling even after small meals with tendency to nausea and vomiting
✱ Thin, copious mucus secretion causing indigestion
✱ Common in children
• Digestive Relaxation

Digestive Irritation

Herbs

✓ Catnip
Chamomile
✗ Fennel
✓ ✗ Meadowsweet
Sage
Slippery Elm
Yarrow

Supplements

Calcium

Combinations

Catnip & Fennel
✓ Intestinal Soothe and Build
✓ Stomach Comfort

Digestive Depression

Herbs

Capsicum
Chamomile
Dong Quai
✓ Ginger
✗ Myrrh Gum
✓ Peppermint EO
✓ Safflowers

Supplements

✓ Food Enzymes
Hi-Lipase
PDA Combination
✓ Proactazyme Plus
Protease Plus
Protease, High Potency

Combinations

Anti-Gas Formula with Lobelia
✓ GastroHealth
Papaya Mint
✓ Spleen Activator

Digestive Stagnation

Herbs

Dandelion
✓ ✗ Gentian
✓ Goldenseal
✗ Myrrh Gum
✗ Orange Peel

Combinations

✓ Anti-Gas, Chinese
✓ Digestive Bitters
FV
HCP-X

Digestive Atrophy

Herbs

✓ Aloe Vera
✓ Ginseng, American
✓ Licorice ATC Concentrate
Marshmallow
Saw Palmetto
Slippery Elm
St. John's Wort
Wild Yam

Combinations

GastroHealth
✓ Intestinal Soothe and Build
Kudzu/St. John's Wort
✓ Spleen Activator
Trigger Immune

Digestive Constriction

Herbs

Blue Vervain
Catnip
Chamomile
Dandelion
✓ Lobelia
St. John's Wort

Combinations

Gall Bladder Formula
✓ CLT-X
Intestinal Soothe and Build
✓ Spleen Activator
Menstrual-Reg

Digestive Relaxation

Herbs

✓ Bayberry
Red Raspberry
Uva Ursi
White Oak bark

Combinations

✓ HCP-X

Important Notice: This chart is a reference tool for the ABC+D approach to nutritional support of each body system. It does not replace the services of qualified practitioners. For serious issues and emergencies, please seek the advice of a certified or licensed health professional.

Key: ✓-Top choices ✗-Not available as a single through NSP
EO-Essential Oil (topical use) ✻-Toxic botanical for professional use only

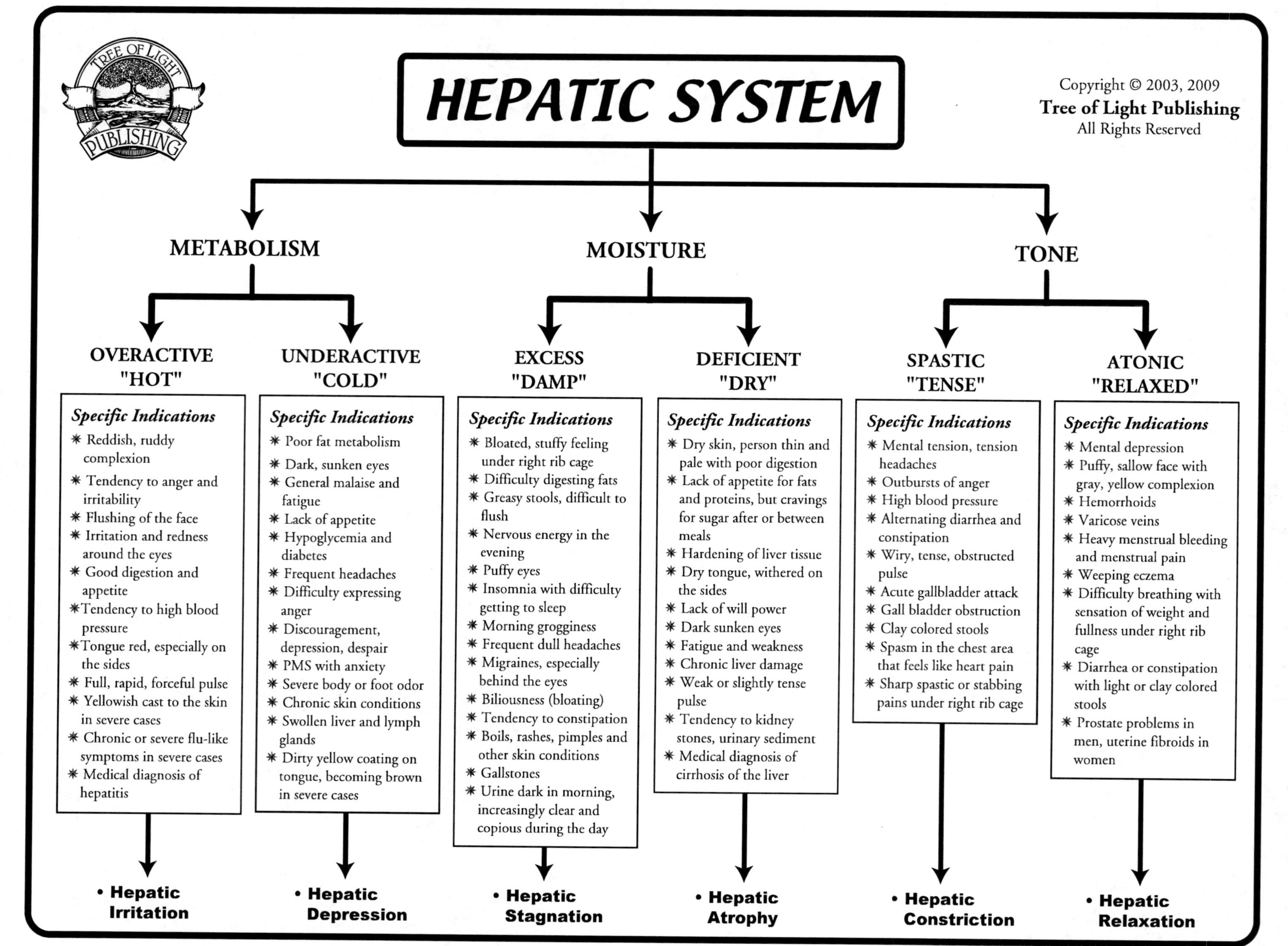

TREE OF LIGHT PUBLISHING
HEPATIC SYSTEM
Copyright © 2003, 2009
Tree of Light Publishing
All Rights Reserved
METABOLISM
MOISTURE
TONE
OVERACTIVE "HOT"
Specific Indications
* Reddish, ruddy complexion
* Tendency to anger and irritability
* Flushing of the face
* Irritation and redness around the eyes
* Good digestion and appetite
* Tendency to high blood pressure
* Tongue red, especially on the sides
* Full, rapid, forceful pulse
* Yellowish cast to the skin in severe cases
* Chronic or severe flu-like symptoms in severe cases
* Medical diagnosis of hepatitis
• Hepatic Irritation
UNDERACTIVE "COLD"
Specific Indications
* Poor fat metabolism
* Dark, sunken eyes
* General malaise and fatigue
* Lack of appetite
* Hypoglycemia and diabetes
* Frequent headaches
* Difficulty expressing anger
* Discouragement, depression, despair
* PMS with anxiety
* Severe body or foot odor
* Chronic skin conditions
* Swollen liver and lymph glands
* Dirty yellow coating on tongue, becoming brown in severe cases
• Hepatic Depression
EXCESS "DAMP"
Specific Indications
* Bloated, stuffy feeling under right rib cage
* Difficulty digesting fats
* Greasy stools, difficult to flush
* Nervous energy in the evening
* Puffy eyes
* Insomnia with difficulty getting to sleep
* Morning grogginess
* Frequent dull headaches
* Migraines, especially behind the eyes
* Biliousness (bloating)
* Tendency to constipation
* Boils, rashes, pimples and other skin conditions
* Gallstones
* Urine dark in morning, increasingly clear and copious during the day
• Hepatic Stagnation
DEFICIENT "DRY"
Specific Indications
* Dry skin, person thin and pale with poor digestion
* Lack of appetite for fats and proteins, but cravings for sugar after or between meals
* Hardening of liver tissue
* Dry tongue, withered on the sides
* Lack of will power
* Dark sunken eyes
* Fatigue and weakness
* Chronic liver damage
* Weak or slightly tense pulse
* Tendency to kidney stones, urinary sediment
* Medical diagnosis of cirrhosis of the liver
• Hepatic Atrophy
SPASTIC "TENSE"
Specific Indications
* Mental tension, tension headaches
* Outbursts of anger
* High blood pressure
* Alternating diarrhea and constipation
* Wiry, tense, obstructed pulse
* Acute gallbladder attack
* Gall bladder obstruction
* Clay colored stools
* Spasm in the chest area that feels like heart pain
* Sharp spastic or stabbing pains under right rib cage
• Hepatic Constriction
ATONIC "RELAXED"
Specific Indications
* Mental depression
* Puffy, sallow face with gray, yellow complexion
* Hemorrhoids
* Varicose veins
* Heavy menstrual bleeding and menstrual pain
* Weeping eczema
* Difficulty breathing with sensation of weight and fullness under right rib cage
* Diarrhea or constipation with light or clay colored stools
* Prostate problems in men, uterine fibroids in women
• Hepatic Relaxation

Hepatic Irritation

Important Notice:

In severe cases (hepatitis), seek medical attention

Herbs

Barley Juice Powder
✓ ✗ Black Cherry Juice
✓ ✗ Lemon Juice
✓ ✗ Lycium (Wolfberry)
✓ ✗ Mangosteen fruit
✓ Milk Thistle
✗ Peach Leaf
✗ Schizandra
Yarrow
Yellow Dock

Supplements

Bioflavionoids
✗ Lactic acid
MSM
✓ SAM-e
Vitamin C

Combinations

Carotenoid Blend
I-X
✓ Milk Thistle Combination
Nervous Fatigue Formula
✓ Super ORAC
✓ Thai-Go

Topical Remedies

✓ Helicrysum EO

Hepatic Depression

Herbs

Chamomile
Chickweed
✗ Fennel
✗ Fenugreek
✓ Feverfew
Garlic
Ginger
✗ Rosemary
✗ Turmeric
✗ Watercress

Supplements

B-Complex
Hi-Lipase
Indole-3 Carbynol
MSM
N-Acetyl Cysteine
SAM-e

Combinations

✓ Blood Build
Liver Cleanse Formula
Milk Thistle Combination
✓ Mood Elevator
SF

Hepatic Stagnation

Herbs

✓ ✗ Artichoke
✗ Barberry
✗ Blue Flag
✗ Celandine
✗ Culver's Root
Dandelion Root
✓ ✗ Fringetree Bark
Ho Shou Wu
Oregon Grape Root
✗ Toadflax
✗ Turmeric

Supplements

Lecithin

Combinations

All Cell Detox
BP-X
Cellular Energy
✓ Enviro-Detox
✓ Liver Balance
LIV-J
Lymphatic Drainage Formula
✓ Milk Thistle Combination
SF

Other Therapies

Coffee Enema
✓ Gall Bladder Flush

Hepatic Atrophy

Important Notice:

In severe cases (cirrhosis of the liver), seek medical attention

Herbs

✓ ✗ Beets
Burdock
Chickweed
✗ Ganoderma
Licorice
Milk thistle
✗ Schizandra
✗ Shiitake mushrooms
Wild Yam
Wood betony

Supplements

✓ N-Acetyl Cysteine
Super GLA

Combinations

✓ Blood Build
Milk Thistle Combination
LIV-J

Other Therapies

Fresh vegetable juices (especially beet roots and greens, celery, chard)

Hepatic Constriction

Important Notice:

In severe cases (gallbladder obstruction), seek medical attention

Herbs

✗ Agrimony
Blessed Thistle
Blue Vervain
Lobelia
✓ Wild Yam

Supplements

✓ Magnesium Complex
SAM-e

Combinations

✓ CLT-X
✓ Gall Bladder Formula
Stress-J
SUMA Combination

Other Therapies

Epsom Salt Baths
Gall Bladder Flush

Avoid

Caffeine (coffee, tea, guarana, energy drinks, etc.)
Capsicum
Garlic

Hepatic Relaxation

Herbs

Bayberry
Butcher's Broom
✗ Horse Chestnut
✓ Pau d'Arco
Sage
Schizandra
✓ White Oak Bark

Supplements

✓ Bioflavinoids
Green Tea Extract

Combinations

Blood Build
Menstrual Reg
Mood Elevator
✓ Vari-Gone
Vari-Gone Cream

Important Notice: This chart is a reference tool for the ABC+D approach to nutritional support of each body system. It does not replace the services of qualified practitioners. For serious issues and emergencies, please seek the advice of a certified or licensed health professional.

Key: ✓-Top choices ✗-Not available as a single through NSP
EO-Essential Oil (topical use) ❋-Toxic botanical for professional use only

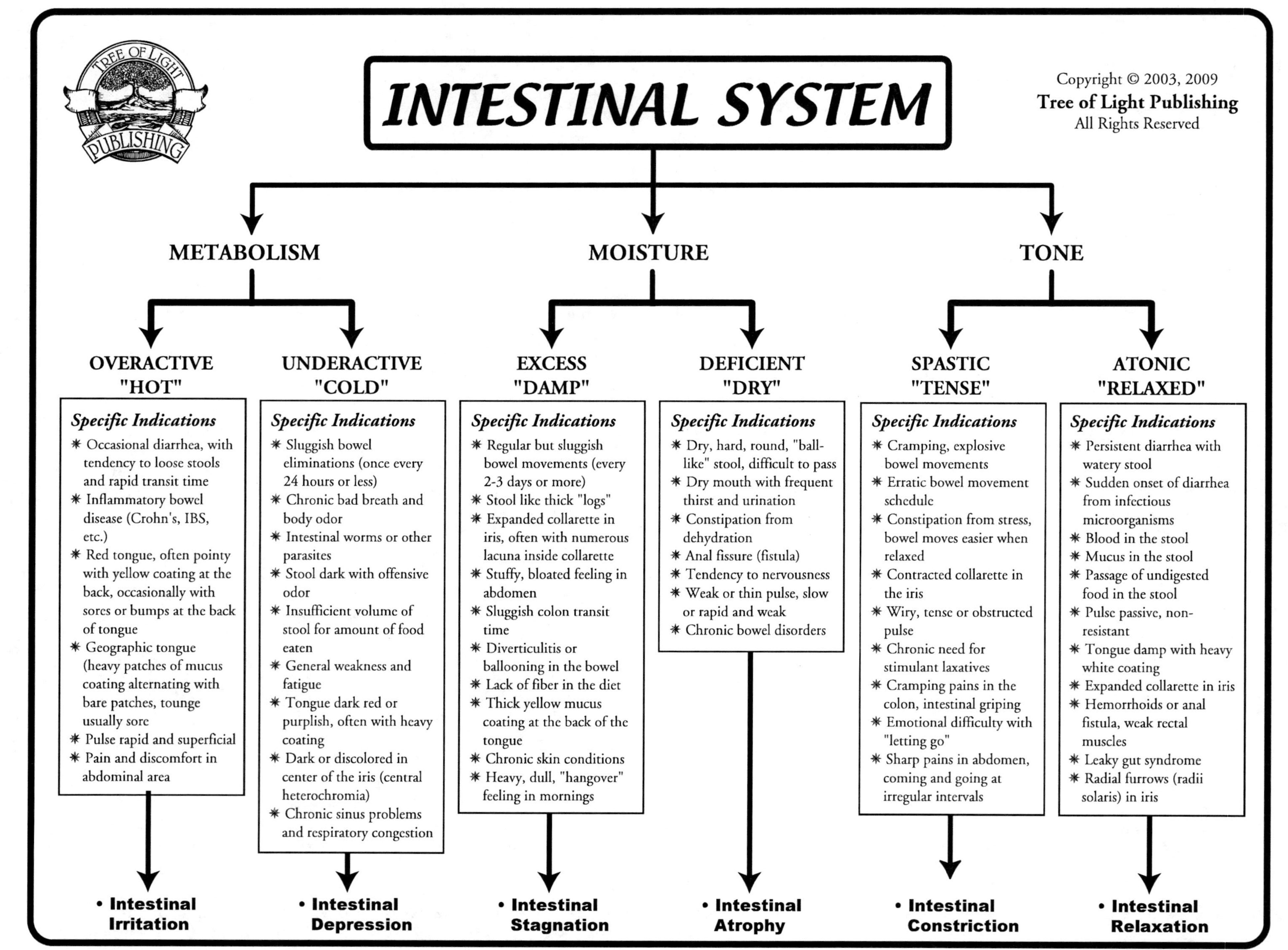
INTESTINAL SYSTEM
TREE OF LIGHT
PUBLISHING
Copyright © 2003, 2009
Tree of Light Publishing
All Rights Reserved
METABOLISM
MOISTURE
TONE
OVERACTIVE "HOT"
Specific Indications
* Occasional diarrhea, with tendency to loose stools and rapid transit time
* Inflammatory bowel disease (Crohn's, IBS, etc.)
* Red tongue, often pointy with yellow coating at the back, occasionally with sores or bumps at the back of tongue
* Geographic tongue (heavy patches of mucus coating alternating with bare patches, tounge usually sore
* Pulse rapid and superficial
* Pain and discomfort in abdominal area
• Intestinal Irritation
UNDERACTIVE "COLD"
Specific Indications
* Sluggish bowel eliminations (once every 24 hours or less)
* Chronic bad breath and body odor
* Intestinal worms or other parasites
* Stool dark with offensive odor
* Insufficient volume of stool for amount of food eaten
* General weakness and fatigue
* Tongue dark red or purplish, often with heavy coating
* Dark or discolored in center of the iris (central heterochromia)
* Chronic sinus problems and respiratory congestion
• Intestinal Depression
EXCESS "DAMP"
Specific Indications
* Regular but sluggish bowel movements (every 2-3 days or more)
* Stool like thick "logs"
* Expanded collarette in iris, often with numerous lacuna inside collarette
* Stuffy, bloated feeling in abdomen
* Sluggish colon transit time
* Diverticulitis or ballooning in the bowel
* Lack of fiber in the diet
* Thick yellow mucus coating at the back of the tongue
* Chronic skin conditions
* Heavy, dull, "hangover" feeling in mornings
• Intestinal Stagnation
DEFICIENT "DRY"
Specific Indications
* Dry, hard, round, "ball-like" stool, difficult to pass
* Dry mouth with frequent thirst and urination
* Constipation from dehydration
* Anal fissure (fistula)
* Tendency to nervousness
* Weak or thin pulse, slow or rapid and weak
* Chronic bowel disorders
• Intestinal Atrophy
SPASTIC "TENSE"
Specific Indications
* Cramping, explosive bowel movements
* Erratic bowel movement schedule
* Constipation from stress, bowel moves easier when relaxed
* Contracted collarette in the iris
* Wiry, tense or obstructed pulse
* Chronic need for stimulant laxatives
* Cramping pains in the colon, intestinal griping
* Emotional difficulty with "letting go"
* Sharp pains in abdomen, coming and going at irregular intervals
• Intestinal Constriction
ATONIC "RELAXED"
Specific Indications
* Persistent diarrhea with watery stool
* Sudden onset of diarrhea from infectious microorganisms
* Blood in the stool
* Mucus in the stool
* Passage of undigested food in the stool
* Pulse passive, non-resistant
* Tongue damp with heavy white coating
* Expanded collarette in iris
* Hemorrhoids or anal fistula, weak rectal muscles
* Leaky gut syndrome
* Radial furrows (radii solaris) in iris
• Intestinal Relaxation

Intestinal Irritation

Important Notice:

In severe cases seek medical attention

Herbs

✓ Aloe Vera
✓ Chamomile
✓ Licorice
Marshmallow
Rose Hips
✓ Slippery Elm
✓ Una De Gato (Cat's Claw)
Yarrow
Yellow Dock

Supplements

Probiotics

Combinations

CLT-X
Everybody's Fiber
Gentle Move
✓ Intestinal Soothe and Build
IF Relief
PLS II
✓ Stress-J
Super ORAC
Thai-Go
Una De Gato Combination

Intestinal Depression

Herbs

✗ Butternut bark
Capsicum
✓ Chamomile
✗ Cloves
Garlic
✓ Ginger
Horseradish
✓ ✗ Peppermint herb
✗ Tansy
✗ Wormwood

Supplements

Chlorophyll Capsules

Combinations

✓ Artemesia Combination
Capsicum, Garlic and Parsley
Cellular Energy
✓ Gentle Move
Herbal Pumpkin
HCP-X
Para-Cleanse

Topical Remedies

Deep Relief massaged into abdomen

Intestinal Stagnation

Herbs

✗ Barberry
Black Walnut
✗ Buckthorn
✗ Butternut
Cascara Sagrada
✗ Senna
✗ Turkey Rhubarb
Yellow Dock

Supplements

Hydrated Bentonite

Combinations

✓ All Cell Detox
Bowel Detox
Clean Start
Cellular Energy
Gentle Move
LB Extract
✓ LBS II
LB-X
Lymphatic Drainage
Psyllium Hulls Combination
Senna Combination
✓ Tiao He Cleanse

Intestinal Atrophy

Herbs

✓ Aloe Vera
✓ ✗ Flaxseed
Licorice
Marshmallow
✓ Psyllium Hulls
Slippery Elm

Supplements

Flax Seed Oil
Liquid B-12 Complete
Omega 3
✓ Probiotics
Vitamin B-12

Combinations

Everybody's Fiber
Fat Grabbers
✓ Intestinal Soothe and Build
LOCLO
Nature's Three
Psyllium Hulls Combination
✓ Spleen Activator
SUMA Combination

Other Therapies

Drink lots of water

Intestinal Constriction

Herbs

✗ Agrimony
Blue Vervain
Catnip
✗ Crampbark
Lobelia
✓ Wild Yam
Wood Betony

Supplements

✓ Magnesium

Combinations

✓ CLT-X
Cramp Relief
Everybody's Fiber
✓ Gall Bladder Formula
✓ Gentle Move
Nutri-Calm
Stress-J

Intestinal Relaxation

Herbs

✓ Bayberry
✓ ✗ Blackberry root
✗ Collinsonia (Stone Root)
✓ Goldenseal
Una De Gato (Cat's Claw)
Uva Ursi
White Oak Bark

Supplements

✓ Charcoal, Activated
Mineral Chi Tonic
Skeletal Strength

Combinations

CA, Herbal
CLT-X
Herbal Trace Minerals
✓ Kudzu/St. John's Wort
Vari-Gone
✓ Una De Gato Combination

Important Notice: This chart is a reference tool for the ABC+D approach to nutritional support of each body system. It does not replace the services of qualified practitioners. For serious issues and emergencies, please seek the advice of a certified or licensed health professional.

Key: ✓-Top choices ✗-Not available as a single through NSP
EO-Essential Oil (topical use) ✻-Toxic botanical for professional use only

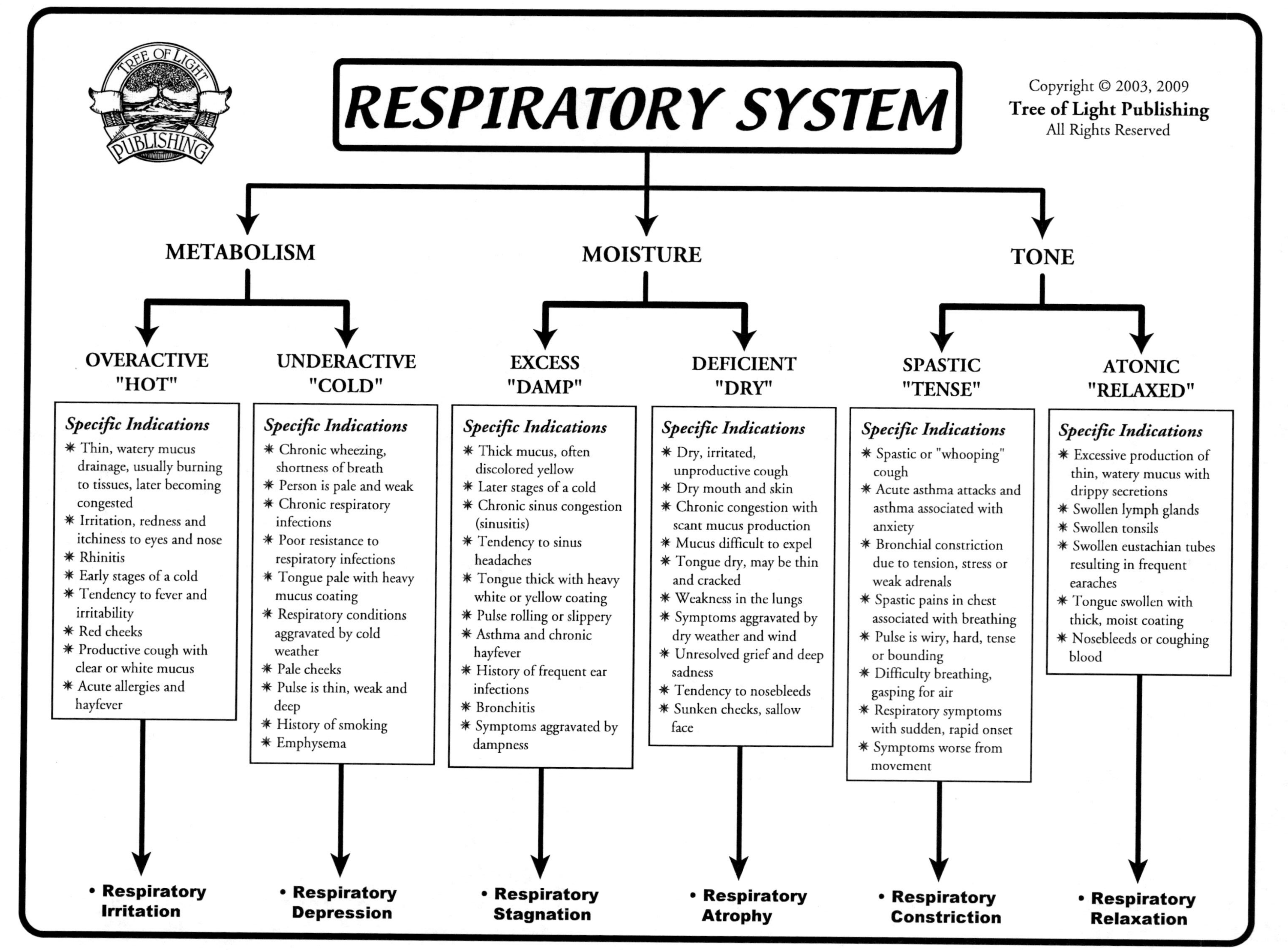

TREE OF LIGHT PUBLISHING
RESPIRATORY SYSTEM
Copyright © 2003, 2009
Tree of Light Publishing
All Rights Reserved
METABOLISM
MOISTURE
TONE
OVERACTIVE "HOT"
Specific Indications
* Thin, watery mucus drainage, usually burning to tissues, later becoming congested
* Irritation, redness and itchiness to eyes and nose
* Rhinitis
* Early stages of a cold
* Tendency to fever and irritability
* Red cheeks
* Productive cough with clear or white mucus
* Acute allergies and hayfever
• Respiratory Irritation
UNDERACTIVE "COLD"
Specific Indications
* Chronic wheezing, shortness of breath
* Person is pale and weak
* Chronic respiratory infections
* Poor resistance to respiratory infections
* Tongue pale with heavy mucus coating
* Respiratory conditions aggravated by cold weather
* Pale cheeks
* Pulse is thin, weak and deep
* History of smoking
* Emphysema
• Respiratory Depression
EXCESS "DAMP"
Specific Indications
* Thick mucus, often discolored yellow
* Later stages of a cold
* Chronic sinus congestion (sinusitis)
* Tendency to sinus headaches
* Tongue thick with heavy white or yellow coating
* Pulse rolling or slippery
* Asthma and chronic hayfever
* History of frequent ear infections
* Bronchitis
* Symptoms aggravated by dampness
• Respiratory Stagnation
DEFICIENT "DRY"
Specific Indications
* Dry, irritated, unproductive cough
* Dry mouth and skin
* Chronic congestion with scant mucus production
* Mucus difficult to expel
* Tongue dry, may be thin and cracked
* Weakness in the lungs
* Symptoms aggravated by dry weather and wind
* Unresolved grief and deep sadness
* Tendency to nosebleeds
* Sunken checks, sallow face
• Respiratory Atrophy
SPASTIC "TENSE"
Specific Indications
* Spastic or "whooping" cough
* Acute asthma attacks and asthma associated with anxiety
* Bronchial constriction due to tension, stress or weak adrenals
* Spastic pains in chest associated with breathing
* Pulse is wiry, hard, tense or bounding
* Difficulty breathing, gasping for air
* Respiratory symptoms with sudden, rapid onset
* Symptoms worse from movement
• Respiratory Constriction
ATONIC "RELAXED"
Specific Indications
* Excessive production of thin, watery mucus with drippy secretions
* Swollen lymph glands
* Swollen tonsils
* Swollen eustachian tubes resulting in frequent earaches
* Tongue swollen with thick, moist coating
* Nosebleeds or coughing blood
• Respiratory Relaxation

Respiratory Irritation

Herbs

- ✗ Elderberry
- ✓ Eyebright
- Lemon Juice
- Plantain
- Rose Hips
- Schizandra Berries
- ✓ ✗ Wild Cherry Bark
- Yarrow

Supplements

- Carotenoid Blend
- Vitamin C

Combinations

- CC-A
- ✓ Elderberry Defense
- Elderberry Plus
- EW
- ✓ Hista-Block
- Seasonal Defense
- Super ORAC
- Thai-Go

Topical Remedies

- Eucalyptus EO
- Tei Fu Oil

Respiratory Depression

Herbs

- ✗ Elecampane
- ✓ Garlic
- Ginger
- ✗ Gumweed (Grindelia)
- ✓ ✗ Horseradish
- ✗ Onion
- ✗ Rosemary
- ✗ Thyme
- ✓ ✗ White Pine Bark
- ✓ ✗ Yerba Santa

Combinations

- ✓ AL-J
- Breathe EZ
- Fenugreek and Thyme
- ✓ Garlic, High Potency
- HCP-X

Topical Remedies

- Breathe Free EO Blend
- Eucalyptus EO
- ✓ Pine Needle EO
- ✓ Rosemary EO
- Thyme EO

Respiratory Stagnation

Herbs

- Burdock
- Dandelion
- ✓ Echinacea
- ✓ Goldenseal
- ✗ Gumweed
- ✗ Horehound
- Oregon Grape

Combinations

- AL-J
- Bronchial Formula
- ✓ Echinacea/Goldenseal
- ✓ Four
- Hista-Block
- ✓ Sinus Support

Topical Remedies

- Frankincense EO
- Myrrh EO
- Tei Fu Oil or Massage Lotion

Respiratory Atrophy

Herbs

- Chickweed
- ✓ Cordyceps
- ✗ Irish Moss
- Horsetail
- ✓ Licorice
- ✓ Mullein
- ✓ Marshmallow
- ✗ Plantain
- Slippery Elm

Combinations

- ✓ Lung Support
- Marshmallow and Fenugreek

Respiratory Constriction

Herbs

- Black Cohosh
- Licorice
- ✓ Lobelia
- ✗ Pleurisy Root

Supplements

- Magnesium Complex
- Pantothenic Acid

Combinations

- ✓ Adrenal Support
- Breathe EZ
- ✓ Hista-Block
- IN-X
- Seasonal Defense

Topical Remedies

- ✓ Lavender EO
- ✓ Rosemary EO

Respiratory Relaxation

Herbs

- Bayberry
- Eyebright
- Horsetail
- Sage
- White Oak Bark
- ✗ Yerba Santa

Combinations

- CC-A with Yerba Santa
- ✓ EW
- ✓ HCP-X
- HSN-W

Important Notice: This chart is a reference tool for the ABC+D approach to nutritional support of each body system. It does not replace the services of qualified practitioners. For serious issues and emergencies, please seek the advice of a certified or licensed health professional.

Key: ✓-Top choices ✗-Not available as a single through NSP
EO-Essential Oil (topical use) ✻-Toxic botanical for professional use only

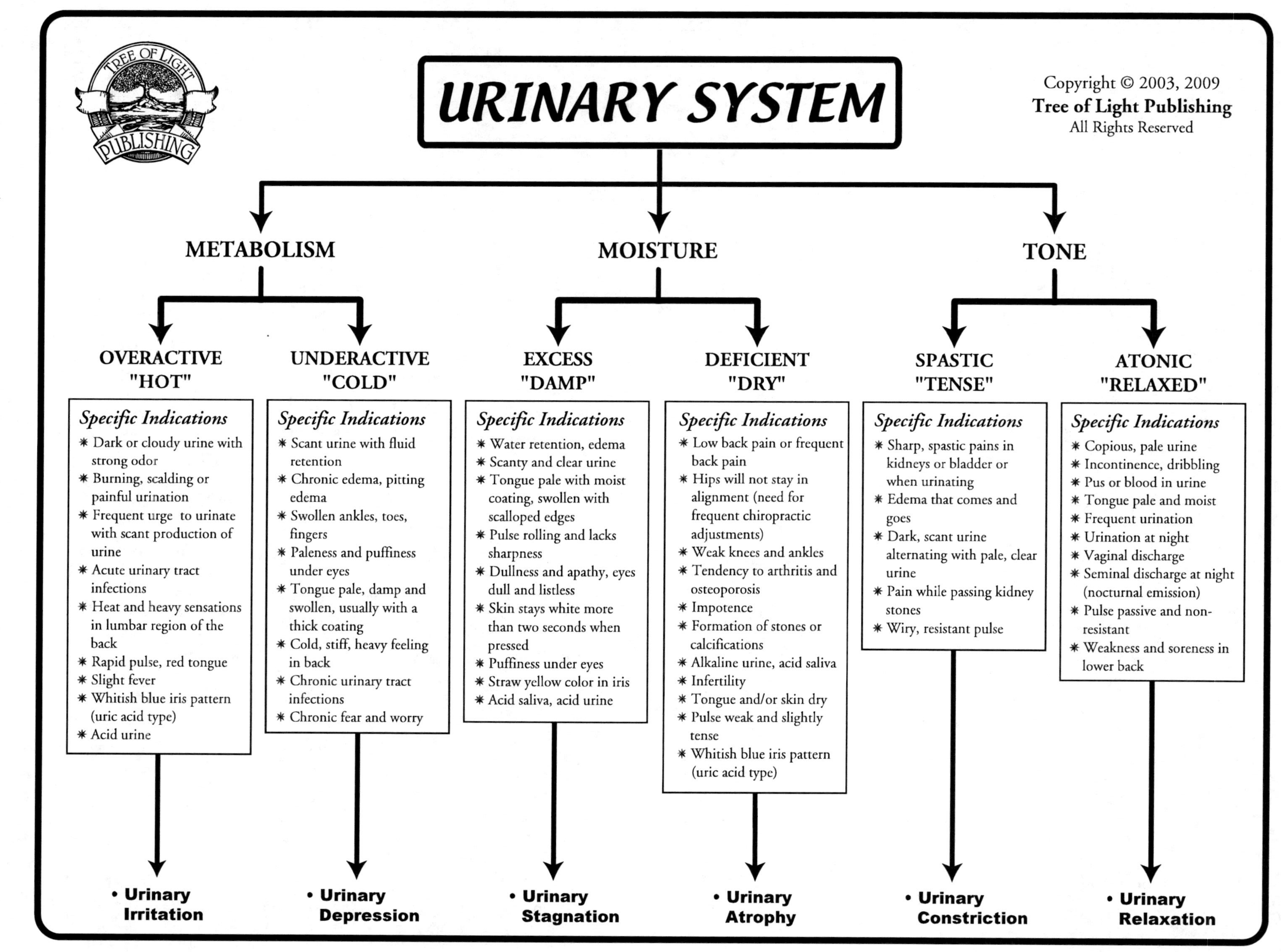

TREE OF LIGHT PUBLISHING
URINARY SYSTEM
Copyright © 2003, 2009
Tree of Light Publishing
All Rights Reserved
METABOLISM
MOISTURE
TONE
OVERACTIVE "HOT"
Specific Indications
* Dark or cloudy urine with strong odor
* Burning, scalding or painful urination
* Frequent urge to urinate with scant production of urine
* Acute urinary tract infections
* Heat and heavy sensations in lumbar region of the back
* Rapid pulse, red tongue
* Slight fever
* Whitish blue iris pattern (uric acid type)
* Acid urine
• Urinary Irritation
UNDERACTIVE "COLD"
Specific Indications
* Scant urine with fluid retention
* Chronic edema, pitting edema
* Swollen ankles, toes, fingers
* Paleness and puffiness under eyes
* Tongue pale, damp and swollen, usually with a thick coating
* Cold, stiff, heavy feeling in back
* Chronic urinary tract infections
* Chronic fear and worry
• Urinary Depression
EXCESS "DAMP"
Specific Indications
* Water retention, edema
* Scanty and clear urine
* Tongue pale with moist coating, swollen with scalloped edges
* Pulse rolling and lacks sharpness
* Dullness and apathy, eyes dull and listless
* Skin stays white more than two seconds when pressed
* Puffiness under eyes
* Straw yellow color in iris
* Acid saliva, acid urine
• Urinary Stagnation
DEFICIENT "DRY"
Specific Indications
* Low back pain or frequent back pain
* Hips will not stay in alignment (need for frequent chiropractic adjustments)
* Weak knees and ankles
* Tendency to arthritis and osteoporosis
* Impotence
* Formation of stones or calcifications
* Alkaline urine, acid saliva
* Infertility
* Tongue and/or skin dry
* Pulse weak and slightly tense
* Whitish blue iris pattern (uric acid type)
• Urinary Atrophy
SPASTIC "TENSE"
Specific Indications
* Sharp, spastic pains in kidneys or bladder or when urinating
* Edema that comes and goes
* Dark, scant urine alternating with pale, clear urine
* Pain while passing kidney stones
* Wiry, resistant pulse
• Urinary Constriction
ATONIC "RELAXED"
Specific Indications
* Copious, pale urine
* Incontinence, dribbling
* Pus or blood in urine
* Tongue pale and moist
* Frequent urination
* Urination at night
* Vaginal discharge
* Seminal discharge at night (nocturnal emission)
* Pulse passive and non-resistant
* Weakness and soreness in lower back
• Urinary Relaxation

Urinary Irritation

Herbs

✓ Cornsilk
✓ ✗ Cranberry
Marshmallow
Noni Juice
✗ Peach (Leaf or Bark)
✓ ✗ Watermelon (Juice and Seeds)
Yarrow
Yellow Dock

Supplements

Magnesium
Vitamin C

Combinations

Kidney Drainage
Potassium Combination
✓ Urinary Maintenance

Urinary Depression

Herbs

✗ Buchu
✓ Juniper Berries
✗ Goldenrod
Parsley
Uva Ursi

Combinations

Cranberry/Buchu
JP-X
✓ Kidney Activator
✓ Kidney Activator, Chinese
Kidney Drainage
Lymphatic Drainage

Urinary Stagnation

Herbs

✗ Boneset
✗ Dandelion Leaf
✓ Goldenseal
Parthenium
✗ Pipsissewa
Safflower

Combinations

IN-X
JP-X
✓ Kidney Activator, Chinese
Kidney Drainage
Lymph Gland Cleanse

Urinary Atrophy

Herbs

✗ Cleavers
✗ Gravel Root
Ho Shou Wu
✓ Horsetail
Hydrangea
Marshmallow
✓ ✗ Nettle Seed

Supplements

Magnesium Complex

Combinations

✓ KB-C

Urinary Constriction

Herbs

✓ ✗ Agrimony
Blue Vervain
✓ Kava Kava
Lobelia

Supplements

✓ Magnesium Complex

Combinations

Adrenal Support
✓ Cramp Relief
Stress-J

Urinary Relaxation

Herbs

✗ Geranium Root
✓ Horsetail
Pau d'Arco
Red Raspberry
Sage
✓ Uva Ursi

Combinations

CA, Herbal
✓ HSN-W
KB-C
Skeletal Strength

Important Notice: This chart is a reference tool for the ABC+D approach to nutritional support of each body system. It does not replace the services of qualified practitioners. For serious issues and emergencies, please seek the advice of a certified or licensed health professional.

Key: ✓-Top choices ✗-Not available as a single through NSP
EO-Essential Oil (topical use) ✻-Toxic botanical for professional use only

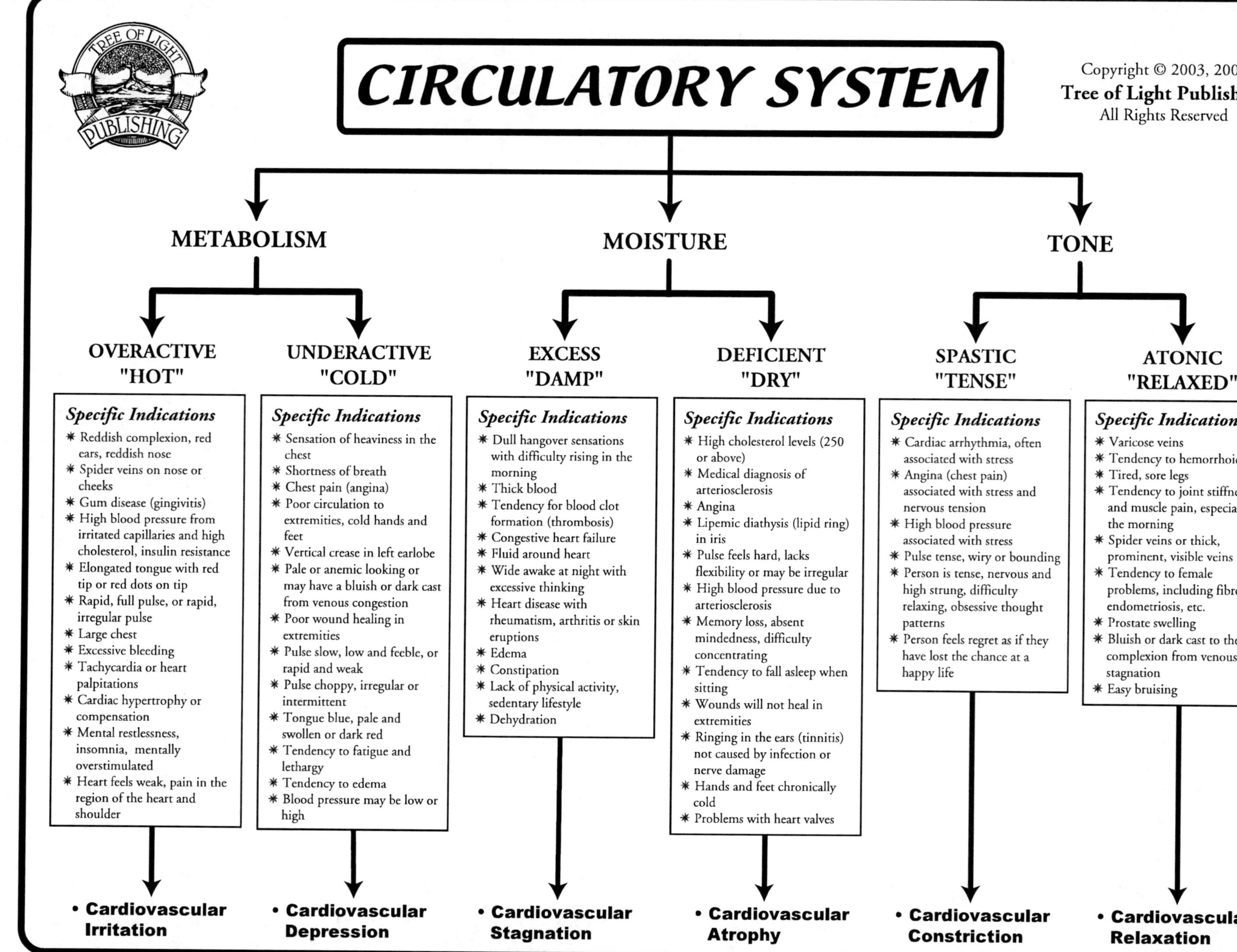

TREE OF LIGHT PUBLISHING
CIRCULATORY SYSTEM
Copyright © 2003, 2009
Tree of Light Publishing
All Rights Reserved
METABOLISM
MOISTURE
TONE
OVERACTIVE "HOT"
Specific Indications
* Reddish complexion, red ears, reddish nose
* Spider veins on nose or cheeks
* Gum disease (gingivitis)
* High blood pressure from irritated capillaries and high cholesterol, insulin resistance
* Elongated tongue with red tip or red dots on tip
* Rapid, full pulse, or rapid, irregular pulse
* Large chest
* Excessive bleeding
* Tachycardia or heart palpitations
* Cardiac hypertrophy or compensation
* Mental restlessness, insomnia, mentally overstimulated
* Heart feels weak, pain in the region of the heart and shoulder
• Cardiovascular Irritation
UNDERACTIVE "COLD"
Specific Indications
* Sensation of heaviness in the chest
* Shortness of breath
* Chest pain (angina)
* Poor circulation to extremities, cold hands and feet
* Vertical crease in left earlobe
* Pale or anemic looking or may have a bluish or dark cast from venous congestion
* Poor wound healing in extremities
* Pulse slow, low and feeble, or rapid and weak
* Pulse choppy, irregular or intermittent
* Tongue blue, pale and swollen or dark red
* Tendency to fatigue and lethargy
* Tendency to edema
* Blood pressure may be low or high
• Cardiovascular Depression
EXCESS "DAMP"
Specific Indications
* Dull hangover sensations with difficulty rising in the morning
* Thick blood
* Tendency for blood clot formation (thrombosis)
* Congestive heart failure
* Fluid around heart
* Wide awake at night with excessive thinking
* Heart disease with rheumatism, arthritis or skin eruptions
* Edema
* Constipation
* Lack of physical activity, sedentary lifestyle
* Dehydration
• Cardiovascular Stagnation
DEFICIENT "DRY"
Specific Indications
* High cholesterol levels (250 or above)
* Medical diagnosis of arteriosclerosis
* Angina
* Lipemic diathysis (lipid ring) in iris
* Pulse feels hard, lacks flexibility or may be irregular
* High blood pressure due to arteriosclerosis
* Memory loss, absent mindedness, difficulty concentrating
* Tendency to fall asleep when sitting
* Wounds will not heal in extremities
* Ringing in the ears (tinnitis) not caused by infection or nerve damage
* Hands and feet chronically cold
* Problems with heart valves
• Cardiovascular Atrophy
SPASTIC "TENSE"
Specific Indications
* Cardiac arrhythmia, often associated with stress
* Angina (chest pain) associated with stress and nervous tension
* High blood pressure associated with stress
* Pulse tense, wiry or bounding
* Person is tense, nervous and high strung, difficulty relaxing, obsessive thought patterns
* Person feels regret as if they have lost the chance at a happy life
• Cardiovascular Constriction
ATONIC "RELAXED"
Specific Indications
* Varicose veins
* Tendency to hemorrhoids
* Tired, sore legs
* Tendency to joint stiffness and muscle pain, especially in the morning
* Spider veins or thick, prominent, visible veins
* Tendency to female problems, including fibroids, endometriosis, etc.
* Prostate swelling
* Bluish or dark cast to the complexion from venous stagnation
* Easy bruising
• Cardiovascular Relaxation

Cardiovascular Irritation

Herbs

Bilberry
Ginkgo
✓ Hawthorn
Ho Shou Wu
✗ Lemon Balm
✗ Linden
✓ ✗ Mangosteen (Fruit and Pericarp)
✗ Motherwort
Rosehips
Yarrow

Supplements

Carotinoid Blend
Citrus Bioflavinoids with Vitamin C
✓ Co-Q 10 75
Vitamin E with Selenium

Combinations

Blood Pressurex
✓ Ginkgo & Hawthorn Combination
HS II
IF Relief
SUMA Combination
✓ Super ORAC
✓ Thai-Go

Cardiovascular Depression

Herbs

✻ ✗ Arnica
✓ Capsicum
✻ ✗ Foxglove
✓ Garlic
✻ ✗ Lily of the Valley
✗ Rosemary Herb
✻ ✗ Scotch Broom

Supplements

✓ CoQ-10 75
✗ D-Ribose
✓ L-Carnatine
✓ Magnesium Complex
Niacin

Combinations

✓ Capsicum, Garlic and Parsley
✓ CardioAssurance
GC-X
HS II
RG-Max

Topical

Bergamot EO
✓ Rose EO
Rosemary EO

Cardiovascular Stagnation

Herbs

Alfalfa
Butcher's Broom
Dandelion
Goldenseal
✗ Nettles
Oregon Grape
✗ Rehmannia

Supplements

✓ Chlorophyll
Nattozimes Plus
Vitamin E

Combinations

✓ I-X
Cellular Energy
Kidney Activator, Chinese
Kidney Drainage
Lymphatic Drainage

Cardiovascular Atrophy

Herbs

✗ Cactus (Night Blooming Cereus)
✗ Coleus
Fenugreek
✓ Guggul Lipid
✓ Ho Shou Wu
Hydrangea
Korean Ginseng
✻ ✗ Lily of the Valley
✗ Maitake Mushrooms
✗ Tulip Poplar

Supplements

Co-Q 10 75
Flax Seed Oil
Magnesium Complex
✓ Omega 3
Protease Plus
Vitamin E with Selenium
SAM-e
✓ Super GLA

Combinations

Cholester-Reg II
Fat Grabbers
LOCLO
✓ Mega-Chel
Nature's Three
✓ Red Yeast Rice
SF

Cardiovascular Constriction

Herbs

Black Cohosh
✗ Khella Seed
Lobelia
✗ Skullcap
Valerian
✻ ✗ Mistletoe

Supplements

✓ Magnesium Complex

Combinations

✓ Blood Pressurex
Kidney Drainage
✓ RG-Max
Stress-J
Stress Relief
SUMA Combination

Cardiovascular Relaxation

Herbs

Black walnut
✓ Butcher's Broom
✗ Collinsonia
✗ Horse Chestnut
✗ Lady's Mantle
Rose Hips
Sage
✓ White Oak Bark
✗ Witch Hazel

Supplements

✓ Bioflavinoids
✓ Co-Q 10 75
Collatrim
Vitamin C
Vitamin E

Combinations

Cellu-Tone EO Blend
High Potency Grapine
HSN-W
✓ Mega-Chel
Super Antioxidant
✓ Vari-Gone
Vitamin E with Selenium

Topical

Vari-Gone Cream

Important Notice: This chart is a reference tool for the ABC+D approach to nutritional support of each body system. It does not replace the services of qualified practitioners. For serious issues and emergencies, please seek the advice of a certified or licensed health professional.

Key: ✓-Top choices ✗-Not available as a single through NSP
EO-Essential Oil (topical use) ✻-Toxic botanical for professional use only

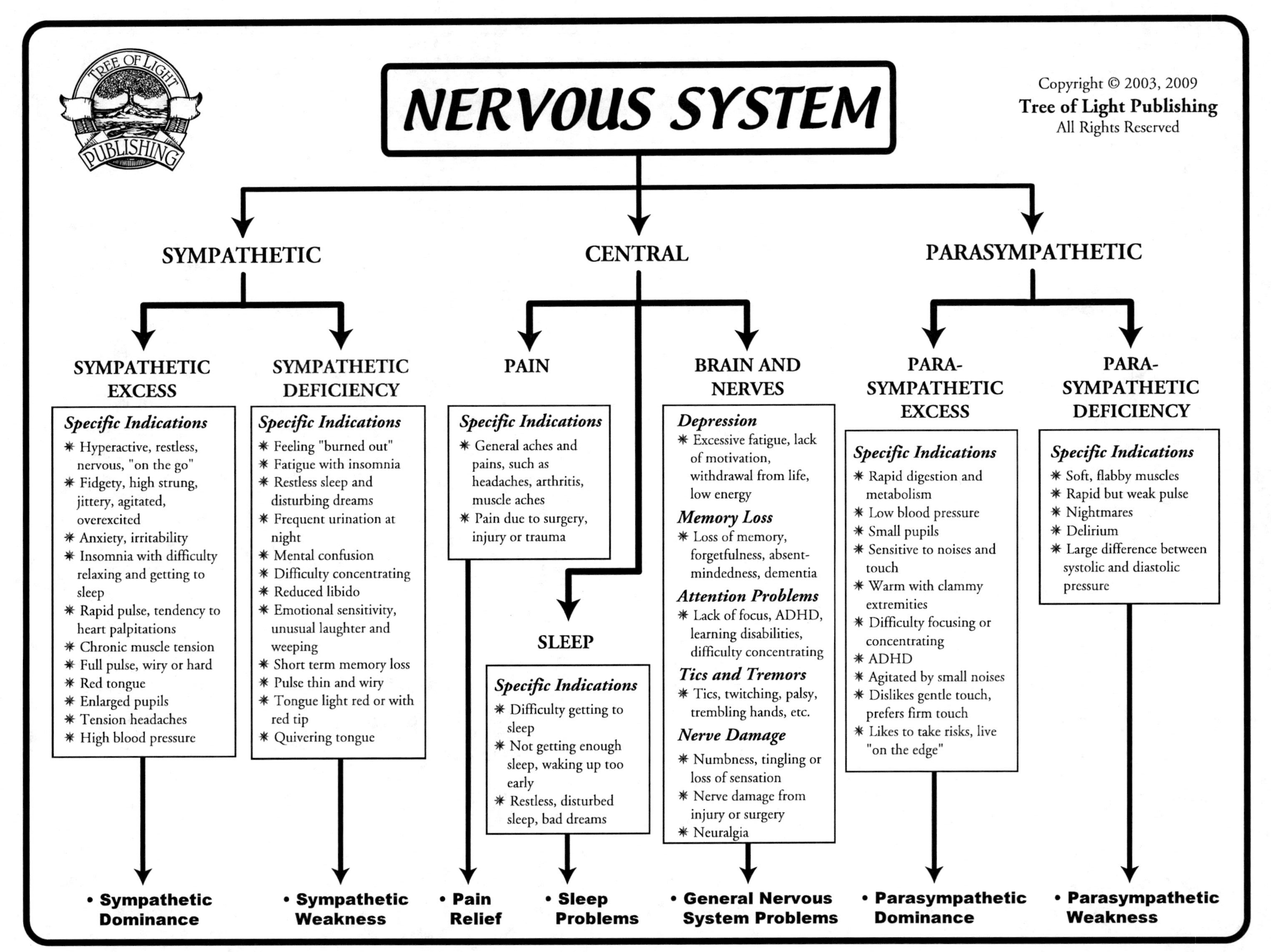

TREE OF LIGHT
PUBLISHING
NERVOUS SYSTEM
Copyright © 2003, 2009
Tree of Light Publishing
All Rights Reserved
SYMPATHETIC
CENTRAL
PARASYMPATHETIC
SYMPATHETIC EXCESS
Specific Indications
* Hyperactive, restless, nervous, "on the go"
* Fidgety, high strung, jittery, agitated, overexcited
* Anxiety, irritability
* Insomnia with difficulty relaxing and getting to sleep
* Rapid pulse, tendency to heart palpitations
* Chronic muscle tension
* Full pulse, wiry or hard
* Red tongue
* Enlarged pupils
* Tension headaches
* High blood pressure
• Sympathetic Dominance
SYMPATHETIC DEFICIENCY
Specific Indications
* Feeling "burned out"
* Fatigue with insomnia
* Restless sleep and disturbing dreams
* Frequent urination at night
* Mental confusion
* Difficulty concentrating
* Reduced libido
* Emotional sensitivity, unusual laughter and weeping
* Short term memory loss
* Pulse thin and wiry
* Tongue light red or with red tip
* Quivering tongue
• Sympathetic Weakness
PAIN
Specific Indications
* General aches and pains, such as headaches, arthritis, muscle aches
* Pain due to surgery, injury or trauma
• Pain Relief
SLEEP
Specific Indications
* Difficulty getting to sleep
* Not getting enough sleep, waking up too early
* Restless, disturbed sleep, bad dreams
• Sleep Problems
BRAIN AND NERVES
Depression
* Excessive fatigue, lack of motivation, withdrawal from life, low energy
Memory Loss
* Loss of memory, forgetfulness, absent-mindedness, dementia
Attention Problems
* Lack of focus, ADHD, learning disabilities, difficulty concentrating
Tics and Tremors
* Tics, twitching, palsy, trembling hands, etc.
Nerve Damage
* Numbness, tingling or loss of sensation
* Nerve damage from injury or surgery
* Neuralgia
• General Nervous System Problems
PARA-SYMPATHETIC EXCESS
Specific Indications
* Rapid digestion and metabolism
* Low blood pressure
* Small pupils
* Sensitive to noises and touch
* Warm with clammy extremities
* Difficulty focusing or concentrating
* ADHD
* Agitated by small noises
* Dislikes gentle touch, prefers firm touch
* Likes to take risks, live "on the edge"
• Parasympathetic Dominance
PARA-SYMPATHETIC DEFICIENCY
Specific Indications
* Soft, flabby muscles
* Rapid but weak pulse
* Nightmares
* Delirium
* Large difference between systolic and diastolic pressure
• Parasympathetic Weakness

Sympathetic Excess

Herbs

Blue Vervain
Chamomile
Hops
✓ Eleuthero (Siberian Ginseng)
✓ Kava Kava
✗ Lemon Balm
Lobelia
Passion Flower
✗ Skullcap
Valerian
Eleuthero
Wood Betony

Supplements

IGF-1
GABA Plus
Balanced B-Complex
Free Amino Acids

Combinations

Herbal Sleep
Nerve Control
✓ Nutri-Calm
✓ Stress-J
Stress Relief
Adrenal Support
HSN-Complex
Fatigue/Exhaustion Homeopathic
Focus Attention
Kudzu/St. John's Wort
✓ Nervous Fatigue Formula
SUMA Combination
Super Algae

Topical Remedies

Lavender EO
Roman Chamomile EO
Ylang Ylang EO
✓ Epsom Salt Baths
✓ Lavender EO
Rose EO
Bergamot EO

Sympathetic Deficiency

Herbs

✗ Agrimony
✓ Licorice
✗ Milky Oat Seed

Pain Relief

Herbs

✗ Agrimony
Black Cohosh
✗ California Poppy
Chamomile
Feverfew
✓ Kava Kava
Licorice
✓ Lobelia
✗ White Willow
Wild Yam
✓ Wood Betony
Yucca

Supplements

5-HTP Power
Magnesium Complex (under the tongue)

Combinations

APS II w/White Willow
✓ IF Relief
Nature's Phenyltol with NEM
Nerve Eight
Triple Relief

Topical Remedies

Deep Relief
MSM Cream
Tei Fu Oil or Massage Lotion

Sleep

5-HTP Power
Adrenal Support
✓ Herbal Sleep
✓ Kava Kava
Melatonin Extra
Nerve Eight
Nervous Fatigue Formula

General Nervous System Problems

Depression

5-HTP Power
Bergamot EO
Black Cohosh
✓ Chinese Mood Elevator
Damiana
Depressaquel Homeopathic
Kava Kava
✗ Lemon Balm
Liquid Dulse
St. John's Wort
✓ SAM-e
Thyroid Activator

General Brain Tonic and Memory

✗ Ashwaganda
Ginkgo
✓ Ginkgo/Gotu Kola with Bacopa
Gotu Kola
HSN-Complex
HSN-W
Rosemary EO
Sage EO

Attention Problems

Energ-V
✓ Focus Attention
GABA Plus
✗ Lemon Balm
Nature's Chi

Tics and Tremors

✓ Black Cohosh
Kava Kava
IGF-1
✓ Lobelia
Magnesium Complex
Nervous Fatigue Formula
✗ Skullcap
St. John's Wort

Nerve Damage

DHA
✓ HSN-W
Omega-3
✓ St. John's Wort
✓ Super GLA
Wood Betony

Parasympathetic Dominance

Herbs

✗ Coffee
✓ ✗ Guarana
✗ Kola Nuts
Licorice
Liquid Dulse
✗ Yerba Mate

Supplements

GABA
Iodine

Combinations

✓ ENERG-V
Focus Attention
Nature's Chi
SUMA Combination
Super Algae

Topical Remedies

Lemon EO
Pine EO
Rosemary EO
Sage EO
Thyme EO

Parasympathetic Deficiency

Herbs

✗ Calendula
✗ California Poppy
Chamomile
✗ Shepherd's Purse
✓ Valerian
✓ Wood Betony

Supplements

IGF-1

Combinations

Herbal Sleep (HVP)
Nervous Fatigue Formula
✓ SUMA Combination

Topical Remedies

Oregano EO
Marjoram EO
Rosemary EO

Important Notice: This chart is a reference tool for the ABC+D approach to nutritional support of each body system. It does not replace the services of qualified practitioners. For serious issues and emergencies, please seek the advice of a certified or licensed health professional.

Key: ✓-Top choices ✗-Not available as a single through NSP
EO-Essential Oil (topical use) ❊-Toxic botanical for professional use only

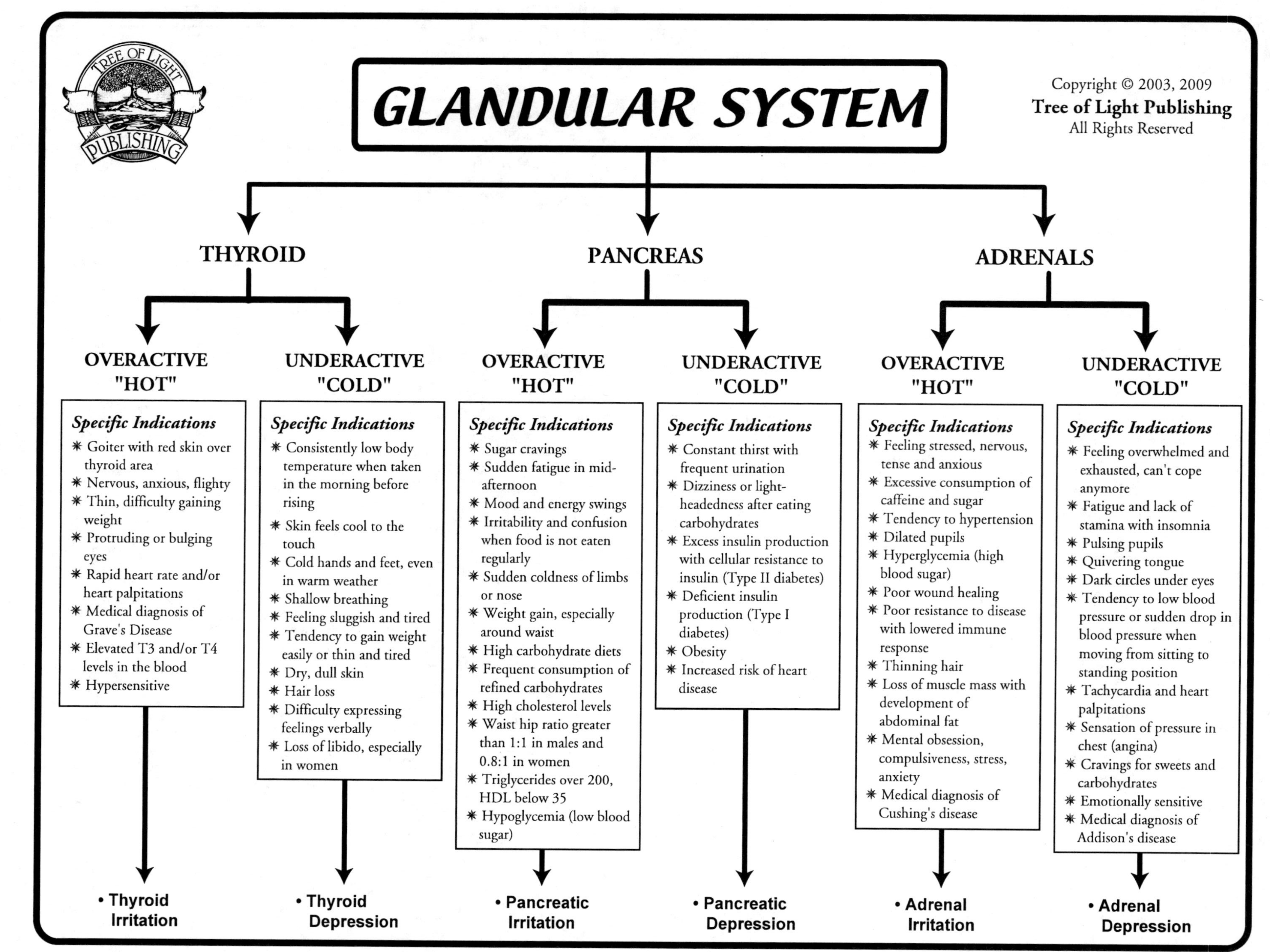

TREE OF LIGHT PUBLISHING
GLANDULAR SYSTEM
Copyright © 2003, 2009
Tree of Light Publishing
All Rights Reserved
THYROID
PANCREAS
ADRENALS
OVERACTIVE "HOT"
Specific Indications
✱ Goiter with red skin over thyroid area
✱ Nervous, anxious, flighty
✱ Thin, difficulty gaining weight
✱ Protruding or bulging eyes
✱ Rapid heart rate and/or heart palpitations
✱ Medical diagnosis of Grave's Disease
✱ Elevated T3 and/or T4 levels in the blood
✱ Hypersensitive
• Thyroid Irritation
UNDERACTIVE "COLD"
Specific Indications
✱ Consistently low body temperature when taken in the morning before rising
✱ Skin feels cool to the touch
✱ Cold hands and feet, even in warm weather
✱ Shallow breathing
✱ Feeling sluggish and tired
✱ Tendency to gain weight easily or thin and tired
✱ Dry, dull skin
✱ Hair loss
✱ Difficulty expressing feelings verbally
✱ Loss of libido, especially in women
• Thyroid Depression
OVERACTIVE "HOT"
Specific Indications
✱ Sugar cravings
✱ Sudden fatigue in mid-afternoon
✱ Mood and energy swings
✱ Irritability and confusion when food is not eaten regularly
✱ Sudden coldness of limbs or nose
✱ Weight gain, especially around waist
✱ High carbohydrate diets
✱ Frequent consumption of refined carbohydrates
✱ High cholesterol levels
✱ Waist hip ratio greater than 1:1 in males and 0.8:1 in women
✱ Triglycerides over 200, HDL below 35
✱ Hypoglycemia (low blood sugar)
• Pancreatic Irritation
UNDERACTIVE "COLD"
Specific Indications
✱ Constant thirst with frequent urination
✱ Dizziness or light-headedness after eating carbohydrates
✱ Excess insulin production with cellular resistance to insulin (Type II diabetes)
✱ Deficient insulin production (Type I diabetes)
✱ Obesity
✱ Increased risk of heart disease
• Pancreatic Depression
OVERACTIVE "HOT"
Specific Indications
✱ Feeling stressed, nervous, tense and anxious
✱ Excessive consumption of caffeine and sugar
✱ Tendency to hypertension
✱ Dilated pupils
✱ Hyperglycemia (high blood sugar)
✱ Poor wound healing
✱ Poor resistance to disease with lowered immune response
✱ Thinning hair
✱ Loss of muscle mass with development of abdominal fat
✱ Mental obsession, compulsiveness, stress, anxiety
✱ Medical diagnosis of Cushing's disease
• Adrenal Irritation
UNDERACTIVE "COLD"
Specific Indications
✱ Feeling overwhelmed and exhausted, can't cope anymore
✱ Fatigue and lack of stamina with insomnia
✱ Pulsing pupils
✱ Quivering tongue
✱ Dark circles under eyes
✱ Tendency to low blood pressure or sudden drop in blood pressure when moving from sitting to standing position
✱ Tachycardia and heart palpitations
✱ Sensation of pressure in chest (angina)
✱ Cravings for sweets and carbohydrates
✱ Emotionally sensitive
✱ Medical diagnosis of Addison's disease
• Adrenal Depression

High Thyroid (Hyperthyroidism)

Herbs

- ✗ Broccoli
- ✓ ✗ Bugleweed
- Eleuthero
- Hops
- Licorice Root
- ✓ ✗ Lemon Balm
- ✓ ✗ Motherwort
- ✗ Radishes
- ✗ Self-Heal
- ✗ Watercress

Supplements

- Magnesium Complex
- Pantothenic Acid
- Selenium

Combinations

- Adrenal Support
- ✓ IF-C
- ✓ Nervous Fatigue Formula
- TS II with Hops

Low Thyroid (Hypothyroidism)

Herbs

- ✓ Black Walnut
- ✓ Dulse
- ✗ Irish Moss
- ✓ Kelp

Supplements

- ✗ Coconut Oil
- ✗ Iodine
- Natural Salt
- MSM
- SAM-e

Combinations

- Master Gland
- Super Algae
- Target TS II
- ✓ Thyroid Activator
- ✓ Thyroid Support
- TS II w/Hops

Pancreatic Irritation

Herbs

- Bee Pollen
- Bilberry
- ✓ Licorice Root
- Red Beet Root Formula
- Spirulina
- Stevia

Supplements

- Chromium GTF
- ✓ ✗ Coconut Oil
- Free Amino Acids
- IGF-1
- L-Glutamine
- SynerProtein
- Xylitol

Combinations

- ✓ HY-A
- HY-C
- ✓ Super Algae
- SugarReg

Pancreatic Depression

Herbs

- Anamu
- Fenugreek
- ✓ Goldenseal
- Noni (Morinda)
- ✓ Nopal
- Stevia

Supplements

- Alpha Lipoic Acid
- Chromium GTF
- Omega-3
- Xylitol
- Zinc

Combinations

- Blood Sugar Formula, Ayurvedic
- PBS
- Pro-Pancreas
- ✓ Target P-14
- ✓ SugarReg

Adrenal Stress

Herbs

- ✓ Blue Vervain
- ✗ Borage
- Chamomile
- ✓ Eleuthero
- ✓ Kava Kava
- Passion Flower
- ✗ Rhodiola
- ✗ Schizandra
- Valerian

Supplements

- Balanced B-Complex
- ✓ Nutri-Calm

Combinations

- Adaptamax
- Adrenal Support
- Focus Attention
- ✓ Stress-J
- Stress Relief
- SUMA Combination
- Nerve Eight

Topical Remedies

- Lavender EO
- Lemon EO
- Roman Chamomile EO
- Rose EO

Adrenal Fatigue

Herbs

- Ho Shou Wu
- Korean Ginseng
- ✓ Licorice
- Eleuthero
- Wild American Ginseng

Supplements

- Balanced B-Complex
- ✓ Pantothenic Acid
- Vitamin C

Combinations

- Adrenal Support
- ✓ Mineral-Chi Tonic
- Energ-V
- Focus Attention
- ✓ Nervous Fatigue Formula
- ✓ SUMA Combination
- Target Endurance

Topical Remedies

- Roman Chamomile EO

Important Notice: This chart is a reference tool for the ABC+D approach to nutritional support of each body system. It does not replace the services of qualified practitioners. For serious issues and emergencies, please seek the advice of a certified or licensed health professional.

Key: ✓-Top choices ✗-Not available as a single through NSP
EO-Essential Oil (topical use) ✻-Toxic botanical for professional use only

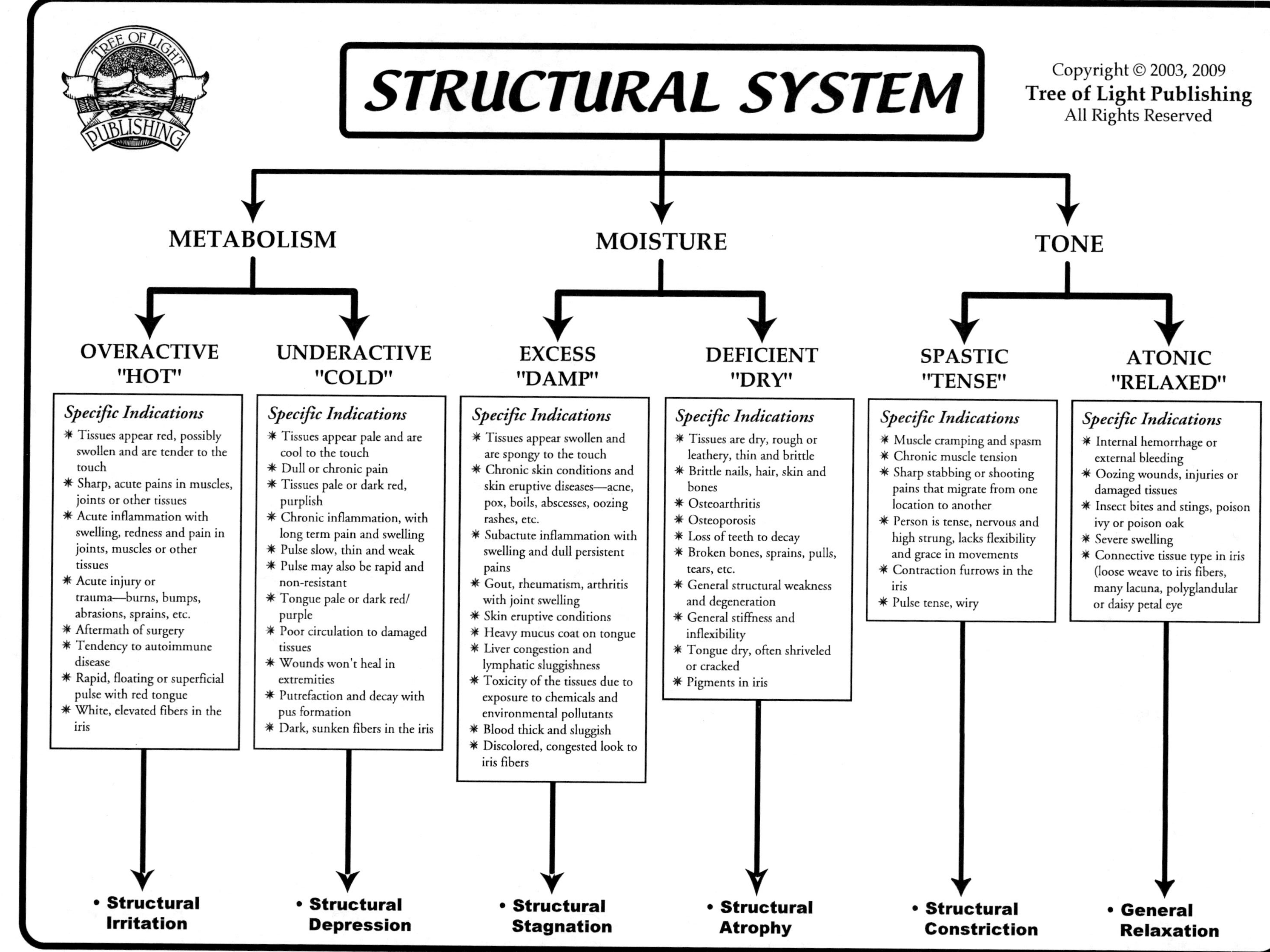

TREE OF LIGHT PUBLISHING
STRUCTURAL SYSTEM
Copyright © 2003, 2009
Tree of Light Publishing
All Rights Reserved
METABOLISM
MOISTURE
TONE
OVERACTIVE "HOT"
Specific Indications
* Tissues appear red, possibly swollen and are tender to the touch
* Sharp, acute pains in muscles, joints or other tissues
* Acute inflammation with swelling, redness and pain in joints, muscles or other tissues
* Acute injury or trauma—burns, bumps, abrasions, sprains, etc.
* Aftermath of surgery
* Tendency to autoimmune disease
* Rapid, floating or superficial pulse with red tongue
* White, elevated fibers in the iris
• Structural Irritation
UNDERACTIVE "COLD"
Specific Indications
* Tissues appear pale and are cool to the touch
* Dull or chronic pain
* Tissues pale or dark red, purplish
* Chronic inflammation, with long term pain and swelling
* Pulse slow, thin and weak
* Pulse may also be rapid and non-resistant
* Tongue pale or dark red/ purple
* Poor circulation to damaged tissues
* Wounds won't heal in extremities
* Putrefaction and decay with pus formation
* Dark, sunken fibers in the iris
• Structural Depression
EXCESS "DAMP"
Specific Indications
* Tissues appear swollen and are spongy to the touch
* Chronic skin conditions and skin eruptive diseases—acne, pox, boils, abscesses, oozing rashes, etc.
* Subactute inflammation with swelling and dull persistent pains
* Gout, rheumatism, arthritis with joint swelling
* Skin eruptive conditions
* Heavy mucus coat on tongue
* Liver congestion and lymphatic sluggishness
* Toxicity of the tissues due to exposure to chemicals and environmental pollutants
* Blood thick and sluggish
* Discolored, congested look to iris fibers
• Structural Stagnation
DEFICIENT "DRY"
Specific Indications
* Tissues are dry, rough or leathery, thin and brittle
* Brittle nails, hair, skin and bones
* Osteoarthritis
* Osteoporosis
* Loss of teeth to decay
* Broken bones, sprains, pulls, tears, etc.
* General structural weakness and degeneration
* General stiffness and inflexibility
* Tongue dry, often shriveled or cracked
* Pigments in iris
• Structural Atrophy
SPASTIC "TENSE"
Specific Indications
* Muscle cramping and spasm
* Chronic muscle tension
* Sharp stabbing or shooting pains that migrate from one location to another
* Person is tense, nervous and high strung, lacks flexibility and grace in movements
* Contraction furrows in the iris
* Pulse tense, wiry
• Structural Constriction
ATONIC "RELAXED"
Specific Indications
* Internal hemorrhage or external bleeding
* Oozing wounds, injuries or damaged tissues
* Insect bites and stings, poison ivy or poison oak
* Severe swelling
* Connective tissue type in iris (loose weave to iris fibers, many lacuna, polyglandular or daisy petal eye
• General Relaxation

Structural Irritation

Herbs

- Aloe Vera
- ✓ ✗ Arnica (used homeopathically)
- Black Cohosh
- ✗ Calendula
- ✗ Meadowsweet
- ✗ Plantain
- Rose Hips
- ✗ White Willow Bark
- ✗ Wintergreen
- Yarrow

Supplements

- Bioflavinoids
- Grapine
- Vitamin C

Combinations

- APS II with White Willow
- ✓ IF-C
- ✓ IF Relief
- Joint Support
- Nerve Eight
- Super ORAC
- Thai-Go

Topical Remedies

- ✓ Distress Remedy
- Golden Salve
- ✓ Healing AC Cream
- ✓ Nature's Fresh
- Tei Fu Oil

Structural Depression

Herbs

- ✓ Anamu
- ✓ Capsicum
- ✗ Clove
- Ginger
- ✓ ✗ Prickly Ash

Supplements

- Chondroitin
- Glucosamine Sulfate
- ✓ MSM

Combinations

- Capsicum, Garlic and Parsley
- ✓ EverFlex w/ Hyaluronic Acid
- Mega-Chel
- Skeletal Strength

Topical Remedies

- ✓ Deep Relief
- Everflex Pain Cream
- Herbal Trim Skin Treatment
- Nature's Fresh
- ✓ Tei Fu Massage Lotion
- Tei Fu Oil
- Tea Tree Oil

Structural Stagnation

Herbs

- Alfalfa
- Anamu
- ✓ Black Walnut
- ✓ Burdock
- Chickweed
- ✗ Devil's Claw
- Oregon Grape
- Pau D'Arco
- Safflowers
- Yellow Dock
- ✓ Yucca

Combinations

- All Cell Detox
- BP-X
- ✓ Eight
- Herbal Trace Minerals
- ✓ Joint Support
- Joint Health
- Kidney Drainage
- Lymphatic Drainage
- Nerve Eight
- Skin Detox, Ayurvedic

Topical Remedies

- Black Ointment
- Nature's Fresh
- Pau D'Arco Lotion

Structural Atrophy

Herbs

- Alfalfa
- Aloe Vera
- Chickweed
- ✗ Comfrey
- Dulse
- Licorice Root
- ✗ Plantain
- Slippery Elm

Supplements

- Collatrim
- Evening Primrose Oil
- Flax Seed Oil
- Chondroitin
- Glucosamine Sulfate
- Omega-3 EPA

Combinations

- ✓ Bone/Skin Poultice
- CA, Herbal
- Herbal Trace Minerals
- ✓ HSN-W
- ✓ KB-C
- Mineral Chi Tonic
- PLS II
- Super GLA
- V-X

Topical Remedies

- Aloe Vera Gel
- EverFlex Pain Cream
- Herbal Trim Skin Treatment
- Nature's Fresh

Structural Constriction

Herbs

- ✗ Agrimony
- Black Cohosh
- ✓ Kava Kava
- ✓ Lobelia
- ✗ Skullcap
- Wild Yam

Supplements

- ✓ Magnesium Complex

Combinations

- ✓ Cellular Energy
- ✓ Cramp Relief
- Fibralgia
- PLS II
- Stress Relief
- Target Endurance

Structural Relaxation

Herbs

- Bayberry
- ✗ Calendula
- Horsetail
- Uva Ursi
- ✓ White Oak Bark
- ✓ Yarrow

Supplements

- Calcium
- Chlorophyll Capsules
- IGF-1
- Mineral Chi Tonic

Combinations

- CA, Herbal
- Hista-Block
- ✓ HSN-W
- ✓ Menstrual-Reg
- Skeletal Strength

Important Notice: This chart is a reference tool for the ABC+D approach to nutritional support of each body system. It does not replace the services of qualified practitioners. For serious issues and emergencies, please seek the advice of a certified or licensed health professional.

Key: ✓-Top choices ✗-Not available as a single through NSP
EO-Essential Oil (topical use) ❊-Toxic botanical for professional use only

IMMUNE SYSTEM

FOUR STAGES OF DISEASE (INFECTION)

ACUTE (IRRITATION)

Specific Indications

* Initial stages of colds, flu, fevers and other contagious diseases (first 24-48 hours)
* Redness, fever
* Thin, watery or white mucus
* Pulse rapid, superficial
* Tongue bright red, often elongated
* Person generally healthy, but may be run down due to stress
* Active discharge from the body (watery eyes, sinus drainage, coughing, sneezing, sweating, diarrhea, etc.)

→ **• Acute Infection**

SUBACUTE (STAGNATION)

Specific Indications

* Later stages of colds, flu, fevers and other contagious diseases (after 24-48 hours)
* Mucus thick, discolored
* Low grade fever
* Swelling
* Sluggish, worn out feeling
* Tongue damp, swollen
* Pulse slippery or rolling
* Swampy, boggy condition
* Lymphatic congestion, swollen lymph nodes
* Frequent sore throats and earaches
* Reduced discharge
* Congestion in lungs and sinuses
* Loss of appetite
* Slow recovery from colds and flu

→ **• Subacute Infection**

CHRONIC (DEPRESSION)

Specific Indications

* Lingering, low grade, chronic infections
* Weakness, long term sickness, slow convalescence
* Tongue dark red or purplish, often with heavy yellow coating
* Tongue sometimes pale with small red spots
* Pulse rapid, but feeble
* Chronic, low grade fever or alternating fever with chills
* Parasitic infections
* Chronic candida or yeast infections
* Chronic cough or sinusitis
* Wounds that won't heal

→ **• Chronic Infection**

DEGENERATIVE (ATROPHY)

Specific Indications

* Chronic immune weakness with frequent infections and fatigue
* Chronic respiratory weakness, frequent infections, low resistance
* Decaying flesh, oozing sores, pus filled wounds
* Bad body odor, may smell like rotten meat or ammonia
* Constantly sickly
* Acquired immune deficiency syndrome (AIDS)
* Antibiotic resistant infections
* Flesh-eating bacteria
* Skin is dry and rough
* Pulse is weak, thin and deep
* Tongue dry, often shriveled or cracked, black in severe cases
* Person may be near death in severe cases

→ **• Degeneration/ Petrification**

IMMUNE ACTIVITY

OVERACTIVE (AUTOIMMUNE)

Specific Indications

* Conditions where the body's immune system is attacking its own tissues, such as:
* Multiple Sclerosis (MS)
* Rheumatoid Arthritis
* Fibromyalgia
* Myasthenia Gravis
* Chronic Fatigue Syndrome
* Amyotropic Lateral Sclerosis -ALS (Lou Gerhig's Disease)
* Lupus (Systemic Lupus Eyrthematosus)
* Scleroderma
* Hashimoto's Thyroiditis
* Graves' Disease
* Autoimmune Hepatitis
* Type I Diabetes
* May also involve vaccine reactions

→ **• Overactive Immune System**

UNDERACTIVE (ABNORMAL CELLS)

Specific Indications

* Medical diagnosis of cancer
* Cysts
* Abscesses
* Growths and lumps
* Chronic swollen lymph nodes or severe lymphatic swelling
* Changes in moles
* Warts
* Endometriosis
* Psoriasis

→ **• Underactive Immune System**

Acute Infection

- ✓ AL-J (V)
- Astragalus (V)
- Capsicum (V)
- CC-A with Yerba Santa (V)
- CC-A (V)
- Elderberry Defense (V)
- Elderberry Plus (V)
- Garlic, Raw or High Potency (B, V, F, P)
- Ginger (V)
- ✓ HCP-X (B, V)
- ✓ IF-C (B, V)
- Immune Stimulator (B, V)
- Rose Hips (V)
- ✓ Silver Shield (B, V, F)
- Tea Tree EO (B, V, F)
- Vitamin C (V)
- ✓ Yarrow (B, V)
- Zinc Lozenges (V)

Subacute Infection

- ✓ AL-J (B, V)
- ✗ Boneset (V)
- Blue Vervain
- Burdock
- Echinacea (B, V)
- ✓ Echinacea/ Goldenseal (B)
- Goldenseal (B)
- IN-X (B, V)
- Lobelia
- Lymph Gland Cleanse (B)
- Lymph Gland Cleanse HY (B)
- ✓ Lymphatic Drainage Formula
- ✓ Lymphomax (B, V)
- Oregon Grape (B)
- Pau D'Arco (F)
- Protease (B, V, F, P)
- Red Clover
- ✗ Red Root (B)
- ✓ Silver Shield (B, V, F)
- ✓ Ultimate Echinacea (B, V)
- ✓ VS-C (B, V)
- Yarrow (V)

Chronic Infection

- Astragalus (V)
- ✓ Candida Clear (F)
- Caprylic Acid Combo
- ✓ Cordyceps
- ✓ Garlic, Raw or High Potency (B, V, F)
- ✓ GastroHealth (B, P)
- Guardian EO Blend (B, V, F)
- ✓ Immune Stimulator (V, B)
- Lavender EO (F)
- Olive Leaf (B, V)
- Oregano, Wild EO (B, F)
- Paw Paw Cell Reg (V, F, P)
- Pau d'Arco (F)
- Rosemary EO (B, V)
- Seasonal Defense (B, V)
- ✓ Silver Shield (B, V, F)
- St. John's Wort (V)
- THIM-J
- Thyme EO (B, V, F)
- Trigger Immune
- Ultimate Echinacea
- ✓ Una De Gato Combination
- Viral Recovery Homeopathic (V)
- VS-C (B, V)
- ✓ Yeast/Fungal Detox (F)

Degeneration/ Petrification

- ✓ Seek Medical Assistance
- ✓ ✗ Baptista (B)
- Echinacea (B)
- ✓ Garlic, Raw (B, P)
- Guardian EO Blend (B, P)
- Lobelia (B)
- ✗ Red Root (B)
- ✓ Silver Shield (B, V, P)
- Tea Tree EO (B)
- Ultimate Echinacea (B)

Autoimmune Disorders

- Antioxidant Arsenal
- Antioxidants with Grapine
- ✓ Black Walnut
- Carotenoid Blend
- Chlorophyll
- ✓ Colostrum
- Co-Q10
- Eleuthero
- Evening Primrose Oil
- ✓ Fibralgia
- Flax Seed Oil
- Food Enzymes
- Licorice Root
- ✓ Mineral-Chi Tonic
- N-Acetyl Cysteine
- ✓ Omega-3 EPA
- Proactazyme Plus
- ✓ Protease Plus
- Silver Shield
- Suma Combination
- Super Antioxidant
- Thai-Go
- ✓ Yucca

Avoid Immune Stimulants Like:

- Colostrum Plus
- Echinacea
- Ginseng, Korean or Wild American
- Immune Stimulator
- Trigger Immune
- Ultimate Echinacea

Underactive Immune System

- Adaptamax
- Anamu
- Astragalus
- Burdock
- Cordyceps
- Defense Maintenance
- ✓ E-Tea
- ✓ Immune Stimulator
- ✗ Maitake Mushrooms
- ✓ Mineral-Chi Tonic
- ⁂ ✗ Mistletoe
- Germanium Combination
- Pau d'Arco
- ✓ Paw Paw Cell Reg
- Protease Plus and High Potency Protease
- Red Clover
- ✗ Red Root
- SC Formula
- SUMA Combination
- Trigger Immune
- ✓ Una De Gato Combination

Letters indicate which types of infection the remedy works best for: B = Bacterial Infections V = Viral Infections F = Fungal Infections P = Parasites

Important Notice: This chart is a reference tool for the ABC+D approach to nutritional support of each body system. It does not replace the services of qualified practitioners. For serious issues and emergencies, please seek the advice of a certified or licensed health professional.

Key: ✓-Top choices ✗-Not available as a single through NSP
EO-Essential Oil (topical use) ⁂-Toxic botanical for professional use only

REPRODUCTIVE SYSTEM

FEMALE REPRODUCTIVE PROBLEMS

ESTROGEN

High Estrogen
* PMS with irritability, anxiety, mood swings
* Low sex drive
* Uterine fibroids
* Heavy menstrual bleeding
* Headaches, migraines
* Nervous tension, panic attacks
* Increased risk of breast cancer
* Cystic breast disease

Low Estrogen
* PMS with depression, mental confusion, sadness
* Insomnia
* Thinning skin
* Low sex drive
* Vaginal dryness, painful intercourse
* Forgetfulness, mental confusion, brain fog
* Wrinkles, acne, oily skin
* Hot flashes, night sweats
* Decreased breast size

• **Estrogen Imbalance**

PROGESTERONE

High Progesterone
* PMS with depression, sadness and mental confusion
* Fatigue, drowsiness
* Forgetfulness
* Increased appetite
* Painful swollen breasts
* Cystitis, bladder infections
* Yeast infections
* Taking birth control pills

Low Progesterone
* PMS with irritability, anxiety and mood swings
* Nervous tension
* Low bone density
* Tendency to miscarriage
* Irregular menstrual cycle
* Infertility
* Vaginal dryness and painful intercourse
* Low sex drive
* Uterine fibroids
* Fibrocystic breasts

• **Progesterone Imbalance**

OTHER HORMONES

Oxytocin
* Causes uterine contractions during childbirth, expels placenta after childbirth
* Excess contributes to painful menstrual cramps
* Promotes bonding between mother and child and between couples engaging in intercourse
* Causes milk letdown in breast feeding

Prolactin
* Assists in initiating and sustaining breast milk production after childbirth

Follicle-Stimulating Hormone (FSH)
* Promotes follicle development, egg maturation and fertility
* Stimulates estrogen secretion

• **Other Female Hormones**

OTHER FEMALE ISSUES

Menstrual Irregularities
* Painful periods (dysmenorrhea) with sharp, cramping pains
* Painful periods with dull congestive pain
* Heavy menstrual bleeding
* Lack of periods (amenorrhea)

Menopausal Issues
* Hot flashes
* Depression, anxiety or mood changes
* Vaginal dryness
* Osteoporosis
* Loss of libido (sex drive)
* Cession of periods or irregular periods

• **Other Female Issues**

MALE REPRODUCTIVE PROBLEMS

TESTOSTERONE

High Testosterone
* Excessive aggression and irritability
* Unlikely to be a problem in men
* In women, high testosterone can cause facial hair, baldness and aggression

Low Testosterone
* Loss of energy and stamina
* Depression
* Loss of muscle mass and weight gain
* Erectile dysfunction or lack of firm erections
* Loss of libido (sex drive)
* Loss of emotional drive
* Increased risk of heart disease
* Low self esteem
* Decreased athletic performance
* Declining health with age, andropause

• **Testosterone Imbalance**

OTHER MALE ISSUES

Prostatitis & BPH
* Decreased urine flow
* Misaim, difficulty controlling flow of urine stream
* Frequent nighttime urination
* Difficulty urinating
* Painful urination
* Prostate swelling or enlargement
* Medical diagnosis of prostatitis or benign prostate hyperplasia (BPH)

Erectile Dysfunction
* Difficulty getting or maintaining an erection
* Lack of "hardness" with erections
* Decreased sexual performance

• **Other Male Issues**

Estrogen Imbalance

High Estrogen

C-X
✓ ✗ Chaste Tree
Enviro-Detox
Folic Acid
✓ Indol-3-Carbynol
✓ Liver Balance
✓ Magnesium Complex
Monthly Maintenance
✗ Pasque Flower
✓ Pro-G-Yam Cream
SAM-e
Sarsaparilla
✓ Vitamin B6
✓ Wild Yam and Chaste Tree

Low Estrogen

✓ Black Cohosh
B-Complex
✓ Breast Assured
✓ Breast Enhance
Clary Sage EO
✓ DHEA-F
Dong Quai
FC with Dong Quai
✓ Flash Ease
Geranium EO
Heavy Metal Detox
Hops
Licorice
Phyto-Soy
Pink Grapefruit (fruit and EO)
Pregnenolone
Red Clover

Progesterone Imbalance

High Progesterone

Bergamot EO
✓ Black Cohosh
Clary Sage EO
Female Comfort
Geranium EO
✗ L-Tyrosine
✓ Magnesium Complex
Nutri-Calm
✓ Mood Elevator
Phyto-Soy
Rose EO
Stress-J
✓ Vitamin B6

Low Progesterone

Blue Cohosh
C-X
✓ False Unicorn
Magnesium Complex
Monthly Maintenance
✓ Pro-G-Yam Cream
Sarsaparilla
Vitamin B6
Wild Yam and Chaste Tree

Other Female Hormones

Oxytocin Enhancers

✓ Blue Cohosh
✓ Chocolate
Clove EO
✗ Lady's Mantle
Goldenseal
✗ Scotch Broom

Oxytocin Inhibitors

✗ Cramp Bark
Jasmine EO
Lavender EO
✓ Menstrual Reg
✗ White Willow

Prolactin Enhancers (Herbs that enhance breast milk)

Alfalfa
Blessed Thistle
✗ Fennel
Marshmallow
Milk Thistle
✗ Nettle Leaf

Prolactin Inhibitors (Herbs that dry up breast milk)

Parsley
Sage

FSH Inhibitors

Hops
Sage
✗ White Willow
✓ Wild Yam and Chaste Tree

Other Female Issues

Heavy Menstrual Bleeding

✓ Capsicum
Bayberry
Menstrual Reg
Sarsaparilla
Yarrow

Painful Periods (Dysmenorrhea with sharp cramps)

✓ Cramp Relief
Lobelia
✓ Magnesium
Menstrual Reg
Wild Yam and Chaste Tree

Painful Periods (Dysmenorrhea with dull pains)

Caster Oil Packs
✓ Dong Quai
✓ FCS II
Ginger
✓ Monthly Maintenance
Niacin
Vari-Gone

Menopausal Issues (Remedies that aid menopause)

✓ Adrenal Support
Black Cohosh
✓ C-X
Eleuthero
Evening Primrose Oil
False Unicorn
✓ Flash-Ease
Dong Quai
✗ Kudzu
Licorice Root
NF-X
✓ Omega 3
Pantothenic Acid
✓ Pro G-Yam Cream
✓ Skeletal Strength
Super GLA
Vitamin B-Complex
Vitamin E

Testosterone Imbalance

High Testosterone

C-X
✓ Liver Balance
Milk Thistle Combination

Low Testosterone

✓ DHEA-M
Eleuthero
✓ Ginseng, Korean
✗ Horny Goat Weed
✗ Muira Puama
✓ ✗ Pine Tree Pollen
Pregnenolone
Sarsaparilla
✗ Tienchi ginseng
✗ Tribulus (Puncture Vine)
✓ X-Action for Men
Zinc

Other Male Issues

Prostatitis & BPH

Eleuthero
✓ KB-C
MACA
✓ Men's Formula
✗ Nettle Root
Omega 3
P-X
✓ PS II
✗ Pygeum
Saw Palmetto
✗ White Sage
Zinc

Erectile Dysfunction

✓ Damiana
Eleuthero
✓ Ginseng, Korean
KB-C
MACA
✓ Mega-Chel
Men's Formula
✓ RG-Max
X-Action for Men
✓ ✗ Yohimbe

Important Notice: This chart is a reference tool for the ABC+D approach to nutritional support of each body system. It does not replace the services of qualified practitioners. For serious issues and emergencies, please seek the advice of a certified or licensed health professional.

Key: ✓-Top choices ✗-Not available as a single through NSP
EO-Essential Oil (topical use) ❋-Toxic botanical for professional use only

Section Two

Products

Section Two

Product Guide

This section allows you to look up products alphabetically and learn about their historical uses, the body systems they affect, their properties, how to use them, any warnings or contraindications, and what specific product and dosage forms are available. The best uses for each product, as well as its most important properties and affected body systems are highlighted in bold type.

~5-A~

5-HTP

Product Type: Formula

Properties: Antidepressant, Appetite Suppressant

Systems Affected: Brain, Digestive System, Dopamine, Hypothalamus, Nerves, Pineal, **Pituitary (infundibulum), Serotonin,** Weight Loss

Conditions: Addictions (sugar or food), Anxiety (Panic Attack), Appetite (deficient), Appetite (excessive), Defensiveness, **Depression, Insomnia,** Pregnancy (herbs and supplements to avoid during), Stress, Weight Loss (aids for)

Usage: Hydroxytryptophan is a precursor to serotonin, a neurotransmitter that regulates appetite, mood and sleep. This product can be used for depression involving low serotonin levels. If a person is taking an SSRI (selective serotonin reuptake inhibitor), 5-HTP can be used to help with the side effects of discontinuing it as 5-HTP will have a similar effect. Never discontinue any SSRI medication all at once and always adjust these medications while being monitored by the prescribing physician or a health professional. 5-HTP Power can be used with St. John's wort or Mood Elevator for depression. A great remedy for O blood types because it raises dopamine levels, which are usually low when O blood types feel depressed. Take 1 capsule three times daily with a meal for depression. 5-HTP can also help with carbohydrate cravings. Low serotonin levels are sometimes the cause of carbohydrate cravings. Because it reduces these cravings, 5-HTP can also be combined with weight loss products to support weight loss. As a sleep aid, take 3 capsules of 5-HTP Power with an evening meal. Before bedtime, turn the lights down low or otherwise darken the room. 5-HTP is converted to serotonin and when it is dark, the pineal gland converts the serotonin to melatonin a hormone that induces sleep.

Warnings: NSP recommends using no more than 3 capsules per day. If taking a prescription medication, consult a health-care practitioner before taking this product. Pregnant or nursing women should seek the advice of a health-care practitioner before using this supplement. Avoid with ADHD since high dopamine levels are associated with hyperactivity.

5-HTP - Capsule (60)

Available in: US - Stock #2806-4

Ingredients: 5-Hydroxytryptophan, eleuthero root (Siberian ginseng), ashwaganda and suma, vitamin B6 and zinc

HTP Power - Capsule (100)

Available in: Canada - Stock #2806-4

Ingredients: Same as Stock #2806-4

5-W

Product Type: Formula

Properties: Antispasmodic, Glandular, Parturient, Tonic, Uterine Tonic

Systems Affected: Female Reproductive

Conditions: Labor and Delivery, Pregnancy (herbs and supplements to avoid during), Varicose Veins

Usage: This formula is used during the last five weeks of pregnancy to help induce labor and make childbirth easier. It promotes uterine contractions, tones the uterus and helps improve circulation. Take 2 capsules three times per day only during the last 5 weeks of pregnancy. Consume with a minimum of 12 ounces of water. Can be useful with hormonally induced vasoconstrictive headaches that occur in conjunction with menses. Take 2 capsules as needed every 3-4 hours as needed during a headache and 2-4 capsules two times per day several days prior to onset of menses if headaches occur regularly.

Warnings: Do not use where vasodilative migraine headaches are present as black cohosh exacerbates this type of migraine. Some midwives caution that using black cohosh during the last 5 weeks of pregnancy may increase bleeding during labor in some women. If a tendency towards heavy bleeding exists do not use this formula. Avoid use in the first and second trimesters of pregnancy.

5-W - Capsule (100)

Available in: US - Stock #1120-5

Ingredients: Black cohosh, squaw vine, dong quai, butcher's broom, red raspberry

7-Keto

Product Type: Nutrient

Properties: **Anti-obesic**, Immune Stimulant

Systems Affected: Liver, Parathyroid, Thyroid, Weight Loss

Conditions: Acquired Immune Deficiency Syndrome (AIDS/HIV), Alzheimer's Disease, Appetite (excessive), Arthritis, Blood Pressure (high), Diabetes, Grave's Disease, Hypothyroid, Lupus, Parkinson's Disease, **Pregnancy (herbs and supplements to avoid during)**, Weight Loss (aids for)

Usage: This supplement stimulates the conversion of T4, the inactive form of the thyroid hormone, to T3 the active form. This increases the metabolic rate and causes the body to burn fat at an accelerated rate. Clinical trials also showed this supplement to enhance the immune system, increase energy, and enhance memory. 7-KetoTM does not convert into the sex steroids, testosterone and estrogen, as does DHEA, so it may provide some of the same benefits as DHEA, but without the hormonal side effects. As an aid to weight loss, take 1-2 capsule daily with a meal.

Warnings: Those with thyroid problems should consult their health care professional prior to use. Use of this product "burns up" T4 reserves in the blood and long term use may affect thyroid function. This product is not recommended for people suffering from low thyroid. Prolonged use may cause some symptoms of low thyroid.

7-Keto™ - Capsule (30)

Available in: US - Stock #2922-4

Ingredients: 75 mg. of 7-Keto (a metabolite of dehydroepiandrosterone or DHEA), chickweed

Acidophilus

See *Probiotics*

Acne

Product Type: Homeopathic

Properties: Vulnerary

Systems Affected: Skin

Conditions: Acne (Pimples, Blackheads)

Usage: Used for the treatment and prevention of pimples. Taken internally to clear and heal skin, dry up pimples, and prevent future outbreaks. Take 10-15 drops under tongue four-six times daily, decreasing to three times daily after improvement.

Warnings: NSP does not recommend this product for children under the age of 12 except on the advice of a health professional. Other than the alcohol content, which is not good for infants, we consider this product safe for children.

Acne (1 fl. oz.) - Homeopathic (Liquid)

Available in: US - Stock #8890-9

Ingredients: Echinacea angustifolia (Cone Flower) 4x, Kali Bromatum (Potassium Bromide) 4x, Mezereum (Spurge Olive) 4x, Belladonna (Nightshade) 6x, Ledum palustre (Wild Rosemary) 6x, Graphites (Graphite) 10x, Carbo Vegetabilis (Vegetable Carbon) 12x, Hepar Sulfuris Calcareum (Calcium Sulfide) 12x, purified water and 20% USP alcohol

Acne Treatment Gel

Product Type: Household

Properties: Anti-inflammatory, Antiseptic, Astringent, Cosmetic

Systems Affected: Skin

Conditions: **Acne (Pimples, Blackheads)**, Skin (infections)

Usage: A topical gel for the treatment of acne. It helps to fight infection, reduce inflammation, tone skin pores and heal blemishes. Apply topically following directions on the product. Acne is a whole body condition that should be addressed by supplements to cleanse the colon and liver and otherwise improve general health.

Warnings: No known warnings.

Acne Treatment Gel (1 fl. oz.) - Topical

Available in: US - Stock #6017-8

Ingredients: Salicylic acid, water, butylene glycol, Artemisia princeps, algae extract, alpha-glucan oligosaccharide, Glycyrrhiza glabra (licorice) root extract, retinyl palmitate (Vitamin A), panthenyl triacetate (Vitamin B), tetrahexyldecyl ascorbate (vitamin C), tocopheryl acetate (vitamin E), Aloe barbadensis, Camellia oleifera (green tea), Chamomilla recutita flower extract, Ginkgo biloba extract, Panax ginseng root extract, Ulva lactuca (sea lettuce) extract, Vitis vinifera (grape) seed extract, ergothioneine, sclerotium gum, PEG-8 dimethicone, phenoxyethanol, methylparaben, butylparaben, ethylparaben, propylparaben

AD-C

See *Mood Elevator*

AdaptaMax

Product Type: Formula

Properties: Adaptagen, Anti-aging, Anxiolytic, Glandular, Immune Amphoterics, Tonic

Systems Affected: Glandular System, Immune System, Liver, Nerves

Conditions: Acquired Immune Deficiency Syndrome (AIDS/HIV), Addison's Disease, **Adrenals (exhaustion, weakness or burnout), Aging (prevention), Anxiety (Panic Attack)**, Anxiety Disorders, Attention Deficit Disorder (ADD, ADHD), Autoimmune Disorders, Blood Pressure (high), Circulation (poor), Convulsions, Cushing's Disease, Depression, Diabetes, Energy (lack of), Exercise, **Fatigue**, Free Radical Damage, Hashimoto's Disease (Thyroiditis), Insomnia, Pregnancy (herbs and supplements to avoid during), **Stress**, Thinking (cloudy)

Usage: AdaptaMax is an adaptagenic formula, meaning it helps the body deal better with stress. It helps with fatigue, immune weakness, and symptoms of aging brought on by excessive stress. It works best with A blood types. Blood type O does better with Suma as an adaptagenic formula. This blend also has antioxidant properties. Take 2 capsules with a meal two to three times daily.

Warnings: Pregnant or lactating women should consult their health care professional prior to taking this supplement.

AdaptaMax - Capsule (100)

Available in: US - Stock #872-9

Ingredients: Chromium (amino acid chelate), Korean ginseng extract, rhodiola extract, eleuthero root, gynostemma extract, ashwaganda, suma, alfalfa, astragalus, kelp, reishi mushroom, rosemary extract, ginkgo concentrate, broccoli, carrot, red beet, rosemary, tomato, turmeric, cabbage, Chinese cabbage, schizandra, grapefruit bioflavonoid, hesperidin, and orange bioflavonoid

Adrenal Support

Product Type: Formula

Properties: Adrenal Tonic, Antilipemic, Anxiolytic, **Glandular, Hypolipidemic**, Mineralizer, Stimulant

Systems Affected: Adrenal Glands, Pituitary (general)

Conditions: Addictions (alcohol), Addictions (coffee, caffeine), Addictions (drugs), Addictions (sugar or food), Addison's Disease, Adrenals (exhaustion, weakness or burnout), Allergies (food), **Anxiety (Panic Attack), Anxiety Disorders, Asthma, Attention Deficit Disorder** (ADD, ADHD), **Autoimmune Disorders, Bedwetting, Blood Pressure (low)**, Body Building, **Cholesterol (low), Confusion**, Congestion (bronchial), Cough (dry), Dehydration, Depression, Diabetes, **Eczema, Emotional Sensitivity**, Endurance (lack of), **Energy (lack of), Fatigue, Fear (excessive)**, Fibromyalgia Syndrome (FMS), Grave's Disease, **Hot Flashes, Hypochondria**, Hypothyroid, Insomnia, **Jet Lag, Lupus, Menopause, Mental Illness**, Nervous Exhaustion (Enervation), **Nervousness, Nightmares, Obsessive Compulsive Disorder, Phobias**, PMS Type S, **Post Traumatic Stress Disorder, Restless Dreams, Sleep (restless and disturbed)**, Tremors, Triglycerides (high), Triglycerides (low), Weight Loss (aids for), **Worry**

Usage: This is a highly effective product for treating adrenal weakness. It strengthens the adrenal glands, increasing energy and stamina. Helps to maintain normal blood sugar levels, blood pressure and increase resistance to stress. Adrenal weakness is often involved in autoimmune disorders, hypoglycemia, fatigue, food cravings for sweets and alcohol, depression, poor digestion and poor bowel function, making this product potentially helpful in all of these conditions. To strengthen under active adrenal function take 1 capsule one to three times daily. This product is ideally used for a short period of time (2-3 months) to rebuild severely depleted adrenals. Follow the use of Adrenal Support by using an herbal formula that supports the adrenal glands, such as Nervous Fatigue Formula, Suma Combination or Energ-V.

Warnings: Not recommended for long-term use.

Adrenal Support - Capsule (60)

Available in: US & Canada - Stock #1507-0

Ingredients: Bovine adrenal glandular substance from New Zealand, vitamins B1, B2, B6, pantothenic acid and vitamin C, plus zinc, potassium, magnesium, protease, borage oil powder, licorice root, and schizandra fruit

AG-C

See *Anti-Gas (Chinese)*

AL-C

See *Breathe EZ*

Alfalfa

Latin Name: *Medicago sativa*

Product Type: Single Herb

Properties: Alterative (Blood Purifier), Anti-arthritic, **Anticarious, Anticoagulant (Blood Thinner)**, Antilipemic, **Aperitive**, Appetite Stimulant, Bitter, Digestive Tonic, **Galactagogue**, Mineralizer, Stomachic

Systems Affected: Bones, Cardiovascular System, Digestive System, Nerves, Pineal, **Pituitary (anterior)**, **Pituitary (general)**, **Pituitary (posterior)**, Red Blood Cells, Skeletal System, Structural System

Conditions: Allergies (food), **Anemia**, Aneurysm, Anorexia, Appetite (deficient), Arthritis, Asthma, Bell's Palsy, Blood Clots (prevention of), Breast Milk (increase or enrich), Bursitis, Cancer (natural therapy for), Cholesterol (high), Convalescence, Cystic Fibrosis, Dyspepsia, Endometriosis, **Gout**, Indigestion, Inflammatory Bowel Disorders (Colitis, IBS), Jaundice (adults), Morning Sickness, Osteoporosis, Overacidity, Oxygen Deficiency, Pernicious Anemia, **Pregnancy (herbs and supplements for)**, Rheumatoid Arthritis (Rheumatism), Rosacea, Thrombosis, **Tooth Decay (prevention)**, Ulcers, Uric Acid Retention

Usage: Alfalfa has been called the king of herbs and has been used since ancient times. Alfalfa is a rich source of many vitamins, minerals, amino acid and other nutrients; particularly trace minerals. This is due to its roots growing 30-60 feet deep to pick up minerals and water other plants cannot reach. Its trace mineral content and amino acids are what makes it valuable for the pituitary because trace mineral and amino acid deficiencies adversely affect this gland. It acts as a mild alterative and blood purifier and has been used for arthritis (due to its blood thinning qualities), poor appetite, general weakness and mineral deficiencies. Use 2 capsules three or four times daily. Add powder from capsules to cool or room temperature drinks for mineralization of tissues and as a tonic for poor digestion. Use in a tea with red raspberry for morning sickness and with peppermint for poor appetite.

Warnings: Large amounts can thin the blood. Should not be taken with blood thinners. Should not be used as a single herb in cases of pernicious anemia or when drugs that increase viscosity of blood are being used. Should not be used the last month of pregnancy and only after delivery. Do not use when signs of hemorrhaging are present.

Alfalfa (100) - Capsule (100)

Available in: US - Stock #30-2

Ingredients: Medicago sativa (alfalfa)

Alfalfa (270) - Capsule (270)

Available in: US & Canada - Stock #32-7

Ingredients: Same as Stock #30-2

Algin

Product Type: Nutrient

Properties: Anticholesteremic, Chelating

Systems Affected: Immune System, Intestinal System, Small Intestines

Conditions: Cholesterol (high), **Heavy Metal Poisoning, Lead Poisoning, Mercury Poisoning, Radiation Sickness, Vaccines (detoxification from)**

Usage: Algin is a mucilage derived from kelp. It absorbs toxins in the digest tract, particularly heavy metals like mercury. Best used in conjunction with Heavy Metal Detox. Take 2-3 capsules three times daily with plenty of water.

Warnings: Like other mucilaginous substances, algin absorbs nutrients. Separate any vitamins and minerals you are taking by at least one hour. Drink plenty of water because fibers like algin can bind the colon.

Algin - Capsule (100)

Available in: US - Stock #675-1

Ingredients: Sodium alginate

ALJ

Product Type: Formula

Properties: **Anti-allergenic, Anticatarrhal, Decongestant, Expectorant**, Lung Tonic, Pectoral, Pulmonary, Stomachic

Systems Affected: Bronchials, Digestive System, **Lungs**, Lymphatic System, Mucus Membranes, Respiratory System, Sinuses, Small Intestines, Thymus

Conditions: Abdominal Pain and Inflammation, Allergies (food), **Allergies (respiratory), Asthma, Bronchitis**, Children's Remedy, **Colds (decongestant), Colds (general remedies for), Congestion (bronchial)**, Congestion (general), Congestion (lungs), Congestion (lymphatic), Congestion (sinus), **Cough (damp), Cough (general)**, Diarrhea, Dyspepsia, Ear Infection or Earache, **Eyes (red or itching)**, Glands (swollen lymph), Grief (excessive), **Headache (sinus)**, Indigestion, **Lungs (fluid in), Pneumonia, Rhinitis, Snoring**

Usage: One of the best 'all round' formulas for almost every type of respiratory problem, including sinus problems, allergies, hay fever, pneumonia, coughs and chronic sinus congestion. For chronic conditions take 2 capsules three times daily with meals. For acute problems take 2-4 capsules every two to four hours. For liquid ALJ use 1/2-1 teaspoon every two to four hours. Works well combined with garlic, Immune Stimulator or Ultimate Echinacea for respiratory infections.

Warnings: No known warnings.

ALJ Capsules - Capsule (100)

Available in: US - Stock #774-3

Ingredients: Boneset, horseradish, mullein, fennel, fenugreek

ALJ Vegitabs - Vegitab (100)

Available in: US - Stock #767-4

Ingredients: Same as Stock #774-3

ALJ Vegitabs - Vegitab (270)

Available in: US - Stock #768-6

Ingredients: Same as Stock #774-3

ALJ Vegitabs - Vegitab (100)

Available in: Canada - Stock #778-8

Ingredients: Horseradish root, mullein leaves, fenugreek seeds, and fennel seeds

ALJ Liquid (2 fl. oz.) - Liquid (Glycerite)

Available in: US & Canada - Stock #3166-5

Ingredients: Glycerine, water, boneset, horseradish, mullein, fennel, fenugreek

All Cell Detox

Other Names: Special Formula #1

Product Type: Formula

Properties: **Alterative (Blood Purifier)**, Hepatic, Laxative (general), Parasiticide, Vermifuge

Systems Affected: Digestive System, Female Reproductive, **Intestinal System**, Liver, Mitochondria, Skin

Conditions: **Acne (Pimples, Blackheads)**, Addictions (drugs), Anemia, Arthritis, Blood Pressure (high), Body Odor, **Boils**, **Breast Lumps**, Cancer (natural therapy for), Cancer (prevention), Cellulite, Chemical Poisoning, Congestion (general), **Constipation (adults)**, **Cysts**, Digestion (poor), Dyspepsia, Fatty Tumors or Deposits, **Fibroids (uterine)**, Lou Gehrig's Disease, **Pap Smear (abnormal)**, Parasites (general), **Polyps**, Psoriasis, **Toxemia**

Usage: This is a general cleansing formula. It contains herbs that stimulate digestive organs, cleanse the bowel, strengthen the liver, acts as blood purifiers and glandular tonics. It is an excellent basis for a colon-cleansing program. May also be taken for endometriosis, gelatinous breast lumps, and for sebaceous cysts. Recommended use is 2 capsules with each meal and works best when used with psyllium hulls or Nature's Three. It can be used externally as a poultice for cysts. This formula is also included in the Tiao He Cleanse.

Warnings: No known warnings.

All Cell Detox - Capsule (100)

Available in: US - Stock #1072-4

Ingredients: Gentian, Irish moss plant, cascara sagrada, fenugreek seeds, golden seal, slippery elm bark, safflower, black walnut hulls, myrrh gum, parthenium root, yellow dock root, dandelion root, Oregon grape root, uva ursi leaves, chickweed herb, catnip, cyani flowers.

All Cell Detox - Capsule (100)

Available in: Canada - Stock #1078-2

Ingredients: Gentiana lutea L. (Gentian) root and rhizome, Chondrus crispus L. (Irish moss) plant, Trigonella foenum-graecum L. (Fenugreek), Hydrastis Canadensis L. (Goldenseal) root and rhizome, Ulmus fulva Michaux (slippery elm) bark, Carthamus tinctorius L. (Safflower) flowers, Juglans nigra L. (Black walnut) hulls, Parthenium integrifolium L. (Parthenium) root, Commiphora molmol Engl. Ex Tschirch (Myrrh gum) stem, Rumex crispus L. (Yellow dock) root, Taraxacum officinale Wiggers (Dandelion) root, Berberis aquifolium L. (Oregon grape) root and rhizome, Arctostaphylos uva-ursi L. (Uva ursi) leaves, Stellaria media (L.) Vill. (Chickweed) herb, Nepeta cataria L. (Catnip) herb, Centaurea cyanus L. (Cyani) flowers

Allergies-Hayfever/Pollen

Product Type: Homeopathic

Properties: **Anti-allergenic**

Systems Affected: Immune System, Respiratory System

Conditions: **Allergies (respiratory)**, Congestion (sinus), Eyes (red or itching)

Usage: Used for the relief of minor symptoms of hay fever and pollen allergy, including runny nose, sneezing, itchy eyes, watery eyes and respiratory congestion. Take 10-15 drops under the tongue every three hours for acute symptoms. For chronic allergies, take 10-15 drops three times daily for at least 30 days.

Warnings: NSP does not recommend this product for children under the age of 12 except on the advice of a health professional. Other than the alcohol content, which is not good for infants, we consider this product safe for children.

Allergies-Hayfever/Pollen (1 fl. oz.) - Homeopathic (Liquid)

Available in: US - Stock #8925-0

Ingredients: Active ingredients: Adrenalinum (Adrenalin) 6x, Allium cepa (Red Onion) 6x, Euphrasia (Eyebright) 6x, Lycopodium clavatum (Club Moss) 6x, Sabadilla (Cevadilla) 6x, Silicea (Silica) 6x, Histaminum (Histamine Hydrochloride) 12x, Mixed Pollen Allersodes: Amaranthus, Chenopodium, Cockleburr, Daisy, Dandelion, Goldenrod, Honeysuckle, Marsh Elder, Mugwort, Ragweed, Timothy Grass, all at 12x. Other ingredients: Purified water and 20% USP alcohol.

Allergies-Mold/Yeast/Dust

Product Type: Homeopathic

Properties: Anti-allergenic

Systems Affected: Immune System, Respiratory System

Conditions: **Allergies (respiratory)**, Congestion (general), Eyes (red or itching), Itching, Sore Throat

Usage: Used for the relief of minor symptoms related to allergies to mold, yeast and dust, including congestion, headaches, sore throat, watery eyes, sneezing and itching. Take 10-15 drops under the tongue every three hours for acute symptoms. For chronic allergies, take 10-15 drops three times a day for at least 30 days.

Warnings: NSP does not recommend this product for children under the age of 12 except on the advice of a health professional. Other than the alcohol content, which is not good for infants, we consider this product safe for children.

Allergies-Mold/Yeast/Dust (1 fl. oz.) - Homeopathic (Liquid)

Available in: US - Stock #8920-8

Ingredients: Active ingredients: Adrenalinum (Adrenalin) 6x, Lycopodium clavatum (Club Moss) 6x, Natrum Sulfuricum (Sodium Sulfate) 6x, Silicea (Silica) 6x, Phosphorus (Phosphorus) 8x, Histaminum (Histamine Hydrochloride) 12x, Mixed Mold/ Yeast/Dust Allersodes: Aspergillus flavus, A. fumigatus, A. glaucus, A. nidulans, A. niger, Candida albicans, Curvularia spicifera, Farinae (mite), Fusarium moniliforme, Geotrichum candidum, House Dust, Mucor plumbeus, Penicillinum camemberti, P. chrysogenum, P. digitatum, P. notatum, P. roqueforti, Pullularia pullulans, Rhizopus nigricans, Saccharomyces (bakers and brewers), Trichophyton mentagrophytes, all at 12x.

Other ingredients: Purified water and 20% USP alcohol.

Allergy

Product Type: Homeopathic

Properties: Anti-allergenic

Systems Affected: Respiratory System, Skin

Conditions: **Allergies (respiratory)**, Bites and Stings, Eyes (red or itching), Eyes (red or itching), Itching, Poison Ivy or Oak

Usage: Used for the relief of common allergy symptoms, including runny nose, watery eyes, cough and itching associated with hay fever, certain foods, poison ivy and insect bites. Take 10-15 drops under the tongue every 10-15 minutes or as needed until symptoms improve; then decrease to every one to two hours, then to four times daily until symptoms are relieved.

Warnings: NSP does not recommend this product for children under the age of 12 except on the advice of a health professional. Other than the alcohol content, which is not good for infants, we consider this product safe for children.

Allergy (1 fl. oz.) - Homeopathic (Liquid)

Available in: US - Stock #8714-3

Ingredients: Active ingredients: Arnica montana (Mountain Arnica) 6x, Euphrasia officinalis (Eyebright) 6x, Ignatia amara (St. Ignatius' Bean) 6x, Lycopodium clavatum (Club Moss) 6x, Pothos foetidus (Skunk Cabbage) 6x, Thuja occidentalis (Tree of Life) 6x, Antimonium Crudum (Antimonious Sulfide) 10x, Histaminum (Histamine Hydrochloride) 12x. Other ingredients: Purified water and 20% USP alcohol.

Aloe Vera

Latin Name: *Aloe vera*

Product Type: Single Herb

Properties: Anti-arthritic, Anti-inflammatory, Aperient, **Balm, Balsamic, Demulcent (Mucilant), Emollient,** Laxative (bulk), Moistening, Tonic, Vermifuge, **Vulnerary**

Systems Affected: Bronchials, Ears, Gall Bladder, Intestinal System, Large Intestine (Colon), Mucus Membranes, **Nipples**, Ovaries, Skin, Small Intestines, Stomach, Structural System, Throat, Vagina

Conditions: Abrasions, Acid Indigestion (Heartburn, Acid Reflux), Acne (Pimples, Blackheads), Acquired Immune Deficiency Syndrome (AIDS/HIV), Allergies (food), Amenorrhea, Arthritis, Asthma, Bites and Stings, Bleeding (internal), Blood in Stool, Blood Pressure (high), Boils, Bruises (healing), **Burns and Scalds**, Bursitis, Cancer (natural therapy for), Canker Sores (Mouth Ulcers), Carbuncles, **Chemotherapy (reducing side effects)**, Chicken Pox, Children's Remedy, Cholesterol (high), Colds (general remedies for), Colic (children), **Colitis**, Constipation (adults), Convalescence, Convulsions, Cough (dry), Cradle Cap, Crohn's Disease, Dandruff, **Denture Sores**, Dermatitis, Diabetes, **Diaper Rash, Duodenal ulcers**, Dyspepsia, Eczema, Epstein Barr

Virus (Chronic Fatigue Syndrome, CFS), Flu, **Gastritis**, Gingivitis (Bleeding Gums, Gum Disease, Pyorrhea), Gray Hair, Hair (loss or thinning), Hemorrhoids, Hepatitis, Herpes, Hysteria, Ileocecal Valve, Indigestion, Infection (viral), Inflammation, **Inflammatory Bowel Disorders (Colitis, IBS)**, Irritability, Itching, Leukemia, Lupus, Malaria, Menopause, Nausea and Vomiting, Pain (general remedies for), Parasites (general), Parasites (tapeworm), **Poison Ivy or Oak**, Poisoning (food), Polyps, Psoriasis, **Radiation Sickness**, Rashes and Hives, Scars / Scar Tissue, Scratches and Abrasions, Seborrhea, Skin Care (general), Sprains, Staph Infections, **Sunburn**, Tendonitis, Tinnitus (Ringing in the Ears), Toxemia, Ulcerations (external), **Ulcers**, Vaginal Dryness, Vaginitis, Varicose Veins, Warts, Wounds and Sores

Usage: Aloe vera juice is extremely soothing to irritated skin and mucous membranes. Take 1 tablespoonful in a glass of water or juice. Sip slowly to relieve heartburn and inflammation of the esophagus. Sip during the day for ulcers and intestinal inflammation. Gel may be applied topically for burns and skin irritations. Apply liberally and keep the skin moist for best results. Whole leaf aloe vera juice also builds the immune system to help fight arthritis, AIDS, cancer and other degenerative illnesses. Freeze-dried aloe vera juice is available in capsules. It has similar properties.

Warnings: Some herbalists suggest that children, the elderly and pregnant women should not drink aloe vera juice. However, this may apply only to the green leaf portion (which is strongly cathartic) or to aloe vera concentrates. The green part of the leaf is filtered out in the juice and gel. The diluted juice is a mild, harmless remedy.

Aloe Vera Juice (32 fl. oz.) - Liquid

Available in: US & Canada - Stock #1680-4

Ingredients: Aloe Vera juice, citric acid, potassium sorbate, sodium benzoate

Aloe Vera, Whole Leaf (32 fl. oz.) - Liquid

Available in: US - Stock #1693-4

Ingredients: Juice of the whole aloe vera leaf with laxative compounds removed, citric acid, sodium benzoate, potassium sorbate

Aloe Vera Gel (8 fl. oz) - Topical

Available in: US & Canada - Stock #1679-2

Ingredients: Aloe vera gel, citric acid, potassium sorbate, sodium benzoate, Irish Moss extract, xanthan gum, carrageenan

Aloe Vera Freeze Dried - Capsule (64)

Available in: US - Stock #1686-1

Ingredients: Freeze dried Aloe vera gel, glycerin

Alpha Lipoic Acid

Product Type: Nutrient

Properties: Antioxidant, Detoxifying, Hypolipidemic

Systems Affected: Brain, Cardiovascular System, Female Reproductive, Immune System, Liver, Mitochondria, Nerves

Conditions: Aging (prevention), Alzheimer's Disease, Blood Pressure (high), Cardiovascular Disease (Heart Disease), Chills, Congestion (general), **Diabetes**, Environmental Pollution (protection from), Heavy Metal Poisoning, Hypoglycemia, Mercury Poisoning, Parkinson's Disease, Triglycerides (high)

Usage: Alpha lipoic acid helps neutralize free radical damage by enhancing Vitamin C, Vitamin E and glutathione, an intracellular antioxidant. Normally once an antioxidant has eliminated a free radical, it is lost forever. Alpha lipoic acid is the only antioxidant with the unique ability to regenerate/recycle itself, and other antioxidants such as vitamins C & E, so that they can continue destroying free radicals. It crosses cell membranes to protect cellular integrity, increase energy production and improve removal of toxins by activating the mitochondria of the cell. It helps remove heavy metals and environmental toxins such as organophosphates, latex, and pesticides from the body. Alpha lipoic acid helps cells take up glucose from the blood stream more quickly, which makes it useful for hypoglycemia, syndrome X and diabetes. It also helps to redirect calories away from fat production and towards energy production, which can help with weight loss. This nutrient also protects the liver, supports myocardial function and increases energy flow to the brain and muscles during exercise. Alpha lipoic acid is one of the few antioxidants that can cross the blood/brain barrier to increase levels of glutathione in the brain. This is important as glutathione protects the brain from free radical damage. Low levels of glutathione in the brain are associated with brain disorders such as stroke, dementia, Parkinson's and Alzheimer's disease. Use 1 capsule two times daily.

Warnings: No known warnings

Alpha Lipoic Acid - Capsule (60)

Available in: US - Stock #1505-6

Ingredients: Alpha lipoic acid, turmeric

Anamu

Latin Name: *Petiveria alliacea*

Product Type: Single Herb

Properties: **Abortifacient**, Analgesic (Anodyne), Anti-arthritic, Anti-inflammatory, Anticancer, Antiparasitic, Antispasmodic, Cytotoxic, Diuretic, Immune Stimulant, Insecticide

Systems Affected: Circulation, Digestive System, Immune System, Liver, Structural System, Thymus

Conditions: Amenorrhea, Arthritis, Blood Clots (prevention of), Cancer (natural therapy for), Colds (general remedies for), Congestion (general), Cough (damp), Diabetes, Diarrhea, Flu, Gastritis, Pain (general remedies for), Parasites (general), **Pregnancy (herbs and supplements to avoid during)**, Rheumatoid Arthritis (Rheumatism)

Usage: Anamu has a long history of use in all of the tropical countries where it grows. Modern research shows that it increases the weight of the thymus gland and stimulates white blood cell production, thus enhancing the immune system. It also reduces the clumping of platelets. Its anti-inflammatory action makes it helpful for arthritis, rheumatism and inflammatory skin conditions. It is also reported to possess pain-relieving properties. It appears to be a COX-1 inhibitor. Central American women use it to ease birthing pains. It helps overcome insulin resistance in cells making it useful for type II diabetes. The plant has also been used for colds, fever, flu, respiratory infections and digestive problems. Suggested use is 1 capsule with a meal three times daily.

Warnings: Pregnant women should not use this product because of its aborfacient properties. Women planning to become pregnant should not use this product as research shows that it inhibits fertilization in rats.

Anamu - Capsule (100)

Available in: US - Stock #39-8

Ingredients: Standardized extract of Petiveria alliacea (anamu) leaf

Anti-Gas (Chinese)

Other Names: Xiao Dao, AG-C

Product Type: Formula

Properties: Appetite Stimulant, Aromatic, **Carminative**, Diuretic, **Stomachic**

Systems Affected: **Digestive System**, Intestinal System, Pancreas, **Stomach**, Urinary System

Conditions: Appetite (deficient), **Belching**, Circulation (poor), Cold Hands and Feet, Congestion (general), Constipation (adults), Diarrhea, Digestion (poor), **Dyspepsia**, Fatigue, **Gas and Bloating**, **Halitosis (Bad Breath)**, **Indigestion**, Sugar Cravings, Worry

Usage: In Chinese medicine, the earth element is associated with the digestive system. Earth energy has traditionally been seen as nurturing, nourishing and caregiving-think "mother earth." Sometimes, however, this energy is excessive and needs to be calmed down, as when we have upset stomach and acid indigestion. This often comes because we have overloaded our digestive organs and they become congested. The result is stomach pain, bloating, gas, indigestion, sore stomach, foul belching, a heavy feeling in our stomach and loss of appetite.

The nurturing, mothering energy can also become excessive on the emotional side, too. Excess mothering quickly becomes smothering, plagued by constant fears, worries and a general sense of being "off balance."

Anti-Gas (Chinese) is a traditional Chinese formula for reducing excessive "earth" energy. It decongests the stomach and gastrointestinal tract, helping to relieve acute indigestion. It can ease nausea, expel excess gas, restore appetite, ease sensations of fullness after eating and ease stomach and intestinal pain. However, it can also be used when there is a general sluggish, heavy feeling in the body or a tendency to sugar cravings and weight problems.

This combination works to eliminate stagnant and undigested food in the stomach and intestines. It also enhances digestion by increasing the production of digestive fluids and enzymes. Chinese Anti-Gas detoxifies the body, soothes inflammation and acts as a laxative. It also increases blood flow and the production of urine.

To strengthen digestion and relieve problems with gas, take 4 capsules of Chinese Anti-Gas with a meal 2 times daily. For acute indigestion, take 4 capsules at the first sign of indigestion. Anti-Gas Formula is also available in a TCM concentrate, where the dose is 1 capsule instead of four.

Warnings: No known warnings.

Anti-Gas, Chinese - Capsule (100)

Available in: US - Stock #1869-9

Ingredients: Agastache tops (Agastache rugosa), crataegus fruit (Crataegus pinnatifida), hoelen sclerotium (Poria cocos), magnolia bark (Magnolia liliflora and officinalis), oryza fruit (Oryza sativa), shen-chu whole plant (Xanthium stramonium), citrus peel (Citrus aurantium), gastrodia rhizome (Gastrodia elata), ginseng root (Panax ginseng), typhonium rhizome (Typhonium flagelliforme), atractylodes rhizome (Atractylodes lancea), cardamon fruit (Amomum villosum), platycodon root (Platycodon grandiflorum), ginger rhizome (Zingiber officinale), licorice root (Glycyrrhiza uralensis)

Anti-Gas TCM Conc. - Capsule (30)

Available in: US - Stock #1018-9

Ingredients: Same as Stock #1869-9

AG-C - Capsule (100)

Available in: Canada - Stock #1870-4

Ingredients: Agastache herb, crataegus fruit, hoelen plant, magnolia bark, oryza seed, shenqu tea, citrus peel, cyperus rhizome, Panax ginseng root, pinellia rhizome, uncaria rhynchophylla stem, atractylodes rhizome, cardamon fruit, inula flower, ginger rhizome, and licorice root.

Anti-Gas Formula

Other Names: AG-X

Product Type: Formula

Properties: Antacid, Appetite Stimulant, Aromatic, **Carminative, Stomachic**

Systems Affected: **Digestive System**, Intestinal System, Liver, Pancreas, **Stomach**

Conditions: Abdominal Pain and Inflammation, Acid Indigestion (Heartburn, Acid Reflux), Allergies (food), Anorexia, Appetite (deficient), Asthma, Belching, Bell's Palsy, Bursitis, Digestion (poor), Dyspepsia, **Gas and Bloating**, Halitosis (Bad Breath), Hiatal Hernia, **Indigestion**, Morning Sickness, Rheumatoid Arthritis (Rheumatism)

Usage: This is the Western formula for indigestion, gas, belching, bloating and other acute digestive disturbances. It supplements digestive enzymes with papaya. The formula stimulates production of digestive secretions with various aromatics and soothes inflammation of the digestive tract with wild yam and dong quai. Use 2 capsules after meals as a digestive aid or 2-4 capsules as needed to relieve gas, bloating and indigestion. It can also be used in an enema solution for food allergies.

Warnings: No known warnings.

Anti-Gas Formula (with Lobelia) - Capsule (100)

Available in: US - Stock #1198-4

Ingredients: Papaya fruit, ginger, peppermint, wild yam, fennel seed, dong quai, spearmint, catnip, lobelia

Appetite Control

Product Type: Homeopathic

Properties: Appetite Suppressant

Systems Affected: Digestive System, Nerves

Conditions: Appetite (excessive), Weight Loss (aids for)

Usage: This formula assists in the control of appetite as an aid to weight loss. Take 10-15 drops under the tongue 30 minutes prior to meals.

Warnings: NSP does not recommend this product for children under the age of 12 except on the advice of a health professional. Other than the alcohol content, which is not good for infants, we consider this product safe for children.

Appetite Control (1 fl. oz.) - Homeopathic

Available in: US - Stock #8722-4

Ingredients: Active ingredients: Antimonium Crudum (Antimonious Sulfide) 6x, Fucus vesiculosus (Kelp) 6x, Graphites (Graphite) 6x, Iodium (Iodine) 6x, Pulsatilla (Wind Flower) 6x, Aurum Metallicum (Gold) 12x, Sulphur (Sulfur) 12x. Other ingredients: Purified water and 20% USP alcohol.

APS II

Product Type: Formula

Properties: Analgesic (Anodyne), Anti-inflammatory, **Anticephalalgic, Antirheumatic**, Antispasmodic, Febrifuge, Nervine, Sympatholytic

Systems Affected: Muscles, Nerves, Pituitary (anterior), Prostaglandins, Structural System

Conditions: Addictions (tobacco smoking or chewing), Afterbirth Pain, Angina, **Arthritis**, Backache (Back Pain, Lumbago), Colds (general remedies for), Dysmenorrhea, Fever, Fibromyalgia Syndrome (FMS), Fibrosis, Flu, Hashimoto's Disease (Thyroiditis), **Headache (general)**, Headache (tension), Hysteria, Inflammation, Labor and Delivery, Lupus, **Pain (general remedies for)**, Tension, Tooth Extraction

Usage: This is a mild pain-reliever when compared to its drug equivalents. Until synthetic aspirin could be produced in large quantities, white willow bark was the treatment of choice for reducing fevers, relieving headache and arthritis pain and controlling swelling. Native Americans, Europeans and others have a tea of willow bark as a medicinal remedy for everything from pain relief to fevers. The bark of the white willow tree is a source of salicin and other salicylates - compounds that are similar in structure to aspirin (acetyl salicylic acid). Today, white willow bark is still used as a natural alternative to aspirin. It this formula it is combined with other nervines (valerian and wild lettuce) and capsicum (a circulatory stimulant). The formula helps relax nerves, reduce inflammation and swelling and ease pain. Take 1-2 capsules

Warnings: Not for emaciation or weak digestion (however, does not irritate stomach like synthetic aspirin).

APS-II (w/ White Willow Bark) - Capsule (100)

Available in: US - Stock #780-8

Ingredients: White willow bark, lettuce leaves, valerian, capsicum

Artemisia Combination

Product Type: Formula

Properties: Antifungal, **Antiparasitic**, **Parasiticide**, Stomachic, **Vermifuge**

Systems Affected: Digestive System, Female Reproductive, Intestinal System, Respiratory System

Conditions: Amenorrhea, Constipation (adults), Digestion (poor), Dysentery, Fingernail Biting, Inflammatory Bowel Disorders (Colitis, IBS), **Malaria**, Menstruation (scant), **Parasites (general)**, Parasites (nematodes, worms), **Parasites (tapeworm)**, Pets (supplements for), **Pregnancy (herbs and supplements to avoid during)**

Usage: Used to aid in the expulsion of worms and other parasites in the body. This formula can be used for amoebas, Candida (yeast infections), ringworm, and tapeworm. Use 2 capsules three times per day. This product is also included in the ParaCleanse with Paw Paw package. The complete ParaCleanse with Paw Paw will be more effective than this formula taken by itself.

Warnings: Drink plenty of water when on a cleanse of any kind. Does not work on all parasites. May induce menstruation so it is not recommended for pregnancy. Wormwood and mugwort may have side effects like dizziness and nausea when taken in large doses.

Artemisia Combination - Capsule (100)

Available in: US & Canada - Stock #787-6

Ingredients: Wormwood, mugwort, elecampane, clove, garlic, ginger, spearmint, turmeric, olive leaf

Arthritis

Product Type: Homeopathic

Properties: Analgesic (Anodyne), Anti-arthritic, Anti-inflammatory

Systems Affected: Structural System

Conditions: Arthritis

Usage: Used for the relief of minor arthritis symptoms, such as pain, stiffness and inflammation of joints. Take 10-15 drops under the tongue every 10-15 minutes or as needed until symptoms improve; then decrease to every one or two hours, then to four times daily until symptoms are relieved. Best used as part of a complete program for arthritis.

Warnings: NSP does not recommend this product for children under the age of 12 except on the advice of a health professional. Other than the alcohol content, which is not good for infants, we consider this product safe for children.

Arthritis (1 fl. oz.) - Homeopathic

Available in: US - Stock #8800-4

Ingredients: Active ingredients: Bryonia (White Bryony) 4x, Cimicifuga racemosa (Black Cohosh) 4x, Rhus toxicodendron (Poison Ivy) 4x, Apis mellifica (Honeybee) 6x, Arnica montana (Mountain Arnica) 6x, Berberis vulgaris (Barberry) 8x, Causticum (Hahnemann's Causticum) 8x, Calcarea Carbonica (Calcium Carbonate) 9x. Other ingredients: Purified water and 20% USP alcohol.

Asthma

Product Type: Homeopathic

Properties: Antispasmodic, Antitussive, Bronchial Dilator

Systems Affected: Bronchials, Lungs

Conditions: Asthma, Cough (general), Wheezing

Usage: A homeopathic remedy used for control of the symptoms of asthma, including difficulty in breathing, shortness of breath, tightness of chest, wheezing or coughing. Take 10-15 drops under the tongue every 10-15 minutes or as needed until symptoms improve; then decrease to every one or two hours, then to four times daily until symptoms are relieved. Best used as part of an overall program for asthma.

Warnings: NSP recommends this product for asthma sufferers only. NSP does not recommend this product for children under the age of 12 except on the advice of a health professional. Other than the alcohol content, which is not good for infants, we consider this product safe for children.

Asthma (1 fl. oz.) - Liquid (Tincture)

Available in: US - Stock #8704-4

Ingredients: Active ingredients: Zingiber officinale (Ginger) 3x, Aralia racemosa (Spikenard) 6x, Eucalyptus globulus (Eucalyptus) 6x, Grindelia (Gum plant) 6x, Hypericum perforatum (St. John's Wort) 6x, Lobelia inflata (Indian Tobacco) 6x, Sanguinaria canadensis (Blood Root) 6x, Cinchona officinalis (Peruvian Bark) 12x. Other ingredients: Purified water and 20% USP alcohol.

Astragalus

Latin Name: *Astragalus membranaceous*

Product Type: Single Herb

Properties: Adaptagen, Antioxidant, Antiviral, Cardiac, Diuretic, Immune Stimulant, Tonic, Vasodilator

Systems Affected: Heart, **Immune System, Lungs,** Respiratory System, Urinary System, Uterus

Conditions: Acquired Immune Deficiency Syndrome (AIDS/HIV), Adrenals (exhaustion, weakness or burnout), Anorexia, Attention Deficit Disorder (ADD, ADHD), Blood Clots (prevention of), Blood Pressure (high), Bronchitis, Cancer (natural therapy for), Cancer (prevention), **Chemotherapy (reducing side effects),** Circulation (poor), Colds (antiviral), Colds (general remedies for), Convalescence, **Cough (dry),** Diabetes, Digestion (poor), Fatigue, **Hashimoto's Disease (Thyroiditis),** Hepatitis, Infection (viral), Lupus, Mononucleosis, Nephritis, Night Sweating, Prolapsed Uterus, Ulcerations (external), Ulcers

Usage: Astragalus has been used for centuries in Chinese traditional medicine. The Chinese use it as a tonic for persistent infections and chronic ulcerations. It helps with night sweats, chi deficiency (fatigue, weakness, loss of appetite) and diarrhea. It helps to balance and increase immune function and has been used for viral diseases, frequent colds, AIDS, and cancer. It stimulates the body's natural production of interferon. Research indicates that astragalus enhances the production of immunoglobulin, stimulates macrophages, and helps to activate T cells and NK cells. It increases the number of stem cells in the bone marrow and lymph tissue and encourages them to develop into active immune cells. It has a more balancing effect on the immune system than Echinacea. One Chinese trial found that astragalus could balance an overactive immune function in people with lupus, an autoimmune disease. Long-term use (for 35 days) heightens the activity of spleen cells. Astragalus also alleviates the adverse effects of steroids. An interesting use for astragalus is to reduce blood agglutination when eating foods that are an "avoid" for a person's blood type. The suggested dosage is 2 capsules with meals three times a day.

Warnings: No known warnings.

Astragalus - Capsule (100)

Available in: US & Canada - Stock #40-1

Ingredients: Astragalus membranaceous (astragalus) root

Barley Grass

Latin Name: *Hordeum vulgare*

Product Type: Single Herb

Properties: Alkalinizer, Anti-arthritic, Antilipemic, Appetite Suppressant, Digestive Tonic, Mineralizer, Nutritive

Systems Affected: Digestive System, Immune System, Mouth, Mucus Membranes, Pituitary (posterior), Structural System, Throat, Thymus, Uterus, Whole Body

Conditions: Acquired Immune Deficiency Syndrome (AIDS/HIV), Aging (prevention), Appetite (excessive), Convalescence, Cysts, Electromagnetic Pollution, Energy (lack of), Exercise, Free Radical Damage, Lupus, Muscular Dystrophy, Polyps, Radiation Sickness, Triglycerides (high)

Usage: This product is best used in the morning prior to a meal as a dense whole food supplement during weight control, to alkalize the body, to provide a very mild residual laxative action, and as a general immune builder, especially with adaptagenic herbs. Chemical analysis of barley grass has revealed it to be an excellent source of many essential nutrients. Use as a nutritional supplement to strengthen the immune system and improve general health. It is rich in magnesium, potassium and bioflavonoids. It can also be used for constipation, inflammation and as a digestive aid. It also helps to reduce lactic acid buildup in muscle tissues. Barley Juice Powder provides more vitamin B1 and calcium than milk, more vitamin C than several oranges, and more iron than found in a serving of spinach. Use 2-3 capsules, three times daily. While recovering from illness, increase to 6-8 capsules, three times daily. Externally, it can be used in a foot soak with Vitamin C Ascorbate powder for gout and edema. It can also be applied as a poultice for skin sores. A facial mask for refining and softening skin can also be made from the barley juice powder.

See also *Nature's Gold.*

Warnings: No known warnings.

Barley Juice Powder Concentrate - Bulk Powder

Available in: US - Stock #55-1

Ingredients: Dehydrated Hordeum vulgare (Barley) juice

Bayberry

Latin Name: *Myrica cerifera*

Product Type: Single Herb

Properties: **Astringent**, **Coagulant**, Expectorant, **Hemostatic**, Insecticide, Stimulant, **Styptic**, Vermifuge, Vulnerary

Systems Affected: Gums, Mucus Membranes, Respiratory System, Sinuses, Skin, Small Intestines, Stomach, Structural System, Vagina

Conditions: Addictions (drugs), Bites and Stings, **Bleeding (external)**, **Bleeding (internal)**, **Blood in Stool**, Blood in Urine, Chills, Colds (general remedies for), Congestion (lungs), **Congestion (sinus)**, Cough (damp), Cuts, **Diarrhea**, **Dysentery**, Endometriosis, Flu, Gingivitis (Bleeding Gums, Gum Disease, Pyorrhea), Injuries, **Labor and Delivery**, **Menorrhagia (Heavy Menstrual Bleeding)**, **Nose Bleeds**, **Polyps**, Prolapsed Uterus, **Sinus Infection**, Smell (loss of sense of), **Sneezing**, **Sore Throat**, Ulcerations (external), Ulcers, Vaginitis, Varicose Veins, Wounds and Sores

Usage: Used as an astringent for internal and external bleeding and as an aid for dispelling mucus. Works better in liquid form for internal bleeding or topical application. Drink an infusion (tea) for fevers. For a decoction, boil 1 teaspoon of powdered root bark in a pint of water for 10-15 minutes. The decoction may be used topically as a fomentation or compress or used as a gargle. The gargle is helpful for sore throat or the early stages of a cold. It is also very effective for post-nasal drip. Bayberry has been combined with capsicum, honey and apple cider vinegar as a drink to stop internal hemorrhage. It can also help to stop a nosebleed. Bayberry is less useful in capsule form for bleeding because it is slow acting. Open capsules and dump directly into the mouth to stop nosebleed or internal bleeding (1-2 capsules per dose). The powder can be used as snuff to shrink nasal polyps and stop sinus drainage. This procedure can be uncomfortable, but it is very effective at shrinking nasal polyps, opening clogged sinus passages and healing chronic sinus infection. It typically causes sneezing and copious drainage that clears the sinuses. Snuffing the powder can help nosebleeds, too. The tincture can take down the swelling of insect bites or bee stings. It has been used as a tonic and stimulant to support the body's defense against a range of ailments such as coughs, colds, flu, fevers, headache, and sore throat. It was also considered an effective remedy for diarrhea, bloody stools, and excessive menstrual bleeding. As an astringent, this herb helps to dry up and protect exposed membranes, and is often applied to the skin as a poultice to heal boils, cankers and skin ulcers.

Warnings: In large doses bayberry is emetic (induces vomiting). It contains tannins, which sometimes have negative effects on people with cancer. Adding milk to bayberry helps neutralize the tannins. Bayberry should not be given to children under age 2. For older children and people over 65, start with a low-strength preparation and increase strength if necessary.

Bayberry - Capsule (100)

Available in: US - Stock #60-6

Ingredients: Myrica cerifera (bayberry) rootbark

Bedwetting

Product Type: Homeopathic

Properties: Antidiuretic

Systems Affected: Urinary System

Conditions: Children's Remedy, Children's Remedy

Usage: A homeopathic remedy used to assist a child's body in overcoming nighttime bedwetting. For children 2 or older, take 5-8 drops under the tongue four times daily, for at least one month or as needed, reducing upon improvement to once or twice daily for an additional one to three months as maintenance.

Warnings: No known warnings. Homeopathic remedies are very safe.

Bedwetting (1 fl. oz.) - Homeopathic (Liquid)

Available in: US - Stock #8830-1

Ingredients: Active ingredients: Ammonium Carbonicum(Ammonium Carbonate) 6x, Cina (Wormseed) 6x, Equisetum hyemale (Scouring Rush) 6x, Benzoicum Acidum (Benzoic Acid) 8x, Causticum (Hahnemann's Causticum) 8x, Verbascum thapsus (Mullein) 8x. Other ingredients: Purified water and glycerin.

Bee Pollen

Product Type: Nutrient

Properties: Adrenal Tonic, Appetite Stimulant, Corrects Polarity, Glandular, Stimulant

Systems Affected: Bones, Cardiovascular System, Hypothalamus, Hypothalamus, Liver, Muscles, Nerves, Pituitary (anterior), Pituitary (general), Respiratory System

Conditions: Addictions (sugar or food), Aging (prevention), Allergies (respiratory), Anorexia, Appetite (deficient), Capillary Weakness, Children's Remedy, Cholesterol (high), Convalescence, Endurance (lack of), Energy (lack of), Epstein Barr Virus (Chronic Fatigue Syndrome, CFS), Exercise, Fatigue, Hypoglycemia, Mercury Poisoning, Pregnancy (herbs and supplements for), **Reversed Polarity**, Weight Loss (aids for)

Usage: Bee pollen contains every known nutrient in trace amounts, which is why some have dubbed it nature's most complete food. It contains all 22 amino acids and has the highest amino acid content of any known food. Highly energizing, it is used to improve energy, stamina and endurance. For this reason it is widely used by athletes. It also supports the glands and aids the immune system. Bee pollen can be used in many ways. Add it to nutritive drinks for energy and general nutrition. Sprinkle it on cereal or oatmeal for use as a tonic with the young or elderly that aren't able to swallow capsules well. Use 1-2 capsules one to three times daily for energy. It combines well with spirulina and licorice for this purpose, and helps to stabilize blood sugar levels. It can also be used to overcome allergies to pollen. In overcoming allergies it is best to get pollen from local beekeepers, but capsules may work as well. Start with a small amount (just a few grains from a capsule) and gradually increase the dose over a period of several weeks to develop a tolerance to pollen and improve immune function. Bee pollen is also an aid for the pituitary and for balancing reversed polarity in muscle response testing.

Warnings: A few allergic attacks have been reported from use of bee pollen. Symptoms of allergy include itching, dizziness and difficulty swallowing. If you have allergies, be sure to start with a small amount (a few grains).

Bee Pollen - Capsule (100)

Available in: US - Stock #70-9
Ingredients: Bee pollen

Bentonite (Hydrated)

Product Type: Nutrient

Properties: Adsorbant, Antidiarrheal

Systems Affected: Intestinal System, Skin

Conditions: Acne (Pimples, Blackheads), Allergies (food), Constipation (adults), Diarrhea, **Diverticulitis (Diverticuli)**, Heavy Metal Poisoning, **Itching**, **Lead Poisoning**, **Mercury Poisoning**, **Poisoning (food)**

Usage: Take 1 tablespoon in a large glass of water on an empty stomach. As an aid to colon cleansing use 1 tablespoon with a glass of water one or two times daily. It can absorb some heavy metals and help shrink diverticulii. Use one-half of a bottle in a bath to help with rashes, psoriasis, eczema and to detoxify the skin.

Warnings: Use only with an herbal laxative. Drink 6-8 glasses of water per day when taking bentonite to flush it from the system. Otherwise, it may get stuck in the bowel, causing constipation and intestinal blockage. If sensitivity occurs, discontinue use. Not to be taken on a continual basis.

Bentonite, Hydrated (32 fl. oz.) - Liquid (Other)

Available in: US & Canada - Stock #1725-9
Ingredients: Purified water, USP-grade bentonite clay

Bergamot

Product Type: Essential Oil

Properties: Anti-inflammatory, Antidepressant, Antiparasitic, Antiseptic, Appetite Stimulant, Aromatic, Carminative, Febrifuge

Systems Affected: Digestive System, Female Reproductive, Liver, Lymphatic System, Mucus Membranes, Nerves, Respiratory System, Skin, Structural System, Thymus, Urinary System

Conditions: Acne (Pimples, Blackheads), Addictions (alcohol), Anger (excessive), Anorexia, Anxiety (Panic Attack), Apathy, Bronchitis, Burns and Scalds, Chicken Pox, Cold Sores (Fever Blisters), Colic (adults), Colic (children), Concentration (poor), Confusion, Cystitis, Depression, Digestion (poor), **Eczema**, Fear (excessive), Fever, Gas and Bloating, Grief (excessive), Herpes, Hysteria, Leucorrhea, Menopause, Mood Swings, Nervous Exhaustion (Enervation), Nervousness, PMS Type D, Shingles, Skin (oily), Tonsillitis (Adenoids), Tuberculosis (Consumption, Scrofula), Urinary Tract Infections, Vaginitis, Varicose Veins, Wounds and Sores

Usage: Bergamot is a valuable antiseptic and anti-inflammatory. It aids urinary conditions like cystitis and respiratory problems like tonsillitis, bronchitis and tuberculosis when inhaled. Combined with tea tree oil and a carrier oil, it is useful for cold sores, chicken pox, herpes, eczema, and shingles. Mixed with a carrier oil, it can also be applied topically to treat wounds, herpes, acne and oily skin conditions. In douches and baths, bergamot has proven helpful in gonococcal infections, leucorrhoea, vaginal pruritis and urinary infections. It also makes a refreshing bath during late pregnancy. It acts as a carminative on the digestive system, making it useful for relieving colic, flatulence and indigestion. The oil soothes anger and frustration by decreasing the action on the sympathetic nervous system. It helps with self-confidence and exhaustion from physical or psychological illnesses. An effective antidepressant, it uplifts and refreshes the spirit, evokes joy, aids self-confidence and warms the heart. It also relieves excess anger and frustration and helps overcome emotions like bitterness, fear, grief, helplessness, and loneliness.

Warnings: Avoid exposure to the sun after using bergamot in a massage or bath.

Bergamot Essential Oil (5ML) - Essential Oil

Available in: US - Stock #3900-5
Ingredients: Essential oil of Citrus bergamia (bergamot)

Berry Healthy

Product Type: Formula

Properties: Anti-inflammatory, Antioxidant, Nutritive

Systems Affected: Immune System

Conditions: Children's Remedy, Colds (general remedies for), Colds (prevention), Inflammation

Usage: This is an antioxidant beverage mix, an alternative to less healthy beverage drinks. Mix 1 scoop with 6 oz. water, shake and drink.

Warnings: No known warnings.

Berry Healthy (9.8 oz.) - Bulk Powder

Available in: US - Stock #3206-8

Ingredients: Vitamin C, fructose, chokeberry fruit, blueberry fruit, raspberry fruit, malic acid, citric acid, blackberry fruit, cranberry fruit, stevia

Bifidophilus

See *Probiotics*

Bilberry Fruit

Latin Name: *Vaccinium myrtillus*

Product Type: Single Herb

Properties: Antidiabetic, Antioxidant, Diuretic, Nutritive, Opthalmicum, Tonic

Systems Affected: Blood Vessels, Digestive System, **Eyes**, Structural System, Urinary System, Veins

Conditions: Bruises (prevention), **Capillary Weakness**, Cartilage Damage, Circulation (poor), Diarrhea, **Eye Problems (general)**, Eyesight (to improve), Glaucoma, Inflammatory Bowel Disorders (Colitis, IBS), Macular Degeneration, **Night Blindness**, Nose Bleeds, **Spider Veins**, Urinary Tract Infections, Varicose Veins

Usage: Bilberry is famous as a remedy for the eyes. It helps reduce eye irritation and improve night vision. It can also be used in conjunction with other nutritive herbs such as slippery elm and marshmallow for wasting conditions or with cranberry for strengthening the urinary tract. Take 2 capsules with each meal. It needs to be used over time to be beneficial. It can be used as a decoction to help balance blood sugar levels (although it must be taken for longer periods of time). The infusion also helps with chronic diarrhea, vomiting, and nerves as well as an antiseptic gargle for sore throats.

Externally, bilberry is use as an eyewash to replenish the retina, helping to reduce visual fatigue, and as a mouthwash for ulcers and gum inflammation. Add with Herbal Trim skin treatment for sunburn and skin inflammations.

Warnings: No known warnings.

Bilberry Concentrate - Tablet (60)

Available in: US - Stock #74-8

Ingredients: 40 mg Vaccinium myrtillus (bilberry) fruit standardized to 25% anthocyanidins, calcium, prosphorus

Black Cohosh

Latin Name: *Cimicifuga racemosa*

Product Type: Single Herb

Properties: Acrid, **Alexipharmic**, Alterative (Blood Purifier), Analgesic (Anodyne), Anti-arthritic, **Antidepressant**, Antihypertensive, Antirheumatic, **Antispasmodic**, **Anti-toxic**, **Antivenomous**, Aphrodisiac, **Bronchial Dilator**, **Cephalalgic**, Emmenagogue, **Estrogenic**, Expectorant, Glandular, **Hypotensive**, **Insecticide**, Nervine, Parturient, **Phytoestrogen**, Relaxant

Systems Affected: Bronchials, **Estrogen**, Heart, Lungs, Nerves, Ovaries, Pineal, Prostaglandins, Throat, Uterus

Conditions: Angina, Arrhythmia, Bedwetting, **Bites and Stings**, Blood Poisoning, Blood Pressure (high), Breasts (swelling and tenderness), Bronchitis, Cancer (prevention), Colon (spastic), Cough (spastic), **Cramps (menstrual)**, Cramps and Spasms (general), **Depression**, Diphtheria, Dizziness (Vertigo), Dysmenorrhea, **Estrogen (low)**, Gonorrhea, Hashimoto's Disease (Thyroiditis), **Headache (tension)**, **Hysteria**, Insects, Labor and Delivery, Measles, **Menopause**, Menstrual Irregularity, Neuralgia and Neuritis, **Osteoporosis**, Pertussis (Whooping Cough), PMS Type D, **Post Partum Depression**, **Pregnancy (herbs and supplements to avoid during)**, Sciatica, Sex Drive (low), **Snake Bite**, Tension, Tinnitus (Ringing in the Ears), Vaginal Dryness, Whiplash

Usage: This herb is typically used for its estrogenic effect. It does not actually contain any of the estrogens produced in the human body. Instead, it contains isoflavones, compounds that mimic the body's estrogens, but in a weak fashion. It has long been used to regulate the female cycle and help the body adjust during menopause. Use 1-2 capsules per day or less for this purpose. Black Cohosh works when given consistently over a period of time. It has also been used during childbirth to stimulate contractions, help control hemorrhage, nervousness and afterbirth pains. Black cohosh, however, is much more than a female herb. It neutralizes venom from black widow spiders, scorpions and rattlesnakes. For treating venomous bites, take 3-6 capsules of black cohosh with 6,000 mg. of vitamin C immediately after the bite. Do not take in this quantity more than once. Seek medical attention as necessary. Black cohosh can also be applied topically as a poultice for bites. Black cohosh acts as an antispasmodic and nervine. It also has anti-inflammatory and pain relieving qualities, but is best in liquid form (small doses) or combined with other herbs for these purposes. It is helpful for rheumatic neuralgia, fibromyalgia and muscle pain where the muscles feel bruised. Use in a fomentation or compress with decoction for muscle spasms, neck and shoulder aches and tension headaches.

Black cohosh is also helpful for certain types of depression where the person feels like a "black cloud" is hanging over them and they feel "trapped". It is especially helpful for post partum depression.

Warnings: Should not be used for vasodilative headaches because it increases blood flow to the brain. Black cohosh stimulates uterine contractions. It is contraindicated in early pregnancy but can be used (especially as part of a formula) during the last weeks (usually 4 weeks) of pregnancy or during labor. This herb has negative effects in large doses. It can cause headaches, vertigo (dizziness), irritation of the central nervous system, nausea, diarrhea, abdominal pain, visual dimness, tremors, joint pains, depressed heart rate, and vomiting. Pregnant women should not take estrogenic herbs like black cohosh (until it is time to give birth, if needed). For some, these effects (especially dizziness and headaches) may develop at a relatively low dosage. Regular usage should not exceed 1-2 capsules per day for most people. Most people cannot tolerate even this low dose. If headache or dizziness occurs, reduce dose or discontinue use. Using black cohosh as part of a combination of herbs lessens these effects. If you have a problem with taking black cohosh straight, consider NSP's timed-release black cohosh found in Flash-Ease.

Black Cohosh - Capsule (100)

Available in: US - Stock #80-3

Ingredients: Cimicifuga racemosa (black cohosh) root

Black Currant Oil

Product Type: Nutrient

Properties: Analgesic (Anodyne), Immune Amphoterics, Immunomodulator, Nutritive

Systems Affected: Cardiovascular System, Immune System, Liver, Prostaglandins, Structural System, Thymus

Conditions: Addictions (alcohol), Allergies (food), Allergies (respiratory), Angina, Arteriosclerosis (Atherosclerosis, Hardening of the Arteries), Arthritis, Asthma, Attention Deficit Disorder (ADD, ADHD), Breast Lumps, Breasts (swelling and tenderness), Cancer (natural therapy for), Cholesterol (high), Cystic Fibrosis, Diabetes, Down Syndrome, Eczema, Epilepsy, Fat Cravings, Fear (excessive), Fibrosis, Hangover, Lupus, Multiple Sclerosis (MS), PMS (general), Weight Loss (aids for)

Usage: In general, fatty acids are important for heart, nerve and immune function. Black currant oil is an alternative source of GLA (gamma-linolenic acid), a fatty acid also found in evening primrose oil. Black currant oil contains 16 to 18% GLA (more than evening primrose oil) plus two essential fatty acids-linoleic and alpha-linolenic. Oils containing GLA have been used to aid temperature regulation, cell construction, energy, low immune response and inflammation. They also help to relieve some PMS symptoms. GLA is frequently deficient in people with eczema, atherosclerosis and diabetes mellitus. As a nutritional supplement use 1-3 capsules per day. This oil may help to lower cholesterol, tonify the cardiovascular tissues, knit nerve tissue (myelin sheath insulation) and reduce inflammation in arthritis. It may also be helpful in eczema, dry skin, PMS, multiple sclerosis and obesity.

Warnings: No known warnings.

Black Currant Oil - Softgel (90)

Available in: US - Stock #1810-9

Ingredients: Black currant oil, glycerine

Black Ointment

Product Type: Formula

Properties: Antiseptic, **Drawing**, **Escharatic**, Rubefacient

Systems Affected: Skin, Structural System

Conditions: Abrasions, Abscesses, Infection (bacterial), Injuries, Melanoma (Skin Cancer), **Slivers**, Ulcerations (external), **Warts**, Wounds and Sores

Usage: This product is for topical use only. Apply a thick layer over the desired area and cover with a bandage. It has a drawing action, along with an antiseptic quality that helps pull pus and infection out of wounds. It can be helpful for skin blemishes, acne, arthritis, boils, cysts, warts, and dermatitis (inflammation of the skin). It may also be helpful for slivers. It can be applied to the rectum to help heal hemorrhoids. It should be followed with the golden salve for a healing action. To make an escharatic ointment for moles and skin cancers, mix the contents of one capsule of Paw Paw Cell-Reg with enough Black Ointment to form a paste and apply topically.

Warnings: Don't overuse. Once the offending material is drawn to the surface, discontinue use and use a Golden Salve or an astringent herb to promote healing.

Black Ointment (1oz.) - Salve

Available in: US & Canada - Stock #1696-9

Ingredients: Chaparral, chickweed, comfrey, golden seal, lobelia, marshmallow, mullein, myrrh, plantain, red clover, in a base of beeswax, olive oil, pine tar and vitamin E oil

Black Walnut

Latin Name: *Juglans nigra*

Product Type: Single Herb

Properties: **Anticarious**, Antifungal, Antiparasitic, Antiseptic, Antiviral, Astringent, Bitter, **Dentifrice**, **Immune Amphoterics**, **Immunomodulator**, Insecticide, Laxative (general), Parasiticide, Vermifuge

Systems Affected: **Gums**, Immune System, Intestinal System, **Mouth**, Nails, Skin, Structural System, **Teeth**, Testes, **Thyroid**

Conditions: Appetite (excessive), Athlete's Foot, Athletic Performance or Exercise (aids to), Boils, Breasts (swelling and tenderness), Canker Sores (Mouth Ulcers), Chicken Pox, **Cold Sores (Fever Blisters)**, Cradle Cap, Dandruff, Depression, Diverticulitis (Diverticuli), Eczema, Epstein Barr Virus (Chronic Fatigue Syndrome, CFS), **Fibromyalgia Syndrome (FMS)**, **Gingivitis (Bleeding Gums, Gum Disease, Pyorrhea)**, **Goiter**, **Herpes**, **Hypothyroid**, **Impetigo**, Inflammatory Bowel Disorders (Colitis, IBS), **Itching (rectal)**, Lice, Lupus, Malaria, Pancreatitis, **Parasites (general)**, Parasites (nematodes, worms), Parasites (tapeworm), Poison Ivy or Oak, Prolapsed Uterus, Rashes and Hives, **Teeth (loose)**, Tooth Decay (prevention), Wounds and Sores

Usage: Internally it helps get rid of ringworm, amoebas, protozoa, pinworms and fungus. It also acts as a blood purifier for skin eruptive diseases. Take 1-2 capsules two to three times daily for fungus, parasites and as a blood purifier. It helps oxygenates the blood. Extract or tincture can be applied topically to canker sores, cold sores, fungal infections, athlete's foot, cradle cap, dandruff, eczema, herpes sores, impetigo and poison ivy. Also contains iodine for the thyroid. Black walnut powder combined with white oak bark powder makes an excellent tooth powder for bleeding gums. Black Walnut is also used to balance sugar levels and burn up excessive toxins and fatty materials.

Warnings: If excessive dryness of the mouth persists after a time, then reduce intake by half.

Black Walnut - Capsule (100)

Available in: US & Canada - Stock #90-8

Ingredients: Juglans nigra (Black Walnut) hulls

Black Walnut ATC Concentrated - Capsule (50)

Available in: US - Stock #93-3

Ingredients: Same as Stock #90-8

Black Walnut Extract (2 fl. oz.) - Liquid (Tincture)

Available in: US & Canada - Stock #1755-7

Ingredients: Juglans nigra (Black Walnut) hulls, alcohol, water

Blessed Thistle

Latin Name: *Cnicus benedictus*

Product Type: Single Herb

Properties: Alterative (Blood Purifier), Cephalic, Cholagogue, Digestive Tonic, Emmenagogue, **Galactagogue**, Hepatoprotective, Nervine, Stomachic, Tonic

Systems Affected: Blood, Breasts, Circulation, Gall Bladder, Heart, Intestinal System, Liver, Ovaries, Pituitary (posterior), Stomach, Uterus

Conditions: Acid Indigestion (Heartburn, Acid Reflux), Angina, Anorexia, Appetite (deficient), Bleeding (internal), **Breast Milk (increase or enrich)**, Calcium Deposits (Calcification), Circulation (to the brain), Concentration (poor), Gas and Bloating, Hiatal Hernia, Hiccups, Indigestion, Memory and Brain Function, Menstrual Irregularity, Nausea and Vomiting, Pregnancy (herbs and supplements for)

Usage: Used to strengthen the liver and digestive system, 2 capsules two times a day. Also taken by nursing mothers to enrich breast milk (combine with marshmallow for this purpose). It has similar properties to the widely publicized milk thistle. It helps improve liver function and fat metabolism. Another use for this herb is in settling digestive problems such as hiatus hernia, regurgitation of bile and hiccups. This herb has hepatoprotective and detoxifying qualities similar to the more popular milk thistle. It has been used to treat jaundice and hepatitis. A poultice of blessed thistle can help with breast inflammation and swelling, or clogged ducts in nursing mothers.

Warnings: No known warnings.

Blessed Thistle - Capsule (100)

Available in: US - Stock #100-1

Ingredients: Cnicus benedictus (blessed thistle) root

Blood Build

Other Names: Bu Xue

Product Type: Formula

Properties: Alterative (Blood Purifier), **Blood Building**, Hepatic, Immunomodulator, Tonic

Systems Affected: **Blood**, Eyes, **Female Reproductive**, **Liver**, Muscles, **Red Blood Cells**, Skin

Conditions: **Addictions (alcohol)**, Allergies (food), **Anemia**, Appetite (excessive), Athletic Performance or Exercise (aids to), **Blood Pressure (high)**, Cancer (natural therapy for), Canker Sores (Mouth Ulcers), Capillary Weakness, Cataracts, Cholesterol (high), Cholesterol (low), Cirrhosis of the Liver, Cold Hands and Feet, Convalescence, Cramps and Spasms (general), Dehydration, Depression, Dizziness (Vertigo), Dysmenorrhea, Energy (lack of), Eye Problems (general), Fat Cravings, Fatigue, Flu, Gall Bladder (sluggish), Hepatitis, **Hypochondria**, Hypoglycemia, Itching, Jaundice (adults), Leukemia, Lupus, **Menorrhagia (Heavy Menstrual Bleeding)**, **Menstrual Irregularity**, Mood Swings, Morning Sickness, Myasthenia Gravis, Nausea and Vomiting, PMS (general), PMS Type A, Poisoning (food), Polyps, Radiation Sickness, Rashes and Hives, Surgery (healing from)

Usage: Blood Build is a traditional Chinese herbal combination that enhances immunity and strengthens the liver by nourishing the blood and nurturing the Yin. It increases deficient "wood." A person lacking in wood energy will have a tendency to feel frustrated, depressed, discouraged and indecisive with occasional bouts of sudden anger.

The wood element is associated with the liver, which is said to "build the blood" in Chinese medicine. Bu Xue, the Chinese name for this formula, means "build the blood." Bu Xue or Blood Build helps increase blood volume and flow, combating anemia, scanty menstruation and fatigue. Enhancement of blood circulation also helps to resolve menstrual problems, PMS, high blood pressure, blood sugar and cholesterol levels. It also enhances immunity, aids in chronic liver problems, softens hard masses and strengthens immunity.

It works with congested fat in the liver and high cholesterol. It is useful for anemia, dizziness, general weakness and pallor. Aids numerous liver-related problems including hypoglycemia, PMS, migraine headaches, skin eruptions, erratic health problems, depression, mood swings, digestive disturbances, food allergies and immune system weakness. It has an immune-balancing action, making it potentially useful in autoimmune disorders, particularly myasthenia gravis and systemic lupus erythematosus. This is a good formula for people who have hypochondriac feelings or suppressed anger.

To build the liver and the blood take 3 capsules of Chinese Blood Build three times daily. The formula is also available as a TCM concentrate. The concentrated Blood Build formula requires only 2 capsules per day.

Warnings: Use with caution where colitis is present.

Blood Build, Chinese - Capsule (100)

Available in: US - Stock #1881-9

Ingredients: Dang gui, ganoderma, lycium, peony, bupleurum, cornus, curcuma, salvia, achyranthes, alisma, astragalus, atractylodes, ho shou wu, ligusticum, ligustrum, rehmannia, cyperus, panax ginseng, cnidium fruit.

Blood Build TCM Conc. - Capsule (30)

Available in: US - Stock #1005-9

Ingredients: Same as Stock #1881-9

BP-C - Capsule (100)

Available in: Canada - Stock #1882-1

Ingredients: Ho shoou wu root, ligustrum fruit, rehmannia root, ganoderma plant, lycium fruit, dangqui root, bupleurum root, cornus fruit, curcuma root, salvia root, achyranthes root, alisma rhizome, astragalus root, atractylodes rhizome, cnidium fruit, cyperus rhizome, panax ginseng root.

Blood Pressurex

Product Type: Formula

Properties: **Antihypertensive**, Antioxidant, Cardiac, **Hypotensive**, Relaxant, **Vasodilator**

Systems Affected: Arteries, Cardiovascular System, Circulation, **Cyclic AMP (cAMP)**

Conditions: Blood Clots (prevention of), **Blood Pressure (high)**, Cardiovascular Disease (Heart Disease), Circulation (poor), Glaucoma

Usage: This formula works on the problem of high blood pressure from several angles. It reduces arterial inflammation and helps dilate peripheral blood vessels. It won't work in all cases of high blood pressure, but is helpful in a majority of cases. Take 1 capsule with a meal three times daily.

Warnings: No known warnings.

Blood Pressurex - Capsule (60)

Available in: US & Canada - Stock #554-8

Ingredients: Coleus, olive leaf extract, hawthorn berries, golden rod, l-arginine, Vitamin E, grape seed extract

Blood Sugar Formula

Other Names: NBS-AV

Product Type: Formula

Properties: Antidiabetic, Glandular

Systems Affected: Pancreas, Small Intestines

Conditions: Diabetes, Thinking (cloudy)

Usage: Diabetes has long been recognized and treated herbally in India. This Ayurvedic formula is built around gymnema leaves, which block sugar absorption in the intestines. It is also helpful where there is a tendency towards dry mouth and cravings for sugar. Use 2 capsules three times daily for diabetes.

Warnings: Since gymnema blocks sugar absorption, this formula could aggravate hypoglycemia.

Blood Sugar Formula - Capsule (100)

Available in: US - Stock #1298-8

Ingredients: Gymnema sylvestre, Momordica charantia, Pterocarpus marsupium, Aegle marmelos, Enicostemma littorale, Andrographis paniculata, Curcuma longa, Syzygium cumini, Azadirachta indica, Picrorhiza kurroa, trigonella foenumgraecum seed, Cyperus rotundus

Blue Cohosh

Latin Name: *Caulophyllum thalictroides*

Product Type: Single Herb

Properties: Abortifacient, Anti-epileptic, Anti-inflammatory, Antirheumatic, **Antispasmodic**, Diaphoretic, Diuretic, **Emmenagogue, Oxytocic, Parturient**, Uterine Tonic

Systems Affected: Female Reproductive, Nerves

Conditions: Amenorrhea, Arthritis, Asthma, Colon (spastic), **Cramps (menstrual)**, Dysmenorrhea, Epilepsy, Hysteria, Inflammation, **Labor (to induce)**, **Labor and Delivery**, Menorrhagia (Heavy Menstrual Bleeding), **Pregnancy (herbs and supplements to avoid during)**, Progesterone (low), Rheumatoid Arthritis (Rheumatism)

Usage: Blue cohosh has a strong oxytocic effect, meaning it stimulates uterine contractions. In an odd paradox, it also relaxes muscle spasms. This makes it very useful during childbirth because it increases the strength of the contractions while relaxing the pelvic floor. During labor, take 1-2 capsules every two hours. It can also be used to induce labor. Use about 2 capsules three times daily after due date is past. Combine with Master Gland for this purpose. Blue cohosh has also been used for ovarian pain and the pain associated with endometriosis. The antispasmodic action of blue cohosh can be helpful for asthma, colic and nervous coughs. It may also be helpful in epilepsy.

Warnings: This herb is an abortifacient and should not be used during pregnancy, except during the last six weeks. It is dangerous to use this herb as an abortifacient because if it does not work it may damage the fetus by reducing the supply of oxygen to the growing baby. Hence, we do not recommend that anyone use this herb in an attempt to expel a fetus.

Blue Cohosh - Capsule (100)

Available in: US - Stock #110-0

Ingredients: Caulophyllum thalictroides (blue cohosh) root

Blue Vervain

Latin Name: *Verbena hastata*

Product Type: Single Herb

Properties: Alterative (Blood Purifier), Detoxifying, Diaphoretic, Emetic, Emmenagogue, Expectorant, Nervine

Systems Affected: Immune System, Liver, Lymph Nodes, Nerves, Respiratory System

Conditions: Anemia, Anxiety (Panic Attack), Asthma, Bronchitis, **Chicken Pox**, Children's Remedy, Circulation (poor), Colds (general remedies for), Congestion (general), **Convulsions, Cough (spastic)**, Crohn's Disease, **Croup**, Cushing's Disease, Cystic Fibrosis, Digestion (poor), Diphtheria, Edema (Dropsy, Water Retention, Swelling), Fatigue, Fever, Indigestion, Insomnia, Irritability, **Nervous Exhaustion (Enervation)**, Neuralgia and Neuritis, Parasites (nematodes, worms), **Pertussis (Whooping Cough)**, Poisoning (food), Tonsillitis (Adenoids)

Usage: A cerebro/nervous tonic and systemic relaxant (gentle), blue vervain lightly relaxes muscles along the neck and spinal column. Take 1/2 to 1 teaspoonful with water three times daily and just before retiring at night or apply topically to the spine and massage in with a small amount of rubbing alcohol. Use blue vervain internally as a milder alternative to lobelia. It helps with colds and fevers and especially with congestion in the throat and chest. It is a good remedy for children with the flu who are nervous and irritable. Blue vervain has been called a natural tranquilizer, and may be helpful for insomnia or nervous conditions that involve people who are fanatical and "hard driving." Take 1/2 to 1 teaspoonful as needed for congestion, mild pain, nervousness and related conditions.

Warnings: Should not be taken with antidepressant or sedative drugs.

Blue Vervain (2 fl. oz.) - Liquid (Glycerite)

Available in: US - Stock #3160-8

Ingredients: Blue vervain (Verbena officinalis) in vegetable glycerine.

Bone/Skin Poultice

Other Names: BON-C

Product Type: Formula

Properties: Anti-inflammatory, Astringent, Cell Proliferant, Soothing, Tonic, **Vulnerary**

Systems Affected: Bones, Lungs, Mucus Membranes, Muscles, **Skeletal System**, Skin, **Structural System**

Conditions: Arthritis, Backache (Back Pain, Lumbago), Bleeding (external), Bleeding (internal), Broken Bones, Bruises (prevention), Bursitis, Cartilage Damage, Cuts, Disks (spinal - bulging or slipped), Dislocation, Hernias, Inflammation, **Injuries, Ligaments (torn or injuried), Sprains, Surgery (healing from)**, Surgery (preparation for), Tendonitis, Tooth Extraction, Wounds and Sores

Usage: This combination is used to speed healing of damaged tissues. It can be taken internally: 2-3 capsules three or four times daily or used externally as a poultice. Poultice can be mixed with Golden Salve for healing or Black Ointment for soothing. This formula can be helpful for healing any damaged tissues, including healing after surgery, arthritis, bursitis, joint weaknesses, broken bones, damaged cartilage and herniated disks.

Warnings: No known warnings.

Bone/Skin Poultice - Capsule (100)

Available in: US - Stock #1248-2

Ingredients: Yarrow, mullein, plantain, rehmannia

Bowel Detox

Other Names: Bowel Build

Product Type: Formula

Properties: Deodorant, Laxative (general)

Systems Affected: Intestinal System, Large Intestine (Colon), Stomach

Conditions: Cholesterol (high), Colitis, Constipation (adults), Crohn's Disease, Depression, Digestion (poor), Diverticulitis (Diverticuli), Gas and Bloating, Heavy Metal Poisoning, Inflammatory Bowel Disorders (Colitis, IBS), **Prolapsed Colon**

Usage: This formula consists mostly of substances that absorb various types of toxins and irritants. Psyllium is a bulk laxative. Algin absorbs heavy metals. Bentonite clay pulls irritants from bowel pockets and astringes the bowel. Apple pectin lowers cholesterol. Charcoal absorbs most poisons and gas from the bowel. Chlorophyll is deodorizing. Take 2-6 capsules per day (1-2 capsules two to three times daily) with a large glass of water. Best taken 20-30 minutes before breakfast and at bedtime. Drink 6-8 glasses of water per day when taking this formula for best results.

Warnings: If you do not drink enough water with this formula, it may cause constipation.

Bowel Detox - Capsule (120)

Available in: US - Stock #3020-8

Ingredients: Betaine HCl, pepsin, pancreatin, bile salts, vitamins A, C, and E, zinc, selenium, psyllium hulls, algin, cascara sagrada, bentonite clay, apple pectin, marshmallow, parthenium, charcoal, ginger, sodium copper chlorophylln, beta carotene, kelp, d.alphatocopheryl acetate

BWL-BLD - Capsule (120)

Available in: Canada - Stock #2855-3

Ingredients: Vitamin C (10.0 mg), zinc gluconate (1.0 mg), selenium H.V.P. chelate (5.3 mcg),hydrolyzed vegetable protein, psyllium seed hulls, kelp plant powder, cascara sagrada bark, bentonite clay, apple pectin, marshmallow root, parthenium root, charcoal powder, ginger root, pepsin, betaine hydrochloride, bile salts, pancreatin, sodium copper chlorophyllin, d-alpha tocopheryl acetate, beta carotene

BP-C

See *Blood Build*

BP-X

Product Type: Formula

Properties: Alterative (Blood Purifier), Bitter, Cholagogue, Diuretic, Hepatic

Systems Affected: Liver, Skin, Structural System

Conditions: **Abscesses, Acne (Pimples, Blackheads)**, Acquired Immune Deficiency Syndrome (AIDS/HIV), Addictions (drugs), Adenitis, Blood Poisoning, Boils, Cancer (natural therapy for), Chicken Pox, Cystic Breast Disease, Dermatitis, Eczema, Gangrene, Impetigo, Itching, Jaundice (adults), Mononucleosis, Myasthenia Gravis, PMS Type A, **PMS Type S**, Poison Ivy or Oak, Polyps, Psoriasis, Puberty (hormone balancer), **Rashes and Hives**, Sickle Cell Anemia, Skin Care (general), Surgery (healing from), Toxemia, Vaccines (detoxification from)

Usage: A formula containing many blood-purifying herbs, as well as some laxatives and diuretics. It helps skin eruptive diseases and general liver and gall bladder problems. It is primarily cleansing. It also helps to prevent bile duct cirrhosis produced by free radical damage and liver-detoxifying foreign compounds. Use 2 capsules with two meals as a blood purifier. This formula can also be used as part of a general cleansing program.

Warnings: Use with caution where colitis is present.

BP-X - Capsule (100)

Available in: US - Stock #803-2

Ingredients: Burdock, pau d'arco, red clover tops, sarsaparilla, yellow dock, dandelion, buckthorn bark, cascara sagrada bark, yarrow flowers, Oregon grape, prickly ash bark

BP - Capsule (100)

Available in: Canada - Stock #799-7

Ingredients: Burdock, yellow dock, dandelion, licorice, red clover, Turkey rhubarb, goldenseal, yarrow and sarsaparilla

Brain-Protex

Product Type: Formula

Properties: **Anti-aging**, Antioxidant, **Cerebral Tonic**

Systems Affected: **Acetylcholine, Brain**, Liver, Nerves

Conditions: Aging (prevention), **Alzheimer's Disease**, Circulation (to the brain), Dementia, Free Radical Damage, **Memory and Brain Function, Mental Illness, Parkinson's Disease, Senility**

Usage: This formula contains herbs and nutrients that enhance acetylcholine, a neurotransmitter involved in memory and muscle coordination. It also contains antioxidants that prevent free radical damage to brain tissue. It also contains ingredients that inhibit acetylcholinesterase, the enzyme that breaks down acetylcholine. Take 2 capsules with a meal twice daily.

Warnings: No known warnings.

Brain-Protex w/ Huperzine - Capsule (60)

Available in: US - Stock #3114-1

Ingredients: Chinese Club Moss (1% Huperzine A), Ginkgo biloba, lycopene, alpha lipoic acid, soybean lecithin complex, Rhododendron caucasicum.

Breast Assured

Product Type: Formula

Properties: **Phytoestrogen**

Systems Affected: **Breasts**, Glandular System, Lymphatic System, Uterus

Conditions: Benign Prostate Hyperplasia (BPH), **Breast Lumps**, Breasts (swelling and tenderness), **Cancer (prevention)**, Cystic Breast Disease, Estrogen (low)

Usage: This formula contains nutrients that have been shown to have a protective affect against breast cancer. Women with a family history of breast cancer may wish to use this formula as an aid to cancer prevention. Take 2 to 3 capsules daily with a meal.

Warnings: No known warnings.

Breast Assured - Capsule (60)

Available in: US - Stock #1122-4

Ingredients: Flax meal, ellagic acid from pomegranate extract powder, kudzu extract, maitake mushroom, lutein, calcium glucanate.

Breast Enhance

Product Type: Formula

Properties: Glandular, **Phytoestrogen**

Systems Affected: **Breasts**, Glandular System

Conditions: Breasts (enhance size), Menopause

Usage: This formula contains estrogenic substances that stimulate breast tissue and may help to enlarge breast tissue and/or enhance the firmness of breast tissue. Use 1 capsule with a meal three times daily.

Warnings: May cause weight gain.

Breast Enhance - Capsule (90)

Available in: US - Stock #1107-3

Ingredients: Kudzu root extract, saw palmetto extract, dong quai extract, alfalfa

Breathe EZ

Other Names: Xuan Fei

Product Type: Formula

Properties: **Antitussive**, Decongestant, Diuretic, **Expectorant**, Pectoral, Pulmonary

Systems Affected: Bronchials, Lungs, Urinary System

Conditions: **Asthma**, Bronchitis, Colds (general remedies for), **Congestion (bronchial)**, Congestion (lungs), **Cough (damp)**, **Grief (excessive)**, Lungs (fluid in), Pneumonia

Usage: The Chinese name for this formula, "Xuan Fei", means "ventilate the lungs." This formula opens respiratory passages. It is an excellent remedy for asthma, bronchial congestion, damp cough with water retention and fluid in the lungs. For respiratory congestion take 2-4 capsules every four hours. For asthma attack take four capsules every 15 minutes until the attack subsides.

Warnings: Do not take with any products containing caffeine.

Breathe E-Z TCM Conc. - Capsule (30)

Available in: US - Stock #1036-3

Ingredients: Citrus peel, typhonium rhizome, bamboo sap, bupleurum root, fritillaria bulb, hoelen selerotium, perilla leaves, platycodon root, xingren apricot seed, ophiopogon root tuber, tussilago flower buds, ginger rhizome, schizandra fruit, licorice root.

AL-C - Capsule (100)

Available in: Canada - Stock #1864-0

Ingredients: Fructus aurantia immaturi, citrus peel, pinellia rhizome, hoelen plant, bamboo sap, fritillaria bulb, bupleurum root, inula flower, xingren, magnolia bark, morus root bark, ophiopogon root, ginger rhizome, schizandra fruit, and licorice root

Breathe Free

Product Type: Essential Oil

Properties: Antiseptic, Antispasmodic, Antitussive, Decongestant, Expectorant, Pectoral

Systems Affected: Bronchials, Lymphatic System, **Respiratory System**

Conditions: Asthma, Bronchitis, Colds (general remedies for), Concentration (poor), **Congestion (general)**, Cough (damp), Endometriosis, Flu, Grief (excessive), Memory and Brain Function

Usage: Breathe Free is stimulating and promotes healing of pulmonary infections. This blend is helpful during cold and flu season to keep airways open so a person can breathe freely. It is a refreshing and uplifting scent. Use it as an inhalant or in a diffuser or massage lotion for asthma. Use as an inhalant for sinus infections and congestion. Emotionally, Breath Free helps with exhaustion, memory, indecision, clarity, lack of concentration, negativity and grief.

Warnings: No known warnings.

Breathe Free Essential Oil Blend (5ML) - Essential Oil

Available in: US - Stock #3919-1

Ingredients: Geranium, Niaouli Bio, Peppermint, Rosemary

Bronchial Formula

Other Names: BRN-AV

Product Type: Formula

Properties: Anti-inflammatory, Antispasmodic, Bitter, Decongestant, Expectorant, Immune Stimulant, Pectoral, Pulmonary, Stimulant

Systems Affected: Bronchials, Digestive System, Lungs

Conditions: Allergies (respiratory), Asthma, **Bronchitis,** Colds (general remedies for), **Congestion (bronchial),** Congestion (lungs), Cough (damp), Pneumonia

Usage: This is a general respiratory remedy from India for bronchitis, asthma, cough and other forms of respiratory congestion. Use 2 capsules three times daily.

Warnings: No known warnings.

Bronchial Formula - Capsule (100)

Available in: US - Stock #1297-2

Ingredients: Adhatoda vasica, Glycyrrhiza glabra (licorice), Verbascum thapsus, Alpina galanga, Clerodendrum indicum, Inula racemosa, Myrica nagi, Phyllanthus emblica, Hedychium spicatum, Picrorhiza kurroa, Pimpinella anisum, Pistacia integerrima, Zingiber officinale (ginger), Ocimum sanctum, Tylophora asthmatica, Abies webbina, Elettaria cardamomum, Ferula assafoetida

Burdock

Latin Name: *Arctum lappa*

Product Type: Single Herb

Properties: Alterative (Blood Purifier), Anti-allergenic, Anticancer, Antimutagenic, **Antipruritic,** Aperient, Bitter, Cholagogue, Detoxifying, Food, Hepatic, Hypoglycemic, **Mast Cell Stabilizer,** Stomachic

Systems Affected: Blood, Digestive System, Gall Bladder, Intestinal System, Liver, Lymphatic System, Skin, Structural System

Conditions: Acne (Pimples, Blackheads), Allergies (food), Allergies (respiratory), Arthritis, Blood Poisoning, **Boils,** Bruises (healing), Burns and Scalds, Bursitis, Cancer (prevention), Canker Sores (Mouth Ulcers), Chicken Pox, Congestion (lymphatic), **Eczema,** Fat Metabolism (poor), Fatty Liver Disease, **Fatty Tumors or Deposits,** Gall Bladder (sluggish), Glands (swollen lymph), Gout, Herpes, Hypoglycemia, **Itching,** Measles, Mumps, PMS Type S, **Poison Ivy or Oak, Psoriasis, Rashes and Hives, Seborrhea,** Skin (infections), **Skin (oily),** Surgery (healing from)

Usage: Burdock is one of the safest and most popular alteratives or blood purifiers. It has been used to help clear up morbid conditions like cancer and skin eruptive diseases. It stimulates bile production, strengthens the liver and also improves lymphatic and kidney function. For skin conditions and general liver problems, take 1-2 capsules up to three times daily with water. This herb is very helpful for both acute chronic skin conditions (boils, bruises, canker sores, chicken pox, measles, herpes, eczema, psoriasis, poison ivy, etc.). It works best when used both internally and topically. Take larger amounts, 2 capsules three to four times daily, for these problems. Empty the capsules and make a paste to apply as a poultice to infected sores. It can also be made into a strong decoction and applied topically to reduce itching.

Warnings: No known warnings.

Burdock - Capsule (100)

Available in: US & Canada - Stock #140-2

Ingredients: Arctum lappa (burdock) root

Butcher's Broom

Latin Name: *Ruscus aculeatus*

Product Type: Single Herb

Properties: Anticoagulant (Blood Thinner), Antithrombolytic, Cardiac, Deobstruent, Vascular Tonics

Systems Affected: Blood Vessels, Peripheral Blood Vessels, **Veins**

Conditions: Aneurysm, Arteriosclerosis (Atherosclerosis, Hardening of the Arteries), **Blood Clots (prevention of),** Bruises (healing), **Bruises (prevention),** Capillary Weakness, Hemorrhoids, Inflammation, Jaundice (adults), **Phlebitis,** Raynaud's Disease, **Spider Veins, Strokes,** Surgery (preparation for), **Thrombosis,** Urine (scant), **Varicose Veins**

Usage: This herb has been used for hemorrhoids, phlebitis, varicose veins and post-operative thrombosis. Butcher's broom has anti-inflammatory properties and tones (tightens and strengthens) the walls of blood vessels. Research in Europe suggests this herb may help prevent blood clots from forming in the circulatory system without thinning the blood, as well as helps disperse blood clots. Butcher's broom is used in Europe to prevent post-operative thrombosis (blood clots forming in the body after surgery). It is considered the specific herb for unremitting leg pains, cramps and the heavy feeling in the legs of elderly people. Butcher's broom improves blood flow to the brain and may be useful when taken with ginkgo and gotu kola for poor memory, depression and similar conditions. This herb is used as a suppository for hemorrhoids and has been applied externally to shrink varicose veins. For internal use, take 1-3 capsules three times per day. For external use, make this herb into a tea (2-3 capsules per cup), soak a washcloth or other cloth in warm tea and apply to affected area. Change cloth frequently.

Warnings: Because butcher's broom tightens or tones blood vessels, this herb may be contraindicated in cases of high blood pressure. No other warnings.

Butcher's Broom - Capsule (100)

Available in: US & Canada - Stock #135-5

Ingredients: Ruscus aculeatus (butcher's broom) herb

C-X

Product Type: Formula

Properties: Aphrodisiac, Emmenagogue, **Female Tonic**, Glandular

Systems Affected: Adrenal Glands, Female Reproductive, Liver, **Ovaries**, Testes

Conditions: Acne (Pimples, Blackheads), Asthma, Broken Bones, Cramps (menstrual), Diabetes, Disks (spinal- bulging or slipped), Erectile Dysfunction, Estrogen (low), Fatigue, Hot Flashes, Infertility, Menopause, Mood Swings, Osteoporosis, Pregnancy (herbs and supplements to avoid during), Progesterone (low), Sex Drive (low)

Usage: This combination is used as a general female corrective, primarily for symptoms associated with menopause. It helps balance male and female hormones. Work up gradually to about 2 capsules three times daily.

Warnings: Not recommended during pregnancy or lactation.

C-X - Capsule (100)

Available in: US - Stock #1230-6

Ingredients: Black cohosh, licorice, Eluthero, sarsaparilla, squaw vine, blessed thistle, false unicorn

C-X - Capsule (100)

Available in: Canada - Stock #1203-2

Ingredients: Blessed thistle, false unicorn, black cohosh, licorice, squaw vine, sarsaparilla, Siberian ginseng

CA, Herbal

See *Herbal CA*

Caffeine Detox

Product Type: Homeopathic

Properties: Detoxifying, Nervine

Systems Affected: Adrenal Glands, Nerves

Conditions: Acid Indigestion (Heartburn, Acid Reflux), **Addictions (coffee, caffeine)**, Anger (excessive), **Cystic Breast Disease**, Headache (general), Insomnia, **Irritability**

Usage: Used to relieve symptoms of withdrawal from caffeine, including irritability, impatience, anger, headache, insomnia, upset stomach, heartburn and nausea. For adults and children over 12, 20-30 drops under tongue every two hours for the first day. On the second day take 15 drops every two hours, 15 drops every three hours on the third day and 10-30 drops every 4-6 hours thereafter.

Warnings: NSP does not recommend this product for children under the age of 12 except on the advice of a health professional. Other than the alcohol content, which is not good for infants, we consider this product safe for children.

Caffeine Detox (1 fl. oz.) - Homeopathic (Liquid)

Available in: US - Stock #8727-8

Ingredients: Chamomilla 3x, coffea cruda 6x, Gratiola officinalis 6x, Ignatia amara 6x, Nux vomica 6x, Thuja occidentalis 6x, water glycerine

Calcium

Product Type: Nutrient

Properties: Alkalinizer, Anti-arrhythmic

Systems Affected: Bones, Heart, Muscles, Nerves, Skeletal System, Structural System

Conditions: **Broken Bones**, Calcium Deficiency, Heart Fibrillation or Palpitations, Insomnia, Mental Illness, Narcolepsy, Neurosis, **Overacidity**, Pregnancy (herbs and supplements for), Prolapsed Colon, Prolapsed Uterus, Tachycardia, Teeth (grinding), Teething, Tics, Twitching

Usage: Calcium is the most important macromineral in the body, being found in higher concentrations than any other nutrient. It is not only important for bones and other structures, it is also used in muscular contractions, nerves and other tissues. Calcium cannot be properly utilized unless it is bound to proteins. It also requires other trace minerals and vitamins for proper utilization. For this reason, plant based sources of calcium often have more tissue healing ability than calcium supplements. Consider HSN-W or Herbal CA. If you want to take a calcium supplement, Skeletal Strength is the best choice because it combines calcium with other nutrients needed for absorption and utilization.

For a temporary alkalizing effect take one capsule or mix one rounded scoop of the coral calcium powder with 8 ounces of water or other beverage once or twice daily, as needed.

Warnings: Many people take too much calcium. This can raise the free calcium index in the blood which contributes to calcifications, kidney stones and other health problems. Always take calcium with magnesium. Use magnesium instead of calcium when there are problems with muscle cramps.

Coral calcium and Sea Calcium are primarily calcium carbonate, a form of calcium that suppresses hydrochloric acid in the stomach and is very difficult for bone and tissues to assimilate and utilize. Calcium must be bound to protein to be utilized by tissues. Coral calcium tends to raise the free (or unbound) calcium level in the blood. A high level of free calcium is associated with increased risk of calcifications and kidney stones as well as an increased risk of cancer. Because it is not effectively utilized by the tissues it is passed out of the body through the urine, which creates a more alkaline reading in the urine pH. This can be useful as a temporary measure to buffer excess acids in the body, but should not be done on a regular basis. For long-term calcium supplementation we recommend Skeletal Strength, Herbal CA and HSN-W.

Calcium Plus Vitamin D - Tablet (200)

Available in: US - Stock #1675-0

Ingredients: Per Serving: Calcium (amino acid chelate, citrate, bone meal, di-calcium phosphate) 250 mg., magnesium (amino acid, chelate, oxide, stearate) 125 mg., vitamin D, phosphorus, alfalfa

Cal-Mag Plus D - Tablet (200)

Available in: Canada - Stock #1758-6

Ingredients: Calcium (amino acid chelate, citrate, bone meal, di-calcium phosphate) 250 mg., magnesium (amino acid, chelate, oxide, stearate) 125 mg., vitamin D, phosphorus, alfalfa

Calcium-Magnesium, SynerPro - Tablet (240)

Available in: US - Stock #3194-9

Ingredients: Calcium (amino acid chelate, di-calcium phosphate) 200 mg., magnesium 100 mg., phosphorus, zinc, copper, boron, and a blend of broccoli, carrot, red beet, rosemary, tomato, turmeric, cabbage, Chinese cabbage, grapefruit bioflavinoid, orange bioflavinoid, hesperidin

Calcium Magnesium Synerpro - Tablet (240)

Available in: US & Canada - Stock #4041-3

Ingredients: Calcium (amino acid chelate, di-calcium phosphate) 400 mg., magnesium 200 mg., phosphorus, zinc, copper, boron, and a blend of broccoli, carrot, red beet, rosemary, tomato, tumeric, cabbage, Chinese cabbage, grapefruit bioflavinoid, orange bioflavinoid, hesperidin

Calcium, Liquid (16 oz.) - Liquid

Available in: US & Canada - Stock #3191-6

Ingredients: Per Serving: Calcium (phosphate, citrate, lactate) 500 mg. Also contains magnesium (oxide) 200 mg., zinc (glucanate) 1.5 mg., vitamin D, filtered water, fructose, glycerine, natural flavors, xanthan gum, lecithin, potassium sorbate, sodium benzoate

Sunshine Heroes™ Calcium Plus D3 - Tablets (90)

Available in: US - Stock #3343-1

Ingredients: See listing under Sunshine Heroes Calcium Plus D3

Nature's Sea Calcium Capsules - Capsule (120)

Available in: US - Stock #1577-5

Ingredients: Same as Nature's Sea Calcium Powder (75 g)

Nature's Sea Calcium Powder (75g) - Bulk Powder

Available in: US - Stock #1576-9

Ingredients: Sea vegetation calcium, magnesium, fructooligosaccharides, silicon dioxide

Coral Calcium Capsules - Capsule (90)

Available in: US - Stock #1899-7

Ingredients: Coral calcium, montmorilonite clay, magnesium.

Coral Calcium (75 g) - Bulk Powder

Available in: US - Stock #1873-7

Ingredients: Coral calcium, magnesium, montmorillonite

Calming

Product Type: Homeopathic

Properties: Analgesic (Anodyne), Calmative, Nervine, Relaxant, Sedative

Systems Affected: Nerves

Conditions: Children's Remedy, Insomnia, Irritability, Teething

Usage: A homeopathic formula for children experiencing restless sleep, fussiness and irritability accompanying minor illness and teething.

For children and infants over 4 months, take 3-5 drops under the tongue every 15-20 minutes until symptoms improve, then decrease to every two-four hours until symptoms are relieved.

Warnings: NSP recommends not administering this remedy to children under 4 months old except on the advice of a health care professional.

Calming (1 fl. oz.) - Homeopathic (Liquid)

Available in: US - Stock #8870-5

Ingredients: Active ingredients: Gelsemium semper-virens (Yellow Jessamine) 4x, Hyo-scyamus niger (Henbane) 6x, Saccharum officinale (Cane Sugar) 6x, Hepar Sulfuris Calcareum (Calcium Sulfide) 8x, Lycopodium clavatum (Club Moss) 8x, Ambra grisea (Ambergris) 10x. Other ingredients: Purified water, glycerin, and potassium benzoate

Candida

Product Type: Homeopathic

Properties: Antifungal

Systems Affected: Immune System, Intestinal System

Conditions: Fungal Infections (Yeast Infections, Candida albicans)

Usage: Used for relief from itching, burning and other symptoms associated with candida yeast infections. Take 10-15 drops under the tongue four times daily for at least one month, then take half the dosage for an additional three months as maintenance. It is also very important to take probiotics and antifungal remedies (such as Yeast/Fungal Detox or pau d'arco) when dealing with yeast infections.

Warnings: NSP does not recommend this product for children under the age of 12 except on the advice of a health professional. Other than the alcohol content, which is not good for infants, we consider this product safe for children.

Candida (1 fl. oz.) - Homeopathic (Liquid)

Available in: US - Stock #8713-7

Ingredients: Active ingredients: Baptisia tinctoria (Wild Indigo) 4x, Bryonia (White Bryony) 4x, Echinacea angustifolia (Cone Flower) 4x, Eupatorium perfoliatum (Boneset) 4x, Hydrastis canadensis (Golden Seal) 4x, Thuja occidentalis (Tree of Life) 4x, Viscum album (Mistletoe) 6x, Candida albicans (Candida Yeast Extract) 12x. Other ingredients: Purified water and 20% USP alcohol.

Candida Clear

Product Type: Formula

Properties: Antifungal, Immune Stimulant

Systems Affected: Digestive System, Immune System, Large Intestine (Colon)

Conditions: Addictions (sugar or food), Antibiotics (side effects of), **Athlete's Foot, Fungal Infections (Yeast Infections, Candida albicans), Itching Ears,** Jock Itch, Jock Itch, Leaky Gut Syndrome, **Leucorrhea,** Mental Illness, **Sugar Cravings**

Usage: Candida Clear Pack is an excellent and highly effective cleanse for yeast and fungal infections. This pack contains three antifungal products that reduce yeast overgrowth and enzymes which help break down dead yeast. Combined with some dietary changes, Candida Clear can reduce yeast overgrowth and improve overall health.

There are two packets in the Candida Clear program-the Candida Clear Combo (containing Pau d'arco, Yeast/Fungal Detox and Caprylic Acid Combination) and the Candida Clear Enzymes. Take 1 Candida Clear Combo pack three times daily with meals and 1 Candida Clear Enzyme Packet three times daily between meals on an empty stomach. It is also important to make dietary modifications in order for this program to work effectively.

Warnings: No known warnings.

Candida Clear - Packets

Available in: US - Stock #958-7

Ingredients: Pau D'Arco, Caprylic Acid Combination, Yeast Fungal Detox, Candida Cleanse Enzymes

Caprylic Acid Combination

Product Type: Formula

Properties: Antifungal, Antiparasitic

Systems Affected: Immune System, Large Intestine (Colon)

Conditions: Fungal Infections (Yeast Infections, Candida albicans), Leaky Gut Syndrome

Usage: To help rid the body of yeast, take 2 capsules two times per day. Works best in combination with a general yeast control program. For most people, Yeast/Fungal Detox is a better antifungal formula.

Warnings: If bloating or gas gets too uncomfortable (due to candida kill off) reduce the amount to 1 capsule with morning and evening meals or 1 capsule with morning meal.

Caprylic Acid Combination - Capsule (90)

Available in: US & Canada - Stock #1808-2

Ingredients: Caprylic acid, elecampane, black walnut, red raspberry

Caprylimune

Product Type: Formula

Properties: Antibacterial, Antifungal

Systems Affected: Cardiovascular System, Intestinal System, Liver

Conditions: Fungal Infections (Yeast Infections, Candida albicans)

Usage: Helps to destroy yeast and other unfriendly micro-organisms in the colon. Use 2 tablets three times a day. For most people, Yeast/Fungal Detox is a better antifungal formula.

Warnings: No known warnings.

Caprylimune - Tablet (150)

Available in: US - Stock #2808-5

Ingredients: Caprylic acid, vitamins A, C, E, pantothenic acid, calcium, phosphorus, biotin, zinc, selenium, pau d' arco, garlic, golden seal, yucca root, lemon grass, rose hips concentrate, hesperidin, lemon bioflavonoid, rutin, guar gum

Capsicum

Latin Name: *Capsicum minimum, C. frutescens*

Product Type: Single Herb

Properties: Analgesic (Anodyne), **Anesthetic,** Anti-arthritic, Anticoagulant (Blood Thinner), Aperient, Cardiac, Carminative, **Catalyst (Synergist), Coagulant,** Condiment, Counterirritant, **Diaphoretic, Hemostatic, Hypotensive,** Panacea, Rubefacient, Sialogogue, Stimulant, **Stimulant (Circulatory),** Stomachic, **Styptic,** Vulnerary

Systems Affected: Bronchials, Cardiovascular System, **Circulation,** Gums, Heart, Hypothalamus, Mucus Membranes, Nerves, Pancreas Tail, **Peripheral Blood Vessels,** Skin, Stomach, **Sweat Glands,** Throat

Conditions: Abdominal Pain and Inflammation, **Aneurysm,** Arthritis, **Bleeding (external), Bleeding (internal),** Blood Pressure (high), **Blood Pressure (low),** Bronchitis, **Cardiac Arrest (Heart Attack), Chills, Circulation (poor), Cold Hands and Feet, Colds (general remedies for),** Congestion (general), **Cuts,** Dysmenorrhea, Fainting, Fatigue, Fever, Frostbite (prevention), Gangrene, Hangover, Heart Fibrillation or Palpitations, Indigestion, Infection (viral), **Labor and Delivery,** Laryngitis (Hoarseness), Lungs (fluid in), Menorrhagia (Heavy Menstrual Bleeding), Miscarriage (prevention), **Nose Bleeds,** Numbness, Paralysis, **Perspiration (deficient),** Phlebitis, **Raynaud's Disease,** Senility, **Shock, Sore Throat,** Strokes, Surgery (preparation for), Sweat Baths (herbs for), Ulcers, Varicose Veins, Wounds and Sores

Usage: This common spice herb contains the active constituent capsacian, which affects its action by irritating (stimulating) tissues it comes into contact with. Capsicum is a major stimulant for the circulatory system. It increases circulation to every area of the body that it comes in contact with, internally or externally. Capsicum also strengthens the heartbeat. Because adequate blood supply is necessary for all tissues to heal, capsicum has earned a reputation in the West as a kind of "cure-all." As a tonic and aid to circulation, digestion and general health, take 1-2 capsules with meals. Capsicum is also useful for shock, heart attack and trauma. Use 1-2 droppers full of the extract or open a capsule and empty contents in mouth. Repeat in 15-20 minutes if necessary. Gargle with the tea for sore throats. Apply powder to cuts and wounds to stop bleeding. Sprinkle a small amount of powder onto on socks to prevent frostbite. Applying the extract to skin encourages circulation to that body part. Capsicum can also be used in food as a replacement for black pepper.

Warnings: Due to its irritating nature some people have a hard time taking capsicum. Large doses can be irritating to the stomach and cause painful bowel eliminations. Although capsicum stops bleeding and has been used to heal ulcers, this herb can cause pain when used for these purposes, and hence, should be used with caution. It is best to start with extremely small doses to build up tolerance. Capsicum causes burning sensations in sensitive areas, such as genitals, sinuses, etc. Not recommended for hemorrhoids, anal fissures or sensitive or spastic colons.

Capsicum - Capsule (100)

Available in: US & Canada - Stock #160-5

Ingredients: Capsicum annuum (cayenne) fruit

Capsicum Extract (2 fl. oz.) - Liquid (Tincture)

Available in: US - Stock #1782-0

Ingredients: Capsicum annuum (cayenne) fruit, alcohol, water

Capsicum Shaker (2-3 oz.) - Bulk Powder

Available in: US & Canada - Stock #166-7

Ingredients: Same as Capsicum

Capsicum & Garlic w/Parsley

Product Type: Formula

Properties: Antibacterial, **Antihypertensive**, Antiseptic, Carminative, **Hypotensive**, **Stimulant (Circulatory)**, Stomachic

Systems Affected: **Circulation**, Heart, Mucus Membranes

Conditions: Arteriosclerosis (Atherosclerosis, Hardening of the Arteries), **Blood Pressure (high)**, Blood Pressure (low), Bronchitis, Chills, **Circulation (poor)**, Cold Hands and Feet, Colds (general remedies for), Congestion (general), Congestion (lymphatic), Cough (damp), Emphysema, Fatigue, Flu, Infection (viral), Kidney Infection, Pneumonia, Tonsillitis (Adenoids)

Usage: This formula is typically used for high blood pressure. The parsley in the formula not only helps counteract some of the garlic odor, it acts as a mild diuretic to stimulate kidney function. The formula can also be used for general circulatory disorders, including cold hands and feet, arteriosclerosis and cholesterol problems. It is an excellent aid for certain types of infection, including bronchitis, flu, lymphatic congestion and pneumonia. Use 1-2 capsules three to four times daily for circulatory problems. For infection use 2-4 capsules every 2-4 hours with plenty of water.

Warnings: Capsicum and garlic may cause some digestive upset in people. Do not use where gastric ulcers are present. Otherwise, it is completely safe for long-term use when used in moderation.

Capsicum & Garlic with Parsley - Capsule (100)

Available in: US - Stock #832-3

Ingredients: Capsicum, garlic, parsley

Capsicum Garlic & Parsley - Capsule (100)

Available in: Canada - Stock #829-9

Ingredients: Same as Stock #832-3

Carbo-Grabbers

Product Type: Nutrient

Properties: Anti-obesic, Hypolipidemic

Systems Affected: Digestive System, Weight Loss

Conditions: Diabetes, Triglycerides (high), Weight Loss (aids for)

Usage: Take 1-2 capsules before a meal high in carbohydrates. Use with Fat Grabbers to maximize weight loss.

Warnings: Do not take in combination with any protease enzyme-containing product.

Carbo Grabbers with Chromium - Capsule (60)

Available in: US - Stock #3070-7

Ingredients: White Kidney Bean Extract (Phaseolus vulgaris), chromium

Carbo Grabbers - Capsule (60)

Available in: US - Stock #2954-6

Ingredients: Northern white kidney bean extract

Carbo-Grabbers - Capsule (60)

Available in: Canada - Stock #2958-0

Ingredients: Same as Stock #2954-6

Carbo Grabbers Trial Pack - Pack (20)

Available in: US - Stock #2789-9

Ingredients: Same as Stock #2954-6

Cardio Assurance

Product Type: Formula

Properties: Anti-arrhythmic, Antioxidant, **Cardiac**, Stimulant (Circulatory), Tonic

Systems Affected: Arteries, Cardiovascular System, Circulation

Conditions: Angina, Arrhythmia, Arteriosclerosis (Atherosclerosis, Hardening of the Arteries), Blood Clots (prevention of), Bruises (prevention), Cardiac Arrest (Heart Attack), Cardiovascular Disease (Heart Disease), Cholesterol (high), Free Radical Damage, Heart (weakness), **Rheumatic Fever**

Usage: This formula was designed to help support the health of the heart. It helps strengthen and regulate the heartbeat and increase peripheral circulation. It also helps prevent and possibly reduce arterial plaque. Take 1 capsule with a meal three times a day.

Warnings: No known warnings.

Cardio Assurance - Capsule (60)

Available in: US & Canada - Stock #553-2

Ingredients: Polygonum cuspidatum (Resveratrol), Hawthorn Berries (Crataegus laevigata), Vitamin B12 (as cyanocobalamin), Folic Acid, Vitamin B6 (as pyridioxine HCL), Vitamin K2 (MK-7) (contains milk derived)

Carotenoid Blend

Product Type: Formula

Properties: Antioxidant, Nutritive

Systems Affected: Liver

Conditions: Aging (prevention), Arteriosclerosis (Atherosclerosis, Hardening of the Arteries), Cancer (natural therapy for), Cancer (prevention), Cataracts, Convalescence, Croup, Emphysema, Eye Problems (general), Free Radical Damage, Hashimoto's Disease (Thyroiditis), Macular Degeneration, Radiation Sickness, Sprains

Usage: Carotenoids have a variety of nutritional benefits in the body, helping to protect various organs from environmental damage. For general nutrition take 1 capsule three times daily.

Warnings: No known warnings.

Carotenoid Blend - Capsule (60)

Available in: US - Stock #4073-3

Ingredients: Beta-carotene, alpha-carotene, lutein, lycopene, zeaxanthin, cryptoxanthin, phytoene, phytofluence, astaxanthin, hibiscus.

Cascara Sagrada

Latin Name: *Rhamnus purshiana*

Product Type: Single Herb

Properties: Bitter, Cholagogue, **Laxative (stimulant), Purgative (Cathartic)**

Systems Affected: Gall Bladder, Intestinal System, Large Intestine (Colon), Liver

Conditions: Colon (atonic), Congestion (general), Constipation (adults), Gall Bladder (sluggish), Parasites (general), Parasites (nematodes, worms), Pregnancy (herbs and supplements to avoid during)

Usage: To stimulate bowel elimination for occasional constipation, take 1 to 2 capsules at bedtime with a full 8 oz. glass of water. Increases bile flow and stimulates natural peristalsis of the colon. Can also be used to help gallstones and liver problems, but is best used in combination with other herbs for this purpose.

Warnings: Not recommended for use during pregnancy (at least as a single herb) or for weak, feeble or debilitated persons. Avoid prolonged use of this or any stimulant laxative. If a person appears dependent on this or other stimulant laxatives or if cramps or griping become a problem, use nervines or magnesium to counteract bowel spasms. In case of diarrhea caused by excessive use, use charcoal or mucilants. Long-term use will darken the tissue color of the bowel. Habitual use drains the body's energies.

Cascara Sagrada Capsules - Capsule (100)

Available in: US & Canada - Stock #170-8

Ingredients: Rhamnus purshiana (cascara sagrada) aged bark

Cascara Sagrada Vegitabs - Vegitab (100)

Available in: US - Stock #172-1

Ingredients: Same as Stock #170-8

Cat's Claw Combination

See *Uña de Gato Combination*

Catnip

Latin Name: *Nepeta cataria*

Product Type: Single Herb

Properties: Analgesic (Anodyne), Antacid, Antispasmodic, Aromatic, Carminative, Diaphoretic, Nervine, Sedative, Stimulant, Stomachic

Systems Affected: Bronchials, Digestive System, Nerves, Pituitary (posterior), Respiratory System, **Stomach**, Sweat Glands

Conditions: Acid Indigestion (Heartburn, Acid Reflux), Addictions (tobacco smoking or chewing), Anorexia,

Bronchitis, Chicken Pox, Children's Remedy, Colds (general remedies for), Colic (children), Convulsions, Croup, Dyspepsia, Fever, Flu, Gas and Bloating, Inflammatory Bowel Disorders (Colitis, IBS), Insomnia, Measles, Mumps, Pain (general remedies for)

Usage: Soothing and settling to the stomach. Use for colic, gas, indigestion, acid stomach. Dosage: 1-2 capsules every 15-20 minutes with glass of water in acute cases. Works better in liquid form, especially when combined with fennel. Make into tea (2-3 capsules per cup) and sweeten for babies and children. Catnip tea is often used in enemas to bring down fevers. It helps produce perspiration without increasing the heat in the body. Catnip can also be used for nervousness and stress; dosage 2 capsules three times daily. Can also be taken at bedtime (2-4 capsules) as a sleep aid. Excellent herb for children and babies because it is mild and extremely safe. Good cold remedy for children, used in the same manner as yarrow. Use tea as an enema for fevers. Combine with peppermint and yarrow to make tea for colds. Drink warm. See also *Catnip and Fennel.*

Warnings: In extremely large doses, it can cause vomiting. Can cause headaches and mild hallucinations if smoked (used in some medicinal Native American tobaccos). No withdrawal symptoms.

Catnip - Capsule (100)

Available in: US - Stock #180-4

Ingredients: Nepeta cataria (catnip) herb

Catnip & Fennel

Product Type: Formula

Properties: **Antacid**, Appetite Stimulant, Carminative, Diaphoretic, Digestive Tonic, Nervine, Stomachic

Systems Affected: Digestive System, Intestinal System, Large Intestine (Colon), Nerves, Stomach

Conditions: **Abdominal Pain and Inflammation, Acid Indigestion (Heartburn, Acid Reflux)**, Appetite (excessive), Belching, Children's Remedy, Cholesterol (low), **Colic (children)**, Diarrhea, Dyspepsia, **Gas and Bloating, Indigestion**, Insomnia, Irritability, Teething

Usage: Liquid remedy primarily used for infants and children. For colic, diarrhea, teething, fussiness in infants, 3-10 drops straight or diluted in water for tea. For indigestion, nervousness, insomnia in children, take 1/2 teaspoon or more as needed. For same conditions in adults use 1 teaspoon three times daily or as needed, although other digestive formulas may be stronger for adult needs. Although we use catnip and fennel a lot for colic in infants, many of the aromatic qualities have been destroyed in this particular preparation, which renders it ineffective in some cases. If this formula doesn't do the trick, try making a simple catnip and fennel tea.

Warnings: No known warnings, no irritant effect, and a safe remedy for children and infants.

Catnip & Fennel Extract (2 fl. oz.) - Liquid (Glycerite)

Available in: US & Canada - Stock #3195-3

Ingredients: Catnip and fennel, glycerine

CBG Extract

Product Type: Formula

Properties: Antispasmodic, Immune Amphoterics, Nervine, Refrigerant

Systems Affected: Digestive System, **Ears**, Nerves

Conditions: Adenitis, Children's Remedy, Colds (general remedies for), **Deafness, Dizziness (Vertigo), Ear Infection or Earache**, Kidney Infection, Pneumonia, Sore Throat, **Tinnitus (Ringing in the Ears)**, Tonsillitis (Adenoids), **Vaccines (detoxification from)**

Usage: This is a very unusual combination of herbs. It has been used for nervous disorders (particularly muscle spasms) and as a remedy for chronic ear problems. It can also be used to reduce inflammation and irritation in tissues. Recommended dosage is 1/2 to 1 teaspoon with a glass of water 3 times a day between meals. As a mouthwash, gargle with 1 teaspoon in water. For use in the ears, warm the liquid to body temperature and then use 5-10 drops in the ear.

Warnings: Hypoglycemics should use with caution. No other known warnings.

Combination CBG Extract (2 fl. oz.) - Liquid (Tincture)

Available in: US & Canada - Stock #1751-2

Ingredients: Black cohosh, chickweed, golden seal, desert tea (brigham tea), licorice, passion flower, valerian

CC-A

Product Type: Formula

Properties: Anti-inflammatory, Antiviral, Decongestant, Expectorant, Nervine, Stimulant, Stomachic, Tonic

Systems Affected: Mucus Membranes, Nerves

Conditions: Bronchitis, Children's Remedy, Colds (antiviral), Colds (general remedies for), Congestion (general), Flu, Infection (viral)

Usage: General remedy for colds and respiratory congestion. At first sign of a cold or congestion take 2 capsules every hour with a large glass of water until symptoms are relieved. Also available in liquid form. Use 1/4-1/2 teaspoonful with water every hour.

Warnings: No known warnings.

CC-A - Capsule (100)

Available in: US - Stock #840-5

Ingredients: Rose hips, chamomile, slippery elm, yarrow, capsicum, golden seal, myrrh gum, peppermint, sage, lemon grass

CC-A with Yerba Santa (2 fl. oz.) - Liquid (Tincture)

Available in: US - Stock #3165-0

Ingredients: Rose hips, chamomile, mullein, yarrow, yerba santa, golden seal, myrrh gum, peppermint, sage, astragalus, slippery elm, lemon grass, capsicum

Cellu-Smooth

Product Type: Formula

Properties: Antioxidant

Systems Affected: Circulation, Liver, Structural System, Urinary System, Weight Loss

Conditions: Cartilage Damage, **Cellulite**, Circulation (poor), Cramps and Spasms (general), Pregnancy (herbs and supplements to avoid during), Weight Loss (aids for)

Usage: This formula helps to reduce fatty deposits around the thighs, hips, buttocks and upper arms. It helps to improve circulation to these areas of the body and has a protective effect against free radical damage. Use one capsule two to three times daily with plenty of water. Use the Cellu-Tone essential oil blend topically at the same time to enhance the effectiveness of this formula.

Warnings: Not recommended for pregnant or nursing women or children under age 6.

Cellu-Smooth with Coleus - Capsule (90)

Available in: US - Stock #926-0

Ingredients: Bladderwrack, rhodiola extract, milk thistle concentrate, ginkgo concentrate, rhododendron root, coleus

Cellu-Tone

Product Type: Essential Oil

Properties: Detoxifying, Stimulant

Systems Affected: Capillaries, Skin, Structural System, Vocal Cords, Weight Loss

Conditions: Cellulite, Laryngitis (Hoarseness)

Usage: Formulated for topical application to help reduce cellulite. Apply topically over affected areas. The formula can also be applied topically to the throat to help with laryngitis.

Warnings: Recommended for topical use only. Do not use prior to sun exposure or on those with kidney disorders.

Cellu-Tone Essential Oil Blend (5ML) - Essential Oil

Available in: US - Stock #3927-3

Ingredients: Cypress, Geranium, Juniper, Niaouli Bio, Pink Grapefruit Bio, Rosemary, Thyme Linalol Bio, Vetiver

Cellular Energy

Product Type: Formula

Properties: Antilipemic, Antioxidant, Detoxifying, Laxative (general), Mineralizer, Moistening, Stimulant

Systems Affected: Glandular System, Mitochondria, Whole Body

Conditions: Acquired Immune Deficiency Syndrome (AIDS/HIV), Addictions (coffee, caffeine), Addictions (sugar or food), Appendicitis, Appetite (deficient), Blood Pressure (high), Body Odor, Boils, Cholesterol (high), Colon (spastic), Congestion (general), Constipation (adults), Convalescence, Dehydration, Digestion (poor), Exercise, Fatigue, Fibromyalgia Syndrome (FMS), Nervous Exhaustion (Enervation), Overalkalinity, **Post Partum Weakness**, Tics, **Tremors**, Triglycerides (high)

Usage: This product contains nutrients that enhance the production of energy (ATP) inside cells. This helps cells detoxify. It also helps muscles to relax and enhances overall energy levels. Use 1 capsule twice daily.

Warnings: No known warnings.

Cellular Energy - Capsule (60)

Available in: US - Stock #1879-6

Ingredients: Vitamins B1, B2, E, niacin, pantothenic acid, zinc, manganese, magnesium, ferulic acid, alpha lipoic acid, alpha-keto glutaric acid, l-Carnitine, coenzyme-Q10 and dimethyl glycine HCl.

Chamomile

Latin Name: *Matricaria chamomilla*

Product Type: Single Herb

Properties: Analgesic (Anodyne), Anti-inflammatory, Anti-smoking, Antiphlogistic, Antispasmodic, Antiviral, Appetite Stimulant, Aromatic, Bitter, **Calmative**, Carminative, Diaphoretic, Digestive Tonic, Febrifuge, **Nervine**, Sedative, Stomachic, Vermifuge, Vulneraries (for intestinal system)

Systems Affected: Bronchials, Circulation, Eyes, Intestinal System, Liver, Muscles, Nerves, Skin, Solar Plexus, Spleen, Stomach, Sweat Glands, Uterus

Conditions: Abdominal Pain and Inflammation, Abscesses, Addictions (alcohol), Addictions (drugs), Addictions (tobacco smoking or chewing), Anorexia, Appetite (deficient), Attention Deficit Disorder (ADD, ADHD), **Bloodshot Eyes**, Children's Remedy, Circulation (poor), Colic (adults), **Colic (children)**, Colitis, **Conjunctivitis (Pink Eye)**, Corns, Cushing's Disease, Cystic Breast Disease, Defensiveness, Denture Sores, Digestion (poor), Diverticulitis (Diverticuli), Dysentery, Dyspepsia, Emotional Sensitivity, Eye Infections, Eye Problems (general), Fever, Floaters, Gas and Bloating, Hair Care (general), **Hysteria**, **Indigestion**, Inflammation, Inflammatory Bowel Disorders (Colitis, IBS), Insomnia, **Irritability**,

Labor and Delivery, Leaky Gut Syndrome, **Nervousness**, Psoriasis, Rashes and Hives, Skin Care (general), Stye, **Teething**, Tension, Toothache, Ulcers, Worry

Usage: Chamomile has many uses. It is an excellent remedy for colic, fevers, colds and other childhood ailments, especially where heat (inflammation or fever) is found in combination with tension and irritability. It is useful for cases where the client's irritability is out of proportion to the nature of the illness or injury. Herbalist Matthew Wood says that chamomile is for "babies of any age." Use as a tea for children, 2-3 capsules per cup of boiling water. Steep 3-5 minutes and sweeten. Can be combined with peppermint to improve taste. Tea is also useful for settling stomach in adults and in enemas for irritable bowel syndrome and Crohn's disease. Tea can also be applied to injuries as a compress or fomentation. Chamomile contains a volatile oil that contains a powerful anti-inflammatory agent (azulene), making it a good herb to use with all types of inflammatory conditions. Take 1-2 capsules 3-4x daily. Combine with ginger for cramps, nausea, digestive pains, gas and so forth. Combines well with marshmallow and other mucilant herbs to soothe intestinal irritation or as a poultice. Tea used as a hair rinse improves golden highlights. Chamomile has properties similar to catnip and peppermint.

Warnings: There have been some reports of allergic reactions to chamomile, but these are extremely rare. As with yarrow, chamomile may cause severe reactions in persons allergic to ragweed, asters or chrysanthemums.

Chamomile - Capsule (100)

Available in: US - Stock #190-7

Ingredients: Matricaria chamomilla (chamomile) flowers

Chamomile (Roman)

Product Type: Essential Oil

Properties: Adrenal Tonic, Anti-inflammatory, Antiseptic, Antispasmodic, Aromatic, Calmative, Cicatrisant, Parasympatholytic (Anticholinergic), Sympatholytic

Systems Affected: Adrenal Glands, Digestive System, Female Reproductive, Nerves

Conditions: Abdominal Pain and Inflammation, Abscesses, Acne (Pimples, Blackheads), Anger (excessive), Anxiety (Panic Attack), Asthma, Attention Deficit Disorder (ADD, ADHD), Colic (children), Cushing's Disease, Dermatitis, Dysmenorrhea, Emotional Sensitivity, Fear (excessive), Fever, Gas and Bloating, Grief (excessive), Hysteria, Inflammation, **Irritability**, Menopause, Nervousness, PMS Type A, Skin Care (general), Sunburn, **Teething**, Worry, Wounds and Sores

Usage: Chamomile has anti-inflammatory and antispasmodic effects. It helps with irregular periods and PMS, especially when used in a bath or as a massage oil. To ease gall bladder problems, sore throats, abdominal pain, and colic in children it can be applied topically in a warm, moist compress. As an inhalant, it helps with emotional anxiety and tension associated with asthma, hay fever, and other allergies. Roman Chamomile is used when someone feels grumpy, morose, discontented or impatient, short-tempered, self-involved, overly sensitive or rarely satisfied. It is a remedy for children who are feeling impatient, disagreeable or tense. The oil is beneficial for teething pain, colic or flatulence as the underlying cause for emotional distresses. It has mildly sedating properties, calming but not depressing effect.

Warnings: When applying topically, always use in a carrier oil as it may cause contact dermatitis. Should not be used where there is prolonged exposure to direct sunlight. Also contraindicated in chronic anxiety and narcolepsy.

Chamomile Essential Oil, Roman (5ML) - Essential Oil

Available in: US - Stock #3901-6

Ingredients: Roman Chamomile (Chamaemelum nobile) oil

Charcoal (Activated)

Product Type: Nutrient

Properties: **Adsorbant, Antidiarrheal, Antidote, Antitoxic, Antivenomous**

Systems Affected: Digestive System, Intestinal System, Large Intestine (Colon), Skin

Conditions: Belching, **Bites and Stings**, **Blood Poisoning**, **Chemical Poisoning**, Children's Remedy, Cholera, **Cholesterol (high)**, **Diarrhea**, **Dysentery**, **Gas and Bloating**, Giardia, **Jaundice (adults)**, **Jaundice (infants)**, Motion Sickness, **Poisoning (general)**, **Poisoning (food)**

Usage: Charcoal absorbs irritants in the digestive tract, making it useful for diarrhea, intestinal gas, indigestion and poisoning. Use 2 capsules with meals or at the first sign of intestinal gas or distress. Repeat after 2 hours if needed, up to a maximum of 6-8 capsules per day. For severe diarrhea, combine with psyllium seed capsules or bulk slippery elm. For severe gas, combine with carminative herbs. Absorbs many types of poisonous substances. For treating poisoning, contact the Poison Control Center and follow their recommendations. First-aid application may require 15 to 20 capsules, depending on quantity and toxicity of material ingested. Can be applied as a poultice to spider bites. It is even effective with brown recluse bites. Change poultice every hour for maximum benefit.

Warnings: Charcoal can interfere with assimilation of medicines and nutrients. Do not take with prescription medications. Do not take for more than a few days at a time. Large doses may cause constipation.

Charcoal (Activated) - Capsule (100)

Available in: US - Stock #366-2

Ingredients: Charcoal powder

Chickweed

Latin Name: *Stellaria media*

Product Type: Single Herb

Properties: Alterative (Blood Purifier), Anti-obesic, **Antipruritic**, Antitussive, Demulcent (Mucilant), Emollient, **Emulsifier**, Expectorant, Laxative (bulk), **Lipotropic**, Mineralizer

Systems Affected: Bladder (Urinary), Blood, Eyes, Gall Bladder, Intestinal System, Liver, Pancreas Tail, Skin, Spleen, Stomach, Structural System, Uro-genital Tract, Uterus

Conditions: Abscesses, Allergies (food), Anemia, Appetite (excessive), Asthma, Blood Poisoning, Boils, Burns and Scalds, Cancer (natural therapy for), Cataracts, Cellulite, Chicken Pox, Cholesterol (high), Congestion (bronchial), Cysts, Diabetes, Diaper Rash, Eczema, Eye Problems (general), Eyes (red or itching), Fat Cravings, **Fat Metabolism (poor), Fatty Liver Disease, Fatty Tumors or Deposits**, Gout, Hemorrhoids, Inflammation, **Itching**, Psoriasis, Rashes and Hives, Rheumatoid Arthritis (Rheumatism), Skin Care (general), Sprains, Triglycerides (low), Weight Loss (aids for), Wounds and Sores

Usage: Chickweed is a soothing herb that is reported to help break down fats and fatty tumors in the body. Soothes irritated mucous membranes. For internal use take 2-4 capsules two or three times daily with water. Acts as an appetite suppressant and aid to weight loss when taken 1 hour before mealtimes. Can be used in poultices for skin irritations. The tea can be used as an eyewash for soothing irritated eyes. The British herbalists use chickweed in healing salves to relieve itching.

Warnings: No known warnings, a safe food herb.

Chickweed - Capsule (100)

Available in: US - Stock #220-2

Ingredients: Stellaria media (chickweed) herb

Chinese Packs

Product Type: Pack

Usage: NSP's Chinese herb line consists of pairs of formulas. One formula in each pair reduces some kind of excess in the body. In effect, it cleanses away some kind of overbuild up of energy. As this built up energy dissipates, congestion is relieved and toxins are eliminated.

The second formula in each pair strengthens some kind of deficiency. In other words, it builds up or increases an area that is deficient in energy. As energy flow is restored, strength and vitality return to that area of the body.

Five of the pairs of formulas in the line are built around the Chinese five element system of wood, fire, earth, metal and water. One formula increases each of these elements (or energies), the other formula decreases one of these elements (or energies). Of the remaining four formulas, one set has a formula that regulates qi (chi or energy) and another formula that builds qi. The other set has a formula for reducing heat and a formula for moistening dryness.

Due to the requests of some NSP Managers, NSP packaged all the building formulas in one pack and called it the Yin Pack. The Yin Pack has a minus sign (-) associated with it because all the formulas in this pack are used to build deficiencies. The other pack is the Yang Pack, which is designed to reduce excesses in the system. This is why it is designated with a plus (+) sign.

The formulas in the Yang Pack are generally detoxifying and are for people who are suffering from acute illnesses and excesses in their life. These people will tend to be warm, have reddish complexions, red tongues and rapid pulse rates.

The formulas in the Yin Pack are generally tonifying or nourishing and are for people suffering from chronic and degenerative illnesses and deficiencies in their life. These people will generally be weak, cold and pale and have pale tongues and feeble pulses.

Because a person can be excessive in one element and deficient in another, we generally do not recommend taking all the herbs in either pack at the same time, although many people have done so and have reported good effects. Instead, we suggest you look carefully at each formula to see if it matches your particular needs.

Warnings: No known warnings, although if you are taking either the Yin Pack or the Yang Pack and notice any unusual symptoms such as anxiety, heart palpitations, insomnia or other negative changes in your health, we recommend you discontinue taking the pack. These are not healing crisis symptoms, but rather signs that the herbs are throwing your body out of balance.

Yang Pack (+) TCM - Pack

Available in: US - Stock #13343-2

Ingredients: Anti-Gas TCM, IF-C TCM, Kidney Activator TCM, Liver Balance TCM, Stress Relief TCM, Mood Elevator TCM, Breathe EZ TCM

Yin Pack (-) TCM - Pack

Available in: US - Stock #13344-8

Ingredients: Lung Support TCM,Blood Build TCM, HY-C TCM, KB-C TCM, Nervous Fatigue TCM, Spleen Activator TCM, Trigger Immune TCM

Chlorophyll

Product Type: Nutrient

Properties: **Alkalinizer**, Anti-emetic (Antinauseous), Anticoagulant (Blood Thinner), **Antimutagenic**, Blood Building, Carminative, **Deodorant**

Systems Affected: **Circulation**, Digestive System, Hypothalamus, Intestinal System, Large Intestine (Colon), Liver, **Parotids**, **Red Blood Cells**, Spleen

Conditions: **Anemia**, Bleeding (internal), **Body Odor**, Breast Milk (increase or enrich), Bunions, Cancer (natural therapy for), **Cancer (prevention)**, Cholesterol (low), Cold Sores (Fever Blisters), Constipation (adults), Diaper Rash, Disks (spinal - bulging or slipped), Emphysema, Energy (lack of), Fatigue, **Foot Odor**, Free Radical Damage, Gout, **Halitosis (Bad Breath)**, Heart Fibrillation or Palpitations, Hemochromatosis, Labor and Delivery, **Overacidity**, **Oxygen Deficiency**, Pain (general remedies for), **Pets (supplements for)**, Poisoning (general), Post Partum Weakness, **Pregnancy (herbs and supplements for)**, Raynaud's Disease, Rosacea, **Sickle Cell Anemia**, Toxemia, Vaccines (detoxification from)

Usage: Chlorophyll is the molecule that makes leaves green. It captures the sun's energy in photosynthesis. Chlorophyll has an affinity for the blood. It prevents the clumping of red blood cells and increases the oxygen-carrying capacity of the blood. It also seems to nutritionally build the blood when taken regularly. Chlorophyll as it is found in plants is a fat-soluble substance (think of grass stains on clothes), with a magnesium molecule in the center. The chlorophyll in NSP's gel caps is this magnesium-rich, fat- soluble chlorophyll. It acts as a gentle bowel detoxifier in person's with spastic colons due to its magnesium content.

Liquid chlorophyll is a different substance. It is water-soluble chlorophyll, in which the magnesium has been displaced by copper and sodium. Thus, liquid chlorophyll is a good source of copper, but not of magnesium. There are some common misconceptions about chlorophyll. Some people incorrectly teach that chlorophyll is converted directly to hemoglobin in the blood, because the molecules are almost identical except that chlorophyll contains magnesium and hemoglobin contains iron. If you consult any chemistry text that shows the chemical structure of these two substances, you will find this is simply incorrect. The two molecules are considerably different.

Furthermore, some people incorrectly teach that liquid chlorophyll is a source of iron and other trace minerals. This is because NSP derives their chlorophyll from alfalfa, an herb rich in iron and trace minerals. However, chlorophyll is not alfalfa juice-all the iron and trace minerals were removed in the process of isolating the chlorophyll. This does not discount the fact that chlorophyll does improve oxygen uptake in the blood, making it very useful for general weakness and blood disorders, and so forth.

Use 1-3 tablespoons of liquid chlorophyll one to three times daily for most problems or use 1-2 capsules daily with a meal. It can also be taken with powdered vitamin C for a pick-up during labor or for times of fatigue. Another important use is to reduce body odor while a person is detoxifying. Some people drink chlorophyll strait to help restore blood after severe bleeding. Liquid chlorophyll also helps expel gas because it contains spearmint oil, making it a good remedy for bloating and indigestion.

Warnings: Non-toxic even in large quantities. However, liquid chlorophyll contains copper, which may depress zinc levels in the body if taken excessively. Encapsulated chlorophyll acts as a mild laxative in some people.

Chlorophyll, Liquid (16 fl. oz.) - Liquid

Available in: US - Stock #1683-7

Ingredients: Sodium Copper Chlorophlyin, purified water, spearmint oil, propylparaben, methylparaben

Chlorophyll, Liquid (32 fl. oz.) - Liquid

Available in: US & Canada - Stock #1689-6

Ingredients: Same as Stock #1683-7

Liquid Chlorophyll Sample (2 fl. oz.) - Liquid

Available in: Canada - Stock #1832-2

Ingredients: Same as Stock #1683-7

Chlorophyll Capsules - Softgel (60)

Available in: US & Canada - Stock #1690-7

Ingredients: Chlorophyllin complex, soybean oil, glycerin.

Cholester-Reg II

Product Type: Formula

Properties: Anticholesteremic, Cholagogue

Systems Affected: Heart, Liver

Conditions: Blood Clots (prevention of), Cholesterol (high)

Usage: This formula is designed to support cholesterol levels that are already in the normal range. Artichoke leaves have a history of helping to maintain healthy circulation. Research on phytosterol and policosanol indicates they are helpful in maintaining a healthy cholesterol level. The suggested dosage is 1 capsule three times daily.

Warnings: No known warnings.

Cholester-Reg II - Capsule (90)

Available in: US & Canada - Stock #557-7

Ingredients: Artichoke Leaves (Cynara scolymus), Phytosterol (contains beta-sitosterol, campesterol, and stigmasterol), Inositol Nicotinate, Resveratrol (Japanese Knotweed Root), Policosanol.

Chondroitin

Product Type: Nutrient

Properties: Anti-arthritic

Systems Affected: Structural System

Conditions: Arthritis

Usage: Chondroiton helps build cartilage and connective tissue to help in the repair of damaged joints in arthritis and other joint disorders. Use 2 capsules with a large glass of water two times daily.

Warnings: No known warnings.

Chondroitin - Capsule (60)

Available in: US - Stock #1811-5

Ingredients: Chondroitin sulfate

Chromium GTF

Product Type: Nutrient

Properties: Antidiabetic, **Hypoglycemic**, Immune Amphoterics

Systems Affected: Cardiovascular System, Liver, **Pancreas**, Pancreas Head, Weight Loss

Conditions: Arteriosclerosis (Atherosclerosis, Hardening of the Arteries), Bedwetting, Cardiovascular Disease (Heart Disease), Cholesterol (high), **Diabetes**, Hyperinsulinemia (Syndrome X), **Hypoglycemia**, Mental Illness, Narcolepsy, **PMS Type C**, Psoriasis, Sugar Cravings

Usage: Chromium is essential for maintenance of normal blood sugar levels. It is helpful for both hypoglycemia and diabetes, especially when used with other blood sugar regulating herbs and formulas. Chromium is also helpful for preventing heart disease and for enhancing the immune system. Take 1 tablet daily with a meal (preferably lunch).

Warnings: No known warnings.

Chromium GTF/500 mcg - Tablet (90)

Available in: US - Stock #1801-6

Ingredients: Chromium amino acid chelate, horsetail, red clover, yarrow, calcium, phosphorus.

Chromium GTF - Capsule (90)

Available in: Canada - Stock #1802-4

Ingredients: Chromium amino acid chelate (500 mcg.), red clover, yarrow

Cinnamon

Product Type: Essential Oil

Properties: Antimicrobial, Antiparasitic, Antiseptic, Aromatic, Carminative, Glandular, Parturient, Preservative, Stimulant

Systems Affected: Digestive System, Glandular System, Sweat Glands, Testosterone

Conditions: Anorexia, Chills, Circulation (poor), Colds (general remedies for), Colitis, Congestion (general), Depression, Diarrhea, Energy (lack of), Flu, Lice, Nausea and Vomiting, Parasites (general), Shock, **Testosterone (low)**, Warts

Usage: Cinnamon oil is a very warming essential oil that stimulates digestion and circulation. It is helpful for easing digestive upset such as colitis, diarrhea, nausea and vomiting. The smell has a testosterone-enhancing action, which makes it useful for inhaling to stimulate contractions during childbirth or increase sexual desire in men and women. It is also antiseptic and anti-parasitic, so it can be used topically to destroy harmful organisms. It has a calming effect on the nerves, but helps improve energy while overcoming feelings of depression and anxiety. Apply diluted oil topically or smell.

Warnings: Always dilute before putting on the skin. Do not add to bath water. Do not use internally.

Cinnamon Leaf Essential Oil (5ML) - Essential Oil

Available in: US - Stock #3898-6

Ingredients: Cinnamon (Cinnamonum zeylanicum) oil

CLA

Product Type: Nutrient

Properties: Anti-inflammatory, Antioxidant, Nutritive

Systems Affected: Adrenal Glands, Female Reproductive, Liver, Pancreas, Prostaglandins, Weight Loss

Conditions: Allergies (food), Allergies (respiratory), Cholesterol (high), Depression, Dizziness (Vertigo), Exercise, Hyperinsulinemia (Syndrome X), Inflammation, Triglycerides (high), Weight Loss (aids for)

Usage: CLA is conjugated linoleic acid. Linoleic acid is the principle essential fatty acid needed by the body. Part of the omega-6 group of fatty acids, linoleic acid is converted into GLA (gamma linolenic acid) and eventually into specialized messenger chemicals such as leukotrienes and prostaglandins which control the body's inflammatory responses. CLA is a special form of linoleic acid, which has been shown to have very beneficial effects in human health. It occurs naturally in meat and dairy products, when animals are allowed to graze and eat grass. However, it is not present in significant amounts in animals fed on commercial feed or silage. CLA is a powerful antioxidant. In the body, it is taken up by phospholipids, a class of fats that serve as the principal structural components of cell membranes. CLA enhances the cell membrane's defense mechanism against attack by free radicals. CLA also stimulates the production of key immune system cells and inhibits the release of an immunoglobulin associated with allergies. CLA influences the PPAR-gamma agonist, a class of internal cell receptors that are capable of suppressing the manifestations of inflammatory conditions. Research has isolated PPAR-gamma as the specific cell receptor involved in chronic inflammatory responses. It also produces important immune-suppressing compounds such as leukotrienes and prostaglandins, which help reduce the body's inflammation response. In contrast, linoleic acid can increase the body's immune response. Other possible benefits for CLA include the following: It increases metabolic rate and may help thyroid patients. It decreases abdominal fat, helping to balance adrenal hormones. It enhances muscle development and lowers cholesterol and triglycerides. It reduces insulin resistance, enhances the immune system, and reduces food-induced allergic reactions. Take 1 capsule with a meal three times daily

Warnings: No known warnings.

CLA (75) - Softgel (75)

Available in: US - Stock #3010-1

Ingredients: CLA 750 mg (Conjugated Linoleic Acid), safflower oil, glycerin.

Clary Sage

Product Type: Essential Oil

Properties: Antidepressant, Antigalactagogue, Antispasmodic, Aphrodisiac, Aromatic, Emmenagogue, Estrogenic, Parturient

Systems Affected: Breasts, Estrogen, Female Reproductive, Hair, Hypothalamus, Nerves, Pineal, Reproductive Glands, Skin

Conditions: Addictions (alcohol), Alzheimer's Disease, Amenorrhea, Asthma, Attention Deficit Disorder (ADD, ADHD), Breast Milk (dry up), Cramps (menstrual), Depression, Dizziness (Vertigo), Dysmenorrhea, Dyspepsia, Eczema, Erectile Dysfunction, Fear (excessive), Guilt, Hair (loss or thinning), Hair Care (general), **Hot Flashes**, Irritability, Menopause, Menstrual Irregularity, Menstruation (scant), Nervous Exhaustion (Enervation), Nervousness, Night Sweating, PMS Type D

Usage: Clary Sage is an antispasmodic and emmenagogue. Helps with uterine problems such as PMS, regulating scanty periods and easing cramps in the lower back area. It encourages labor by helping the mother to relax and also eases postnatal depression. It promotes estrogen secretion by acting with the pituitary. It inhibits prolactin, which tends to dry up breast milk. It works during menopause to relieve hot flashes, dizziness, headaches, night sweats, palpitations and irritability. Clary Sage can be used for asthma to relax spasms of the bronchial tubes and ease emotional tension found in asthma sufferers. Its effects on the central nervous system make it a good choice for impotency, mid-life crisis, frigidity, and as an aphrodisiac. It can be used topically to reduce excessive sebum production in oily hair and skin. Emotionally, it regenerates energy and inspires both mind and spirit. It is used for nervousness, weakness, fear, paranoia and depression. It brings a long-lasting inner, tranquility and helps to remedy melancholy. It diverts one from negative thoughts and helps guide energies.

Warnings: It is contraindicated in states of high anxiety.

Clary Sage Essential Oil (5ML) - Essential Oil

Available in: US & Canada - Stock #3902-4

Ingredients: Clary Sage (Salvia sclarea) oil

CleanStart

Product Type: Pack

Properties: **Detoxifying**, Laxative (bulk), Laxative (stimulant)

Systems Affected: Intestinal System, Large Intestine (Colon), Liver, Lungs, Skin, Weight Loss

Conditions: Allergies (food), Allergies (respiratory), Constipation (adults), Toxemia

Usage: Clean Start is a convenient cleansing program. It is used for cleansing the colon and liver and for general detoxification. Take the contents of one cleanse packet 15-30 minutes before breakfast and 15-30 minutes before dinner. Take contents of capsule packet with breakfast and dinner. Mix the cleanse packet in 8 ounces of juice or water. Shake, blend, or stir vigorously. Drink immediately. It is recommended that an additional glass of water be consumed immediately after drink. Drink plenty of water throughout the day.

Warnings: Generally speaking, cleansing programs like this are for occasional use only. CleanStart may aggravate spastic or irritable bowel conditions. Use gentler cleansing programs or products like Intestinal Soothe and Build if this happens. See also warnings for cascara sagrada and hydrated bentonite.

CleanStart Apple/Cinnamon (14 days) - Packets

Available in: US - Stock #3992-6

Ingredients: Clean-Start Packet: Psyllium hulls, modified maltodextrin, bentonite, natural apple cinnamon flavor, potassium citrate, freeze dried aloe, stevia extract, sodium copper chlorophyllin. Program also contains LBS II and Enviro-Detox formula.

CleanStart Wild Berry (14 days) - Packets

Available in: US - Stock #3993-8

Ingredients: Same as Apple-Cinnamon CleanStart with black currant and raspberry flavors

Clove Bud

Product Type: Essential Oil

Properties: Analgesic (Anodyne), **Anesthetic**, Anti-emetic (Antinauseous), Anti-inflammatory, Antibacterial, Antidepressant, Antioxidant, Antiparasitic, Antiseptic, Antispasmodic, Antiviral, Aphrodisiac, Aromatic, Carminative, Disinfectant, Expectorant, Stimulant, Stomachic, Vermifuge

Systems Affected: Circulation, Digestive System, Gums, Heart, Immune System, Intestinal System, Nipples, Teeth, Uterus

Conditions: Abdominal Pain and Inflammation, Asthma, Belching, Blood Pressure (low), Bronchitis, Colds (general remedies for), Cough (damp), Denture Sores, Diarrhea, Fungal Infections (Yeast Infections, Candida albicans), Gas and Bloating, Halitosis (Bad Breath), Hiccups, Indigestion, Laryngitis (Hoarseness), Leucorrhea, Mumps, Nausea and Vomiting, Pap Smear (abnormal), Rheumatoid Arthritis (Rheumatism), Sex Drive (low), **Teething, Toothache**, Warts

Usage: Clove is an excellent antiseptic because of its high eugenol content. It is useful in the prevention of viral diseases. It is an anti-parasitic. It helps to stimulate digestion, restores appetite and relieves flatulence. It has a mild anesthetic effect and is commonly used for toothaches. It triggers the release of anti-inflammatory substances. It inhibits the prostaglandins type 2 which are well-known mediators of inflammatory processes. Clove is beneficial for rheumatoid arthritis and other musculoskeletal conditions. Topically it can be used for leg ulcers. Emotionally lifts depression and used when feeling weak and lethargic. It has a positive stimulating effect on the mind. Clove oil applied topically has a numbing effect on the nerves. It is used by dentists to numb the gums prior to giving shots. It has been mixed with olive oil (1 part clove to 10-20 parts olive oil) and applied to the gums for teething babies or toothache. Mix it with a carrier oil and apply topically over the abdominal area to help make a parasite cleanse more effective.

Warnings: Can be irritating in large quantities. Clove oil should not be used internally except by trained professionals.

Clove Bud Essential Oil (5ML) - Essential Oil

Available in: US - Stock #3903-1

Ingredients: Clove Bud (Eugenia caryophyllata) oil

CLT-X

Product Type: Formula

Properties: Anti-inflammatory, Antispasmodic, Laxative (bulk)

Systems Affected: Intestinal System, Large Intestine (Colon), Mucus Membranes, Stomach

Conditions: Abdominal Pain and Inflammation, Colic (adults), Colitis, Constipation (adults), Convalescence, Cramps and Spasms (general), Debility, Diarrhea, Digestion (poor), Dyspepsia, Inflammation, **Inflammatory Bowel Disorders (Colitis, IBS)**, Ulcers

Usage: This formula contains soothing herbs that ease intestinal irritation in colitis, Crohn's, ulceration, irritable bowel syndrome, adult colic, diverticulitis and any other inflammation of the digestive tract. For soothing irritated digestive linings, take 2 capsules with each meal. Can also be made into a bolus (for colitis) and inserted rectally before retiring for the night. This formula is used in a similar manner to Intestinal Sooth and Build. This formula has more antispasmodic and pain-relieving qualities, while Intestinal Sooth and Build has more astringent and anti-inflammatory properties.

Warnings: No known warnings.

CLT-X - Capsule (100)

Available in: US - Stock #1207-7

Ingredients: Slippery elm, marshmallow, ginger, dong quai, wild yam, lobelia

CLT-X - Capsule (100)

Available in: Canada - Stock #1209-1

Ingredients: Slippery elm, marshmallow, wild yam, ginger

Co-Q10

Product Type: Nutrient

Properties: Anti-aging, Anti-allergenic, Anti-inflammatory, **Antioxidant, Cardiac**, Hypotensive, Tonic

Systems Affected: Gums, Heart, Liver, Muscles, Pituitary (anterior), Respiratory System, Thyroid

Conditions: Aging (prevention), Allergies (food), Allergies (respiratory), Alzheimer's Disease, **Angina**, Asthma, Blood Pressure (high), Cancer (natural therapy for), **Cardiac Arrest (Heart Attack), Cardiovascular Disease (Heart Disease)**, Chemotherapy (reducing side effects), **Congestive Heart Failure**, Debility, Diabetes, Down Syndrome, Edema (Dropsy, Water Retention, Swelling), Emphysema, Energy (lack of), Fatigue, Free Radical Damage, **Gingivitis (Bleeding Gums, Gum Disease, Pyorrhea)**, Grave's Disease, Hair (loss or thinning), Hashimoto's Disease (Thyroiditis), **Heart (weakness), Heart Fibrillation or Palpitations**, Incontinence (urinary), **Inflammation**, Muscle Tone (lack of), Muscular

Dystrophy, Narcolepsy, Parkinson's Disease, **Raynaud's Disease**, Rheumatic Fever, **Spider Veins, Teeth (loose)**

Usage: When it comes to protecting your heart, one of the best antioxidants is Co-Q10. If you have gum disease, take statin drugs or already have heart problems, you should definitely use Co-Q10. Over 4,000 scientific studies have been done on Coenzyme Q-10, known as Co-Q10 for short. Research shows that this antioxidant helps reduce cardiac inflammation and both prevents and helps to reverse heart disease. It also reduces inflammation in the gums to help heal gum disease.

The beneficial effects of Co-Q10 are numerous. It reduces blood pressure, aids recovery from heart attacks, keeps LDL cholesterol from oxidizing and improves energy production in the heart muscle. Statin drugs deplete Q-10, so this supplement should always be taken by people using statin drugs to lower cholesterol.

Co-Q10 75 is a very powerful Co-Q10 supplement. For prevention of heart disease consider 1 or 2 gel-caps per day. For angina, arrhythmia, high blood pressure and gum diseases, 2 to 5 gel-caps per day is recommended. For serious heart problems, recovery from heart disease or heart failure, it is safe to use between 4 and 8 caps per day.

Besides helping the heart and circulation, Co-Q10 can help heal bleeding gums, reduce the side effects of chemotherapy and reduce allergic reactions. It also helps cellular energy production and detoxification.

Warnings: No known warnings.

Co Q10 Softgel (75 mg) - Softgel (60)

Available in: US - Stock #1895-5

Ingredients: Co-Q 10 (75 mg.) lecithin, betacarotene, soybean oil, beeswax, glycerin.

Co Q10 Plus - Capsule (60)

Available in: US - Stock #1796-4

Ingredients: Co-enzyme Q10 (10 mg.), copper, iron, magnesium, zinc, glycerin, capsicum, hawthorn.

Co Q10 (30 mg) - Capsule (30)

Available in: US - Stock #4089-6

Ingredients: 30 mg of Co-Q10 per capsule, along with zinc (30 mg), copper (3.9 mg), magnesium (100 mg), capsicum, hawthorn and ginkgo

Co-Q 10 (30 mg.) - Capsule (60)

Available in: Canada - Stock #4034-2

Ingredients: 30 mg of Co-Q10 per capsule, along with zinc (30 mg), copper (3.9 mg), magnesium (100 mg), capsicum and ginkgo

Cold

Product Type: Homeopathic

Properties: Antiviral, Decongestant

Systems Affected: Immune System, Respiratory System

Conditions: Colds (general remedies for), Congestion (sinus), Sneezing, Sore Throat

Usage: Cold is used for the relief of symptoms of the common cold, including runny nose, sneezing, headache, sore throat and difficult breathing. For the adult formula take 10-15 drops under the tongue every 10-15 minutes until symptoms improve, then decrease to hourly, then four times daily until symptoms are relieved. For the children's product take 3-5 drops under the tongue every 15-20 minutes until symptoms improve, then every two to four hours until symptoms are relieved.

Warnings: The adult product isn't recommended for children under 12, while the children's product isn't recommended for children under 4 months.

Cold (1 fl. oz.) - Homeopathic (Liquid)

Available in: US - Stock #8765-2

Ingredients: Aconitum (Aconite) 4x, Euphrasia officinalis (Eyebright) 4x, Sanguinaria canadensis (Blood Root) 4x, Allium cepa (Onion) 6x, Kali Iodatum (Potassium Iodide) 6x, Pulsatilla (Wind Flower) 6x, Sulphur (Sulfur) 8x, Hepar Sulphuris Calcareum (Calcium Sulfide) 10x, water, alcohol.

Cold (1 fl. oz.) (Children's) - Homeopathic (Liquid)

Available in: US - Stock #8865-1

Ingredients: Eupatorium perfoliatum (Boneset) 3x, Sanguinaria canadensis (Blood Root) 3x, Gelsemium sempervirens (Yellow Jessamine) 4x, Aconitum napellus (Aconite) 5x, Euphrasia officinalis (Eyebright) 5x, Pulsatilla (Wind Flower) 5x, Kali Bichromicum (Potassium Dichromate) 6x, Purified water, glycerin, and potassium benzoate

Collatrim

Product Type: Formula

Properties: Nutritive, Vulnerary

Systems Affected: **Collagen**, Structural System

Conditions: Acquired Immune Deficiency Syndrome (AIDS/HIV), Arthritis, Backache (Back Pain, Lumbago), Bedwetting, **Body Building**, Broken Bones, Burning Feet or Hands, Bursitis, Cardiovascular Disease (Heart Disease), Carpal Tunnel Syndrome, **Cartilage Damage**, Cystitis, Diabetes, Disks (spinal - bulging or slipped), Injuries, Knees (weak), **Ligaments (torn or injuried), Weight Loss (aids for)**

Usage: Both Collatrim products contain collagen, which is used to build cartilage, tendons, heart muscle tissue, skin and lean muscle mass. Both can be used for weight loss or for helping to repair damaged tissues. Liquid: Take 1 tablespoon of Collatrim with an 8-ounce glass of Nature's Spring water right before going to bed for weight management. Do not eat or snack for at least three hours before taking Collatrim. For bone and joint support take 1 tablespoon three times daily with water. Capsules: Recommended use is 3 capsules with a meal three times a day for joint-function benefits. For weight management benefits, take 9 capsules of the product on an empty stomach with a glass of water just before bedtime.

Warnings: No known warnings.

Collatrim Plus - Bulk Powder

Available in: US - Stock #3062-0

Ingredients: Xylitol, Protein, Collagen (hydrolyzed), citric acid, natural citrus flavor, silicon dioxide(powdered silica), malic acid, and concentrated aloe vera (Aloe barbadensis)

Collatrim Capsules - Capsule (180)

Available in: US & Canada - Stock #3065-4

Ingredients: Collagen powder

Colloidal Minerals

Product Type: Formula

Properties: **Mineralizer**, Nutritive

Systems Affected: Liver, Skeletal System, Structural System, Teeth, Whole Body

Conditions: Amenorrhea, Appetite (excessive), Arthritis, Bipolar Mood Disorder (Manic Depressive Disorder), **Birth Defects (prevention)**, Broken Bones, Children's Remedy, Digestion (poor), Eczema, Fingernail Biting, Gingivitis (Bleeding Gums, Gum Disease, Pyorrhea), **Gray Hair**, Hiccups, **Infertility**, Lou Gehrig's Disease, Menorrhagia (Heavy Menstrual Bleeding), Oral Surgery, Pets (supplements for), Post Partum Weakness, **Pregnancy (herbs and supplements for)**, Strokes, Tachycardia, **Tooth Decay (prevention)**

Usage: Colloidal minerals are concentrated mineral waters derived from leeching minerals into water from deposits of clay from ancient sea beds. They supply trace amounts of sixty plus mineral elements and are used as a general nutritional aid to supply trace elements that may be missing in the diet. Colloidal minerals can help a wide range of conditions, including cardiovascular problems, joint or structural weakness, urinary problems, digestive problems and glandular imbalances. Use 1/2 to 1 ounce twice daily with a meal. Can be mixed in juice or even mixed with a vinaigrette salad dressing. Can also be added to a bowl of soup just prior to serving.

Warnings: No known warnings.

Colloidal Minerals (32 fl. oz.) - Liquid

Available in: US - Stock #4013-6

Ingredients: 60+ trace elements, purified water, vegetable glycerin, citric acid, sodium benzoate, a natural flavor blend

Essentials Liquid Minerals - Liquid

Available in: Canada - Stock #4037-1

Ingredients: Liquid mineral concentate, purified water, vegetable glycerine, sodium benzoate, citric acid, natural flavor blend

Colostrum

Product Type: Nutrient

Properties: **Immune Amphoterics, Immunomodulator**

Systems Affected: Immune System, Liver

Conditions: Aging (prevention), Autoimmune Disorders, Cancer (natural therapy for), Children's Remedy, Cystitis, Debility, Down Syndrome, Epstein Barr Virus (Chronic Fatigue Syndrome, CFS)

Usage: When a baby is born, the first "milk" produced by the mammary glands is called colostrum. It contains immune-boosting factors that help build the infant's resistance to disease. Colostrum is made from bovine colostrum and is used to help balance the immune system. Because it doesn't over stimulate the immune system, it is very suitable for autoimmune disorders. It contains naturally occurring IGF-1. I have no experience with this product. Use as directed on the bottle. Take 1 capsule four times daily, one hour before meals and bedtimes with 8 ounces of water.

Warnings: No known warnings.

Colostrum - Capsule (90)

Available in: US & Canada - Stock #1828-7

Ingredients: Colostrum Powder (from milk)

Colostrum with Immune Factors

Other Names: Colostrum Plus

Product Type: Formula

Properties: Immune Stimulant

Systems Affected: Immune System, Liver

Conditions: Allergies (food), Allergies (respiratory), Broken Bones, Cancer (natural therapy for), Colds (general remedies for), Convalescence, Digestion (poor)

Usage: This blend mixes the colostrum with herbs that stimulate the immune system. To stimulate the immune system take 1 capsule three times daily.

Warnings: Not for use with autoimmune disorders.

Colostrum w/Immune Factors - Capsule (60)

Available in: US - Stock #1587-3

Ingredients: Colostrum, IP-6 (inositol), maitake mushroom, shiitake mushroom, astragalus

Coral Calcium

See *Calcium*

Cordyceps

Latin Name: *Cordyceps sinensis*

Product Type: Single Herb

Properties: Antifungal, Antiviral, Bronchial Dilator, Kidney Tonic, **Lung Tonic**, Tonic

Systems Affected: Immune System, Lungs

Conditions: Aging (prevention), Anemia, Asthma, **Athletic Performance or Exercise (aids to)**, Autoimmune Disorders, Backache (Back Pain, Lumbago), Bleeding (internal), **Body Building**, **Bronchitis**, Chicken Pox, Cholesterol (high), Congestion (bronchial), Congestion (lungs), **Convalescence**, Cough (damp), **Cough (dry)**, Cough (general), Cystic Fibrosis, Diphtheria, **Emphysema**, Epstein Barr Virus (Chronic Fatigue Syndrome, CFS), Erectile Dysfunction, Menopause, Pregnancy (herbs and supplements to avoid during), **Wheezing**

Usage: Cordyceps is a fungus product that also contains the larval remains from the caterpillar larva on which it grows. Like reishi and shiitake mushrooms it strengthens the immune system. It also helps build strength and endurance. It is used in Chinese medicine for sore and weak lower back and extremities, bleeding, wheezing, chronic cough, coughing of blood, impotence and menopause. Take 2 capsules with a meal three times daily.

Warnings: Cordyceps is safe for adults, but people using immune-suppressing drugs, anticoagulant drugs, or bronchodilators should consult their healthcare practitioners before using this product. Pregnant or lactating women should avoid using this product. Because it tonifies both the yin and yang, the Chinese consider this a very safe substance that can be taken over long periods of time.

Cordyceps - Capsule (90)

Available in: US - Stock #1240-5

Ingredients: Cnicus benedictus (cordyceps) fungus

Cornsilk

Latin Name: *Zea mays*

Product Type: Single Herb

Properties: Demulcent (Mucilant), Diuretic, Soothing

Systems Affected: Urinary System

Conditions: Bedwetting, **Bladder (irritable)**, Blood Pressure (high), **Cystitis**, Cysts, Diaper Rash, Edema (Dropsy, Water Retention, Swelling), **Nephritis**, **Urethritis**, **Urination (burning or painful)**, **Urination (frequent)**

Usage: A mild, soothing diuretic agent. Take 2 capsules three times daily with a full glass of water. Used originally by the Native Americans as a urinary aid for water retention and edema. Best used where a non-irritating/stimulating diuretic is called for (hot or inflamed urinary/bladder conditions). Specifically useful for kidney inflammation, painful urination, cystitis and urethritis.

Warnings: No known warnings, very mild and safe remedy.

Cornsilk - Capsule (100)

Available in: US & Canada - Stock #235-3

Ingredients: Zea mays (cornsilk)

Cough Syrup (Childrens)

Product Type: Homeopathic

Properties: Antitussive, Decongestant, Expectorant

Systems Affected: Lungs

Conditions: Children's Remedy, Cough (general)

Usage: This homeopathic cough syrup is for relief of coughs in children due to colds or minor throat or bronchial irritation. Take 1 or 2 teaspoons as needed.

Warnings: Not recommended for children under four months except under the advice of a health care professional.

Cough Syrup (4 fl. oz.) (Children's) - Homeopathic (Liquid)

Available in: US - Stock #8855-0

Ingredients: Active ingredients: Chamomilla (chamomile) 3x, Thymus (wild thyme) 3x, Trifolium pratense (red clover) 3x, Drosera rotundifolia (sundew) 4x, Ipecacuanha (ipecac) 4x, Bryonia (white bryony) 6x, Spongia tosta (sponge) 6x. Other ingredients: Honey, jujube extract, purified water, and potassium benzoate

Cough Syrup-DH

Product Type: Homeopathic

Properties: Decongestant, Expectorant, Moistening

Systems Affected: Lungs, Throat

Conditions: Allergies (respiratory), Cough (dry), Cough (general), Laryngitis (Hoarseness)

Usage: Used for the relief of dry, hoarse coughs accompanying minor illness, allergic responses or throat irritations. Adults, take 2-4 teaspoons as required. Children 4-12, take 1-3 teaspoons as required.

Warnings: NSP does not recommend this product for children under 4 except on the advice of a health care professional.

Cough Syrup-DH (4 fl. oz.) - Homeopathic (Liquid)

Available in: US - Stock #8780-3

Ingredients: Active ingredients: Allium cepa (onion) 4x, Bryonia (white bryony) 4x, Drosera rotundifolia (sundew) 4x, Mentha piperita (peppermint) 4x, Spongia tosta (sponge) 6x, Corallium rubrum (red coral) 12x, Bromium (bromine) 12x. Other ingredients: Honey, purified water, corn syrup, molasses, jujube extract, citric acid, potassium benzoate, and caramel color

Cough Syrup-LP

Product Type: Homeopathic

Properties: Decongestant, Expectorant

Systems Affected: Lungs

Conditions: Congestion (bronchial), Congestion (lungs), Cough (damp), Cough (general)

Usage: Used for the relief of coughs with respiratory congestion, loose phlegm and difficult expectoration. Adults, take 2-4 teaspoons as required. Children 4-12, take 1-3 teaspoons as required.

Warnings: NSP does not recommend this product for children under 4 except on the advice of a health care professional.

Cough Syrup-LP (4 fl. oz.) - Homeopathic (Liquid)

Available in: US - Stock #8785-5

Ingredients: Active ingredients: Anisum (star anise) 3x, Ipecacuanha (ipecac) 3x, Stachys betonica (betony) 4x, Antimonium Tartaricum (tartar emetic) 6x, Coccus cacti (cochineal) 6x, Scilla maritima (sea onion) 6x, Hepar Sulfuris Calcareum (calcium sulfide) 10x. Other ingredients: Honey, jujube extract, purified water, and potassium benzoate

Cough Syrup-NT

Product Type: Homeopathic

Properties: Antitussive, Decongestant, Expectorant

Systems Affected: Bronchials, Lungs

Conditions: Cough (general), Cough (spastic)

Usage: Used for the relief of persistent nighttime coughing that disturbs sleep. Adults, take 2-4 teaspoons as required. Children 4-12, take 1-3 teaspoons as required.

Warnings: NSP does not recommend this product for children under 4 except on the advice of a health care professional.

Cough Syrup-NT (4 fl. oz.) - Homeopathic (Liquid)

Available in: US - Stock #8790-8

Ingredients: Active ingredients: Aconitum napellus (aconite) 4x, Conium maculatum (hemlock) 6x, Hyoscyamus niger (henbane) 6x, Pulsatilla (wind flower) 6x, Sticta pulmonaria (lungwort) 6x, Calcarea Carbonica (calcium carbonate) 10x, Silicea (silica) 10x. Other ingredients: Honey, jujube extract, purified water, and potassium benzoate

Cramp Relief

Product Type: Formula

Properties: Analgesic (Anodyne), **Antispasmodic**, Relaxant

Systems Affected: Intestinal System, Large Intestine (Colon), Lungs, Muscles, Nerves, Structural System

Conditions: Colon (spastic), **Cramps (menstrual)**, **Cramps and Spasms (general)**, **Dysmenorrhea**, Nervous Exhaustion (Enervation), **PMS Type P**

Usage: This formula contains several antispasmodic herbs that relax muscle cramps and ease pain and tension. Although labeled as a female product, it can also be used for muscle cramps, spastic bowel conditions with painful gripping, and general tension. Use 2 capsules with a meal two or three times daily, or for severe cramping, take two capsules every hour for up to six hours.

Warnings: No known warnings. However, the product would not be advisable in cases of low blood pressure.

Cramp Relief - Capsule (100)

Available in: US - Stock #1124-9

Ingredients: Cramp bark, wild yam root, black cohosh, lobelia, plantain.

Cranberry & Buchu

Product Type: Formula

Properties: Antibacterial, **Diuretic**

Systems Affected: **Bladder (Urinary)**, Pancreas, Prostate, **Urinary System**

Conditions: Bedwetting, **Bladder Infection**, Cystitis, Edema (Dropsy, Water Retention, Swelling), **Kidney Infection**, Prolapsed Uterus, **Urinary Tract Infections**, Urine (scant)

Usage: Cranberry makes the lining of the urinary tract inhospitable for bacteria. It also stops smelly urine in incontinent people. Buchu is a powerful diuretic and improves digestion. It also helps with diabetes. This combination is a specific diuretic formula for preventing urinary tract infections. Take 1-2 capsules three times daily. When treating acute urinary tract infections combine with JP-X, goldenseal, uva ursi or Oregon grape.

Warnings: Buchu may be contraindicated with severe kidney inflammation.

Cranberry & Buchu - Capsule (100)

Available in: US & Canada - Stock #834-5

Ingredients: Cranberry, buchu

Damiana

Latin Name: *Turnera diffusa*

Product Type: Single Herb

Properties: Antidepressant, Aperient, **Aphrodisiac**, Cardiac, Corrects Polarity, Diuretic, Euphoretic, Glandular, Nervine, Stimulant

Systems Affected: Nerves, Ovaries, Parathyroid, **Reproductive Glands**

Conditions: Amenorrhea, Anxiety (Panic Attack), Benign Prostate Hyperplasia (BPH), Calcium Deficiency, Cystitis, Depression, **Erectile Dysfunction**, Fatigue, Hot Flashes, Infertility, Lou Gehrig's Disease, **Premature Ejaculation**, Reversed Polarity, **Sex Drive (low)**, Testosterone (low), Urethritis

Usage: As a sexual stimulant use 2 capsules with two meals daily. The herb has a testosterone-like influence on the glandular system. It may either lower or raise blood pressure, depending on which direction it is out of balance. The herb is a mild antidepressant and aids digestive function, too.

Warnings: Not recommended when a person is emaciated, weak or experiencing inflammation.

Damiana - Capsule (100)

Available in: US - Stock #240-9

Ingredients: Turnera diffusa (damiana)

Dandelion

Latin Name: *Taraxacum officinal*

Product Type: Single Herb

Properties: Alterative (Blood Purifier), Antacid, Appetite Stimulant, Bitter, **Cholagogue**, Diuretic, Food, Hepatic, Lipotropic

Systems Affected: Gall Bladder, Intestinal System, Liver, Pancreas, Solar Plexus, Spleen, Spleen, Stomach, **Tongue**, Urinary System

Conditions: Acne (Pimples, Blackheads), Appetite (deficient), Arthritis, Blood Poisoning, Blood Pressure (low), Cirrhosis of the Liver, Defensiveness, Diabetes, Digestion (poor), Eczema, Edema (Dropsy, Water Retention, Swelling), Gall Bladder (sluggish), Gout, Hepatitis, Hiatal Hernia, Jaundice (adults), **Nephritis**, PMS Type S, Rashes and Hives, Sore or Geographic Tongue

Usage: This common garden weed has been a trusted liver and kidney remedy for centuries. It acts as a mild diuretic for water retention, but the leaf (not available through NSP) works better as a diuretic than the root. It contains potassium, so it replaces the potassium that is often depleted by drug diuretics. Used as a general aid for digestive problems (liver, stomach and pancreas). It has some of the same properties as burdock, but is a milder remedy. Take 1-3 capsules two or three times per day with plenty of water. Aids arthritic conditions. Use as tea for stomach aches, 1/2 cup every half hour. As a diuretic, use 1-2 capsules three times daily with plenty of pure water. Works better in combination with stronger diuretics.

Warnings: Dandelion is entirely harmless even if taken for extended periods. It is not recommended for conditions involving fluid deficiency or dryness, however.

Dandelion - Capsule (100)

Available in: US & Canada - Stock #250-4

Ingredients: Taraxacum officinal (dandelion) roott

Deep Relief Oil

Product Type: Essential Oil

Properties: **Analgesic (Anodyne)**, **Anesthetic**, Anti-arthritic, Antimicrobial, Aromatic, **CNS Depressant**, Counterirritant, Narcotic, **Rubefacient**, Stimulant

Systems Affected: Female Reproductive, Intestinal System, Large Intestine (Colon), Muscles, Skin, Stomach

Conditions: Abdominal Pain and Inflammation, Arthritis, Asthma, Backache (Back Pain, Lumbago), Bronchitis, **Bursitis**, Carpal Tunnel Syndrome, Cramps (menstrual), Cramps and Spasms (general), Disks (spinal - bulging or slipped), Dislocation, Dysmenorrhea, Exercise, Inflammation, Injuries, Ligaments (torn or injuried), Neuralgia and Neuritis, Oral Surgery, **Pain (general remedies for)**, **Stiff Neck**, **Tendonitis**, Tooth Extraction

Usage: Apply over painful areas to relief pain and inflammation. Helps relieve intestinal pain and cramping. Works with female cramping and imbalanced menstrual flow.

Warnings: Not recommended for internal use.

Deep Relief Essential Oil Blend (5ML) - Essential Oil

Available in: US - Stock #3926-2

Ingredients: Ginger, nutmeg, clove.

Defense Maintenance

Product Type: Formula

Properties: Antioxidant, Immune Stimulant

Systems Affected: Digestive System, Immune System, Thymus

Conditions: Acquired Immune Deficiency Syndrome (AIDS/HIV), Cancer (natural therapy for)

Usage: Contains antioxidant vitamins and minerals. Also contains cruciferous vegetables, which have been shown to have a protective effect against cancer, although the amounts of these vegetables are obviously small compared to actual dietary needs. Immune-stimulant herbs are also part of this formula. As a general aid to building the immune system, use 2 capsules two times a day with meals.

Warnings: No known warnings.

Defense Maintenance - Capsule (120)

Available in: US - Stock #1654-5

Ingredients: Vitamins A, C, and E, Zinc, Selenium, barley grass juice powder, asparagus powder, astragalus, broccoli powder, cabbage powder, ganoderma, parthenium, pau d' arco, schizandra, eleuthero, wheat grass juice powder, myrrh gum

Depressaquel

Product Type: Homeopathic

Properties: Antidepressant

Systems Affected: Nerves

Conditions: Apathy, Depression

Usage: Used to help alleviate melancholy, apathy and listlessness by lifting the mood and mental outlook. Take 10-15 drops under the tongue every hour until symptoms improve, then four times daily until symptoms are relieved.

Warnings: NSP does not recommend this product for children under the age of 12 except on the advice of a health professional. Other than the alcohol content, which is not good for infants, we consider this product safe for children.

Depressaquel (1 fl. oz.) - Homeopathic (Liquid)

Available in: US - Stock #8755-1

Ingredients: Active ingredients: Pulsatilla (Wind Flower) 4x, Phosphoricum Acidum (Phosphoric Acid) 6x, Cimicifuga racemosa (Black Cohosh) 6x, Conium maculatum (Hemlock) 6x, Kali Bromatum (Potassium Bromide) 6x, Ignatia amara (St. Ignatius' Bean) 8x, Lycopodium clavatum (Club Moss) 8x, Calcarea Carbonica (Calcium Carbonate) 10x, Silicea (Silica) 10x, Selenium Metallicum (Selenium) 12x. Other ingredients: Purified water and 20% USP alcohol.

Detoxification

Product Type: Homeopathic

Properties: Antinauseous, Detoxifying

Systems Affected: Immune System

Conditions: Heavy Metal Poisoning, Mercury Poisoning, Pain (general remedies for), Poisoning (general)

Usage: Used to ease symptoms such as headache, vomiting, nausea, aches and joint discomfort associated with exposure to pollutants and toxins such as mercury and arsenic. Take 10-15 drops under the tongue three times daily for 60 days. Drink plenty of water and take algin or other fiber products to absorb toxins and carry them out of the body.

Warnings: NSP does not recommend this product for children under the age of 12 except on the advice of a health professional. Other than the alcohol content, which is not good for infants, we consider this product safe for children.

Detoxification (1 fl. oz.) - Homeopathic (Liquid)

Available in: US - Stock #8940-3

Ingredients: Active ingredients: Cinchona officinalis (China) 3x, Iris versicolor (Blue Flag) 3x, Ledum palustre (Wild Rosemary) 3x, Chelidonium majus (Celandine) 4x, 12x, 30x, Mezereum (Spurge Olive) 4x, Thuja occidentalis (Tree of Life) 4x, Berberis vulgaris (Barberry) 6x, 12x, 30x, Gelsemium sempervirens (Yellow Jessamine) 6x, 12x, 30x, Staphysagria (Stavesacre) 6x, Hepar Sulfuris Calcareum (Calcium Sulfide) 12x. Other ingredients: Purified water and 20% USP alcohol.

Devil's Claw

Latin Name: *Harpagophytum procumbens*

Product Type: Single Herb

Properties: Analgesic (Anodyne), **Anti-arthritic**, **Anti-inflammatory**, Bitter, Digestive Tonic

Systems Affected: Gall Bladder, Immune System, Intestinal System, Lungs, Lungs, Pituitary (anterior), Spleen, Structural System

Conditions: Acid Indigestion (Heartburn, Acid Reflux), Adrenals (exhaustion, weakness or burnout), Allergies (food), Allergies (respiratory), **Arthritis**, Backache (Back Pain, Lumbago), Boils, Bursitis, Cholesterol (high), Crohn's Disease, Diabetes, Fever, Fibrosis, **Gout**, Indigestion, **Inflammation**, **Neuralgia and Neuritis**, **Rheumatoid Arthritis (Rheumatism)**, **Uric Acid Retention**, Wounds and Sores

Usage: Devil's claw is a powerful anti-inflammatory with pain relieving properties. It is primarily of value for rheumatic conditions and joint problems, but also stimulates appetite and digestive secretions. It may also be helpful for back pain and headaches. It was applied topically by African natives for boils, sores and ulcers. For arthritic conditions use 1 capsule three times daily.

Warnings: Contraindicated with gastric and duodenal ulcers.

Devil's Claw - Capsule (100)

Available in: US - Stock #255-9

Ingredients: Harpagophytum procumbens (Devil's claw) root

DHA

Product Type: Nutrient

Properties: Anti-inflammatory, Nutritive

Systems Affected: Brain, Circulation, Eyes, Nerves, Skin

Conditions: Arthritis, Cholesterol (high), Circulation (poor), Diabetes, Eye Problems (general), Gout, **Memory and Brain Function**, **Mental Illness**, Multiple Sclerosis (MS), **Nerve Damage**, **Neuralgia and Neuritis**, Triglycerides (high)

Usage: DHA is an essential fatty acid, and the most abundant fatty acid in the brain. It is essential for the myelin sheath and is needed for development, growth and maintenance of the brain. It may be helpful for brain function, memory, visual function, healthy cholesterol levels, and neurological conditions. Take one capsule with a meal three times daily

Warnings: No known warnings.

DHA - Capsule (60)

Available in: US - Stock #1513-5

Ingredients: Eicosapentaenoic acid (EPA) 50 mg, Docosahexaenoic acid (DHA) 250 mg.

DHEA-F

Product Type: Formula

Properties: Anti-aging, **Aphrodisiac**, **Glandular**

Systems Affected: Adrenal Glands, Female Reproductive, Progesterone

Conditions: Acquired Immune Deficiency Syndrome (AIDS/HIV), Aging (prevention), Alzheimer's Disease, Cramps (menstrual), Diabetes, **Lupus**, Menopause, Menstrual Irregularity, **Pregnancy (herbs and supplements to avoid during)**

Usage: DHEA is the most abundant of all hormones in the body. It is produced by the adrenals. DHEA levels usually drop about 80% between age 20 and age 65. The level of DHEA in the bloodstream has been shown to be a good barometer of cancer risk. Take 1 capsule daily with a meal.

Warnings: Because DHEA is a naturally occurring hormone, long term supplementation may alter the body's ability to produce sufficient amounts of it on its own. DHEA is best used for short-term treatment hormone imbalance. Excessive levels of DHEA can cause acne, breast swelling and tenderness and hormonal imbalances like those found in teenagers.

DHEA-F (Women's) - Capsule (100)

Available in: US - Stock #4202-2

Ingredients: 25 mg DHEA, wild yam, false unicorn, chaste tree berries.

DHEA-M

Product Type: Formula

Properties: Anti-aging, **Aphrodisiac**, **Glandular**

Systems Affected: Adrenal Glands, **Male Reproductive**, **Testosterone**

Conditions: Acquired Immune Deficiency Syndrome (AIDS/HIV), Aging (prevention), Alzheimer's Disease, Benign Prostate Hyperplasia (BPH), Diabetes, **Erectile Dysfunction**, **Lupus**, **Pregnancy (herbs and supplements to avoid during)**, **Testosterone (low)**

Usage: DHEA is the most abundant of all hormones in the body. It is produced by the adrenals. DHEA levels usually drop about 80% between age 20 and age 65. Take 1 capsule daily with a meal.

Warnings: Because DHEA is a naturally occurring hormone, long term supplementation may alter the body's ability to produce sufficient amounts of it on its own. DHEA is best used for short-term treatment hormone imbalance. Excessive levels of DHEA can cause acne and hormonal imbalances like those found in teenagers.

DHEA-M (Men's) - Capsule (100)

Available in: US - Stock #4200-7

Ingredients: 25 mg DHEA, sarsaparilla, saw palmetto, damiana, pumpkin seed, Panax ginseng.

Dieter's Cleanse

Other Names: Bod-E Klenz

Product Type: Pack

Properties: Detoxifying, Hepatic, Laxative (stimulant)

Systems Affected: Large Intestine (Colon), Weight Loss

Conditions: Halitosis (Bad Breath), **Weight Loss (aids for)**

Usage: As a cleanse to begin a weight loss program, take the contents of one (1) AM packet with breakfast, one (1) PM packet with evening meal with 8 oz. of Nature's Spring water daily for seven days.

Warnings: Do not use if diarrhea, loose stools, or abdominal pain are present or develop. Not intended for prolonged use.

Dieter's Cleanse (14 Day) - Pack

Available in: US - Stock #3220-4

Ingredients: Bowel Detox, Master Gland Formula, Enviro-Detox, Liver Cleanse Formula, LBS II, SF, Chromium.

Bod-E Klenz - Pack

Available in: Canada - Stock #4082-5

Ingredients: BWL-BLD, Master G Formula, Enviro D-T-X, LIV-A, LBS II, SF, GTF Chromium.

Digestive Bitters Tonic

Product Type: Formula

Properties: **Aperitive, Appetite Stimulant**, Bitter, **Digestive Tonic**, Hepatic, Pancreatic Tonics, Tonic

Systems Affected: Digestive System, Gall Bladder, Intestinal System, Liver, Pancreas, Small Intestines

Conditions: Acid Indigestion (Heartburn, Acid Reflux), **Anorexia, Appetite (deficient), Bulimia**, Cholesterol (low), Colic (adults), Convalescence, Debility, **Digestion (poor), Dyspepsia**, Failure to Thrive, Fat Metabolism (poor), Migraine, **Protein Digestion (poor)**, Rosacea, **Sore or Geographic Tongue**

Usage: This formula strengthens the digestive system. Improves appetite and tones digestive membranes. May help stimulate appetite. Acts as a carminative to relieve gas and digestive distress. Adults should take one teaspoon 15 minutes before meals. For children over six, use 1/2 teaspoon 15 minutes before meals.

Warnings: No known warnings.

Digestive Bitters Tonic (4 fl. oz.) - Liquid (Glycerite)

Available in: US - Stock #3113-9

Ingredients: Gentian, cardamom, orange peel tincture, dandelion, red raspberry syrup, stevia, glycerine, phosphoric acid, potassium sorbate

Digestive Enzymes

See *Food Enzymes*

Distress Remedy

Product Type: Homeopathic

Properties: Anxiolytic

Systems Affected: Nerves

Conditions: Anxiety (Panic Attack), Bulimia, Burns and Scalds, Children's Remedy, **Defensiveness**, Failure to Thrive, **Fainting**, Fear (excessive), Fingernail Biting, **Guilt**, Inflammation, **Injuries**, Labor and Delivery, **Mental Illness, Nervous Exhaustion (Enervation), Neurosis, Obsessive Compulsive Disorder, Phobias**, Post Traumatic Stress Disorder, Scratches and Abrasions, Seizures, **Shock**

Usage: Distress Remedy is a modified version of Dr. Edward Bach's flower essence blend known as Rescue Remedy. Rescue Remedy is used for shock, both physically and emotionally. Distress Remedy contains the same flower essences as Rescue Remedy, but adds several that also promote tissue healing, including comfrey, arnica and calendula. The blend is also in glycerine instead of alcohol, making it easier to administer to children. Take 10-15 drops under the tongue every 10-15 minutes for relief of emotional shock due to injury, illness, or emotional distress. Apply topically to minor injuries to reduce swelling, ease pain and promote healing. Add 15-20 drops to a spray bottle filled with 4-8 oz. of purified water and shake vertically 10-20 times to create a spray form of the product which can be sprayed over larger injuries to promote healing. The spray can also be used to "mist" a room to clear emotional tension or negative energy. The remedy can be used as a general activator to stimulate emotional healing and improve attitude in anyone who is having difficulty healing.

Warnings: No known warnings.

Distress Remedy (Flower Remedy) (1 fl. oz.) - Homeopathic

Available in: US - Stock #8975-1

Ingredients: Arnica montana (Leopard's Bane) 3x, Calendula officinalis (Garden Marigold) 3x, Symphytum officinale (Comfrey) 3x, Belladonna (Nightshade) 6x, Clematis vitalba, Flos (Traveler's Joy) 6x, Helianthemum nummularium (Rock Rose) 6x, Impatiens glandulifera, Flos (Impatiens) 6x, Orinthogalum umbellatum (Star of Bethlehem) 6x, Prunus cerasifera, Flos (Cherry-Plum) 6x, purified water, glycerin, and potassium benzoate

Dong Quai

Latin Name: *Angelica sinensis*

Product Type: Single Herb

Properties: Antispasmodic, Blood Building, Deobstruent, Diuretic, Emmenagogue, Female Tonic, Glandular, Uterine

Systems Affected: Blood, **Circulation**, Muscles, Ovaries, Pineal, Uterus

Conditions: Amenorrhea, **Anemia**, Bites and Stings, Breasts (swelling and tenderness), Circulation (poor), Cramps and Spasms (general), Fibroids (uterine), Infertility, Labor and Delivery, Menopause, Menstrual Irregularity, Migraine, Pregnancy (herbs and supplements to avoid during), Prolapsed Uterus

Usage: Dong quai has been used extensively in Chinese medicine for treating feminine problems. It acts as a tonic to the blood and helps overcome the anemia caused by monthly blood loss. It is also a blood-moving herb. It helps to keep the circulation flowing freely in the abdominal area. Dong quai is a relative of the Western herb angelica, and can be used in a similar manner to treat digestive problems. It is not just a female herb, however, as men can also use it for anemia and general weakness. Generally speaking, use 1-2 capsules two to three times daily as a female tonic and menstrual regulator. Dong quai can also be taken for PMS and other female troubles over a 2-week period beginning 10 days prior to the start of menstruation.

Warnings: Not recommended during pregnancy nor with excessive menstrual flow. Not to be used during the period as it increases menstrual bleeding.

Dong Quai - Capsule (100)

Available in: US - Stock #258-8

Ingredients: Angelica sinensis (dong quai) root

Dulse

Latin Name: *Rhodymenia palmata*

Product Type: Single Herb

Properties: Glandular, Mineralizer, Thyrotropic

Systems Affected: Bones, Hair, Nails, Nerves, Skin, Structural System, Thyroid

Conditions: Cholesterol (low), Depression, **Dermatitis**, Fatigue, **Goiter**, Hair (loss or thinning), Hair Care (general), Hashimoto's Disease (Thyroiditis), **Hypothyroid**, Pregnancy (herbs and supplements for), **Radiation Sickness**, Weight Loss (aids for)

Usage: Dulse is a seaweed that is rich in iodine, making it a valuable remedy for the thyroid gland. It is also rich in silica, a mineral that promotes healthy skin, hair, nails, nerves and bones. Take 15-20 drops one to two times daily.

Warnings: Not recommended for hyperactive thyroid.

Dulse Liquid (2 fl. oz.) - Liquid (Glycerite)

Available in: US - Stock #3156-6

Ingredients: Phodymenia-palmata (dulse) herb, glycerin.

E-Tea

Product Type: Formula

Properties: Alterative (Blood Purifier), Anticancer, Immune Stimulant, Tonic

Systems Affected: Immune System, Liver, Spleen, Thymus

Conditions: Acne (Pimples, Blackheads), Acquired Immune Deficiency Syndrome (AIDS/HIV), **Cancer (natural therapy for)**, Chicken Pox, Fatty Tumors or Deposits, **Leukemia**, Measles, Melanoma (Skin Cancer), Rashes and Hives, Surgery (healing from)

Usage: This is a native American formula for cancer that was given to Rene Caisse of Canada, a nurse who used it to help many people with cancer. She prepared it in a fresh liquid form. It was called Essiac (Caisse spelled backwards). Take 2-4 capsules at bedtime or 2 capsules three times daily. Best used in liquid form. Use 2 capsules in four ounces of hot water to make tea, or take capsules with warm water. This formula would also be helpful for eruptive skin diseases or other degenerative diseases.

Warnings: Cancer is a serious illness. Do not rely on one herbal formula as your sole treatment. Consult a qualified practitioner and develop a complete health program.

E-Tea - Capsule (100)

Available in: US & Canada - Stock #1360-4

Ingredients: Burdock, sheep sorrel, slippery elm, turkey rhubarb.

Echinacea Purpurea

Latin Name: *Echinacea purpurea*

Product Type: Single Herb

Properties: Alterative (Blood Purifier), Anti-allergenic, Anti-inflammatory, Antibacterial, Antiseptic, Antitoxic, Antivenomous, **Antiviral**, Detoxifying, Febrifuge, **Immune Stimulant**, Lymphatic, Tonic, Vulnerary

Systems Affected: Capillaries, Ears, Immune System, Liver, Lymph Nodes, Lymphatic System, Pancreas Tail, Prostate, Respiratory System, Skin, Thymus, **Tonsils**

Conditions: Acne (Pimples, Blackheads), Bites and Stings, **Blood Poisoning**, Boils, Cancer (natural therapy for), Canker Sores (Mouth Ulcers), Colds (antiviral), Colds (general remedies for), Colds (prevention), Congestion (lymphatic), Contagious Diseases, Diphtheria, Ear Infection or Earache, Eczema, Emphysema, Epstein Barr Virus (Chronic Fatigue Syndrome, CFS), Fever, Gangrene, Gingivitis (Bleeding Gums, Gum Disease, Pyorrhea), Infection (bacterial), **Infection (viral)**, Malaria, Meningitis, Mercury Poisoning,

Nephritis, Pets (supplements for), **Snake Bite**, Strep Throat, Syphilis, **Tonsillitis (Adenoids)**, **Typhoid**, Wounds and Sores

Usage: Echinacea has become an extremely popular herb because of its immune-enhancing qualities. It aids the process of antibody formation and stimulates production of white blood cells. It also helps to strengthen and clear lymph nodes. It inhibits enzymes produced by bacteria to break down compounds that bind cells together. Thus it inhibits spread of infection. It also helps the body fight viral infections by tricking the body into thinking it is under a massive viral attack. As an alternative to antibiotics for acute infection or inflammation, take in frequent small doses, 1-2 capsules every 1-2 hours; then taper off as condition improves. Many people use it at the onset of cold or flu to arrest the illness. It is also useful externally on infected wounds to prevent spread of infection. Has also been used in cancer therapy. For prevention take 1-2 capsules 1-2x daily. The herb is a blood purifier and works well in combination with other blood purifiers for lymphatic congestion, skin eruptive diseases, boils and wounds. Adults: take 15-20 drops (1 ml) in water with a meal three times daily. Children age 6 or older: take approximately 5-10 drops in water with a meal three times daily. Ultimate Echinacea is an excellent product containing three different species of the herb: Echinacea purpurea, Echinacea angustifolia and Echinacea pallid.

Warnings: Non-toxic and harmless, but in excessive amounts it can cause excessive salivation and scratchy, tingling sensation in throat. Do not use with autoimmune disorders where the immune system is over-active (i.e. MS, Lupus, Hodgkin's, etc.). No known toxic effects, but some herbalists feel that because it stimulates the immune response it is best to use it for short periods, then take a break and allow the immune system to rest before resuming ingestion. General recommendation is two weeks on, two weeks off. This may also be wise because the body tends to adapt to remedies taken over a long period of time so they become less effective.

Echinacea Purpurea - Capsule (180)

Available in: US - Stock #263-8

Ingredients: Echinacea purpurea root

Echinacea Purpurea - Capsule (50)

Available in: US & Canada - Stock #262-6

Ingredients: Same as Stock #263-8

Echinacea/Golden Seal

Product Type: Formula

Properties: Antibacterial, Antimicrobial, Antiviral, Hepatic, **Immune Stimulant**

Systems Affected: Immune System, Liver, Pineal, Thymus

Conditions: Antibiotics (alternatives to), Bronchitis, Colds (general remedies for), Ear Infection or Earache, Flu, Gangrene, Giardia, **Impetigo, Infection (bacterial), Sinus Infection, Skin (infections), Sore Throat, Strep Throat, Tonsillitis (Adenoids), Urinary Tract Infections**

Usage: Historically used when someone has colds and flu-like conditions, but is more helpful in the later stages of a cold than in the early stages. It helps fight infection on the mucus membranes and reduces "hot" (discolored) mucus. These herbs combine well with garlic, or can be rotated with garlic for treating respiratory infection. This formula is also useful for treating infection in the gastrointestinal tract or topically on the skin. Use 1-2 capsules or 1/2 teaspoon of the extract three times daily with meals. For children, use 1/8-1/4 teaspoon every two to four hours.

Warnings: See warnings for goldenseal and echinacea. No other known warnings.

Echinacea/Golden Seal - Capsule (100)

Available in: US & Canada - Stock #835-2

Ingredients: Echinacea purpurea, echinacea angustifolia, goldenseal root

Echinacea/Golden Seal (2 fl. oz.) - Liquid (Glycerite)

Available in: US & Canada - Stock #3180-1

Ingredients: Same as Stock #835-2

Eczema/Psoriasis

Product Type: Homeopathic

Properties: Anti-inflammatory, Vulnerary

Systems Affected: Skin, Structural System

Conditions: Eczema, Psoriasis

Usage: Used for the treatment of eczema, dermatitis and psoriasis. Take 10-15 drops every two to four hours or as needed. Decrease to three or four times daily upon improvement and continue until the skin clears.

Warnings: NSP does not recommend this product for children under the age of 12 except on the advice of a health professional. Other than the alcohol content, which is not good for infants, we consider this product safe for children.

Eczema/Psoriasis Remedy (1 fl. oz.) - Homeopathic (Liquid)

Available in: US - Stock #8900-2

Ingredients: Active ingredients: Cicuta virosa (Water Hemlock) 6x, Iris versicolor (Blue Flag) 6x, Oleander (Rose Laurel) 6x, Bovista (Puff Ball) 10x, Graphites (Graphite) 10x, Pix liquida (Pine Tar) 10x, Sulphur (Sulfur) 30x. Other ingredients: Purified water and 20% USP alcohol

Eight Combination

See *Nerve Eight*

Elderberry Defense

Product Type: Formula

Properties: **Antiviral**, Decongestant

Systems Affected: Immune System, Liver, Respiratory System

Conditions: **Colds (antiviral)**, **Colds (general remedies for)**, Colds (prevention), Congestion (general), Contagious Diseases, Fever, Flu, **Infection (viral)**, Inflammation

Usage: This formula helps fight viral infections such as colds, flu, fever and sore throat. It also helps stimulate the immune system. Take 2 capsules with a meal three times daily, plus 2 capsules at bedtime.

Warnings: No known warnings.

Elderberry Defense - Capsule (100)

Available in: US - Stock #868-5

Ingredients: Elderberry extract, Echinacea purpurea, royal jelly, olive leaf.

Elderberry Plus

Product Type: Formula

Properties: **Antiviral**, Decongestant, Febrifuge, Immune Stimulant

Systems Affected: Immune System, Liver, Respiratory System, Sinuses

Conditions: Allergies (respiratory), Bronchitis, Children's Remedy, **Colds (antiviral)**, **Colds (general remedies for)**, Congestion (general), **Flu**, **Infection (viral)**, Vaccines (detoxification from)

Usage: This is a chewable tablet for fighting viral disorders in children. It can help boost the immune system and aid colds, allergies, infection, congestion, flu, scratchy throat and bronchitis. Chew 1 to 2 tablets one to two times daily with a meal.

Warnings: No known warnings.

Elderberry Plus, Chewable - Tablet (60)

Available in: US - Stock #3300-9

Ingredients: Elderberry, reishi mushroom, astragalus.

Herbasaurs Elderberry Plus - Tablet (60)

Available in: Canada - Stock #3300-9

Ingredients: Same as Stock #3300-9

Eleuthero

Other Names: Ginseng, Siberian

Latin Name: *Eleuthrococcus senticosus*

Product Type: Single Herb

Properties: **Adaptagen**, Antirheumatic, Aphrodisiac, Sympatholytic, Tonic

Systems Affected: Adrenal Glands, Cardiovascular System, Immune System, Reproductive Glands, Testosterone

Conditions: Addictions (coffee, caffeine), Addison's Disease, **Adrenals (exhaustion, weakness or burnout)**, Aging (prevention), **Anxiety Disorders**, Arrhythmia, Bedwetting, Benign Prostate Hyperplasia (BPH), **Blood Pressure (high)**, Breasts (swelling and tenderness), Cancer (natural therapy for), Chemotherapy (reducing side effects), Chills, Circulation (poor), Colds (general remedies for), Colds (prevention), **Cushing's Disease**, Depression, Dizziness (Vertigo), Endurance (lack of), Epstein Barr Virus (Chronic Fatigue Syndrome, CFS), Erectile Dysfunction, **Exercise**, Fatigue, Hair (loss or thinning), **Hashimoto's Disease (Thyroiditis)**, Headache (tension), Heart (weakness), Heart Fibrillation or Palpitations, Hypoglycemia, Infertility, Insomnia, Mercury Poisoning, Radiation Sickness, Sex Drive (low), **Stress**, Testosterone (low)

Usage: This herb does not have the wide-ranging tonic actions of the true ginsengs, even though it was sold under the name Siberian Ginseng for a long time. It is primarily an adaptagen for helping the body cope with stress. It also increases stamina and endurance, stimulates the brain to improve concentration and stimulates male hormone production. It has been proven to aid the immune response. As a general tonic take 1-2 capsules two to three times daily. Combines well with other tonics, such as gotu kola, licorice, kelp, etc. to improve energy.

Warnings: Not recommended for acute diseases, high fever, severe inflammations, hyperactivity or extreme, nervous anxiety.

Eleuthero - Capsule (100)

Available in: US - Stock #660-9

Ingredients: Eleuthrococcus senticosus (Eleuthero) root

Energ-V

Product Type: Formula

Properties: Adaptagen, Adrenal Tonic, **Adrenergic**, Anticoagulant (Blood Thinner), Aphrodisiac, Glandular, Stimulant, **Sympathomimetic**

Systems Affected: Circulation, Glandular System, Liver, Nerves

Conditions: Addictions (coffee, caffeine), Adrenals (exhaustion, weakness or burnout), Allergies (food), **Attention Deficit Disorder (ADD, ADHD)**, Circulation (poor), Depression, Eczema, Endurance (lack of), Energy (lack of), Epstein Barr Virus (Chronic Fatigue Syndrome, CFS), **Exercise**, **Fatigue**, Post Partum Weakness, Sex Drive (low)

Usage: This formula was designed to address various problems about why a person might be feeling fatigued. It contains adaptogens (eleuthero, gotu kola, schizandra), yellow dock to supply iron to combat anemia, herbs for the thyroid (kelp) and adrenals (licorice root), high-energy foods (bee pollen and barley greens), herbs for immune system weakness (eleuthero, barley greens, rose hips) and capsicum to stimulate circulation. As a general tonic for more energy, take 2 capsules with breakfast and lunch, or whenever extra energy is needed.

Warnings: No known warnings.

Energ-V - Capsule (100)

Available in: US - Stock #875-8

Ingredients: Bee pollen, gotu kola, kelp, licorice, eleuthero, yellow dock, barley grass, rose hips, schizandra, capsicum

ENERG-V - Capsule (100)

Available in: Canada - Stock #859-7

Ingredients: Bee pollen, ginkgo leaves, kelp plant, licorice root, siberian ginseng root, yellow dock root, barley grass herb, rosehips, schizandra fruit, and capsicum fruit

Enviro D-T-X

See *Enviro-Detox*

Enviro-Detox

Product Type: Formula

Properties: **Alterative (Blood Purifier)**, Bitter, **Hepatic**, Tonic

Systems Affected: Digestive System, Immune System, Intestinal System, Large Intestine (Colon), **Liver**, Lungs, Pineal, Skin, Structural System

Conditions: Acne (Pimples, Blackheads), Addictions (drugs), Addictions (tobacco smoking or chewing), Blood Poisoning, Body Odor, Cancer (prevention), Chemical Poisoning, Chicken Pox, **Cysts**, Eczema, **Environmental Pollution (protection from)**, Fibroids (uterine), Fibromyalgia Syndrome (FMS), Infection (bacterial), Insomnia, Irritability, Memory and Brain Function, Pap Smear (abnormal), Poisoning (general), Psoriasis, Radiation Sickness, Rashes and Hives, **Surgery (healing from)**, Thinking (cloudy), Tooth Extraction, **Toxemia**, **Vaccines (detoxification from)**

Usage: This formula strengthens all eliminative organs to help the body expel toxins of all sorts. It is a stronger alterative than BP-X. Use it as part of a periodic cleanse to remove the effects of environmental pollution. It is also a good formula to take if you have been exposed to chemicals on a regular basis, such as painting, manufacturing, lab work, farm chemicals, etc. Take 3 capsules one time daily. Drink plenty of water to help flush the body of toxins.

Warnings: Individuals with colon problems (especially spastic or colitis) should use this formula carefully as it will increase bile flow and may further irritate the colon.

Enviro-Detox (Improved) - Capsule (100)

Available in: US - Stock #874-4

Ingredients: Burdock, dandelion, fenugreek, ginger, marshmallow, pepsin, red clover, sarsaparilla, yellow dock, echinacea purpurea, lactobacillus sporogenes, cascara sagrada, milk thistle

Enviro D-T-X - Capsule (100)

Available in: Canada - Stock #874-4

Ingredients: Same as Stock #874-4

Essential Liquid Minerals

See *Colloidal Minerals*

Eucalyptus

Product Type: Essential Oil

Properties: Analgesic (Anodyne), Antibacterial, Antifungal, Antimicrobial, Antiseptic, Antiviral, Aromatic, Decongestant, Disinfectant, Expectorant, Preservative, Vulnerary

Systems Affected: Immune System, Lungs, Pancreas, Respiratory System

Conditions: Addictions (alcohol), Blisters, Colds (decongestant), Colds (general remedies for), Concentration (poor), Concentration (poor), Congestion (general), Cough (damp), Cough (general), Croup, Cuts, Diabetes, Fear (excessive), Fleas, Infection (bacterial), **Insects**, Neuralgia and Neuritis, Sinus Infection, Sore Throat, Thinking (cloudy), Wounds and Sores

Usage: Eucalyptus is best known for its expectorant properties. It reduces the swelling of mucous membranes and loosens phlegm to make breathing easier, while increasing oxygen supply to the body's cells. It can help with airborne staphylococci bacteria and is stimulating to the immune system. These properties make it very useful for colds and flu as well as sinus problems and throat infections. It supports the pancreas and lowers blood sugar. Eucalyptus is also helpful for pain of a cold cramping nature like muscle pain and neuralgia. Emotionally, eucalyptus restores vitality and positive outlook and assists concentration. It dispels stagnant feelings that can keep one bound to a limiting environment. Relieving a sense of emotional suffocation, it helps one to achieve freedom and reduce fear and excessive caution.

Warnings: Not to be taken internally and should always be used in a carrier oil when applied topically.

Eucalyptus Essential Oil (5ML) - Essential Oil

Available in: US & Canada - Stock #3904-9

Ingredients: Eucalyptus (Eucalyptus globulus) oil

Evening Primrose Oil

Product Type: Nutrient

Properties: Analgesic (Anodyne), Anti-arthritic, Anti-inflammatory, Anticholesteremic, Anticoagulant (Blood Thinner), Immune Amphoterics, Immunomodulator, Nutritive

Systems Affected: Gall Bladder, Immune System, Prostaglandins, Structural System

Conditions: Addictions (alcohol), Arthritis, Blood Pressure (high), Breast Lumps, Cardiovascular Disease (Heart Disease), Cystic Breast Disease, Cystic Fibrosis, Dandruff, Down Syndrome, Eczema, Epilepsy, Fat Cravings, Fibrosis, Hangover, Menopause, Menstrual Irregularity, Multiple Sclerosis (MS), Neuralgia and Neuritis, Parkinson's Disease, PMS (general), PMS Type C, PMS Type H, Rosacea, Schizophrenia, Vitiligo, Weight Loss (aids for)

Usage: Evening primrose oil is a source of gamma-linoleic acid (GLA), an important essential fatty acid that helps the immune system. It has been used for weight loss, PMS, arthritis, multiple sclerosis, eczema, heart disease and schizophrenia. Both evening primrose oil and vitamin E affect the circulatory system. As a nutritional supplement use 1-2 capsules daily.

Warnings: No known warnings.

Evening Primrose Oil - Capsule (90)

Available in: US & Canada - Stock #1787-7

Ingredients: Evening primrose oil, vitamin E, linoleic acid, gamma-linolenic acid.

EverFlex

Other Names: Glucosamine Combo

Product Type: Formula

Properties: Anti-arthritic, Anti-inflammatory, Detoxifying

Systems Affected: Bones, Connective Tissue, Muscles, Skeletal System, Skin, Structural System

Conditions: Arthritis, Backache (Back Pain, Lumbago), **Cartilage Damage**, Dysmenorrhea, Exercise, Injuries, Knees (weak), Lupus, Rheumatoid Arthritis (Rheumatism)

Usage: Millions of Americans suffer from arthritis, inflammation of the joints that makes movement painful and difficult. Unfortunately, the only relief most people experience for this problem is the reduction of inflammation and the temporary relief of pain. Modern medicine offers nothing that actually helps inflamed joints to heal.

For those who are willing to take a more holistic approach and seek real healing, Everflex can contribute to restoring health to their joints. The ingredients in Everflex encourage joint health by helping to lubricate the joints, increase shock absorption in the joints, improving flexibility, and promoting tissue and cartilage repair.

For starters, joints need shock absorbing power, which is provided by aggrecan. Glycosaminoglycans or GAGs are a critical part of aggrecan. Glucosamine derivatives are found in both hyaluronic acid and chondroitin, which are GAGs. In the first stages of degenerative joint disease, the production of aggrecan increases but in later stages it decreases and eventually leads to the loss of cartilage resiliency and the breakdown of joints characteristic of osteoarthritis.

Chondroitin sulfate has been discovered in the human body in cartilage, bone, cornea, skin, and the arterial wall. Chondroitin sulfate is often taken in combination with other products, including glucosamine hydrochloride, for relief of pain and inflammation in osteoarthritis. It may be helpful for rebuilding joints and easing pain in osteoarthritis.

MSM is an organic sulfur-containing compound that is used orally and topically for pain relief in relation to arthritis, chronic pain, joint inflammation, rheumatoid arthritis, osteoporosis, stretch marks, muscle repair, and a variety of wounds.

Reduction of inflammation is also important and Devil's claw is an herb in this formula that is a powerful natural anti-inflammatory and pain reliever and may also be helpful for osteoarthritis, rheumatoid arthritis, gout, myalgia, fibrositis, lumbago, tendonitis, gastrointestinal upset, fever and migraine headache.

Hyaluronic Acid adds significantly in cell proliferation (which is necessary for tissue healing) and is a key element of articular cartilage. The acid also plays a part in some of the primary surface receptors of the body. In medicine, hyaluronic acid has been injected into the knee for treatment of osteoarthritis. It is also used during cataract surgery, cases of detached retina, corneal transplants, and other eye traumas.

The recommended dose of Everflex for adults is two tablets two times daily with a meal.

Warnings: No known warnings.

EverFlex w/Hyaluronic Acid - Tablets (60)

Available in: US - Stock #948-4

Ingredients: Glucosamine Hydrochloride (from crab shells), Methylsulfonylmethane (MSM), Devil's Claw Root (Harpagophytum procumbens), Chondroitin Sulfate, Hyaluronic Acid

EverFlex Tablets - Tablet (60)

Available in: Canada - Stock #925-5

Ingredients: Glucosamine, MSM, chondroitin, devil's claw.

Everflex Pain Cream

Product Type: Formula

Properties: **Analgesic (Anodyne), Anti-arthritic**

Systems Affected: Muscles, Structural System

Conditions: Arthritis, Backache (Back Pain, Lumbago), Exercise, Pain (general remedies for)

Usage: Everflex Pain Cream is designed as a topical cream to apply to painful joints and muscles. It has a penetrating action and helps to ease pain and stiffness. It also reduces inflammation and promotes healing.

Warnings: No known warnings.

EverFlex Pain Cream (2 oz.) - Topical

Available in: US & Canada - Stock #3535-6

Ingredients: Menthol, benzyl alcohol, butylparaben, carbormer, cetyl laurate, cetyl myristate, cetyl myristoleate, cetyl oleate, cetyl palmitate, cetyl palmitoleate, ethylparaben, glycerin, glycerl stearate, isobuylparaben, lecithin, methylparaben, methylsulfonylmethane (MSM), olive oil, PEG-100 stearate, phenoxyethanol, potassium hydroxide, propylparaben, tocopheryl acetate, water

Every Body's Fiber

Product Type: Formula

Properties: Absorbant, Anti-inflammatory, **Laxative (bulk)**, Laxative (general), **Vulneraries (for intestinal system)**

Systems Affected: Digestive System, Intestinal System, Large Intestine (Colon), **Rectum**, Weight Loss

Conditions: Anal Fistula or Fissure, Celiac Disease, **Colitis**, Colon (spastic), **Constipation (adults)**, **Crohn's Disease**, Diarrhea, Dysentery, Enteritis, Hemorrhoids, Inflammatory Bowel Disorders (Colitis, IBS), Leaky Gut Syndrome

Usage: This is a soft fiber blend suitable for use with inflammatory conditions in the intestinal tract such as Chron's disease, Celiac's disease, colitis, spastic colon and leaky gut syndrome. It feeds the friendly bacteria in the colon, reduces inflammation and absorbs toxins, while acting as a very gentle bulk laxative. Mix one scoop in water or juice and take before meals 2-4 times daily.

Warnings: Take with plenty of water.

Everybody's Fiber (133 g) - Bulk Powder

Available in: US - Stock #1336-6

Ingredients: Short Chain Fructooligosaccharides (prebiotic dietary fiber), chamomile flowers (Matricaria recutita), flax meal (Linum usitatissimum), fennel seed (Foeniculum vulgare), malic acid, marshmallow root (Althaea officinalis), peppermint leaves (Mentha piperita), stevia leaves (Stevia rebaudiana), asparagus stem (Asparagus officinalis), cat's claw inner bark (Uncaria tomentosa) and natural peach, apricot, and plum flavors

EveryBody's Formula

Product Type: Formula

Properties: Nutritive

Systems Affected: Structural System, Weight Loss

Conditions: Body Building, Exercise, Hypoglycemia, Weight Gain (aids for), Weight Loss (aids for)

Usage: EveryBody's Formula is a soy-based protein powder. Add approximately 2 level tablespoons (20g) of Everybody's Formula to 8 ounces of your favorite juice, water, or skim milk.

Warnings: No known warnings.

Everybody's Formula (17 oz.) - Bulk Powder

Available in: US - Stock #2939-8

Ingredients: Isolated soy protein, calcium, sodium caseinate, fructose, dried whey, natural and artificial flavors, lactalbumin, dl-methionine, magnesium oxide, soybean oil, lecithin, ascorbic acid, ferrous fumarate, niacinamide, vitamin A palmitate, dried honey, licorice extract, thiamine hydrochloride, riboflavin, kelp, bromelain, papain, potassium iodide.

EW

Product Type: Formula

Properties: Anti-allergenic, Anticatarrhal, Astringent, Decongestant, **Opthalmicum**, Tonic

Systems Affected: **Eustachian Tubes**, **Eyes**, Sinuses

Conditions: **Allergies (respiratory)**, **Bloodshot Eyes**, **Cataracts**, Colds (general remedies for), Congestion (bronchial), **Conjunctivitis (Pink Eye)**, Cough (damp), Diarrhea, Dizziness (Vertigo), Ear Infection or Earache, **Eye Infections**, **Eye Problems (general)**, **Eyes (red or itching)**, Eyesight (to improve), Floaters, Glaucoma, **Rhinitis**, Sinus Infection, **Sneezing**, Snoring, **Stye**

Usage: This formula was designed for use as an herbal eye wash for poor vision, cataracts, eye infections, such as pink eye and other eye problems. For eyewash, use 2-3 capsules per cup of purified, boiling water. Use glass or stainless steel pans. Strain carefully through a fine cloth and refrigerate. Make fresh tea every three to four days. Put drops of tea into eyes like eye drops or use an eyewash cup to rinse eyes. Tea can also be soaked in cotton balls and used as a compress over the eyelids. It may be used as a gargle for sore throats. Internally, the formula has some effect on the eyes, but not anywhere near as effective as its external use. For internal help for the eyes, see *Perfect Eyes* or *Bilberry*. The formula also has a strong affinity for the respiratory system and mucous membranes. It contains herbs that may relieve itchy, red eyes, itchy nose with watery sinus drainage, swollen Eustachian tubes that can cause frequent ear infections and itchy ears. For these types of upper respiratory problems and allergic reactions, take internally 2 capsules three to four times daily.

Warnings: Use as an eyewash may cause mucus drainage from eyes and temporary irritation. This is considered a "healing crisis" by most herbalists and usually subsides after a time. If irritation persists, discontinue use. A tea made of straight marshmallow may be used as an eyewash to soothe irritation. Make sure all powder is thoroughly removed before using tea in eyes to avoid herb grains irritating the eyes. Be sure to make fresh tea every few days and keep tea refrigerated to avoid bacterial contamination.

E-W - Capsule (100)

Available in: US - Stock #861-3

Ingredients: Bayberry, golden seal, eyebright, red raspberry

Eyebright

Latin Name: *Euphrasia officinalis*

Product Type: Single Herb

Properties: Anti-allergenic, Anti-inflammatory, Anticatarrhal, Expectorant, **Opthalmicum**

Systems Affected: **Eustachian Tubes**, Eyes, Hypothalamus, Mucus Membranes, Sinuses

Conditions: Allergies (respiratory), Bloodshot Eyes, Cataracts, Colds (decongestant), Colds (general remedies for), Conjunctivitis (Pink Eye), **Ear Infection or Earache**, Eye Infections, Eye Problems (general), Eyes (red or itching), Eyesight (to improve), Glaucoma, Nose Bleeds, Rhinitis, Sneezing, Stye

Usage: Eyebright is commonly used to strengthen the eyes and treat eye infections. Although it is commonly taken internally for these conditions, it works best when used as an eyewash. For eyewash use 2-3 capsules per cup of boiling water. Use glass or stainless steel pans when making eyewash. Strain carefully through a fine cloth and refrigerate. Make fresh tea every three to four days. Put drops of tea into eyes like eye drops or use an eyewash cup to rinse eyes. Tea can also be soaked in cotton and used as a compress over the eyelids. See also *EW* formula, which is used in a similar manner.

The best use of Eyebright as an internal remedy is for upper respiratory congestion involving acute irritation of the sinuses and eyes with thin, watery mucus and itching eyes and ears. A tincture made from the fresh plant will open the Eustachian tubes in children, allowing the inner ear to drain and thus preventing earaches. The dried plant found in the capsules is less effective, but it does seem to work for this purpose if taken in large enough doses (say 4 capsules every 2 hours). General internal dose is 2 capsules two to three times per day with water.

Warnings: May cause mucus drainage from eyes. Make sure all powder is thoroughly filtered out before using tea in eyes.

Eyebright - Capsule (100)

Available in: US - Stock #260-3

Ingredients: Euphrasia officinalis (eyebright) herb

False Unicorn

Other Names: Helonias root

Latin Name: *Chamaelirium luteum*

Product Type: Single Herb

Properties: **Anti-abortive**, Diuretic, Female Tonic, Kidney Tonic, Uterine Tonic

Systems Affected: Digestive System, Female Reproductive

Conditions: Amenorrhea, Benign Prostate Hyperplasia (BPH), Bladder (irritable), Breast Lumps, Cystic Breast Disease, Dysmenorrhea, **Endometriosis**, Erectile Dysfunction, Infertility, Leucorrhea, Lice, Menopause, **Menstrual Irregularity**, **Miscarriage (prevention)**, Morning Sickness, Nocturnal Emission (Wet Dreams), **Ovarian Pain**, Parasites (nematodes, worms), PMS Type A, **Progesterone (low)**

Usage: Helps in some cases of threatened miscarriage, especially when associated with nervous irritability and periodic contractions. For threatened miscarriage, take 2 capsules every two to four hours. If spotting is occurring, add 1 capsule of capsicum. If cramping is present add one capsule of lobelia. It can also be taken in small doses during the first trimester to reduce the risk of miscarriage (1-2 capsules per day). The indications for this plant otherwise are a sensation of sluggishness in the lower pelvic area and pressure on the bladder. The person may feel like they are sitting on a ball. It improves circulation and decongests this region in both men and women. Use 1-2 capsules two times daily.

Warnings: No known warnings.

False Unicorn - Capsule (100)

Available in: US - Stock #267-1

Ingredients: Chamaelirium luteum (false unicorn) roots

Fat Grabbers

Product Type: Formula

Properties: Absorbant, Anti-obesic, Anticholesteremic, Antilipemic, Appetite Suppressant, Emulsifier, Hypolipidemic, Laxative (bulk)

Systems Affected: Digestive System, Large Intestine (Colon), Pancreas Tail, Small Intestines, Spleen, Thyroid, Weight Loss

Conditions: **Cholesterol (high)**, **Fat Cravings**, Gall Bladder (sluggish), Inflammatory Bowel Disorders (Colitis, IBS), Weight Loss (aids for)

Usage: Psyllium and guar gum are both mucilants, which means they increase in size when combined with water. They also absorb toxins and fats. Lecithin is an emulsifier or in other words, it can convert fat into water. Chickweed nourishes the thyroid, which regulates metabolism. To prevent fat absorption and reduce fat deposits, take 4 capsules three times daily, drinking one glass of water before taking them with another glass of water.

Warnings: Mucilants like psyllium and guar gum can absorb a large amount of water. If you don't drink sufficient amounts of water they can cause constipation.

Fat Grabbers - Capsule (120)

Available in: US - Stock #3035-9

Ingredients: Guar gum, psyllium hulls, lecithin, chickweed

Fat Grabbers - Capsule (360)

Available in: US - Stock #3030-4

Ingredients: Same as Stock #3035-9

Fat Grabbers Trial Pack - Packets (20)

Available in: US - Stock #2485-4

Ingredients: Same as Stock #3035-9

Fat Grabbers - Capsule (120)

Available in: Canada - Stock #2946-3

Ingredients: Psyllium hulls, chickweed herb, lecithin, and guar gum

Fatigue/Exhaustion

Product Type: Homeopathic

Properties: Stimulant

Systems Affected: Whole Body

Conditions: Endurance (lack of), Energy (lack of), Fatigue, Insomnia, Nervous Exhaustion (Enervation), Stress

Usage: Used to alleviate the symptoms of minor fatigue and exhaustion due to insomnia, illness, overwork and stress. Take 10-15 drops under the tongue four to six times daily until symptoms improve, then decrease to twice daily and continue for two to four weeks or as needed.

Warnings: NSP does not recommend this product for children under the age of 12 except on the advice of a health professional. Other than the alcohol content, which is not good for infants, we consider this product safe for children.

Fatigue/Exhaustion (1 fl. oz.) - Homeopathic (Liquid)

Available in: US - Stock #8840-0

Ingredients: Active ingredients: Cinchona officinalis (China) 4x, Cocculus indicus (Indian Cockle) 4x, Helonias dioica (False Unicorn) 4x, Picricum Acidum (Picric Acid) 6x, Kali Phosphoricum (Potassium Phosphate) 6x, Zincum Metallicum (Zinc) 10x, Sepia (Sepia) 12x. Other ingredients: Purified water and 20% USP alcohol

FCS II

Product Type: Formula

Properties: **Emmenagogue, Female Tonic,** Glandular

Systems Affected: Female Reproductive, **Ovaries,** Uterus

Conditions: Acne (Pimples, Blackheads), Addictions (alcohol), Amenorrhea, Anemia, Anxiety (Panic Attack), Birth Control (countering side effects), Breasts (swelling and tenderness), Cramps (menstrual), Dysmenorrhea, Edema (Dropsy, Water Retention, Swelling), Fibroids (uterine), Hot Flashes, **Infertility,** Menopause, Nephritis, PMS (general), Post Partum Depression, Pregnancy (herbs and supplements to avoid during), Puberty (hormone balancer)

Usage: This formula is used as a general female corrective for PMS, menstrual irregularity and menopause. More medicinal than Female Comfort (formerly FC w/Dong Quai.) Start with small dose (1-2 capsules a day) and work up to find best dose. Typical use is 2 capsules three times daily.

Warnings: Not recommended during pregnancy or lactation.

FCS II (with Lobelia) - Capsule (100)

Available in: US - Stock #878-7

Ingredients: Red raspberry, black cohosh, golden seal, blessed thistle, dong quai, queen of the meadow, marshmallow, lobelia, ginger, capsicum

FCS II - Capsule (100)

Available in: Canada - Stock #879-5

Ingredients: Raspberry, blessed thistle, cramp bark, squaw vine, black cohosh, ginger, marshmallow, queen of the meadow, capsicum

Female Comfort

Other Names: FC with Dong Quai

Product Type: Formula

Properties: **Emmenagogue, Female Tonic,** Glandular

Systems Affected: Female Reproductive, **Ovaries, Uterus**

Conditions: Acne (Pimples, Blackheads), Amenorrhea, Breasts (swelling and tenderness), Cramps (menstrual), Dysmenorrhea, Endometriosis, Fibroids (uterine), Hot Flashes, Infertility, Menopause, PMS (general), Pregnancy (herbs and supplements to avoid during)

Usage: Used as a general female corrective for PMS, menstrual irregularity and menopause. More nutritive than FCS II, this combination is a better choice if a female corrective is needed for long periods of time. Use 2 capsules three times daily. This combination has more tonic herbs and less black cohosh than FCS II

Warnings: Not recommended during pregnancy or lactation, however, could be taken in small amounts.

Female Comfort - Capsule (100)

Available in: US - Stock #882-2

Ingredients: Red raspberry, dong quai, ginger, licorice, black cohosh, blessed thistle, marshmallow, queen of the meadow

Women's Formula - Capsule (100)

Available in: Canada - Stock #891-0

Ingredients: Red raspberry leaves, dong quai root, black cohosh root, ginger root, licorice root, blessed thistle herb, marshmallow root,and queen of the meadow herb

Feminine Tonic

Product Type: Homeopathic

Properties: Glandular

Systems Affected: Ovaries, Uterus

Conditions: Menstrual Irregularity, Sex Drive (low)

Usage: Used to support the female glands or reproductive organs. Take 10-15 drops under the tongue three times daily.

Warnings: Not recommended for children under 12.

Feminine Tonic (1 fl. oz.) - Homeopathic (Liquid)

Available in: US - Stock #8701-1

Ingredients: Active ingredients: Viburnum opulus (High Bush Cranberry) 4x, Pulsatilla (Wind Flower) 4x, Helonias dioica (False Unicorn) 4x, Cimicifuga racemosa (Black Cohosh) 6x, Ignatia amara (St. Ignatius' Bean) 6x, Adrenal Tissue 6x, Ovary Tissue 6x, Uterus Tissue 12x. Other ingredients: Purified water and 20% USP alcohol.

Fenugreek & Thyme

Product Type: Formula

Properties: **Anticatarrhal**, Antiseptic, Antitussive, Carminative, **Decongestant**, Diaphoretic, Expectorant, Nervine, Parasympathomimetic

Systems Affected: Mucus Membranes, Nerves, **Respiratory System**, **Sinuses**, Thymus

Conditions: Allergies (respiratory), Asthma, Colds (decongestant), Colds (general remedies for), **Congestion (general)**, **Congestion (sinus)**, Croup, Cystic Fibrosis, Diabetes, Dyspepsia, Emphysema, Gas and Bloating, **Headache (sinus)**, Indigestion, **Sinus Infection**, Smell (loss of sense of), Sneezing

Usage: This formula is particularly effective for sinus headaches, although it can be used as a general decongestant and expectorant. Take 2 capsules three times a day or 6 capsules every two to three hours for sinus congestion and sinus headaches.

Warnings: No known warnings.

Fenugreek & Thyme - Capsule (100)

Available in: US - Stock #885-1

Ingredients: Fenugreek, thyme

Fenugreek & Thyme - Capsule (100)

Available in: Canada - Stock #884-6

Ingredients: Fenugreek, thyme and wood betony

Feverfew

Latin Name: *Tanacetum parthenium*

Product Type: Single Herb

Properties: Analgesic (Anodyne), Anti-inflammatory, Antibacterial, **Anticephalalgic**, Bitter, Carminative, Diuretic, Emmenagogue, Febrifuge, Insecticide, Nervine, Stomachic, Vermifuge

Systems Affected: Cyclic AMP (cAMP), Mucus Membranes, Nerves, Skin, Uro-genital Tract

Conditions: Amenorrhea, Arthritis, Bites and Stings, Fever, Inflammation, **Migraine**, **Psoriasis**, Psoriasis

Usage: Used to relieve vasodilative migraine headaches and inflammatory conditions. For best results, don't wait for a migraine, take 4-6 capsules daily as a preventative. Has been used as an insect repellent, also.

Warnings: Because of its emmenagogue properties, some herbalists recommend that it should not be used in pregnancy. If mouth soreness or ulcerations develop reduce dosage or discontinue use. Does not work on migraine headaches caused by weakness or deficiency. Is not recommended for vasoconstrictive headaches (tension, cluster headaches, etc.).

Feverfew Concentrate - Capsule (100)

Available in: US - Stock #288-1

Ingredients: Tanacetum parthenium (feverfew) herb

Feverfew, High Potency - Capsule (100)

Available in: Canada - Stock #288-1

Ingredients: Same as Stock #288-1

Fibralgia

Product Type: Formula

Properties: Analgesic (Anodyne), Mineralizer

Systems Affected: **Muscles**, Structural System

Conditions: **Autoimmune Disorders**, Cramps and Spasms (general), Energy (lack of), Epstein Barr Virus (Chronic Fatigue Syndrome, CFS), Fatigue, **Fibromyalgia Syndrome (FMS)**, **Tremors**

Usage: Both Chronic Fatigue Syndrome and Fibromyalgia seem to be caused when the body fails to produce ATP (adenosine triphosphate). ATP is generally formed from the energy in foods. When foods are broken down, the vitamins, minerals and coenzymes provide electrical charges to carry out enzymatic reactions. Once this step occurs, ATP is made and energy is released. Malic acid is a fruit acid mostly found in apples and it plays an important role in ATP production in low oxygen conditions. When malic acid is combined with magnesium, it provides fuel that generates energy to operate the body. Magnesium also aids absorption of numerous vitamins and minerals, encourages bone strength, and regulates nerve impulses. Recommended dosage is 2 capsules twice per day, with meals.

Warnings: No known warnings.

Fibralgia - Capsule (90)

Available in: US & Canada - Stock #4061-6
Ingredients: Malic acid and magnesium

Fizz Active-Immune

Product Type: Formula

Properties: Immune Stimulant

Systems Affected: Immune System

Conditions: Antibiotics (alternatives to), Colds (general remedies for), Colds (prevention), **Flu**, Infection (bacterial), Infection (viral), Sore Throat

Usage: Fizz Active Immune is a formula designed to boost your immune system. You can take it whenever illness is going around to help your body resist getting sick in the first place. It is a blend of antioxidants and immune stimulants that put your body's immune system on "red alert" so they can fight off whatever comes your way. It can also be used in the early stages of colds or flu to boost the immune system for a more rapid recovery.

Fizz Active Immune a great remedy to take when traveling or when exposed to crowds or sick people. The fizzy action of Fizz Active-Immune allows the body to quickly absorb the herbs and nutrients for a rapid effect.

Simply drop two tablets to a half cup (about 4-6 ounces) of water and let the tablets dissolve for 2-4 minutes before drinking. You can repeat this dose every four hours as needed.

Warnings: Immune stimulants like Fizz Active-Immune are contraindicated in autoimmune disorders.

Fizz Active-Immune - Tablets (20)

Available in: US - Stock #3045-5
Ingredients: Zinc, Potassium (bicarbonate), Beta Glucans 70% (Saccharomyces cerevisiae), Arabinogalactan (Larix spp), Echinacea purpurea Root Extract, Elder Berries Extract (Sambucus nigra), Korean Ginseng Root Extract (Panax ginseng), Citric acid, sodium bicarbonate, sorbitol, natural flavors, cellulose (plant fiber), stearic acid (vegetable), stevia leaf extract (Stevia rebaudiana), mineral oil and magnesium sterate (vegetable)

Flash-Ease T/R

Product Type: Formula

Properties: Estrogenic, Nervine, Phytoestrogen

Systems Affected: Female Reproductive

Conditions: Breasts (swelling and tenderness), **Estrogen (low)**, **Hot Flashes**, **Menopause**, PMS Type D, **Pregnancy (herbs and supplements to avoid during)**, Vaginal Dryness

Usage: This herbal combination is designed to reduce or alleviate the hot flashes and other menopausal symptoms many women experience. The black cohosh found in Flash-Ease is a standardized extract of the rhizome containing 2.5% triterpene glycosides. The ingredients are also time-released so that they are introduced into the system gradually. This is a big plus, since large doses of black cohosh sometimes cause headaches and dizziness. By releasing the herb slowly, these undesirable effects are avoided. This formula also contains dong quai, a Chinese tonic herb frequently used for female problems. Dong quai nourishes the blood and supports the female organs. Take one tablet in the morning and at bed time. Flash-Ease provides time-released coverage offering a full 10 hours of benefits.

Warnings: Pregnant or lactating women should consult their health care providers before using this product.

Flash-Ease T/R - Tablets (60)

Available in: US - Stock #81-4
Ingredients: Black cohosh, dong quai.

F.E. Formula - Tablets (60)

Available in: Canada - Stock #81-4
Ingredients: Same as Stock #81-4

Flax Seed Oil

Product Type: Nutrient

Properties: Anticoagulant (Blood Thinner), Antirheumatic, Antitussive, Food, Moistening, Nutritive

Systems Affected: Female Reproductive, Large Intestine (Colon), Nerves, Prostaglandins, Structural System, Whole Body

Conditions: Abscesses, Amenorrhea, Appetite (excessive), Arteriosclerosis (Atherosclerosis, Hardening of the Arteries), Arthritis, Attention Deficit Disorder (ADD, ADHD), Autism, Bipolar Mood Disorder (Manic Depressive Disorder), Blood Clots (prevention of), Boils, Burns and Scalds, Cancer (prevention), Cardiovascular Disease (Heart Disease), Chemical Poisoning, Children's Remedy, Cholesterol (high), Circulation (poor), Cold Hands and Feet, Constipation (adults), Constipation (children), Cough (dry), Cystic Fibrosis, Dandruff, Depression, Dermatitis, **Eczema**, Epilepsy, **Failure to**

Thrive, Fat Cravings, Hashimoto's Disease (Thyroiditis), Heavy Metal Poisoning, Lupus, Multiple Sclerosis (MS), Muscular Dystrophy, Nerve Damage, **Nervous Exhaustion (Enervation)**, Pain (general remedies for), **Pets (supplements for)**, Pregnancy (herbs and supplements for), Psoriasis, Reye's Syndrome, Schizophrenia, Senility, **Skin (dry and/or flaky)**, Skin Care (general), Strokes

Usage: Flax seed oil is a vegetarian source of Omega-3 and Omega-6 fatty acids that help lower cholesterol. It is also is a source of lignan which is linked to lowering the incidence of breast and colon cancer. It is one of the most nutritious oils on the earth. Most people in our society are getting too much of the wrong kind of fats, and hence, are winding up deficient in essential fatty acids. As a nutritional supplement, take 1-2 soft gels three times daily with meals.

Warnings: No known warnings.

Flax Seed Oil Liquid (8 fl. oz.) - Liquid

Available in: US - Stock #3162-1

Ingredients: Flax (Linum sp.) seed oil

Flax Seed Oil w/Lignans - Softgel (60)

Available in: US - Stock #1583-6

Ingredients: Same as Stock #3162-1

Flax Seed Oil - Softgel (60)

Available in: US & Canada - Stock #1770-3

Ingredients: Same as Stock #3162-1

Focus Attention

Product Type: Formula

Properties: Anti-epileptic, Nervine

Systems Affected: Brain, Nerves

Conditions: Alzheimer's Disease, **Attention Deficit Disorder (ADD, ADHD)**, Children's Remedy, **Concentration (poor)**, Confusion, **Epilepsy**, Memory and Brain Function, **Thinking (cloudy)**

Usage: This formula provides nutrients that help to reduce over activity in the brain's neurotransmitter system. The supplement was designed to help hyperactivity and attention deficit disorder. It also helps protect the brain from toxic chemicals. For best results, take with NSP flax seed oil. Adults or children over age 12: take 2 capsules twice daily with meals, or as needed, or take 3/4 teaspoon of the powder in 2 ounces of water twice daily. Children age 6-12: take 1 capsule twice daily with a meal or take 1/2 teaspoon of the powder in 1 1/2 ounces of water twice daily. Children age 3-6: take 1/4 teaspoon of the powder in 1 ounce of water twice daily. If using the powder, mix well and drink immediately. For optimum utilization, take capsules with Flax Seed Oil.

Warnings: No known warnings, but NSP recommends that a health care provider be consulted before giving this product to children under 6 years.

Focus Attention - Capsule (90)

Available in: US - Stock #1833-4

Ingredients: Slippery elm, l-Glutamine DMAE (dimethylamino- ethanol), lemon balm, grape seed extract, ginkgo, cellulose, magnesium stearate

Focus ATN - Capsule (90)

Available in: Canada - Stock #1833-4

Ingredients: Same as Stock #1833-4

Focus Attention (Powder) - Bulk Powder

Available in: US - Stock #1843-0

Ingredients: Vitamins B1, B2, B6, B12, Niacine, Pantothenic acid, folic acid, biotin, flax seed powder, DMAE, lemon balm, bacopa, ginko, grape seed extract, natural raspberry flavor, grape skin extract, citric acid, lohan fruit concentrate, eleuthero, malic acid, stevia leaf extract, inosiaol, choline.

Folic Acid Plus

Product Type: Formula

Properties: Nutritive

Systems Affected: Brain, DNA, Red Blood Cells, Spleen

Conditions: Angina, Arteriosclerosis (Atherosclerosis, Hardening of the Arteries), Birth Defects (prevention), Cancer (natural therapy for), Cataracts, Cholesterol (low), Depression, Hemochromatosis, **Pregnancy (herbs and supplements for)**, Restless Leg Syndrome, Schizophrenia

Usage: Used in red blood cell formation, metabolism of proteins and growth and division of cells. It is necessary for fetal development. Deficiencies in pregnant women may cause birth defects. RDA infants: 3 - 4.5 mcg, RDA children: 100 - 400 mcg, RDA adults: 400 mcg, dietary levels: 400 mcg, therapeutic range: 400-2000 mcg. Take 1 tablet daily with a meal. During pregnancy, take 2 tablets daily with a meal. Works best with vitamin B-12, 400 mcg. Recommended dosages of Vitamin B-Complex and Balanced B-Complex each supply 400 mcg of folic acid.

Warnings: No toxic effects. Oral contraceptives may increase the need for folic acid. Avoid high doses in hormone-related cancer or convulsive disorders.

Folic Acid Plus - Tablet (90)

Available in: US & Canada - Stock #1585-8

Ingredients: Each tablet contains 400 mcg folic acid plus vitamin C and bee pollen.

Food Enzymes

Product Type: Formula

Properties: Antilipemic, Carminative, **Digestant**, Pancreatic Tonics

Systems Affected: Digestive System, Gall Bladder, Intestinal System, **Liver**, **Pancreas**, **Pancreas Head**, **Stomach**

Conditions: Acid Indigestion (Heartburn, Acid Reflux), Allergies (food), **Belching**, Calcium Deficiency, Cancer (natural therapy for), Cystic Fibrosis, Debility, **Digestion (poor)**, Dyspepsia, Fat Metabolism (poor), Fatigue, Fibromyalgia Syndrome (FMS), **Gas and Bloating**, **Halitosis (Bad Breath)**, Hypoglycemia, Indigestion, Inflammatory Bowel Disorders (Colitis, IBS), Lyme Disease, Overacidity, **Overalkalinity**, Pancreatitis, **Protein Digestion (poor)**, Triglycerides (high), Triglycerides (low), Ulcerations (external), **Wasting**

Usage: Supplies hydrochloric acid, pancreatic enzymes and bile for aiding in the digestion of fats, proteins and carbohydrates. Used for digestive system weakness. Take 1-2 capsules during or after meals as an aid to digestion. Enzyme supplements may also be helpful for cancer if taken between meals or in the early hours of the morning (between midnight and 3 AM) on an empty stomach.

Warnings: Do not take if you have stomach ulcers. May cause stomach irritation in some persons. Always take with food. We recommend that this product not be used continually, except in the case of very elderly persons, as supplementing these digestive secretions will tend to atrophy the body's ability to produce them. Use to help rebuild a person while supplementing their diet with herbs and nutrients that rebuild the digestive organs. For regular long-term use, we recommend Proactazyme Plus as that product does not duplicate the body's natural digestive secretions.

Food Enzymes Capsules - Capsule (120)

Available in: US - Stock #1836-9

Ingredients: Mycozyme (alpha-amylase), betaine HCl, bile salt, bromelain, lipase, pancreatin, papain, pepsin.

Food Enzymes Trial Pack - Packets (20)

Available in: US - Stock #2494-1

Ingredients: Same as Stock #1836-9

Digestive Enzymes - Capsule (120)

Available in: Canada - Stock #1761-1

Ingredients: Pancreatin (180 mg), mycozyme (90 mg), bromelain (50 mg), papain (45 mg), lipase (15 mg), pepsin (200 mg), betaine HCl (162 mg), and bile salt (40 mg).

Four

Product Type: Formula

Properties: Antihistamine, Decongestant, Nervine

Systems Affected: Lungs, Mucus Membranes, Respiratory System, Sinuses

Conditions: **Allergies (respiratory)**, Asthma, Bronchitis, Congestion (general), Cough (general), Headache (sinus), Lungs (fluid in), **Pleurisy**, Pneumonia

Usage: This is an excellent formula for hay fever and allergies. Blessed thistle helps detoxify the liver, which is usually involved in allergies. The formula also has anti-inflammatory and nervine action. Take 2-3 capsules three times per day during allergy season.

Warnings: No known warnings.

Four - Capsule (100)

Available in: US - Stock #892-5

Ingredients: Blessed thistle, catnip, pleurisy root, yerba santa

Frankincense

Product Type: Essential Oil

Properties: Antidepressant, Antimicrobial, Antiseptic, Aromatic, Astringent, Bitter, Calmative, Disinfectant, Immune Stimulant, Preservative, Uterine Tonic

Systems Affected: Breasts, Immune System, Immune System, Lungs, Nerves, Ovaries, Uterus

Conditions: Asthma, Autism, Breasts (swelling and tenderness), Cancer (natural therapy for), Confusion, Depression, Fear (excessive), Grief (excessive), Inflammation, Insomnia, Irritability, Laryngitis (Hoarseness), Menorrhagia (Heavy Menstrual Bleeding), Nightmares, Ovarian Pain, PMS Type H, Tension, Worry

Usage: Frankincense is effective for respiratory catarrhal discharge and respiratory congestion. It is helpful for asthma sufferers as it eases shortness of breath and increases the amplitude of the breath. It is also useful for both sinusitis and laryngitis. Its astringent properties may relieve uterine hemorrhages as well as heavy periods and it acts as a general tonic to the uterus. It helps to relax and revitalize the nervous system and is good for nervous tension and nervous exhaustion, reducing stress levels that lead to irritability, restlessness, and insomnia. It's uplifting energy gives it an antidepressant quality. It is an immune strengthener, especially in cases where nervous depression is involved. Frankincense has been applied topically to help shrink tumors, especially in the breasts. It has also been massaged over the ovaries to ease ovarian inflammation and pain during menses.

Emotionally it slows down breathing and produces feelings of calm. Possessing an elevating and soothing effect on the mind, frankincense allows the consciousness to expand. It works well with anxious and obsessive states linked to worrying about the past and is a strong support tool for agitation, worry or conditions where the mind is distracted and overwhelmed by a multitude of thoughts. Releasing feelings of being tied down or restricted from over-attachment to a thought, idea or thing, it helps a person gain insight, spiritual self-discipline and tranquility.

Warnings: Recommended for topical use only.

Frankincense Essential Oil (5ML) - Essential Oil

Available in: US - Stock #3899-9

Ingredients: Frankincense (Boswellia carterii) essential oil

Free Amino Acids

Product Type: Formula

Properties: Nutritive

Systems Affected: Structural System, Whole Body

Conditions: Athletic Performance or Exercise (aids to), Attention Deficit Disorder (ADD, ADHD), Bedwetting, **Body Building**, Broken Bones, Canker Sores (Mouth Ulcers), Cataracts, Coordination, Diabetes, Digestion (poor), Down Syndrome, Exercise, Failure to Thrive, Muscular Dystrophy, Narcolepsy, Nervous Exhaustion (Enervation), Reye's Syndrome, Schizophrenia, Thinking (cloudy), Triglycerides (high), Wasting

Usage: Free amino acids can be a very beneficial way to rebuild the health of very weak individuals with poor muscle tone who suffer from digestive problems. They can also be helpful for body builders and athletes. Take 2 tablets with a meal three times daily in addition to your daily multiple vitamin and mineral supplement.

Warnings: No known warnings.

Free Amino Acids - Tablet (60)

Available in: US - Stock #3664-6

Ingredients: l-Lysine, l-Histidine, l-Arginine, l-Aspartic acid, l-Threonine, l-Serine, l-Proline, l-Alanine, Glycine, l-Glutamic Acid, l-Cystine, l-Valine, l-Methionine, l-Isoleucine, l-Leucine, l-Tryosine, l-Phenylalanine, l-Tryptophan, l-Carnitine, Magnesium aspartate, magnesium oxide.

FV

Product Type: Formula

Properties: Bitter, Stimulant, Stomachic, Vulnerary

Systems Affected: Adrenal Glands, Mucus Membranes, Pancreas, Stomach

Conditions: Colds (general remedies for), Cramps and Spasms (general), Diarrhea, Fever, **Flu**, Indigestion, Motion Sickness, **Nausea and Vomiting**

Usage: Use 2-4 capsules every hour to settle stomach during acute flu, nausea, vomiting. Helps to settle the stomach and fight infection. This combination for flu and vomiting also makes an effective cold remedy. Use 2 capsules every two hours with a large glass of water.

Warnings: Capsicum and ginger may be irritating to sensitive stomachs.

FV - Capsule (100)

Available in: US - Stock #900-7

Ingredients: Ginger, golden seal, capsicum, licorice

GABA Plus

Product Type: Formula

Properties: **Anti-epileptic**, Nervine

Systems Affected: Brain, **GABA**, Nerves

Conditions: Addictions (drugs), Anxiety (Panic Attack), **Attention Deficit Disorder (ADD, ADHD)**, Concentration (poor), **Epilepsy**, Parkinson's Disease, Pregnancy (herbs and supplements to avoid during), **Schizophrenia**, **Seizures**, Senility

Usage: This formula is designed to help prevent nerve cells from over-firing by making them less susceptible to stimuli. It helps balance neurotransmitters, improve circulation to the brain and calm brain function. Use as directed on the bottle, 1-2 capsules two times daily with meals.

Warnings: Do not exceed directed amount, do not combine with prescription drugs or use with pregnancy or nursing.

GABA Plus - Capsule (60)

Available in: US - Stock #1823-6

Ingredients: GABA, glutamine, taurine, spirulina, passion flower.

Gall Bladder Formula

Other Names: BLG-X

Product Type: Formula

Properties: Carminative, Cholagogue, Hepatic, Stomachic

Systems Affected: Digestive System, **Gall Bladder**, Liver

Conditions: **Abdominal Pain and Inflammation**, Cholesterol (high), Cholesterol (low), **Colic (adults)**, Constipation (adults), Digestion (poor), Fat Cravings, Fat Metabolism (poor), Gall Bladder (sluggish), Gall Stones, **Indigestion**, Jaundice (adults)

Usage: A Western formula for indigestion, gas and general digestive problems. Barberry stimulates bile production while aromatics stimulate other digestive secretions. Use 2 capsules with meals as an aid to liver congestion, jaundice, sluggish gall bladder production or frequent bloating or stuffiness in the abdomen.

Warnings: If diarrhea ensues, discontinue use for a period of time, then continue again at 1/2 dosage.

Gall Bladder Formula - Capsule (100)

Available in: US - Stock #1202-0

Ingredients: Oregon grape root, ginger root, cramp bark, fennel seeds, peppermint leaves, wild yam root, catnip herb

Garcinia Combination

Product Type: Formula

Properties: Anti-obesic, Anticholesteremic, Antilipemic, Appetite Suppressant

Systems Affected: Cardiovascular System, Digestive System, Heart, Intestinal System, Liver

Conditions: Appetite (excessive), Cardiovascular Disease (Heart Disease), Cholesterol (high), Fatty Tumors or Deposits, Triglycerides (high), **Weight Loss (aids for)**

Usage: This formula is used as an aid to weight loss. It helps to control appetite and aid the body in burning fat. Garcinia has also been shown to reduce LDL cholesterol and triglycerides and promote HDL cholesterol. This formula may help to prevent the formation of arterial plaque and be of some benefit in cardiovascular disease and fatty congestion of the liver. Take two capsules 30 minutes before a meal three times daily.

Warnings: No known warnings.

Garcinia Combination - Capsule (100)

Available in: US - Stock #906-9

Ingredients: Garcinia, chickweed, l-carnatine

Garden Essence

See *Proactazyme Plus*

Garlic

Latin Name: *Allium sativum*

Product Type: Single Herb

Properties: **Antibacterial**, Anticholesteremic, Anticoagulant (Blood Thinner), Antidiabetic, Antifungal, Antihypertensive, Antimicrobial, Antiparasitic, **Antiseptic**, Antiviral, Cardiac, Carminative, Cholagogue, Condiment, Decongestant, **Diaphoretic**, Expectorant, Febrifuge, Hypotensive, Lung Tonic, Lymphatic, **Parasiticide**, Pectoral, Preservative, Pulmonary, Stimulant, Stomachic, Vasodilator, Vermifuge

Systems Affected: Appendix, Arteries, Digestive System, **Ears**, Eustachian Tubes, Hypothalamus, Immune System, Intestinal System, Liver, **Lungs**, Lymphatic System, Peripheral Blood Vessels, Respiratory System, Sinuses, Stomach, Throat, Thyroid, Tonsils

Conditions: **Acquired Immune Deficiency Syndrome** (AIDS/HIV), **Antibiotics (alternatives to)**, Antibiotics (side effects of), **Appendicitis**, **Arteriosclerosis (Atherosclerosis, Hardening of the Arteries)**, Arthritis, Asthma, Athlete's Foot, **Blood Poisoning**, Blood Pressure (high), Blood Pressure (low), Bronchitis, Cancer (natural therapy for), **Cholera**, Circulation (poor), Cold Sores (Fever Blisters), Colds (antiviral), Colds (decongestant), Colds (general remedies for), Colds (prevention), Colds (with fever), Congestion (lungs), **Contagious Diseases**, Corns, **Cough (damp)**, **Cough (general)**, Cystitis, Diabetes, Diarrhea, **Diphtheria**, Dysentery, Dyspepsia, **Ear Infection or Earache**, Fever, Flu, Fungal Infections (Yeast Infections, Candida albicans), Gangrene, Gas and Bloating, Gonorrhea, **Infection (bacterial)**, Infection (viral), Insects, Itching Ears, Jock Itch, Lead Poisoning, Leucorrhea, **Lungs (fluid in)**, Lupus, Lyme Disease, Malaria, Mercury Poisoning, Mumps, **Parasites (general)**, Parasites (nematodes, worms), **Pertussis (Whooping Cough)**, Pets (supplements for), **Pneumonia**, Poisoning (food), Rheumatic Fever, Scabies, Sore Throat, **Staph Infections**, **Strep Throat**, Tick, **Tinnitus (Ringing in the Ears)**, Tonsillitis (Adenoids), **Toothache**, **Tuberculosis (Consumption, Scrofula)**, Urinary Tract Infections, Vaginitis, Warts

Usage: Garlic has been called nature's penicillin. Garlic is also one of the most powerful circulatory remedies. This herb is as effective as many drugs in lowering high blood pressure, but needs to be used regularly for long periods of time (3-6 months and longer). Garlic may also help to chelate (remove) plaque from arterial walls, thus lowering cholesterol levels. Garlic destroys many types of parasites and infections. This aromatic herb is a powerful decongestant and expectorant.

Raw garlic works well for serious lung problems and infections, but for less serious problems take 2 capsules regular garlic or 1 of high potency garlic tablet every three to four hours until relief is experienced. For more serious respiratory conditions use 2 capsules or 1 tablet every one to two hours along with 4 capsules of ALJ. For an even more powerful decongestant and expectorant action, add 1/2 dropperful of lobelia extract every one to two hours. Garlic has anti-viral properties and is helpful for colds and congestion. Use 1-4 capsules up to every 2 hours during a cold. Tea used as an enema helps break fevers, especially in children. For prevention take 2-3 capsules or 1 high potency tablet daily with meals. For infection (bacterial, viral or yeast) or respiratory congestion, take up to 2 capsules every two hours or 1 high potency tablet every two hours. For bacterial infection, use in combination with golden seal, echinacea, Lymph Gland Cleanse HY or Lymph Gland Cleanse, or IN-X. For viral infection, use in combination with Una D'Gato Combination, yarrow or Elderberry Defense. For yeast infection, combine with pau d'arco or Yeast/Fungal Detox. For respiratory congestion, combine with ALJ, mullein or lobelia. For swollen lymph nodes combine with lobelia and mullein. For parasites take eight to twelve capsules per day. Combine with Herbal Pumpkin and black walnut.

For hypertension (high blood pressure) and/or high cholesterol, take 6-10 capsules or 2-3 high potency tablets per day for three to six months. The fresh bulb has the best anti-hypertensive properties. Always use for at least three to six months for circulation conditions before expecting to see real results. Can be taken continually for prevention after improvement. Eat 1-3 cloves the size of almonds daily. (Cut up small and eat on bread.) Larger doses need to be taken for arterial effects.

Using Garlic Oil: For lung congestion, rub liberally on chest and back. For earache, use 3 drops (warm) in ear, rub on ear and rub on neck beneath ear. Constituents of garlic penetrate right through skin. Even the fumes kill harmful micro-organisms. Rub on throat for sore throat.

Raw Garlic: Eat 1-2 cloves every morning and regularly at meals for general health. For warts and cysts, rub with small slices of raw garlic throughout day. Follow with clay packs. Use 3-5 cloves in a decoction or eaten raw, three to six times daily for worms. Enema: For a garlic enema, simmer one chopped clove for 10 minutes in 1 pint of water or use 6-8 capsules and steep as tea. Use garlic enema for fevers, earache, infection and worms.

Warnings: Gastric irritation is possible; eat with food to lessen this effect. In rare instances may cause gastrointestinal upset de to stimulation of digestive secretions, changes in intestinal flora or allergic reactions. These reactions may be decreased by taking garlic with food.

Garlic is not recommended during nursing because it imparts an off flavor to the milk. Garlic also increases the effects of anticoagulant drugs like warfarin. Because of its blood thinning effect, garlic should be avoided prior to surgery. Also, even though High Potency Garlic is odorless and tasteless when ingested, the odor of garlic may pervade the breath and skin as it is secreted.

Garlic is stimulating and may not be the best choice of remedy for people who are feeble, emaciated or suffering from wasting conditions. Tonic herbs are better for these individuals.

Garlic - Capsule (100)

Available in: US - Stock #290-0

Ingredients: Allium sativum (garlic) bulb

Garlic Oil Capsules - Softgel (60)

Available in: US - Stock #1694-6

Ingredients: Garlic oil, soybean oil, glycerin.

Garlic, High Potency, SynerPro - Tablet (60)

Available in: US & Canada - Stock #292-9

Ingredients: 400 mg. standardized garlic extract, broccoli, carrot, red beet, rosemary, tomato, tumeric, cabbage, Chinese cabbage, grapefruit and orange bioflavinoids, hesperidin, calcium, phosphorus.

Gastro Health

Other Names: Herbal H-p Fighter

Product Type: Formula

Properties: Antibacterial, Antiparasitic, Soothing, Vulneraries (for intestinal system)

Systems Affected: Digestive System, Immune System, Large Intestine (Colon), **Mucus Membranes**, Pancreas Head, Small Intestines

Conditions: Acid Indigestion (Heartburn, Acid Reflux), Addictions (alcohol), Angina, Belching, Bleeding (external), Cancer (natural therapy for), Cholesterol (low), Colitis, Congestion (bronchial), Crohn's Disease, Cystic Fibrosis, **Diarrhea**, Digestion (poor), **Duodenal ulcers, Gastritis**, Inflammation, Leaky Gut Syndrome, Parasites (general), Ulcerations (external), **Ulcers**

Usage: Recent studies suggest that bacteria, Helicobater pylori, is the real cause of ulcers. This combination combines digestive stimulants, antibacterial and anti-ulcer herbs to fight against H. pylori bacteria and make the stomach lining inhospitable to infection. To combat ulcers take 2 capsules four times daily with plenty of water.

Warnings: No known warnings.

Gastro Health Concentrate - Capsule (60)

Available in: US - Stock #917-9

Ingredients: Pau d'arco, cloves, inula racemosa, licorice (deglycyrrhizinated), capsicum, lecithin

ULC-R+ - Capsule (60)

Available in: Canada - Stock #917-9

Ingredients: Same as Stock #917-9

GC-X

Product Type: Formula

Properties: Antibacterial, Hypotensive, Stimulant, Stomachic

Systems Affected: Circulation, Digestive System, Heart, Intestinal System, Mucus Membranes

Conditions: Arteriosclerosis (Atherosclerosis, Hardening of the Arteries), Asthma, Blood Pressure (high), Bronchitis, Dyspepsia, Energy (lack of)

Usage: This formula has been used to treat arteriosclerosis as well as high and low blood pressure. It's stimulating effect on circulation, heart, mucus membranes, kidneys, immune system, and the digestive system have long been known. This formula can also be used as an antibiotic for infection as well as an antiseptic. Take 2 capsules with meals.

Warnings: No known warnings. Completely safe for long-term use.

GC-X - Capsule (100)

Available in: US - Stock #1212-1

Ingredients: Garlic, capsicum, parsley, ginger, golden seal, eleuthero.

Gentle Move

Product Type: Formula

Properties: Laxative (general), Tonic

Systems Affected: **Intestinal System**, Large Intestine (Colon)

Conditions: **Colitis**, Colon (spastic), **Colon (spastic), Constipation (adults), Constipation (children), Hemorrhoids**, Inflammatory Bowel Disorders (Colitis, IBS), **Leaky Gut Syndrome**

Usage: Gentle move is an effective alternative to herbal stimulant laxatives. It helps to hydrate the colon and improve bowel tone to promote natural elimination. It can be used to wean someone off of stimulant laxatives and restore normal tone to the colon.

The key ingredient in Gentle Move is magnesium hydroxide, a salt of magnesium that attracts water to the bowel. This helps hydrate the stool and make it easier to pass. Magnesium hydroxide is found in milk of magnesia, an over the counter laxative.

Straight magnesium hydroxide can cause diarrhea and loss of potassium. It can also interfere with the absorption of iron and folic acid. However, in Gentle Move, a smaller amount is used along with a blend of herbs that help to tonify and rebuild the colon.

Gentle Move contains a blend of three herbs used in Ayurvedic medicine in India called triphalia. Triphalia is a blend of three fruits that has been used as a gentle laxative, bowel tonic and blood purifier in India for thousands of years.

The triphalia blend contains all the flavors (or tastes) used in Ayurvedic medicine, so it is considered a harmonizing or balancing remedy and is very safe for long term use. Besides normalizing colon function, it improves liver function, protects the liver against environmental toxins and improves digestion.

Triphalia's benefits extend far beyond the digestive tract. It is antioxidant and anti-inflammatory, so it slows aging and protects the body from degenerative disease. It enhances circulation, lowers blood pressure and protects the heart. It helps expel mucus from the respiratory passages and fights infection. It also supports healthy adrenal function.

In Ayurvedic medicine, triphalia is used for constipation, indigestion, flatulence, poor appetite, digestive headaches, sinus congestion, joint pain and general toxicity. It is also used for eye problems, general weakness and lethargy.

When you put all the benefits of Gentle Move together, you can see that the formula provides a mild laxative action while tonifying and improving colon health. It reduces intestinal inflammation and helps normalize bowel function in people who have become dependent on stimulant laxatives.

It is a safe bowel formula for people who are suffering from inflammatory diseases of the digestive tract. It is also safe for pregnancy.

Gentle Move may also help to relieve skin conditions, sinus problems and respiratory congestion. It also aids and protects the liver.

A safe long-term dose of Gentle Move (for maintenance of normal bowel function is 2 capsules once or twice daily. People who have serious constipation problems or have become dependent on stimulant laxatives for normal bowel function will need to start with a higher dose, about 6-9 capsules daily. This could be 2-3 capsules three times daily or 3-4 capsules twice daily (morning and evening are best). As bowel function improves, the dose can be reduced. Taking Gentle Move with plenty of water and some fiber for the colon will greatly enhance its effects.

Warnings: No known warnings.

Gentle Move - Capsule (90)

Available in: US - Stock #952-9

Ingredients: Magnesium (hydroxide), Triphala Extract, (Ayurvedic fruit blend: Terminalia chebala, Terminalia belerica, Embilica officinalis),Yellow Dock Root (Rumex crispus),Ginger Rhizome (Zingiber officinale),Marshmallow Root Extract (Althaea, officinalis), Slippery Elm Bark (Ulmus rubra)

Geranium

Product Type: Essential Oil

Properties: Analgesic (Anodyne), Anti-inflammatory, Antispasmodic, Antiviral, Anxiolytic, Aromatic, Decongestant, Dermatic, Nervine, Relaxant, Stimulant, Uterine Tonic

Systems Affected: Adrenal Cortex, Estrogen, Female Reproductive, Gums, Lymphatic System, Nerves, Nerves, Rectum, Vagina, Veins

Conditions: Acquired Immune Deficiency Syndrome (AIDS/HIV), Adrenals (exhaustion, weakness or burnout), Amenorrhea, Anorexia, Apathy, Breasts (swelling and tenderness), Confusion, Congestion (general), Congestion (lymphatic), Convulsions, Cuts, Depression, Dermatitis, Eczema, Edema (Dropsy, Water Retention, Swelling), Hemorrhoids, Hot Flashes, Infertility, Insects, Menopause, Menorrhagia (Heavy Menstrual Bleeding), Mood Swings, Nose Bleeds, Pain (general remedies for), PMS Type A, Skin (oily), **Vaginal Dryness**, Varicose Veins

Usage: Geranium has a stimulating effect on the adrenal cortex and a regulatory effect on the hormonal system. It helps with PMS and menopausal problems such as depression, lack of vaginal secretion and heavy periods. It may even be helpful with infertility. As a lymphatic stimulant it helps relieve congestion, fluid retention and swollen extremities. Geranium also helps with nose bleeds, bleeding gums, hemorrhoids, varicose veins, excessive menstrual bleeding and internal bleeding. It clears the heat from inflammation, relaxes nerves and calms feelings of anxiety. With both analgesic and antispasmodic properties, it is useful for nerve, eye and joint pain. Emotionally, geranium works for chronic and acute anxiety where there is stress due to feeling overworked. It is ideal for workaholics and those with perfectionist attitudes. Ideal for someone who has forgotten imagination, intuition and sensory experience, it reconnects us to a feeling for life.

Warnings: Contraindications: Skin that copiously secretes oils (chamomile is best for this condition). Estrogenic (over production of estrogen) conditions. Anxiety conditions or overly Type A individuals.

Geranium Essential Oil (5ML) - Essential Oil

Available in: US - Stock #3905-3

Ingredients: Geranium (Pelargonium graveolens) Essential Oil

Germanium Combination

Product Type: Formula

Properties: Antioxidant, Immune Stimulant

Systems Affected: Brain, Eyes, Heart, Immune System, Intestinal System, Liver, Muscles, Skeletal System, Skin, Thymus

Conditions: Arthritis, Autism, Backache (Back Pain, Lumbago), Cancer (natural therapy for), Cardiovascular Disease (Heart Disease), Cataracts, Chemotherapy (reducing side effects), Cirrhosis of the Liver, Corns, Endurance (lack of), Epstein Barr Virus (Chronic Fatigue Syndrome, CFS), Eye Problems (general), Fatigue, Glaucoma, Hepatitis, Infection (viral), Leukemia, Meningitis, Mercury Poisoning, Muscle Tone (lack of), Osteoporosis, Oxygen Deficiency, Pain (general remedies for), Toxemia, Warts

Usage: For boosting the immune system against normal infections, colds, etc. use 1-3 tablets per day. For AIDS and serious immune problems use 3-10 tablets per day. Germanium boosts the body's natural production of interferon.

Warnings: Not to be used for autoimmune disorders.

Germanium Combination - Tablet (30)

Available in: US - Stock #1652-3

Ingredients: Germanium sesquioxide, Echinacea purpurea

Ginger

Latin Name: *Zingiber officinale*

Product Type: Single Herb

Properties: Analgesic (Anodyne), Antacid, **Anti-emetic (Antinauseous)**, Antihistamine, **Antinauseous**, Antioxidant, Aperient, Aromatic, Carminative, Catalyst (Synergist), Condiment, **Diaphoretic**, Nervine, Sialogogue, Stimulant, Stomachic

Systems Affected: Circulation, Intestinal System, Liver, Pineal, Pineal, Reproductive Glands, Small Intestines, Sweat Glands

Conditions: Amenorrhea, Bronchitis, Circulation (poor), Colds (general remedies for), Cough (damp), Cramps and Spasms (general), Diarrhea, Dizziness (Vertigo), Dysmenorrhea, **Dyspepsia**, Fatigue, Fever, **Flu**, Gas and Bloating, Indigestion, Migraine, **Morning Sickness, Motion Sickness, Nausea and Vomiting**, PMS Type P, Shock, Sweat Baths (herbs for), Toothache, Ulcers

Usage: Helps allay nausea, vomiting and motion sickness. Take 2 capsules before traveling to prevent motion sickness. Use 4-6 capsules for nausea, vomiting and flu. Take 1 to 2 capsules at mealtime to stimulate digestive secretions or make into tea for better results. Take 2 capsules every hour for colds or chills. Used in herb formulas to transport ingredients to the abdominal areas and to increase circulation in digestive organs and pelvic region. Ginger is used worldwide as a spice.Helps allay nausea, vomiting and motion sickness. Take 2 capsules before traveling to prevent motion sickness. Use 4-6 capsules for nausea, vomiting and flu. Take 1 to 2 capsules at mealtime to stimulate digestive secretions or make into tea for better results. Take 2 capsules every hour for colds or chills. Used in herb formulas to transport ingredients to the abdominal areas and to increase circulation in digestive organs and pelvic region. Ginger is used worldwide as a spice.

Warnings: No known warnings. Although ginger is often recommended for alleviating the nausea of morning sickness, some herbalists believe it should be used with caution during pregnancy. We do not consider this caution valid, as it would suggest that ginger ale and ginger cookies should be avoided during pregnancy.

Ginger - Capsule (100)

Available in: US & Canada - Stock #300-6

Ingredients: Zingiber officinale (ginger) root

Ginkgo & Hawthorn

Product Type: Formula

Properties: **Anti-aging**, Anti-arrhythmic, Cardiac, Cerebral Tonic, Nervine

Systems Affected: Adrenal Glands, Brain, Cardiovascular System, **Circulation**, Ears, **Heart**, Nerves

Conditions: Age Spots, Angina, Blood Pressure (low), Cardiovascular Disease (Heart Disease), **Circulation (poor)**, Cold Hands and Feet, Concentration (poor), Dizziness (Vertigo), Fainting, Heart (weakness), **Heart Valves**, Memory and Brain Function, **Numbness, Raynaud's Disease**

Usage: In addition to the use of hawthorn berries as a cardiac tonic, this formula contains gingko, a nervine and brain tonic. Ginkgo & Hawthorn improves circulation to the heart and brain. It is an excellent general tonic for the elderly. Take 1-2 capsules with each meal.

Warnings: No known warnings.

Ginkgo & Hawthorn - Capsule (100)

Available in: US & Canada - Stock #909-3

Ingredients: Ginkgo, hawthorn, olive leaf

Ginkgo Biloba

Latin Name: *Ginkgo biloba*

Product Type: Single Herb

Properties: **Anti-aging**, Anti-allergenic, Anti-inflammatory, **Anticoagulant (Blood Thinner)**, Antioxidant, **Cephalalgic**, Tonic, Vascular Tonics, Vasodilator

Systems Affected: **Brain**, Nerves, Serotonin

Conditions: Aging (prevention), Alzheimer's Disease, Arteriosclerosis (Atherosclerosis, Hardening of the Arteries), Asthma, Autism, Bruises (prevention), Circulation (poor), Circulation (to the brain), Concentration (poor), Dementia, **Dizziness (Vertigo)**, Erectile Dysfunction, Glaucoma, Irritability, Macular Degeneration, Memory and Brain Function, Migraine, Multiple Sclerosis (MS), Narcolepsy, Parkinson's Disease, Raynaud's Disease, **Senility**, Strokes, Thinking (cloudy), **Tinnitus (Ringing in the Ears)**, Tremors, Wheezing

Usage: Ginkgo enhances circulation to the brain, enhancing memory and brain function. It also enhances peripheral circulation. NSP's ginkgo is a timed-release concentrate. Take one tablet with breakfast.

:Ginkgo Biloba Extract - Tablet (30)

Available in: US - Stock #898-8

Ingredients: Ginkgo biloba concentrate

Ginkgo Biloba, Slow Release - Tablet (30)

Available in: US & Canada - Stock #898-8

Ingredients: Ginkgo biloba concentrate

Ginkgo/Gotu Kola

Product Type: Formula

Properties: Adaptagen, Anti-aging, Anticoagulant (Blood Thinner), Antioxidant, **Cerebral Tonic**, Hypotensive, Nervine, Vasodilator

Systems Affected: Arteries, **Brain**, Capillaries, Circulation, Ears, Nerves, Peripheral Blood Vessels, Uterus, Veins

Conditions: **Aging (prevention)**, **Alzheimer's Disease**, Circulation (poor), **Circulation (to the brain)**, **Concentration (poor)**, **Dementia**, Depression, Dizziness (Vertigo), Fainting, **Memory and Brain Function**, **Senility**, Thinking (cloudy), Tinnitus (Ringing in the Ears)

Usage: To improve memory and aid in brain function, use 1 tablet three times daily.

Warnings: No known warnings.

Ginkgo/Gotu Kola w/Bacopa - Capsule (60)

Available in: US - Stock #899-6

Ingredients: Ginkgo biloba, gotu kola, bacopa

Ginseng (Korean)

Latin Name: *Panax ginseng*

Product Type: Single Herb

Properties: Adaptagen, **Androgenic**, Anti-aging, Antioxidant, Aphrodisiac, Blood Building, Cardiac, Nervine, Panacea, Stimulant, Tonic

Systems Affected: Adrenal Glands, Adrenal Medulla, Cardiovascular System, Central Nerves, Circulation, Cyclic AMP (cAMP), Heart, Immune System, Intestinal System, Liver, Lungs, Nerves, Prostate, **Reproductive Glands**, Respiratory System, Spleen, **Testes**, **Testosterone**

Conditions: Addictions (drugs), **Aging (prevention)**, Anemia, Anxiety (Panic Attack), Arrhythmia, Arteriosclerosis (Atherosclerosis, Hardening of the Arteries), Asthma, Benign Prostate Hyperplasia (BPH), Blood Pressure (high), Blood Pressure (low), Cancer (natural therapy for), Cardiovascular Disease (Heart Disease), Cataracts, Debility, Depression, Diabetes, Dysmenorrhea, Endurance (lack of), Energy (lack of), **Erectile Dysfunction**, **Fatigue**, Infertility, Neurosis, Poison Ivy or Oak, **Premature Ejaculation**, Radiation Sickness, **Sex Drive (low)**, Stress, **Testosterone (low)**

Usage: Korean ginseng is a general tonic for the entire body. It helps to strengthen the immune system, improve digestive and respiratory function, overcome anemia and fatigue and aids the body's ability to adapt to stress. It warms the body and helps strengthen people who are cold, pale and weak. As a daily tonic for strengthening sexual hormones and improving resistance and stamina, use 1-2 capsules per day. For acute stress take 2-4 capsules per day. For chronic weakness and immune deficiency, use 2-5 capsules per day. Combines well with other sweet tonics, such as gotu kola, licorice, kelp, etc.

Warnings: Has no toxic effects but is contraindicated in acute diseases, high fevers, severe inflammation, hyperactivity or extreme, nervous anxiety. Ginseng should not be used in conjunction with caffeine. Generally speaking, ginseng should not be used by children, teenagers or robust athletic types. It is more for the elderly, weak, feeble or sickly. Taking too much ginseng when young can "burn-up" your chi or vital energy.

Ginseng, Korean - Capsule (100)

Available in: US & Canada - Stock #665-4

Ingredients: Panax ginseng (Korean ginseng) root

Ginseng (Wild American)

Latin Name: *Panax quinquefolium*

Product Type: Single Herb

Properties: Adaptagen, Androgenic, Antioxidant, Aphrodisiac, Cardiac, Nervine, Panacea, Spleen Chi Tonic, Stimulant, Tonic

Systems Affected: Adrenal Glands, Adrenal Medulla, Cardiovascular System, Central Nerves, Circulation, Heart, Intestinal System, Liver, Lungs, Nerves, Prostate, **Reproductive Glands**, Respiratory System, Spleen, Testes

Conditions: **Aging (prevention)**, Anemia, Anxiety (Panic Attack), Arrhythmia, Arteriosclerosis (Atherosclerosis, Hardening of the Arteries), Asthma, Blood Pressure (high), Blood Pressure (low), Cancer (natural therapy for), Cardiovascular Disease (Heart Disease), Debility, Depression, Diabetes, Digestion (poor), Endurance (lack of), Energy (lack of), Erectile Dysfunction, **Fatigue**, Fear (excessive), Infertility, Neurosis, Poison Ivy or Oak, Protein Digestion (poor), Radiation Sickness, **Sex Drive (low)**, Stress, Testosterone (low), Wasting, Weight Gain (aids for)

Usage: Wild American Ginseng is a bitter-sweet tasting tonic recommended for daily use to improve vital energy and strengthen sexual hormones. By improving resistance and stamina it works well for acute stress, anxiety, depression, chronic weakness, fatigue, immune deficiency, asthma, cancer, anemia, heart disease, diabetes, menstrual problems, hypertension, hypotension, radiation sickness and poison ivy prevention. Use 1-3 capsules per day. For acute stress take 3-6 capsules per day. For chronic weakness and immune deficiency, use 2-5 capsules per day. Combines well with other sweet tonics, such as gotu kola, licorice, kelp, etc. Can be used by women as well as men. American ginseng is an excellent tonic for individuals with poor digestion. It is less heating than Korean ginseng.

Warnings: Has no toxic effects but is contraindicated in acute diseases, high fevers, severe inflammation, hyperactivity or extreme, nervous anxiety. Ginseng should not be used in conjunction with caffeine. Generally speaking, ginseng should not be used by children, teenagers or robust athletic types. It is more for the elderly, weak, feeble or sickly. Taking too much ginseng when young can "burn-up" your chi or vital energy.

Ginseng, Wild American - Capsule (50)

Available in: US - Stock #725-8

Ingredients: Panax quinquefolium (wild American ginseng) root

GlucoReg

See *SugarReg*

Glucosamine

Product Type: Formula

Properties: Anti-arthritic, Anti-inflammatory, Tonic, Vulnerary

Systems Affected: Connective Tissue, Structural System

Conditions: Arthritis, Cartilage Damage, Inflammation, Rheumatoid Arthritis (Rheumatism)

Usage: Glucosamine is an amino acid/sugar substance used by the body to produce connective tissues. It is derived from crab shells. Uña de gato increases the anti-inflammatory action of glucosamine helping to relieve joint pain. Take 4 capsules daily with a meal.

Warnings: May irritate ulcers if not taken with food.

Glucosamine - Capsule (60)

Available in: US & Canada - Stock #903-4

Ingredients: Glucosamine hydrochloride, uña de gato

Glyco Essentials

Product Type: Formula

Properties: Immune Amphoterics, Nutritive

Systems Affected: Immune System

Conditions: Autoimmune Disorders, Cancer (natural therapy for), Infection (viral), Pets (supplements for)

Usage: When most people think of sugar, they think of sucrose or fructose, the kinds of sugar we eat for energy. However, there are many different types of sugar. Glyco Essentials contains eight sugars that are essential for the normal function of the body. These sugars aid in cellular communication and play a role in building enzymes, hormones and antibodies.

Because the body uses the sugars coating cell membranes to help distinguish "friend and foe" these sugars are very helpful in regulating the immune response. This product contains a number of herbs rich in these sugars that have been used to modulate the immune response.

For maintenance, take 1 capsule three times daily. For diseases that involve immune stress, take 1 capsule five times a day.

Warnings: No known warnings.

Glyco Essentials - Capsule (90)

Available in: US - Stock #876-5

Ingredients: Glyco Essentials contains fucose, glucose, galactose, N-acetylgalactosamine, N-acetylglucosamine, N-acetylneuraminic acid, mannose and xylose in a phytonutrient base of short-chain fructo-oligosaccharides, aloe vera extract, tragacanth gum, arabinogalactan, beta-glucan, shiitake mushroom extract, cordyceps mushroom extract, maitake mushroom extract, glucosamine sulfate, ghatti gum, guar gum and rice starch

Golden Salve

Product Type: Formula

Properties: Antiseptic, Astringent, **Balm**, Emollient, **Vulnerary**

Systems Affected: Digestive System, Eyes, Mucus Membranes, **Nipples**, Pancreas, **Skin**, Spleen, Stomach, Structural System, Uro-genital Tract, Uterus, Vagina

Conditions: Abrasions, Abscesses, Anal Fistula or Fissure, Bites and Stings, Blisters, Burns and Scalds, Cold Sores (Fever Blisters), Cuts, **Diaper Rash, Hemorrhoids,** Inflammation, Injuries, Itching (rectal), Scratches and Abrasions, **Wounds and Sores**

Usage: A healing salve with disinfectant properties that can be applied to minor injuries to promote healing.

Warnings: Not recommended for deep wounds or cuts.

Golden Salve (1 oz.) - Salve

Available in: US & Canada - Stock #1698-0

Ingredients: Chickweed, comfrey leaf, golden seal, yarrow, white oak, black walnut, marshmallow, lobelia, scullcap, myrrh, mullein, wheat germ, olive oil, beeswax, vitamin E, eucalyptus oil

Golden Seal

Latin Name: *Hydrastis canadensis*

Product Type: Single Herb

Properties: Alterative (Blood Purifier), **Antacid**, Antibacterial, **Antidiabetic**, Antimicrobial, Antiparasitic, Antipruritic, Antiseptic, Antiviral, **Aperitive, Appetite Stimulant**, Astringent, Bitter, Cholagogue, Detergent, Emmenagogue, Hemostatic, **Hypoglycemic, Insulinomimetic**, Nervine, Oxytocic, Stomachic

Systems Affected: Appendix, Digestive System, Eyes, Female Reproductive, Gall Bladder, Gums, Immune System, Intestinal System, Lungs, **Mouth, Mucus Membranes**, Pancreas Tail, Pineal, Respiratory System, Skin, **Spinal Disks**, Structural System, **Tongue, Urinary System**

Conditions: Acid Indigestion (Heartburn, Acid Reflux), Adenitis, Aneurysm, Anorexia, Antibiotics (alternatives to), **Appendicitis**, Appetite (deficient), Bladder (irritable), **Bladder**

(ulcerated), **Bladder Infection**, Bleeding (external), Bleeding (internal), Bloodshot Eyes, Bronchitis, **Canker Sores (Mouth Ulcers)**, Chicken Pox, **Cholera**, Colds (general remedies for), Colitis, Conjunctivitis (Pink Eye), Contagious Diseases, Cuts, **Diabetes**, **Diarrhea**, Digestion (poor), **Disks (spinal - bulging or slipped)**, **Dysentery**, Ear Infection or Earache, Edema (Dropsy, Water Retention, Swelling), **Eye Infections**, Eye Problems (general), Eyes (red or itching), Gangrene, **Giardia**, Gingivitis (Bleeding Gums, Gum Disease, Pyorrhea), Gonorrhea, Hemorrhoids, **Infection (bacterial)**, Inflammation, Injuries, Itching, **Kidney Infection**, Lyme Disease, Measles, Meningitis, Morning Sickness, Nephritis, **Oral Surgery**, **Pancreatitis**, Parasites (general), Pets (supplements for), Poisoning (food), Pregnancy (herbs and supplements to avoid during), Prostatitis, Scabies, Scratches and Abrasions, **Sinus Infection**, Skin (infections), Smell (loss of sense of), **Sore or Geographic Tongue**, Sore Throat, **Staph Infections**, Syphilis, Tonsillitis (Adenoids), **Tooth Extraction**, Tuberculosis (Consumption, Scrofula), Typhoid, **Ulcerations (external)**, **Ulcers**, **Urinary Tract Infections**, Urination (burning or painful), **Vaginitis**, Wounds and Sores

Usage: Golden seal has been considered by many to be one of those "cure-all" herbs. This has resulted in its being over-harvested and the plant is rapidly becoming endangered. Hence, where other herbs can be substituted, they should be used and golden seal saved for those things it does best. One of golden seal's primary benefits is for acute inflammation of the mucus membranes lining the respiratory and digestive system. It tonifies, astringes and cools this first line of immune defense in the body. It is not very useful for the early stages of a cold where the mucus is watery. It works better in cases where the mucus is thick and discolored. Golden seal acts as a tonic to improve digestion and neutralize waste acids in the system, especially when a very small amount of the powder is "sucked" on so that the bitter taste is experienced.

Golden seal lowers blood sugar levels. Some borderline diabetics have successfully used golden seal instead of insulin. Some diabetics have weaned themselves off of insulin with this herb. For this reason it is often avoided in hypoglycemia. This herb also heals injuries, both on the mucous membranes and on the skin. It has been used as a wash for sore, red eyes. It is a very good treatment for amoebic dysentery (giardia). For giardia use 10 grams daily for 10 days. Suck on golden seal powder to relieve canker sores. To fight acute infection take 2-3 capsules three times daily. Do not exceed 10 days at this dose. For chronic problems, take 1 capsule one to three times daily for up to 4 weeks. It works even better when combined with other immune-stimulant and antibiotic herbs and combinations such as Lymph Gland Cleanse, Lymph Gland Cleanse HY, garlic, etc. For blood sugar regulation individuals need to experiment and monitor their blood sugar levels to determine how much golden seal to use.

Warnings: Not recommended for autoimmune disorders where the immune system is over active (i.e. M.S. Lupus, Hodgkins, etc.). Safe when used in recommended dosages and times, but not to be used for long periods. It can cause malabsorption of vitamin B, resulting in fatigue and listlessness, when used for long periods. Contraindicated for hypoglycemics because it lowers blood sugar levels. Hypoglycemics should use Oregon grape, myrrh gum or other herbs instead of golden seal. Not for emaciated persons, weak digestion or high blood pressure. Goldenseal is reported to have an oxytocic effect, and many herbalists recommend avoiding it during pregnancy. We have not found it to be a problem, at least for temporary use in lower doses, but large doses could be a problem during pregnancy.

Golden Seal - Capsule (100)

Available in: US - Stock #340-7

Ingredients: Hydrastis canadensis (goldenseal) root

Golden Seal/Parthenium

Product Type: Formula

Properties: Antiseptic, Astringent, Diuretic

Systems Affected: Immune System, Mucus Membranes, Urinary System

Conditions: Bladder Infection, **Urinary Tract Infections**

Usage: This product helps the body fight bacterial infection and has some ability to reduce inflammation on mucus membranes. Parthenium does not have the wide immune-stimulating action of Echinacea, but it does have some affinity for the urinary tract, making this a potentially useful product for urinary tract infections. For other applications use Echinacea/Golden Seal, which is a superior formula for most infection fighting purposes. Use 1/2 to 1 teaspoon two to four times per day.

Warnings: See warnings for goldenseal. No known warnings.

Golden Seal/Parthenium Extract (2 fl. oz.) - Liquid (Tincture)

Available in: US - Stock #1781-5

Ingredients: Goldenseal, parthenium

Gotu Kola

Latin Name: *Centella asiatica syn. Hyrocotyle asiatica*

Product Type: Single Herb

Properties: Adaptagen, Alterative (Blood Purifier), Anti-aging, Anti-epileptic, Anti-inflammatory, Antibacterial, Aphrodisiac, Cephalic, Cerebral Tonic, Hypotensive, Insecticide, Nervine, Vulnerary

Systems Affected: **Brain**, Central Nerves, Circulation, Connective Tissue, Eyes, Heart, Nerves, Pineal, Pituitary (posterior), Skin

Conditions: Aging (prevention), Anxiety (Panic Attack), Arthritis, Blisters, Burns and Scalds, Circulation (to the brain), Concentration (poor), Dementia, Depression, Dermatitis, Eczema, Epilepsy, Fatigue, Hashimoto's Disease (Thyroiditis), Inflammation, Jaundice (adults), Leprosy, Memory and Brain Function, Meningitis, Myasthenia Gravis, Narcolepsy, Psoriasis, Senility, Sex Drive (low), Skin (infections), Skin Care (general), Stress, Ulcers

Usage: Neutralizes the blood acids and cools the blood, removes excess toxic fat. Gotu kola has a reputation for improving memory and brain function. It is a general tonic for the body, helping it adapt to stress. As a general tonic use 2 capsules two to three times daily for short-term use. For long term use 1 capsule two times daily. Gotu kola, or Indian pennyworth, is used extensively in Indian Ayurvedic medicine as an important healing and toning herb. The herb is still relied on in promoting the healing of ulcerations and wasting diseases, such as leprosy. Modern research has begun to, validate these traditional uses as well as showing important, adaptagenic properties.

Warnings: Overdose may cause dizziness.

Gotu Kola - Capsule (100)

Available in: US - Stock #360-0
Ingredients: Centella asiatica (gotu kola) herb

Gotu Kola - Capsule (100)

Available in: Canada - Stock #8005-3
Ingredients: Same as Stock #360-0

Grapefruit (Pink)

Product Type: Essential Oil

Properties: Alterative (Blood Purifier), Analgesic (Anodyne), Anti-obesic, Antidepressant, Appetite Suppressant, Aromatic, Detoxifying, Diuretic, Nervine

Systems Affected: Circulation, Digestive System, Female Reproductive, Intestinal System, Liver, Lymphatic System, Nerves, Skin

Conditions: Abdominal Pain and Inflammation, Anorexia, Appetite (excessive), Arteriosclerosis (Atherosclerosis, Hardening of the Arteries), Blood Pressure (high), Cellulite, Confusion, Depression, Edema (Dropsy, Water Retention, Swelling), Herpes, **Hot Flashes**, Irritability, Mood Swings, Mood Swings, Nausea and Vomiting, Nervous Exhaustion (Enervation), Rheumatoid Arthritis (Rheumatism), Skin (oily), Stretch Marks, Tension

Usage: Pink grapefruit is beneficial for an overheated liver and a sluggish lymphatic system. It acts as a blood purifier, helping the body to eliminate toxins, which makes it useful in drug withdrawal. It's diuretic properties and ability to help break down fats make it useful for treating water retention and cellulite. It can also help to prevent and reverse arteriosclerosis and hypertension. Having a stimulating effect on the digestive system, it helps with abdominal distention, constipation and nausea. Grapefruit oil helps with rheumatic pain of a hot nature when the joints feel warm and swollen, mixed with a burning sensation. Topically it is good for oily skin, acne and stretch marks. Emotionally, pink grapefruit helps with feelings of tension, frustration, irritability and moodiness. It is an uplifting antidepressive and helps with nervous exhaustion. It helps obesity in those who, under pressure and tension, resort to comfort eating to deal with emotions, because it eases the hunger for immediate satisfaction or the desperate need to be full. It also helps those who have high expectations of life, other people, and themselves, and feel let-down when their expectations have not been met. When people react with anger, blame and self-criticism, often followed by feelings of guilt and depression, grapefruit oil helps clear the emotional congestion with its cleansing, clarifying and refreshing effects.

Warnings: Always use topically diluted in a carrier oil. Since it can be phototoxic, avoid direct exposure to sunlight after use.

Grapefruit, Pink Essential Oil (5ML) - Essential Oil

Available in: US & Canada - Stock #3906-7
Ingredients: Pink Grapefruit (Citrus paradisi) essential oil

Grapine

Product Type: Formula

Properties: Anti-inflammatory, Antioxidant

Systems Affected: Bones, Liver, Pancreas Tail

Conditions: Aging (prevention), Arthritis, Bruises (prevention), Cancer (natural therapy for), Cancer (prevention), Diabetes, Environmental Pollution (protection from), Free Radical Damage, Inflammation, Lupus, Macular Degeneration, Meningitis, Mental Illness, Pain (general remedies for), Parkinson's Disease, Surgery (healing from), Vaccines (detoxification from)

Usage: Proanthocyanidins are powerful antioxidants, 50 times more potent than vitamin E. Antioxidants help prevent cell damage believed to be responsible for cancer, hardening of the arteries and aging. Helps inflammatory diseases such as arthritis. For maintenance: use 1-2 tablets of Grapine with Protectors three times a day. For saturation: take one 60 mg High-Potency tablet for every 60 lbs of body weight daily for two weeks. For example, if you weigh 180 lbs., take 3 tablets daily. Children can chew 1-2 Herbasaurs Chewable Antioxidants with Grapine tablets daily.

Warnings: No known warnings.

Grapine with Protectors, SynerPro - Capsule (90)

Available in: US - Stock #1750-1

Ingredients: Grapine (maritime pine bark extract and grape seed extract which supply proanthocyanidins), vitamin C, broccoli, cruciferous vegetables, turmeric, rosemary, carrot, tomato and others nutrients

Grapine, High Potency - Tablet (60)

Available in: US & Canada - Stock #1699-3

Ingredients: Grapine (maritime pine bark extract and grape seed extract)

Antioxidants with Grapine, Chewable - Tablet (120)

Available in: US - Stock #3301-5

Ingredients: Grapine (Maritime pine bark extract and grape seed extract), beta-carotene, vitamin C, vitamin E, licorice root extract, sorbitol, mannitol, fructose, stearic acid (vegetable), natural lemon and lime flavor, silicon dioxide (powdered silica), citric acid

Green Tea Extract

Latin Name: *Camellia sinensis*

Product Type: Single Herb

Properties: **Antioxidant**, Astringent

Systems Affected: Liver

Conditions: Aging (prevention), Alzheimer's Disease, Anorexia, Arteriosclerosis (Atherosclerosis, Hardening of the Arteries), Autism, Bedwetting, Blood Clots (prevention of), Blood in Stool, Cancer (natural therapy for), Cholesterol (high), Convalescence, Diarrhea, Free Radical Damage, Inflammation, Pregnancy (herbs and supplements to avoid during), **Psoriasis**

Usage: Green tea decaffeinated extract is a standardized extract of green tea. Known to be 200 times stronger than vitamin E in neutralizing free radicals, green tea also helps normalize vascular blood clotting and total cholesterol levels and has antimicrobial properties as well. Take 1 capsule three times daily with a meal; 3 capsules equals 10 cups of liquid green tea.

Warnings: Not recommended for children under 6 or pregnant or nursing women.

Green Tea Extract - Capsule (60)

Available in: US & Canada - Stock #1096-6

Ingredients: Standardized extract of Camellia sinensis (green tea)

GreenZone

Product Type: Formula

Properties: Food, Low Glycemic, **Nutritive**

Systems Affected: Pituitary (anterior), **Whole Body**

Conditions: Acquired Immune Deficiency Syndrome (AIDS/HIV), Aging (prevention), Anemia, **Autoimmune Disorders**, Birth Defects (prevention), Blood Pressure (low), Body Building, Body Odor, Cancer (natural therapy for), Cold Sores (Fever Blisters), **Convalescence**, Corns, **Debility**, Depression, Digestion (poor), Disks (spinal - bulging or slipped), Failure to Thrive, **Infertility**, **Overacidity**, **Post Partum Weakness**, Psoriasis, Triglycerides (high)

Usage: GreenZone is a convenient whole food supplement that can be used to improve general nutrition or as a meal replacement. Mix 2 scoops (21g) in 8 to 12 ounces of water or juice once daily. (You may want to start with 1/2 scoop and work up to 2 scoops, as this product can be very detoxifying.)

Warnings: No known warnings.

GreenZone, pH, Powder (131g) - Bulk Powder

Available in: US - Stock #1091-5

Ingredients: Spirulina (Spirulina platensis), amaranth (Amaranthus spp.), natural lemon juice, lecithin, chlorella (Chlorella regularis), sprouted kamut whole leaf 5:1 extract

(Triticum durhum egyptum), alfalfa juice concentrate (Medicago sativa), sprouted barley grass herb (Hordeum vulgare), apple pectin, lemongrass (Cymbopogon citratus), bee pollen, acerola cherry fruit extract (Malpighia glabra), brown rice, spinach (Spinacia oleracea), ginger rhizome (Zingiber officinale), astragalus root (Astragalus membranaceus), Echinacea purpurea root, milk thistle seed (Silybum marianum), asparagus stem (Asparagus officinalis), broccoli flowers (Brassica oleracea botrytis), kale whole leaf (Brassica oleracea acephala), flaxseed (Linum usitatissimum), beet extract (Beta vulgaris), stevia leaf (Stevia rebaudiana), orange bioflavonoid, royal jelly, bladderwrack plant (Fucus vesiculosus), elderberry 5:1 extract (Sambucus nigra), hawthorn berries (Crataegus oxycantha and monogyna), red grape skin extract (Vitis vinifera), eleuthero root (Eleutherococcus senticosus), Ginkgo biloba leaf concentrate 24%, licorice root (Glycyrrhiza glabra), polyphenol catechins (from green tea extract), rhodenol root (Rhododendron caucasicum), sodium copper chlorophyllin.

GreenZone, Ultimate (483g) - Bulk Powder

Available in: US - Stock #1103-6

Ingredients: Amaranth seed (Amaranthus cruentus), brown rice (Oryza sativa), flaxseed (Linum usitatissimum), spirulina algae (Spirulina platensis), fructooligosaccharides, quinoa (Chenopodium quinoa), flaxseed hull lignans (Linum usitatissimum), chia seed (Salvia hispanica), chlorella algae (Chlorella pyrenoidosa), millet seed (Panicum millaceum), alfalfa juice concentrate (Medicago sativa), licorice root (Glycyrrhiza glabra), lecithin-from soy (Glycine max), carrot root (Daucus carota), lemon grass aerial parts (Cymbopogon citratus), papaya fruit extract (Carica papaya), artichoke leaves (Cynara scolymus), spinach leaves and stems (Spinacia oleracea), broccoli flowers (Brassica oleracea), kale leaves (Brassica oleracea acephala), asparagus stems (Asparagus officinalis), red beet root extract (Beta vulgaris), bromelainpineapple extract (Ananas comosus), chicory root (Cichorium intybus), acerola fruit extract (Malpighia punicifolia), horsetail stems and strobilus (Equisetum arvense), lemon bioflavonoid extract (Citrus limon), sodium copper chlorophyllin, parsley leaves (Petroselinum crispum), and pau d'arco bark concentrate (Tabebuia avellanedae)

GreenZone, Ultimate, Capsules - Capsule (180)

Available in: US - Stock #1104-4

Ingredients: Same as Stock #1103-6

GreenZone Powder - Bulk Powder

Available in: Canada - Stock #1095-4

Ingredients: Spirulina (blue-green algae), amaranth, lemon juice powder, lecithin, chlorella, wheat grass, alfalfa herb, apple pectin, barley grass herb, lemon grass, acerola cherry extract, bee pollen, spinach, ginger rhizome, astragalus, echinacea root, milk thistle seed, papaya fruit, shiitake mushroom, beet extract, flaxseed, stevia leaf, orange bioflavonoid, royal jelly, rosemary leaf, hawthorn berries, elder berry extract, grape skin extract, brown rice, bladderwrack, Siberian ginseng root, sodium copper chlorophyllin, rhodenol root, Ginkgo biloba leaf concentrate, polyphenol catechins, and licorice root

GreenZone Capsules - Capsule (360)

Available in: Canada - Stock #1097-7

Ingredients: Same as Stock #1095-4

Guardian

Product Type: Formula

Properties: Antifungal, Antimicrobial, Aromatic

Systems Affected: Immune System, Lungs, Skin

Conditions: Acquired Immune Deficiency Syndrome (AIDS/HIV), Anger (excessive), Antibiotics (alternatives to), Athlete's Foot, Defensiveness, Diphtheria, Fungal Infections (Yeast Infections, Candida albicans), Infection (bacterial), Infection (viral), Inflammation, Irritability, **Skin (infections)**, Worry, Wounds and Sores

Usage: This blend can be applied topically or diffused into the air to prevent or help to fight bacterial infections.

Warnings: See warnings for individual oils.

Guardian Essential Oil Blend (5ML) - Essential Oil

Available in: US - Stock #3922-9

Ingredients: Lavender, ravensara, Roman chamomile, tea tree

Guggul

Latin Name: *Commiphora mukul*

Product Type: Single Herb

Properties: Anticholesteremic, **Anticoagulant (Blood Thinner)**, Antirheumatic, Hepatic

Systems Affected: Circulation

Conditions: Blood Pressure (high), Cholesterol (high), Circulation (poor), Cold Hands and Feet, Pregnancy (herbs and supplements to avoid during), Triglycerides (high)

Usage: Research has shown guggul to be one of the most powerful cholesterol-lowering herbs known. This herb lowers unhealthy low density lipo-proteins (LDLs) and raises the healthy high density lipoproteins (HDLs). Guggul also lowers triglycerides, decreases the stickiness of blood platelets, stimulates weight loss and improves the thyroid gland's ability to absorb iodine. Use 1-2 tablets of guggul lipids three times daily.

Warnings: Because it thins the blood, it should not be used in persons who bleed easily or during pregnancy.

Guggul Lipid (Concentrate) - Tablet (120)

Available in: US - Stock #904-6

Ingredients: Commiphora mukul (guggul) extract

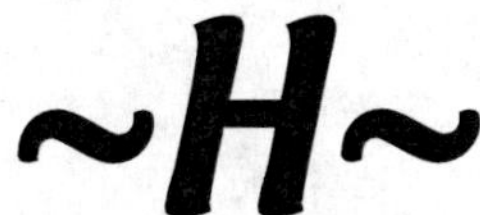

Hawthorn Berries

Latin Name: *Crataegus oxycanthia*

Product Type: Single Herb

Properties: **Anti-arrhythmic**, Antihypertensive, Antilipemic, Antioxidant, **Cardiac**, Food, Hypertensive, Hypotensive, Sympatholytic, Vasodilator

Systems Affected: Adrenal Glands, Arteries, Capillaries, Cardiovascular System, Circulation, Cyclic AMP (cAMP), **Heart**, Peripheral Blood Vessels, Thyroid

Conditions: Angina, **Arrhythmia**, Arteriosclerosis (Atherosclerosis, Hardening of the Arteries), Blood Pressure (high), Bruises (prevention), **Cardiac Arrest (Heart Attack)**, Cardiovascular Disease (Heart Disease), **Congestive Heart Failure**, Diabetes, Edema (Dropsy, Water Retention, Swelling), Grave's Disease, **Heart (weakness)**, **Heart Fibrillation or Palpitations**, Heart Valves, Memory and Brain Function, Rheumatic Fever, Shock, **Tachycardia**

Usage: Studies around the world have confirmed that hawthorn berries improve the tone of the cardiac muscle, improve oxygen uptake by the heart, improve circulation in the heart, energize the heart cells and dilate blood vessels in the extremities to reduce strain on the heart. Thus, hawthorn berries are an excellent herbal food for building up the heart muscle. However, these berries need to be taken on a regular basis for best results. Take 2-4 capsules three times daily with water at mealtime. When hawthorn berries are used for high blood pressure, other herbs (such as garlic, valerian, chamomile and capsicum) should be used as well.

Warnings: No known warnings. Completely safe for long-term use. However, hawthorn potentiates digitalis (a heart drug) which can lead to dangerous side effects if the dose of digitalis is not carefully adjusted. If individuals are using digitalis, they will need to consult with their physician to reduce the dose if they start using hawthorn.

Hawthorn Berries - Capsule (100)

Available in: US & Canada - Stock #370-3

Ingredients: Crataegus oxycanthia (hawthorn) berries

Hawthorn Berries Extract (2 fl. oz.) - Liquid (Tincture)

Available in: US - Stock #1760-0

Ingredients: Crataegus oxycanthia (hawthorn) berries, alcohol, water

HCP-X

Product Type: Formula

Properties: Astringent, Decongestant, Diaphoretic, **Expectorant**, Stimulant

Systems Affected: Digestive System, Immune System, Lungs, **Mucus Membranes**, Respiratory System, Small Intestines, Sweat Glands, Thymus

Conditions: **Colds (decongestant)**, **Colds (general remedies for)**, Congestion (general), **Cough (damp)**, Cough (general), Fever, **Flu**, Perspiration (deficient), **Polyps**, Sinus Infection, Sneezing, Sweat Baths (herbs for)

Usage: This is a traditional formula that dates back to the pioneer herbalist Samuel Thomson. He used it to "scour the bowels and remove the canker [mucus and/or toxins]." It was originally intended to be made as a tea, and sipped. Useful for inducing perspiration as part of a sweat bath to quickly rid the body of colds, flu and fevers. It can also be taken as capsules with warm water. For tea, use 2-4 capsules per cup of water. Use the tea as a gargle for sore throats. When using as capsules, take 1-2 capsules every two to four hours with plenty of fluids until symptoms subside.

Warnings: This is a spicy combination, so use caution if spicy foods irritate your stomach.

HCP-X - Capsule (100)

Available in: US - Stock #1216-5

Ingredients: Bayberry, ginger, white pine, cloves, capsicum

Healing AC Cream

Product Type: Formula

Properties: **Anti-inflammatory, Vulnerary**

Systems Affected: Bones, Skin, Structural System

Conditions: **Abrasions**, **Bruises (healing)**, Bruises (prevention), Burns and Scalds, Carpal Tunnel Syndrome, Children's Remedy, Concussions, Cradle Cap, **Dislocation**, **Inflammation**, **Injuries**, **Ligaments (torn or injuried)**, **Scratches and Abrasions**, **Sprains**, **Tendonitis**, **Varicose Veins**, **Wounds and Sores**

Usage: Healing AC Cream combines two homeopathic remedies used to treat injuries-arnica and calendula. The cream can be applied topically to bruises, bumps, sprains, pulls, swellings, rashes, etc., as long as the skin is not punctured or the wound open. The product reduces swelling and prevents bruising when applied immediately after an injury has occurred. Apply generously and massage into affected areas.

Warnings: Do not apply to open wounds, cuts or puncture wounds. Avoid contact with the eyes.

Healing AC Cream (2 oz.) - Topical

Available in: US - Stock #8723-1

Ingredients: Arnica montana 1x, Calendula officinalis 1x, medium chain triglycerides, glycerine, isopropyl myristate, hydrogenated vegetable oils, steryl ethers, cetyl alcohol, carbomer, mixed parabens

Heavy Metal Detox

Product Type: Formula

Properties: Antioxidant, **Chelating**, Detoxifying

Systems Affected: Digestive System, Immune System, Intestinal System, Liver, Nerves

Conditions: **Alzheimer's Disease**, Attention Deficit Disorder (ADD, ADHD), Autism, Autism, Autoimmune Disorders, Body Odor, Bronchitis, Cataracts, Chemical Poisoning, Cholesterol (low), Copper Toxicity, Disks (spinal - bulging or slipped), Epstein Barr Virus (Chronic Fatigue Syndrome, CFS), Floaters, Grave's Disease, **Heavy Metal Poisoning, Lead Poisoning**, Memory and Brain Function, **Mental Illness, Mercury Poisoning, Multiple Sclerosis** (**MS**), PMS Type D, Pregnancy (herbs and supplements to avoid during), Thinking (cloudy), **Vaccines (detoxification from)**

Usage: This formula aids the body in detoxifying and expelling mercury and other heavy metals. It helps the liver bind and eliminate these toxins. It is a good product to take after removal of amalgam fillings. Take 1 capsule with a meal twice daily. Combines well with algin, which bonds to heavy metals to remove them from the system.

Warnings: Do not exceed the recommended dosage as this can create too rapid of a detoxification process. If headache, nausea or diarrhea develop, reduce dosage. You may need to take extra fiber (algin is recommended) in order to bind heavy metals in the gut. Baths with clay and Epsom salts may also aid the detoxification process.

Heavy Metal Detox - Capsule (90)

Available in: US & Canada - Stock #507-1

Ingredients: Cilantro, N-Acetyl-Cysteine, Vitamin B6, apple pectin, sodium alginate, kelp algae, l-methionine, alpha lipoic acid, magnesium citrate.

Helichrysum

Product Type: Essential Oil

Properties: Anti-inflammatory, Antibacterial, Anticoagulant (Blood Thinner), Antiseptic, Antispasmodic, Antiviral, Aromatic, **Cicatrisant**, Hepatic, **Vulnerary**

Systems Affected: Breasts, Circulation, Digestive System, Ears, Liver, Lymphatic System, Respiratory System, Skin, Structural System

Conditions: Addictions (alcohol), Allergies (respiratory), Anger (excessive), Arthritis, Asthma, Bronchitis, **Bruises (healing)**, Cholesterol (low), **Cirrhosis of the Liver**, Colds (general remedies for), Confusion, Cuts, Cysts, Deafness, Depression, Digestion (poor), Diphtheria, Disks (spinal - bulging or slipped), Ear Infection or Earache, Grief (excessive), Headache (general), **Hepatitis**, Inflammation, **Injuries**, Migraine, Mood Swings, **Nerve Damage**, Pertussis (Whooping Cough), Rashes and Hives, **Scars / Scar Tissue**, Seizures, Skin (infections), Skin Care (general), Sprains, Stretch Marks, **Wounds and Sores**

Usage: Helichrysum, also known as immortelle or everlasting, has been a preferred remedy for chronic ailments of the skin and lymphatic system. It helps with respiratory complaints such as asthma, chronic bronchitis and whooping cough and is helpful in allergic conditions that involve nasal catarrh, sneezing and itchy skin rashes. It is also effective for headaches, migraines and liver disorders. For earaches, it can be diluted in a carrier oil and dropped into the ear to reduce inflammation and fight infection. It has anticoagulant actions making it useful for any severe bruising that results in clotted blood. Topically it speeds cellular growth, eases eczema and dermatitis and assists in wound healing of all types. It reduces inflammation, helps heal old scar tissue and stretch marks, and loosens adhesions after injuries. Emotionally, helichrysum alleviates the tension that arises from issues involving excessive effort and over-control. Easing long-standing frustrations and helping a person break through negative "stuck" emotions, it is highly effective on stubborn, negative attitudes. These may be blockages in expressing anger and despair or even in admitting the depth of one's emotional wounds. Though secretly despairing, they cannot bear to see others admit vulnerability and may feel rage when they do so. Helichrysum can loosen the hardest of attitudes, restoring compassion for self and others. It also increases dream activity and awareness to stimulate the right side of the brain.

Warnings: Recommended for topical use only.

Helichrysum Essential Oil (5ML) - Essential Oil

Available in: US - Stock #3943-0

Ingredients: Helichrysum (Helichrysum italicum) Essential Oil

Herbal Beverage

Product Type: Formula

Properties: Hepatic, Nutritive

Systems Affected: Digestive System, Liver

Conditions: Addictions (coffee, caffeine), Adrenals (exhaustion, weakness or burnout), Anxiety (Panic Attack)

Usage: Herbal Beverage is a coffee substitute, made naturally with healthy roasted grains and herbal flavorings. It has a mild liver-cleansing effect. For a hot beverage add 1 teaspoon of Herbal Beverage to a cup of hot water. Add milk and honey (or another natural sweetener) as desired. It can also be made into a cold drink by preparing a stronger drink and pouring it over crushed ice.

Warnings: No known warnings.

Herbal Beverage (3.5 oz.) - Bulk Powder

Available in: US - Stock #1600-1

Ingredients: Roasted barley, malt, chicory, rye, and red beet root

Herbal CA

Product Type: Formula

Properties: Anti-arthritic, Demulcent (Mucilant), **Mineralizer**, Soothing, **Vulnerary**

Systems Affected: Bones, Hypothalamus, Nerves, **Parathyroid, Skeletal System, Structural System**

Conditions: Arthritis, Backache (Back Pain, Lumbago), Bedwetting, Breast Milk (increase or enrich), **Broken Bones**, Bursitis, **Calcium Deficiency**, Cartilage Damage, Cramps and Spasms (general), **Dislocation**, Eczema, Fingernail Biting, Fingernails (weak or brittle), Gout, Hernias, **Ligaments (torn or injuried)**, Menopause, Multiple Sclerosis (MS), Sciatica, Shingles, **Surgery (healing from)**, Teeth (grinding), Teeth (loose), Teething, Tendonitis, Tooth Decay (prevention)

Usage: This combination consists primarily of vulnerary (tissue-healing) mucilants. It does not contain a large amount of calcium but does help bones and injured tissues to heal. It contains many trace elements that would help the body utilize calcium and other major minerals better. It is commonly used for problems like arthritis, back ache, broken bones, damaged cartilage and teeth grinding. It would also help enrich breast milk and make an excellent tonic for pregnant women in combination with red raspberry. For injuries take 2-4 capsules 3-6x daily with a large glass of water. This formula can also be used externally as a poultice for swelling and injuries.

Warnings: No known warnings.

Herbal CA - Capsule (100)

Available in: US - Stock #823-5

Ingredients: Alfalfa, horsetail, oatstraw, plantain, marshmallow, wheat grass, hops

Herbal CA - Capsule (100)

Available in: Canada - Stock #819-2

Ingredients: Alfalfa herb, marshmallow root, plantain herb, Irish moss plant, oatstraw stem, wheatgrass herb, and hops flowers

CA, ATC Concentrate - Capsule (50)

Available in: US - Stock #826-3

Ingredients: Alfalfa, marshmallow, slippery elm, oatstraw, Irish moss, passion flower.

CA Concentrate, ATC - Capsule (50)

Available in: Canada - Stock #827-4

Ingredients: Same as Stock #826-3

Herbal Pumpkin

Product Type: Formula

Properties: Antiparasitic, Bitter, Laxative (general), **Parasiticide, Vermifuge**

Systems Affected: Intestinal System, **Prostate, Testes**

Conditions: Benign Prostate Hyperplasia (BPH), Fingernail Biting, Inflammatory Bowel Disorders (Colitis, IBS), **Parasites (general)**, Parasites (nematodes, worms), Parasites (tapeworm), Pets (supplements for), Prostatitis

Usage: Used to aid in the expulsion of worms and other parasites in the body. Take 2 capsules one to three times per day. Works best when used with garlic, black walnut and a stimulant laxative formula. A sample program would be: 2 herbal pumpkin, 1 black walnut, 2 garlic and 2 LBS-II every morning and evening. Should be taken for two to three weeks. Then discontinue for a week or two and start again for an additional two to three weeks. Drink plenty of water when on a cleanse of any kind. Pumpkin seeds are a rich source of zinc and are helpful for the prostate gland in men. Many men have found this formula helpful in reducing prostate enlargement.

Warnings: Does not work on all parasites.

Herbal Pumpkin - Capsule (100)

Available in: US - Stock #915-2

Ingredients: Pumpkin seeds, black walnut, cascara sagrada, violet leaves, chamomile flowers, mullein leaves, marshmallow, slippery elm bark

Herbal Pumpkin - Capsule (100)

Available in: Canada - Stock #914-5

Ingredients: Pumpkin seeds, black walnut, Turkey rhubarb, violet leaves, chamomile flowers, mullein leaves, marshmallow, slippery elm bark

Herbal Punch

Product Type: Formula

Properties: Nutritive

Systems Affected: Whole Body

Conditions: Children's Remedy, Energy (lack of)

Usage: Herbal Punch is a natural alternative to sugar-laden beverages. For a healthy drink add 1 ounce of Herbal Punch concentrate to 7 ounces of water and serve chilled. Add some liquid chlorophyll and/or vitamin C powder for a quick energy pick-me-up.

Warnings: This product contains honey and there are some potential problems with giving honey to infants, so this drink should probably be given to children over the age of one.

Herbal Punch (16 fl. oz.) - Liquid

Available in: US - Stock #1610-0

Ingredients: Natural honey, natural herbal extracts, gum acacia, ascorbic acid with rose hips, niacinamide, vitamin A (fish liver oil), pyridoxine, riboflavin, thiamine, vitamin D (fish liver oil), cyanocobalamin.

Herbal Sleep

Other Names: HVP

Product Type: Formula

Properties: Analgesic (Anodyne), Antispasmodic, Aromatic, **Hypnotic**, Hypotensive, Nervine, **Sedative**, **Sympatholytic**, Tranquilizer

Systems Affected: Brain, Digestive System, Nerves

Conditions: Addictions (alcohol), Addictions (drugs), Addictions (tobacco smoking or chewing), Anxiety (Panic Attack), Attention Deficit Disorder (ADD, ADHD), Bell's Palsy, Blood Pressure (high), Headache (tension), **Insomnia**, Nocturnal Emission (Wet Dreams), Pain (general remedies for), Parkinson's Disease, Restless Leg Syndrome, Stress

Usage: This formula was created primarily as a sleep aid. Originally, it contained skullcap instead of passion flower. The new version is less effective, but can be improved by adding kava kava. Take 2-4 capsules of Herbal Sleep, or 1 capsule of Herbal Sleep concentrate with 1 kava kava capsule before bedtime to aid in falling asleep. Herbal Sleep can also be used to calm restlessness and for nervous stress. It is helpful for stress, high blood pressure, anxiety, and as an aid to help quit smoking. Take 2 capsules three times daily for stress.

Warnings: See cautions for valerian.

Herbal Sleep - Capsule (100)

Available in: US - Stock #940-8

Ingredients: Valerian root, passion flower, hops

HVP - Capsule (100)

Available in: US & Canada - Stock #940-8

Ingredients: Same as Stock #940-8

Herbal Trace Minerals

Other Names: Three

Product Type: Formula

Properties: Alterative (Blood Purifier), Bitter, Diuretic, Mineralizer

Systems Affected: Blood, Bones, Liver, Skeletal System, Structural System, Teeth, Thyroid, Weight Loss

Conditions: Acne (Pimples, Blackheads), Anemia, Cholesterol (high), Dermatitis, Digestion (poor), Dyspepsia, Eczema, Endurance (lack of), Fingernails (weak or brittle), Hair Care (general), **Pets (supplements for)**, Pregnancy (herbs and supplements for)

Usage: This combination is designed to supply various trace elements to the body in herbal form. It is an excellent supplement for animals. Mix it with their food. It is also a general tonic for children and adults. Take 1-2 capsules three times daily.

Warnings: No known warnings. Safe for long-term use.

Herbal Trace Minerals - Capsule (100)

Available in: US - Stock #980-5

Ingredients: Kelp, dandelion, alfalfa

Three - Capsule (100)

Available in: Canada - Stock #980-5

Ingredients: Same as Stock #980-5

Herbal Trim Skin Treatment

Product Type: Formula

Properties: **Balm**, **Demulcent (Mucilant)**, Emollient, **Vulnerary**

Systems Affected: Cuticle, Ears, **Skin**, Structural System

Conditions: **Burns and Scalds**, Children's Remedy, Corns, **Cradle Cap**, Cystic Breast Disease, Dandruff, Inflammation, Itching, **Itching Ears**, **Jock Itch**, **Rashes and Hives**, Rosacea, Scratches and Abrasions, **Skin Care (general)**, Stretch Marks, **Sunburn**, Weight Loss (aids for), **Wounds and Sores**, **Wrinkles**

Usage: Apply externally for rashes, skin irritations, bites, stings, sunburn, scrapes, bruises and other injuries. It speeds healing. Massage into skin after bath to condition skin and to aid in toning skin during weight loss.

Warnings: Avoid contact with eyes and genitals.

Herbal Trim Skin Treatment (8 fl. oz.) - Topical

Available in: US - Stock #1669-0

Ingredients: Water, aloe vera, pau d'arco, lobelia, collagen, elastin, citric acid, xanthan gum, Tei Fu essential oil blend (eucalyptus oil, wintergreen oil, menthol, camphor, clove oil), allantoin, lavendar essential oil, carrageenan, safflower seed oil

Hi Lipase

Product Type: Formula

Properties: Digestant, **Emulsifier**

Systems Affected: Digestive System, Hypothalamus, Liver, Small Intestines, Spleen, Weight Loss

Conditions: Acne (Pimples, Blackheads), Eczema, **Fat Cravings, Fat Metabolism (poor), Fatty Liver Disease**, Fatty Tumors or Deposits, **Gall Bladder (sluggish)**, Gall Stones, Psoriasis, **Seborrhea, Skin (dry and/or flaky)**, Triglycerides (low)

Usage: This formula helps break down fats in the body. It is useful for anyone who is having problems with poor fat metabolism such as indigestion after eating fats, dry skin or gall bladder problems. It is very helpful after a person has had their gallbladder surgically removed to help the body break down fats more effectively. Take 1-2 capsules before consuming foods high in fat.

Warnings: No known warnings.

Hi Lipase (120 LU) - Capsule (100)

Available in: US - Stock #1528-4

Ingredients: Lipase, potassium citrate, caraway, ginger, genitian, fennel, dandelion, red beet

HistaBlock

Product Type: Formula

Properties: Anti-allergenic, Antihistamine, Antioxidant, Bronchial Dilator, Decongestant, **Mast Cell Stabilizer**

Systems Affected: Bronchials, Digestive System, Eustachian Tubes, Immune System, Liver, Lungs, **Respiratory System**, Skin

Conditions: Allergies (food), Allergies (respiratory), Attention Deficit Disorder (ADD, ADHD), Bites and Stings, Bronchitis, Colds (decongestant), Colds (general remedies for), Congestion (bronchial), **Dermatitis**, Disks (spinal - bulging or slipped), **Eczema**, Eye Problems (general), **Eyes (red or itching), Itching**, Migraine, **Rashes and Hives**, Snoring

Usage: This formula blocks histamine reactions in the body. It helps to stabilize mast cells, which reduces allergic reactions. It is not only helpful for hayfever, it can also help with food allergies and leaky gut syndrome. It can also help reduce allergic reactions in the skin in eczema and dermatitis. Take 2 capsules with a meal twice daily.

Warnings: Nursing mothers should not take this formula as it could decrease milk production.

HistaBlock - Capsule (90)

Available in: US & Canada - Stock #776-1

Ingredients: Stinging nettles, bioflavonoid, fructus aurantia immaturi, bromelain, di-calcium phosphate.

Ho Shou Wu

Latin Name: *Polygonum multiflorum*

Product Type: Single Herb

Properties: Anti-aging, **Anticholesteremic**, Blood Building, Kidney Tonic, Tonic

Systems Affected: Hair, Liver, Reproductive Glands

Conditions: Abscesses, **Aging (prevention)**, Boils, Bunions, Carbuncles, Cataracts, **Cholesterol (high)**, Dizziness (Vertigo), Glands (swollen lymph), **Gray Hair**, Hair (loss or thinning), **Hashimoto's Disease (Thyroiditis)**, Hypoglycemia, **Hypothyroid**, Infertility, Insomnia, Malaria, Nocturnal Emission (Wet Dreams), Numbness, Tuberculosis (Consumption, Scrofula), Wounds and Sores

Usage: Used in Chinese medicine as a liver and kidney tonic, for yin and blood deficiency. Has been reported to restore normal color to dark hair. This herb is an excellent general tonic for slowing the aging process. Take 2 capsules with a meal two times daily.

Warnings: No known warnings.

Ho Shou Wu - Capsule (100)

Available in: US - Stock #375-5

Ingredients: Polygonum multiflorum (ho shu wu) root

Hops

Latin Name: *Humulus lupulus*

Product Type: Single Herb

Properties: Analgesic (Anodyne), **Anaphrodisiac**, Antacid, Anti-epileptic, Antispasmodic, Aromatic, Bitter, Carminative, Cholagogue, **CNS Depressant**, Diaphoretic, Diuretic, **Estrogenic**, Febrifuge, Hypnotic, **Narcotic, Nervine, Phytoestrogen, Sedative, Sympatholytic**, Tonic, Tranquilizer, Vermifuge

Systems Affected: Central Nerves, Estrogen, GABA, Liver, Muscles, Nerves, Pituitary (posterior), Stomach

Conditions: Addictions (alcohol), Anxiety (Panic Attack), Colic (adults), Colon (spastic), Cramps (leg), Cramps and Spasms (general), Epilepsy, Estrogen (low), Grave's Disease, Headache (tension), Insomnia, Pain (general remedies for), Parkinson's Disease, **Sex Drive (excessive)**, Shock

Usage: This herb has a cooling and slightly drying effect with strong relaxing and sedative qualities. It acts primarily on the nervous system and glands. Effective for relief from alcoholism, anxiety, cramps, fever, insomnia, leg cramps, nervous disorders, pain, acute pain, restlessness, sex drive decrease, and shock. It also affects liver, kidney, male and female organs, muscles, and stomach functions Use: This herb is a powerful nervine and sleep aid, and can be combined with other carminatives for settling a nervous, acidic stomach. Take 2 capsules 2x daily as a general nervine. Take 1-2 capsules before bedtime as an aid to sleep. Hops has a stronger estrogenic effect than black cohosh and can be used to calm-down an overactive sex drive in men. It is also reported to help increase sex drive in women.

Warnings: Hops contains a central nervous system depressant. It is also possible that hops can be mildly addictive. Research done by beer companies showed that increasing the amount of hops in beer increased consumption. Although it is depressant and estrogenic actions are mild, some herbalists feel that the herb should not be given to small children.

Hops - Capsule (100)

Available in: US - Stock #380-9

Ingredients: Humulus lupulus (Hops) flowers

Horsetail

Latin Name: *Equisetum arvense*

Product Type: Single Herb

Properties: Astringent, Diuretic, Hemostatic, Kidney Tonic, Mineralizer, Parasiticide, Parasympatholytic (Anticholinergic), Styptic, Vermifuge, Vulnerary

Systems Affected: Bladder (Urinary), Central Nerves, Connective Tissue, Cuticle, Eyes, Gall Bladder, Hair, Intestinal System, **Kidneys**, Lungs, **Nails**, Nerves, Parathyroid, Parotids, Pineal, Skeletal System, Skin, Structural System, Urinary System, Uterus

Conditions: Anal Fistula or Fissure, Arthritis, Bell's Palsy, Bleeding (external), **Bleeding (internal)**, **Blood in Urine**, Broken Bones, Cystitis, Edema (Dropsy, Water Retention, Swelling), **Emphysema**, Fingernail Biting, **Fingernails (weak or brittle)**, Hair Care (general), Hemorrhoids, Inflammatory Bowel Disorders (Colitis, IBS), Nephritis, Nose Bleeds, Osteoporosis, Pancreatitis, Parasites (general), Parasites (nematodes, worms), Parkinson's Disease, Stye, Urethritis, Wheezing, Wounds and Sores

Usage: Horsetail is used primarily as a nutritive to supply silica, an element essential to healthy hair, skin, nails, bones and nerves. Take 2 capsules with three meals. This herb has also been used externally as a poultice for hemorrhoids and anal fistula. It is also a kidney remedy and helps to expel parasites from the colon. It is helpful for bleeding in the lungs or urinary tract. The herb may also help to repair damaged tissues in arthritis, osteoporosis and emphysema. The silica content is also reported to help expel parasites, especially worms.

Warnings: While this herb is traditionally and safely used to treat kidney-related ailments, its long-term use may prove damaging to kidney function. Very excessive doses have led to poisoning in grazing animals, though not in humans. Care should also be taken to avoid any exaggerated usage that could result in intestinal irritations due to rough silica scraping. Make sure to achieve a high B vitamin intake to accompany a course of horsetail, since the herb encourages B vitamin breakdown.

Horsetail - Capsule (100)

Available in: US & Canada - Stock #390-2

Ingredients: Equisetum arvense (horsetail) herb

HRP-C

See *VS-C*

HS II

Product Type: Formula

Properties: Antiseptic, **Cardiac**, Hypertensive, Hypotensive, Stimulant, Stimulant (Circulatory), Stomachic

Systems Affected: Cardiovascular System, **Circulation**, Heart

Conditions: Angina, Anorexia, Arrhythmia, Blood Pressure (high), Blood Pressure (low), Cardiac Arrest (Heart Attack), **Cardiovascular Disease (Heart Disease)**, Cholesterol (high), Circulation (poor), Colds (decongestant), Fatigue, Flu, **Heart (weakness)**, Thrombosis

Usage: This formula combines cardiac tonic hawthorn berries, with two other key herbs for circulation, garlic and capsicum. It is an excellent general aid for strengthening the heart and the entire circulatory system. It may help angina, arrhythmia, high blood pressure, high cholesterol, poor circulation, dropsy, thrombosis and cardiac weakness. Use 1-2 capsules three times daily.

Warnings: Capsicum and garlic may cause some digestive upset in some people. No other warnings. Completely safe for long-term use.

HS II - Capsule (100)

Available in: US - Stock #930-9

Ingredients: Hawthorn, capsicum, garlic

HS-C

See *Nervous Fatigue Formula*

HSN Complex

Product Type: Formula

Properties: Mineralizer, Vulnerary

Systems Affected: **Hair**, **Nails**, **Skin**, Structural System

Conditions: Arthritis, Broken Bones, Dandruff, Eczema, Fingernail Biting, Fingernails (weak or brittle), Hair (loss or thinning), Hair Care (general), Pregnancy (herbs and supplements for), Rosacea, Skin Care (general), Wrinkles

Usage: A formula to strengthen hair, skin and nails. Take 2 or 3 capsules with a meal twice daily.

Warnings: No known warnings.

HSN Complex, Natria - Capsule (90)

Available in: US - Stock #3505-8

Ingredients: Horsetail, hyssop, rosemary, Ginkgo biloba concentrate, rhododendron, freeze-dried aloe vera

HSN-W

Product Type: Formula

Properties: **Anti-arthritic**, Glandular, **Mineralizer**, Vulnerary

Systems Affected: Acetylcholine, **Bones, Central Nerves, Connective Tissue**, Cuticle, Gall Bladder, **Hair**, **Legs**, **Nails**, Parathyroid, Parotids, **Pineal**, **Skeletal System**, **Skin**, Stomach, **Structural System**, Sweat Glands

Conditions: Anxiety (Panic Attack), **Arthritis**, Autoimmune Disorders, Bell's Palsy, **Bleeding (internal)**, **Blood in Urine**, Body Odor, **Broken Bones**, Burning Feet or Hands, Burning Feet or Hands, **Calcium Deficiency**, **Calcium Deposits (Calcification)**, Canker Sores (Mouth Ulcers), Capillary Weakness, Copper Toxicity, Cramps (leg), Cystitis, Dandruff, **Dislocation**, Dizziness (Vertigo), Eczema, **Fingernail Biting**, **Fingernails (weak or brittle)**, **Hair (loss or thinning)**, **Hair Care (general)**, Hemochromatosis, Injuries, **Memory and Brain Function**, Menopause, **Multiple Sclerosis (MS)**, **Nerve Damage**, **Osteoporosis**, Parkinson's Disease, Pregnancy (herbs and supplements for), Rosacea, Scoliosis, **Skin Care (general)**, Surgery (healing from), **Surgery (preparation for)**, **Teeth (grinding)**, Vitiligo, Wrinkles

Usage: This formula contains two of the richest herbs in silica, dulse and horsetail. It is designed to supply nutrients for healthy hair, skin and fingernails. The silica in this formula helps prevent split ends in hair, keeps nails strong and flexible and promotes a healthy luster to the skin. It also helps make bones and joints strong and flexible. Silica is also a critical element in helping the nerves stand up to stress. The dulse aids the thyroid, which indirectly helps the skin. Use 1-2 capsules two to three times per day.

Warnings: No known warnings.

HSN-W - Capsule (100)

Available in: US - Stock #945-0

Ingredients: Dulse, horsetail, sage, rosemary

HSN-W - Capsule (100)

Available in: Canada - Stock #944-7

Ingredients: Dulse, ginger, sage, rosemary

HTP Power

See *5-HTP*

HVP

See *Herbal Sleep*

HY-A

Product Type: Formula

Properties: Adaptagen, Adrenal Tonic, Glandular, Stomachic

Systems Affected: Adrenal Glands, Aldosterone, Liver, Pancreas, Pancreas Tail, Stomach

Conditions: Addictions (alcohol), **Anorexia**, Bedwetting, Confusion, Dizziness (Vertigo), **Fainting**, Fatigue, **Hypoglycemia**, Mental Illness, **Mood Swings**, Phobias, Schizophrenia, Sickle Cell Anemia, **Sugar Cravings**

Usage: This combination was designed to aid hypoglycemia. It contains licorice to stabilize the adrenals, dandelion for the liver plus safflowers and horseradish to improve digestion. Use 2 capsules with meals to aid in correcting blood sugar imbalances. Often, extra licorice root is needed when taking this formula.

Warnings: No known warnings.

HY-A - Capsule (100)

Available in: US - Stock #950-0

Ingredients: Licorice, dandelion, safflowers, horseradish

HY-C

Other Names: Bu Yin

Product Type: Formula

Properties: Antidiuretic, Febrifuge, **Moistening**, **Refrigerant**, Tonic

Systems Affected: Adrenal Medulla, Aldosterone, Brain, Capillaries, Ears, Eyes, Female Reproductive, Glandular System, Intestinal System, Pancreas Tail, Throat

Conditions: **Burning Feet or Hands**, Cardiovascular Disease (Heart Disease), Chills, Constipation (adults), Cough (dry), **Dehydration**, Diabetes, Eye Problems (general), Fever, Hot Flashes, **Hyperinsulinemia (Syndrome X)**, Inflammation, Memory and Brain Function, **Night Sweating**, Numbness, PMS Type C, **Skin (dry and/or flaky)**, Sore Throat, Tinnitus (Ringing in the Ears), **Urination (frequent)**

Usage: This formula is indicated in conditions of excessive dryness in the body and thirst. The symptoms are dry skin, dry mouth, dry cough, dry eyes, hot flashes, night sweats and burning sensations in the hands and feet. This is accompanied by constant thirst and frequent urination. These may be indications of diabetes, which this formula can aid. It helps the tissues retain moisture. This formula helps to overcome insulin resistance and is useful for syndrome X. Use 3 capsules three times daily.

Warnings: No known warnings.

HY-C, Chinese - Capsule (100)

Available in: US - Stock #1886-5

Ingredients: Dendrobium, eucommia, rehmannia, ophiopogon, pueraria, trichosanthes, achyranthes, alisma, amemarrhna, asparagus, hoelen, moutan, cornus, licorice, phellodendron, schizandra

HY-C TCM Concentrate - Capsule (30)

Available in: US - Stock #1006-1

Ingredients: Same as Stock #1886-5

HY-C - Capsule (100)

Available in: Canada - Stock #1885-0

Ingredients: Dendrobium, eucommia, rehmannia, ophiopogon, pueraria, trichosanthes, achyranthes, alisma, amemarrhna, asparagus, hoelen, moutan, cornus, licorice, phellodendron, schizandra, inula.

Hydrangea

Latin Name: *Hydrangea arborescens*

Product Type: Single Herb

Properties: **Anti-urolithic**, **Antilithic**, Diuretic, **Lithotriptic**, Sialogogue

Systems Affected: Bladder (Urinary), Gall Bladder, Skeletal System, Structural System, Structural System, Urinary System

Conditions: Arteriosclerosis (Atherosclerosis, Hardening of the Arteries), Arthritis, Backache (Back Pain, Lumbago), Breast Lumps, **Calcium Deposits (Calcification)**, Cystic Breast Disease, Cystitis, Edema (Dropsy, Water Retention, Swelling), **Kidney Stones**, Urethritis

Usage: An excellent stone and calcium solvent. Useful for arthritis, bone spurs, calcifications, kidney stones and gallstones. Take 2 capsules three times per day with plenty of water. Best used as part of an overall program.

Warnings: Not recommended in cases of fluid deficiency, wasting or dryness. Overdose can cause vertigo and stuffiness of the chest.

Hydrangea - Capsule (100)

Available in: US & Canada - Stock #395-6

Ingredients: Hydrangea arborescens (Hydrangea) root

Hydrated Bentonite

See *Bentonite (Hydrated)*

I-X

Product Type: Formula

Properties: Alterative (Blood Purifier), **Blood Building**, Hepatic, Mineralizer

Systems Affected: **Blood**, Liver, Lymphatic System, Red Blood Cells, Stomach

Conditions: **Anemia**, Fatigue, Menorrhagia (Heavy Menstrual Bleeding), Oxygen Deficiency, Pernicious Anemia, Post Partum Weakness, Pregnancy (herbs and supplements for), **Sickle Cell Anemia**

Usage: This formula supplies iron in plant form for helping to build up the blood in cases of anemia and general weakness. Take 2 capsules three times daily with meals. Works better when taken with extra yellow dock.

Warnings: No known warnings. Safe for pregnancy and long term use.

I-X - Capsule (100)

Available in: US & Canada - Stock #1218-4

Ingredients: Red beets, yellow dock, red raspberry, chickweed, burdock, nettle, mullein

IF Relief

Product Type: Formula

Properties: **Analgesic (Anodyne)**, Anodyne, **Anti-inflammatory**, Antioxidant, Antiphlogistic, **Febrifuge**

Systems Affected: Cardiovascular System, Eustachian Tubes, Gums, Respiratory System, Structural System

Conditions: Abdominal Pain and Inflammation, Afterbirth Pain, **Allergies (respiratory)**, Appendicitis, Arthritis, Arthritis, Asthma, Backache (Back Pain, Lumbago), Bronchitis, Bursitis, Cancer (natural therapy for), Cancer (prevention), Cardiovascular Disease (Heart Disease), Colitis, Cystitis, Dermatitis, Disks (spinal - bulging or slipped), Ear Infection or Earache, Eczema, Enteritis, Eyes (red or itching), Gastritis, Gingivitis (Bleeding Gums, Gum Disease, Pyorrhea), Hashimoto's Disease (Thyroiditis), **Headache (general)**, Hepatitis, **Inflammation**, Inflammatory Bowel Disorders (Colitis, IBS), **Injuries**, Laryngitis (Hoarseness), Leaky Gut Syndrome, Mastitis, Meningitis, Nephritis, **Neuralgia and Neuritis**, **Oral Surgery**, **Pain (general remedies for)**, Pancreatitis, Phlebitis, Prostatitis, Rhinitis, Sore Throat, **Surgery (healing from)**, **Tendonitis**, **Tooth Extraction**

Usage: IF Relief is an anti-inflammatory formula, which can help reduce inflammation and ease chronic or acute pain. It contains herbs that act as natural COX-2 inhibitors, making it a viable alternative to over-the-counter pain medications.

Warnings: No known warnings.

IF Relief - Capsule (90)

Available in: US - Stock #1175-4

Ingredients: Turmeric root extract, mangosteen pericarp extract, andrographis paniculata, boswellia gum extract, white willow bark extract

IF-C

Other Names: Qing Re

Product Type: Formula

Properties: Alterative (Blood Purifier), Analgesic (Anodyne), **Anti-inflammatory, Antiphlogistic**, Bitter, **Febrifuge**

Systems Affected: Eyes, Gums, Immune System, Intestinal System, Nerves, Structural System

Conditions: Abrasions, Acquired Immune Deficiency Syndrome (AIDS/HIV), Anal Fistula or Fissure, Arthritis, Asthma, Autoimmune Disorders, Bleeding (external), Bleeding (internal), Blood Poisoning, **Burning Feet or Hands**, Bursitis, Chicken Pox, **Chills, Colds (general remedies for), Colds (with fever)**, Colitis, Congestion (bronchial), Crohn's Disease, Cystitis, Dermatitis, Disks (spinal - bulging or slipped), Diverticulitis (Diverticuli), Dizziness (Vertigo), **Ear Infection or Earache, Endometriosis, Fever, Flu**, Gingivitis (Bleeding Gums, Gum Disease, Pyorrhea), **Grave's Disease, Hashimoto's Disease (Thyroiditis)**, Headache (general), **Inflammation**, Inflammatory Bowel Disorders (Colitis, IBS), Injuries, Nephritis, Pain (general remedies for), Rashes and Hives, Rosacea, **Sore Throat**, Tachycardia, Tinnitus (Ringing in the Ears), TMJ, Tonsillitis (Adenoids)

Usage: This Chinese formula is designed for earaches, sore throats, bleeding gums, red eyes, fever and conditions involving heat and inflammation. Use 4 capsules two times daily or take 4 capsules every two hours for severe acute conditions.

Warnings: Not for cold conditions. Thins the blood, should probably be avoided if taking blood thinners.

IF-C, Chinese - Capsule (100)

Available in: US - Stock #1875-9

Ingredients: Lonicera flowers, forsythia fruit, chrysanthemum flowers, gardenia fruit, ligusticum rhizome, peony root, platycodon root, schizonepeta, scute root, arctium seed, bupleurum root, dang gui root, philodendron bark, siler root, vitex fruit, carthamus flowers, coptis rhizome, licorice

IF-C TCM Concentrate - Capsule (30)

Available in: US - Stock #1007-2

Ingredients: Lonicera flowers, forsythia fruit, chrysanthemum flowers, gardenia fruit, ligusticum rhizome, peony root, platycodon root, schizonepeta, scute root, arctium seed, bupleurum root, dang gui root, philodendron bark, siler root, vitex fruit, carthamus flowers, coptis rhizome, licorice

IF-C - Capsule (100)

Available in: Canada - Stock #1874-3

Ingredients: Lonicera flowers, forsythia fruit, chrysanthemum flowers, cnidum fruit, gardenia fruit, inula flower, ho shou wu root, schizonepeta herb, scute root, arctium seed, bupleurum root, phellodendron bark, siler root, dang gui root, vitex fruit, carthamus flowers, coptis rhizome, licorice root

IGF-1

Product Type: Nutrient

Properties: Anti-aging, **Cell Proliferant**, Glandular, Tonic

Systems Affected: Cardiovascular System, Liver, Reproductive Glands, Structural System

Conditions: Aging (prevention), Arthritis, Athletic Performance or Exercise (aids to), Blood Pressure (high), Body Building, Broken Bones, Bursitis, Cardiovascular Disease (Heart Disease), Carpal Tunnel Syndrome, Cartilage Damage, Debility, Diabetes, Disks (spinal - bulging or slipped), Exercise, Fatigue, Fibromyalgia Syndrome (FMS), Multiple Sclerosis (MS), Pregnancy (herbs and supplements to avoid during), Sex Drive (low), Weight Loss (aids for)

Usage: IGF-1 is a hormone released from the liver and other tissues in response to growth hormone released by the pituitary. It has many functions in the body, but in general, it stimulates cellular growth and repair. This product is derived from deer antler velvet, which has been used in Chinese medicine as a sexual stimulant and anti-aging product. IGF-1 may be beneficial for athletes as it increases energy, enhances athlete performance and improves muscle strength. It improves circulatory and sexual function and enhances immunity. It is anti-inflammatory and may improve joint health. It has also been demonstrated to improve blood pressure in hypotensive persons. Another use for the product is to speed recovery from illness or injury. Spray twice under your tongue three times daily. Hold for 20 seconds before swallowing.

Warnings: There are no known warnings for this product. In short-term studies it has shown no toxicity. However, since it is a hormone, use of this product constitutes a form of hormone replacement therapy. The long-term effects of this are unknown, so we recommend it only as a temporary supplement to use as needed and not as a product to use on a continual basis.

IGF-1 (1 fl. oz.) - Liquid

Available in: US - Stock #966-1

Ingredients: IGF-1 (insulin-like growth factor 1) derived from deer antler velvet, stevia, lecithin.

IGS II

See *Lymph Gland Cleanse*

IMM-C

See *Trigger Immune*

Immune Stimulator

Product Type: Formula

Properties: **Anticancer**, Antioxidant, **Antiviral**, **Immune Stimulant**

Systems Affected: **Immune System**, Lungs, Lymphatic System, Spleen

Conditions: Acquired Immune Deficiency Syndrome (AIDS/HIV), Antibiotics (alternatives to), Antibiotics (side effects of), **Cancer (natural therapy for)**, Chemotherapy (reducing side effects), **Colds (antiviral)**, Colds (general remedies for), Colds (prevention), **Contagious Diseases**, **Epstein Barr Virus (Chronic Fatigue Syndrome, CFS)**, Fever, Free Radical Damage, **Infection (bacterial)**, **Infection (viral)**, **Leukemia**, **Meningitis**, **Mononucleosis**, **Pneumonia**, Poisoning (food), **Staph Infections**, **Strep Throat**, Weight Loss (aids for)

Usage: This formula contains a number of ingredients, which stimulate the immune system. It can be used to strengthen a weakened immune system by taking 1 capsule between meals two or three times daily. For cancer or fighting acute bacterial infections larger doses are needed. Use 1-2 capsules every two hours for a total of six to ten capsules per day.

Warnings: Not for use with autoimmune diseases. Not recommended for long term use.

Immune Stimulator - Capsule (90)

Available in: US - Stock #1839-3

Ingredients: Beta glucans, arabinogalactan, colostrum, reishi mushroom, maitake mushroom, cordyceps.

IN-X

Product Type: Formula

Properties: Antibacterial, Bitter, Bitter, Immune Stimulant, **Lymphatic**

Systems Affected: Ears, Immune System, Liver, Lymph Nodes, Lymphatic System, Respiratory System, **Tonsils**

Conditions: Adenitis, Bites and Stings, **Blood Poisoning**, Bronchitis, Cold Hands and Feet, Colds (general remedies for), Congestion (lymphatic), Ear Infection or Earache, Emphysema, Epstein Barr Virus (Chronic Fatigue Syndrome, CFS), Fever, Flu, Glands (swollen lymph), Impetigo, **Infection (bacterial)**, Kidney Infection, Staph Infections, Strep Throat, **Tonsillitis (Adenoids)**

Usage: For relieving all kinds of inflammation in the digestive tract, including Crohn's disease, colitis, hemorrhoids, irritable bowel syndrome and ulcerations; it contains both mucilants for soothing (slippery elm and marshmallow) and astringents for toning (plantain, rose hips and bugleweed). Chamomile is both nervine and anti-inflammatory. Take 3 capsules with each meal. Drink plenty of water when taking this formula. Works in a similar manner to Lymph Gland Cleanse (formerly IGS II) and Lymph Gland Cleanse-HY (formerly HIGS II), but is somewhat stronger.

Warnings: No known warnings.

IN-X - Capsule (100)

Available in: US - Stock #1220-1

Ingredients: Goldenseal, black walnut hulls, althea, parthenium, plantain, bugleweed

Indole 3 Carbinol

Product Type: Formula

Properties: Anticancer, Detoxifying

Systems Affected: Breasts, Liver, Prostate

Conditions: Benign Prostate Hyperplasia (BPH), Cancer (natural therapy for), Cancer (prevention), Cystic Breast Disease, Fibroids (uterine), **PMS Type A**

Usage: Indole 3 Carbinol helps with both Phase 1 and Phase 2 detoxification in the liver. It helps to break down excess estrogen compounds, including xenoestrogens, which may contribute to breast, prostate and uterine cancers. This also makes it useful for PMS symptoms brought on by estrogen dominance. Recommended dose is two capsules daily with a meal.

Warnings: No known warnings.

Indole 3 Carbinol - Capsule (60)

Available in: US - Stock #1506-4

Ingredients: Indole-3-Carbinol, broccoli flowers, brocoli sprouts

Inflammation

Product Type: Homeopathic

Properties: Anti-inflammatory

Systems Affected: Immune System, Nerves

Conditions: Inflammation, Pain (general remedies for)

Usage: Used as a remedy for general and localized inflammation with its accompanying heat, swelling, redness and pain. Helpful for conditions like arthritis, rheumatism, minor injuries and acute infections. Take 10-15 drops under the tongue every hour until symptoms improve, then four times daily until symptoms are relieved.

Warnings: Do not give this product for pain for more than five days. If pain is persistent or gets worse, seek professional assistance. NSP does not recommend this product for children under the age of 12 except on the advice of a health professional. Other than the alcohol content, which is not good for infants, we consider this product safe for children.

Inflammation (1 fl. oz.) - Homeopathic (Liquid)

Available in: US - Stock #8810-3

Ingredients: Active ingredients: Arnica montana (Mountain Arnica) 3x, Echinacea angustifolia (Cone Flower) 3x, Eupatorium perfoliatum (Boneset) 3x, Apis mellifica (Honeybee) 4x, Baptisia tinctoria (Wild Indigo) 4x, Bryonia (White Bryony) 4x, Colchicum autumnale (Meadow Saffron) 4x, Dulcamara (Bittersweet) 4x, Lycopersicum esculentum (Tomato) 4x, Rhus toxicodendron (Poison Ivy) 6x. Other ingredients: Purified water and 20% USP alcohol.

Nature's Phenyltol w/NEM - Capsule (60)

Available in: US - Stock #1180-3

Ingredients: DL-Phenylalanine, natural eggshell membrane (NEM), white willow bark, wood betony, morinda root

Influenza Remedy

Product Type: Homeopathic

Properties: Antiviral

Systems Affected: Immune System

Conditions: Abdominal Pain and Inflammation, Bronchitis, Chills, Fever, Flu

Usage: Used to relieve minor influenza symptoms, including stomach and abdominal pain, body aches, chills, fever, bronchitis, cough and thirst. Take 10-15 drops under the tongue every 10-15 minutes, or as needed until symptoms improve, then decrease to hourly, then to 4 times daily until symptoms are relieved.

Warnings: NSP does not recommend this product for children under the age of 12 except on the advice of a health professional. Other than the alcohol content, which is not good for infants, we consider this product safe for children.

Influenza Remedy (1 fl. oz.) - Homeopathic (Liquid)

Available in: US - Stock #8770-9

Ingredients: Active ingredients: Eupatorium perfoliatum (Boneset) 3x, Aconitum (Aconite) 4x, Baptisia tinctoria (Wild Indigo) 4x, Bryonia (White Bryony) 4x, Phosphorus (Phosphorus) 6x, Influenzinum (Influenza) 12x, Lachesis mutus (Lachesis) 12x. Other ingredients: Purified water and 20% USP alcohol.

Intestinal Soothe & Build

Other Names: UC3-J

Product Type: Formula

Properties: Absorbant, Astringent, Demulcent (Mucilant), Emollient, Soothing, **Vulneraries (for intestinal system)**

Systems Affected: **Appendix, Large Intestine (Colon), Mucus Membranes, Small Intestines**, Stomach, Thyroid

Conditions: **Abdominal Pain and Inflammation**, Abscesses, Acid Indigestion (Heartburn, Acid Reflux), Anal Fistula or Fissure, Bites and Stings, **Blood in Stool**, Burns and Scalds, **Celiac Disease**, Chemical Poisoning, Chemotherapy (reducing side effects), Cold Sores (Fever Blisters), **Colitis**, Colon (spastic), Copper Toxicity, **Crohn's Disease**, Cystitis, Denture Sores, Diarrhea, **Diverticulitis (Diverticuli)**, Duodenal ulcers, Dysentery, **Enteritis**, Grave's Disease, **Hemorrhoids**, Hiatal Hernia, **Ileocecal Valve, Inflammatory Bowel Disorders (Colitis, IBS)**, Itching, Leaky Gut Syndrome, Slivers, Tooth Extraction, Ulcerations (external), Ulcers

Usage: This is an excellent formula for relieving all kinds of inflammation in the digestive tract. It contains both mucilants for soothing (slippery elm and marshmallow) and astringents for toning (plantain, rose hips and bugleweed). Chamomile is both nervine and anti-inflammatory. Take 3 capsules with each meal. Works well with Stress-J (formerly STR-J.) Drink plenty of water when taking this formula for best results.

Warnings: No known warnings.

Intestinal Soothe & Build - Capsule (100)

Available in: US - Stock #1106-2

Ingredients: Chamomile, marshmallow, plantain, rosehips, slippery elm, bugleweed

UC3-J - Capsule (100)

Available in: Canada - Stock #1105-8

Ingredients: Chamomile flowers, marshmallow root, plantain herb, rose hips, slippery elm bark, and bugleweed herb

Ionic Minerals

Product Type: Nutrient

Properties: Antioxidant, Mineralizer, Nutritive

Systems Affected: Bones, Immune System, Structural System

Conditions: Arthritis, Bipolar Mood Disorder (Manic Depressive Disorder), Birth Defects (prevention), Broken Bones, Children's Remedy, Fingernail Biting, Fingernails (weak or brittle), Gray Hair, Pregnancy (herbs and supplements for), Scoliosis, Surgery (preparation for), Tachycardia, Tooth Decay (prevention)

Usage: Ionic Minerals provides varying amounts of 70 trace minerals. The ionic form makes them easy to absorb and utilize. The blend also contains the popular antioxidant acai berry and fulvic acid to aid absorption. Recommended dose is 1 tablespoon twice daily with meals.

Warnings: No known warnings.

Ionic Minerals - Liquid (Other)

Available in: US - Stock #310-5

Ingredients: 70 trace minerals plus purified water, fulvic acid, vegetable glycerin, açai berry concentrate, red grape skin extract, tartaric acid, xanthan gum, sodium benzoate, and citric acid

Iron

Product Type: Nutrient

Properties: Blood Building, Nutritive

Systems Affected: Blood, Heart, Liver, Skin

Conditions: Anemia, Fatigue, Hair (loss or thinning), Heart Fibrillation or Palpitations, Irritability, Muscle Tone (lack of), Pregnancy (herbs and supplements for)

Usage: Take 1 tablet of the iron supplement daily with a meal. Because iron must be bound to organic compounds to be utilized we find I-X or herbs like alfalfa or yellow dock are often more effective at raising iron levels in the blood.

Warnings: Store out of the reach of children.

Iron Chelated (25 mg) - Tablet (180)

Available in: US - Stock #1784-8

Ingredients: 25 mg iron (ferrous gluconate), vitamin C, calcium (di-calcium phosphate), rose hips, chickweed, mullein, thyme, yellow dock

I-X [Herbal Iron Formula] - Capsule (100)

Available in: US & Canada - Stock #1218-4

Ingredients: Red beets, yellow dock, red raspberry, chickweed, burdock, nettle, mullein

Iron with Vitamin C - Tablet (120)

Available in: US & Canada - Stock #1785-4

Ingredients: 5 mg. iron (ferrous gluconate) per tablet, vitamin C, rose hips, chickweed, mullein, thyme, yellow dock

Jasmine Absolute

Product Type: Essential Oil

Properties: Antidepressant, Aphrodisiac, Aromatic, Calmative, Expectorant, Febrifuge, Glandular, Parturient, **Perfume**, Relaxant, Sedative

Systems Affected: Hypothalamus, Liver, Pineal, Reproductive Glands, Respiratory System, Skin

Conditions: Afterbirth Pain, Apathy, Blood Pressure (low), Colds (general remedies for), Cough (damp), Cramps (menstrual), Depression, Dermatitis, Eczema, Emotional Sensitivity, Fatigue, Fear (excessive), Fever, Flu, Guilt, Headache (tension), Hypochondria, Insomnia, Labor and Delivery, Laryngitis (Hoarseness), Mood Swings, Post Partum Depression, Pregnancy (herbs and supplements to avoid during), Sex Drive (low), Skin (dry and/or flaky), Sprains

Usage: Jasmine is an exquisite floral scent that is sensual and luxuriant. It is often referred to as the King of flower oils. A single kilogram is extracted from about 8 million flowers picked before dawn. It is an exotic aphrodisiac with a powerful fragrance and only small quantities are needed. It has beneficial effects on the reproductive system and can be used to relieve uterine pain, childbirth pain, and to strengthen contractions during childbirth. It is also used to create a sensual, romantic mood. Jasmine is beneficial for respiratory problems such as colds, cough and congestion. It is useful for inflammatory conditions of the skin. A powerful mood enhancer, jasmine helps overcome feelings of anxiety, bitterness, burnout, dejection, depression, guilt, indifference, jealousy, negativity and repression. It brings feelings of confidence, creativity, happiness, intuition, peace and self-awareness.

Warnings: For topical use only. Do not take jasmine internally. Do a patch test first before using on skin. Avoid in cases of epilepsy. Keep away from eyes. Not to be used on babies. Do not use during the first trimester of pregnancy.

Jasmine Absolute Essential Oil - Essential Oil

Available in: US - Stock #3890-1

Ingredients: Jasmine (Jasminum officinalis) Essential Oil

Joint Health

Other Names: JNT-AV

Product Type: Formula

Properties: Anti-arthritic, Anti-inflammatory, Antirheumatic, Vulnerary

Systems Affected: Liver, Skeletal System, Structural System

Conditions: Arthritis, Cartilage Damage, Disks (spinal - bulging or slipped), **Gout**, Inflammation, **Rheumatoid Arthritis (Rheumatism)**

Usage: An Ayurvedic formula for arthritis and joint problems. Used for similar problems as Joint Support. Take 2 capsules with a meal three times daily.

Warnings: No known warnings.

Joint Health - Capsule (100)

Available in: US - Stock #1296-1

Ingredients: Withania somnifera, Commiphora mukul, Smilax china, Boswellia serrata, Holarrbena antidysenterica, Paederia foetida, Vitex negundo, Apium graveolens, Boerhaavia diffusa, Trachyspermum ammi, Tribulus terrestris, Trigonella foenum-graeceum

Joint Support

Other Names: JNT-A

Product Type: Formula

Properties: Alterative (Blood Purifier), Analgesic (Anodyne), **Anti-arthritic**, Anti-inflammatory, **Antirheumatic**, Diuretic, Lithotriptic, Nervine

Systems Affected: Liver, Lymphatic System, Muscles, Nerves, Skeletal System

Conditions: **Arthritis**, **Bursitis**, Calcium Deposits (Calcification), Cartilage Damage, **Fibrosis**, **Gout**, Lupus, **Neuralgia and Neuritis**, Overacidity, **Rheumatoid Arthritis (Rheumatism)**, **Uric Acid Retention**

Usage: This formula was designed for arthritis but also works as a general anti-inflammatory and pain reliever. Do not expect instant results. Takes effect gradually over a period of about 1-2 weeks. Take 2 capsules three times daily. For better results, combine with alfalfa, aloe vera, yucca, Herbal CA and/or HSN-W.

Warnings: No known warnings.

Joint Support - Capsule (100)

Available in: US - Stock #810-8

Ingredients: Hydrangea, yucca, horsetail, celery seed, alfalfa, black cohosh, bromelain, catnip, yarrow, capsicum, valerian, white willow, burdock, slippery elm, sarsaparilla

ART-A with Devil's Claw - Capsule (100)

Available in: Canada - Stock #798-9

Ingredients: Bromelain, white willow bark, yucca, devil's claw, marshmallow, celery seed, alfalfa, yarrow, capsicum, burdock, nettle, rosemary

Jojoba Oil

Product Type: Household

Properties: Cosmetic, **Emollient**, Moistening

Systems Affected: **Scalp**, Skin

Conditions: **Dandruff**, **Hair (loss or thinning)**, **Hair Care (general)**, Itching, **Seborrhea**, Skin Care (general)

Usage: Massage into scalp for relief from itchy, flaky scalp and dandruff. Can also be applied to moisten dry skin. It is a good carrier oil for essential oils.

Warnings: No known warnings.

Jojoba Oil (0.5 fl. oz.) - Oil

Available in: US - Stock #1695-1

Ingredients: Jojoba oil

JP-X

Product Type: Formula

Properties: Antiseptic, **Diuretic**

Systems Affected: Bladder (Urinary), Urinary System

Conditions: Backache (Back Pain, Lumbago), Bedwetting, Bladder Infection, Cystitis, Edema (Dropsy, Water Retention, Swelling), Gas and Bloating, Kidney Infection, Pregnancy (herbs and supplements to avoid during), Urinary Tract Infections, **Urine (scant)**

Usage: Diuretic formula with some soothing and antiseptic qualities. Better for inflamed kidneys and infections than Kidney Activator (formerly K.) Take 2 capsules three times per day with water. For bedwetting, take 1 hour before bedtime.

Warnings: This formula should be avoided when the kidneys are inflamed. It also contains herbs that should be avoided in the early stages of pregnancy.

JP-X - Capsule (100)

Available in: US - Stock #1222-7

Ingredients: Dong quai, golden seal, juniper, uva ursi, parsley, ginger, marshmallow

Juniper Berries

Latin Name: *Juniperus communis*

Product Type: Single Herb

Properties: Antifungal, Antiseptic, Aromatic, Astringent, Carminative, Diaphoretic, Diuretic, Fumigant, Lipotropic, Preservative, Stimulant, Stomachic, Vermifuge

Systems Affected: Bladder (Urinary), Gums, Kidneys, Nerves, Pancreas, Skin, Stomach, Urinary System, Uro-genital Tract

Conditions: Addictions (coffee, caffeine), Bedwetting, Cystic Fibrosis, Diabetes, Edema (Dropsy, Water Retention, Swelling), Hypoglycemia, Incontinence (urinary), Kidney Infection, Kidney Stones, PMS Type H, **Urine (scant)**

Usage: For fluid retention, take 2 capsules, two to three times daily with meals. Juniper contains an essential oil that strongly stimulates kidney function. It is an excellent diuretic with some antiseptic properties for infection. It also stimulates hydrochloric acid production and aids digestion (especially proteins) and has been used for parasites. If used as a diuretic, or to treat the inflammation associated with arthritis, or attempt to bring on a period, take juniper as an infusion. For an infusion, use one teaspoon of bruised berries per cup of boiling water. Steep 10 to 20 minutes. Drink up to two cups per day for no more than 6 weeks.

Warnings: Aromatic qualities can be irritating to kidneys (with lesions) and nervous system with long-term use. Not recommended when kidneys are inflamed or in cases of nephritis and nephrosis. Never take juniper for longer than six weeks. Overdose symptoms include diarrhea, intestinal pain, kidney pain, protein in the urine, blood in the urine, purplish urine, rapid heartbeat and elevated blood pressure. Up to one-third of hay fever sufferers develop allergy symptoms from exposure to juniper. Juniper should not be given to children under age two. For older children and people over 65, start with low-strength preparations and increase strength if necessary.

Juniper Berries - Capsule (100)

Available in: US - Stock #400-4

Ingredients: Juniperus communis (juniper) berries

K

See *Kidney Activator*

K-C

See *Kidney Activator*

Kava Kava

Latin Name: *Piper methysticum*

Product Type: Single Herb

Properties: Acrid, **Analgesic (Anodyne)**, **Anesthetic**, Antidepressant, Antiseptic, **Antispasmodic**, **Anxiolytic**, Aphrodisiac, Diuretic, **Euphoretic**, **Hypnotic**, **Narcotic**, **Relaxant**, **Sedative**

Systems Affected: Central Nerves, GABA, Kidneys, Muscles, Nerves, Urinary System

Conditions: Addictions (drugs), **Anxiety (Panic Attack)**, **Anxiety Disorders**, Asthma, **Backache (Back Pain, Lumbago)**, Bedwetting, Bladder (irritable), Bronchitis, **Convulsions**, **Cramps and Spasms (general)**, Cushing's Disease, **Depression**, Dysmenorrhea, Fatigue, Fibromyalgia Syndrome (FMS), Gas and Bloating, Glaucoma, Gonorrhea, Gout, Headache (tension), **Insomnia**, Kidney Infection, Kidney Stones, Laryngitis (Hoarseness), Leucorrhea, Nervous Exhaustion (Enervation), **Nervousness**, Neurosis, **Pain (general remedies for)**, PMS Type P, **Post Traumatic Stress Disorder**, **Restless Leg Syndrome**, Rheumatoid Arthritis (Rheumatism), Sex Drive (low), **Stiff Neck**, **Stress**, **Tension**, TMJ, Toothache, Urinary Tract Infections, **Urination (burning or painful)**, **Vaginitis**

Usage: For chronic tension, anxiety or muscle spasms, a recommended dosage would be 1,500 to 3,000 mg of dried root daily or 3-6 herbal capsules of 500 mg each of kava kava daily for short periods of time. NSP's product is a concentrate and contains 200 mg of kava kava concentrate, so the dosage is lower. Take 1 capsule with a meal twice daily. Take 1-2 capsules at night in preparation for bedtime. Do not exceed 4 capsules per day of the kava kava concentrate. For anxiety attacks, use in tincture or liquid form, give one dropperful (about 30 drops) every minute or two until anxiety subsides. Because kava kava is a strong relaxant that helps to induce sleep when tired, a good way to use the herb is to take 1-2 capsules with dinner or before going to bed. It works best for insomnia caused by muscle tension, but is not helpful for insomnia associated with worry and excessive mental activity.

Warnings: Do not use kava kava while driving, operating heavy machinery or other tasks where alcohol should be avoided. Large doses of kava kava can interfere with muscle coordination. Kava kava should only be used in small doses for a short period. Continuous overdose of this herb may cause loss of balance and appetite, chronic "drunk" feeling, diarrhea and skin problems. Do not take kava kava daily for longer than three months. Large doses over a long period of time may cause liver problems and skin eruptions. Fortunately, the condition disappears once the person quits taking the herb. In large doses, some people have experienced becoming so relaxed that they could not move (This symptom wears off quickly without side effects). Again, use common sense as you would with any herb. Reports in Europe suggested kava kava has hepatotoxicity (toxicity to the liver), but these reports appear to be based on using highly concentrated extracts of the stems, which were never used in traditional medicine. No reports of kava toxicity to the liver are reported in the areas where it has been traditionally used. Kava may aggravate alcohol's toxic effect on the liver and should not be consumed with alcoholic beverages.

Kava Kava Concentrate - Capsule (60)

Available in: US - Stock #405-9

Ingredients: Standardized extract of Piper methysticum (kava kava) root

KB-C

Other Names: Jian Gu

Product Type: Formula

Properties: Analgesic (Anodyne), **Anti-arthritic**, Diuretic, **Kidney Tonic**

Systems Affected: Bladder (Urinary), Bronchials, Ears, Hair, **Legs**, Male Reproductive, Muscles, Reproductive Glands, Skeletal System, Spinal Disks, **Urinary System**, **Uro-genital Tract**

Conditions: Arthritis, **Backache (Back Pain, Lumbago)**, Benign Prostate Hyperplasia (BPH), Bladder Infection, Calcium Deposits (Calcification), Cystitis, **Disks (spinal - bulging or slipped)**, Dizziness (Vertigo), Erectile Dysfunction, Fatigue, **Fear (excessive)**, **Gout**, **Gray Hair**, **Incontinence (urinary)**, Infertility, Kidney Infection, **Knees (weak)**, **Nocturnal Emission (Wet Dreams)**, **Osteoporosis**, **Overacidity**, **Premature Ejaculation**, **Prostatitis**, **Sciatica**, Scoliosis, Tinnitus (Ringing in the Ears), **Uric Acid Retention**, **Urination (frequent)**

Usage: When we think of the characteristics of water, we can readily see that water is fluid, flexible, flowing and lubricating. So, when these qualities are lacking in the body, the Chinese would say that one is deficient in the water element (or energy). This lack of water energy would make a person stiff, brittle, rigid and dry.

KB-C is a Chinese formula that corrects this deficiency of water energy. It strengthens the organs associated with water in the body, the kidneys and bladder and it also strengthens the bones. In Chinese medicine, the kidneys help "build the bones." Although this may seem like a strange concept to Western minds, the kidneys also regulate the balance of mineral electrolytes and flush acid waste to help maintain pH balance, which maintains the health of the structural system. Acid pH contributes to a loss of bone density.

KB-C can be helpful for broken bones, arthritis, backache, weak and brittle bones (osteoporosis), constipation, frequent urination (irritable bladder), chronic back problems, weak knees and ankles, kidney inflammation, kidney stones, prostate swelling, impotence, fatigue and insomnia due to waking up frequently to urinate.

Fear is the emotion associated with the kidneys, and KB-C can also be helpful for a person who is fearful and timid and lacks "back bone." It rejuvenates and strengthens the bones (especially the spine), kidneys, connective tissues and sexual organs. This leads to a healthier urogenital system and more energy and sexual vitality.

To strengthen weak bones and kidneys take 3 capsules 3 times daily. Also available as a TCM concentrate where the dose is 1 capsule two to three times daily.

Warnings: No known warnings.

KB-C, Chinese - Capsule (100)

Available in: US & Canada - Stock #1883-3

Ingredients: Eucommia, cistanche, achyranthes, dipsacus, drynarea, hoelen, morinda, rehmannia, astragalus, cornus, dang gui, dioscorea, epimedium, ligustrum, liquidambar, lycium, panax ginseng, atractylodes

KB-C TCM Conc. - Capsule (30)

Available in: US - Stock #1016-0

Ingredients: Same as Stock #1883-3

KC-X

See *Thyroid Activator*

Kelp

Latin Name: *Laminaria sp., Fucus sp*

Product Type: Single Herb

Properties: Antacid, Condiment, Demulcent (Mucilant), Emollient, Mineralizer, Stimulant, Thyrotropic

Systems Affected: Gall Bladder, Glandular System, Hair, Lymphatic System, Nails, Parathyroid, Pineal, Pituitary (anterior), Pituitary (general), Pituitary (posterior), Reproductive Glands, Skin, Thyroid

Conditions: Acne (Pimples, Blackheads), Adrenals (exhaustion, weakness or burnout), Anxiety (Panic Attack), Arteriosclerosis (Atherosclerosis, Hardening of the Arteries), Birth Defects (prevention), Calcium Deficiency, Cancer (natural therapy for), Cardiovascular Disease (Heart Disease), Cramps and Spasms (general), Cysts, Dandruff, Down Syndrome, Eczema, Fatigue, Fingernails (weak or brittle), Goiter, Hair (loss or thinning), **Hypothyroid**, Lead Poisoning, Pregnancy (herbs and supplements for), Radiation Sickness, Skin (dry and/or flaky), Weight Loss (aids for)

Usage: Kelp is a rich source of iodine and other trace elements and is most frequently used to aid thyroid function. It is also a source of sodium and potassium for maintaining fluid balance in the body. Two capsules typically provide the U.S. RDA of iodine. Use 1-2 capsules with two meals or use as seasoning by sprinkling on food. Should be taken consistently for best effect. Kelp also contains algin, a mucilage that absorbs heavy metals.

Warnings: Contraindicated in cases of overactive thyroid (Grave's disease). Safe food otherwise.

Kelp - Capsule (100)

Available in: US & Canada - Stock #410-3

Ingredients: Fucus sp. (kelp) herb

Kidney Activator

Other Names: K

Product Type: Formula

Properties: Diuretic

Systems Affected: Bladder (Urinary), Kidneys, **Urinary System**

Conditions: Backache (Back Pain, Lumbago), Bedwetting, Blood in Urine, Blood Pressure (low), Breasts (swelling and tenderness), Cystitis, Cysts, **Edema (Dropsy, Water Retention, Swelling)**, Gonorrhea, Hypochondria, Incontinence (urinary), Kidney Stones, PMS Type H, **Urine (scant)**

Usage: Diuretic formula for water retention. Take 2 capsules one to four times daily. For bedwetting, take 1 hour before bedtime.

Warnings: Not recommended when kidneys are inflamed. See warnings under Juniper.

Kidney Activator - Capsule (100)

Available in: US - Stock #970-2

Ingredients: Juniper, parsley, uva ursi, dandelion and chamomile

Kidney Activator ATC Concentrate - Capsule (50)

Available in: US - Stock #973-9

Ingredients: Same as Stock #970-2

K - Capsule (100)

Available in: Canada - Stock #959-5

Ingredients: Juniper, parsley, uva ursi, dandelion

Kidney Activator (Chinese)

Other Names: Qu Shi

Product Type: Formula

Properties: Anti-inflammatory, **Diuretic**, Lymphatic

Systems Affected: Bladder (Urinary), Kidneys, Lymphatic System, Nerves, Stomach, Testes, **Urinary System**

Conditions: Abdominal Pain and Inflammation, Backache (Back Pain, Lumbago), Blood Pressure (high), Breasts (swelling and tenderness), Dizziness (Vertigo), **Edema (Dropsy, Water Retention, Swelling)**, Fatigue, Fear (excessive), Indigestion, Kidney Stones, Pain (general remedies for), **PMS Type H**, Prostatitis, Urethritis, Urinary Tract Infections, **Urine (scant)**, Weight Loss (aids for)

Usage: An excess of water is a pretty obvious symptom in both Chinese and Western medicine. We call it edema or water retention. It is characterized by swelling of the tissues due to retained fluids. Chinese Kidney Activator is an herbal formula designed to reduce this excess of water.

Logically, the Chinese associate the water element with the kidneys and bladder, which filter the body's fluids and flush excess moisture from the body. However, the kidneys aren't the only consideration with water retention. The lymphatics must be functioning properly in order to flush fluid from the tissues and inflammation (an underlying cause of fluid retention) must be reduced. So, Kidney Activator not only acts as a diuretic, stimulating the kidneys to flush more fluid from the body, it also works with other body systems and processes that flush excess fluid from the tissues.

Kidney Activator can be helpful for frequent urination, scanty urination, kidney stones, abdominal inflammation and swelling, kidney inflammation, burning urination, backache, leg pain, hip pain, swelling of the legs, ankles, fingers, etc., and high blood pressure due to fluid retention. Qu Shi can also help feelings of late afternoon sluggishness, heavy feelings in the body and dizziness.

Chinese herbal formulas can also help with related emotional problems; the emotional issues associated with excess water energy include fear and a "wishy-washy" personality. Kidney Activator can improve these emotional issues by redistributing and eliminating excess moisture. It relieves pain and inflammation, especially in the joints and back, as well as reducing abdominal pressure and swelling in the breasts.

To build the kidneys and eliminate excess moisture from the body, it is suggested to take 4 capsules of Chinese Kidney Activator 3 times daily. A TCM concentrate is also available. One capsule of the TCM concentrate will replace 4 capsules of the regular blend.

Warnings: Not recommended for conditions of dryness or fluid deficiency.

Kidney Activator, Chinese - Capsule (100)

Available in: US - Stock #1872-5

Ingredients: Hoelen, siler root, chaenomeles fruit, morus root bark, astragalus, psyllium seed, alisma rhizome, peony root without bark, atractylodes rhizome, magnolia bark, polyporus, cinnamon, citrus peel, ginger, typhonium rhizome, licorice

Kidney Activator TCM Conc. - Capsule (30)

Available in: US - Stock #1040-0

Ingredients: Hoelen sclerotium, siler root, chaenomeles fruit, morus root bark, astragalus root, psyllium seed, alisma rhizome, peony root without bark, atractylodes rhizome, magnolia bark, polyporus sclerotium, cinnamon twig, citrus peel, ginger rhizome, typhonium rhizome, licorice root

K-C - Capsule (100)

Available in: Canada - Stock #1872-5

Ingredients: Sileris root, hoelen plant, morus root bark, chaenomeles fruit, astragalus root, atractylodes rhizome, alisma rhizome, magnolia bark, polyporus plant, areca peel, psyllium seed, cinnamon twig, pinellia rhizome, ginger rhizome, citrus peel, and licorice root

Kidney Drainage

Product Type: Formula

Properties: **Diuretic**

Systems Affected: Kidneys, **Urinary System**

Conditions: Adenitis, Angina, Arthritis, Bedwetting, Bladder (irritable), Blood in Urine, Blood Pressure (high), Calcium Deposits (Calcification), Congestive Heart Failure, **Cystitis**, Diabetes, **Edema (Dropsy, Water Retention, Swelling)**, Gout, Pain (general remedies for), PMS Type H, Urination (burning or painful), Urine (scant)

Usage: This formula helps the kidneys filter wastes out of the blood stream more efficiently. It also has mild kidney stimulating effects. Take approximately 15-20 drops or 1/4 teaspoon in water twice daily.

Warnings: Not recommended with kidney inflammation.

Kidney Drainage (2 fl. oz.) - Liquid (Glycerite)

Available in: US - Stock #3168-4

Ingredients: Asparagus, plantain, juniper, goldenrod, vegetable glycerine

Korean Ginseng

See *Ginseng (Korean)*

Kudzu/St.John's Wort

Product Type: Formula

Properties: Antidepressant, Detoxifying, **Vulneraries (for intestinal system)**

Systems Affected: Large Intestine (Colon), Nerves, **Peyer's Patches, Small Intestines**

Conditions: **Addictions (alcohol)**, Belching, Depression, **Hangover**, Inflammation, Inflammatory Bowel Disorders (Colitis, IBS), Irritability, **Leaky Gut Syndrome**, Stiff Neck, Stress

Usage: In laboratory tests, kudzu was shown to lessen alcohol craving in hamsters. This only verifies Chinese medicine's use of kudzu to lessen the extreme effects of alcohol consumption. St. John's wort soothes irritability and depression while alfalfa provides trace minerals and detoxifies the body. Take 1 capsule three times daily. Add Milk Thistle Combination (formerly LIV-Guard) and milk thistle to increase the body's ability to cleanse the liver.

Warnings: No known warnings.

Kudzu/St. John's Wort (Concentrate) - Capsule (100)

Available in: US - Stock #975-6

Ingredients: Kudzu, St. John's wort, alfalfa

L-Carnitine

Product Type: Nutrient

Properties: Nutritive

Systems Affected: **Heart**, Thyroid, Weight Loss

Conditions: Angina, Cataracts, Cellulite, **Cholesterol (low)**, **Congestive Heart Failure**, Diabetes, Fatigue, Grave's Disease, **Heart (weakness)**, Heart Fibrillation or Palpitations, Reye's Syndrome, Rheumatic Fever, **Triglycerides (low)**, Weight Loss (aids for)

Usage: Carnitine is an amino acid that helps transport long chain fatty acids. Symptoms of carnitine deficiency include progressive muscle weakness and severe hypoglycemia. Carnitine is not found in vegetable foods but is mostly found in animal muscle tissue. Used for heart problems, weight control, lack of energy, Reye's Syndrome and as a weight loss aid. Take 1-3 capsules daily with meals.

Warnings: No known warnings.

l-Carnitine - Capsule (30)

Available in: US - Stock #1632-6

Ingredients: 250 mg. of l-carnatine

L-Glutamine

Product Type: Nutrient

Properties: Nutritive, **Vulneraries (for intestinal system)**

Systems Affected: Brain, Nerves

Conditions: Addictions (alcohol), Addictions (drugs), Addictions (sugar or food), Aging (prevention), Anemia, Bruises (healing), Colitis, Depression, Dizziness (Vertigo), Ear Infection or Earache, Fat Metabolism (poor), Fatigue, Gout, Hypoglycemia, **Leaky Gut Syndrome**, Menopause, Motion Sickness, Narcolepsy, Schizophrenia, Senility, Senility, Ulcers

Usage: Important along with glucose in supplying the brain with energy. Used for aging prevention, treatment of alcoholism, anemia, bruises, colitis, dizziness, drug withdrawal, ear infections, poor fat metabolism, fatigue, gout, hemorrhoids, hypoglycemia, menopause, motion sickness, senility, sore throat and ulcers of the stomach. Take 1-2 capsules daily with a meal.

Warnings: No known warnings.

L-Glutamine - Capsule (30)

Available in: US - Stock #1776-0

Ingredients: 500 mg. of L-glutamine

L-Lysine

Product Type: Nutrient

Properties: Antiherpetic, Nutritive

Systems Affected: Bones, Immune System, Mouth, Tongue

Conditions: Canker Sores (Mouth Ulcers), Cold Sores (Fever Blisters), Epstein Barr Virus (Chronic Fatigue Syndrome, CFS), Herpes, Mononucleosis, Shingles, Sprains

Usage: Helps ensure adequate absorption of calcium and the formation of collagen for bone, cartilage and connective tissue. Lysine strengthens circulation and helps the immune system manufacture antibodies. It also helps control the body's acid/alkaline balance, influences the pineal and mammary glands and plays a role in gallbladder function. It is necessary for all amino acid assimilation and assists in the storage of fats. Used for treating canker sores, cold sores, genital herpes, fever blisters, mononucleosis, shingles and to promote growth in children. Take 1 capsule with a meal three times daily.

Warnings: No known warnings.

L-Lysine - Capsule (100)

Available in: US - Stock #1631-4

Ingredients: 475 mg of l-lysine monohydrochloride

L. Reuteri

See *Probiotics*

Lactase Plus

Product Type: Formula

Properties: Digestant

Systems Affected: Digestive System, Liver

Conditions: **Allergies (food)**, Cataracts, Digestion (poor), Failure to Thrive, Gas and Bloating, Lactose Intolerance

Usage: Contains enzymes that help digest milk and dairy products. It is used specifically for lactose intolerance and dairy allergies. Take 1-2 capsules to aid in digestion of milk, ice cream and other dairy foods.

Warnings: No known warnings.

Lactase Plus - Capsule (100)

Available in: US - Stock #1655-2

Ingredients: Lactase, protease, lipase, beet root, potassium citrate, caraway, dandelion, fennel, gentian, ginger

Lavender

Product Type: Essential Oil

Properties: Analgesic (Anodyne), Antibacterial, Antifungal, Antiseptic, Aromatic, Carminative, Cholagogue, **Cicatrisant**, CNS Depressant, Narcotic, Nervine, Parasympatholytic (Anticholinergic), **Perfume**, **Relaxant**, Stimulant (Circulatory), **Sympatholytic**, Tonic

Systems Affected: Central Nerves, **Ears**, Nerves, Pineal, Skin, Vagina

Conditions: Abrasions, **Afterbirth Pain**, Anger (excessive), Anxiety (Panic Attack), Autism, **Burns and Scalds**, Colic (children), Confusion, Cushing's Disease, Cuts, Dysmenorrhea, Ear Infection or Earache, Edema (Dropsy, Water Retention, Swelling), Fainting, Fear (excessive), Fever, Halitosis (Bad Breath), **Headache (tension)**, **Hysteria**, Indigestion, Irritability, Nausea and Vomiting, Nervous Exhaustion (Enervation), **Nervousness**, Neuralgia and Neuritis, Paralysis, PMS Type A, PMS Type D, **Scars / Scar Tissue**, **Sunburn**, Teeth (loose), **Thrush**, Worry

Usage: Use the undiluted essential oil as an antiseptic, mild sedative and painkiller, particularly on insect bites, stings and small (cooled) burns. Add six drops to hot bath water to calm irritable children, and place one drop on the temple for headache relief. For enervation, nervous exhaustion, anxiety, excess stress and even cancer, lavender oil baths with Epsom salts (2 cups per tub) are very beneficial. Lavender baths can also help fight fungal infections in children and adults. Lavender oil can be blended for use as a massage oil in aromatherapy. The diluted oil can be applied topically for throat infections, skin sores, inflammation, rheumatic aches, anxiety, insomnia and depression. It can also be massaged into the chest, with a little chamomile oil added, for asthmatic and bronchial spasm. Other uses for lavender oil include the following: Add a few drops of oil to a little water for sunburn or scalds. Massage diluted oil into the temples and nape of the neck for tension headaches or at the first hint of a migraine. Apply undiluted oil to insect bites and stings. Dilute 10 drops oil in 25 ml carrier oil for sunstroke (Note: this is not an effective sun block). Add a few drops of the oil to a chamomile-based cream for eczema. Dilute in water and apply rinse to hair for lice.

Warnings: Not recommended in cases of depression. Skin to which lavender oil or salves/lotions containing lavender has been applied should not be exposed to sunlight, because lavender is photo reactive and will bleach and blotch the skin.

Lavender Fine AOC (5ML) - Essential Oil

Available in: US & Canada - Stock #3907-8

Ingredients: Lavender Fine (Lavandula angustifolia) Essential Oil

LB Extract

Product Type: Formula

Properties: Bitter, **Laxative (general)**, **Laxative (stimulant)**, **Purgative (Cathartic)**, Vermifuge

Systems Affected: Intestinal System, Large Intestine (Colon), Liver

Conditions: Colon (atonic), Constipation (adults), Constipation (children), Digestion (poor), Diverticulitis (Diverticuli), Parasites (nematodes, worms), Weight Loss (aids for)

Usage: Liquid version of LBS II. To cleanse and tone the colon, take 1/2-1 teaspoon in water or juice. Can be given in smaller doses to children for occasional problems with constipation. Disguise taste in fruit juice for children.

Warnings: Same as for cascara sagrada, except that a combination such as this one is gentler and safer for pregnancy.

LB Extract (2 fl. oz.) - Liquid (Tincture)

Available in: US & Canada - Stock #1794-1

Ingredients: Cascara sagrada, senna, buckthorn, alfalfa, psyllium hulls, licorice, turkey rhubarb, barberry, ginger, slippery elm, water, alcohol

LB-X

Product Type: Formula

Properties: Bitter, **Laxative (general)**, **Laxative (stimulant)**, **Purgative (Cathartic)**, Vermifuge

Systems Affected: Intestinal System, Large Intestine (Colon), Liver

Conditions: Colon (atonic), **Constipation (adults)**

Usage: Formula similar to LBS II only milder. Take 2-4 capsules before bedtime or 1-2 capsules three times per day.

Warnings: Same as for LBS II, except that a combination such as this one is gentler and safer for pregnancy.

LB-X - Capsule (100)

Available in: US - Stock #1226-3

Ingredients: Cascara sagrada, dong quai, turkey rhubarb, capsicum fruit, golden seal, ginger, Oregon grape, fennel, red raspberry, lobelia.

LB-X - Capsule (100)

Available in: Canada - Stock #1227-4

Ingredients: Buckthorn, wild yam, yellow dock, Turkey rhubarb, ginger, fennel, red raspberry, peppermint, capsicum, marshmallow.

LBS II

Product Type: Formula

Properties: Bitter, **Laxative (general)**, **Laxative (stimulant)**, **Purgative (Cathartic)**, Vermifuge

Systems Affected: **Intestinal System**, **Large Intestine (Colon)**, Liver

Conditions: **Colon (atonic)**, **Constipation (adults)**, Diverticulitis (Diverticuli), Leaky Gut Syndrome, Lou Gehrig's Disease, Parasites (general), Parasites (nematodes, worms), Weight Loss (aids for)

Usage: This is an old standby laxative formula. To cleanse and tone the colon, take 2-4 capsules before bedtime or 1-2 capsules with meals. Best taken with lots of water and with bulking herbs like psyllium.

Warnings: The constituents of stimulant laxatives such as cascara sagrada can be transferred to babies through the milk of nursing mothers, so use with caution when nursing. Long-term use of stimulant laxatives can drain the energy of the colon. Generally speaking, these laxatives should be necessary for occasional use only. If bowel does not return to normal function, look deeper into the underlying causes of the constipation and work with those. Typical problems include lack of digestive enzymes, poor liver function, not drinking enough water, lack of dietary fiber, magnesium deficiency, sedentary lifestyle (insufficient exercise), and stress (muscle spasms). See also warnings for cascara sagrada.

LBS II Capsules - Capsule (100)

Available in: US - Stock #990-1

Ingredients: Cascara sagrada, buckthorn, licorice, capsicum, ginger, Oregon grape, turkey rhubarb, couch grass, red clover

LBS II - Capsule (100)

Available in: Canada - Stock #994-0

Ingredients: Cascara sagrada (bark), buckthorn (bark), licorice (root), Turkey rhubarb (root), ginger root, capsicum fruit, couch grass herb, Oregon grape root, and red clover tops

LBS II Vegitabs - Vegitab (100)

Available in: US - Stock #993-5

Ingredients: Same as Stock #994-0

LBS II Vegitabs - Vegitab (270)

Available in: US - Stock #998-6

Ingredients: Same as Stock #994-0

LBS II Trial Pack - Packets (20)

Available in: US - Stock #2497-0

Ingredients: Same as Stock #994-0

Lecithin

Product Type: Nutrient

Properties: Antioxidant, **Emulsifier**, Nervine

Systems Affected: Acetylcholine, Brain, Circulation, Gall Bladder, Heart, Liver, Muscles, Nerves, Pineal, Spleen, Testes

Conditions: Alzheimer's Disease, Arteriosclerosis (Atherosclerosis, Hardening of the Arteries), Cardiovascular Disease (Heart Disease), Cholesterol (high), Cholesterol (low), Down Syndrome, Fibroids (uterine), Hashimoto's Disease (Thyroiditis), Memory and Brain Function, Myasthenia Gravis, Narcolepsy, Neuralgia and Neuritis, Pancreatitis, Parkinson's Disease, Phlebitis, Reye's Syndrome, Skin (dry and/or flaky), Strokes

Usage: Helps emulsify fats in the body. Used with cholesterol in cell membranes. May help high cholesterol levels. Works well with ginkgo, garlic, hawthorn and capsicum for circulatory problems. May also aid brain function. Use 1-3 capsules with meals.

Warnings: No known warnings.

Lecithin - Capsule (270)

Available in: US - Stock #1660-5

Ingredients: 520 mg. of lecithin, glycerin.

Lecithin - Softgel (170)

Available in: US & Canada - Stock #1661-6

Ingredients: Same as Stock #1660-5

Lemon Oil

Product Type: Essential Oil

Properties: Antibacterial, Aromatic, Condiment, Decongestant, Febrifuge, Hypotensive, Stomachic, Sympathomimetic

Systems Affected: Circulation, Immune System, Intestinal System, Liver, Testosterone

Conditions: Abscesses, Acid Indigestion (Heartburn, Acid Reflux), Addictions (alcohol), Addictions (tobacco smoking or chewing), Aging (prevention), Alzheimer's Disease, Apathy, Appetite (deficient), Arthritis, Asthma, Bites and Stings, Blood Pressure (high), Boils, Breasts (swelling and tenderness), Bronchitis, Canker Sores (Mouth Ulcers), Capillary Weakness, Cellulite, Children's Remedy, Cholesterol (high), Colds (decongestant), Colds (general remedies for), Concentration (poor), Confusion, Conjunctivitis (Pink Eye), Cough (dry), Dermatitis, Diabetes, Diarrhea, Digestion (poor), Dysentery, Eczema, Fatigue, Fear (excessive), Fever, Fleas, Flu, Gall Stones, Gas and Bloating, Gingivitis (Bleeding Gums, Gum Disease, Pyorrhea), Gonorrhea, Gout, Headache (general), Herpes, Hot Flashes, Infection (viral), Insects, Jaundice (adults), Kidney Stones, Malaria, Mood Swings, Nausea and Vomiting, Pap Smear (abnormal), Parasites (general), Phlebitis, PMS Type H, Pneumonia, Rashes and Hives, Rheumatoid Arthritis (Rheumatism), Scabies, Seborrhea, Senility, Syphilis, Teeth (loose), Thrombosis, Tinnitus (Ringing in the Ears), Tonsillitis (Adenoids), Tuberculosis (Consumption, Scrofula), Typhoid, Varicose Veins, Weight Loss (aids for), Wounds and Sores, Wrinkles

Usage: As a preventative remedy for liver and urinary problems, many people have found that the juice of 1/2 lemon in water taken each morning upon arising is very beneficial. To cleanse the system, lemon can be taken as part of a fast using lemon water (sweetened with a small amount of pure maple syrup, preferably grade B). The juice of four lemons dissolved in a gallon of distilled water and consumed over the course of a day while fasting has been known to help dissolve kidney stones. The lemon peel can be used as an immune tonic and helps to strengthen the capillaries and weak tissues (similar to rose hips). The essential oil is a disinfectant (topical) and immune booster. It also helps to move lymph and lift mood. When using the essential oil internally or externally dilute 5-10 drops in a half pint of warm water.

Warnings: Overuse of lemon oil internally may cause nausea. Externally, lemon oil may cause dermal irritation. Lemon oil is phototoxic, so avoid using on skin and then exposing to the sun.

Lemon BIO Essential Oil (5ML) - Essential Oil

Available in: US & Canada - Stock #3908-2

Ingredients: Lemon (Citrus limon) Essential Oil

LH-C

See *Lung Support*

Licorice Root

Latin Name: *Glycyrrhiza glabra*

Product Type: Single Herb

Properties: **Adrenal Tonic**, **Adrenergic**, **Anti-inflammatory**, Antidiuretic, Antilipemic, **Antitussive**, Antiviral, Aperient, Aphrodisiac, Appetite Stimulant, Bronchial Dilator, **Catalyst (Synergist)**, Emollient, Estrogenic, Expectorant, **Hypertensive**, **Immune Amphoterics**, Immunomodulator, **Moistening**, Nutritive, **Parasympatholytic (Anticholinergic)**, Phytoestrogen, Refrigerant, Sialogogue, Stimulant, Sweetener, **Sympathomimetic**, Vulneraries (for intestinal system)

Systems Affected: Adrenal Cortex, Adrenal Glands, Aldosterone, Bladder (Urinary), Cortisol, Estrogen, Intestinal System, Liver, Lungs, Mucus Membranes, Pancreas, Pituitary (anterior), Pituitary (posterior), Reproductive Glands, Spleen, Stomach, **Throat**, Uro-genital Tract

Conditions: Addictions (alcohol), Addictions (coffee, caffeine), Addictions (drugs), **Addison's Disease**, **Adrenals (exhaustion, weakness or burnout)**, Allergies (respiratory), Anorexia, Appetite (deficient), Arthritis, **Asthma**, Attention Deficit Disorder (ADD, ADHD), Autoimmune Disorders, **Bedwetting**, Bronchitis, Bursitis, **Chills**, Cirrhosis of the Liver, Confusion, Constipation (children), **Cough (dry)**, **Dehydration**, **Dizziness (Vertigo)**, Duodenal ulcers, Eczema, Emphysema, Endurance (lack of), Epstein Barr Virus (Chronic Fatigue Syndrome, CFS), Estrogen (low), **Fainting**, **Fatigue**, Fear (excessive), Fibromyalgia Syndrome (FMS), Fibrosis, **Gastritis**, Grave's Disease, Halitosis (Bad Breath), **Hangover**, Hot Flashes, Hyperinsulinemia (Syndrome X), **Hypoglycemia**, Hypothyroid, Inflammation, **Inflammatory Bowel Disorders (Colitis, IBS)**, Jet Lag, Leaky Gut Syndrome, **Lupus**, **Mental Illness**, **Mood Swings**, Muscular Dystrophy, Myasthenia Gravis, **Pancreatitis**, Parkinson's Disease, Pertussis (Whooping Cough), Phobias, **PMS Type C**, Psoriasis, **Schizophrenia**, Sex Drive (low), Shingles, Sleep (restless and disturbed), **Sore Throat**, Stress, **Sugar Cravings**, Tendonitis, Thrush, **Tickle in Throat**, Triglycerides (high), **Ulcers**, **Weight Gain (aids for)**, Weight Loss (aids for)

Usage: Licorice root helps to stabilize blood sugar levels and is useful in treating both hypoglycemia and diabetes. It improves stamina and endurance. Licorice has anti-inflammatory properties and can be used in place of steroids. It increases energy in the body but is not highly stimulating. This herb has helped in cases of hyperactivity and bedwetting in children. The Chinese seldom use licorice alone but include it as a catalyst in many of their formulas. Licorice contains a glycoside that is many times sweeter than sugar and can be used in herbal teas for children to mask the taste of other, more disagreeable, herbs.

For bedwetting, take before bed. For blood sugar problems, stress or weak adrenals, take with breakfast, lunch and with a mid-afternoon snack if fatigued in the afternoon. This also reduces sugar cravings. As an anti-inflammatory, 2 capsules three times daily. Empty capsules and suck on powder for relief of sore or irritated throat and coughs. For ulcers, use 1 capsule of ATC concentrated licorice two to three times per day after meals for 3-4 weeks

Warnings: Although licorice is a safe herb, some cautions are necessary when taking large doses of licorice for long periods of time. Licorice should be avoided in cases of high blood pressure or when taking digitalis. It causes retention of water and sodium, and excretion of potassium, which can cause edema (water retention), high blood pressure, heart palpitations or a slowing of the heartbeat. Vertigo (dizziness) and headaches are early symptoms of overuse of licorice. Taking potassium supplement with licorice can help counteract some of these effects. These effects are seen most in individuals using licorice extracts or licorice derived drug isolates rather than taking the whole licorice root.

Licorice Root - Capsule (100)

Available in: US & Canada - Stock #420-6

Ingredients: Glycyrrhiza glabra (licorice) root

Licorice Root, ATC Concentrated - Capsule (50)

Available in: US & Canada - Stock #424-5

Ingredients: Concentrated extract of Glycyrrhiza glabra (licorice) root

Licorice Concentrate - Capsule (50)

Available in: Canada - Stock #420-6

Ingredients: Glycyrrhiza glabra (licorice) root

Licorice Root Extract (2 fl. oz.) - Liquid (Tincture)

Available in: US & Canada - Stock #1780-9

Ingredients: Licorice root (Glycyrrhiza glabra) in water and alcohol

Liquid Calcium

See *Calcium*

Liquid Chlorophyll

See *Chlorophyll*

Liquid Cleanse

Product Type: Formula

Properties: Digestive Tonic, Laxative (general), Purgative (Cathartic)

Systems Affected: Digestive System, Large Intestine (Colon)

Conditions: Colon (atonic), Constipation (adults), Gas and Bloating

Usage: A laxative formula in liquid form, Liquid Cleanse can be used like LBS II, LB-X or Senna Combination as a stimulant laxative to promote bowel evacuations. It also has some digestive tonic properties.

Warnings: This product contains senna, which is a harsh laxative that can cause dependancy with long term use. Avoid during pregnancy or while nursing unless directed otherwise by a competent health practitioner. Do not use if diarrhea, loose stools or abdominal pain are present or develop.

Liquid Cleanse (16 fl. oz.) - Liquid (Other)

Available in: US - Stock #3193-1

Ingredients: Aloe vera inner leaf (Aloe barbadensis), red raspberry fruit concentrate (Rubus idaeus), senna leaves (Cassia senna), cinnamon bark (Cinnamomum cassia), trace minerals, barberry root bark (Berberis vulgaris), xanthan gum, fennel seeds (Foeniculum vulgare), cornsilk (Zea mays), ginger rhizome (Zingiber officinalis), bitter orange (Citrus Aurantium), dandelion root (Taraxacum officinale), capsicum fruit (Capsicum annuum)

LIV-A

See *Liver Cleanse Formula*

LIV-C

See *Liver Balance*

LIV-GD

See *Milk Thistle Combination*

LIV-J

Product Type: Formula

Properties: Alterative (Blood Purifier), Bitter, Hepatic, Stomachic

Systems Affected: Digestive System, Gall Bladder, Liver, Pineal

Conditions: Addictions (alcohol), Allergies (food), Anger (excessive), Blood Poisoning, Boils, Cholesterol (low), Constipation (adults), Copper Toxicity, Fat Metabolism (poor), Gall Bladder (sluggish), Hyperinsulinemia (Syndrome X), Indigestion, Psoriasis

Usage: A general liver and digestive formula. Designed to work well with other J-formulas, such as Stress-J and Intestinal Soothe & Build. Primarily, LIV-J is a liver cleansing formula. Use 2-4 capsules with each meal and before bed.

Warnings: No known warnings.

LIV-J - Capsule (100)

Available in: US & Canada - Stock #1011-4

Ingredients: Dandelion, barberry, rose hips, fennel, horseradish, parsley, red beet

Liver Balance

Other Names: Tiao He

Product Type: Formula

Properties: **Alterative (Blood Purifier)**, **Cholagogue**, Detoxifying, Digestive Tonic, Galactagogue, **Hepatic**, Nervine

Systems Affected: Blood, Cyclic AMP (cAMP), Digestive System, Gall Bladder, **Liver**, Muscles, Ovaries, Pancreas

Conditions: Abdominal Pain and Inflammation, Allergies (food), **Anger (excessive)**, Backache (Back Pain, Lumbago), Breast Lumps, Chemical Poisoning, **Defensiveness**, Diarrhea, Dysmenorrhea, Dyspepsia, Eczema, Eye Infections, Fat Cravings, Fat Metabolism (poor), **Gall Bladder (sluggish)**, Gall Stones, **Hangover**, Hemochromatosis, **Hepatitis**, **Hypochondria**, Hypoglycemia, Inflammatory Bowel Disorders (Colitis, IBS), **Insomnia**, **Irritability**, Leaky Gut Syndrome, **Migraine**, **Mood Swings**, **Morning Sickness**, Nightmares, Pap Smear (abnormal), **PMS Type A**, **PMS Type S**, Poisoning (food), **Rashes and Hives**, Skin (infections), Skin Care (general), Sore Throat, Stress, Tonsillitis (Adenoids), Toxemia, **Triglycerides (low)**

Usage: The Chinese name for this formula, Tiao He, means "mediate harmony" because this formula helps harmonize the internal workings of the body. Liver Balance is a traditional Chinese formula for reducing excess "wood" energy in the body. In Chinese medicine, wood is associated with the liver and gallbladder. A person with excess "wood" is not only prone to a congested liver and gallbladder, they are also prone to anger, irritability and frustration, aggression and depression. Symptoms of excess "wood" energy could include hypoglycemia, migraine headaches, allergies, poor fat metabolism, abdominal pain and distention and skin problems. It may also be helpful for colitis, sore throat, tonsillitis, gallstones, eye infections, insomnia, digestive upset, addictions, PMS and breast lumps. Liver Balance helps cleanse the liver, nourish the blood, improve digestive function and promote detoxification and elimination. It is an excellent addition to a general cleansing program. Use 4 capsules two times daily.

Warnings: No known warnings.

Liver Balance [Wood Reducing] Tiao He - Capsule (100)

Available in: US - Stock #1860-1

Ingredients: Bupleurum, peony, thyphonium, cinnamon twig, dang gui, fushen, scute, zhishi fruit, atractylodes rhizome, panax ginseng, ginger rhizome, licorice

Liver Balance TCM Conc. - Capsule (30)

Available in: US - Stock #1008-8

Ingredients: Same as Stock #1860-1

LIV-C - Capsule (100)

Available in: Canada - Stock #1862-8

Ingredients: Bupleurum root, ho shou wu root, pinellia rhizome, cinnamon twig, dang gui root, fushen plant, zhishi fruit, scute root, atractylodes rhizome, Panax ginseng root, ginger rhizome, and licorice root

Liver Cleanse Formula

Other Names: LIV-A

Product Type: Formula

Properties: Alterative (Blood Purifier), Bitter, Cholagogue, Hepatic

Systems Affected: Gall Bladder, **Liver**, Pancreas

Conditions: Addictions (alcohol), Appetite (deficient), Constipation (adults), Gall Bladder (sluggish), Gall Stones, Hangover, Hepatitis, Jaundice (adults), Mononucleosis, Morning Sickness

Usage: A general liver and digestive formula with additional herbs for building up the blood and strengthening the kidneys. This formula is an old standby for basic liver related problems. Use 1-2 capsules two to three times per day.

Warnings: No known warnings.

Liver Cleanse Formula - Capsule (100)

Available in: US - Stock #1010-3

Ingredients: Red beet, dandelion, parsley, horsetail, yellow dock, black cohosh, birch leaves, blessed thistle, angelica, chamomile flowers, gentian, golden rod

LIV-A - Capsule (100)

Available in: Canada - Stock #1013-8

Ingredients: Dandelion, turmeric, parsley, blessed thistle, angelica, chamomile flowers, gentian root, Goldenrod

Lobelia

Latin Name: *Lobelia inflata*

Product Type: Single Herb

Properties: Acrid, **Alexipharmic**, Analgesic (Anodyne), Anti-abortive, Anti-epileptic, **Anti-smoking**, Anticephalalgic, **Antidote**, **Antihypertensive**, **Antispasmodic**, **Antitussive**, **Antivenomous**, **Anxiolytic**, **Bronchial Dilator**, Catalyst (Synergist), Decongestant, **Deobstruent**, Diaphoretic, **Emetic**, Expectorant, **Hypotensive**, Lymphatic, **Nauseant**, **Nervine**, Panacea, Relaxant, Sedative, Stimulant

Systems Affected: Bronchials, Circulation, **Ears**, Eyes, Lungs, Lymph Nodes, Lymphatic System, Mucus Membranes, Muscles, Nerves, Respiratory System, Stomach, Sweat Glands, Tonsils, Urinary System, **Vocal Cords**

Conditions: Abscesses, **Addictions (tobacco smoking or chewing)**, Allergies (respiratory), Angina, **Anxiety (Panic Attack)**, Anxiety Disorders, Arrhythmia, **Asthma**, Attention Deficit Disorder (ADD, ADHD), **Backache (Back Pain, Lumbago)**, **Bites and Stings**, **Blood Poisoning**, **Blood Pressure (high)**, **Bronchitis**, Bruises (healing), Bunions, Burning Feet or Hands, Canker Sores (Mouth Ulcers), **Cardiac Arrest (Heart Attack)**, Chicken Pox, Colds (decongestant), Colds (general remedies for), **Colic (adults)**, Colic (children), **Colon**

(spastic), Congestion (general), Congestion (lymphatic), Contagious Diseases, **Convulsions**, **Cough (general)**, **Cough (spastic)**, **Cramps and Spasms (general)**, **Croup**, Diphtheria, Dislocation, Dysmenorrhea, **Ear Infection or Earache**, Emphysema, **Epilepsy**, Fever, Fibromyalgia Syndrome (FMS), **Glands (swollen lymph)**, Glaucoma, Hashimoto's Disease (Thyroiditis), **Headache (tension)**, Hepatitis, Hiatal Hernia, **Hiccups**, Insomnia, **Kidney Stones**, Labor and Delivery, Laryngitis (Hoarseness), **Lead Poisoning**, Lymph Nodes or Glands (swollen), Mastitis, Meningitis, Mercury Poisoning, Migraine, **Miscarriage (prevention)**, **Mumps**, **Pain (general remedies for)**, Paralysis, **Pertussis (Whooping Cough)**, **Pleurisy**, PMS Type P, Pneumonia, Poison Ivy or Oak, **Poisoning (general)**, **Poisoning (food)**, Raynaud's Disease, Restless Leg Syndrome, Seizures, **Sore Throat**, Sprains, **Stiff Neck**, **Tachycardia**, **Teething**, **Tension**, Tetanus, **Tics**, **Tinnitus (Ringing in the Ears)**, **TMJ**, Toothache, Tuberculosis (Consumption, Scrofula), Ulcers, **Urination (burning or painful)**, Vaccines (detoxification from), Wheezing

Usage: Lobelia has earned a reputation with capsicum as a cure-all. Small doses are stimulating, large doses cause a profound state of relaxation. Taken internally in small doses, it can clear lymphatic channels, open the air passages and relieve mild pain. Larger doses may cause nausea, profuse sweating, vomiting and even a coma-like state of deep relaxation. Can slow the heartbeat. Use with capsicum or peppermint to counteract this effect. Extract: Internally, use 15 drops as an antispasmodic for asthma attack, muscle spasms, etc. Externally, apply lobelia extract combined with capsicum extract to relax muscle spasms, reduce pain and inflammation. Follow with Tei Fu oils. Capsicum and lobelia extracts may be applied to back and followed with Tei Fu oils for an herbal back adjustment. Use 1 teaspoonful in pint of water for an enema to relieve spastic colon. Capsules: Take 1 capsule as needed. Take larger doses with lots of warm water to induce vomiting. Can be used as an aid to quit smoking. Use 2 capsules of chamomile and 1 capsule of lobelia three times daily for this purpose. Comments: Small doses stimulate; large doses relax every muscle.

Warnings: Lobelia is often accused of being poisonous, although there is no record of it causing convulsions, coma or death in people as has been claimed. Although lobelia can produce severe symptoms, they pass quickly. Not recommended for weak, debilitated persons or persons who are deeply relaxed. Not recommended for long term use.

Lobelia - Capsule (100)

Available in: US - Stock #430-1

Ingredients: Lobelia inflata herb

Lobelia Essence (2 fl. oz.) - Liquid (Tincture)

Available in: US & Canada - Stock #1765-8

Ingredients: Lobelia inflata herb.

LOCLO

Product Type: Formula

Properties: Absorbant, **Anticholesteremic**, Demulcent (Mucilant), **Laxative (bulk)**, Laxative (general), Soothing

Systems Affected: Gall Bladder, Large Intestine (Colon), Liver, Weight Loss

Conditions: Chemical Poisoning, Cholesterol (high), Cholesterol (high), Constipation (adults), Diarrhea, Hemorrhoids, Inflammatory Bowel Disorders (Colitis, IBS), Multiple Sclerosis (MS), Weight Loss (aids for)

Usage: A bulk-forming laxative that cleanses the colon, helps diarrhea, hemorrhoids and high cholesterol. Add 1 tablespoon to 8 ounces of liquid and stir. Drink immediately. Drink 6-8 glasses of water per day when taking a fiber supplement like LOCLO.

Warnings: If you do not drink enough water with this formula, it may cause constipation. Long-term use can weaken digestion unless you also take some ginger, capsicum or other aromatic stomachic herbs occasionally.

LOCLO High Fiber Supplement (12 oz.) - Bulk Powder

Available in: US - Stock #1348-4

Ingredients: Psyllium hulls, apple fiber, acacia gum, guar gum, oat bran, broccoli, carrot, red beet, rosemary, tomato, turmeric, cabbage, Chinese cabbage, grapefruit and orange bioflavonoids, hesperidin, stevia, cinnamon bark, flax seed, fructose, citric acid, malic acid, natural apple flavor

LOCLO - Bulk Powder

Available in: Canada - Stock #1347-6

Ingredients: Psyllium seed hulls, fructose, apple fibre, oat bran, citric acid, cinnamon, acacia gum, guar gum, natural apple flavour, and SynerPro¨ concentrate (broccoli, carrot, red beet, rosemary, tomato, turmeric, cabbage, Chinese cabbage, grapefruit and orange bioflavonoids, and hesperidin)

Love and Peas

Product Type: Formula

Properties: Antioxidant, Nutritive

Systems Affected: Kidneys, Whole Body

Conditions: Addictions (sugar or food), Blood Pressure (high), Body Building, Children's Remedy, **Hypoglycemia**, Overalkalinity, Pregnancy (herbs and supplements for), **Weight Loss (aids for)**

Usage: Love and Peas is a protein powder made from yellow split peas. It also contains whole foods with antioxidant properties and enzymes for better digestion and absorption. It is a great alternative to whey and soy proteins, which cause allergies in many people. Love and Peas is dairy-free, lactose-free and gluten-free and Vegan Certified. Studies suggest pea protein may support kidney function and the maintenance of normal blood pressure. Mix 2 level scoops (45 g) of Love and Peas powder with approximately 12-16 ounces of cold water, or mix to taste.

Warnings: No known warnings.

Love and Peas - Bulk Powder

Available in: US - Stock #3082-9

Ingredients: 20 grams protein per serving from split, yellow peas, antioxidant blend of grape skin and grape seed extract, whole food complex (oat, beta glucans, blueberries, cranberries, pomegranate, carrots, broccoli), flax seed and borage oil, vitamin blend,vanilla flavor

Lung Support

Other Names: Fu Lei

Product Type: Formula

Properties: Antibacterial, Bitter, Bronchial Dilator, Decongestant, Expectorant, **Lung Tonic**, Pectoral, Pulmonary, Tonic

Systems Affected: Bronchials, **Lungs**, Muscles, **Respiratory System**, Testes

Conditions: Addictions (tobacco smoking or chewing), Asthma, Bronchitis, Cancer (natural therapy for), Cold Hands and Feet, Colds (prevention), Convalescence, Cough (dry), **Cough (dry)**, Dizziness (Vertigo), **Emphysema**, Flu, Grief (excessive), Grief (excessive), Muscle Tone (lack of), Pain (general remedies for), Perspiration (excessive), Tuberculosis (Consumption, Scrofula), Urine (scant), **Weight Gain (aids for)**, **Wheezing**

Usage: Metal is the stuff swords and shields are made of, so the element of metal in Chinese medicine is associated with the immune system and the ability of the body to protect itself from infection. The primary line of immune defense is the mucous membranes lining our lungs and digestive tract, as most infections enter the body through these membranes. This helps explain why the Chinese metal element is also associated with the lungs and colon.

Lung Support is a Chinese formula designed to enhance deficient metal energy. It enhances the function of the mucous membranes lining the lungs and the colon to improve immune response. Chinese Lung Support is useful for people who suffer from chronic infections of the lungs, frequent colds and flu, dry cough, tightness in the chest, shortness of breath, general weakness and fatigue. It can also be helpful for chronic lung diseases, including emphysema, tuberculosis, chronic asthma or bronchitis and lungs that have been damaged or weakened by smoking.

The deficiency of metal energy can also be associated with deep-seated grief and sadness that can make a person aloof, cold and withdrawn. This formula can help to resolve this deep-seated grief. People who are unnaturally thin, suffer from weak muscles and energy loss or who are pale with flushed cheeks may also find this formula helpful.

Lung Support stimulates the immune system and increases the body's stamina during times of weakness. It is a decongestant and expectorant. By dilating bronchioles and increasing blood circulation, it restores deep breathing and strengthens the lungs.

For respiratory weakness it is suggested to take 3 capsules 3 times daily. Also available in a TCM concentrate where the dose is 1 capsule twice daily.

Warnings: No known warnings.

Lung Support [Metal Supporting] Fu Lei - Capsule (100)

Available in: US - Stock #1887-6

Ingredients: Astragalus root, Aster root, Qinjiao root, Platycodon root, Anemarrhena rhizome, Bupleurum root, Dang gui root, Lycium bark, Ophiopogon root, Panax ginseng root, Atractylodes rhizome, Blue citrus peel, Citrus peel, Typhonium flagelliforme, Schizandra fruit, licorice.

Lung Support TCM Concentrate - Capsule (30)

Available in: US - Stock #1004-3

Ingredients: Same as Stock #1887-6

LH-C - Capsule (100)

Available in: Canada - Stock #1890-3

Ingredients: Same as Stock #1887-6

Lutein

Product Type: Formula

Properties: Antioxidant

Systems Affected: Breasts, Eyes

Conditions: Cancer (natural therapy for), Macular Degeneration

Usage: Lutein is a carotenoid that helps protect the macula lutea from oxidative damage. It may also help to improve vision. Lutein also has a protective effect against breast cancer. Use 1 capsule daily with a meal.

Warnings: No known warnings.

Lutein (10 mg) - Capsule (60)

Available in: US - Stock #1855-6

Ingredients: Lutein, zeaxanthin, hibiscus.

LYM-MX

See *Lymphomax*

Lymph Gland Cleanse

Other Names: IGS II

Product Type: Formula

Properties: Antibacterial, Antiscrofulous, **Lymphatic**

Systems Affected: Ears, Immune System, Liver, **Lymph Nodes**, Lymphatic System, Tonsils

Conditions: Abscesses, **Adenitis**, Blood Poisoning, Colds (with fever), Ear Infection or Earache, Epstein Barr Virus (Chronic Fatigue Syndrome, CFS), Eyes (red or itching), Fever, Flu, Gastritis, Gingivitis (Bleeding Gums, Gum Disease, Pyorrhea), Glands (swollen lymph), Infection (bacterial), Inflammation, Lymph Nodes or Glands (swollen), Mastitis, Rashes and Hives, Staph Infections, TMJ, Tonsillitis (Adenoids)

Usage: This formula was designed primarily for swollen lymph nodes and low-grade infections in the body. Use 2 capsules three times daily or every 2 hours for serious infection. Can be applied externally as a poultice.

Warnings: Not for hypoglycemics, use Lymph Gland Cleanse-HY (formerly HIGS-II) instead.

Lymph Gland Cleanse - Capsule (100)

Available in: US - Stock #960-4

Ingredients: Parthenium, golden seal, yarrow, capsicum

IGS II - Capsule (100)

Available in: Canada - Stock #964-3

Ingredients: Same as Stock #960-4

Lymph Gland Cleanse-HY

Other Names: HIGS II

Product Type: Formula

Properties: Antibacterial, Antiscrofulous, Bitter, Immune Stimulant, **Lymphatic**

Systems Affected: Ears, Immune System, Liver, **Lymph Nodes, Lymphatic System**, Tonsils

Conditions: **Adenitis**, Bites and Stings, Blood Poisoning, Bronchitis, Colds (general remedies for), Contagious Diseases, Ear Infection or Earache, Emphysema, Fever, Flu, **Glands (swollen lymph)**, Infection (bacterial), Inflammation, Lymph Nodes or Glands (swollen), **Mastitis**, Psoriasis, Staph Infections, Tonsillitis (Adenoids), Wounds and Sores

Usage: An alternative formula to Lymph Gland Cleanse. Substitutes myrrh gum for golden seal. Used internally for lymphatic swellings and infections. It is very helpful for low-grade infections, tonsillitis, bronchitis and ear infections. May also help psoriasis and breast swelling. Use 2 capsules 3x daily or every 2 hours for serious infection. Can be applied externally as a poultice to sores, bites and stings. We prefer this formula to Lymph Gland Cleanse.

Warnings: No known warnings.

Lymph Gland Cleanse-HY - Capsule (100)

Available in: US - Stock #920-3

Ingredients: Parthenium, yarrow, myrrh gum, capsicum

Lymphatic Drainage

Product Type: Formula

Properties: Alterative (Blood Purifier), Antiscrofulous, Decongestant, **Deobstruent**, Diuretic, **Lymphatic**, Stimulant

Systems Affected: Appendix, Breasts, Circulation, Eustachian Tubes, Immune System, Kidneys, Liver, Lungs, **Lymph Nodes**, **Lymphatic System**, Tonsils, Urinary System

Conditions: Abrasions, Abscesses, Acquired Immune Deficiency Syndrome (AIDS/HIV), Addictions (sugar or food), **Adenitis**, Antibiotics (side effects of), Appendicitis, Appetite (deficient), Arthritis, Bladder (irritable), Blood Pressure (high), Body Odor, Boils, Breast Lumps, Breasts (swelling and tenderness), Congestion (general), **Congestion (lymphatic)**, Congestive Heart Failure, Cough (damp), Cystic Breast Disease, Cysts, Diabetes, Digestion (poor), Disks (spinal - bulging or slipped), Ear Infection or Earache, Edema (Dropsy, Water Retention, Swelling), Epstein Barr Virus (Chronic Fatigue Syndrome, CFS), Eyes (red or itching), **Glands (swollen lymph)**, Hodgkin's Disease, Lymph Nodes or Glands (swollen), **Mastitis**, Mumps, Oral Surgery, Overacidity, Pain (general remedies for), PMS Type H, **Tonsillitis (Adenoids)**, Urine (scant), Vaccines (detoxification from)

Usage: This formula helps stimulate lymphatic drainage and clear toxins from tissue spaces. Use approximately 15-20 drops or 1/4 teaspoon, in water twice daily. A good practice to improve lymphatic drainage is to put some Lymphatic Drainage and Kidney Drainage formulas in a quart of water and sip the water frequently throughout the day.

Warnings: No known warnings.

Lymphatic Drainage (2 fl. oz.) - Liquid (Glycerite)

Available in: US - Stock #3171-7

Ingredients: Cleavers, red clover, stillingia, prickly ash bark, vegetable glycerin

Lymphomax

Product Type: Formula

Properties: Antioxidant, Antiscrofulous, Immune Stimulant, Lymphatic

Systems Affected: Appendix, Immune System, Liver, **Lymph Nodes**, Lymphatic System, **Tonsils**

Conditions: **Acquired Immune Deficiency Syndrome (AIDS/HIV)**, **Congestion (lymphatic)**, Contagious Diseases, **Glands (swollen lymph)**, **Hodgkin's Disease**, **Leukemia**, **Mumps**, **Tonsillitis (Adenoids)**

Usage: This is an improved version of the Lymphomax formula. It helps improve lymphatic drainage, shrink swollen lymph nodes, support the immune system in fighting chronic infections and purifying the blood. It is useful for problems involving lymphatic congestion such as earaches, sore throat, tonsillitis and swollen lymph nodes. It may also help Hodgkin's disease. Use 2 capsules two to three times daily.

Warnings: No known warnings.

Lymphomax - Capsule (100)

Available in: US - Stock #4077-6

Ingredients: Mullein, lobelia, plantain, cleavers, red root, bayberry, alfalfa, chamomile, echinacea, yarrow, garlic, chlorophyll

LYM-MX - Capsule (100)

Available in: Canada - Stock #4036-0

Ingredients: Beta-carotene, vitamin C, vitamin B6, vitamin D3, magnesium, cleavers, alfalfa, chamomile, Echinacea purpurea, parthenium, Siberian ginseng, yarrow, lecithin, garlic, Co-Q10, chlorophyll

Lymphostim

Product Type: Homeopathic

Properties: Lymphatic

Systems Affected: Immune System, Lymph Nodes, Lymphatic System

Conditions: Congestion (lymphatic)

Usage: Used to stimulate the lymphatic system and to boost the immune system. It helps to relieve infection and treat both mental and physical fatigue. Take 10-15 drops under the tongue three to four times daily.

Warnings: NSP does not recommend this product for children under the age of 12 except on the advice of a health professional. Other than the alcohol content, which is not good for infants, we consider this product safe for children.

Lymphostim (1 fl.oz.) - Homeopathic (Liquid)

Available in: US - Stock #8710-4

Ingredients: Active ingredients: Echinacea angusti-folia (Cone Flower) 4x, Aralia quinquefolia (Ginseng) 4x, Arnica montana (Mountain Arnica) 4x, Baptisia tinctoria (Wild Indigo) 4x, Adrenal Tissue 6x, Spleen Tissue 6x, Thymus Tissue 6x, Selenium (Selenium) 10x, Zincum Muriaticum (Zinc Chloride) 10x, Sulphur (Sulfur) 30x. Other ingredients: Purified water and 20% USP alcohol

Maca

Latin Name: *Lepidium meyennii*

Product Type: Single Herb

Properties: Adaptagen, Anti-aging, **Aphrodisiac**, Stimulant, Tonic

Systems Affected: Adrenal Cortex, Glandular System, **Reproductive Glands, Testes**

Conditions: Adrenals (exhaustion, weakness or burnout), Circulation (poor), Energy (lack of), Epstein Barr Virus (Chronic Fatigue Syndrome, CFS), Erectile Dysfunction, Fatigue, Hair (loss or thinning), **Sex Drive (low)**, Stress, Testosterone (low)

Usage: This product is a standardized extract of maca root, a plant from the Andean mountains of South America. The plant has a reputation as an aphrodisiac, strengthening sexual desire and performance in both men and women. It has also been called Peruvian ginseng, because it is a general tonic and adaptagen, helping to reduce stress and increase general health and energy. It also helps mental concentration. Dosage is 1 capsule three times daily.

Warnings: No known warnings, although tonic herbs are generally contraindicated with acute inflammation.

Maca - Capsule (90)

Available in: US & Canada - Stock #1117-2

Ingredients: Standardized extract of Lepidium meyennii (maca) root

Magnesium

Other Names:

Product Type: Nutrient

Properties: Alkalinizer, Anti-arrhythmic, Anti-epileptic, Anti-urolithic, Antihypertensive, Antilithic, Antispasmodic, Aperient, Lithotriptic, Vasodilator

Systems Affected: Bones, Brain, Gall Bladder, **Heart**, Intestinal System, Large Intestine (Colon), Mitochondria, **Muscles**, Nerves, Ovaries, Thyroid, Urinary System

Conditions: Abdominal Pain and Inflammation, Addictions (alcohol), Adrenals (exhaustion, weakness or burnout), Afterbirth Pain, Anal Fistula or Fissure, Aneurysm, **Angina, Anxiety (Panic Attack), Arrhythmia**, Arthritis, Attention Deficit Disorder (ADD, ADHD), Autism, **Backache (Back Pain, Lumbago), Bedwetting**, Birth Control (countering side effects), **Blood Pressure (high), Breasts (swelling and tenderness)**, Bursitis, Calcium Deficiency, **Calcium Deposits (Calcification), Cardiac Arrest (Heart Attack)**, Cardiovascular Disease (Heart Disease), Celiac Disease, **Colon** (spastic), Confusion, **Congestive Heart Failure, Constipation** (adults), **Constipation** (children), **Cramps** (leg), **Cramps** (menstrual), **Cramps and Spasms** (general), Crohn's Disease, Cystic Breast Disease, Depression, Diarrhea, Digestion (poor), **Dysmenorrhea**, Epilepsy, Epstein Barr Virus (Chronic Fatigue Syndrome, CFS), Fatigue, Fibromyalgia Syndrome (FMS), Floaters, Gall Stones, **Glaucoma**, Grave's Disease, **Heart (weakness), Heart Fibrillation or Palpitations, Heart Valves**, Inflammatory Bowel Disorders (Colitis, IBS), **Insomnia, Irritability, Kidney Stones**, Labor and Delivery, Mental Illness, Migraine, **Myasthenia Gravis**, Narcolepsy, **Neurosis, Overacidity, Pain (general remedies for)**, PMS (general), PMS Type A, **PMS Type C**, PMS Type D, **PMS Type H, PMS Type P**, Pregnancy (herbs and supplements for), Progesterone (low), Prolapsed Uterus, **Restless Leg Syndrome, Schizophrenia, Stress**, Strokes, **Tachycardia, Teeth** (grinding), **Tension, Tics**, TMJ, **Tremors, Twitching**

Usage: Magnesium is extremely important to many body functions. It acts as a catalyst in the utilization of carbohydrates, fats, protein, calcium, phosphorus and possibly potassium. This vital mineral helps produce energy inside cells. Magnesium works hand in hand with calcium in the body. Calcium ions make muscles contract, while magnesium ions help muscles relax. So magnesium helps relieve muscle cramping and spasm, relieves colic and spastic bowel conditions and helps prevent heart attacks. Magnesium, along with vitamin B6, is very helpful for relieving most PMS symptoms. With vitamin B6, it also helps dissolve calcium oxalate kidney stones. Spastic muscles, nervous twitching, addiction to stimulant laxatives, spastic colon, hypersensitivity to noises and calcium deposits are all symptoms of magnesium deficiency, which is far more common than calcium deficiency. In fact, over consumption of calcium as a supplement actually contributes to the development of magnesium deficiency. Magnesium, along with potassium, are also the first mineral buffers to be lost in an over acid condition in the body. Alcohol, diuretics, diarrhea, fluoride, zinc and vitamin D deplete the body's supply of magnesium. Foods high in oxalic acid also inhibit absorption of magnesium. The magnesium complex is a better form of calcium for most people, as it is generally better absorbed. Take 1-2 capsules one to three times daily. For the Magnesium tablets, use 1-2 daily.

Warnings: There are no known warnings for the therapeutic doses listed above.

Magnesium Complex - Capsule (100)

Available in: US - Stock #1859-8

Ingredients: 100 mg each of magnesium citrate and magnesium malate.

Magnesium (250 mg) - Tablet (180)

Available in: US & Canada - Stock #1786-6

Ingredients: 250 mg. of magnesium oxide, magnesium amino acid chelate, magnesium stearate, calcium, phosphorus, licorice root, kelp plant, peppermint leaves, white willow bark.

Mandarin (Red)

Product Type: Essential Oil

Properties: Antioxidant, Antiseptic, Antispasmodic, Aromatic, Carminative, Cholagogue, Cicatrisant, Diuretic, Tonic

Systems Affected: Digestive System, Intestinal System, Liver, Nerves, Skin

Conditions: Acne (Pimples, Blackheads), Asthma, Belching, Children's Remedy, Colic (children), Constipation (adults), Cramps (menstrual), Depression, Digestion (poor), Dyspepsia, Gas and Bloating, Hiccups, Insomnia, Scars / Scar Tissue, Stretch Marks

Usage: Mandarin oil is gentle, smoothing, and has strengthening properties. The aroma is sweet and tangy. The name comes from the mandarins of Cochin China where it originates and to whom the fruit was offered as a gift. In France, it has long been regarded as a safe children's remedy because of its gentle action. Mandarin can be used in a massage oil for nervousness and tension and lymphatic stimulation. Massaged over the digestive organs, it helps with the breakdown of fats in the gallbladder and in relieving colic, hiccups and intestinal gas. It can also be applied topically as a skin toner, an acne treatment, and to prevent stretch marks and scaring. Emotionally, it relieves depression, having an uplifting and freeing energy. It is calming and cheering, easing stress, grief and hysteria. It promotes self-awareness and an improved self-image.

Warnings: Do not take internally. Do a patch test first before using on skin. Avoid sun exposure after topical use. Always mix with a carrier and never apply directly on the skin.

Mandarin, Red Essential Oil (5ML) - Essential Oil

Available in: US - Stock #3897-4

Ingredients: Mandarin (Citrus reticulata) Essential Oil

Marjoram (Sweet)

Product Type: Essential Oil

Properties: Analgesic (Anodyne), Anaphrodisiac, Antibacterial, Antiseptic, Antispasmodic, Antiviral, Aromatic, Carminative, Cephalic, Diaphoretic, Diuretic, Emmenagogue, Expectorant, Hypotensive, Nervine, Parasympathomimetic, Sedative, Stomachic, Vasodilator, Vulnerary

Systems Affected: Digestive System, Female Reproductive, Intestinal System, Muscles, Respiratory System, Skin, Structural System

Conditions: Addictions (alcohol), Blood Pressure (high), Bronchitis, Bruises (healing), Colds (general remedies for), Colic (children), Congestion (general), Diphtheria, Grief (excessive), Hypochondria, Insomnia, Irritability, Menstruation (scant), Migraine, Pregnancy (herbs and supplements to avoid during), Rheumatoid Arthritis (Rheumatism), Sex Drive (excessive), Sprains, Stress

Usage: Marjoram originates from Greek meaning 'joy of the mountains.' It is a warm woody scent. In ancient times it was often referred to for longevity. It warms both the mind and the body. Marjoram can be used in massages and baths for insomnia, PMS, scanty or painful periods. A compress can be used for vasoconstrictive migraines, rheumatic pains, sprained or strained muscles, bruises and sore throat. Emotionally, marjoram brings a sense of peace, calmness, balance and self-assurance. It reduces stress and anxiety, and eases grief, hostility, anger and irritability. It promotes courage and confidence in those with a weak will.

Warnings: Do not take internally. Do a patch test first before using on skin. Avoid sun exposure after topical use. Avoid using topically if pregnant. Do not use with low blood pressure. Do not overuse.

Marjoram, Sweet Essential Oil (5ML) - Essential Oil

Available in: US - Stock #3896-3

Ingredients: Marjoram (Origanum majorana) Essential Oil

Marshmallow

Latin Name: *Althaea officinalis*

Product Type: Single Herb

Properties: Absorbant, Antacid, Balsamic, **Demulcent (Mucilant)**, Diuretic, Emollient, Expectorant, **Galactagogue**, Mineralizer, Moistening, Soothing, Vulnerary

Systems Affected: Bladder (Urinary), Intestinal System, Lungs, Lymphatic System, Mucus Membranes, Pancreas Head, Respiratory System, Skin, Small Intestines, Urinary System

Conditions: Abscesses, Anemia, Bedwetting, **Bladder (irritable)**, Bladder (ulcerated), Blood in Urine, Boils, **Breast Milk (increase or enrich)**, Bronchitis, Burns and Scalds, Cholera, Colds (general remedies for), Colitis, Convalescence, **Cough (dry)**, Cough (general), Cystic Fibrosis, **Cystitis**, Diarrhea, Diverticulitis (Diverticuli), Emphysema, Eyes (red or itching), Failure to Thrive, Gastritis, Gonorrhea, Hemorrhoids, Inflammation, Inflammatory Bowel Disorders (Colitis, IBS), Injuries, Kidney Stones, **Nephritis**, Pertussis (Whooping Cough), Rashes and Hives, Sprains, Teething, Tickle in Throat, Ulcerations (external), **Urethritis**, **Urination (burning or painful)**, Urination (frequent), Wasting, Wounds and Sores

Usage: For intestinal irritation, take 1-3 capsules up to four times daily. Can be used as a suppository for hemorrhoids. Make into tea for gargle to help relieve mouth and throat irritations. Can be used externally as a poultice to heal minor irritations. Useful to soothe burning urination, inflamed kidneys, and eases the passing of kidney stones. Use 2 capsules every two hours with lots of water. Combines well with cornsilk, golden seal and uva ursi for urinary problems. May also ease respiratory congestion and dry cough. It is also useful as a mild food for nourishing weak and debilitated persons. Tea is a good mild remedy for children. Combined with blessed thistle, it is excellent for nursing mothers. Use 1-2 capsules three times daily for nursing.

Warnings: Very mild and safe remedy for children, infants and elderly persons. There are no known toxic effects of marshmallow, but if any diarrhea occurs from its use, cut back or discontinue use.

Marshmallow - Capsule (100)

Available in: US & Canada - Stock #440-0

Ingredients: Althaea officinalis (marshmallow) root.

Marshmallow & Fenugreek

Product Type: Formula

Properties: Absorbant, Demulcent (Mucilant), Emollient, Expectorant, Soothing

Systems Affected: Lungs, Respiratory System, Stomach

Conditions: Allergies (respiratory), Angina, Asthma, Breast Milk (increase or enrich), Bronchitis, Congestion (general), Cough (dry), Cough (general), Emphysema, Laryngitis (Hoarseness), Pleurisy, Pneumonia

Usage: This is one of the few formulas that is moistening to the lung tissue, rather than drying. It helps where there is a deficient production of mucus, resulting in a dry cough. For soothing irritated respiratory passages, dry cough, tickle in throat, take 2 capsules every hour with large glass of water. For chronic lung conditions, take 2 capsules two to three times daily. Taking this formula with mullein can enhance its effectiveness.

Warnings: No known warnings.

Marshmallow & Fenugreek - Capsule (100)

Available in: US - Stock #843-1

Ingredients: Fenugreek, marshmallow, slippery elm

Marshmallow & Pepsin

See *Small Intestine Detox*

Master G

See *Master Gland*

Master Gland

Product Type: Formula

Properties: Adaptagen, **Glandular**

Systems Affected: **Glandular System**, Liver, **Pituitary (anterior)**, Urinary System, Weight Loss

Conditions: Acne (Pimples, Blackheads), Adrenals (exhaustion, weakness or burnout), Age Spots, Aging (prevention), Anxiety (Panic Attack), Attention Deficit Disorder (ADD, ADHD), Bedwetting, Breast Lumps, Cholesterol (low), Cystic Breast Disease, Depression, Disks (spinal - bulging or slipped), Energy (lack of), Hot Flashes, **Infertility**, Labor (to induce), Myasthenia Gravis, Triglycerides (low)

Usage: This is a general glandular supplement containing herbs for each gland in a base of vitamins and minerals. This formula is useful as an overall glandular tonic. Take 1-2 capsules three times daily.

Warnings: No known warnings.

Master Gland Capsules - Capsule (120)

Available in: US - Stock #3040-3

Ingredients: Vitamins A, C, E, zinc, pantothenic acid, manganese, potassium, lecithin, licorice, alfalfa, asparagus, black walnut, kelp, parsley, parthenium, thyme, dandelion, dong quai, lemon bioflavonoids, schizandra, eleuthero, marshmallow root, uva ursi

Master G - Capsule (120)

Available in: Canada - Stock #2874-6

Ingredients: Vitamin A (palmitate), beta carotene, vitamin D3 (cholecalciferol), vitamin E (d-alpha tocopheryl succinate), vitamin B1 (thiamine mononitrate), vitamin B2 (riboflavin), niacinamide 8.3, vitamin B6 (pyridoxine hydrochloride), vitamin B12 (cyanocobalamin), folic Acid, pantothenic acid (calcium pantothenate), biotin 8.3, vitamin C (ascorbic acid), choline bitartrate, inositol, calcium (dicalcium phosphate, amino-acid chelate), magnesium (oxide, amino acid chelate), iron (gluconate), copper (gluconate), manganese (amino acid chelate), potassium (citrate), iodine (potassium), zinc (gluconate, oxide), selenium (amino acid chelate), chromium (amino acid chelate), p-aminobenzoic acid, citrus bioflavonoids, rutin, freeze-dried bovine adrenal, spleen, thymus substances, butcher's broom, and ginkgo biloba.

MC

See *Mega-Chel*

Mega-Chel

Product Type: Formula

Properties: Cardiac, **Chelating**, Deobstruent, Opthalmicum, Stimulant

Systems Affected: Arteries, **Blood Vessels**, Brain, Capillaries, Cardiovascular System, **Circulation**, Eyes, **Heart**, **Veins**

Conditions: Aging (prevention), **Alzheimer's Disease**, **Arteriosclerosis (Atherosclerosis, Hardening of the Arteries)**, Blood Clots (prevention of), **Blood Pressure (high)**, **Burning Feet or Hands**, **Calcium Deposits (Calcification)**, **Cardiovascular Disease (Heart Disease)**, Cataracts, Chills, Cholesterol (high), **Circulation (poor)**, **Circulation (to the brain)**, **Cold Hands and Feet**, Concentration (poor), **Dementia**, Depression, Diabetes, Erectile Dysfunction, Eye Problems (general), **Eyesight (to improve)**, Floaters, Free Radical Damage, **Gangrene**, Glaucoma, **Heavy Metal Poisoning**, **Lead Poisoning**, **Macular Degeneration**, **Memory and Brain Function**, Mental Illness, Mercury Poisoning, **Numbness**, Parkinson's Disease, PMS Type D, **Senility**, Spider Veins, **Strokes**, **Thinking (cloudy)**, Tinnitus (Ringing in the Ears), **Varicose Veins**

Usage: This is an oral chelation formula, used to strip plaque from artery walls to improve circulation. It aids symptoms of hardening of the arteries such as poor circulation, absent-mindedness, wounds that will not heal in extremities, fatigue and heart disease. It can also be used for Alzheimer's disease. Another benefit of Mega-Chel is that it removes some heavy metals from the system. The full dose is 5-6 tablets two times per day, mornings and evenings. However, it is best to start slowly with 1 tablet morning and evening and gradually increase the dose. It is also wise to taper off slowly after having been on this formula. The effect is enhanced by taking, 1 ounce of Colloidal Minerals morning and evening. Besides reducing arterial plaque and heavy metals, this formula is helpful for diabetes, failing eyesight, macular degeneration, varicose veins and as a vitamin/mineral supplement for elderly people. For these purposes use 1-2 tablets two times daily. Use with Perfect Eyes for eye problems and combine with butcher's broom for varicose veins.

Warnings: Starting with a full dose may cause dizziness, headache, skin rashes, kidney stress and intestinal gas. May temporarily elevate blood pressure and cholesterol levels when first starting use. These problems subside with continued use. Mega-Chel often requires supplements to support liver detoxification and kidney function, as these organs are used to flush the toxins released during oral chelation.

Mega-Chel - Tablet (90)

Available in: US - Stock #4050-6

Ingredients: Vitamins A, B-1, B-2, B-6, B-12, C, D, E, niacin, calcium, iron, folic acid, iodine, magnesium, zinc, copper, biotin, pantothenic acid, phosphorus, l-cystine HCL, potassium, p-aminobenzoic acid (PABA), l-methionine, citrus bioflavonoids, rutin, adrenal substance, spleen substance, thymus substance, inositol, ginkgo biloba, hawthorn berries, coenzyme Q10, manganese, selenium, chromium.

Mega-Chel - Tablet (180)

Available in: US - Stock #1611-1

Ingredients: Same as Stock #4050-6

MC - Tablet (180)

Available in: Canada - Stock #4002-8

Ingredients: Vitamin A, Beta Carotene, vitamin D3, vitamin E, vitamin B1, vitamin B2, niacinamide, vitamin B6, vitamin B12, Folic Acid, pantothenic acid, biotin, vitamin C, p-aminobenzoic acid, citrus bioflavonoids, rutin, butcher's broom, ginkgo biloba, freeze-dried bovine adrenal, spleen, thymus substances, choline bitartrate, inositol, calcium, magnesium, iron, copper, manganese, potassium, iodine, zinc, selenium, chromium

Melatonin Extra

Product Type: Formula

Properties: Antioxidant, Sedative

Systems Affected: **Pineal, Pituitary (infundibulum)**

Conditions: Aging (prevention), Cataracts, Depression, Fatigue, **Insomnia, Jet Lag,** Pregnancy (herbs and supplements to avoid during), Triglycerides (high)

Usage: Melatonin is a hormone that is produced naturally in the pineal gland. Its production is linked to the amount of sunlight received by the eye. Take 1 capsule 30 minutes prior to bedtime

Warnings: Because melatonin is a naturally occurring hormone, long-term supplementation may alter the body's ability to produce sufficient amounts of it on its own. Melatonin is best used for short-term treatment of insomnia and jet lag. To help the body produce more melatonin, spend more time in the sunlight. Melatonin is not recommended for use by children, adolescents, pregnant or lactating women.

Melatonin Extra (3 mg) - Capsule (60)

Available in: US - Stock #2830-4

Ingredients: 3 mg melatonin, ginkgo biloba, eleuthero, Vitamin E.

Men's Formula

Product Type: Formula

Properties: Diuretic, Glandular

Systems Affected: **Male Reproductive, Prostate,** Urinary System

Conditions: Asthma, **Benign Prostate Hyperplasia (BPH),** Circulation (poor), Erectile Dysfunction, Prostatitis, Testosterone (low), Urine (scant)

Usage: This formula improves kidney and prostate function in males. It has a protective effect against prostate cancer and helps reduce prostate swelling. It may also help with impotency. For intensive use: take 3 capsules twice daily with morning and evening meals. For maintenance: take 2 capsules daily with a meal.

Warnings: No known warnings.

Men's Formula with Lycopene - Capsule (60)

Available in: US - Stock #3112-7

Ingredients: Saw palmetto extract, pygeum extract, lycopene, stinging nettle extract, gotu kola, zinc

Men's Formula Minerals - Capsule (60)

Available in: Canada - Stock #3111-3

Ingredients: Saw palmetto extract, pygeum extract, stinging nettle extract, gotu kola, zinc

Menopause

Product Type: Homeopathic

Properties: Glandular

Systems Affected: Female Reproductive, Glandular System

Conditions: Hot Flashes, Menopause

Usage: Used for the relief of symptoms associated with menopause, including nervous irritability, bloating, cravings, discomfort and hot flashes. Take 10-15 drops under the tongue every two hours, until symptoms improve or as needed. Then take every four hours until symptoms are relieved.

Warnings: NSP recommends not taking this product for longer than five days.

Menopause (1 fl. oz.) - Homeopathic (Liquid)

Available in: US - Stock #8728-2

Ingredients: Active ingredients: Sanguinaria canadensis (Blood Root) 3x, Cimici-fuga racemosa (Black Cohosh) 6x, Helonias dioica (False Unicorn) 6x, Salix nigra (Black Willow) 6x, Sepia (Cuttle Fish) 10x, Lachesis (Viper Venom) 12x. Other ingredients: Purified water and 20% USP alcohol

Menstrual

Product Type: Homeopathic

Properties: Analgesic (Anodyne), Glandular

Systems Affected: Female Reproductive, Glandular System

Conditions: Dysmenorrhea, Menstrual Irregularity

Usage: Used for the relief of the minor pain and discomfort of premenstrual syndrome (PMS) and menstrual periods. Take 10-15 drops under tongue every 10-15 minutes or as needed until symptoms improve. Then decrease to every one or two hours, then to four times daily until symptoms are relieved.

Warnings: No known warnings.

Menstrual (1 fl. oz.) - Homeopathic (Liquid)

Available in: US - Stock #8702-9

Ingredients: Active ingredients: Viburnum opulus (High Bush Cranberry) 4x, Cimicifuga racemosa (Black Cohosh) 4x, Caulophyllum thalictroides (Blue Cohosh) 6x, Cuprum Sulphuricum (Cupric Sulfate) 6x, Magnesia Phosphorica (Magnesium Phosphate) 6x, Platinum Metallicum (Metallic Platinum) 12x. Other ingredients: Purified water and 20% USP alcohol

Menstrual-Reg

Product Type: Formula

Properties: Astringent, Glandular, **Hemostatic**, Styptic, Vasoconstrictor

Systems Affected: Blood, Female Reproductive, Intestinal

Conditions: Bleeding (external), **Bleeding (internal), Blood in Stool**, Dysmenorrhea, Fibroids (uterine), Hemorrhoids, **Menorrhagia (Heavy Menstrual Bleeding), Nose Bleeds**

Usage: This formula contains herbs that reduce heavy menstrual bleeding through a variety of mechanisms. It helps balance hormones, as well as slow blood flow. The herbs in this formula can also arrest bleeding in the gastrointestinal tract in cases of ulceration or hemorrhoids. To reduce heavy menstrual bleeding take 2-3 capsules with a meal three times daily. For best results, this formula should be taken over a period of several months. For a more immediate effect while bleeding is actually in progress, it can be taken in larger amounts for short periods of time, such as 2-3 capsules every two hours for a period of several days.

Warnings: No known warnings.

Menstrual-Reg - Capsule (100)

Available in: US - Stock #1125-3

Ingredients: Yarrow, shepherd's purse, lady's mantle, black haw root, sarsaparilla root, false unicorn, stinging nettle leaves, chaste tree concentrate

MetaboMax

Product Type: Formula

Properties: Anti-obesic, Decongestant, Stimulant

Systems Affected: Circulation, Digestive System, Liver, Respiratory System

Conditions: Appetite (excessive), Cellulite, Energy (lack of), Fat Metabolism (poor), Fatigue, Pregnancy (herbs and supplements to avoid during), **Weight Loss (aids for)**

Usage: MetaboMax is a product designed to boost metabolism, increasing thermogenisis and metabolic fat burn. The new formula is called MetaboMax Plus. It contains caffeine-bearing herbs like yerba mate and gurana, along with stimulants like capsicum and bitter orange extract. It helps increase metabolism, stimulant energy and lower appetite to aid weight loss. There is also a caffeine-free version, MetaboMax Free. Recommended dose is 2 capsules two to three times daily.

Warnings: MetaboMax Plus contains caffeine and should be avoided by people suffering from anxiety, nervous exhaustion and adrenal burnout. Not recommended for pregnant or nursing mothers.

MetaboMax Plus - Capsule (60)

Available in: US - Stock #3072-2

Ingredients: Capsicum, green tea, bitter orange fruit (synephrine), yerba mate, guarana, ginger, chickweed, guggulsterones.

MetaboMax Free - Capsule (60)

Available in: US - Stock #3074-6

Ingredients: Capsicum, green tea, bitter orange fruit (synephrine), ginger, chickweed, guggulsterones

MetaboMax EF (Ephedra-Free) - Capsule (90)

Available in: US - Stock #3008-5

Ingredients: Lotus leaf extract, green tea extract, Fructus aurantia, garcina, l-carnatine, cordyceps, eleuthero root, chromium, bee pollen, spirulina, vitamin E, kelp

MetaboStart

Product Type: Formula

Properties: Anti-obesic, Stimulant

Systems Affected: Digestive System, Glandular System, Thyroid, Weight Loss

Conditions: Pregnancy (herbs and supplements to avoid during), **Weight Loss (aids for)**

Usage: This 14-day weight-loss program is designed to increase thermogenisis, decrease fat deposits, boost metabolism and reduce the body's ability to absorb fats and carbohydrates from certain foods. It contains 7-Keto, which increases T4 to T3 conversion from the thyroid, which stimulates fat burning. It contains substances which slow or block absorption of fat and carbohydrates. Recommended use is to take the contents of one AM packet before breakfast, one Noon packet before lunch and one PM packet before dinner.

Warnings: Please note that permanent weight loss involves changing one's diet and lifestyle and that no weight loss product will work long term without these changes. Protease enzymes should not be taken with this product as they will neutralize the effect of the Carbo Grabbers. Not recommended for people with thyroid disorders because it contains 7-Keto. (See warnings for 7-Keto.) This program would also be contraindicated in pregnancy or while nursing.

MetaboStart Plus - Packets

Available in: US - Stock #3076-9

Ingredients: AM packet contains: 1 MetaboMax Plus capsule, 1 7-Keto™ capsule, 3 Fat Grabbers capsules,1 Carbo Grabbers capsule. Noon packet contains 2 MetaboMax capsules, 3 Fat Grabbers capsules, 1 Carbo Grabbers capsule. PM packet contains 3 Carbo Grabbers capsules, 2 CLA gel capsules.

MetaboStart EF (Ephedra-Free) (14 day) - Packets

Available in: US - Stock #3018-6

Ingredients: Metabomax EF, 7-Keto, Fat Grabbers, Carbo Grabbers

Migraquel

Product Type: Homeopathic

Properties: Analgesic (Anodyne)

Systems Affected: Nerves

Conditions: Migraine

Usage: Used for the relief of pain and other symptoms associated with throbbing, debilitating cluster headaches. Take 10-15 drops under the tongue every 10-15 minutes or as needed until symptoms improve. Then decrease to hourly, then to four times daily until symptoms are relieved.

Warnings: NSP recommends not taking this product for more than five days.

Migraquel (1 fl. oz.) - Homeopathic (Liquid)

Available in: US - Stock #8745-7

Ingredients: Active ingredients: Gelsemium sempervirens (Yellow Jessamine) 4x, Spigelia anthelmia (Pink Root) 4x, Phosphoricum Acidum (Phosphoric Acid) 6x, Natrum Sulfuricum (Sodium Sulfate) 6x, Secale cornutum (Rye Ergot) 6x, Carbo Vegetabilis (Vegetable Carbon) 10x, Silicea (Silica) 10x, Bryonia (White Bryony) 30x. Other ingredients: Purified water and 20% USP alcohol

Milk Thistle

Latin Name: *Silybum marianum*

Product Type: Single Herb

Properties: Alterative (Blood Purifier), Anti-inflammatory, Anticholesteremic, Antidiabetic, Antioxidant, Bitter, **Cholagogue**, Emmenagogue, **Galactagogue**, Hepatic, **Hepatoprotective**, Immune Stimulant, Tonic

Systems Affected: Gall Bladder, **Liver**, Spleen

Conditions: **Breast Milk (increase or enrich)**, Cancer (prevention), Chemical Poisoning, Cholesterol (high), Cirrhosis of the Liver, Constipation (adults), Defensiveness, Environmental Pollution (protection from), **Gall Bladder (sluggish)**, Gall Stones, Hangover, Hashimoto's Disease (Thyroiditis), Hemorrhoids, Hepatitis, Jaundice (adults), Nightmares, **Poisoning (general)**, Poisoning (food)

Usage: This herb is very helpful for hepatitis. It protects the liver against chemical poisoning. It has been clinically proven to prevent liver cell death from carbon tetrachloride and even the toxic effects of amanita (death cap) mushrooms. It accomplishes this feat in very small doses of 1-3 capsules/tablets per day or 1/2 to 1 teaspoon of tincture. Recommended use for this product is 1 tablet with a morning meal and 1 tablet at dinner.

Warnings: No toxic effects. Milk thistle may work as a mild laxative in some people. This may be due to the increase in bile secretion and flow into the intestinal tract.

Milk Thistle T/R - Capsule (60)

Available in: US & Canada - Stock #4071-9

Ingredients: Standardized extract of Silybum marianum (milk thistle) seed (80% silymarin)

Milk Thistle Combination

Product Type: Formula

Properties: Bitter, Cholagogue, Detoxifying, **Hepatic**, **Hepatoprotective**

Systems Affected: **Liver**

Conditions: Acquired Immune Deficiency Syndrome (AIDS/HIV), **Addictions (alcohol)**, **Addictions (drugs)**, Antibiotics (side effects of), Bronchitis, Cancer (prevention), **Chemical Poisoning**, **Cirrhosis of the Liver**, Dermatitis, Dysmenorrhea, **Environmental Pollution (protection from)**, Fatigue, Fatty Liver Disease, Gall Bladder (sluggish), Gall Stones, **Hangover**, Heavy Metal Poisoning, **Hepatitis**, **Jaundice (adults)**, Menstrual Irregularity, Mercury Poisoning, Nightmares, **Poisoning (general)**, Poisoning (food), Psoriasis, Varicose Veins

Usage: This product contains milk thistle, which has an hepatoprotective effect, in combination with other ingredients that help to protect and heal the liver. This is a very valuable formula for protecting the liver against chemical poisoning, or helping the liver to heal after exposure to chemicals. The formula is also beneficial for inflammatory and degenerative liver diseases such as hepatitis and cirrhosis of the liver. Recommended use is 2 tablets twice daily.

Warnings: See information under Milk Thistle T/R. Rapid detoxification of the liver will occasionally result in a "healing crisis," often involving flushing of the face and upper body, headaches or skin eruptions. If these symptoms occur, reduce dosage or discontinue supplement.

Milk Thistle Combination - Tablet (90)

Available in: US - Stock #4076-5

Ingredients: Milk thistle seed concentrate, N-Acetyl-Cysteine, dandelion, choline, inositol, vitamins A and C

LIV-GD - Tablet (50)

Available in: Canada - Stock #4015-8

Ingredients: Beta-carotene, vitamin C, iron (ferrus gluconate), lipotropic factors, choline, inositol, milk thistle extract, dandelion

Mineral-Chi Tonic

Product Type: Formula

Properties: **Adaptagen, Adrenal Tonic, Anti-aging,** Anti-arrhythmic, Bitter, Chelating, Digestive Tonic, **Immune Amphoterics, Immunomodulator, Mineralizer, Tonic**

Systems Affected: Adrenal Glands, Cuticle, Immune System, Liver, Nerves, Pancreas Tail, **Structural System, Whole Body**

Conditions: **Adrenals (exhaustion, weakness or burnout), Aging (prevention), Appetite (excessive),** Arrhythmia, **Autoimmune Disorders, Bipolar Mood Disorder** (Manic Depressive Disorder), **Blood Pressure (low)**, Broken Bones, Calcium Deficiency, Cartilage Damage, Children's Remedy, Concentration (poor), Constipation (adults), **Convalescence**, Coordination, Cramps (leg), Cramps (menstrual), Cramps and Spasms (general), Crohn's Disease, Dandruff, **Debility**, Depression, Diabetes, Diarrhea, Disks (spinal - bulging or slipped), Electromagnetic Pollution, **Energy (lack of)**, Fatigue, Fibromyalgia Syndrome (FMS), **Fingernail Biting**, Fingernails (weak or brittle), Grave's Disease, **Hair (loss or thinning), Infertility, Post Partum Weakness**, Pregnancy (herbs and supplements for), Scoliosis, Sugar Cravings, **Surgery (preparation for)**, Tachycardia, Tooth Decay (prevention), Triglycerides (high)

Usage: This product is a good overall tonic for people with severe fatigue, chronic illness and autoimmune disorders. As a general tonic for energy and overall health, take 1 ounce one to two times daily.

Warnings: Not recommended for acute or "hot" conditions like fever and acute inflammation.

Mineral-Chi Tonic (32 fl. oz.) - Liquid

Available in: US & Canada - Stock #1818-3

Ingredients: Colloidal minerals, gynostemma herb, lycium fruit, schizandra berry, eleuthero, astragalus root, deglycyrrhizinated licorice root, reishi mushroom, ginger rhizome, Ginkgo biloba leaf, potassium, purified water, white grape juice, glycerin, sodium benzoate, citric acid, natural apple and cherry flavors.

Monthly Maintenance

Product Type: Formula

Properties: Alterative (Blood Purifier), Emmenagogue, Glandular, Nervine

Systems Affected: Adrenal Glands, Glandular System, Liver, Nerves, Ovaries

Conditions: **Breasts (swelling and tenderness), Dysmenorrhea**, Electromagnetic Pollution, PMS (general), PMS Type A, Progesterone (low), **Puberty (hormone balancer)**

Usage: This is a combination of herbs and nutrients used to help prevent and relieve the symptoms of PMS. Take 6 capsules three times daily for 10 days prior to onset of menses (last 10 days of the menstrual cycle).

Warnings: No known warnings.

Monthly Maintenance - Capsule (180)

Available in: US - Stock #1812-0

Ingredients: Dong quai, peony, bupleurum, hoelen, atractylodes, codonopsis, alisma, licorice, magnolia, ginger, peppermint, moutan, gardenia, cyperus, vitamins A, B1, B2, B6, B12, C, D, E, biotin, calcium, choline, chromium, folic acid, iodine, iron, magnesium, manganese, niacin, pantothenic acid, potassium, selenium, zinc, copper, bioflavonoids, inositol, PABA, phosphorus

Mood Elevator

Other Names: Jie Yu

Product Type: Formula

Properties: **Antidepressant**, Corrects Polarity, Nervine, Sedative, Stimulant

Systems Affected: Brain, Liver, Nerves, Pituitary (infundibulum), Thyroid

Conditions: Addictions (alcohol), **Bipolar Mood Disorder (Manic Depressive Disorder)**, Breast Lumps, Concentration (poor), Cramps and Spasms (general), Cystic Breast Disease, **Depression**, Dizziness (Vertigo), Emotional Sensitivity, Energy (lack of), Fatigue, Hypothyroid, Insomnia, Menopause, **Mental Illness**, Neurosis, PMS Type D, **Post Partum Depression, Prolapsed Colon, Prolapsed Uterus, Reversed Polarity, Seasonal Affective Disorder**, Tension, Thinking (cloudy)

Usage: Chinese Mood Elevator is a Traditional Chinese Medicine (TCM) formula that is used to relieve sagging energy (chi). It is helpful for sadness, depression, fatigue, insomnia and anxiety. Unlike modern Western approaches to depression which focus on the neurotransmitter serotonin, Mood Elevator works in a holistic manner to relieve depression and sadness by balancing liver, digestive, intestinal and nervous functions.

The formula supports the liver, expels mucus and toxins from the liver and digestive tract, relaxes muscle spasms, stimulates circulation and energy and eases indigestion. In TCM terms, it relieves sagging chi (or energy) and disperses stagnant chi. It also expels wind (constriction and spasm) and dampness (excess fluid). So, it is useful not only for depression, but for congestion and sluggishness of the liver, abdominal bloating and pain, lymphatic congestion, anxiety and nervous tension and muscle spasms.

For depression and nervous problems, recommended dose is 4 capsules twice daily. For digestive upset, liver problems, etc., use 2-3 capsules three times daily. Many people have been able to get off antidepressant drugs with this formula, taking the formula along with their medication, until they begin to feel better. They then gradually reduce the dose of their medication. Never discontinue antidepressant medications abruptly. Ideally, this process should be done under professional supervision.

Warnings: Do not take a person off prescription medications for depression. Leave this up to the individual and his/her doctor.

Mood Elevator [Regulate Chi] Jie Yu - Capsule (100)

Available in: US - Stock #1878-8

Ingredients: Perilla leaves, aurantium peel, bamboo sap, bupleurum, cyperus rhizome, gambir twig, ophiopogon, typhonium flageliforme, hoelen, ligusticum, panax ginseng, platycodon, coptis rhizome, ginger rhizome, licorice, tang-kuei.

AD-C - Capsule (100)

Available in: Canada - Stock #1877-2

Ingredients: Perilla leaves, saussurea, aurantium peel, bamboo sap, bupleurum, cyperus rhizome, gambir twig, ophiopogon, pinellia rhizome, zhishi fruit, dang gui, hoelen, ligusticum rhizome, panax ginseng, platycodon, coptis rhizome, ginger rhizome, licorice

Mood Elevator TCM Conc. - Capsule (30)

Available in: US - Stock #1035-7

Ingredients: Same as Stock #1878-8

Morinda

See *Noni (Morinda)*

MSM

Product Type: Nutrient

Properties: Analgesic (Anodyne), Anti-arthritic, Anti-inflammatory, Lipotropic, Vulnerary

Systems Affected: Connective Tissue, Hair, Hypothalamus, Immune System, Muscles, Nails, Skin, Structural System

Conditions: Acid Indigestion (Heartburn, Acid Reflux), Acne (Pimples, Blackheads), Allergies (food), Allergies (respiratory), Alzheimer's Disease, Arthritis, Asthma, Blood Poisoning, Bronchitis, Carpal Tunnel Syndrome, Cartilage Damage, Cataracts, Cholesterol (low), Cirrhosis of the Liver, Confusion, Congestion (bronchial), Copper Toxicity, Defensiveness, Dermatitis, Diabetes, Disks (spinal - bulging or slipped), Dislocation, Epstein Barr Virus (Chronic Fatigue Syndrome, CFS), Eyesight (to improve), Fibromyalgia Syndrome (FMS), Fungal Infections (Yeast Infections, Candida albicans), Hypothyroid, Inflammation, Injuries, Itching, Memory and Brain Function, Psoriasis, Rashes and Hives, Scars / Scar Tissue, Skin (dry and/or flaky), Toxemia

Usage: MethylSulfonylMethane (MSM) is an organic sulfur compound found in vegetables, fruit, meat and diary products. It is a crystalline derivative of DMSO, a product that was popular in the alternative health community during the previous decade. MSM is one of hundreds of naturally occurring sulfur compounds in foods. Sulfur is crucial in the process of maintaining a vital healthy body and mind. It is part of the cellular structure and necessary for effecting repairs in the body. It promotes the health of hair, skin, nails and joints. Sulfur is also part of our immune systems. It has natural disinfectant properties. It aids liver detoxification. Taking about 1000 mg. (1 gram) of MSM for every 30 pounds of body weight (taken two or three times throughout the day) is a good amount, although you may wish to start with a smaller dose and work up. Take 2 tablets with a meal three times daily to supply 4,500 mg. (4.5 grams) per day. For best results, add Vitamin C/Citrus Bioflavonoids or Timed-Release Vitamin C to your MSM regimen.

Warnings: No known warnings.

MSM - Tablet (90)

Available in: US & Canada - Stock #4059-4

Ingredients: Methylsulfonylmethane (MSM) 750 mg

MSM/Glucosamine Cream

Product Type: Formula

Properties: Analgesic (Anodyne), Anti-arthritic, Anti-inflammatory

Systems Affected: Muscles, Skin, Structural System

Conditions: Cartilage Damage, Inflammation, Injuries

Usage: Massage liberally in areas where there is joint or tissue inflammation and pain.

Warnings: No known warnings.

MSM/Glucosamine Cream (2 oz. tube) - Topical

Available in: US - Stock #3522-4

Ingredients: Purified water, caprylic/capric/stearic triglycerides, glyceryl stearate, stearyl alcohol, PEG 100 stearate, glucosamine sulfate 4%, dimenthicone, butylene glycol, methylsulphonylmethane (MSM) 2%, phenoxyethanol, hydrogenated lecithin, tocopheryl acetate, methylparaben, ethylparaben, propylparaben, butylparaben, cholesterol.

Mullein

Latin Name: *Verbascum thapsus*

Product Type: Single Herb

Properties: Absorbant, Demulcent (Mucilant), Emollient, Expectorant, Lymphatic, Mineralizer, Vermifuge, Vulnerary

Systems Affected: Bones, Bronchials, Ears, Gall Bladder, Intestinal System, Lungs, Lymph Nodes, **Lymphatic System**, Mucus Membranes, Nerves, Nipples, Respiratory System, Structural System, Throat, Tonsils

Conditions: Allergies (respiratory), Asthma, Breasts (swelling and tenderness), Bronchitis, Congestion (bronchial), Congestion (general), Congestion (lymphatic), Cough (dry), Cough (general), Croup, Dislocation, Ear Infection or Earache, **Emphysema**, Epstein Barr Virus (Chronic Fatigue Syndrome, CFS), Fibrosis, **Glands (swollen lymph)**, Hashimoto's Disease (Thyroiditis), Hodgkin's Disease, Injuries, Lymph Nodes or Glands (swollen), **Mastitis, Mumps**, Rashes and Hives, Tuberculosis (Consumption, Scrofula), Vaccines (detoxification from), **Wheezing**

Usage: A very powerful lung remedy. Heals weakened lung tissue. Useful for chronic respiratory congestion as it is soothing and lubricating. Take 1-2 capsules two to three times daily. A decoction is more astringent and anodyne, use 3-4 capsules per cup of water and simmer for 20 minutes. Has been smoked to relieve lung congestion. Can also be applied externally as a poultice. An oil from the flowers is an excellent earache remedy.

Warnings: No known warnings.

Mullein - Capsule (100)

Available in: US & Canada - Stock #460-7

Ingredients: Verbascum thapsus (mullein) leaf

Multiple Vitamins & Minerals

See *Vitamins & Minerals, Multiple*

Myrrh

Product Type: Essential Oil

Properties: Anti-inflammatory, Antibacterial, Antimicrobial, **Antiseptic**, Antiviral, Aromatic, Bitter, Disinfectant, Expectorant, **Perfume**, Preservative

Systems Affected: Digestive System, Gums, Immune System, Respiratory System, Skin, Throat, Vagina

Conditions: Abdominal Pain and Inflammation, Amenorrhea, Appetite (deficient), Athlete's Foot, Autism, Bronchitis, Canker Sores (Mouth Ulcers), Cholesterol (high), Colds (general remedies for), Cough (general), Diarrhea, Dysmenorrhea, Eczema, Gas and Bloating, Grief (excessive), Hashimoto's Disease (Thyroiditis), Infection (bacterial), Inflammation, Oral Surgery, Pregnancy (herbs and supplements to avoid during), Sore Throat, **Ulcerations (external)**, Ulcers, Worry

Usage: Myrrh prevents infection, clears toxins and promotes tissue repair. It acts as an excellent expectorant and helps with coughs, bronchitis, colds and any condition involving excessive thick mucus. Warming and stimulating to the stomach, it helps diarrhea, flatulence, abdominal distention, poor appetite and a weakened, damp spleen. It can be used in conjunction with a Candida program for vaginal discharge. Clinical research shows that it helps lower blood cholesterol levels helping in obesity and ischemic heart disease. Traditionally, it has been used for mouth, gum and throat infections. Myrrh promotes menstruation and helps relieve painful periods. Topically, myrrh is used for chronic wounds and ulcers. It can be used for wounds that are slow to heal and for weepy eczema and athlete's foot. It works to heal deep cracks on the heels and hands. Emotionally, myrrh is valuable for those that feel stuck emotionally or spiritually and want to move forward in their lives. It is calming to the nervous system and instills a deep sense of mental tranquility. It is a principle oil for those who over think things, worry and have mental distractions. The sense of peace it imparts helps to ease sorrow and grief. Myrrh helps with a connection of inner self so that dreams can be brought to reality.

Warnings: Highly toxic in concentration, not recommended for internal use. It is a uterine stimulant and not for use with pregnancy.

Myrrh Essential Oil (5ML) - Essential Oil

Available in: US - Stock #3939-3

Ingredients: Myrrh (Commiphora myrrha) Essential Oil

N-acetyl Cysteine

Product Type: Nutrient

Properties: Antioxidant, Detoxifying, Hepatic

Systems Affected: Liver

Conditions: Acquired Immune Deficiency Syndrome (AIDS/HIV), Alzheimer's Disease, Bell's Palsy, Blood Pressure (high), Cancer (natural therapy for), Cataracts, **Chemical Poisoning**, Cholesterol (high), Congestion (general), Cystic Fibrosis, Environmental Pollution (protection from), **Glaucoma**, **Heavy Metal Poisoning**, Inflammation, Lead Poisoning, **Mercury Poisoning**, **Multiple Sclerosis** (MS), Parkinson's Disease

Usage: The cells of our body contain glutathione, an important substance that aids in cellular detoxification and recycles antioxidants to protect the cell from damage. Raising levels of glutathione can help autoimmune diseases, AIDS and other degenerative diseases. It can also help the body eliminate heavy metals and other toxic substances. N-Acetyl Cysteine is the precursor to glutathione. It can aid a wide variety of conditions, including helping liver detoxification, maintaining blood pressure and protecting the skin, eyes and other tissues against oxidative stress. Suggested use is 1 tablet twice daily with meals.

Warnings: No known warnings.

N-Acetyl Cysteine (250 mg) - Tablet (60)

Available in: US - Stock #509-7

Ingredients: 250 mg of N-Acetyl Cysteine, tumeric

Nattozimes Plus

Product Type: Formula

Properties: Anticoagulant (Blood Thinner)

Systems Affected: Blood, Circulation

Conditions: Blood Clots (prevention of), **Blood Clots (prevention of)**, Cardiovascular Disease (Heart Disease), Strokes, **Thrombosis**

Usage: Nattokinase is a protease (protein digesting) enzyme formed during the soybean fermentation process that produces natto. Natto has been consumed as a food in Japan for thousands of years. Recent research shows that nattokinase helps dissolve fibrin the the blood. Fibrin "thickens" the blood and contributes to the formation of blood clots, which can cause heart attack, stroke and other clotting disorders.

Nattozimes Plus contains a blend of fungal enzymes (nattozimes) that help dissolve fibrin like nattokinase. It may be helpful for people who want an alternative to aspirin or blood thinners to help prevent cardiovascular disease. To get these circulatory benefits, take 1 capsule between meals twice daily on an empty stomach.

Warnings: Do not use or handle this product if you are taking blood thinning medications, have a bleeding disorder or if are allergic to Aspergillus.

Nattozimes Plus - Capsule (60)

Available in: US - Stock #520-7

Ingredients: Aspergillus orgyzae, Aspergillus melleus, hawthorn berries (Crataegus laevigata), dandelion leaf (Taraxacum officinale), capsicum fruit (Capsicum annuum), resveratrol (from red grape skin Vitis vinifera)

Natural Changes

Product Type: Pack

Properties: Glandular

Systems Affected: Glandular System, Nerves, Structural System

Conditions: Hot Flashes, Menopause

Usage: A convenient pack of supplements to aid with menopausal symptoms. Take the contents of 1 packet in the morning and in the evening, for three weeks.

Warnings: No known warnings.

Natural Changes (42 packets) - Packet (42)

Available in: US - Stock #4055-2

Ingredients: C-X, Skeletal Strength, Wild Yam with Chaste Tree, Flax Seed Oil, Nutri-Calm

Nature's Chi

Product Type: Formula

Properties: Adaptagen, Adrenal Tonic, Anti-aging, Appetite Suppressant, Blood Building, Immune Amphoterics, Kidney Tonic, Tonic

Systems Affected: Circulation, Immune System, Respiratory System

Conditions: Addictions (drugs), Debility, **Energy (lack of)**, Exercise, Fatigue, **Fatigue**, Surgery (healing from), Weight Loss (aids for)

Usage: Nature's Chi is an energy tonic that nourishes both blood and chi according to the principles of traditional Chinese medicine. Instead of giving you an energy "buzz" followed by a "crash," Nature's Chi gradually builds energy reserves in the body. It helps increase metabolism and reduce appetite, which may be helpful for weight loss. It also helps overcome anemia and supports a healthy immune system. It is a great general tonic to help counteract the effects of aging.

As an energy tonic take 1 capsules once or twice daily. For appetite reduction take 2 capsules in between breakfast and lunch.

Warnings: No known warnings.

Nature's Chi TCM Concentrate - Capsule (30)

Available in: US - Stock #836-8

Ingredients: Astragalus root, tang-kuei root, fo-ti (ho shou wu) root, eleuthero root, cinnamon twig, peony root without bark, forsythia fruit, gardenia fruit, panax ginseng root, hoelen sclerotium, mint leaves, schizonepeta flower, scute root, siler root, licorice root

Nature's Cortisol

Product Type: Formula

Properties: Adaptagen, Anti-obesic

Systems Affected: Adrenal Glands, Weight Loss

Conditions: **Cushing's Disease**, Stress, Weight Loss (aids for)

Usage: Cortisol is a hormone produced by the adrenal glands when the body is under stress. It has the positive effect of reducing inflammation, but too much cortisol also leads to a breakdown of lean muscle tissue and an accumulation of fat. Nature's Cortisol contains herbs that reduce the output of stress hormones, including cortisol, from the adrenals. It also contains substances which boost metabolism. It can be used as an aid to weight loss when weight problems are related to stress. It can also reduce overactive adrenal function.

Recommended dose is one capsule with a meal three times daily.

Warnings: Should not be used when the adrenals are underactive or exhausted or when there is a lot of chronic inflammation present, as is usually the case in autoimmune disorders.

Nature's Cortisol Formula™ - Capsule (90)

Available in: US - Stock #3209-4

Ingredients: Vitamin C, chromium, magnolia bark, phellodendron bark, holy basil leaves, green tea, bababa leaf, l-theanine, DHEA, vanadium

Nature's Fresh

Product Type: Formula

Properties: Analgesic (Anodyne), **Anti-inflammatory**, **Antivenomous**, **Balm**, Cosmetic, **Deodorant**, Panacea, **Vulnerary**

Systems Affected: **Cuticle**, **Nipples**, **Skin**, **Spinal Disks**

Conditions: Abrasions, Acne (Pimples, Blackheads), Adenitis, **Age Spots**, Arthritis, **Backache (Back Pain, Lumbago)**, **Bites and Stings**, **Body Odor**, Boils, **Breast Lumps**, Broken Bones, Bruises (healing), **Burns and Scalds**, Cartilage Damage, Concussions, Conjunctivitis (Pink Eye), **Corns**, Cramps (leg), Cramps (menstrual), Cramps and Spasms (general), Cuts, **Cystic Breast Disease**, **Cysts**, Dandruff, **Denture Sores**, Dermatitis, **Diaper Rash**, **Disks (spinal - bulging or slipped)**, Eczema, **Fleas**, **Foot Odor**, **Inflammation**, **Injuries**, **Itching**, **Pets (supplements for)**, **Rashes and Hives**, **Rhinitis**, Sciatica, Scoliosis, **Skin Care (general)**, Spider Veins, **Sprains**, Stiff Neck, **Sunburn**, TMJ, Whiplash, **Wounds and Sores**, **Wrinkles**

Usage: While this formula was originally designed to break down odors without harming fabrics and to help remove stains, some innovative NSP managers discovered it was also therapeutic for a wide variety of conditions. It can be sprayed topically on the body to reduce pain and inflammation for a wide variety of injuries and ailments. It has proven helpful in repairing damaged disks in the spine. Applying a goldenseal tincture to the areas where the damaged disks are, along with and following up with Nature's Fresh can be very helpful. The product also makes an excellent deodorizer, spot remover, and cleaner.

Warnings: No known warnings.

Nature's Fresh Enzyme Spray - Liquid

Available in: US - Stock #9834-7

Ingredients: Water, enzymes: oxidoreductases, transferases, lyases, hydrolases, isomerases, ligases

Nature's Gold

Latin Name:

Product Type: Single Herb

Properties: Alkalinizer, Anti-arthritic, Anti-inflammatory, Anti-obesic, Antioxidant, Detoxifying, Food, Galactagogue, Immunomodulator, Mineralizer, Nutritive, Tonic, Vulneraries (for intestinal system), Vulnerary

Systems Affected: Cardiovascular System, Digestive System, Immune System, Skeletal System, Skin, Structural System, Weight Loss, Whole Body

Conditions: Aging (prevention), Allergies (food), Allergies (respiratory), Arthritis, Bipolar Mood Disorder (Manic Depressive Disorder), Blood Pressure (high), Convalescence, Depression, Diverticulitis (Diverticuli), Eczema, Energy (lack of), Free Radical Damage, Inflammatory Bowel Disorders (Colitis, IBS), Migraine, Pain (general remedies for), Radiation Sickness, Rashes and Hives, Warts, Weight Gain (aids for)

Usage: Nature's Gold is a whole food produced from organic barley seed. It is made by a special process which causes all the seeds to germinate within minutes of each other. The nutritional value is stabilized to maintain nutritional value and enhance nutrient absorption. This gives it the benefits of barley grass, but with an enhanced assimilation of nutrients. It is a highly energized food which is 97% digestible, while ordinary barley grass is only 5% digestible. This is a great product for enhancing general health.

Start with Nature's Gold Level I on the following schedule.

Week 1: 1/2 teaspoon per day
Week 2: 1 heaping teaspoon per day
Week 3: 2 heaping teaspoons per day
Week 4: 1 heaping tablespoon per day
Maintain: 2 heaping tablespoons per day

Proceed with Nature's Gold Level II after maintaining maximum servings of Level I for at least one month. Then take Level II as follows:

Week 1: 1 heaping tablespoon per day
Maintain: 2 heaping tablespoons per day

Nature's Gold can be added to other foods and beverages. It does not have a strong flavor and is a great addition to smoothies. Do not cook Nature's Gold as this will destroy its nutritional value.

Warnings: Follow the instructions carefully and do not exceed the recommended program. It is best to start slowly with Nature's Gold because it can cause intense healing crisis reactions.

Nature's Gold Level I - Bulk Powder

Available in: Canada - Stock #8003-1
Ingredients: Organic barley grass

Nature's Gold Level II - Bulk Powder

Available in: Canada - Stock #8004-9
Ingredients: Same as Nature's Gold Level I

Nature's Hoodia

Product Type: Formula

Properties: Appetite Suppressant

Systems Affected: Weight Loss

Conditions: Appetite (excessive), **Weight Loss (aids for)**

Usage: Nature's Hoodia is a natural appetite suppressant formula. It contains herbs that suppress appetite and provide a feeling of fullness. Take one or two capsules between meals daily.

Warnings: No known warnings.

Nature's Hoodia Formula™ - Capsule (90)

Available in: US - Stock #3009-8
Ingredients: Chromium (amino acid chelate), Hoodia Gordonii aerial parts extract, Caralluma Fimbriata ariel parts extract, Gymnemna leaves, Apple Cider vinegar extract, Garcinia pericarp extract, l-carnitine, marshmallow root, psyllium seed hulls

Nature's Phenyltol with NEM

Product Type: Formula

Properties: Analgesic (Anodyne), Anti-inflammatory, Euphoretic, Vulnerary

Systems Affected: Brain, Nerves, Structural System

Conditions: Arthritis, Bursitis, Dislocation, Emotional Sensitivity, Inflammation, Ligaments (torn or injuried), Mood Swings, Pain (general remedies for), Surgery (healing from), Tooth Extraction

Usage: Nature's Phenyltol with NEM is an anti-inflammatory formula with special properties to support joint health. It is a good choice for pain relief for arthritis and other structural problems. It contains DL-phenylalanine, which helps release chemicals in the brain that improve mood and ease pain. It also contains natural eggshell membrane, which contains glucosamine, chondroitin sulfate, hyaluronic acid and collagen, which support joint health and repair.

Warnings: No known warnings.

Nature's Prenatal

Product Type: Formula

Properties: Nutritive

Systems Affected: Pituitary (anterior), Whole Body

Conditions: **Birth Defects (prevention), Pregnancy (herbs and supplements for)**

Usage: Formulated for pregnant and lactating women with 800 mcg of folic acid and ginger to soothe the stomach. Take 1 tablet daily with a meal.

Warnings: No known warnings.

Nature's Prenatal - Tablet (180)

Available in: US - Stock #4917-8
Ingredients: Vitamins A (beta-carotene), C (ascorbic acid), D, E (d-alpha tocopherol), B1 (thiamine mononitrate), B2 (riboflavin), B6 (pyridoxine hydrochloride), folic acid (folacin), B12 (cyanocobalamin), niacin (niacinamide), biotin, pantothenic acid (d-calcium pantothenate), iron (ferrous fumarate), iodine (potassium iodide), magnesium (oxide and stearate), zinc (gluconate), copper (gluconate), ginger

Nature's Sea Calcium

See *Calcium*

Nature's Three

Product Type: Formula

Properties: **Absorbant**, Anticholesteremic, Demulcent (Mucilant), Emollient, **Laxative (bulk)**, Laxative (general), Soothing

Systems Affected: Large Intestine (Colon), Liver, Small Intestines, Weight Loss

Conditions: Cholesterol (high), **Constipation (adults)**, **Diarrhea**, Gall Stones, Heavy Metal Poisoning, Hemorrhoids, Inflammatory Bowel Disorders (Colitis, IBS), Lead Poisoning, Multiple Sclerosis (MS), Psoriasis, Weight Loss (aids for)

Usage: This is a bulk laxative formula that also contains substances known to lower cholesterol. Take 1 teaspoon to 1 tablespoon and mix in a small glass of water or juice. Drink as soon as powder is stirred into the liquid. Do not let stand or it will form a slimy, mucilaginous mass that is impossible to drink. Follow with a tall glass of water. Can be taken one to three times daily in this manner. Drink 6-8 glasses of water per day when taking this formula for best results.

Warnings: If you do not drink enough water with this formula, it may cause constipation. Long-term use can weaken digestion unless you also take some ginger, capsicum or other aromatic stomachic herbs occasionally.

Nature's Three (12 oz.) - Bulk Powder

Available in: US - Stock #1345-0

Ingredients: Psyllium hulls, apple fiber, oat bran

Neroli

Product Type: Essential Oil

Properties: Antibacterial, Antidepressant, Antiseptic, Antispasmodic, Anxiolytic, Aromatic, Carminative, Deodorant, **Perfume**, Relaxant, Stimulant, Vulnerary

Systems Affected: Capillaries, Cardiovascular System, Intestinal System, Muscles, Nerves, Skin, Structural System

Conditions: Angina, Anxiety (Panic Attack), Athlete's Foot, Body Odor, Bronchitis, Circulation (poor), Colic (children), Depression, Diarrhea, Dyspepsia, Fear (excessive), Fungal Infections (Yeast Infections, Candida albicans), Gas and Bloating, Headache (general), Heart Fibrillation or Palpitations, Hysteria, Insomnia, Mood Swings, Neuralgia and Neuritis, Parasites (general), Shock, Skin (dry and/or flaky), Varicose Veins

Usage: Neroli is extracted from the flowers of the bitter orange tree and has a soft delicate aroma. It takes one thousand pounds of blossoms to make one pound of oil. Traditionally, it is associated with marriage and purity. It helps to relax the nerves and can be used in massages or baths for stress, depression, anxiety, shock and insomnia. It can be inhaled for headaches, neuralgia, heart palpitations or vertigo. It has affinity for the circulatory system and can be used in a massage oil for broken capillaries, constricted circulation, and varicose veins. It can be massaged on the abdomen for chronic diarrhea, gas, colic, intestinal spasms and parasites. It can also be used in a soak for nail fungus and athlete's foot. Emotionally, neroli helps a person to have positive, confident feelings and to believe in him or herself. It can help overcome moodiness and nervous exhaustion, irritability, grief and fear and inspire feelings of hope and joy. Neroli may also help to overcome amnesia, phobias, panic, hysteria and withdrawal.

Warnings: Do not take internally. Do a patch test first before using on skin.

Neroli Essential Oil (2ML) - Essential Oil

Available in: US - Stock #3891-2

Ingredients: Neroli (Citrus aurantium) Essential Oil

Nerve Control

Other Names: RE-X

Product Type: Formula

Properties: Analgesic (Anodyne), Antispasmodic, Emmenagogue, Nervine

Systems Affected: Female Reproductive, Nerves

Conditions: Addictions (alcohol), Addictions (drugs), Addictions (tobacco smoking or chewing), Anxiety (Panic Attack), Headache (tension), Insomnia, Schizophrenia, Shingles, Stress, Triglycerides (high), Vaccines (detoxification from)

Usage: Take 2 capsules two to three times daily with meals for nervous stress and other nervous disorders.

Warnings: No known warnings.

Nerve Control - Capsule (100)

Available in: US - Stock #1242-4

Ingredients: Black cohosh, valerian, capsicum, passion flower, catnip, hops, wood betony

RE-X - Capsule (100)

Available in: Canada - Stock #1241-6

Ingredients: Valerian root (Valeriana officinalis), passion flower herb (Passiflora incarnata), black cohosh root (Cimicifuga racemosa), hops flowers (Humulus lupulus), catnip leaves and flowers (Nepeta cataria), wood betony leaves (Betonica officinalis)

Nerve Eight

Other Names: Stress Relief

Product Type: Formula

Properties: **Analgesic (Anodyne)**, Anti-arthritic, Anti-arthritic, Anti-inflammatory, **Anticephalalgic**, **Antirheumatic**, Antispasmodic, Nervine, Sedative

Systems Affected: Digestive System, Nerves, Prostaglandins, Structural System

Conditions: Addictions (alcohol), Addictions (drugs), Addictions (tobacco smoking or chewing), Anxiety (Panic Attack), Appetite (excessive), Arthritis, Asthma, Attention Deficit Disorder (ADD, ADHD), **Bell's Palsy**, Burning Feet or Hands, Chicken Pox, Cramps (menstrual), Defensiveness, Dyspepsia, Epstein Barr Virus (Chronic Fatigue Syndrome, CFS), **Headache (general)**, **Headache (tension)**, **Hysteria**, **Inflammation**, Injuries, Lupus, Multiple Sclerosis (MS), **Neuralgia and Neuritis**, Sciatica, Stress

Usage: A Western nervine formula for mild inflammation and pain, nervous tension, headaches and other nervous system disorders. Take 2 capsules three times daily or as needed.

Warnings: No known warnings.

Nerve Eight - Capsule (100)

Available in: US - Stock #873-6

Ingredients: White willow, black cohosh, capsicum fruit, valerian, ginger, hops, wood betony, devil's claw

Eight Combination - Capsule (100)

Available in: Canada - Stock #849-0

Ingredients: White willow bark, hops flowers, wood betony herb, passion flower herb, ginger root, chamomile flowers, capsicum fruit, and schizandra fruit

Nervous Fatigue Formula

Other Names: Yang Xin

Product Type: Formula

Properties: **Adaptagen**, **Adrenal Tonic**, **Anxiolytic**, Glandular, Hepatoprotective, Nervine

Systems Affected: Adrenal Glands, Adrenal Medulla, Brain, Heart, Liver, **Nerves**

Conditions: Addictions (coffee, caffeine), Addictions (drugs), **Adrenals (exhaustion, weakness or burnout)**, Anemia, **Angina**, **Anxiety (Panic Attack)**, Anxiety Disorders, Backache (Back Pain, Lumbago), **Concentration (poor)**, **Confusion**, Constipation (adults), Coordination, Dehydration, Depression, Diabetes, **Emotional Sensitivity**, **Energy (lack of)**, Erectile Dysfunction, Fatigue, **Grave's Disease**, Heart (weakness), Hot Flashes, **Hysteria**, **Insomnia**, Jet Lag, **Memory and Brain Function**, **Mental Illness**, **Nervous Exhaustion (Enervation)**, **Nervousness**, **Night Sweating**, **Nightmares**, Obsessive Compulsive Disorder, Perspiration (excessive), Phobias, PMS Type S, **Post Traumatic Stress Disorder**, **Restless Dreams**, **Sex Drive (low)**, **Sleep (restless and disturbed)**, **Stress**, **Thinking (cloudy)**, Tremors, Urination (frequent)

Usage: Sometimes the stress of life just gets to be too much and a person starts to feel "burned-out." This shows the subconscious association of fire with passion, excitement and motivation. This same association exists in Chinese medicine. When a person becomes deficient in fire energy, they become tired and nervous, feel overwhelmed, vulnerable and "broken hearted" and lose sexual desire.

Nervous Fatigue Formula is a Chinese herbal formula that nourishes this deficient fire energy. It is a tonic for the heart, nerves and glands, particularly the adrenal glands, which become exhausted under constant stress. A key indication that a person could benefit from this blend is fatigue during the day coupled with disturbed and restless sleep patterns at night. Disturbing dreams are often an early warning sign that one is approaching "burn-out." As the problem becomes more severe, a person wakes up frequently at night or suffers from night sweats.

This state of chronic stress may also involve sensations of pressure or pain on the left side of the chest, loss of sexual desire, nervous exhaustion, trembling, and burning sensations in the palms of the hands, the souls of the feet or over the heart area. The fire-weakened person has muddled thoughts, mental confusion, hypersensitive emotions and a loss of short term memory. Frequently, the person suffering from this burnout has dark circles under their eyes and a quivering tongue.

Sold under NSP's trade name Nervous Fatigue Formula, Yang Xin helps to improve mental functions, strengthen the adrenals, increase circulation, improve digestion and reduce the body's response to stress. Other conditions it may help include frequent urination, scanty menstruation, constipation, excessive perspiration, heart disease, leg pain, impotence and diabetes.

For those who are "burned-out" and need to strengthen the heart, nerves and glands take 3-4 capsules Nervous Fatigue Formula of 3 times daily with meals. It is also available in a TCM concentrate where the dose is 1 capsule two to three times daily. It can also be taken when a person wakes up at night and can't seem to go back to sleep.

Warnings: No known warnings.

Nervous Fatigue Formula - Capsule (100)

Available in: US - Stock #1884-7

Ingredients: Schizandra fruit, biota seed, cistanches, cuscuta seed, dang gui, lycium fruit, ophiopogon, succinum, acorus rhizome, astragalus, dioscorea, hoelen, lotus seed, panax ginseng, polygala, polygonum rhizome, zizyphus seed, rehmannia

Nervous Fatigue Form. TCM Conc. - Capsule (30)

Available in: US - Stock #1017-1

Ingredients: Same as Stock #1884-7

HS-C - Capsule (100)

Available in: Canada - Stock #1884-7

Ingredients: Same as Stock #1884-7

Nervousness

Product Type: Homeopathic

Properties: Anxiolytic, Calmative

Systems Affected: Nerves

Conditions: Irritability, Nervous Exhaustion (Enervation)

Usage: Used for the relief of occasional nervousness and irritability caused by simple nervous tension and stress. Take 10-15 drops under the tongue every 10-15 minutes or as needed until symptoms improve. Then decrease to every one or two hours, then to four times daily.

Warnings: NSP does not recommend this product for children under the age of 12 except on the advice of a health professional. Other than the alcohol content, which is not good for infants, we consider this product safe for children.

Nervousness (1 fl. oz.) - Liquid (Tincture)

Available in: US - Stock #8711-0

Ingredients: Active ingredients: Passiflora incarnata (Passion Flower) 4x, Valeriana officinalis (Valerian) 4x, Acidum Phosphoricum (Phosphoric Acid) 6x, Agnus castus (Chaste Tree) 6x, Ferrum Phosphoricum (Ferric Phosphate) 6x, Ignatia amara (St. Ignatius' Bean) 6x, Moschus (Musk) 6x, Silicea (Silica) 6x, Arnica montana (Mountain Arnica) 12x. Other ingredients: Purified water and 20% USP alcohol

NF-X

Product Type: Formula

Properties: Emmenagogue, **Female Tonic**, Glandular, Stimulant (Circulatory)

Systems Affected: Digestive System, **Female Reproductive**, Urinary System

Conditions: Acne (Pimples, Blackheads), Birth Control (countering side effects), Breasts (swelling and tenderness), Cramps (menstrual), Edema (Dropsy, Water Retention, Swelling), Fibroids (uterine), Hot Flashes, Infertility, **Menopause**, Nephritis, Post Partum Depression, Pregnancy (herbs and supplements to avoid during), Puberty (hormone balancer), Puberty (hormone balancer)

Usage: This formula is used as a general female corrective for PMS, menstrual irregularity and menopause. More medicinal than Female Comfort (formerly FC with Dong Quai.) Start with small dose (1-2 capsules a day) and work up to find best dose. Typical use is 2 capsules three times daily.

Warnings: Not recommended during pregnancy or lactation.

NF-X - Capsule (100)

Available in: US - Stock #1232-3

Ingredients: Golden seal, capsicum fruit, ginger, uva ursi, cramp bark, squaw vine, blessed thistle, red raspberry, false unicorn

Niacin

Product Type: Nutrient

Properties: **Anticholesteremic**, Nutritive, **Stimulant (Circulatory)**, Stomachic

Systems Affected: Brain, **Circulation**, Liver, Nerves, Skin, Tongue

Conditions: Addictions (alcohol), Addictions (drugs), Backache (Back Pain, Lumbago), **Cholesterol (high)**, **Cold Hands and Feet**, Depression, Diarrhea, Fatigue, Halitosis (Bad Breath), Migraine, Schizophrenia, Stress

Usage: Also known as B3, niacin is an important nutrient for circulation. It helps lower cholesterol and enhance blood flow to the extremities. It also helps the skin, nervous system, metabolism of nutrients, and the production of hydrochloric acid. Take 1 tablet daily with a meal.

Warnings: A flushing (reddening) can occur when taking niacin, but this is harmless and passes. Avoid high amounts if you have gout, peptic ulcer, glaucoma, liver disease or diabetes or if you are pregnant.

Niacin (250 mg) - Capsule (90)

Available in: US - Stock #1797-0

Ingredients: 250 mg. of niacin in a base of hops flowers and feverfew herb.

Noni (Morinda)

Latin Name: *Morinda citrifolia, M. officinalis*

Product Type: Single Herb

Properties: Alkalinizer, Analgesic (Anodyne), Anti-inflammatory, Anticoagulant (Blood Thinner), Antidiabetic, Cell Proliferant, Emmenagogue, Hypolipidemic, Kidney Tonic, Vulnerary

Systems Affected: Capillaries, Digestive System, Intestinal System, Liver, Nerves, Pancreas Tail, Respiratory System, Skeletal System, Urinary System

Conditions: Aging (prevention), Arthritis, Bruises (prevention), Cancer (natural therapy for), Cancer (prevention), Chemotherapy (reducing side effects), Diabetes, Epstein Barr Virus (Chronic Fatigue Syndrome, CFS), Erectile Dysfunction, Hair (loss or thinning), Hemorrhoids, Incontinence (urinary), Inflammation, Malaria, Nervous Exhaustion (Enervation), **Overacidity**, Pain (general remedies for), Premature Ejaculation, Rheumatoid Arthritis (Rheumatism), Sex Drive (low), Vaccines (detoxification from)

Usage: Noni juice is used as an antioxidant, anti-inflammatory and alkalizing agent. The root is used in Chinese medicine to tonify the kidney and fortify the yang energy. Deficiency of kidney yang is connected to symptoms like impotence, male or female infertility, premature ejaculation, frequent urination, urinary incontinence, irregular menstruation and a sore back. The kidney energy is also thought to strengthen the bones. Noni is used in the Chinese formula KB-C for strengthening the kidneys and the bones. Capsules: Take 1-2 capsules with a meal three times daily. Liquid: Take 2 tablespoons of Noni juice daily preferably before meals.

Warnings: No known warnings for noni juice. The noni in capsules is contraindicated with damp-heat (conditions of excess fluid and inflammation) or heat from yin deficiency (heat with fatigue). It is also contraindicated in people who have a difficult time urinating.

Nature's Noni (16 fl. oz.) - Liquid

Available in: US & Canada - Stock #4066-7

Ingredients: Morinda citrifolia (Noni) fruit juice, agave nectar, grape juice concentrate, grapeskin extract, malic acid, tartatic acid, citric acid, ascorbic acid.

Morinda Fruit Drink (16 fl. oz.) - Liquid

Available in: US & Canada - Stock #4066-7

Ingredients: Same as Stock #4066-7

Nature's Noni (2-32 fl. oz. each) - Liquid

Available in: US - Stock #4042-7

Ingredients: Same as Stock #4066-7

Nature's Noni Capsules - Capsule (100)

Available in: US - Stock #457-6

Ingredients: Morinda citrifolia fruit, Morinda citrifolia root and leaves, Morinda officinalis root

Nature's Noni Capsules - Capsule (100)

Available in: Canada - Stock #456-5

Ingredients: Same as Stock #457-6

Nopal

Latin Name: *Opuntia streptacantha and O. ficus-indica*

Product Type: Single Herb

Properties: Antidiabetic, **Hypoglycemic**, Low Glycemic

Systems Affected: Pancreas, Pancreas Tail, Urinary System

Conditions: Dehydration, Diabetes, Urination (burning or painful)

Usage: Researchers hypothesize that the nopal treatments may improve the ability of insulin to stimulate the glucose movement from the blood and into body cells where it is used as energy. Recommended intake of the capsule form is 2 capsules with meals three times a day.

Warnings: No known warnings.

Nopal - Capsule (100)

Available in: US - Stock #475-3

Ingredients: Opuntia ficus-indica (nopal) leaf

Nutri-Burn

Product Type: Formula

Properties: Anti-obesic, Antidiabetic, Low Glycemic, Nutritive

Systems Affected: Weight Loss, Whole Body

Conditions: Appetite (excessive), Body Building, Diabetes, Diabetes, Exercise, Hypoglycemia, **Weight Loss (aids for)**

Usage: This is a lactose-free whey protein which can aid building muscle and burning fat. It contains enzymes to aid digestion and assimilation, conjugated linoleic acid (CLA) to enhance the buildup of lean muscle mass, prebiotic fiber for intestinal health and green tea and L-carnitine to boost energy and fat burn. Calcium caseinate offers a sustained release of amino acids which may help curb appetite. For best results, you should consume Nutri-Burn within 45 minutes of exercise. Mix 2 rounded scoops of Nutri-Burn powder with approximately 12-16 ounces of cold water, or mix to taste.

Warnings: No known warnings.

Nutri-Burn, Chocolate - Bulk Powder

Available in: US - Stock #3080-0

Ingredients: Protein matrix, fat matrix (including CLA), fructose, cocoa powder (chocolate flavor only), milk calcium, fructo-oligosaccharides, natural flavors, soy lecithin, magnesium amino acid chelate, potassium citrate, xanthan gum, l-carnitine, proteolytic plant enzyme blend, lohan fruit extract, sodium chloride, mono and diglycerides, ascorbic acid, green tea extract, potassium phosphate, calcium phosphate, stevia extract, vitamins and minerals

Nutri-Burn, Vanilla - Bulk Powder

Available in: US - Stock #3078-0

Ingredients: Same as Stock #3080-0

Nutri-Burn (Vanilla) - Bulk Powder

Available in: Canada - Stock #3214-9

Ingredients: Protein blend (whey protein concentrate, whey protein isolate, calcium caseinate), natural French Vanilla powder, sunflower oil creamer (sunflower seed oil, corn syrup solids, sodium caseinate, mono-and diglycerides, dipotassium phosphate, tricalcium phosphate, soy lecithin and natural tocopherol), fructose, maltodextrin, xanthan gum, CLA (conjugated linoleic acid) powder (safflower seed oil, casein), cloves flowers

Nutri-Calm

Product Type: Formula

Properties: Adrenal Tonic, **Anti-smoking**, Anxiolytic

Systems Affected: Adrenal Glands, Nerves, Pancreas Tail, Pineal, Spleen

Conditions: Addictions (alcohol), Addictions (drugs), Addictions (tobacco smoking or chewing), Adrenals (exhaustion, weakness or burnout), Angina, **Anorexia**, Anxiety (Panic Attack), **Anxiety Disorders**, Asthma, Attention Deficit

Disorder (ADD, ADHD), Bell's Palsy, Bipolar Mood Disorder (Manic Depressive Disorder), Burning Feet or Hands, Crohn's Disease, Cushing's Disease, Depression, Dermatitis, Eczema, **Emotional Sensitivity, Fingernail Biting,** Grief (excessive), Inflammatory Bowel Disorders (Colitis, IBS), Insomnia, Memory and Brain Function, **Mental Illness,** Narcolepsy, **Nerve Damage, Nervous Exhaustion (Enervation), Nervousness, Neurosis, Paralysis,** Parkinson's Disease, Post Traumatic Stress Disorder, **Restless Dreams, Schizophrenia,** Sleep (restless and disturbed), **Stress,** Strokes, Teeth (grinding), **Tension,** Thinking (cloudy), Triglycerides (high), Weight Loss (aids for)

Usage: An anti-stress vitamin/mineral supplement that helps calm nervous and high-strung individuals, Nutri-Calm is a good supplement for anyone who feels depleted from nervous stress. It does not cause drowsiness, and can increase energy while maintaining a sense of calm. The formula feeds the nerves and adrenal glands. It is very useful for people who have contraction furrows (stress rings) in their iris. Use 1 tablet three times a day.

Warnings: No known warnings.

Nutri-Calm - Tablet (100)

Available in: US - Stock #1617-3

Ingredients: Vitamins C, B-1 (thiamin), B-2 (riboflavin), B-6, B-12, folic acid, biotin, niacin, pantothenic acid, schizandra fruit, choline bitartrate, wheat germ, inositol, para-aminobenzoic acid (PABA), bee pollen, lemon bioflavonoids, hops, passion flower, valerian root, di-calcium phosphate.

Nutri-Calm - Capsule (60)

Available in: US - Stock #4803-3

Ingredients: Same as Stock #1617-3

Nutri-Calm Trial Pack - Packets

Available in: US - Stock #2493-9

Ingredients: Vitamins C, B-1 (thiamin), B-2 (riboflavin), B-6, B-12, folic acid, biotin, niacin, pantothenic acid, schizandra fruit, choline bitartrate, wheat germ, inositol, para-aminobenzoic acid (PABA), bee pollen, lemon bioflavonoids, hops, passion flower, valerian root, di-calcium phosphate.

Stress Formula - Capsule (100)

Available in: Canada - Stock #1645-6

Ingredients: Vitamin B1 (thiamine mononitrate), vitamin B2 (riboflavin), vitamin B6 (pyridoxine hydrochloride), vitamin B12 (cyanocobalamin), vitamin C (ascorbic acid), biotin 100, folic acid, niacinamide, pantothenic acid, schizandra fruit, choline bitartrate, inositol powder, para-aminobenzoic acid (PABA), and concentrated extracts of hops flowers, passion flower herb, and valerian root.

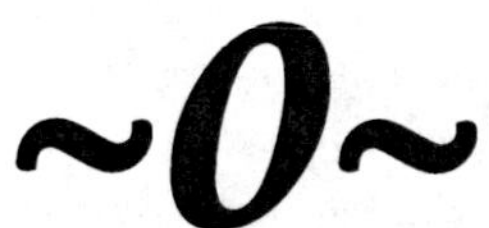

Olive Leaf

Latin Name: *Olea europae*

Product Type: Single Herb

Properties: Antibacterial, Antidiabetic, Antifungal, Antioxidant, Antiviral, Astringent, Diuretic, **Hypoglycemic,** Hypotensive, Vasodilator

Systems Affected: Heart, Immune System, Liver, Nerves

Conditions: Angina, Arteriosclerosis (Atherosclerosis, Hardening of the Arteries), Bladder Infection, Blood Pressure (high), **Chicken Pox,** Circulation (poor), Cold Sores (Fever Blisters), Colds (antiviral), Colds (general remedies for), Contagious Diseases, Diabetes, Epstein Barr Virus (Chronic Fatigue Syndrome, CFS), Fungal Infections (Yeast Infections, Candida albicans), Hashimoto's Disease (Thyroiditis), Herpes, Infection (viral)

Usage: Olive leaf has antimicrobial properties. It is helpful for all kinds of infections, bacterial, viral and fungal. It also reduces oxidative stress and aids circulation. Take 1-3 capsules per day.

Warnings: No known warnings.

Olive Leaf Extract (Concentrate) - Capsule (60)

Available in: US & Canada - Stock #204-7

Ingredients: Standardized extract of Olea europae (olive) leaf (12% oleuropein)

Omega-3

Product Type: Nutrient

Properties: Anti-arthritic, Anti-inflammatory, **Immunomodulator**

Systems Affected: Cuticle, Female Reproductive, **Heart,** Immune System, Liver, **Prostaglandins,** Prostate, Weight Loss

Conditions: Allergies (food), Allergies (respiratory), Appetite (excessive), Arthritis, **Autoimmune Disorders,** Bell's Palsy, **Benign Prostate Hyperplasia (BPH), Blood Clots (prevention of),** Blood Pressure (low), Breasts (swelling and tenderness), Calcium Deficiency, Calcium Deposits (Calcification), Cancer (prevention), **Cardiovascular Disease (Heart Disease), Chemical Poisoning,** Cholesterol (high), Cholesterol (low), Circulation (poor), Cold Hands and Feet, Cold Sores (Fever Blisters), Colitis, Cradle Cap, Croup, **Dementia,** Depression, **Dermatitis,** Diabetes, Dizziness (Vertigo), Down Syndrome, Dyspepsia, **Eczema,** Environmental Pollution (protection from), **Fat Cravings,** Fibroids (uterine), **Heavy Metal Poisoning,** Hyperinsulinemia (Syndrome X), **Inflammation, Lupus, Memory and Brain Function,** Mental

Illness, Mercury Poisoning, **Multiple Sclerosis (MS)**, Narcolepsy, **Nerve Damage, Neuralgia and Neuritis, Pain (general remedies for), Pets (supplements for)**, PMS Type C, **Pregnancy (herbs and supplements for)**, Prostatitis, **Psoriasis**, Schizophrenia, Seborrhea, **Senility, Skin (dry and/or flaky), Skin Care (general)**, Strokes, **Triglycerides (low)**, Vaccines (detoxification from)

Usage: American diets contain too much omega-6 essential fatty acid (EFA) and too little of the omega-3 EFA. Omega-3 helps to reduce inflammation and the risk of cardiovascular disease. It also helps overcome insulin resistance in cells. Omega-3 is also used to help lower cholesterol levels and aid circulation. Take 1-2 capsules with a meal three times daily. Problems with burping a fishy taste from taking the capsules can be reduced or eliminated by freezing the capsules and swallowing the frozen capsules.

Warnings: Not recommended where there is fatty congestion in the liver.

Super Omega-3 EPA - Capsule (60)

Available in: US - Stock #1515-7

Ingredients: eicosapentaenoic acid (EPA), Docosahexaenoic acid (DHA), natural lemon oil, glycerin.

Oregano (Wild)

Product Type: Essential Oil

Properties: Antiviral, Aromatic, Parasympathomimetic, Preservative, Sympatholytic

Systems Affected: Digestive System, Immune System, Respiratory System

Conditions: Arthritis, Asthma, Bronchitis, Fungal Infections (Yeast Infections, Candida albicans), Hypochondria, Infection (bacterial), Infection (viral), Lice, Parasites (general), Pertussis (Whooping Cough), Tuberculosis (Consumption, Scrofula)

Usage: Oregano is known for its use with parasitical conditions. It can help soothe the stomach, stimulate appetite and act as an antispasmodic sedative. It thins bronchial secretions and is an antiseptic to the respiratory tract. It acts as an analgesic, to remove pain. Oregano can be stimulating to a sluggish digestive system. It is highly effective with chronic bronchitis, tickling cough, pulmonary tuberculosis, asthma and whooping cough. Topically it is effective with lice, cellulitis and an arthritic remedy in a poultice. Emotionally, Oregano helps with hypochondria. It can help deflect the sense of dependency on others and shifts responsibility to oneself, which often helps clear up any sense of the problem.

Warnings: Can be a dermal irritant and mucus membrane irritant. Always dilute with a carrier oil for topical use. Not recommended for internal use, as this oil is hepatotoxic, meaning it can damage liver tissue. Symptoms of liver damage do not develop gradually and will show up suddenly, without warning, from using this oil internally.

Oregano, Wild Essential Oil (5ML) - Essential Oil

Available in: US & Canada - Stock #3934-6

Ingredients: Wild Oregano (Origanum campactum) Essential Oil

Oregon Grape

Latin Name: *Berberis aquifolium*

Product Type: Single Herb

Properties: Antipruritic, Bitter, Detergent, Hepatic, Lymphatic

Systems Affected: Appendix, Bladder (Urinary), Bronchials, Gall Bladder, Immune System, Intestinal System, Liver, Lymph Nodes, Lymphatic System, Skin, Thyroid

Conditions: Acid Indigestion (Heartburn, Acid Reflux), Appetite (deficient), Blood Pressure (low), Chicken Pox, Children's Remedy, Cholesterol (low), Colic (children), Congestion (bronchial), Congestion (lymphatic), Congestive Heart Failure, **Defensiveness**, Dermatitis, Digestion (poor), Glands (swollen lymph), Hepatitis, Infection (bacterial), **Itching**, Jaundice (adults), **Measles**, Rashes and Hives, Rheumatic Fever, Staph Infections, Syphilis, Vaccines (detoxification from), Vaginitis

Usage: Alternative to golden seal for hypoglycemics. The alkaloids are similar to golden seal. Helps clear lymphatic channels. Use with dandelion root for hepatitis and jaundice. Stimulates the menses. Works well when applied externally for itching and skin eruptions. Take 1/4 to 1 teaspoon three times daily.

Warnings: Not for use in cases of emaciation or weak digestion. Diabetics should use Oregon grape only in moderate amounts.

Oregon Grape (2 fl. oz.) - Liquid (Glycerite)

Available in: US & Canada - Stock #3395-9

Ingredients: Berberis aquifolium (Oregon grape) root, glycerin, water

P-14

See *Pro-Pancreas*

P-X

Product Type: Formula

Properties: Diuretic

Systems Affected: Liver, Pancreas, Prostate

Conditions: Benign Prostate Hyperplasia (BPH), Diabetes, Prostatitis

Usage: For prostate swelling or blood sugar problems use 2 capsules three times daily with meals.

Warnings: No known warnings.

P-X - Capsule (100)

Available in: US - Stock #1234-5

Ingredients: Juniper berries, golden seal, capsicum, parsley, ginger, eleuthero, uva ursi, queen of the meadow, marshmallow

Pain

Product Type: Homeopathic

Properties: Analgesic (Anodyne)

Systems Affected: Nerves

Conditions: Backache (Back Pain, Lumbago), Neuralgia and Neuritis, Pain (general remedies for), Sciatica, Teething

Usage: Used for the relief of minor pain associated with neuralgia, backache and sciatica. Take 10-15 drops under the tongue every 10-15 minutes or as needed until symptoms improve. Then decrease to every one or two hours, then to four times daily until symptoms are relieved.

Warnings: NSP recommends not taking this product for more than five days. If symptoms do not improve after 10 days, consult your health care professional.

Pain (1 fl. oz.) - Homeopathic (Liquid)

Available in: US - Stock #8709-5

Ingredients: Active ingredients: Colocynthis (Bitter Apple) 4x, Gnaphalium polycephalum (Common Everlasting) 4x, Ammonium Muriaticum (Ammonium Chloride) 6x, Cimicifuga racemosa (Black Cohosh) 6x, Gelsemium sempervirens (Yellow Jessamine) 6x, Spigelia anthelmia (Pink Root) 6x. Other ingredients: Purified water and 20% USP alcohol.

Pantothenic Acid

Product Type: Nutrient

Properties: **Adrenal Tonic**, Nutritive

Systems Affected: Adrenal Glands, Adrenal Medulla, Hair, Muscles, Nerves, Throat

Conditions: **Adrenals (exhaustion, weakness or burnout)**, Aging (prevention), Antibiotics (side effects of), Anxiety (Panic Attack), Asthma, Congestion (bronchial), Cramps and Spasms (general), Dermatitis, Eczema, Fatigue, Gray Hair, **Hot Flashes**, Itching, Nervous Exhaustion (Enervation), Neuralgia and Neuritis, **Post Traumatic Stress Disorder**, **Stress**, Surgery (healing from), Teeth (grinding)

Usage: Aids in the formation of some fats and participates in energy metabolism. Plays a role in formation of antibodies and is important for the production of adrenal hormones. RDA infants: 2-3 mg., RDA children: 3-7 mg., RDA adults: 5-10 mg., dietary levels: 4-7 mg. Therapeutic range: 50-100 mg. Take 1 capsule daily with a meal.

Warnings: No known warnings.

Pantothenic Acid (250 mg) - Capsule (100)

Available in: US & Canada - Stock #1640-2

Ingredients: 250 mg of d-calcium pantothenate

Papaya Mint

Product Type: Formula

Properties: Deodorant, Digestant

Systems Affected: Liver, Small Intestines, Stomach

Conditions: **Acid Indigestion (Heartburn, Acid Reflux)**, **Belching**, Cancer (natural therapy for), Children's Remedy, Cystic Fibrosis, Digestion (poor), Dyspepsia, Gas and Bloating, **Indigestion**, Parasites (nematodes, worms)

Usage: Raw papaya fruit contains the enzyme papain, which helps break down proteins. Peppermint stimulates hydrochloric acid production. Hence, this combination primarily aids in the digestion of proteins. It is useful for people with allergies, immune system weakness and glandular weakness due to poor digestive function. Papain also dissolves outer cuticle of worms, so it aids in the expulsion of parasites. Chew 1-3 tablets with each meal to aid in digestion. Chew 1-2 tablets every half hour in acute indigestion.

Warnings: No known warnings. Mild, safe remedy for children or adults.

Papaya Mint Chewable Tablets - Tablet (70)

Available in: US & Canada - Stock #485-6

Ingredients: Papaya fruit, peppermint leaf and oil.

Para Pack

See *Para-Cleanse*

Para-Cleanse

Product Type: Pack

Properties: Antiparasitic, Vermifuge

Systems Affected: Immune System, Intestinal System, Large Intestine (Colon), Liver

Conditions: Allergies (food), Allergies (respiratory), Constipation (adults), Dysentery, Fingernail Biting, Fungal Infections (Yeast Infections, Candida albicans), Inflammatory Bowel Disorders (Colitis, IBS), **Itching (rectal)**, Leaky Gut Syndrome, **Parasites (general)**, Parasites (nematodes, worms), **Parasites (tapeworm)**, Pets (supplements for), Pregnancy (herbs and supplements to avoid during), Teeth (grinding)

Usage: This is an outstanding and very effective program for removing yeast, worms and other parasites from the intestinal tract. Take the contents of one packet (6 capsules) 15 minutes before breakfast and 15 minutes before dinner daily for 10 days. Drink one glass (8 ounces) of water with the capsules. Continue this program for ten days, wait seven days, and then do a second round of cleansing. The theory is that the first cleanse removes the parasites and the break in between allows any eggs to hatch so that the second cleanse removes any newly hatched parasites. After doing a colon cleansing program like this one it is a good idea to follow up with a round of probiotics.

Warnings: Not for continuous or long-term use. Not recommended for children.

Para-Cleanse (20 packets) - Packet

Available in: US - Stock #4115-7

Ingredients: Herbal Pumpkin Combination, Paw Paw Cell-Reg™, Black Walnut ATC Concentrate, Artemisia Combination

Para Pack - Packet

Available in: Canada - Stock #4021-5

Ingredients: Herbal Pumpkin, Caprilic Acid Combo, Elecampane Combination

Parasites

Product Type: Homeopathic

Properties: Antiparasitic

Systems Affected: Immune System, Large Intestine (Colon)

Conditions: Abdominal Pain and Inflammation, Diarrhea, Gas and Bloating, Parasites (general)

Usage: Used for the relief of minor intestinal symptoms such as bloating, abdominal pain, flatulence and diarrhea that may be associated with parasites. Take 10-15 drops under the tongue four times daily for at least one month. Then take half the dosage for an additional three months as maintenance.

Warnings: NSP does not recommend this product for children under the age of 12 except on the advice of a health professional. Other than the alcohol content, which is not good for infants, we consider this product safe for children.

Parasites (1 fl. oz.) - Homeopathic (Liquid)

Available in: US - Stock #8735-6

Ingredients: Active ingredients: Baptisia tinctoria (Wild Indigo) 4x, Bryonia (White Bryony) 4x, Colocynthis (Bitter Apple) 4x, Nux vomica (Quaker Button) 4x, Cina (Wormseed) 6x, Spigelia anthelmia (Pink Root) 6x, Teucrium marum (Cat Thyme) 6x, Terebinthina (Oil of Turpentine) 8x,. Other ingredients: Purified water and 20% USP alcohol

Parsley

Latin Name: *Petroselium crispum*

Product Type: Single Herb

Properties: **Antigalactagogue**, Bitter, Condiment, Diuretic, Food, Mineralizer, Nutritive

Systems Affected: Digestive System, Kidneys, Pituitary (posterior), Progesterone, Urinary System

Conditions: Adrenals (exhaustion, weakness or burnout), Bed-wetting, Body Odor, **Breast Milk (dry up)**, Breasts (swelling and tenderness), Cancer (prevention), Cystitis, Edema (Dropsy, Water Retention, Swelling), **Halitosis (Bad Breath)**, Kidney Stones

Usage: Parsley is rich in sodium and potassium necessary to regulate fluids in the body. It has a volatile oil that stimulates kidney function and has a mild alkalizing effect on the system. It also helps to lower blood pressure and slow the pulse. An excellent mild remedy for improving kidney function. Use 1-2 capsules with meals. Fresh parsley juice is a non-toxic insect repellent. Parsley also heals sick fish! Chariot horses were fed parsley.

Warnings: Not recommended in cases involving fluid deficiency, wasting or dryness. Avoid when nursing as parsley tends to dry up breast milk.

Parsley - Capsule (100)

Available in: US & Canada - Stock #490-9

Ingredients: Petroselium crispum (parsley) herb

Parthenium

Latin Name: *Parthenium integrifolium*

Product Type: Single Herb

Properties: Antiseptic, Diuretic

Systems Affected: **Bladder (Urinary)**, Urinary System

Conditions: Bladder Infection, Infection (bacterial), Kidney Infection, Urinary Tract Infections

Usage: Has some affinity for the urinary system and has been used to treat urinary system problems. Also may have some immune stimulating properties. Use 1-2 capsules one to three times daily.

Warnings: May cause allergic reactions in some people. Not recommended for long-term use.

Parthenium - Capsule (100)

Available in: US & Canada - Stock #265-5

Ingredients: Parthenium integrifolium root

Passion Flower

Latin Name: *Passiflora incarnata*

Product Type: Single Herb

Properties: Anti-epileptic, Antispasmodic, Anxiolytic, Bronchial Dilator, Calmative, Hypnotic, Nervine, Sedative, Sympatholytic

Systems Affected: Eyes, Heart, Nerves, Serotonin, Skin, Stomach

Conditions: Addictions (alcohol), Anxiety (Panic Attack), **Arrhythmia**, Asthma, Blood Pressure (high), Colic (adults), Convulsions, Cushing's Disease, Depression, Dysmenorrhea, Epilepsy, Headache (tension), Heart (weakness), **Heart Fibrillation or Palpitations**, Hysteria, Insomnia, Neuralgia and Neuritis, Parkinson's Disease, Rheumatic Fever, Teething

Usage: As a general nervine, use 1 capsule two to three times daily. Works best when combined with other nervines. The herb helps people who can't fall asleep because their mind keeps going over and over things (i.e., won't "shut down").

Warnings: No known warnings.

Passion Flower - Capsule (100)

Available in: US & Canada - Stock #500-3

Ingredients: Passiflora incarnata (passion flower) herb

Patchouli

Product Type: Essential Oil

Properties: Anti-inflammatory, Aphrodisiac, Aromatic, Astringent, Deodorant, Insecticide, **Perfume**, Relaxant

Systems Affected: Hypothalamus, Pancreas, Pineal, Scalp, Spleen

Conditions: Aging (prevention), Anger (excessive), Anxiety (Panic Attack), Apathy, Bites and Stings, Cellulite, Confusion, Edema (Dropsy, Water Retention, Swelling), Fungal Infections (Yeast Infections, Candida albicans), Insects, Mood Swings, Nervous Exhaustion (Enervation), Nervousness, Scars / Scar Tissue, Sex Drive (low), Wrinkles

Usage: Patchouli's action is astringing. It helps prevent the formation of scar tissue. Historically, it was used to help prevent the spread of fevers and epidemics. It does so by strengthening the immune system. It is also considered to be an effective insecticide and continues to be one of the most important remedies for snake and insect bites. It has anti-inflammatory action and can give an uplifting feeling along with a synergistic calming effect. It has a wide range of uses for dermal disorders such as eczema, acne, impetigo and herpes. It helps to relieve fluid retention and cellulite and also has a deodorizing effect. Topically, patchouli is used as a tissue regenerator helping to stimulate re-growth of skin cells and forming scar tissue. It helps heal fungal infections and scalp disorders. Patchouli's strongest area of use however is for emotional imbalances. It helps with an energetic imbalance in the spleen/pancreas in Chinese medicine, meaning it helps with chronic anxiety, over thinking and worry. Where excessive mental activity and nervous strain have caused people to lose touch with their body and their sensuality, patchouli is a relaxing aphrodisiac that helps with impotence, frigidity and sexual anxiety. It is an antidepressant. Gently stimulating the senses and uplifting the mind, it expands creative expression and imagination.

Warnings: Not recommended for internal use.

Patchouli Essential Oil (5ML) - Essential Oil

Available in: US & Canada - Stock #3909-0

Ingredients: Patchouli (Pogostemon patchouli) Essential Oil

Pau D'Arco

Latin Name: *Tabebuia impetiginosa, T. avellanedae, T. ipe, T. cassanoides, Tecoma ochracea*

Product Type: Single Herb

Properties: Alterative (Blood Purifier), Anti-inflammatory, Antibacterial, Anticancer, Anticoagulant (Blood Thinner), **Antifungal**, Antiviral, Astringent, Hepatic, Tonic

Systems Affected: Immune System, Large Intestine (Colon), Liver, Parathyroid, Respiratory System, Skin, Spleen

Conditions: Acquired Immune Deficiency Syndrome (AIDS/HIV), Age Spots, Antibiotics (side effects of), Arthritis, Athlete's Foot, Blood Poisoning, Body Odor, Cancer (natural therapy for), Chemotherapy (reducing side effects), Cradle Cap, Crohn's Disease, Dermatitis, Diarrhea, Eczema, Endometriosis, Epstein Barr Virus (Chronic Fatigue Syndrome, CFS), Fever, **Fungal Infections (Yeast Infections, Candida albicans)**, Herpes, Inflammation, **Itching**, Itching Ears, Jock Itch, **Leucorrhea**, Leukemia, Lou Gehrig's Disease, Malaria, Meningitis, Parasites (general), Polyps, Psoriasis, **Rashes and Hives**, Skin (infections), Snake Bite, Sprains, Thrush, Toxemia, **Vaginitis**, Wounds and Sores

Usage: Although these trees grow in tropical rain forests, it is reported that no fungus of any kind grows on them. Used as both an anti-cancer remedy and as an antifungal remedy. Take 2-5 capsules three times a day. For tea, simmer 1 tablespoon in 2 pints of water for 20 minutes, and drink 2-3 cups per day.

Warnings: No known warnings.

Pau D'Arco Capsules - Capsule (100)

Available in: US & Canada - Stock #504-2

Ingredients: Tabebuia sp. (pau d'arco) bark

Pau D'Arco Capsules - Capsule (270)

Available in: US - Stock #505-5

Ingredients: Same as Stock #504-2

Pau D'Arco Bulk (7 oz.) - Bulk Tea

Available in: US - Stock #1372-7

Ingredients: Same as Stock #504-2

Pau D'Arco Extract (2 fl. oz.) - Liquid (Tincture)

Available in: US - Stock #1774-2

Ingredients: Pau d'arco bark (Tabebuia heptaphylla) in water and alcohol

Pau D'Arco Lotion (4 fl. oz.) - Topical

Available in: US & Canada - Stock #1614-4

Ingredients: Purified water, Pau d' arco extract(tabebuia heptaphylla), Ethylhexyl palmitate, Carthamus tinctorius (hybrid safflower) seed oil, glycerin, Helianthus annuus (sunflower) oil, glyceryl sterate, cetyl alcohol, stearic acid, Aloe barbadensis leaf juice, sodium steroyl lactylate, tocopheryl acetate, panthenol, allantion, disodium EDTA, carbomer, aminiomenthyl propanol, dimethicone, polysorbate-20, phenoxyethanol, methylparaben, ethylparaben, propylparaben, butylparaben, isobutylparaben.

Paw Paw

Latin Name: *Asimina triloba*

Product Type: Single Herb

Properties: **Anticancer**, Antifungal, Antimicrobial, **Antiparasitic**, Antiviral, Cytotoxic, **Escharatic**, **Insecticide**, **Nauseant**, Vermifuge, **Virostatic**

Systems Affected: Breasts, Immune System, Large Intestine (Colon), Liver, Mitochondria

Conditions: Athlete's Foot, **Cancer (natural therapy for)**, Chemotherapy (reducing side effects), **Endometriosis**, **Fleas**, Fungal Infections (Yeast Infections, Candida albicans), **Herpes**, Infection (viral), **Insects**, Lice, Melanoma (Skin Cancer), **Pap Smear (abnormal)**, **Parasites (general)**, Parasites (nematodes, worms), **Parasites (tapeworm)**, **Pregnancy (herbs and supplements to avoid during)**, **Psoriasis**, Scabies, **Shingles**, Tick, **Warts**

Usage: The acetogenins in paw paw inhibit the production of ATP in the mitochondria of cells. Since cancer cells have a much faster metabolic rate than normal cells, this induces cancer cells to self-destruct. Scientific research demonstrates that paw paw extract is more effective than many chemotherapy drugs at destroying cancer cells, and yet is completely non-toxic. In a clinical trial involving over 100 people with cancer, Paw Paw was shown to reduce tumor markers and tumor sizes, usually within 1-4 weeks, with virtually no side effects. Paw Paw Cell-Reg may also be helpful in fighting chronic viral conditions such as shingles and herpes, because viruses require ATP to reproduce. These compounds also have antifungal and antibacterial activity. The acetogenins from paw paw are antiparasitic which is why this product is included in the ParaCleanse with Paw Paw pack.

While using this program, products that increase cellular vitality, especially those that increase energy metabolism, should be avoided. Antioxidants also decrease the effectiveness of the paw paw extract. Supplements to be avoided include Co-Q10, SOD, Super Antioxidants, Grapine, Alpha Lipoic Acid, Cellular Energy, N-Acetyl Cysteine, Garlic, IGF-1, 7-Keto, TS II and Thyroid Activator. Because these supplements increase energy production and inhibit cellular damage, they reduce the effectiveness of the paw paw extract. Thyroid Activator, for instance, would stimulate the thyroid gland to secrete hormones that would raise the production of ATP in the cells. However, we have found that alternating taking Paw Paw for two to four weeks, followed by one to two weeks of building with antioxidants and other nutrients may increase the effectiveness of the program.

Dosage is 4 capsules per day, taken as 1 capsule four times daily. Higher amounts can be taken if desired, but nausea and vomiting, dizziness, fatigue, or light headedness can occur. Paw Paw Cell-Reg works better when taken with Immune Stimulator and Protease Plus Enzymes.

It may also be combined with traditional anti-cancer herbs and formulas such as anamu, pau d'arco, and E-tea. Paw Paw Cell-Reg can also be applied topically as a poultice to warts and for fungal, bacterial or viral infections. It can be mixed with Black Ointment for that purpose.

Paw Paw Cell-Reg and tea tree oil can be added to shampoo to help kill head lice. This mixture can also be used as a flea and tic shampoo for dogs and cats. Paw Paw Cell-Reg can also be diluted with water containing a little bit of Sunshine Concentrate (for emulsification) and used as a non-toxic insecticidal spray.

Warnings: The paw paw extract is not toxic to people or animals in normal doses, but it may cause nausea and vomiting when ingested. This product is not for cancer prevention, use antioxidants for that purpose. Do not take antioxidant products in combination with Paw Paw Cell-Reg. Don't use the Para-Cleanse with Paw Paw and the Paw Paw Cell-Reg together, as this is an excessive use of the Paw Paw. Paw Paw is not recommended for healthy people as it will reduce energy levels, causing fatigue. It should only be used by persons with cancer, viral disorders, parasites or other specific health problems. Excessive doses will cause nausea and vomiting. Paw paw may also induce a healing crisis or cleansing reaction if cell die off is too quick, which may require cleansing herbs to help the body detoxify. Cancer is a serious illness, seek professional assistance and advice.

Paw Paw Cell-Reg - Capsule (180)

Available in: US - Stock #515-6

Ingredients: Standardized extract of acetogenins from Asimina triloba (paw paw)

Paw Paw Cell-Reg™ - Capsule (120)

Available in: Canada - Stock #511-3

Ingredients: Same as Stock #515-6

PBS

Product Type: Formula

Properties: Antidiabetic, Bitter, **Hypoglycemic**, Hypolipidemic, Pancreatic Tonics

Systems Affected: Adrenal Glands, Digestive System, Liver, Pancreas, Pancreas Tail

Conditions: Diabetes, Triglycerides (high)

Usage: This formula is also designed for diabetics. Cedar berries help the pancreas produce insulin, while golden seal helps lower blood sugar levels. Burdock root aids liver function and ginseng acts as an overall tonic (adaptogen) for the body. Horseradish aids digestion. Use 2 capsules three times daily.

Warnings: Not for hypoglycemics.

PBS - Capsule (100)

Available in: US - Stock #1054-7

Ingredients: Burdock, cedar berries, horseradish, golden seal, Siberian ginseng

PDA

Product Type: Formula

Properties: Digestant

Systems Affected: Pancreas Head, Stomach

Conditions: Acid Indigestion (Heartburn, Acid Reflux), Appetite (deficient), Belching, Calcium Deficiency, Cancer (natural therapy for), Cystic Fibrosis, Diarrhea, Digestion (poor), Dyspepsia, Failure to Thrive, Fibroids (uterine), Gas and Bloating, **Gout**, **Gray Hair**, Indigestion, Osteoporosis, Poisoning (food), **Protein Digestion (poor)**, Psoriasis, Rosacea, Weight Gain (aids for)

Usage: Supplements hydrochloric acid (HCl) for protein digestion. Take 1-2 capsules with meals as an aid to digestion of proteins.

Warnings: Do not take with ulcers. May cause stomach irritation in some people. Take with food. It is not a good idea to constantly take HCl, as this reduces the need for the body to make it for itself. Use Digestive Bitters to stimulate the body's own production of HCl.

PDA Combination Capsules - Capsule (200)

Available in: US - Stock #1837-5

Ingredients: Betaine hydrochloric acid, pepsin, cellulose, magnesium stearate

Protein Digestive Aid - Capsule (200)

Available in: Canada - Stock #1756-2

Ingredients: Same as Stock #1837-5

Peppermint

Product Type: Essential Oil

Properties: **Anti-emetic (Antinauseous)**, **Antinauseous**, Aromatic, **Carminative**, Deodorant, Sialogogue, Stimulant, Stomachic

Systems Affected: Brain, Digestive System, Intestinal System, Large Intestine (Colon), Mouth, **Stomach**

Conditions: Apathy, Belching, Burns and Scalds, Cold Sores (Fever Blisters), **Colic (children)**, Concentration (poor), Confusion, Constipation (adults), Cysts, Diarrhea, Dizziness (Vertigo), Dysentery, Dysmenorrhea, Energy (lack of), Fever, Flu, Fungal Infections (Yeast Infections, Candida albicans), **Gas and Bloating**, Halitosis (Bad Breath), Hot Flashes, Hypochondria, Hysteria, Impetigo, **Indigestion**, Insects, Menopause, Morning Sickness, **Motion Sickness**, Mumps, **Nausea and Vomiting**, Rheumatic Fever, Shock, **Thinking (cloudy)**, Ulcers

Usage: Apply peppermint oil full strength externally as a mild stimulant or antiseptic for healing minor injuries, bites and stings. Put 1 or 2 drops in a cup of water and sip slowly to relieve gas, heartburn and indigestion. A drop can also be placed on the bottom of the thumb and sucked on or placed on the back of the tongue with the finger. Inhaling the oil helps promote mental alertness for study, late night driving, etc. Also relieves muscle spasms. Inhale for sinus and respiratory conditions. Peppermint leaf is best used as a soothing tea for digestive and nervous system problems.

Warnings: Undiluted oil can cause skin irritation if used excessively. Do not use orally for kids under six years. For young children, make a tea of peppermint leaves and sweeten. Do not use externally in high concentrations.

Peppermint Oil (0.17 fl oz.) - Essential Oil

Available in: US & Canada - Stock #1706-8

Ingredients: Peppermint (Mentha x. piperita) essential oil

Peppermint Essential Oil (5ML) - Essential Oil

Available in: US & Canada - Stock #3910-9

Ingredients: Same as Stock #1706-8

Perfect Eyes

Product Type: Formula

Properties: **Opthalmicum**

Systems Affected: **Eyes**, Liver

Conditions: Cataracts, **Eye Problems (general)**, **Eyesight (to improve)**, Floaters, Glaucoma, **Macular Degeneration**, Night Blindness

Usage: This formula helps protect the eyes from damage due to oxidative stress. It can help diminish the effects of macular degeneration and may help to prevent damage to the eyes from diabetes. It can also help when eyes are irritated from environmental pollutants. Use 2 capsules one time daily.

Warnings: No known warnings.

Perfect Eyes - Capsule (60)

Available in: US & Canada - Stock #4075-0

Ingredients: Lutein, N-acetyl-cysteine, bilberry, curcuma root, eyebright herb, alpha and beta carotene, lycopene, zeaxanthin, cryptoxantin, zinc gluconate, selenium, taurine, quercetin and hesperidin

Phyto-Soy

Product Type: Nutrient

Properties: **Estrogenic**, Nutritive, **Phytoestrogen**

Systems Affected: Female Reproductive, Prostate

Conditions: Arthritis, Broken Bones, Burns and Scalds, Cancer (natural therapy for), Cancer (prevention), Cholesterol (high), Dizziness (Vertigo), **Estrogen (low)**, Grave's Disease, Menopause, Osteoporosis, Perspiration (excessive)

Usage: Take 1 capsule with a meal three times daily. Phyto-Soy is a standardized soy extract powder. Soy contains all eight essential amino acids and is useful in promoting healing. Its genistein and isoflavones have proven to be effective against several forms of cancer. Soy isoflavones are a form of mild plant estrogen that can help relieve the symptoms of menopause and osteoporosis.

Warnings: No known warnings.

Phyto-Soy - Capsule (90)

Available in: US - Stock #4981-5

Ingredients: Soy Bean Extract (standardized to 10% isoflavones- 12 mg isoflavones per dose as genistein, daidzein, and glycitein.)

Pine Needle

Product Type: Essential Oil

Properties: Anti-inflammatory, Antimicrobial, Antiparasitic, Antiseptic, Aromatic, Decongestant, Expectorant, Pectoral, Stimulant, Sympathomimetic

Systems Affected: Adrenal Cortex, Adrenal Medulla, Aldosterone, Epinephrine, Lungs, Reproductive Glands, Respiratory System, Sinuses, Uro-genital Tract

Conditions: Addictions (tobacco smoking or chewing), Appetite (deficient), Asthma, Bronchitis, Circulation (poor), Colds (general remedies for), Concentration (poor), Concentration (poor), Confusion, Congestion (sinus), Cough (dry), **Cough (general)**, Depression, Fatigue, Fleas, Flu, Foot Odor, Gout, Grief (excessive), **Guilt**, Insects, Laryngitis (Hoarseness), Nervous Exhaustion (Enervation), Nervous Exhaustion (Enervation), Parasites (general), Sinus Infection, **Thinking (cloudy)**, Uric Acid Retention

Usage: Pine is an effective expectorant and works well with asthma, bronchitis, laryngitis and influenza. It cleanses the chest and lungs, relieving chronic cough. Pine is also a tonifier that helps overcome mental, physical and sexual fatigue, chills and poor appetite. Stimulating the adrenal glands, both the medulla and the cortex, it can help relieve nervous exhaustion. It also benefits the genito-urinary system with antiseptic and anti-inflammatory properties. Pine reduces uric acid in the blood that can relieve rheumatic symptoms, gout, sciatica and arthritis. Emotionally, Pine helps to restore self-confidence. Promoting feelings of well-being, mental alertness and energy, it disperses melancholy and counteracts pessimism. It reawakens our instinctive connection to life. Cleansing and invigorating, this oil is often chosen to scent cleaning solutions for that sense of "fresh." For those who blame themselves and feel responsible for all the mistakes and sufferings of others, pine can help to establish appropriate boundaries and self-identity, relieving feelings of helplessness and unworthiness. It disperses negative self-images and feelings of remorse, helping one to feel forgiveness and self-acceptance.

Warnings: Pine is recognized as a dermal irritant especially where allergic skin conditions exist. It is also contraindicated in dry conditions in the lungs.

Pine Needle Essential Oil (5ML) - Essential Oil

Available in: US & Canada - Stock #3911-5

Ingredients: Pine needle (Pinus sylvestris) essential oil

PLS II

Product Type: Formula

Properties: Absorbant, Anti-inflammatory, Cell Proliferant, Demulcent (Mucilant), Soothing, Vulnerary

Systems Affected: Bones, Lungs, Mucus Membranes, Respiratory System, Skeletal System, Skin, Structural System

Conditions: Arthritis, Backache (Back Pain, Lumbago), Bites and Stings, Blood Poisoning, Broken Bones, Bursitis, Colitis, Convalescence, Cough (dry), Cuts, Disks (spinal - bulging or slipped), Dislocation, Diverticulitis (Diverticuli), Epstein Barr Virus (Chronic Fatigue Syndrome, CFS), **Hernias**, Hiatal Hernia, Inflammation, **Injuries**, **Ligaments (torn or injuried)**, Sciatica, Ulcers, Wounds and Sores

Usage: This formula was designed primarily as an instant poultice. Open capsules and moisten powder with water or aloe vera to produce a thick paste. Apply paste to swellings, minor injuries, etc. to promote healing. PLS II can also be taken internally to soothe intestinal irritations, or reduce irritation to the respiratory and urinary systems. Use 2-4 capsules two to four times per day with a large glass of water. The ingredients in this formula suggest it would also be beneficial for "hot" respiratory conditions, as it would cool the mucus membranes of the lungs.

Warnings: No known warnings.

PLS II - Capsule (100)

Available in: US - Stock #1029-4

Ingredients: Slippery elm, marshmallow, golden seal, fenugreek

PMS

Product Type: Homeopathic

Properties: Glandular

Systems Affected: Female Reproductive, Glandular System

Conditions: Dysmenorrhea, Hot Flashes, PMS (general)

Usage: Used for the relief of symptoms associated with PMS, including nervous irritability, bloating, cravings, discomfort and hot flashes. Take 10-15 drops under the tongue every two hours, until symptoms improve, or as needed. Then decrease to every four hours until symptoms are relieved.

Warnings: Not recommended for girls under the age of 12.

PMS (1 fl. oz.) - Homeopathic (Liquid)

Available in: US - Stock #8738-5

Ingredients: Active ingredients: Cimicifuga racemosa (Black Cohosh) 6x, Ignatia amara (St. Ignatius' Bean) 6x, Pulsatilla nigricans (Wind Flower) 6x, Magnesia Phosphorica (Magnesium Phosphate) 10x, Phosphorus (Phosphorus) 10x, Sepia (Cuttle Fish) 12x. Other ingredients: Purified water and 20% USP alcohol

Potassium

Product Type: Formula

Properties: Alkalinizer, Glandular, Kidney Tonic, Mineralizer

Systems Affected: Digestive System, Hypothalamus, Parotids, **Pituitary (posterior)**, Structural System, Thyroid, Thyroid, Urinary System

Conditions: Copper Toxicity, **Cramps (leg)**, Cramps (menstrual), Cramps and Spasms (general), Dehydration, Diarrhea, Edema (Dropsy, Water Retention, Swelling), Heart Fibrillation or Palpitations, Overacidity, Paralysis, Tachycardia, Tics, **Tremors**, Triglycerides (high), **Twitching**

Usage: Used as a natural potassium supplement, but it is also high in natural sodium. It is used to benefit the kidneys and help to maintain fluid and electrolyte balance. It also contains many herbs beneficial to the thyroid. Take 2 capsules with a meal three times daily.

Warnings: No known warnings.

Potassium Combination - Capsule (100)

Available in: US - Stock #1673-3

Ingredients: Potassium, kelp, di-potassium phosphate, alfalfa, dulse, white cabbage, horseradish, horsetail

Potas - Capsule (100)

Available in: Canada - Stock #4978-9

Ingredients: Kelp, dulse, alfalfa, horseradish, white cabbage, thyme, potassium

Pregnenolone

Product Type: Nutrient

Properties: Glandular

Systems Affected: Adrenal Cortex

Conditions: Aging (prevention), Memory and Brain Function, Pregnancy (herbs and supplements to avoid during), Stress

Usage: Pregnenolone is a hormone that can be converted into other hormones produced by the glands of the body (especially the adrenals and sex glands.) Take 1-2 capsules daily with a meal.

Warnings: Not for use by persons under the age of 40, persons with severe benign hypertrophy hormone-responsive cancer, or anyone with a medical condition or taking prescription medication. Pregnant or nursing women should consult a health care professional before using this product. Be cautious and observant when taking a hormone replacement product such as this one, as increasing levels of one hormone always affects other hormones. Not recommended for long term use.

Pregnenolone - Capsule (60)

Available in: US - Stock #1824-4

Ingredients: Pregnenolone 10 mg.

Prevention

Product Type: Homeopathic

Properties: Anti-inflammatory, Immune Stimulant

Systems Affected: Immune System

Conditions: Colds (general remedies for), Colds (prevention), Flu, Infection (viral), Inflammation

Usage: A homeopathic which helps to prevent or relieve symptoms associated with minor infections, including inflammation, secondary infections, colds and influenza. Take 10-15 drops under the tongue 1-3 times daily as a preventive health measure, or with other remedies for colds, influenza and minor infections and illness.

Warnings: NSP does not recommend this product for children under the age of 12 except on the advice of a health professional. Other than the alcohol content, which is not good for infants, we consider this product safe for children.

Prevention (1 fl. oz.) - Homeopathic (Liquid)

Available in: US - Stock #8827-9

Ingredients: Active ingredients: Echinacea angustifolia (Cone Flower) 2x, Aconitum (Aconite) 3x, Rhus toxicodendron (Poison Ivy) 4x, Sanguinaria (Blood Root) 4x, Bryonia (White Bryony) 6x, Eupatorium perfoliatum (Boneset) 6x, Thuja occidentalis (Tree of Life) 8x, Hepar Sulfuris Calcareum (Calcium Sulfide) 10x, Sulphur (Sulfur)10x, Influenzinum (Influenza) 12x, Lachesis mutus (Lachesis) 12x, Staph-ylococcinum (Staphylococci) 12x, Streptococcinum (Streptococci) 12x. Other ingredients: Purified water and 20% USP alcohol

Pro-G-Yam Cream

Product Type: Formula

Properties: Antidepressant, Emollient, Glandular

Systems Affected: Female Reproductive, Ovaries, **Progesterone**

Conditions: Breast Lumps, Breasts (swelling and tenderness), Broken Bones, Cancer (natural therapy for), Cardiovascular Disease (Heart Disease), **Cystic Breast Disease**, Depression, Dermatitis, Disks (spinal - bulging or slipped), Epstein Barr Virus (Chronic Fatigue Syndrome, CFS), Fibroids (uterine), Hair (loss or thinning), **Hot Flashes, Menopause, Menstrual Irregularity, Miscarriage (prevention), Osteoporosis, PMS Type A, Progesterone (low), Vaginal Dryness**

Usage: Progesterone is a hormone that is produced by the ovaries. It helps regulate the menstrual cycle and works in opposition to estrogen to regulate its production. During menopause, progesterone levels decrease leading to symptoms such as depression and hot flashes. Pro-G-Yam also contains wild yam, an emollient and antispasmodic agent. To use, simply rub 1/4 teaspoon of cream gently into soft skin regions, such as the inside upper arm, inner thigh or abdomen once or twice daily. Rotate application to a different area of the body each day. Use no more than one tube per month. For external use only.

Warnings: Because progesterone is a naturally occurring hormone, long term supplementation may alter the body's ability to produce sufficient amounts of it on its own. Progesterone is best used for short term treatment of hormonal imbalance.

Pro-G-Yam Cream (2 oz.) - Topical

Available in: US - Stock #4948-0

Ingredients: Water, glycerin, glyceryl stearate, cetearyl alcohol, sodium stearoyl lactylate, dimethicone, didecene, caprylic/capric triglyceride, macadamia ternifolia seed oil, progesterone, dioscorea villosa (wild yam), oleyl alcohol, glycine soja (soybean), sterols root extract, citrus grandis (grapefruit) peel oil, citrus aurantium, bergamia (bergamot) fruit oil, cananga_odorata flower oil, cimicifuga racemosa root extract, caulophyllum thalictroides_root, glycyrrhiza glabra (licorice) root extract, glycine soja (soybean) germ extract, aloe barbanensis leaf juice, panax ginseng root extract, ginkgo biloba_leaf extract, squalane, tocopheryl acetate, tetrahexyldecyl acsorbate, dipotassium glycyrrhizate, menthol, carbomer, cetearyl methicone, tetrasodium EDTA, phenoxyethanol, caprylyl glycol, ethylhexylglycerin, hexylene glycol.

Pro-G-Yam (500 mg) (2 oz. tube) - Topical

Available in: US - Stock #4949-3

Ingredients: Water, glycerin, glyceryl stearate, cetearyl alcohol, sodium stearoyl lactylate, dimethicone, didecene, caprylic/capric triglyceride, progesterone, Macadamia ternifolia seed oil, Dioscorea villosa (wild yam) root extract, oleyl alcohol, Glycine soja (soybean) sterols root extract, Citrus grandis (grapefruit) peel oil, Citrus aurantium bergamia (bergamot) fruit oil, Cananga odorata flower oil, Cimicifuga racemosa root extract, Caulophyllum thalictroides root, Glycyrrhiza glabra (licorice) root extract, Glycine soja (soybean) germ extract, Aloe barbadensis leaf juice, Panax ginseng root extract, Ginkgo biloba leaf extract, squalane, to co pher yl acetate, tetrahexyldecyl ascorbate, dipotassium glycyrrhizate, menthol, carbomer, cetearyl methicone, tetrasodium EDTA, phe nox y eth anol, caprylyl glycol, ethylhexyl glycerin, and hexylene glyco

Pro-Pancreas

Other Names: P-14

Product Type: Formula

Properties: Antidiabetic, Digestive Tonic, Diuretic, Glandular, Hypoglycemic, Pancreatic Tonics, Stomachic

Systems Affected: Adrenal Glands, Intestinal System, Liver, Pancreas

Conditions: Bedwetting, Blood Pressure (high), Cystic Fibrosis, Cystitis, Diabetes, **Hyperinsulinemia (Syndrome X)**, Mental Illness

Usage: This formula was designed to aid the pancreas in cases of diabetes, It contains golden seal to help lower blood sugar levels, diuretics to flush the system through the urine and other herbs to aid the overall body. Take 2 capsules with each meal, three times daily.

Warnings: Not for hypoglycemics.

Pro-Pancreas - Capsule (100)

Available in: US - Stock #1027-9

Ingredients: Golden seal, juniper berries, uva ursi, cedar berries, mullein, garlic bulb, yarrow, slippery elm bark, capsicum fruit, dandelion, marshmallow, nettle, white oak, licorice

P-14 - Capsule (100)

Available in: Canada - Stock #1023-0

Ingredients: Dandelion root, garlic bulb, golden seal root, buchu leaves concentrate, cedar berries, mullein leaves, yarrow flowers, slippery elm bark, capsicum fruit, marshmallow root, nettle herb, white oak bark, and licorice root

Proactazyme Plus

Other Names: Garden Essence

Product Type: Formula

Properties: **Digestant**, Immune Amphoterics

Systems Affected: **Digestive System**, Immune System, Liver, **Stomach**, Weight Loss

Conditions: **Allergies (food)**, Appetite (deficient), **Autoimmune Disorders**, **Belching**, Body Odor, Broken Bones, Convalescence, Diabetes, **Digestion (poor)**, Dyspepsia, Epstein Barr Virus (Chronic Fatigue Syndrome, CFS), **Failure to Thrive**, Fatigue, Fibromyalgia Syndrome (FMS), **Gas and Bloating**, **Halitosis (Bad Breath)**, Ileocecal Valve, **Indigestion**, Inflammatory Bowel Disorders (Colitis, IBS), **Leaky Gut Syndrome**, Migraine, Overacidity, Overalkalinity, Pancreatitis, Rosacea, Sore or Geographic Tongue, Sore or Geographic Tongue, Wasting, **Weight Gain (aids for)**

Usage: Contains enzymes for breaking down fats, proteins and carbohydrates. Take 1-2 capsules with meals as an aid to digesting all types of food.

Warnings: Do not use in cases of ulceration.

Proactazyme Plus - Capsule (100)

Available in: US - Stock #1525-0

Ingredients: Protease, amylase, glucoamylase, lipase, cellulase, hemicellulase, invertase, malt diastase, alpha-galactosidase, peptidase, beet root, potassium citrate, caraway seed, dandelion root, fennel seed, gentian, ginger

Garden Essence Plant Enz. - Capsule (100)

Available in: Canada - Stock #1643-9

Ingredients: Protease, amylase, glucoamylase, lipase, pectinase, cellulase, beet root, caraway seed, ginger, fennel, gentian, dandelion root, potassium citrate

Probiotics

Product Type: Formula

Properties: Antifungal, Detoxifying, Laxative (general)

Systems Affected: Large Intestine (Colon), Liver, Pancreas Head

Conditions: Anorexia, **Antibiotics (side effects of)**, Cancer (natural therapy for), **Celiac Disease**, Chemical Poisoning, Cold Sores (Fever Blisters), Constipation (adults), Contagious Diseases, Cradle Cap, **Crohn's Disease**, Cystic Fibrosis, Cystitis, Diaper Rash, Diaper Rash, **Diarrhea**, Digestion (poor), Dysentery, Dyspepsia, Enteritis, Epstein Barr Virus (Chronic Fatigue Syndrome, CFS), Fibromyalgia Syndrome (FMS), Fungal Infections (Yeast Infections, Candida albicans), Gas and Bloating, **Inflammatory Bowel Disorders (Colitis, IBS)**, Itching Ears, Jock Itch, **Leaky Gut Syndrome**, **Leucorrhea**, Lupus, Malaria, Mononucleosis, Overalkalinity, Pancreatitis, Poisoning (food), Thrush, Vaginitis

Usage: Acidophilus, bifidophilus and other species of Lactobacillis are friendly bacteria necessary for colon health. They aid in digestion and assimilation of some nutrients, and may help produce vitamin B-12 for the body. They protect the body against yeast and other harmful micro-organisms. They can protect against diarrhea when traveling or act as a mild laxative in constipation. They are depleted and destroyed by antibiotics. Hence, probiotics should always be taken after a round of antibiotics. They can be taken orally or used in enemas or douches for yeast infection. They may be sprinkled in the diaper for thrush-related diaper rash. The best product for general use is Probiotic Eleven. Children benefit highly from Bifidophilus. After a round of antibiotics, L. Reuteri is the best choice as it aggressively knocks down yeast and helps other friendly bacteria gain a new foothold. For general use take 1 to 2 capsules before meals on an empty stomach. These products work best when taken first thing in the morning about one hour before breakfast. Even better, empty 3-5 capsules into a cup of room temperature water, dissolve, and inject rectally using a bulb syringe. This implants the bacteria right where they are needed. Probiotics are especially helpful when traveling abroad because they help to prevent traveler's diarrhea. When traveling, double or triple the amount normally consumed, and take them with meals. Also take 1 capsule with any in-between meal snack.

Warnings: No known warnings.

Acidophilus (Milk Free) - Capsule (90)

Available in: US - Stock #1666-7

Ingredients: Lactobacillus acidophilus

Bifidophilus Flora Force - Capsule (90)

Available in: US & Canada - Stock #4080-4

Ingredients: Lactobacillus acidophilus, Bifidobacterium longum, Lactobacillus caseii, Lactobacillus rhamnosus, fructo-oligsaccharides

L. Reuteri, Chewable - Tablet (60)

Available in: US - Stock #1559-0

Ingredients: Lactobacillus reuteri

Probiotic Eleven - Capsule (90)

Available in: US - Stock #1510-1

Ingredients: Bifidobacterium bifidum, Bifidobacterium infantis, Bifidobacterium longum, Lactobacillus bulgaricus, Lactobacillus brevis, Lactobacillus plantarum, Lactobacillus rhamnosus, Lactobacillus salivarius, Streptococccus thermophilus, Lactobacillus acidophilus, Lactobacillus caseii, fructooligsaccharides.

Sunshine Heroes™ Probiotic Power - Bulk Powder

Available in: US - Stock #3346-7

Ingredients: See listing under Sunshine Heroes™ Probiotic Power

Acidophilus Bifidus - Capsule (75)

Available in: Canada - Stock #8002-4

Ingredients: Lactobacillus acidophilus, L. rhamnosus, Bifidobacterium longum, maltodextrin, bacteria culture

Herbasaurs Chewable Bifidophilus - Capsule (90)

Available in: Canada - Stock #3302-0

Ingredients: Bifidobacterium adolescentis, B. infantis, Lactobacillus acidophilus, L. casei subspecies rhamnosus, sorbitol, mannitol, fructose, fructooligosaccharides, natural orange flavor

Protease

Product Type: Nutrient

Properties: Anticancer, Antilipemic, Digestant, Immune Amphoterics

Systems Affected: Digestive System, Immune System, Liver, Pancreas Head, Structural System, Weight Loss

Conditions: Acquired Immune Deficiency Syndrome (AIDS/HIV), Allergies (food), Arteriosclerosis (Atherosclerosis, Hardening of the Arteries), Autism, Autoimmune Disorders, Belching, Blood Clots (prevention of), **Cancer (natural therapy for)**, Celiac Disease, Congestion (general), Crohn's Disease, Cystic Fibrosis, Cysts, Digestion (poor), Fibromyalgia Syndrome (FMS), Hair (loss or thinning), Inflammatory Bowel Disorders (Colitis, IBS), Leukemia, Lupus, Overacidity, Overalkalinity, **Parasites (tapeworm)**, **Protein Digestion (poor)**, Psoriasis, Weight Gain (aids for)

Usage: Take 1 capsule with a meal three times daily for protein digestion. Alternate use: Take 1-3 capsules between meals to break down undigested proteins in the body and enhance immune function.

Warnings: Do not use when ulcers are present. If stomach burning or pain occurs reduce dose or discontinue use.

Protease Plus - Capsule (90)

Available in: US & Canada - Stock #1841-7

Ingredients: Protease (60,000 HUT) , beet root fiber, plant-sourced trace mineral concentrate.

Protease, High Potency - Capsule (60)

Available in: US - Stock #1876-1

Ingredients: Protease (180,000 HUT), trace mineral concentrate

Protector Pak

Product Type: Pack

Properties: Antioxidant

Systems Affected: Whole Body

Conditions: Aging (prevention), Cancer (prevention), Cardiovascular Disease (Heart Disease), Free Radical Damage

Usage: Antioxidant pack for protecting the body from free radical damage. Take the contents of one packet (5 tablets and 1 capsule) twice daily with meals.

Warnings: No known warnings.

Protein Digestive Aid

See *PDA*

PS II

Product Type: Formula

Properties: Anti-inflammatory, Glandular

Systems Affected: Adrenal Glands, Male Reproductive, Nerves, **Prostate**, Spleen

Conditions: Acne (Pimples, Blackheads), **Benign Prostate Hyperplasia (BPH)**, Hair (loss or thinning), Kidney Infection, **Prostatitis**, Urine (scant)

Usage: For prostate enlargement and urinary problems in men, use 2-3 capsules three times daily. This remedy also works on the spleen in men and women.

Warnings: No known warnings.

PS II - Capsule (100)

Available in: US - Stock #1050-8

Ingredients: Black cohosh, licorice, kelp, gotu kola, capsicum fruit, golden seal, ginger, dong quai, lobelia

Psyllium

Latin Name: *Plantago psyllium or P. ovata*

Product Type: Single Herb

Properties: **Absorbant**, Anticholesteremic, Demulcent (Mucilant), Emollient, Laxative (bulk), Laxative (general), Soothing

Systems Affected: Large Intestine (Colon), Liver, Mucus Membranes, Skin

Conditions: Appetite (excessive), **Cholesterol (high)**, Colitis, Colon (spastic), Constipation (adults), Crohn's Disease, Diarrhea, Hemorrhoids, Inflammatory Bowel Disorders (Colitis, IBS), Multiple Sclerosis (MS), Poisoning (food), Weight Loss (aids for)

Usage: Psyllium comes from the seeds of plantain. Highly mucilaginous, the seeds and the seed hulls are used as a bulk laxative to absorb toxins in the colon. Psyllium seeds are milder acting than the hulls, which are mostly mucilage. The seeds are used for irritated intestinal membranes. Psyllium is also available in several bulk laxative formulas. Typical dose on the capsules is 2-4 capsules two to three times daily with plenty of water.

Warnings: If you do not drink enough water with psyllium, it may cause constipation. Long-term use can weaken digestion unless you also take some ginger, capsicum or other aromatic stomachic herbs occasionally.

Psyllium Hulls - Capsule (100)

Available in: US & Canada - Stock #545-9

Ingredients: Pyllium hulls (Plantago ovata)

Psyllium Seed - Capsule (100)

Available in: US & Canada - Stock #540-4

Ingredients: Psyllium seed (Plantago psyllium)

Psyllium Hulls Combination

Product Type: Formula

Properties: **Absorbant, Anticholesteremic**, Antidiarrheal, Demulcent (Mucilant), **Laxative (bulk)**, Laxative (general), Soothing

Systems Affected: **Large Intestine (Colon)**, Liver, Weight Loss

Conditions: **Cholesterol (high), Colon (atonic), Constipation (adults), Diarrhea**, Dysentery, Gall Stones, Heavy Metal Poisoning, Hemorrhoids, Inflammatory Bowel Disorders (Colitis, IBS), Multiple Sclerosis (MS), Psoriasis, Weight Loss (aids for)

Usage: This is a bulk laxative formula with licorice and hibiscus for flavoring. Take 1 teaspoon to 1 tablespoon and mix in a small glass of water or juice. Drink as soon as powder is stirred into the liquid. Do not let stand or it will form a slimy, mucilaginous mass that is impossible to drink. Follow with a tall glass of water. Can be taken one to three times daily in this manner. Drink 6-8 glasses of water per day when taking this formula for best results.

Warnings: If you do not drink enough water with this formula, it may cause constipation. Long-term use can weaken digestion unless you also take some ginger, capsicum or other aromatic stomachic herbs occasionally.

Psyllium Hulls Combination (13 oz.) - Bulk Powder

Available in: US & Canada - Stock #1375-6

Ingredients: Psyllium hulls, hibiscus, licorice root

RE-X

See *Nerve Control*

Recovery

Product Type: Formula

Properties: Nutritive

Systems Affected: Immune System

Conditions: Athletic Performance or Exercise (aids to), **Dehydration**, Diarrhea, Exercise

Usage: Recovery helps the body stay hydrated by replacing nutrients that are lost through perspiration when exercising or engaging in other strenuous activities. It can be used both during and after exercise. It also helps neutralize acid waste in tissues created by exercise. Add one level scoop (15 g) of drink mix powder to one glass (8 ounce) of cold Nature's Spring water.

Warnings: No known warnings.

Recovery (14.8 oz. or 420 G) - Bulk Powder

Available in: US - Stock #3662-2

Ingredients: Fructose, citric acid, potassium citrate, natural orange flavor, tricalcium phosphate, maltodextrin, rice syrup solids, xanthan gum, magnesium oxide, potassium chloride, sodium chloride, inositol, beta-carotene, ascorbic acid, niacinamide, l-carni-tine, glycine, l-taurine, pyridoxine hydrochloride, riboflavin, thiamin hydrochloride, folic acid, biotin, chromium chelate, cyanocobalamin.

Red Beet Formula

Other Names: Fasting Plus

Product Type: Formula

Properties: Antacid, Stomachic, Tonic

Systems Affected: Adrenal Glands, Brain, Muscles, Small Intestines, Stomach, Weight Loss

Conditions: Acne (Pimples, Blackheads), Allergies (food), Appetite (excessive), Arthritis, Cancer (natural therapy for), Constipation (adults), Cysts, Disks (spinal - bulging or slipped), Fatigue, Gout, Hepatitis, Jaundice (adults), Lupus, Rashes and Hives, Toxemia

Usage: Used as an aid to maintain energy and aid the body when fasting. Licorice stabilizes blood sugar levels to take away hunger cravings. Hawthorn supports the heart and red beet, the liver. Use 2 capsules five times daily while fasting. Fasting helps to cleanse the body. Semi-fasting on juices (drinking only fresh, raw fruit or vegetable juice) is best for most people. Consult books on fasting for more information. Fasting is contraindicated for weak persons, pregnancy and nursing.

Warnings: No known warnings.

Red Beet Root Formula - Capsule (100)

Available in: US - Stock #870-0

Ingredients: Licorice, red beet, fennel seeds, hawthorn berries

Red Clover

Latin Name: *Trifolium pratense*

Product Type: Single Herb

Properties: Alkalinizer, Alterative (Blood Purifier), Anticancer, Antiparasitic, Antitussive, Deobstruent, Detergent, Expectorant, Lymphatic, Mineralizer, Tonic

Systems Affected: Bladder (Urinary), Eyes, Immune System, Liver, Lymph Nodes, Lymphatic System, Skin, Spleen

Conditions: Acne (Pimples, Blackheads), Arthritis, Blood Poisoning, Boils, **Cancer (natural therapy for)**, Cancer (prevention), Chemical Poisoning, Cold Sores (Fever Blisters), Congestion (lymphatic), Cysts, Eczema, Estrogen (low), Fibrosis, Gout, Hodgkin's Disease, Leukemia, Mastitis, Melanoma (Skin Cancer), Menopause, Mercury Poisoning, Mononucleosis, Muscle Tone (lack of), Pertussis (Whooping Cough), PMS Type S, Psoriasis, Tuberculosis (Consumption, Scrofula)

Usage: 2 capsules two to three times daily with water or prepared as a tea (2-3 capsules per cup). Use in combination with other cleansing herbs, such as yellow dock, chaparral, burdock and echinacea. Used with chaparral for cancer.

Warnings: No known warnings.

Red Clover - Capsule (100)

Available in: US & Canada - Stock #550-9

Ingredients: Trifolium pratense (red clover) herb

Red Clover Blend

Product Type: Formula

Properties: Alterative (Blood Purifier), Anticancer, Bitter, Deobstruent, Detergent, Diuretic, Lymphatic, Tonic

Systems Affected: Bladder (Urinary), Eyes, Immune System, Liver, Lymph Nodes, Skin

Conditions: Acne (Pimples, Blackheads), Arthritis, Blood Poisoning, Boils, **Cancer (natural therapy for)**, Chemical Poisoning, Children's Remedy, Congestion (lymphatic), Gout, Melanoma (Skin Cancer), Mercury Poisoning, Mononucleosis, Muscle Tone (lack of), Pertussis (Whooping Cough), Psoriasis, Tuberculosis (Consumption, Scrofula)

Usage: Take 1 teaspoon with a meal three times daily or place several drops in water for a tea or take directly under the tongue or in juice.

Warnings: No known warnings.

Red Raspberry

Latin Name: *Rubus idaeus, R. strigosis*

Product Type: Single Herb

Properties: Antacid, **Anti-abortive**, Anti-emetic (Antinauseous), Antidepressant, Antidiarrheal, Astringent, Hemostatic, Parturient, Tonic, Uterine, Uterine Tonic

Systems Affected: Eyes, **Female Reproductive**, Heart, Hypothalamus, Intestinal System, Male Reproductive, Mucus Membranes, Prostate, Structural System, Urinary System, **Uro-genital Tract, Uterus**

Conditions: Acid Indigestion (Heartburn, Acid Reflux), **Afterbirth Pain**, Bedwetting, **Birth Defects (prevention)**, Bleeding (external), Bleeding (internal), Blood in Stool, Chemotherapy (reducing side effects), Colds (general remedies for), Cystitis, Diarrhea, Flu, Gastritis, **Hernias**, **Labor and Delivery**, Menstruation (scant), **Miscarriage (prevention)**, **Morning Sickness**, Myasthenia Gravis, **Pregnancy (herbs and supplements for)**, **Prolapsed Uterus**, Puberty (hormone balancer), **Rosacea**, Vaginitis

Usage: This is an herb rich in manganese, an essential element for oxygenation of the cells. It has been used as a tonic to strengthen the uterine muscles for childbirth. This plant also helps to relieve and prevent morning sickness during pregnancy. It may also act as a tonic to the male reproductive system as well. Raspberry may tone the entire abdominal wall, making it useful for healing hernias. It can be used as a gargle for sore throats, etc. in children. For pregnancy, take 2 capsules three times daily, or better yet, make tea out of equal parts raspberry, oatstraw, alfalfa and peppermint and drink. Use 2-4 capsules or 1 tsp. of herbs per cup of tea.

Warnings: No known warnings.

Red Raspberry - Capsule (100)

Available in: US & Canada - Stock #560-8

Ingredients: Rubus idaeus (raspberry) leaves

Red Raspberry Liquid

Product Type: Formula

Properties: Antacid, Anti-emetic (Antinauseous), Antidiarrheal, Digestive Tonic, Parturient

Systems Affected: Digestive System, Liver, Uterus

Conditions: Colds (general remedies for), Flu, Gas and Bloating, Gastritis, **Morning Sickness**, Nausea and Vomiting, Pregnancy (herbs and supplements for)

Usage: Red raspberry liquid is a blend of red raspberry leaves with herbs that settle the digestive tract. It has been used as a tonic to strengthen the uterine muscles for childbirth and to help settle the stomach during morning sickness.

Warnings: No known warnings.

Red Raspberry Liquid (2 fl. oz.) - Liquid (Glycerite)

Available in: US - Stock #3425-4

Ingredients: Peppermint leaves (Mentha piperita), rose hips (Rosa canina), hibiscus flowers (Hibiscus sabdariffa), red raspberry (Rubus idaeus) leaves, vegetable glycerin.

Red Yeast Rice

Product Type: Nutrient

Properties: **Anticholesteremic, Hypolipidemic**

Systems Affected: Cardiovascular System, Liver

Conditions: Cholesterol (high)

Usage: Red Yeast Rice helps to lower cholesterol production in the liver, which can help to lower blood cholesterol levels. Take two capsules with meals two or three times daily.

Warnings: Do not reduce cholesterol levels below 175 as this is unhealthy. Normal cholesterol should optimally be between 200 and 250.

Red Yeast Rice - Capsule (120)

Available in: US - Stock #558-3

Ingredients: Red Yeast Rice (Monoascus purpureus)

RG-Max

Product Type: Formula

Properties: Hypotensive, Nutritive, Vasodilator

Systems Affected: Circulation, Heart, Male Reproductive

Conditions: Acquired Immune Deficiency Syndrome (AIDS/HIV), **Angina, Angina, Blood Pressure (high), Blood Pressure (high), Cardiovascular Disease (Heart Disease), Congestive Heart Failure,** Cystitis, Energy (lack of), **Erectile Dysfunction,** Migraine

Usage: RG-Max is an l-arginine supplement that is balanced with other amino acids. L-arginine is an amino acid found in both plant and animal proteins. It is a not one of the eight essential amino acids because some can be synthesized in the liver. However, synthesis can be impaired in some people, including people with type II diabetes. Burns, injuries and infections can also deplete stores.

L-arginine is used to make the chemical messenger nitric oxide, which dilates blood vessels. This enlarging of the blood vessels takes stress off the heart, allows blood to flow more easily through the body and reduces blood pressure. This has several positive benefits for cardiovascular health.

For starters, supplementation with l-arginine has proven beneficial in helping angina. Several studies have shown that oral supplementation with l-arginine decreases angina symptoms and improves tolerance to exercise. It also appears to aid congestive heart failure when used as part of a comprehensive treatment plan.

Because of its influence on nitric oxide, l-arginine supplements can be helpful in temporarily reducing high blood pressure, particularly in type II diabetics. It may also improve the effectiveness of angiotensin converting enzyme (ACE) inhibitors used in treating hypertension. It has also been helpful in peripheral arterial disease and improves tolerance to transdermal nitrate medications.

In spite of all these benefits to circulation, research suggests that l-arginine is not helpful in aiding a person's recovery from a heart attack. Experts suggest people who have recently suffered a heart attack should not take l-arginine.

The drug sildenafil citrate (Viagra) affects nitric oxide by blocking an enzyme which breaks down nitric oxide. It was originally developed as a medication for high blood pressure, but was found to be more helpful in erectile dysfunction. Erections are dependent on good blood flow and the ability of the arteries in the penis to dilate.

L-arginine supplements help the body make more nitric oxide and have also proven beneficial in erectile dysfunction. A study of 50 men with erectile dysfunction showed that the group taking 5 grams of L-arginine per day had improvement over those on the placebo. L-arginine must be taken daily to have this affect.

L-arginine has other benefits, too. By increasing blood flow to the brain, L-arginine can help vasoconstrictive migraines. These are migraines where the head feels like it is being squeezed in a vise or by a belt. This is due to a lack of blood flow to the brain.

Other possible benefits of l-arginine include: improving recovery time after surgery, preventing wasting in AIDS patients and helping to recover from recurring interstitial cystitis.

Ingredients in RG-Max

RG-Max contains 5 grams of l-arginine per serving. It also contains other amino acids and nutrients that provide additional health benefits. These include:

L-citrulline is a non-essential amino acid that is a precursor to the production of l-arginine. The amino acid taurine is used in Japan to treat ischemic heart disease as well as arrhythmia. It may be helpful for congestive heart failure, hypertension and high cholesterol. L-isoleucine and l-leucine are used nutritionally to prevent fatigue and enhance exercise performance.

L-methionine is used to produce sam-e and aids in liver detoxification. L-tyrosine is an important amino acid for brain function. L-threonine is a precursor to glycine and may help reduce muscle spasms. N-acetyl-l-cysteine helps the body's detoxification mechanisms. It is a precursor to glutathione an important cellular antioxidant.

L-glutamine is used for a wide variety of health problems. It may be helpful for balancing mood and relieving insomnia. It can be helpful for leaky gut syndrome and enhancing athletic performance and recovery from critical illness. Acetyl-l-carnitine helps to strengthen the heart muscle. It works in the mitochondria to improve energy production.

Suggested Use

Use one scoop mixed in eight ounces of water once daily.

Warnings: High levels of l-arginine can provoke canker sores and herpes outbreaks in some people.

RG-Max - Bulk Powder

Available in: US - Stock #587-4

Ingredients: Xylitol, l-arginine, citric acid, natural strawberry flavor, natural lemon juice, malic acid, silicon dioxide, red grapeskin extract, l-citrulline, taurine, l-isoleucine, l-leucine, l-methionine, l-tyrosine, l-threonine, N-acetyl-l-cysteine, l-glutamine and acetyl-l-carnitine.

Rose Bulgaria

Product Type: Essential Oil

Properties: Antidepressant, Antiseptic, Antispasmodic, Antiviral, Aromatic, Dermatic, Emollient, Moistening, Nervine, **Perfume**, Soothing

Systems Affected: Digestive System, Female Reproductive, Heart, Liver, Skin

Conditions: Addictions (alcohol), Afterbirth Pain, Age Spots, Amenorrhea, Anger (excessive), Anxiety (Panic Attack), Anxiety (Panic Attack), Apathy, Arrhythmia, Blood Pressure (high), Boils, Bruises (healing), Burns and Scalds, Conjunctivitis (Pink Eye), Cushing's Disease, Defensiveness, Depression, Digestion (poor), Eczema, Fever, **Grief (excessive)**, Guilt, Headache (general), Herpes, Hot Flashes, Hot Flashes, Indigestion, Insomnia, Menopause, Nervous Exhaustion (Enervation), Nervousness, PMS Type D, Rashes and Hives, Skin Care (general)

Usage: Rose helps to regulate hormones and menstruation, overcome infertility, stop uterine bleeding, strengthen the uterus and relieves menstrual cramps. Historically, it is used for cooling hot conditions like hot flashes, irritated liver and gall bladder and stressed heart conditions, making it good for irritability, bleeding, fever, cholecystitis, billiary dyspepsia, headaches and gastro-enteritis. Rose can help reduce high blood pressure and correct arrhythmia. Possessing rehydrating and emollient properties, it is applied topically for inflamed or dehydrated skin and is useful for conditions like boils, rashes and sensitive skin. Rose has been confirmed with clinical studies to help with herpes zoster and herpes simplex when applied neat (undiluted) to the afflicted area. Emotionally, rose brings joy to the heart. It has harmonizing, anti-depressive actions and helps to heal grief, sorrow and disappointment. It opens the heart and soothes feelings such as anger, fear and anxiety. It is beneficial for nervous anxiety, insomnia and heart palpitations. Offering comfort to help heal emotional wounds, rose opens the cold heart, rebuilds trust and restores one's capacity for self-love and self-nurturing. It is often used with abused children to help with the deep despair and feeling of unworthiness and violation.

Warnings: Recommended for external use only.

Rose Bulgaria Essential Oil (2ML) - Essential Oil

Available in: US - Stock #3892-8

Ingredients: Rose Bulgaria (Rosa damascena) Essential Oil

Rose Hips

Latin Name: *Rosa sp.*

Product Type: Single Herb

Properties: Antiscorbutic, Astringent, Food, Nutritive, Refrigerant, Tonic, Vascular Tonics

Systems Affected: Blood Vessels, **Capillaries**, Connective Tissue, Liver, Mucus Membranes, **Peripheral Blood Vessels**, Pituitary (posterior), Structural System, Thymus, Veins

Conditions: Aneurysm, Arteriosclerosis (Atherosclerosis, Hardening of the Arteries), Asthma, **Bruises (healing)**, **Bruises (prevention)**, **Capillary Weakness**, Colds (general remedies for), Fever, Inflammation, **Nose Bleeds**, Scurvy, Spider Veins

Usage: The fruit is a rich source of vitamin C and bioflavonoids. However, after drying much of the vitamin C is lost. Rose hips are used to aid assimilation of vitamin C and are useful for persons who cannot tolerate vitamin C supplements. Also take for colds, flu and immune system weakness. Use 1-2 capsules three times daily for prevention, or up to 2 capsules every 2 hours for acute ailments. Rose hips make a pleasant tasting tea.

Warnings: No known warnings.

Rose Hips - Capsule (100)

Available in: US & Canada - Stock #580-1

Ingredients: Rosa sp. fruits

Rosemary

Product Type: Essential Oil

Properties: Analgesic (Anodyne), Antimicrobial, **Antiseptic,** Aromatic, Carminative, Cephalic, Decongestant, Parasympathomimetic, Pectoral, Preservative, **Sympathomimetic**

Systems Affected: Aldosterone, Circulation, Digestive System, Epinephrine, Hair, Immune System, Lungs, Respiratory System, **Scalp**, Skin

Conditions: Apathy, Arthritis, Colds (decongestant), Colds (general remedies for), Concentration (poor), Confusion, Congestion (general), **Cough (general)**, Cough (spastic), **Croup**, Digestion (poor), Fever, Gas and Bloating, Hair (loss or thinning), Hair Care (general), Hyperinsulinemia (Syndrome X), Hypochondria, Hysteria, Indigestion, Infection (bacterial), **Insects**, Memory and Brain Function, Nervous Exhaustion (Enervation), **Pertussis (Whooping Cough)**, Pregnancy (herbs and supplements to avoid during), Rheumatoid Arthritis (Rheumatism), Senility, Skin (infections), Skin (oily), Skin Care (general), **Thinking (cloudy)**

Usage: Rosemary is a valuable oil for respiratory problems, helping conditions like the common cold, catarrh, asthma and sinusitis. It also acts to relieve pain in rheumatism, arthritis and tired, stiff or overworked muscles. Rosemary stimulates blood circulation, improving low blood pressure and helping relieve cold feet, tired or weak legs and circulatory problems in the extremities. In the digestive area it helps to relieve dyspepsia, flatulence and abdominal distention. Rosemary helps lower high blood sugar and aids arteriosclerosis, probably by overcoming insulin resistance. Topically, rosemary has traditionally been used in shampoos and hair treatments to stimulate blood circulation to the scalp to help promote hair growth. Emotionally it renews enthusiasm and bolsters self confidence. In the nervous system it promotes mental clarity and awareness, stimulates the brain and helps improve memory. For those who lack faith in their own potential and tend to doubt their every action, rosemary can help boost their confidence and help develop a sense of self worth. It helps one find his or her own path. It is known to enhance the spiritual dedication of love rather than the experience of romantic ecstasy.

Warnings: Rosemary's stimulating action is not recommended for people with epilepsy or high blood pressure. Avoid during pregnancy or breast feeding.

Rosemary Essential Oil (5ML) - Essential Oil

Available in: US - Stock #3914-8

Ingredients: Rosemary (Rosemarinus officinalis) Essential Oil

S-O-D With Gliadin

Product Type: Formula

Properties: Anti-inflammatory, Antioxidant

Systems Affected: Liver, Structural System

Conditions: Arthritis, Cancer (natural therapy for), Cardiovascular Disease (Heart Disease), Chemotherapy (reducing side effects), Free Radical Damage, Radiation Sickness, Rosacea

Usage: SOD is an enzyme that breaks down one of the major free radicals in the body, the super oxide radical. It may help with any inflammatory condition. SOD helps protect the body against radiation damage. Take 1 tablet on an empty stomach.

Warnings: No known warnings.

Safflowers

Latin Name: *Carthamus tinctorius*

Product Type: Single Herb

Properties: Alterative (Blood Purifier), **Antacid**, Appetite Stimulant, Aromatic, Carminative, Stomachic

Systems Affected: Heart, Muscles, Skin, Stomach

Conditions: Acid Indigestion (Heartburn, Acid Reflux), Amenorrhea, Arthritis, Athletic Performance or Exercise (aids to), Body Building, Bunions, Burning Feet or Hands, Chicken Pox, Cholesterol (high), Colic (children), Diabetes, Digestion (poor), **Exercise**, Gas and Bloating, Gout, Hypoglycemia, Indigestion, Jaundice (adults), **Jaundice (infants)**, Pain (general remedies for), **Tendonitis**, Uric Acid Retention

Usage: Believed to neutralize waste acids and stimulate natural hydrochloric acid production. Take 2 capsules with each meal as a digestive aid. Use 2 capsules with a pint of water to make tea. Drink 1 to 2 pints of tea to relieve sore muscles after over-exertion. Take prior to exercise to avoid sore muscles. Excellent when taken with Energ-V to help promote energy and avoid soreness when doing strenuous work or exercise. Apply tea externally for injuries and bruises. The tea is also helpful for jaundice and colic in infants.

Warnings: Do not use with excessive menstrual bleeding. Some herbalists feel it should also be avoided in pregnancy, but I do not believe this caution is warranted.

Safflowers - Capsule (100)

Available in: US & Canada - Stock #600-2

Ingredients: Carthamus tinctorius (safflower) flowers

Sage

Latin Name: *Salvia officinalis*

Product Type: Single Herb

Properties: Antacid, Antidiarrheal, **Antigalactagogue**, Antioxidant, Antiseptic, **Antisudorific**, Aromatic, Astringent, Carminative, Cephalic, Condiment, Estrogenic, Febrifuge, Fumigant, Insecticide, Nervine, Stimulant, Stomachic, **Sympathomimetic**, Vermifuge, Vulnerary

Systems Affected: Acetylcholine, Epinephrine, Hair, Hypothalamus, Lymphatic System, Pancreas Tail, Parotids, Scalp, Skin, Structural System, Sweat Glands, Throat, Tonsils, **Vocal Cords**

Conditions: Asthma, Bites and Stings, Bleeding (external), Body Odor, **Breast Milk (dry up)**, Canker Sores (Mouth Ulcers), Chills, Diarrhea, Gray Hair, Hair (loss or thinning), Hair Care (general), Hashimoto's Disease (Thyroiditis), Hot Flashes, Indigestion, **Laryngitis (Hoarseness)**, Memory and Brain Function, **Night Sweating**, Parasites (nematodes, worms), **Perspiration (excessive)**, Senility, Sore Throat, Wounds and Sores

Usage: For gas and indigestion, 1-2 capsules two to three times daily before meals. Take 2-4 capsules before bedtime for night sweats. Use as tea for sore throat, laryngitis, hoarseness, 2-4 capsules per cup. Use tea as a mouthwash or gargle for irritation of the throat or mouth.

Warnings: No known warnings.

Sage - Capsule (100)

Available in: US - Stock #610-1

Ingredients: Salvia officinalis (sage) herb

SAM-e

Product Type: Nutrient

Properties: Hepatic, Lipotropic

Systems Affected: Liver, Pituitary (infundibulum)

Conditions: **Addictions (alcohol)**, Blood Poisoning, Cartilage Damage, Cholesterol (low), **Cirrhosis of the Liver**, Defensiveness, Depression, Dysmenorrhea, Energy (lack of), **Hepatitis**, Hypothyroid, Irritability, Mood Swings, Pregnancy (herbs and supplements to avoid during)

Usage: Take 1 tablet once or twice daily on an empty stomach with 8 oz of water. For best results, take SAM-e with folic acid and Vitamin B12.

Warnings: If you are taking prescription antidepressants or have bipolar (manic) depression, consult your physician before or during the use of this product. Pregnant or lactating women should consult their health care provider prior to taking this supplement.

SAM-e (200 mg active) - Tablet (30)

Available in: US - Stock #1845-2

Ingredients: S-Adenosylmethionine 200 mg, methacrylate copolymers, glycerol palmito stearate, sodium starch glycolate, silicon dioxide, microcrystalline cellulose, titanium dioxide, citric acid, mannitol, magnesium stearate, iron oxide.

Sandalwood

Product Type: Essential Oil

Properties: Antiseptic, Antispasmodic, Antiviral, Aphrodisiac, Bronchial Dilator, Calmative, Expectorant, Moistening, Nervine, **Perfume**

Systems Affected: Digestive System, Mucus Membranes, Pineal, Reproductive Glands, Respiratory System, Skin, Vagina

Conditions: Acne (Pimples, Blackheads), Bronchitis, Colic (adults), Cough (dry), Cough (spastic), Cystitis, Eczema, Fear (excessive), Gastritis, Nightmares, Psoriasis, Sex Drive (low), Skin (dry and/or flaky), Skin Care (general), Sore Throat, Stress

Usage: Sandalwood's dominant action is on the mucus membranes. It eases dry cough and is especially effective in deep, painful coughs. It also helps with vaginal discharge. It helps both acute and chronic diarrhea and relieves intestinal spasms and inflammation, making it useful in colic and gastritis. It is also very effective as a gargle for sore throats. Topically, sandalwood works for dryness, irritation and inflammation making it very useful with eczema and psoriasis. It helps both dehydrated and oily skin. Emotionally, sandalwood has a calming, harmonizing effect helping to reduce tension and confusion. It works for nervous depression, fear, stress and a hectic lifestyle. Many traditions consider sandalwood an aphrodisiac because it brings openness, warmth and understanding. It clarifies and stills the mind and refreshes an over-heated body. It instills an inner unity helping with incessant over-thinking. For those who feel a neurotic need for security and manipulation of outcomes, it helps to reestablish acceptance of reality.

Warnings: Not recommended for internal use.

Sandalwood Essential Oil (5ML) - Essential Oil

Available in: US & Canada - Stock #3915-4

Ingredients: Sandalwood (Santalum album) Essential Oil

Sarsaparilla

Latin Name: *Smilax officinalis*

Product Type: Single Herb

Properties: Alterative (Blood Purifier), **Androgenic**, Antacid, Anti-inflammatory, Antirheumatic, Deobstruent, Diuretic, Glandular, Hepatoprotective, Relaxant, Stimulant, Tonic

Systems Affected: Digestive System, Liver, Progesterone, Reproductive Glands, Testes, Testosterone

Conditions: Angina, Arthritis, Blood Poisoning, Burns and Scalds, Bursitis, Colitis, Congestive Heart Failure, Cough (dry), Edema (Dropsy, Water Retention, Swelling), Erectile Dysfunction, Eye Infections, Gout, Hair (loss or thinning), Leaky Gut Syndrome, Leprosy, Lou Gehrig's Disease, Menorrhagia (Heavy Menstrual Bleeding), Mercury Poisoning, Pleurisy, Progesterone (low), Psoriasis, Rheumatoid Arthritis (Rheumatism), Sex Drive (low), Testosterone (low)

Usage: For strengthening the liver and purifying the blood, use 1 or 2 capsules two to three times per day with water. For long-term restorative usage, dosage can be decreased to 1 capsule once or twice per day after several weeks at initial dosage. It works best when combined with other blood purifiers such as yellow dock, burdock, dandelion and red clover. Helps strengthen the male reproductive hormones, relieve arthritic pains and clear skin conditions. Sarsaparilla is a member of the ginseng family. It is the major flavoring found in old-fashioned root beer.

Warnings: No known warnings.

Sarsaparilla - Capsule (100)

Available in: US - Stock #620-8

Ingredients: Smilax officinalis (sarsaparilla) root

Saw Palmetto

Latin Name: *Serenoa repens*

Product Type: Single Herb

Properties: Glandular, Spleen Chi Tonic

Systems Affected: Adrenal Glands, Male Reproductive, Prostate, Respiratory System, Skin, Stomach, Uro-genital Tract

Conditions: Benign Prostate Hyperplasia (BPH), **Breasts (enhance size)**, Diabetes, Digestion (poor), Hashimoto's Disease (Thyroiditis), Hypothyroid, Muscular Dystrophy, Wasting, Weight Gain (aids for)

Usage: For prostate enlargement and urinary problems in men, 2 capsules three times daily with meals; concentrate: take 1 capsule with a meal twice daily. Combines well with echinacea and many kidney herbs for this purpose. The herb was traditionally used for elderly men who were feeble, thin and emaciated. It acts as a digestive tonic and promotes weight gain. It also is reported to help increase bust size in women.

Warnings: None, except that individuals prone to weight gain (especially men) should be aware that saw palmetto does enhance weight gain.

Saw Palmetto - Capsule (100)

Available in: US & Canada - Stock #630-4

Ingredients: Serenoa repens (saw palmetto) berries

Saw Palmetto Concentrate - Capsule (60)

Available in: US & Canada - Stock #635-9

Ingredients: Concentrated extract of Serenoa repens (saw palmetto) berries, olive oil, glycerin.

SC Formula

Product Type: Formula

Properties: Anticancer, Immune Stimulant

Systems Affected: Immune System, Liver

Conditions: Arthritis, Cancer (natural therapy for), Dermatitis

Usage: Shark cartilage is reported to help inhibit tumor growth. Reishi mushrooms are a powerful immune-stimulant. This formula may help cancer patients. Use 3 capsules three times a day with meals. Higher doses may be required for maximum effectiveness.

Warnings: Do not treat cancer without advice from a competent practitioner.

SC Formula (Shark Cartilage) - Capsule (100)

Available in: US - Stock #1602-8

Ingredients: Shark cartilage, reishi mushroom.

Sciatic

Product Type: Homeopathic

Properties: Analgesic (Anodyne)

Systems Affected: Structural System

Conditions: Backache (Back Pain, Lumbago), Sciatica

Usage: Used for the relief of lower back pain due to sciatica or overexertion. Take 10-15 drops under the tongue every 10-15 minutes, or as needed. As symptoms improve, decrease to every one or two hours, then to four times daily until symptoms are relieved.

Warnings: NSP does not recommend this product for children under the age of 12 except on the advice of a health professional. Other than the alcohol content, which is not good for infants, we consider this product safe for children.

Sciatica (1 fl. oz.) - Homeopathic (Liquid)

Available in: US - Stock #8820-6

Ingredients: Active ingredients: Gnaphalium poly-cephalum (Everlasting) 4x, Capsicum (Cayenne) 6x, Colocynthis (Bitter Apple) 6x, Lycopodium clavatum (Club Moss) 6x, Kali Carbonicum (Potassium Carbonate) 8x, Rhus toxicodendron (Poison Ivy) 8x. Other ingredients: Purified water and 20% USP alcohol

Sea Salt

Product Type: Nutrient

Properties: Detoxifying, Nutritive

Systems Affected: Adrenal Glands, Circulation, Thyroid, Whole Body

Conditions: Adrenals (exhaustion, weakness or burnout), Asthma, Blood Pressure (high), Cholesterol (high), Circulation (poor), Cramps and Spasms (general), Diabetes, **Edema (Dropsy, Water Retention, Swelling)**, Exercise, Fatigue, Hypothyroid, Insomnia, Insomnia, Osteoporosis, PMS Type D, Poison Ivy or Oak, **Sinus Infection**

Usage: This natural salt is actually beneficial to the body. Table salt has been stripped of essential trace minerals, bleached and treated with additives. Natural salt is actually beneficial for a wide variety of health conditions. It does not raise blood pressure, and will often help to lower blood pressure. Mixed with pure water can be used as a nasal wash or a sore throat gargle. A pinch of natural salt followed by a glass of water can also be helpful for asthma. A pinch of natural salt taken before meals can also stimulate hydrochloric acid production and digestion. Salt can be used with baking soda or Epsom salt for a detoxification bath. For more information on the uses of salt, see the book *Salt Your Way to Health* by David Brownstein, M.D.

Warnings: Ordinary table salt does not have the health producing benefits of natural salt.

Sea Salt Shaker (2-7.5 oz.) - Bulk Powder

Available in: US - Stock #150-6

Ingredients: Natural salt from Redmond Utah

Seasonal Defense

Product Type: Formula

Properties: Anti-allergenic, Antibacterial, Antifungal, **Antihistamine**, Decongestant

Systems Affected: Immune System, Liver, Lungs, Respiratory System

Conditions: Allergies (respiratory), Colds (decongestant), Colds (general remedies for), Congestion (lungs), Congestion (sinus), Contagious Diseases, Sneezing, Snoring

Usage: This formula helps decongest the sinuses, reduce sinus inflammation, combat allergic responses and fight respiratory infections. Take one capsule with a meal three times daily.

Warnings: No known warnings.

Seasonal Defense - Capsule (90)

Available in: US - Stock #806-6

Ingredients: Andrographis paniculata herb extract, thyme herb, bitter orange fruit, eleuthero root, oregano herb

Senna Combination

Product Type: Formula

Properties: Bitter, **Cathartic, Laxative (stimulant)**

Systems Affected: Intestinal System, Large Intestine (Colon)

Conditions: Colon (atonic), Constipation (adults), Pregnancy (herbs and supplements to avoid during)

Usage: Senna is an extremely strong purgative. Senna is best used with antispasmodic and carminative herbs like fennel, ginger and catnip, as found in this formula. Otherwise, it can cause severe intestinal cramping. Take 2-4 capsules with water and a late evening snack for occasional relief from severe constipation.

Warnings: For occasional use only. Not recommended during convalescence. Do not take while nursing (can cause diarrhea in baby). Overdoses and daily usage can cause dependency, colic, pain, nausea, vomiting. Can cause uterine contractions, so do not take when pregnant or during menstruation.

Senna Combination - Capsule (100)

Available in: US - Stock #650-5

Ingredients: Senna, fennel seed, ginger, catnip

SF

Product Type: Formula

Properties: Alterative (Blood Purifier), **Anti-obesic**, Diuretic, **Emulsifier**, Laxative (general), **Lipotropic**, Parasiticide

Systems Affected: Intestinal System, Liver, Pancreas

Conditions: Acne (Pimples, Blackheads), Adenitis, Anemia, Appetite (excessive), Arteriosclerosis (Atherosclerosis, Hardening of the Arteries), Boils, Calcium Deposits (Calcification), Cardiovascular Disease (Heart Disease), Cataracts, Cellulite, Cholesterol (high), Cholesterol (low), Constipation (adults), Cysts, Diabetes, Edema (Dropsy, Water Retention, Swelling), Fat Cravings, **Fat Metabolism (poor)**, Fatigue, **Fatty Liver Disease, Fatty Tumors or Deposits**, Hair (loss or thinning), Hypothyroid, Hypothyroid, Parasites (nematodes, worms), Seborrhea, Toxemia, **Triglycerides (low)**, Weight Loss (aids for)

Usage: This formula acts as a diuretic and mild laxative to aid the process of weight loss. It also contains herbs that aid digestive and glandular function. Take 2 capsules 1/2 hour prior to meals each day. Begin with smaller doses and work up. Can be taken with extra chickweed for best results.

Warnings: No known warnings.

SF - Capsule (100)

Available in: US - Stock #1067-5

Ingredients: Chickweed, cascara sagrada, licorice, safflower, black walnut hulls, parthenium, gotu kola, hawthorn berries, papaya fruit, fennel seeds, dandelion

Siberian Ginseng

See *Eleuthero*

Silver Shield

Product Type: Nutrient

Properties: **Antibacterial, Antifungal, Antimicrobial,** Antiparasitic, **Antiseptic,** Antiviral, **Disinfectant,** Immune Stimulant

Systems Affected: **Immune System, Lymph Nodes, Nipples,** Structural System, **Tonsils, Vagina**

Conditions: **Abrasions, Abscesses, Acne (Pimples, Blackheads), Acquired Immune Deficiency Syndrome (AIDS/HIV),** Acquired Immune Deficiency Syndrome (AIDS/HIV), **Antibiotics (alternatives to),** Antibiotics (side effects of), **Appendicitis, Athlete's Foot,** Bladder (irritable), **Bladder Infection,** Blisters, **Blisters, Blood Poisoning,** Broken Bones, **Burns and Scalds, Carbuncles, Chicken Pox, Cholera, Cold Sores (Fever Blisters), Colds (general remedies for),** Colds (prevention), Congestion (sinus), **Conjunctivitis (Pink Eye), Contagious Diseases, Cuts,** Cystitis, Denture Sores, **Diarrhea, Diphtheria, Dysentery, Ear Infection or Earache, Eye Infections,** Eye Problems (general), **Fever, Flu, Fungal Infections (Yeast Infections, Candida albicans), Gangrene, Giardia, Gingivitis (Bleeding Gums, Gum Disease, Pyorrhea), Gonorrhea, Herpes, Impetigo, Infection (bacterial), Infection (viral), Jock Itch, Kidney Infection, Leprosy, Leucorrhea,** Lou Gehrig's Disease, **Lyme Disease, Malaria, Measles, Mumps,** Parasites (tapeworm), **Pertussis (Whooping Cough), Pets (supplements for), Pneumonia, Poisoning (food),** Prostatitis, **Rheumatic Fever,** Scabies, **Scratches and Abrasions, Sinus Infection, Skin (infections), Skin Care (general),** Slivers, Smell (loss of sense of), **Staph Infections, Strep Throat, Syphilis, Tetanus, Thrush,** Tick, **Tonsillitis (Adenoids), Toothache, Tuberculosis (Consumption, Scrofula), Typhoid, Urinary Tract Infections,** Urination (burning or painful), **Vaginitis, Wrinkles**

Usage: Silver was employed as a germicide and an antibiotic long before modern antibiotics were developed. In previous centuries, people would shave silver particles or submerge silver coins or articles in drinking water to prevent it from turning bad. This works, because silver inhibits the growth of bacteria and other disease-causing organisms. It seems likely that the practice of eating with silver utensils (i.e., silverware) resulted from a belief in the healthful properties of silver.

As people become increasingly concerned about the overuse of antibiotics and the rise of antibiotic resistant bacteria, the use of silver as a natural antibacterial agent is coming back. While many companies sell what they claim to be "colloidal" silver, many products are actually ionic silver. Ionic silver reacts with the hydrochloric acid in the stomach to form insoluble silver salts, which can deposit in the body causing argyria (a condition that causes a blue/gray discoloration of the skin and inner eyelid inflammation).

Silver Shield is a true, colloidal silver product and contains the only colloidal silver product backed by a patent. It is also the only colloidal silver that has been approved as non-toxic by the Environmental Protection Agency (EPA). The colloidal silver in Silver Shield is antibacterial, antifungal, antimicrobial, antiparasitic, antiviral and antiseptic and has been tested against numerous micro-organisms in doses of only 10 parts per million. The product itself contains 14 ppm and is not only effective in this low concentration, it is also safe to take regularly.

Organisms this silver has been found to be effective against include: the bacteria that cause dental plaque and tooth decay (Streptococcus mutans), ear and sinus infection (Streptococcus pneumonie), eye infections (Staphylococcus aureus), food poisoning (Salmonella arizona, S. tyhimurium, Escherichia coli), bacillary dysentery, (Shigella boydii), diarrhea (Escherichia coli), enteric fever (Salmonella tyhimurium), nosocomial infections in hospitals (Klebsiella pneumoniae, Pseudomonas aeruginosa, Streptococcus pyogenes), strep throat (Streptococcus pyogenes), impetigo (Staphylococcus aureus), septicemia (Enterobacter aerpyogenes), pneumonia (Staphylococcus aureus, Haemophilus influenzae, Pseudomonas aeruginosa, Streptococcus pneumonie), scarlet fever (Streptococcus pyogenes), respiratory tract infections (Streptococcus pyogenes, E. coli, Klebsiella pneumoniae), urinary tract infections (E. coli, Klebsiella pneumoniae, Pseudomonas aeruginosa, Streptococcus faecalis, Enterobacter aerpyogenes), urethritis in men and vaginitis in women (Trichomonas vaginalis), yeast infections (Candida albicans), burn and wound infections (Pseudomonas aeruginosa), skin infections (Staphylococcus aureus, Streptococcus pyogenes), wound infections (Escherichia coli, Enterobacter aerpyogenes, Klebsiella pneumoniae, Pseudomonas aeruginosa, Streptococcus faecalis), endocarditis (Streptococcus faecalis, S. gordonii), meningitis (Pseudomonas aeruginosa, Streptococcus pneumonie, Enterobacter aerogenes , Haemophilus influenzae), suppurative arthritis (Haemophilus influenzae) and osteomyelitis (Staphylococcus aureus).

In fact, colloidal silver has been found to be both bacteriostatic and bactericidal for all gram-negative and gram-positive organisms tested, in vivo and in vitro (in patients and in the lab) all causing infections in humans, including emerging resistant strains. Clinical use of 10 ppm patented colloidal silver has been successful in the treatment of upper respiratory infections, sinusitis and rhinitis, eye, ear, nose infections, bronchitis, pharyngitis, tonsillitis, halitosis and gingivitis, abdominal pain and diarrhea, pelvic inflammatory disease, urinary tract infections, malaria, Staphylococcus skin infections, septic ulcers, infected abscesses, fungal skin infections, vaginal yeast infections, sexually transmitted diseases such as gonorrhea, and retrovirus infection (HIV). The studies have shown that this silver has equal or broader spectrum antibacterial activity than any one antibiotic tested. This makes it a very effective alternative to antibiotics.

Internally, the recommended dose for adults is one teaspoon daily on an empty stomach for prevention and 2-3 teaspoons daily for fighting infection. Children's suggested dose is 1/4-1/2 teaspoon daily. Colloidal silver may be used undiluted on cuts, scrapes and wounds. It can also be used as drops in the eyes or ears or used in a douche for vaginal yeast infections

Warnings: The colloidal silver in silver shield will not cause agyria or a discoloration of the skin. It is approved as non-toxic by the EPA and is safe for regular use.

Silver Shield w/Aqua Sol(18 ppm) (4 fl. oz.) - Liquid

Available in: US - Stock #4274-1

Ingredients: Silver (colloidal and ionic) 18 pppm, deionized water

Liquid Silver - Liquid

Available in: Canada - Stock #4011-1

Ingredients: Same as Stock #4274-1

Silver Shield Gel - Topical

Available in: US & Canada - Stock #4950-1

Ingredients: Silver Shield Solution 24ppm (purified water, silver), TEA, carbomer

Sinus

Product Type: Homeopathic

Properties: Antihistamine, Decongestant

Systems Affected: Sinuses

Conditions: Congestion (sinus), Headache (sinus)

Usage: Used for the relief of stuffy nose, runny nose and headache due to common sinusitis. Take 10-15 drops under the tongue every 10-15 minutes or as needed until symptoms improve. As symptoms improve decrease to hourly, then to four times daily until symptoms are relieved.

Warnings: NSP does not recommend this product for children under the age of 12 except on the advice of a health professional. Other than the alcohol content, which is not good for infants, we consider this product safe for children.

Sinus (1 fl. oz.) - Homeopathic (Liquid)

Available in: US - Stock #8740-1

Ingredients: Active ingredients: Euphorbium officinarum (Gum Euphorbium) 4x, Melilotus officinalis (Yellow Melilot) 4x, Pulsatilla (Wind Flower) 4x, Hydrastis canadensis (Golden Seal) 6x, Kali Bichromicum (Potassium Bichromate) 6x, Thuja occidentalis (Tree of Life) 6x, Phosphorus (Phosphorus) 8x. Other ingredients: Purified water and 20% USP alcohol

Sinus Support

Other Names: SN-X

Product Type: Formula

Properties: Alterative (Blood Purifier), Antibacterial, **Anticatarrhal**, Antihistamine, Decongestant, Expectorant

Systems Affected: Liver, Lungs, Mucus Membranes, **Respiratory System, Sinuses**, Stomach

Conditions: Asthma, Colds (decongestant), Colds (general remedies for), **Congestion (sinus)**, Cough (damp), Eyes (red or itching), Rhinitis, **Sinus Infection**, Wheezing

Usage: An alternative to over-the-counter antihistamines. Helps to dry up excessive sinus drainage while working to improve digestion, expel excess phlegm and generally detoxify the body. Use 2 capsules three to four times daily as needed.

Warnings: No known warnings.

Sinus Support EF - Capsule (100)

Available in: US - Stock #1250-3

Ingredients: Capsicum Fruit (Capsicum annuum), Burdock root (Articum lappa), Goldens Seal root (Hydrastis canadensis), Parsley Herb (Petroselinum crispum), Horehound Herb (Marrubium vulgare), Althea Root (Althea officinalis), Bitter Orange Fruit (Fructus aurantia) (Synephrine 6% extract), Yerba Santa herb.

SN-X - Capsule (100)

Available in: Canada - Stock #1247-8

Ingredients: Fructus aurantia immaturi, burdock, parsley, horehound, Siberian ginseng, marshmallow, goldenseal, capscium, yerba santa

Skeletal Strength

Other Names: SKL Formula

Product Type: Formula

Properties: Alkalinizer, Mineralizer

Systems Affected: Bones, Connective Tissue, Cuticle, Parathyroid, **Skeletal System, Structural System**

Conditions: Arthritis, Backache (Back Pain, Lumbago), **Broken Bones, Calcium Deficiency**, Cramps (leg), Cramps (menstrual), Cramps and Spasms (general), Crohn's Disease, Cystitis, Disks (spinal - bulging or slipped), Menopause, **Osteoporosis**, Overacidity

Usage: Provides nutrients to strengthen the skeletal system. Use for healing broken bones, preventing osteoporosis or aiding bone development. Use 1-2 tablets two times per day.

Warnings: No known warnings.

Skeletal Strength - Tablet (150)

Available in: US - Stock #1806-7

Ingredients: Vitamin A, Vitamin C, Vitamin D, Vitamin B6, Vitamin B12, Calcium (amino acid chelate, di-calcium phosphate, calcium citrate), iron, phosphorus, magnesium (amino acid chelate, magnesium oxide), zinc, copper, manganese, potassium, boron, horsetail, bentaine HCL, papaya, parsley, pineapple, valerian, licorice

Skin Detox

Other Names: SKN-AV

Product Type: Formula

Properties: **Alterative (Blood Purifier)**, Anti-inflammatory, Antifungal, Antiseptic, Antiviral, Bitter, Diuretic, Hepatic, Styptic

Systems Affected: Liver, Skin

Conditions: **Acne (Pimples, Blackheads)**, Bleeding (external), **Boils, Carbuncles, Dermatitis, Eczema**, Inflammation, PMS Type S, Psoriasis, **Rashes and Hives**, Skin (infections), **Skin Care (general)**, Ulcerations (external)

Usage: This is an Ayurvedic formula for skin conditions such as rashes, eczema, dermatitis, psoriasis and acne. Use 2 capsules three times daily.

Warnings: No known warnings.

Skin Detox, Ayurvedic - Capsule (100)

Available in: US - Stock #1299-6

Ingredients: Dandelion (tarasacum), Acacia catechu, Azadirachta indica (neem), Smilax china, Picrorhiza kurroa, Hemidesmus indicus (Indian sarsaparilla), Holarrbena antidysenterica, Rubia cordifolia, Swertia chirata, Caesalpinia crista, Fumaria parviflora, Alstonia scholaris, Tinospora cordifolia, Curcuma longa, Phylilanthus emblica, Terminalia belerica, Terminalia chebula

Slippery Elm

Latin Name: *Ulmus fulva*

Product Type: Single Herb

Properties: Absorbant, Antidiarrheal, Balsamic, **Demulcent (Mucilant)**, Drawing, Emollient, Expectorant, Laxative (bulk), Nutritive, Soothing, Vulneraries (for intestinal system)

Systems Affected: Digestive System, Intestinal System, Mucus Membranes, Nipples, Rectum, Respiratory System, Skeletal System, Skin, **Small Intestines**, Stomach, Structural System, **Throat**, Uro-genital Tract, Vagina

Conditions: Abscesses, Breasts (swelling and tenderness), Broken Bones, Burns and Scalds, **Chemotherapy (reducing side effects)**, Cholera, **Colitis**, Constipation (children), Convalescence, **Crohn's Disease**, Cystitis, Debility, Diaper Rash, **Diarrhea, Diverticulitis (Diverticuli)**, Dysentery, **Enteritis**, Failure to Thrive, **Gastritis**, Hemorrhoids, Hiatal Hernia, Ileocecal Valve, Indigestion, **Inflammatory Bowel Disorders (Colitis, IBS)**, Leprosy, Mastitis, **Sore Throat, Tickle in Throat**, Tonsillitis (Adenoids), Ulcerations (external), **Ulcers**, Vaginal Dryness, **Wasting**, Wounds and Sores

Usage: Used internally for irritation of the stomach and intestines, diarrhea (especially in children) and as a mild, nourishing food for weak and debilitated persons. In capsule form, take 2 to 4 capsules, three to six times daily with plenty of water. It is better, however, to use it in bulk form, especially for throat problems. Mix 1 to 2 teaspoons bulk powder with hot water or juice (mixes best using a blender). Sweeten to taste and drink (may be slightly "thick" depending on how much powder is used). To give to children, combine with applesauce or juice into a porridge consistency, sweeten and eat like cereal (combine with suma, ginseng, almond meal and elder berries for strengthening weak persons). Porridge can be flavored with cinnamon or other spices to improve taste. Or make into a tea using 1/2 teaspoon per cup of boiling water, steep for 1 hour, then sweeten with honey or raw sugar. Can be mixed with hot milk (1/2 teaspoon per cup) and molasses (to taste) for a soothing "mock chocolate" drink. Sip tea or suck on powder in small, frequent doses for relief of sore throat, tickling or irritations of the throat, cough, heartburn and ulcers (combines well with licorice for these conditions). The tea can be used in an enema or rectal injection for hemorrhoids or bowel inflammation.

Externally, slippery elm makes excellent poultice for abscesses, wounds, congestion, eruptions, swollen glands, injuries, burns and swollen, inflamed surfaces. Combine with powdered plantain, comfrey, white oak bark, golden seal, lobelia, myrrh or bayberry, depending on problem. Moisten with aloe vera and/or water to make a thick paste. Cover swelling, injury or other damaged area with paste. Cover paste with plastic wrap to protect it and keep it moist. Poultice can be taped in place and worn to bed.

Warnings: Completely mild and safe remedy for internal use. Not recommended for external ulcers or purulent sores

Slippery Elm - Capsule (100)

Available in: US & Canada - Stock #670-7

Ingredients: Ulmus fulva (slippery elm) bark

Slippery Elm Bulk (7 oz.) - Bulk Powder

Available in: US - Stock #1391-2

Ingredients: Same as Stock #670-7

Small Intestine Detox

Other Names: Marshmallow & Pepsin

Product Type: Formula

Properties: Antacid, Anticatarrhal, Stomachic

Systems Affected: **Intestinal System**, Mucus Membranes, **Small Intestines**, Stomach

Conditions: Acid Indigestion (Heartburn, Acid Reflux), Allergies (food), Allergies (respiratory), Anemia, Cancer (prevention), Colitis, Colon (spastic), Congestion (general), Diarrhea, Digestion (poor), Diverticulitis (Diverticuli), Dyspepsia, Indigestion, Inflammatory Bowel Disorders (Colitis, IBS), Lou Gehrig's Disease

Usage: This formula was designed to break down undigested proteins that may be clogging the small and large intestines. It is designed to improve absorption of nutrients. The mucilage in the marshmallow helps to deliver the pepsin into the small intestines, where it can break down proteins. Works well as part of a colon cleanse. Use 2 capsules three times daily with water.

Warnings: No known warnings.

Small Intestine Detox - Capsule (100)

Available in: US - Stock #848-2

Ingredients: Marshmallow, pepsin (an enzyme)

Marshmallow & Pepsin - Capsule (100)

Available in: US & Canada - Stock #848-2

Ingredients: Same as Stock #848-2

SN-X

See *Sinus Support*

SnorEase

Product Type: Formula

Properties: Anticatarrhal, Antihistamine

Systems Affected: Digestive System, Respiratory System

Conditions: Congestion (sinus), Headache (sinus), Pregnancy (herbs and supplements to avoid during), **Snoring**

Usage: Helps open respiratory passages to inhibit snoring. It can also be used as a natural sinus decongestant in a similar manner to sinus support. Take 1-3 capsules daily 30 minutes before bedtime for snoring. For sinus problems take 1-2 capsules two to three times daily.

Warnings: Do not use during pregnancy.

SnorEase - Capsule (60)

Available in: US - Stock #1815-4

Ingredients: Immature bitter orange, bromelain, Co-Q10

Sore Throat/Laryngitis

Product Type: Homeopathic

Properties: Analgesic (Anodyne), Decongestant

Systems Affected: Throat

Conditions: Laryngitis (Hoarseness), Sore Throat

Usage: Used for the relief of sore throats, hoarseness and laryngitis. Take 10-15 drops under the tongue every 10-15 minutes or as needed until symptoms improve. Then decrease to every one or two hours, then to four times daily until symptoms are relieved.

Warnings: NSP does not recommend this product for children under the age of 12 except on the advice of a health professional. Other than the alcohol content, which is not good for infants, we consider this product safe for children.

Sore Throat/Laryngitis (1 fl. oz.) - Homeopathic (Liquid)

Available in: US - Stock #8795-0

Ingredients: Active ingredients: Arum triphyllum (Indian Turnip) 4x, Causticum (Hahnemann's Causticum) 6x, Phytolacca decandra (Poke Root) 6x, Verbascum thapsus (Mullein) 8x, Ferrum Phosphoricum (Ferrous Phosphate) 10x, Phosphorus (Phosphorus) 10x. Other ingredients: Purified water and 20% USP alcohol

Spirulina

Latin Name:

Product Type: Single Herb

Properties: Antilipemic, Appetite Suppressant, Corrects Polarity, Food, Glandular, Nutritive

Systems Affected: Glandular System, Liver, Pancreas Tail, Pineal, Pituitary (anterior), Pituitary (general)

Conditions: Acquired Immune Deficiency Syndrome (AIDS/HIV), Addictions (alcohol), Addictions (sugar or food), Adrenals (exhaustion, weakness or burnout), Appetite (excessive), Attention Deficit Disorder (ADD, ADHD), Cartilage Damage, Cataracts, Diabetes, Energy (lack of), Fatigue, Glaucoma, Hepatitis, Hypoglycemia, Triglycerides (high), Weight Gain (aids for)

Usage: Spirulina is rich in many nutrients, particularly amino acids. The amino acids are easily assimilated, especially for vegetarians and people who have a hard time digesting animal proteins. It is a high-energy food and may act as an appetite suppressant. It also reduces food cravings. Use 1-2 capsules w/meals or up to 10 capsules per day. For weight management, take 20 to 30 minutes prior to meals and when experiencing food cravings.

Warnings: No known warnings.

Spirulina - Capsule (100)

Available in: US & Canada - Stock #681-1

Ingredients: Spirulina sp.

Spleen Activator

Other Names: Wen Zhong

Product Type: Formula

Properties: Appetite Stimulant, Digestive Tonic, Spleen Chi Tonic, Stomachic

Systems Affected: Intestinal System, Liver, Muscles, Ovaries, Stomach

Conditions: Abdominal Pain and Inflammation, Anorexia, Appetite (excessive), **Bulimia,** Cold Hands and Feet, Convalescence, Cramps (menstrual), Cramps and Spasms (general), Debility, Diarrhea, **Digestion (poor),** Dyspepsia, Energy (lack of), Hemorrhoids, Hernias, Hiatal Hernia, Ileocecal Valve, Indigestion, **Muscle Tone (lack of),** Prolapsed Uterus, **Protein Digestion (poor), Wasting, Weight Gain (aids for),** Worry

Usage: In Chinese medicine, the earth element is associated with digestion, especially the stomach, spleen and pancreas. When a person has an inability to properly digest and metabolize food, they tend to lose muscle mass. They may become very thin and pale, or simply overweight with a lack of muscle tone. The Chinese associate these symptoms with a deficiency of spleen chi (or energy). The Chinese name, Wen Zhong means "warm the center" referring to the power of this formula to warm up or enhance digestive energy.

Spleen Activator is a formula for strengthening the earth element by tonifying the spleen chi. It improves the digestion and metabolism of proteins, minerals and other nutrients to enhance muscle tone, improve physical development, enhance energy and increase appetite for healthy foods.

Emotionally, a lack of earth element energy will lead to chronic worries and fears, a sense of hopelessness, lack of control over one's own life and sluggish thought processes. Physical symptoms may include difficulty gaining muscle weight or losing fat, easy bruising, fatigue, sinus congestion, chronic diarrhea, hemorrhoids, hernia or hiatal hernia, intestinal inflammation, intestinal cramping, menstrual cramping and abdominal bloating.

By restoring the ability to digest food and enhancing circulation and immunity, Chinese Spleen Activator can improve general health. Here is a breakdown of the ingredients in this formula.

To improve chronic digestive system weakness take 3 capsules Spleen Activator (Wen Zhong) 3 times daily. Spleen Activator is also available as a TCM concentrate. Take 1 capsule of the concentrate two to three times daily.

Warnings: No known warnings.

Spleen Activator, Chinese - Capsule (100)

Available in: US - Stock #1880-8

Ingredients: Panax ginseng, astragalus, atractylodes rhizome, hoelen, discorea, lotus seed, chaenomeles fruit, citrus peel, dang gui, dolichos seed, galanga rhizome, ginger rhizome, licorice, magnolia bark, typhonium flageliforme, saussurea, cardamom fruit, zanthoxylum fruit

Spleen Activator TCM Conc. - Capsule (30)

Available in: US - Stock #1070-5

Ingredients: Same as Stock #1880-8

UC-C - Capsule (100)

Available in: Canada - Stock #1891-4

Ingredients: Panax ginseng, astragalus, atractylodes rhizome, hoelen, discorea, lotus seed, chaenomeles fruit, citrus peel, dang gui, dolichos seed, galanga rhizome, ginger rhizome, licorice, magnolia bark, pinellia, saussurea, cardamom fruit, zanthoxylum fruit

Sprains and Pulls

Product Type: Homeopathic

Properties: Anti-inflammatory

Systems Affected: Structural System

Conditions: Injuries, Sprains, Tendonitis

Usage: Used for the relief of pain, stiffness and inflammation caused by muscle or tendon strains, tennis elbow and overexertion during sports activities or workouts. Take 10-15 drops under the tongue every one to two hours following overexertion or appearance of symptoms. Then decrease to three or four times daily until symptoms disappear.

Warnings: NSP does not recommend this product for children under the age of 12 except on the advice of a health professional. Other than the alcohol content, which is not good for infants, we consider this product safe for children.

Sprains and Pulls (1 fl. oz.) - Homeopathic (Liquid)

Available in: US - Stock #8970-7

Ingredients: Active ingredients: Arnica montana (Mountain Arnica) 3x, Bellis perennis (Daisy) 3x, Cinchona officinalis (China) 3x, Gnaphalium polycephalum (Common Everlasting) 4x, Ruta graveolens (Rue) 4x, Colocynthis (Bitter Apple) 6x, Asafoetida (Asafetida) 8x, Kalmia latifolia (Mountain Laurel) 8x

St. John's Wort

Latin Name: *Hypericum perforatum*

Product Type: Single Herb

Properties: Anti-smoking, Antidepressant, Antiviral, Expectorant, Febrifuge, Hypotensive, Nervine, Relaxant, Sedative, Styptic, Tonic, Vermifuge

Systems Affected: Bladder (Urinary), **Central Nerves**, Intestinal System, Liver, Lungs, Nerves, Pituitary (infundibulum), Serotonin, **Solar Plexus**, Spleen, Structural System

Conditions: Acquired Immune Deficiency Syndrome (AIDS/HIV), Addictions (alcohol), Addictions (drugs), Addictions (tobacco smoking or chewing), Anxiety (Panic Attack), Asthma, Autism, Bedwetting, Bell's Palsy, Bites and Stings, Bleeding (external), Bruises (healing), Bunions, Chicken Pox, Cold Sores (Fever Blisters), Colic (children), **Concussions**, Cough (dry), Cuts, Deafness, Depression, Dermatitis, Diarrhea, Dysentery, Ear Infection or Earache, Fatigue, Fear (excessive), Gastritis, Headache (sinus), Hemorrhoids, Incontinence (urinary), Infection (viral), Inflammation, Insomnia, Irritability, Jaundice (adults), **Meningitis**, Menopause, Myasthenia Gravis, Narcolepsy, **Nerve Damage**, **Neuralgia and Neuritis**, Neurosis, **Nightmares**, **Numbness**, Pain (general remedies for), **Paralysis**, PMS Type D, Restless Leg Syndrome, **Seasonal Affective Disorder**, Shingles, Sprains, Tension, Tinnitus (Ringing in the Ears), Tuberculosis (Consumption, Scrofula), Ulcerations (external)

Usage: While having gained recent fame as an antidepressant, St. John's wort has many other uses. It helps to regenerate nerve tissue that has been damaged. In homeopathy, it is considered to be the "arnica" of the nervous system. It helps to heal cuts and injuries when applied topically. It has an antiviral activity, too. General use is one capsule with food three times daily. Time-release: take 1 tablet morning and evening.

Warnings: It has been reported that livestock grazing in pastures with St. John's wort have developed extreme photosensitivity that leads to serious sunburns accompanied by blistering. In more grave cases, this has resulted in the death of some cattle. The remedy is, of course, limiting the exposure to sun either by shading the animal or by painting the animal with dark colored dyes. The long use of St. John's Wort in herbal medicine suggests that these complications are not as serious in humans, although some photosensitivity has been reported by AIDS patients treated with St. John's wort. When taking high doses of this herb, one should avoid excessive exposure to sunlight.

St. John's Wort (Concentrate) T/R - Tablet (60)

Available in: US - Stock #653-1

Ingredients: Same as Stock #

St. John's Wort (Concentrate) - Capsule (100)

Available in: US - Stock #655-3

Ingredients: Standardized extract of Hypericum perforatum (St. John's wort)

St. John's Wort with Passion Flower

Product Type: Formula

Properties: Antidepressant, Nervine, Sedative

Systems Affected: Central Nerves, Nerves, Serotonin, Solar Plexus

Conditions: Anxiety (Panic Attack), Insomnia

Usage: This formula combines the benefits of St. John's wort with the relaxing nervine action of passion flower. It is useful for anxiety, insomnia and mild depression. Take 1 capsules three times daily.

Warnings: No known warnings.

St. John's Wort - Capsule (100)

Available in: US & Canada - Stock #655-3

Ingredients: Same as Stock #655-3

Stevia

Latin Name: *Stevia rebaudiana*

Product Type: Single Herb

Properties: Antidiabetic, Condiment, **Sweetener**

Systems Affected: Liver, Weight Loss

Conditions: Addictions (sugar or food), Blood Pressure (high), Diabetes, Hyperinsulinemia (Syndrome X), **Hypoglycemia**, Schizophrenia, **Sugar Cravings**, Weight Loss (aids for)

Usage: Stevia contains glycosides (steviosides) which are many times sweeter than sugar, but are remarkably stable, making it a natural non-caloric sweetener. It is about 30 times sweeter than sugar. It helps to lower blood sugar levels in hypoglycemic diabetics, but does not have this effect on people with normal blood sugar levels. It is an ideal natural sweetener for weight loss because it appears to help reset the body's appestat (appetite thermostat) and reduce hunger. Stevia also has other health benefits, as well. Research suggests it may help to reduce high blood pressure, fight bacterial infection, aid digestion and tonify the heart. NSP's stevia is not a whole plant extract, but rather a standardized concentrate of the plant's sweet-tasting constituents, which can be used to sweeten beverages or other foods. It comes with a tiny 1/4 teaspoon scoop for convenient measuring. Experimentation will be required to determine how much is needed in various recipes to obtain the desired sweetness.

Warnings: No known warnings. Stevia has a long-term record of safe use in other countries and has been well researched. It is actually a much safer sweetener than the chemical artificial sweeteners currently sold in the marketplace.

Stevia Powder Extract (1.26 oz.) - Bulk Powder

Available in: US - Stock #1386-7

Ingredients: Extract of glycosides from Stevia rebaudiana (stevia), Fructooligosaccharides.

Stevia Powder Packets - Packets (50)

Available in: US - Stock #1381-6

Ingredients: Same as Stock #1386-7

Stomach Comfort

Product Type: Formula

Properties: **Antacid**

Systems Affected: **Stomach**

Conditions: **Acid Indigestion (Heartburn, Acid Reflux),** Dyspepsia

Usage: This natural alternative to antacids contains a combination of natural ingredients that help neutralize excess acid, soothe the stomach, improve digestion and prevent stomach acid from "backwashing" into the esophagus. It soothes burning sensations in the stomach and lessens acidic tastes in the throat and mouth. Use for quick temporary relief by taking 2 tablets when experiencing digestive difficulty.

Warnings: It is important to note that digestive problems may be a sign of more serious health concerns, so if symptoms persist, consult a competent health care provider.

Stomach Comfort - Tablet (60)

Available in: US & Canada - Stock #1820-0

Ingredients: Calcium carbonate, alginic acid, papaya fruit, slippery elm bark, licorice, ginger, guar gum.

Stomach Comfort Trial Pack - Packets

Available in: US & Canada - Stock #2489-1

Ingredients: Same as Stock #1820-0

STR-C

See *Stress Relief*

STR-J

See *Stress-J*

Stress Formula

See *Nutri-Calm*

Stress Relief

Other Names: An Shen

Product Type: Formula

Properties: Alkalinizer, **Nervine**, Relaxant, Sedative, Stomachic

Systems Affected: **Nerves**, Thyroid

Conditions: Abdominal Pain and Inflammation, Angina, Anxiety (Panic Attack), **Anxiety Disorders**, Asthma, Bell's Palsy, Blood Pressure (high), Canker Sores (Mouth Ulcers), Cholesterol (high), Constipation (adults), Convulsions, Cushing's Disease, Depression, Diarrhea, Diverticulitis (Diverticuli), Dizziness (Vertigo), Dysmenorrhea, Emotional Sensitivity, Fear (excessive), Grave's Disease, Headache (tension), Heart Fibrillation or Palpitations, Indigestion, Insomnia, Irritability, Mental Illness, Neurosis, Perspiration (excessive), Stress, Triglycerides (high), Urinary Tract Infections, Varicose Veins

Usage: Stress Relief is a traditional Chinese herb formula for reducing excessive "fire" energy. Fire energy is associated with nervous, glandular and circulatory functions in the body. In Chinese medicine, a person with too much "fire" is prone to excessive excitement, restlessness, anxiety and insomnia, conditions we associate with excessive stress in Western society. The Chinese name for this formula, An Shen, means "pacify the spirit." It helps to calm the nervous system, relieving anxiety and excitability.

An Shen can also be helpful for conditions associated with a high strung and tense nervous system, including heart palpitations, high blood pressure, absent-mindedness, dizziness, convulsions, neurosis, pains in the chest due to stress, emotional instability and excessive perspiration. People who tend to be fast talkers with high strung personalities and vivid dreams will find An Shen helps to balance their nervous system. It can ease feelings of fear and insecurity and reduce conditions involving over acidity, such as urinary tract irritation and mouth sores.

An Shen is designed to help calm the nervous system and relieve anxiety, fright, insomnia and excitability. It is a mild sedative that contains natural calcium to relax the nerves. Since stress often causes digestive problems, this combination also contains herbs that aid digestion by increasing the production of digestive fluids. Chinese Stress Relief re-establishes energy by balancing the normal flow of Chi.

To help nervous problems the suggested dose is 4 capsules of Chinese Stress Relief, once or twice daily. Chinese Stress Relief is also available as a TCM concentrate. The dose of the concentrate would be 1 capsule once or twice daily.

Warnings: No known warnings.

Stress Relief, Chinese - Capsule (100)

Available in: US - Stock #1863-5

Ingredients: Dragon bone, oyster shell, albizzia, polygonum, fushen, acorus rhizome, curcuma, haliotis shell, panax ginseng, polygala, saussurea, zizyphus, coptis, cinnamon, ginger, licorice

Stress Relief TCM Conc. - Capsule (30)

Available in: US - Stock #1033-5

Ingredients: Same as Stock #1863-5

STR-C - Capsule (100)

Available in: Canada - Stock #1871-0

Ingredients: Dragon bone, oyster shell, albizzia bark, polygonum stem, fushen plant, acorus rhizome, curcuma rhizome, cyperus rhizome, haliotis shell, Panax ginseng root, polygala root, zizyphi seed, coptis rhizome, cinnamon twig, ginger rhizome, and licorice root

Stress-J

Other Names: STR-J

Product Type: Formula

Properties: Antispasmodic, **Nervine**, Relaxant, Stomachic

Systems Affected: Digestive System, **Nerves**, Thyroid

Conditions: Addictions (drugs), Addictions (tobacco smoking or chewing), Amenorrhea, Anorexia, Anxiety (Panic Attack), Anxiety Disorders, Asthma, Attention Deficit Disorder (ADD, ADHD), Colic (adults), Colitis, Colon (spastic), Convulsions, Crohn's Disease, Depression, Digestion (poor), Diverticulitis (Diverticuli), Dyspepsia, Fingernail Biting, Grave's Disease, **Inflammatory Bowel Disorders (Colitis, IBS)**, Mental Illness, Nervous Exhaustion (Enervation), **Nervousness**, Pain (general remedies for), **Stress**, Tension, Triglycerides (high)

Usage: This is probably one of the best general nervine formulas for stress and tension. Take 2-4 capsules every two to four hours.

Warnings: No known warnings.

Stress-J - Capsule (100)

Available in: US - Stock #1084-1

Ingredients: Chamomile, passion flower, fennel, feverfew, hops, marshmallow

STR-J - Capsule (100)

Available in: US & Canada - Stock #1087-0

Ingredients: Same as Stock #

Stress-J Liquid Herb (2 fl. oz.) - Liquid (Glycerite)

Available in: US - Stock #3163-3

Ingredients: Same as Stock #

SugarReg

Product Type: Formula

Properties: **Antidiabetic**, Antioxidant, Glandular, Hypoglycemic, **Hypolipidemic**, Tonic

Systems Affected: Glandular System, **Pancreas**, **Pancreas Tail**

Conditions: Cholesterol (high), **Diabetes**, Hyperinsulinemia (Syndrome X)

Usage: This formula helps to overcome cellular resistance to insulin, making it very useful for diabetes. It is very helpful as part of an overall program for Type II diabetics, and can reduce insulin requirements in Type I diabetics. Take 1 capsule three times daily.

Warnings: Since gymnema blocks sugar absorption, this formula could aggravate hypoglycemia. Diabetes is a serious condition and professional assistance should be sought for. Do not stop taking medication without advice of a medical doctor. Type I diabetics must remain on insulin. Check blood sugar levels frequently when taking herbs and supplements for diabetes so that dosages of medications can be adjusted as the situation improves.

SugarReg - Capsule (60)

Available in: US - Stock #927-1

Ingredients: Banaba, gymnema, bitter melon, nopal, fenugreek, chromium, vanadium

GlucoReg - Capsule (60)

Available in: Canada - Stock #927-1

Ingredients: Same as Stock #927-1

SUMA Combination

Product Type: Formula

Properties: Adaptagen, Antioxidant, Anxiolytic, Aphrodisiac, Cerebral Tonic, Glandular, Immune Stimulant, Stimulant

Systems Affected: Adrenal Glands, Brain, Glandular System, Liver, Nerves, Pineal

Conditions: Acquired Immune Deficiency Syndrome (AIDS/HIV), Addison's Disease, Adrenals (exhaustion, weakness or burnout), Aging (prevention), Anxiety Disorders, Attention Deficit Disorder (ADD, ADHD), Convalescence, Crohn's Disease, Cushing's Disease, Depression, Diabetes, Endurance (lack of), Energy (lack of), Epstein Barr Virus (Chronic Fatigue Syndrome, CFS), Fatigue, Glands (swollen lymph), **Grave's Disease**, Senility, Sex Drive (low), Triglycerides (high)

Usage: Suma combination is a formula containing several adaptogens that help the body cope with stress and resist disease. This is a combination of adaptogenic herbs that aid the body's ability to cope with stress. Many of these herbs also enhance circulation to the brain and stimulate the immune response. Use 2-4 capsules three times daily.

Warnings: No known warnings.

SUMA Combination - Capsule (100)

Available in: US - Stock #1088-5
Ingredients: Astragalus, ginkgo, gotu kola, eleuthero, suma

Sunshine Brite Toothpaste

Product Type: Household

Properties: Antiseptic

Systems Affected: Teeth

Conditions: Tooth Decay (prevention)

Usage: This is a natural toothpaste, an alternative to chemical-laden commercial toothpaste. It is fluoride-free (fluoride is a toxic substance). Use as directed on the package.

Warnings: No known warnings.

Sunshine Brite Toothpaste (3.5 fl. oz.) - Topical

Available in: US - Stock #2851-6
Ingredients: Sorbitol, hydrated silica, deionized water, glycerin, sodium bicarbonate, calcium carbonate, sodium lauroyl sarcosinate, natural mint flavors, stevia (Chondrus crispus), corn (Zea mays) oils, unsaponifiables, Astragalus sinicus extract, myrrh gum (Commiphonra myrrha) resin, golden seal (Hydrastis canadensis) root extract, iceland moss (Cetraria islandica) extract, green tea (Camellia sinensis) leaf extract, elderberry (Sambucus nigra) fruit extract, Aloe barbadensis.

Sunshine Concentrate Cleaner

Product Type: Household

Properties: Detergent

Conditions: Chemical Poisoning, **Rhinitis, Skin Care (general)**

Usage: Sunshine Concentrate is a natural, non-toxic, bio-degradable cleaning solution for all household uses. Combined with Nature's Fresh and a few essential oils, it can replace virtually all your toxic household cleaning products. Here are some suggested uses:

- Household Cleaning Spray: Use 1-1/2 teaspoons per quart of water. Add essential oils such as tea tree or lemon for fragrance and disinfectant action. For stains add Nature's Fresh. This spray can be used on appliances, cabinets, walls, woodwork, countertops, etc.
- Dishwashing: Use 2 pumps (about 2 teaspoons) per gallon of hot water. Add more for heavy grease. Use 2 pumps in the dishwasher and fill the dispenser the rest of the way with Nature's Fresh.
- Laundry: Add 5 pumps to washer for a full load of clothes. For grease stains, apply full strength, rub it in, then wash and rinse. Apply full strength to stains along with Nature's Fresh.
- Liquid Hand Soap: Add one part Sunshine Concentrate to seven parts water then add a few drops of Tea Tree Oil or Guardian EO as a natural disinfectant.
- Bathing: Use like bath or shower gel as a soap for bathing.
- Shaving: Use full strength with a little water to make a lather. Use Herbal Trim as aftershave.

Warnings: Avoid contact with eyes.

Sunshine Concentrate Cleaner (32 oz.)

Available in: US - Stock #1551-6
Ingredients: Contains linear alcohol alkoxylates.

Nature's Concentrate (946 mL) - Not Applicable

Available in: Canada - Stock #1551-6
Ingredients: Same as Stock #1551-6

Sunshine Heroes Calcium Plus D3

Product Type: Nutrient

Properties: Calmative, Mineralizer, Nutritive

Systems Affected: Bones, Nerves, Teeth

Conditions: Broken Bones, Insomnia, Nervousness, Teeth (grinding), Tooth Decay (prevention)

Usage: A calcium supplement for children to help ensure they get the calcium they need for healthy bones and teeth. This supplement is in a base of Sunshine Heroes Protector Shield blend, which contains more antioxidants and amino acids. Children should chew from one to three tablets per day. Since this product does not contain magnesium, it may be wise to give children the contents of one capsule of Magnesium Complex with each tablet of this product to ensure they are getting an adequate balance of calcium and magnesium.

Warnings: No known warnings.

Sunshine Heroes™ Calcium Plus D3 - Tablets (90)

Available in: US - Stock #3343-1
Ingredients: Vitamin D (cholecalciferol) (300 IU), calcium (tricalcium phosphate) (300 mg.), phosphorus (tricalcium phosphate) (140 mg), Sunshine Heroes Protector Shield blend (mangosteen fruit, cranberry fruit, broccoli flower, spinach leaves and stems, asparagus stems, carrot root, tomato fruit, açai fruit, pomegranate fruit extract, white grape, apple, pear, orange, pineapple, cherry, strawberry and blueberry fruit juice concentrates, l-leucine, l-lysine, l-valine, choline, nositol, l-isoleucine, l-threonine, l-phenylalanine, l-arginine, l-cysteine, l-methionine, l-tyrosine), organic tapioca syrup, organic evaporated cane juice, pectin, citric acid, natural colors and flavors, sodium citrate and organic sunflower oil

Sunshine Heroes Multiple Vitamin & Mineral

Product Type: Nutrient

Properties: Antioxidant, Nutritive

Systems Affected: Whole Body

Conditions: Addictions (sugar or food), Appetite (excessive), Children's Remedy, Failure to Thrive

Usage: A multiple vitamin and mineral supplement for children that supplies 50-100% of the Recommended Daily Allowance for eleven essential nutrients that may be deficient in modern diets. This supplement is in a base of Sunshine Heroes Protector Shield blend, which contains more antioxidants and amino acids. Children can chew up to four tablets per day.

Warnings: No known warnings.

Sunshine Heroes Multiple Vitamin & Mineral - Tablets (90)

Available in: US - Stock #3341-6

Ingredients: Vitamin A, Vitamin C, Vitamin D3, Vitamin E, Vitamin B6, Folic Acid, Vitamin B12, Biotin, Panothenic Acid, Iodine, Zinc, Sunshine Heroes Protector Shield blend (mangosteen fruit, cranberry fruit, broccoli flower, spinach leaves and stems, asparagus stems, carrot root, tomato fruit, açai fruit, pomegranate fruit extract, white grape, apple, pear, orange, pineapple, cherry, strawberry and blueberry fruit juice concentrates, l-leucine, l-lysine, l-valine, choline, nositol, l-isoleucine, l-threonine, l-phenylalanine, l-arginine, l-cysteine, l-methionine, l-tyrosine), tapioca syrup, organic evaporated cane juice, pectin, citric acid, natural colors and flavors, sodium citrate and organic sunflower oil

Sunshine Heroes Omega 3 with DHA

Product Type: Nutrient

Properties: **Cerebral Tonic**, Nutritive

Systems Affected: **Brain**, Immune System, Nerves, Whole Body

Conditions: Addictions (sugar or food), Allergies (respiratory), Appetite (excessive), Asthma, **Attention Deficit Disorder (ADD, ADHD)**, Autism, **Children's Remedy**, Concentration (poor), **Eczema**, Environmental Pollution (protection from), Epilepsy, Failure to Thrive, Inflammatory Bowel Disorders (Colitis, IBS), **Memory and Brain Function**, Mercury Poisoning, Mood Swings, Skin (dry and/or flaky), Weight Loss (aids for)

Usage: DHA (docosahexaenoic acid) and EPA (eicosapentaenoic acid) are essential fatty acids that are essential to every cell in the body, but particularly the brain and nervous system. These omega-3 fatty acids are found almost exclusively in deep ocean fish, and are lacking in modern diets. Omega-3 fatty acids have been shown to enhance brain development and intelligence in children. The can also be helpful in behavior problems. They also help with the development of the eyes. DHA is the major fatty acid in the brain and in the retina of the eye. This supplement is in a base of Sunshine Heroes Protector Shield blend, which contains more antioxidants and amino acids. Children can chew up to four tablets per day.

Warnings: No known warnings.

Sunshine Heroes Omega 3 with DHA - Tablets (90)

Available in: US - Stock #3342-4

Ingredients: DHA (100 mg.) DHA and EPA (20 mg.) from fish oil, Sunshine Heroes Protector Shield blend (mangosteen fruit, cranberry fruit, broccoli flower, spinach leaves and stems, asparagus stems, carrot root, tomato fruit, açai fruit, pomegranate fruit extract, white grape, apple, pear, orange, pineapple, cherry, strawberry and blueberry fruit juice concentrates, l-leucine, l-lysine, l-valine, choline, nositol, l-isoleucine, l-threonine, l-phenylalanine, l-arginine, l-cysteine, l-methionine, l-tyrosine), organic tapioca syrup, organic evaporated cane juice, pectin, citric acid, natural colors and flavors, sodium citrate and organic sunflower oil

Sunshine Heroes Probiotic Power

Product Type: Formula

Properties: Immune Amphoterics, Nutritive

Systems Affected: Immune System, Intestinal System

Conditions: Abdominal Pain and Inflammation, Acne (Pimples, Blackheads), Addictions (sugar or food), Allergies (food), Antibiotics (side effects of), Children's Remedy, Colds (prevention), Constipation (children), Contagious Diseases, Diarrhea, Eczema, Fungal Infections (Yeast Infections, Candida albicans), Inflammatory Bowel Disorders (Colitis, IBS)

Usage: Probiotics are friendly bacteria that live in the digestive system. They help protect the body from infections from yeast and harmful bacteria. See *Probiotics* for more information on these friendly bacteria. This is a probiotic supplement for children containing 11 strains of probiotic bacteria in a base of Sunshine Heroes Protector Shield blend, which contains more antioxidants and amino acids. Children can chew one or two tablets per day.

Warnings: No known warnings.

Sunshine Heroes™ Probiotic Power - Bulk Powder

Available in: US - Stock #3346-7

Ingredients: Lactobacillus rhamnosus, Bifidobacterium bifidus,Lactobacillus acidophilus, Lactobacillus brevis,Lactobacillus bulcaricus, Lactobacillus plantarum,Streptococcus thermophilus, Bifidobacterium infantis, Bifidobacterium longum, Lactobacillus casei,

Lactobacillus salivarius, Sunshine Heroes Protector Shield blend (mangosteen fruit, cranberry fruit, broccoli flower, spinach leaves and stems, asparagus stems, carrot root, tomato fruit, açai fruit, pomegranate fruit extract, white grape, apple, pear, orange, pineapple, cherry, strawberry and blueberry fruit juice concentrates, l-leucine, l-lysine, l-valine, choline, nositol, l-isoleucine, l-threonine, l-phenylalanine, l-arginine, l-cysteine, l-methionine, l-tyrosine).

Sunshine Heroes Whole Foods Antioxidant

Product Type: Formula

Properties: Anti-inflammatory, Antioxidant, Nutritive, Refrigerant

Systems Affected: Immune System, Intestinal System, Urinary System

Conditions: Adrenals (exhaustion, weakness or burnout), Allergies (respiratory), Anger (excessive), Asthma, Bruises (prevention), Cancer (prevention), Children's Remedy, Colds (prevention), Colic (children), Eczema, Environmental Pollution (protection from), Fever, Flu, Free Radical Damage, Inflammation, Inflammatory Bowel Disorders (Colitis, IBS), Irritability, Urination (burning or painful)

Usage: Free radical damage is believed to be the cause of premature aging and many degenerative diseases. Antioxidants found primarily in fresh fruits and vegetables help neutralize these free radicals. Unfortunately, many children (and adults) today eat almost no fresh fruits and vegetables. This supplement supplies antioxidant nutrients in a base of the Sunshine Heroes Protector Shield blend, which contains more antioxidants and amino acids. This blend also has cooling, soothing anti-inflammatory properties that would be helpful in recovering from acute ailments involving fever and inflammation. Children can chew from one to three tablets per day.

Warnings: No known warnings.

Sunshine Heroes™ Whole Foods Antioxidant - Tablets (90)

Available in: US - Stock #3344-9

Ingredients: Vitamin A, Vitamin C, Vitamin E, Super Fruit Blend (Apple, Grapeseed, Strawberry, Raspberry, Blueberry, Tart Cherry, Pomegranate, Cranberry, Carrot, Tomato, Broccoli, Spinach, Kale), Sunshine Heroes Protector Shield blend (mangosteen fruit, cranberry fruit, broccoli flower, spinach leaves and stems, asparagus stems, carrot root, tomato fruit, açai fruit, pomegranate fruit extract, white grape, apple, pear, orange, pineapple, cherry, strawberry and blueberry fruit juice concentrates, l-leucine, l-lysine, l-valine, choline, nositol, l-isoleucine, l-threonine, l-phenylalanine, l-arginine, l-cysteine, l-methionine, l-tyrosine), organic tapioca syrup, organic evaporated cane juice, pectin, citric acid, natural flavors, natural colors (black carrot juice concentrate, annatto, turmeric), sodium citrate, organic sunflower oil and carnauba wax.

Sunshine Heros Whole Foods Papayazyme

Product Type: Formula

Properties: Antioxidant, Aperient, Decongestant, Digestant, Nutritive

Systems Affected: Digestive System, Immune System, Respiratory System

Conditions: Acid Indigestion (Heartburn, Acid Reflux), Allergies (food), Allergies (respiratory), Asthma, Autism, Children's Remedy, Constipation (children), Cough (damp), Failure to Thrive, Parasites (general)

Usage: Enzymes help break down nutrients in the food we eat. Live foods are rich in enzymes, but cooking deactivates enzymes. This supplement provides digestive enzymes in a chewable form for children (and adults) to offset the lack of enzymes in modern diets. The enzymes in this blend can also help to loosen mucus in respiratory congestion. In addition, the formula contains a blend of amino acids and antioxidant fruits and vegetables to boost brain function, immunity and general health. This supplement is in a base of Sunshine Heroes Protector Shield blend, which contains more antioxidants and amino acids. For children under four, chew one tablet before meals. For children four and over, chew two tablets before meals.

Warnings: No known warnings.

Sunshine Heroes™ Whole Foods Papayazyme - Tablets (90)

Available in: US - Stock #3345-3

Ingredients: Pineapple fruit (Ananas comosus), papaya fruit (Carica papaya), bromelain and papain (proteolytic enzymes), mycozyme (an amylase enzyme), Sunshine Heroes Protector Shield blend (mangosteen fruit, cranberry fruit, broccoli flower, spinach leaves and stems, asparagus stems, carrot root, tomato fruit, açai fruit, pomegranate fruit extract, white grape, apple, pear, orange, pineapple, cherry, strawberry and blueberry fruit juice concentrates, l-leucine, l-lysine, l-valine, choline, nositol, l-isoleucine, l-threonine, l-phenylalanine, l-arginine, l-cysteine, l-methionine, l-tyrosine), sorbitol, xylitol, fructose, stearic acid, natural flavors and magnesium stearate

Super Algae

Product Type: Formula

Properties: Antilipemic, Antiviral, Corrects Polarity, Food, Nutritive

Systems Affected: Brain, Cuticle, Digestive System, **DNA**, Liver, Pituitary (anterior), Pituitary (general), Structural System, Thymus, Weight Loss, Whole Body

Conditions: Acquired Immune Deficiency Syndrome (AIDS/HIV), Addictions (sugar or food), Alzheimer's Disease, Amenorrhea, Anemia, Appetite (excessive), Athletic Performance or Exercise (aids to), Attention Deficit Disorder (ADD, ADHD), Bedwetting, Bell's Palsy, Bipolar Mood Disorder (Manic Depressive Disorder), Calcium Deficiency, Cancer (natural therapy for), Cartilage Damage, Children's Remedy, Cholesterol (high), Cholesterol (low), Confusion, Convalescence, Coordination, Cramps (menstrual), Cramps and Spasms (general), Crohn's Disease, Dehydration, Depression, Disks (spinal - bulging or slipped), Energy (lack of), Epstein Barr Virus (Chronic Fatigue Syndrome, CFS), Exercise, Failure to Thrive, Fatigue, Hyperinsulinemia (Syndrome X), **Hypoglycemia**, **Memory and Brain Function**, Mercury Poisoning, Mood Swings, Phobias, **PMS Type C**, Radiation Sickness, Reversed Polarity, Sugar Cravings, Surgery (healing from), Triglycerides (high), Weight Gain (aids for), Weight Loss (aids for)

Usage: The various forms of algae in this combination provide super concentrated nutrition. They are a source of protein, carbohydrates, carotenoids, amino acids and trace minerals. This makes Super Algae especially useful in recuperating from serious illness or surgery. Take 2-4 capsules one to three times daily.

Warnings: No known warnings.

Super Algae - Capsule (100)

Available in: US - Stock #1056-5

Ingredients: Spirulina, chlorella, Klamath Lake blue-green algae

Super Antioxidant

Product Type: Formula

Properties: Anti-inflammatory, **Antioxidant**, Hepatic, Nutritive

Systems Affected: Immune System, Liver

Conditions: Aging (prevention), Alzheimer's Disease, Autoimmune Disorders, Cancer (prevention), **Cataracts**, **Chemotherapy (reducing side effects)**, Environmental Pollution (protection from), Eye Problems (general), **Free Radical Damage**, Memory and Brain Function, Radiation Sickness, Toxemia

Usage: This formula contains a variety of antioxidants and liver protecting herbs that help protect the body against free radical damage and assist in the process of detoxification. Prevents cellular damage, supports liver, supports nervous system and enhances immunity. Take 1 capsule two times daily.

Warnings: If headache or rapid heart rate occurs then discontinue. No known side effects are reported.

Super Antioxidant - Capsule (60)

Available in: US - Stock #1825-8

Ingredients: Tocotrienols, lycopene, alpha lipoic acid, rose hips, milk thistle, tumeric

Super GLA

Product Type: Formula

Properties: Analgesic (Anodyne), Anti-arthritic, Anti-inflammatory, Antirheumatic, Immune Amphoterics, Immunomodulator, Nutritive

Systems Affected: Central Nerves, Cuticle, Female Reproductive, Immune System, Intestinal System, Liver, Prostaglandins, Reproductive Glands, Structural System, Weight Loss, Whole Body

Conditions: Allergies (food), Allergies (respiratory), Anemia, Appetite (excessive), Arthritis, Asthma, Autoimmune Disorders, Bipolar Mood Disorder (Manic Depressive Disorder), Blood Pressure (high), Breast Lumps, Burning Feet or Hands, Cancer (prevention), Chemical Poisoning, Children's Remedy, Cradle Cap, Cystic Breast Disease, Cystic Fibrosis, Dermatitis, Diabetes, Down Syndrome, Eczema, Endometriosis, Environmental Pollution (protection from), Epilepsy, Failure to Thrive, Fat Cravings, Fibroids (uterine), Hair (loss or thinning), Leucorrhea, Ligaments (torn or injuried), Lupus, Lyme Disease, Memory and Brain Function, **Menopause**, Menstrual Irregularity, Multiple Sclerosis (MS), Nerve Damage, Neuralgia and Neuritis, Neurosis, Parkinson's Disease, Pets (supplements for), PMS (general), Post Partum Weakness, Psoriasis, Restless Leg Syndrome, Seborrhea, Senility, Skin (dry and/or flaky), Triglycerides (low), Vitiligo, Weight Loss (aids for), Wrinkles

Usage: The three oils in this blend are all high in the omega-6 essential fatty acids, linoleic acid and gamma-linolenic acid (GLA). These oils are used in the creation of eicosanoids that play essential roles in maintaining normal blood pressure and body weight, and in reducing platelet aggregation and inflammation. GLA is also very helpful with PMS symptoms like breast swelling and tenderness. Normal dose is 1 capsule three times daily with food.

Warnings: In the presence of high insulin levels and a lack of omega-3 essential fatty acids, however, GLA is converted into eicosanoids that are pro-inflammatory and have opposing effects to the benefits listed above. So, it is also important to obtain adequate omega-3 in the diet and to avoid refined carbohydrates for maximum effects.

Super GLA Oil Blend - Capsule (90)

Available in: US - Stock #1844-5

Ingredients: Evening primrose oil, black currant oil, borage oil, glycerin.

Super Oil - Softgel (90)

Available in: Canada - Stock #8235-8

Ingredients: Linseed oil, fish oil, soya lecithin, evening primrose oil, vitamin E

Super Oil

See *Super GLA*

Super ORAC

Product Type: Formula

Properties: Anodyne, Anti-aging, **Anti-inflammatory, Antioxidant**, Febrifuge, Refrigerant

Systems Affected: Cardiovascular System, Immune System, Liver

Conditions: Abdominal Pain and Inflammation, Aging (prevention), Arthritis, Asthma, Autoimmune Disorders, Cancer (prevention), Cardiovascular Disease (Heart Disease), Environmental Pollution (protection from), Free Radical Damage, Free Radical Damage, **Free Radical Damage, Glaucoma, Inflammation**, Leaky Gut Syndrome, **Macular Degeneration**, Memory and Brain Function, Neuralgia and Neuritis, Oral Surgery, Pain (general remedies for), Radiation Sickness, Rheumatoid Arthritis (Rheumatism), Sore or Geographic Tongue, Urination (burning or painful)

Usage: Super ORAC was formulated to neutralize oxidation and reduce the destructive effects of free radicals in the body. ORAC stands for oxygen radical absorbance capacity and this product has a high ability for absorbing free radicals, which are believed to be the cause of premature aging and many chronic and degenerative diseases such as heart disease, cancer and senility.

Super ORAC contains green tea extract which is rich in protective flavanols, flavandiols, flavonoids, and phenolic acids. It also contains mangosteen pericarp extract, which contains xanthones, known for their antioxidant and anti-inflammatory properties. This fruit is full of antioxidants and more than half of the two hundred know natural xanthones, which help boost the immune system. Mangosteen is also known for stimulating the immune system, anti-inflammatory properties, and anti-microbial abilities. When taken orally, Mangosteen is used to treat dysentery, diarrhea, urinary tract infections, gonorrhea, tuberculosis and cancer.

Another key ingredient in Super ORAC is turmeric. Traditionally, in both Ayurvedic and Chinese medicine, it was used to aid liver action, to treat jaundice, and as also used for digestive problems. Turmeric is antioxidant, antibacterial and anti-inflammatory properties.

Acai berry has received a lot of publicity recently because of its healing abilities. Acai has anti-inflammatory properties, it boosts the immune system, increases energy, aids in good sleep, reduces sores and pain, and overall helps maintain regular function of the body's systems and organs.

Other ingredients in Super ORAC include quercitin, revestratrol and selenium. Quercitin has antioxidant, anti-inflammatory, anticarcinogenic, and cardioprotective properties and works on the function of the immune systems. Resveratrol is a polyphenolic compound with anti-inflammatory action. Selenium is an important mineral for the immune system.

Take Super ORAC for prevention of chronic illness or for reducing inflammation and pain in the body. Also consider Thai-Go, another excellent antioxidant product.

Warnings: No known warnings.

Super ORAC - Capsule (90)

Available in: US - Stock #808-3

Ingredients: Polyphenols, Green Tea Leaves Extract (Camellia sinensis), Mangosteen Pericarp Extract (Garcinia mangostana), Turmeric Root Extract (Curcuma longa), Quercetin Dihydrate (Sophora japonica), Resveratrol (from Japanese Knotweed Root-Polygonum cuspidatum), Apple Fruit Extract (Malus domestica), Açai (Acai) Berry Extract (Euterpe oleracea), Selenium

Super Supplemental

Product Type: Formula

Properties: Nutritive

Systems Affected: Whole Body

Conditions: Infertility

Usage: A vitamin and mineral supplement in an herbal base, available with and without iron. Take 1 tablet three times daily with meals.

Warnings: No known warnings.

Super Trio

Product Type: Pack

Properties: Antioxidant, Nutritive

Systems Affected: Cardiovascular System, Whole Body

Conditions: Aging (prevention), Arthritis, Cancer (prevention), Convalescence, Diabetes, Energy (lack of), Environmental Pollution (protection from), Exercise, Fatigue, Inflammation, Inflammatory Bowel Disorders (Colitis, IBS), Memory and Brain Function, Mental Illness, Pregnancy (herbs and supplements for), Skin Care (general)

Usage: Good nutrition is an important part of staying healthy. Unfortunately, people living in our fast-paced, junk-food world don't often take the time to eat right. They skip meals or grab something on the go.

Super Trio is a convenient way to obtain some of the basic nutrients we need to stay healthy. It provides a basic vitamin and mineral supplement, an antioxidant supplement to reduce free radical damage and the essential good fats we need to reduce inflammation and maintain health. Packaged in convenient packets that can be placed in a purse, briefcase or pocket, Super Trio is an easy way to ensure you are getting important nutrients your body needs for good health.

Warnings: No known warnings.

Super Trio - Packets

Available in: US & Canada - Stock #20-5

Ingredients: Super Omega-3 EPA, Super Supplemental, Super ORAC

SynerProTein

Product Type: Formula

Properties: Nutritive

Systems Affected: Structural System, Weight Loss, Whole Body

Conditions: Attention Deficit Disorder (ADD, ADHD), Coordination, Exercise, Hypoglycemia, Overalkalinity, Triglycerides (high)

Usage: A natural protein powder. Add approximately one heaping scoop (28.4 grams) of powder to 8 ounces of your favorite juice, water, or skim milk.

Warnings: Avoid with allergies to soy.

SynerProTein - Bulk Powder

Available in: US - Stock #3025-0

Ingredients: Soy protein isolate (non-GMO), fructose, fructooligosaccharides, natural flavors, powdered sunflower oil, medium chain triglycerides, potassium citrate, xanthan gum, guar gum, carrageenan, dl-methionine, magnesium oxide, sodium chloride, potassium phosphate, ascorbic acid, soy lecithin, stevia leaf extract, SynerPro Concentrate base (broccoli, carrot, red beet, rosemary, tomato, turmeric, cabbage, grapefruit and orange bioflavonoids and hesperidin), ferrous fumarate (iron), niacin (vitamin B3), vitamin A palmitate, chromium amino acid chelate, calcium d-pantothenate, riboflavin, thiamine hydrochloride (vitamin B1), pyridoxine hydrochloride (vitamin B6), potassium iodide, cyanocobalamin (vitamin B12).

SynerProTein Powder - Bulk Powder

Available in: Canada - Stock #8228-0

Ingredients: Same as SynerProTein #3025-0

Target Endurance

Product Type: Formula

Properties: Stimulant

Systems Affected: DNA, **Mitochondria**, **Muscles**, Structural System

Conditions: **Addictions (coffee, caffeine)**, **Addictions (sugar or food)**, Appendicitis, **Athletic Performance or Exercise (aids to)**, **Endurance (lack of)**, **Energy (lack of)**, **Exercise**, **Fatigue**, **Muscular Dystrophy**, Post Partum Weakness

Usage: Contains minerals chelated to amino acids that deliver them to the energy producing center (mitochondria) in cells. Research with animals showed this product to increase endurance levels by over 400 percent. For athletes or people who need more stamina, energy or endurance, use 1-2 capsules one to three times daily.

Warnings: No known warnings.

Target Endurance - Capsule (90)

Available in: US - Stock #2809-8

Ingredients: Copper, potassium, zinc, vitamins B-6, B-12, C, niacin, folic acid, pantothenic acid, calcium, phosphorus, iodine, bee pollen, Eleuthero, gotu kola, capsicum, licorice, l- glutamine, choline, ginger.

Target P-14

Product Type: Formula

Properties: **Antidiabetic**, Glandular, Hypoglycemic, Mineralizer

Systems Affected: Intestinal System, **Pancreas**, **Pancreas Tail**

Conditions: Adrenals (exhaustion, weakness or burnout), Autism, **Diabetes**, Hypoglycemia

Usage: Same as Pro-Pancreas (formerly P-14) (see *Glandular System-Pancreas*) but with the addition of zinc and chromium chelated to amino acids to deliver them to the pancreas where they aid in insulin production. As an aid to diabetes, use 1 capsule per meal. For hypoglycemia use 1-2 capsules per day with licorice root and/or HY-A.

Warnings: No known warnings.

Target P-14 - Capsule (90)

Available in: US - Stock #2810-1

Ingredients: Chromium, zinc, golden seal, juniper berries, uva ursi, cedar berries, mullein, garlic bulb, yarrow, slippery elm, capsicum fruit, dandelion, marshmallow, nettle, white oak bark, licorice.

Target TS-II

Product Type: Formula

Properties: **Anti-obesic**, Stimulant

Systems Affected: **Pituitary (general)**, **Thyroid**, Weight Loss

Conditions: Fatigue, **Hypothyroid**, Schizophrenia, **Weight Loss (aids for)**

Usage: Same as TS-II with Hops, except it also contains minerals chelated to amino acids of which direct them to the pituitary where they stimulate the hormones which stimulate the thyroid gland. Used to stimulate the thyroid and as an aid in weight loss. Use 2 capsules before breakfast, 1 before lunch. Same as TS-II with Hops, except it also contains minerals chelated to amino acids that direct them to the pituitary where they stimulate the hormones which stimulate the thyroid gland. Used to stimulate the thyroid and as an aid in weight loss. Use 2 capsules before breakfast, 1 before lunch.

Warnings: No known warnings.

Target TS-II - Capsule (90)

Available in: US - Stock #2815-7

Ingredients: Zinc, manganese, hops, parsley, capsicum, Irish moss, kelp

Tea Tree Oil

Product Type: Essential Oil

Properties: Anti-inflammatory, **Antibacterial**, **Antifungal**, **Antimicrobial**, **Antiseptic**, Antivenomous, Antiviral, Aromatic, Cicatrisant, **Disinfectant**

Systems Affected: Ears, Immune System, Structural System

Conditions: Abrasions, Abscesses, **Antibiotics (alternatives to)**, Antibiotics (side effects of), **Athlete's Foot**, **Blisters**, Boils, Burns and Scalds, Canker Sores (Mouth Ulcers), Children's Remedy, Cold Sores (Fever Blisters), Corns, Cuts, **Dandruff**, Denture Sores, **Ear Infection or Earache**, Fleas, **Fungal Infections (Yeast Infections, Candida albicans)**, **Gangrene**, Gingivitis (Bleeding Gums, Gum Disease, Pyorrhea), Hair Care (general), Hypochondria, Impetigo, **Infection (bacterial)**, Injuries, **Insects**, Itching, Itching Ears, Jock Itch, Laryngitis (Hoarseness), Leucorrhea, **Lice**, Poison Ivy or Oak, Rashes and Hives, Rosacea, Scabies, Scars / Scar Tissue, **Scratches and Abrasions**, **Seborrhea**, **Skin (infections)**, Slivers, Staph Infections, Strep Throat, Sunburn, **Tetanus**, **Thrush**, Toothache, **Ulcerations (external)**, Vaginitis, **Wounds and Sores**

Usage: This volatile oil has powerful disinfectant and antifungal properties. Externally, apply full strength to aid healing and fight infection in burns, cuts, wounds, abrasions, bites, stings, canker sores and other injuries. Disinfects injuries and speeds healing of tissues. Usually does not sting. Use 1-2 drops in ear for earaches. Internally use highly diluted, 1-2 drops in glass of water. Also useful as a gargle; add 2-4 drops per pint of warm water. Can be used in an enema or douche; use 2-4 drops per pint of water. Can be used in baths; add 5-20 drops to bath water. Use sparingly in baths as it can irritate the skin. As a skin cleanser and acne cream, combine with jojoba oil. Mix with shampoo for lice, dandruff and itching scalp.

Warnings: Do not take undiluted oil internally. The oil can irritate the skin in some people.

Tea Tree Oil (.5 fl. oz.) - Essential Oil

Available in: US & Canada - Stock #1777-1

Ingredients: Australian Tea Tree (Melaleuca alternifolia).

Teething

Product Type: Homeopathic

Properties: Analgesic (Anodyne)

Systems Affected: Teeth

Conditions: Children's Remedy, **Teething**

Usage: Used for the temporary relief of minor gum soreness due to teething in infants and children. Directions for children over 4 months: Take 3-5 drops under the tongue every 15-20 minutes until symptoms improve, then every two-four hours until symptoms are relieved.

Warnings: Administer to children under 4 months only under the advice of a health care professional.

Teething (1 fl. oz.) - Homeopathic (Liquid)

Available in: US - Stock #8885-6

Ingredients: Active ingredients: Chamomilla (chamomile) 3x, Natrum phosphoricum (sodium phosphate) 3x, Calcarea Carbonica (calcium carbonate) 6x, Kali Phosphoricum (potassium phosphate) 6x, Staphy-sagria (stavesacre) 6x, Kali Sulphuricum (potassium sulfate) 12x. Other ingredients: Purified water, glycerin, and potassium benzoate

Tei Fu Oil

Product Type: Formula

Properties: Anticephalalgic, Aromatic, Counterirritant, Rubefacient, Stimulant (Circulatory)

Systems Affected: Muscles, **Sinuses**, Structural System, Tongue

Conditions: Allergies (respiratory), **Arthritis**, Asthma, Backache (Back Pain, Lumbago), **Bites and Stings**, Bruises (healing), Bruises (prevention), **Bursitis**, **Canker Sores (Mouth Ulcers)**, Carpal Tunnel Syndrome, Cold Sores (Fever Blisters), **Colds (decongestant)**, Concentration (poor), **Congestion (general)**, **Congestion (sinus)**, Cough (damp), Cough (general), Croup, Cuts, Dislocation, Energy (lack of), Exercise, Gas and Bloating, **Headache (general)**, **Headache (sinus)**, **Headache (tension)**, Infection (bacterial), **Injuries**, Ligaments (torn or injuried), **Migraine**, Motion Sickness, **Pain (general remedies for)**, Sinus Infection, **Sprains**, **Stiff Neck**, Stress, **Tendonitis**, Toothache

Usage: Both lotion and oil can be massaged into skin for relief of sore muscles. Massage into throat for sore throats, into chest for congestion. Massage into neck and shoulders to relieve headaches. Also apply Tei Fu Oils to temples for headaches. Tei Fu Oils can be used on insect bites, bee stings, bruises, etc., to aid healing. Inhale Tei Fu Oils for relief of sinus congestion. One drop of oil can be used on the tongue to promote mental alertness, while driving. Apply oil directly to canker sores to promote healing.

Warnings: Do not use more than a drop or two internally and for a very limited period of time. Avoid contact with eyes and genitals. Do not apply after a hot shower (will cause severe burning sensation).

Tei Fu Essential Oil (0.17 fl. oz.) - Essential Oil

Available in: US & Canada - Stock #1618-7

Ingredients: Safflower oil, wintergreen oil, menthol, camphor and other essential oils.

Tei Fu Massage Lotion (4 fl. oz. tube) - Essential Oil

Available in: US & Canada - Stock #3538-5

Ingredients: Water, menthol, wintergreen oil, camphor, eucalyptus oil, glyceryl stearate, PEG-100 stearate, stearic acid, cetearyl alcohol, sweet almond oil, ceteareth-20, butylene glycol, caprylic/capric triglyceride, clove oil, aloe barbadensis, squalane, allantoin, citric acid, dimethicone, xanthan gum, tetrasodium EDTA, acrylates/C10-30 alkyl acrylate crosspolymer, hydroxyethyl acrylate/sodium acryloyldimethyl taurate copolymer, polysorbate 60, phenoxyethanol, methylparaben, propylparaben, ethylparaben, butylparaben

Thai-Go

Other Names: Zambroza

Product Type: Formula

Properties: **Anti-aging**, **Anti-inflammatory**, **Antimutagenic**, **Antioxidant**, Febrifuge, **Immunomodulator**, Low Glycemic, Opthalmicum, **Refrigerant**

Systems Affected: Arteries, Blood Vessels, Capillaries, Cardiovascular System, Immune System, Liver, Structural System, Weight Loss, Whole Body

Conditions: Acquired Immune Deficiency Syndrome (AIDS/HIV), Addictions (alcohol), Adenitis, Adrenals (exhaustion, weakness or burnout), Age Spots, **Aging (prevention)**, Alzheimer's Disease, Anemia, Anorexia, Arteriosclerosis (Atherosclerosis, Hardening of the Arteries), Arthritis, Athletic Performance or Exercise (aids to), Backache (Back Pain, Lumbago), Blood Pressure (high), Broken Bones, Bruises (healing), Bruises (prevention), Cancer (natural therapy for), **Cancer (prevention)**, Capillary Weakness, Cardiovascular Disease (Heart Disease), Carpal Tunnel Syndrome, Cartilage Damage, Cataracts, Children's Remedy, Circulation (poor), Colds (general remedies for), Colds (prevention), **Convalescence**, Crohn's Disease, Cystic Breast Disease, Cystitis, **Debility**, Dermatitis, Disks (spinal - bulging or slipped), Electromagnetic Pollution, Emphysema, Environmental Pollution (protection from), Epstein Barr Virus (Chronic Fatigue Syndrome, CFS), Exercise, Eye Problems (general), Eyes (red or itching), Eyesight (to improve), Fatigue, **Fever**, Fibromyalgia Syndrome (FMS), **Free Radical Damage**, Hair (loss or thinning), **Inflammation**, Injuries, Leaky Gut Syndrome, Lupus, Macular Degeneration, Memory and Brain Function, Meningitis, Mental Illness, Myasthenia Gravis, **Nephritis**, Nervous Exhaustion (Enervation), Neuralgia and Neuritis, Night Blindness, Overacidity, Pain (general remedies for), Phlebitis, Pleurisy, Post Partum Weakness, **Pregnancy (herbs and supplements for)**, Prostatitis, Radiation Sickness, Rhinitis, Rosacea, Senility, Toxemia, Urination (burning or painful), Urination (frequent), Vaccines (detoxification from)

Usage: Research continues to show that free radical damage and inflammation are primary causes of aging and degenerative diseases like cancer and heart disease. Antioxidants are nutrients, primarily found in plants, which prevent free radical damage and reduce inflammation. Research also shows that people who eat lots of fresh fruits and vegetables, known to be loaded with natural antioxidants, have less degenerative disease. As a result, many nutritionists are recommending we should eat 5-7 generous servings of fresh fruits and vegetables every day. Unfortunately, research also shows that very few people actually do this.

For those who are on-the-go and having a hard time getting their fruits and vegetables, a convenient way to get more antioxidants into the diet is to take Thai-Go. Thai-Go is a blend of fruit juices and other botanical ingredients that are rich in naturally occurring antioxidants such polyphenols, flavonoids, xanthones and vitamin C. This powerful formula also reduces inflammation, relieves pain, and enhances energy production. One of the key ingredients in Thai-Go is mangosteen.

Mangosteen is a tropical fruit which is highly praised for its luscious flavor. Mangosteen is very high in compounds called xanthones, some of the most powerful antioxidants discovered to date. Two of the most beneficial xanthones in mangosteen are alpha manostin and gamma mangostin. These compounds have been shown to be antibiotic, antiviral and anti-inflammatory. They have histamine-blocking actions and may help protect arteries from damage. Mangosteen has also been found to have antifungal activity.

Recommended use is to take one ounce of Thai-Go daily-two tablespoons twice daily. It tastes great by itself and can also be combined with water or fruit juice.

Warnings: Thai-Go may cause a "healing crisis" in certain people as it causes their body to detoxify. In these cases, some colon and lymphatic cleansing may be needed.

Thai-Go (2-25 fl. oz. each) - Liquid

Available in: US - Stock #4095-1

Ingredients: Reconstituted Mangosteen Fruit (Garcinia mangostana), Concord Grape Fruit Concentrate (Vitis labrusca), Red Grape Fruit Concentrate (Vitis vinifera), Blueberry Fruit Concentrate (Vaccinium corymbosum), Red Raspberry Fruit Concentrate (Rubus idaeus), Red Grapeskin Extract (Vitis vinifera), Wolfberry Lycium Fruit Extract (Lycium barbarum), Sea Buckthorn Fruit Extract (Hippophae rhamnoides), Red Grapeseed Extract (Vitis vinifera), Green Tea Leaves Extract (Camellia sinensis), Apple Fruit Extract (Malus domestica)

Zambroza - Liquid

Available in: US & Canada - Stock #4104-1

Ingredients: Purified water, Vitis vinifera (red grape) fruit concentrate 1135 mg; Vitis labrusca (concord grape) fruit concentrate 1135 mg; Vaccinium corymbosum (blueberry) fruit concentrate 678 mg; reconstituted Garcinia mangostana (mangosteen) fruit powder, 406 mg of a 50:1 extract; Rubus idaeus (red raspberry) fruit concentrate 205 mg; Vitis vinifera (red grape) skin extract 102 mg; Lycium barbarum (wolfberry) fruit extract, 16 mg of a 10:1 extract standardized to 10% polysaccharides; Hippophae rhamnoides (sea buckthorn) fruit, 16 mg of a 12:1 extract standardized to no less than 8% flavonoids; Camellia sinensis (green tea) leaves, 3 mg of an extract standardized to no less than 80% polyphenols; Vitis vinifera (red grape) seed, 3 mg of a 100:1 extract standardized to no less than 83% polyphenols; Malus domestica (apple) skin and fruit, 3 mg of an extract standardized to no less than 80% polyphenols.

THIM-J

Product Type: Formula

Properties: Corrects Polarity, Immune Stimulant

Systems Affected: Immune System, Liver, Pineal, **Thymus**, Thyroid

Conditions: Acquired Immune Deficiency Syndrome (AIDS/HIV), Addictions (drugs), Cancer (natural therapy for), Colds (general remedies for), Convalescence, Epstein Barr Virus (Chronic Fatigue Syndrome, CFS), Flu, Fungal Infections (Yeast Infections, Candida albicans), Grave's Disease, Hepatitis, Lyme Disease, Myasthenia Gravis, **Reversed Polarity**

Usage: This formula was designed to strengthen the thymus gland, which helps to regulate the immune system. It contains herbs and nutrients known to have a protective effect against cancer. To help improve a weakened immune system or help prevent disease, take 2 capsules with each meal. In acute situations take 2 capsules every two hours.

Warnings: No known warnings.

THIM-J - Capsule (100)

Available in: US - Stock #1089-8

Ingredients: Rosehips, beta-carotene, broccoli, cabbage, eleuthero, parsley, red clover, wheat grass, horseradish

Three

See *Herbal Trace Minerals*

Thyme

Product Type: Essential Oil

Properties: Antibacterial, Anticarious, Antifungal, Antimicrobial, Antioxidant, **Antiseptic**, Antispasmodic, Antiviral, Aromatic, Astringent, Carminative, Counterirritant, Diaphoretic, Disinfectant, Emmenagogue, Expectorant, Parasympatholytic (Anticholinergic), Pectoral, Preservative, Stimulant (Circulatory), Sympathomimetic, Tonic

Systems Affected: Aldosterone, Bronchials, Epinephrine, Gall Bladder, Hypothalamus, Immune System, Lungs

Conditions: Addictions (alcohol), Antibiotics (alternatives to), Bronchitis, Colds (antiviral), Colds (decongestant), Colds (general remedies for), Colic (children), Confusion, Congestion (general), Cough (damp), **Cough (general)**, Cough (spastic), Cramps (menstrual), Croup, Digestion (poor), Disks (spinal - bulging or slipped), Dyspepsia, Ear Infection or Earache, Edema (Dropsy, Water Retention, Swelling), Fever, Gas and Bloating, Headache (general), Hypochondria, Hysteria, Indigestion, Infection (bacterial), Infection (viral), Inflammatory Bowel Disorders (Colitis, IBS), Neuralgia and Neuritis, Nightmares, Parasites (nematodes, worms), Pertussis (Whooping Cough), Scabies, Tetanus, **Thrush**, Toothache

Usage: Thyme was used by the ancient Egyptians for embalming. It is a very good household disinfectant and home deodorizer. It has a sweet, warm and strongly herbal fragrance. The oil of thyme is strongly disinfectant and has an antispasmodic and expectorant action. It can be applied topically to minor injuries as a natural disinfectant. It can be inhaled or diffused for respiratory problems such as colds, sinus congestion, asthma, bronchitis, whooping cough and spastic, dry coughs. It can also help overcome laryngitis, sore throats, strep throat and flu. It can be used topically or in a bath for circulatory problems, indigestion, arthritis pain or to stimulate the immune system. It can also be used for thrush, warts and viral infections. Emotionally, thyme helps mental focus, creating an alert, focused state of mind. It revives low spirits, combats exhaustion and stimulates the mind, while calming mental chatter. Thyme can be used to help overcome confusion, fear and to regulate instinctive reactions.

Warnings: Thyme is a safe herb, but thyme oil may be toxic in large doses. Toxic symptoms include nausea, vomiting, gastric pain, headache, dizziness, convulsions, coma, cardiac and respiratory collapse. For internal use, take the whole herb, not the oil unless under the direction of a professional. Even a few teaspoons of thyme oil can be toxic, causing headache, nausea, vomiting, weakness, thyroid impairment and depression of the heart rate and respiration. One animal study showed thyme suppresses thyroid activity in rats. Those with thyroid conditions should consult their physicians before taking medicinal doses. Thyme and thyme oil may cause a rash in sensitive individuals.

Thyme Linalol BIO (5ML) - Essential Oil

Available in: US - Stock #3913-2

Ingredients: Thyme Linalol Bio (Thymus vulgaris ct. linalol) Essential Oil

Thyroid Activator

Other Names: KC-X

Product Type: Formula

Properties: Glandular, Mineralizer, Thyrotropic

Systems Affected: Parathyroid, **Thyroid**, Weight Loss

Conditions: Depression, Energy (lack of), Epilepsy, Fatigue, Fibrosis, Glands (swollen lymph), Goiter, Hypothyroid, Radiation Sickness, Sex Drive (low), **Skin (dry and/or flaky)**, Weight Loss (aids for)

Usage: To aid thyroid function take 2 capsules one to three times daily. This formula is stronger than TS II for most people.

Warnings: Contraindicated in cases of overactive thyroid.

Thyroid Activator - Capsule (100)

Available in: US - Stock #1224-0

Ingredients: Irish moss, kelp, black walnut, parsley, watercress, sarsaparilla.

KC-X - Capsule (100)

Available in: Canada - Stock #1224-0

Ingredients: Same as Stock #1224-0

Thyroid Support

Product Type: Formula

Properties: **Glandular**, Mineralizer, Stimulant, Thyrotropic

Systems Affected: Pituitary (general), **Thyroid, Weight Loss**

Conditions: Amenorrhea, Anxiety (Panic Attack), Blood Pressure (low), **Cholesterol (low)**, Cirrhosis of the Liver, **Cold Hands and Feet, Depression**, Eczema, **Energy (lack of)**, Fat Metabolism (poor), Fatigue, Fibrosis, **Goiter, Hair (loss or thinning), Hashimoto's Disease (Thyroiditis), Hypothyroid**, Jet Lag, Menstrual Irregularity, Mental Illness, **Sex Drive (low)**, Skin (dry and/or flaky), Skin Care (general), Weight Loss (aids for)

Usage: This product contains nutrients necessary to help the thyroid gland produce thyroxine, the major thyroid hormone. It is the strongest of the thyroid formulas because it contains thyroid glandular. Take 1 capsule twice daily with food for low thyroid.

Warnings: Not recommended for use with hyperthyroid (Grave's disease). This formula will not work if the thyroid gland is missing or is destroyed. Not recommended for long-term use, especially in high doses. Use Thyroid Activator or TS II for long-term thyroid supplementation.

Thyroid Support - Capsule (60)

Available in: US - Stock #1228-6

Ingredients: L-tyrosine, kelp, zinc, copper citrate, B6 (pyridoxal-5-phosphate), protease, stinging nettle, manganese, thyroid glandular substance, brain substance (anterior pituitary and hypothalamus)

Tiao He Cleanse

Product Type: Pack

Properties: Alterative (Blood Purifier), **Detoxifying**, Hepatic, Laxative (general), Laxative (stimulant)

Systems Affected: Digestive System, Immune System, **Intestinal System**, Liver, **Weight Loss**

Conditions: **Chemical Poisoning**, Constipation (adults), Halitosis (Bad Breath), Hypothyroid, Migraine, **Pap Smear (abnormal)**, **Toxemia**

Usage: Tiao He Cleanse is a general cleansing and detoxification program. It stimulates bowel elimination, but more importantly helps to cleanse the liver and the tissues. Take the contents of one packet (6 capsules) 15 minutes before meals up to three times daily for 10 days. Drink one glass (8 ounces) of Nature's Spring water with the capsules, followed by another glass of water. This should produce two or three bowel movements daily. If stools become too loose, reduce the number of packets being used. It can be helpful to add fiber to this cleanse. Take one teaspoon of Nature's Three or LOCLO in a large glass of water or juice in the morning. Be sure to drink plenty of water when cleansing.

Warnings: Cleansing products like this should be avoided in cases of irritable or inflammatory bowel conditions. Do not use if diarrhea, loose stools, or abdominal pain are present or develop. Not for prolonged use.

Tiao He Cleanse (15 day) - Packets (30)

Available in: US - Stock #4092-2

Ingredients: Chinese Liver Balance, All Cell Detox, LBS II, Burdock, Black Walnut, Psyllium Hulls.

Tiao He Pack - Packet

Available in: Canada - Stock #4005-7

Ingredients: LIV-C, SP #1, LBS II, Psyllium Hulls, Red Clover, Black Walnut

TNT

Product Type: Formula

Properties: Nutritive

Systems Affected: Whole Body

Usage: TNT stands for Total Nutrition Today. The product is a nutrient/fiber blend for general nutritional supplementation.

Warnings: No known warnings.

TNT Nutritional Drink - Bulk Powder

Available in: US & Canada - Stock #4300-5

Ingredients: Vitamin A (palmitate, beta-carotene) 4400 IU, calcium (tricalcium phosphate) 120 mg, vitamin D3 (cholecalciferol) 250 IU, vitamin B1 (thiamine mononitrate) 0.75 mg, vitamin B2 (riboflavin) 0.9 mg, vitamin B6 (pyridoxine hydrochloride) 1 mg, vitamin B12 (cyanocobalamin) 3 mcg, pantothenic acid (d-calcium pantothenate) 5 mg, zinc (citrate) 2 mg,copper (gluconate) 0.25 mg, chromium (chloride) 16.5 mcg, vitamin C (ascorbic acid) 30 mg, iron (ferrous fumarate) 2.5 mg, vitamin E (d-alpha tocopherol acetate) 12.5 IU, niacinamide 10 mg, folic acid 0.205 mg, biotin 150 mcg, phosphorus (tricalcium phosphate) 75 mg, magnesium (oxide) 45 mg, manganese (citrate) 0.2 mg, molybdenum (citrate) 9.5 mcg, selenium (HVP* chelate) 8.5 mcg, sodium (sodium chloride) 40 mg, iodine (potassium iodide) 22 mcg, potassium (potassium chloride & potassium citrate) 135 mg, fibre blend (gum arabic, soy fibre, oat fibre, cellulose, sugar beet fibre, wheat bran, oat bran flour, rice bran, corn bran, pea fibre, guar gum, citrus pectin, tomato powder, broccoli powder, cabbage powder, carrot powder, spinach powder, celery powder), crystalline fructose, high oleic sunflower oil blend, corn syrup solids, natural flavours, maltodextrin, citric acid, dipotassium phosphate, and montmorillonite

TNT Bar - Bar

Available in: Canada - Stock #8236-2

Ingredients: Protein Blend (soy protein isolate, whey protein concentrate, calcium caseinate, milk protein isolate), glucose-fructose, yogurt-flavoured coating (sucrose, modified palm kernel and palm oil, yogurt powder [cultured whey protein concentrate, cultured skim milk and yogurt culture], skim milk powder, soy lecithin and natural flavours), maltodextrin, honey, sorbitol, vitamin/mineral premix, soy lecithin, soybean oil, fructo-oligosaccharides, salt, citric acid, malic acid, natural flavours

Tobacco Detox

Product Type: Homeopathic

Properties: Detoxifying

Systems Affected: Lungs, Nerves, Respiratory System

Conditions: Addictions (tobacco smoking or chewing)

Usage: To help reduce the craving for tobacco products, dissolve one tablet or take 10-15 drops every 1-2 hours or as needed.

Warnings: NSP does not recommend this product for children under the age of 12 except on the advice of a health professional. Other than the alcohol content, which is not good for infants, we consider this product safe for children.

Tobacco Detox (1 fl. oz.) - Homeopathic (Liquid)

Available in: US - Stock #8712-5

Ingredients: Calcarea Phosphorica 6x, Cinchona officinalis 6x, Lobelia inflata 10x, Tabacum (Tobacco) 10x, Ignatia amara 12x, Posphorus 12x, Tabacum (Tobacco) 20x, purified water, alcohol

Tobacco Detox Tablet - Homeopathic (Tablet)

Available in: US - Stock #8721-6

Ingredients: Same as Stock #8712-5

Tofu Moo

Product Type: Food

Properties: Food, Nutritive

Systems Affected: Whole Body

Conditions: Allergies (food), Children's Remedy, Failure to Thrive

Usage: An alternative to diary products. For 2 quarts: Add 8 scoops of powder to 2 cups warm water. Stir briskly until dissolved, then add enough water to make 2 quarts. Can be served hot or cold. For single serving: Add 1 scoop of powder to 6 ounces of hot or cold water. Stir briskly until dissolved, then add enough water to make 8 ounces.

Warnings: Avoid with allergies to soy.

Tofu Moo (Natural) (22.5 oz) - Bulk Powder

Available in: US - Stock #1703-0

Ingredients: Maltodextrin, tofu powder, fructose, titanium dioxide (color), calcium carbonate, acacia gum, di-potassium phosphate, sea salt, natural flavor, lecithin, carrageenan, cellulose gum, and xanthan gum.

Tofu Moo (Carob) (25.9 oz) - Bulk Powder

Available in: US - Stock #3207-9

Ingredients: Fructose, sunflower oil (non hydrogenated), corn syrup solids, tofu powder, calcium carbonate, carob powder, natural flavors, natural carmel color, salt, xanthan gum, natural beet color, vitamin A palmitate, vitamin D3

Trace Mineral Maintenance

Product Type: Formula

Properties: Mineralizer

Systems Affected: Liver, Structural System, Whole Body

Conditions: Aging (prevention), Lou Gehrig's Disease, Multiple Sclerosis (MS)

Usage: Take 2 tablets with each meal as a mineral supplement. Must have hydrochloric acid to assimilate. Persons with weak digestion should use Colloidal Minerals.

Warnings: No known warnings.

Trace Mineral Maintenance - Capsule (450)

Available in: US - Stock #1672-1

Ingredients: Contains Montmorillonite, a clay from an ancient sea bed which has randomly-occurring amounts of about 70 trace minerals.

Trigger Immune

Other Names: IMM-C

Product Type: Formula

Properties: Adaptagen, **Blood Building, Corrects Polarity, Digestive Tonic,** Glandular, **Immune Stimulant,** Nervine, **Spleen Chi Tonic, Tonic**

Systems Affected: Cortisol, Gall Bladder, **Immune System,** Liver, Lungs, Nerves, **Reproductive Glands,** Spleen, **Thymus, Whole Body**

Conditions: Aging (prevention), **Anorexia,** Backache (Back Pain, Lumbago), **Cancer (natural therapy for),** Chills, Cold Hands and Feet, **Colds (prevention),** Contagious Diseases, **Convalescence, Debility,** Depression, Diphtheria, Dizziness (Vertigo), **Electromagnetic Pollution,** Emotional Sensitivity, Endurance (lack of), **Energy (lack of), Epstein Barr Virus (Chronic Fatigue Syndrome, CFS),** Erectile Dysfunction, **Fatigue,** Hair (loss or thinning), Memory and Brain Function, Menstruation (scant), **Perspiration (excessive), Reversed Polarity, Wasting,** Weight Gain (aids for)

Usage: This formula is for a general "run-down" condition of the entire body. It combats weak immunity, fatigue and reversed polarity in muscle testing. It's chief ingredient, astragalus, is a powerful tonic with adaptagenic qualities. Anciently, it was referred to as "the superior tonic." Along with ginseng (second ingredient in the formula), this herb is considered to be one of the most important herbs in Chinese herbology. Modern research has shown it to have immune stimulant qualities like echinacea, and that it helps recovery from chemotherapy. For recovery from long-term fatigue, illness and general weakness, use 3 capsules three times daily.

Warnings: This formula was not designed for acute illnesses, high fever or severe inflammation.

Trigger Immune - Capsule (100)

Available in: US - Stock #1889-2

Ingredients: Astragalus, panax ginseng, dang gui, epimedium leaf, eucommia bark, ganoderma, lycium fruit, rehmannia, achyranthes, atractylodes rhizome, citrus peel, hoelen, ligustrum fruit, ophiopogon, peony, polygala, schizandra fruit, licorice

Trigger Immune TCM Conc. - Capsule (30)

Available in: US - Stock #1034-0

Ingredients: Same as Stock #1889-2

IMM-C - Capsule (100)

Available in: Canada - Stock #1892-6

Ingredients: Astragalus root, Panax ginseng root, epimedium leaf, eucommia bark, ganoderma plant, lycium fruit, rehmannia root, dang gui root, achyranthes root, atractylodes rhizome, citrus peel, hoelen plant, ligustrum fruit, ophiopogon root, ho shou wu root, polygala root, schizandra fruit, and licorice root

Triple Effect Age Relief

Product Type: Topical

Properties: Emollient, Tonic

Systems Affected: Skin

Conditions: **Age Spots,** Aging (prevention), Wrinkles

Usage: A topical cream designed to support youthful-looking skin. It helps counteract the effects of aging on the skin.

Warnings: No known warnings.

Triple Effect Age Relief Cream (1 fl. oz. tube) - Topical Preparation

Available in: US - Stock #6013-1

Ingredients: Water, acetyl hexapeptide-3, palmitoyl oligopeptide, palmitoyl tetrapeptide-3, glycerin, bytylene glycol, carbomer, polysorbate 20, glyceryl stearate, cetearyl alcohol, sodium stearoyl lactylate, glucosamine HCI, algae extract, yeast extract, urea, saccharomyces/xylinum black tea ferment, hydroxyethylcellulose, cyclopentasiloxane, hydrolyzed lupine protein, artemisia priceps leaf, triisononanoin, ethylene/acrylic acid copolymer, dimethicone, hydrolyzed hibiscus esculentus extrac, vitamin C, vitamin E, mango, shea butter, sandalwood extract, phellodendron amurense bark extract, barley extract, camelliasinensis leaf extract, ginkgo biloba leaf extract, squalane, fragrance (essential oils), xanthan gum, stearyl glycyrrhetinate, dextrin, isohexadecane, cetyl alcohol, dimethicone/vinyl dimethicone crosspolymer, tetrasodium EDTA, phenoxyethanol, methylparaben, ethylparaben, propylparaben, butylparaben

Triple Relief

Product Type: Formula

Properties: **Analgesic (Anodyne)**, Anti-arthritic, Anti-inflammatory, Anticephalalgic

Systems Affected: Central Nerves, Nerves, **Prostaglandins**, Structural System

Conditions: Backache (Back Pain, Lumbago), Bunions, Burns and Scalds, Disks (spinal - bulging or slipped), Dislocation, Dysmenorrhea, Headache (general), **Pain (general remedies for)**

Usage: Nexrutine is an extract from a plant. It has Cox-2 inhibiting properties, which gives it a similar action to non-steroidal anti-inflammatory drugs (NAISDs), except that the drugs also inhibit the Cox-1 enzyme giving them more side-effects. Willow bark contains salicin, a natural form of aspirin, while boswellia has been used for pain in Ayurvedic medicine. Use 2 capsules as needed up to three times daily for relief from minor pain.

Warnings: No known warnings. The long-term effects of Nexrutine use are unknown.

Triple Relief - Capsule (90)

Available in: US - Stock #1851-3

Ingredients: Nexrutine™ extract from Phellodendron amurense, Boswellia extract (Boswellia serrata) standardized to 25% total boswellic acid, Willow Bark extract (Salix spp.) standardized to 15% total salicins

Triple Relief Trial Pack - Packets (20)

Available in: US - Stock #2786-3

Ingredients: Same as Stock #1851-3

TS II

Product Type: Formula

Properties: Glandular, Mineralizer, Nervine, Thyrotropic

Systems Affected: Nerves, **Thyroid**, Weight Loss

Conditions: Attention Deficit Disorder (ADD, ADHD), Blood Pressure (low), Cold Hands and Feet, Depression, Energy (lack of), Epilepsy, Fatigue, Glands (swollen lymph), Goiter, **Grave's Disease**, Hair (loss or thinning), Hashimoto's Disease (Thyroiditis), Hypothyroid, Weight Loss (aids for)

Usage: To aid thyroid function take 2 capsules one to three times daily. The hops in this formula act to calm the nerves, since many individuals with thyroid problems have nervous stress as well.

Warnings: Contraindicated in cases of overactive thyroid and possibly with high blood pressure.

TS II (with Hops) - Capsule (100)

Available in: US - Stock #1092-0

Ingredients: Kelp, Irish moss, parsley, hops, capsicum fruit.

TS II - Capsule (100)

Available in: US & Canada - Stock #1092-0

Ingredients: Same as Stock #1092-0

UC-C

See *Spleen Activator*

UC3-J

See *Intestinal Soothe & Build*

Ultimate Echinacea

Product Type: Formula

Properties: Acrid, Alterative (Blood Purifier), **Antibacterial**, Antiseptic, **Antitoxic**, **Antivenomous**, **Antiviral**, Detoxifying, Febrifuge, **Immune Stimulant**, Lymphatic, Tonic, Vulnerary

Systems Affected: Capillaries, Ears, **Immune System**, Liver, **Lymph Nodes**, **Lymphatic System**, Peyer's Patches, Prostate, Respiratory System, Skin, Spleen, **Thymus**, **Tonsils**

Conditions: Acne (Pimples, Blackheads), Adenitis, Antibiotics (alternatives to), Bites and Stings, Bladder Infection, **Blood Poisoning**, **Boils**, Bronchitis, Cancer (natural therapy for), Canker Sores (Mouth Ulcers), Carbuncles, Chemotherapy (reducing side effects), Children's Remedy, Colds (antiviral), **Colds (general remedies for)**, **Colds (prevention)**, Congestion (general), **Congestion (lymphatic)**, **Contagious Diseases**, Croup, **Diphtheria**, Ear Infection or Earache, Ear Infection or Earache, Eczema, Emphysema, Epstein Barr Virus (Chronic Fatigue Syndrome, CFS), Fever, Gangrene, Gingivitis (Bleeding Gums, Gum Disease, Pyorrhea), Glands (swollen lymph), Gonorrhea, **Infection (bacterial)**, **Infection (viral)**, Lyme Disease, Lymph Nodes or Glands (swollen), Malaria, Mastitis, **Meningitis**, Mercury Poisoning, **Mononucleosis**, **Mumps**, Pets (supplements for), **Poisoning (food)**, Sinus Infection, Snake Bite, **Sore Throat**, **Staph Infections**, Strep Throat, **Syphilis**, **Tonsillitis (Adenoids)**, Typhoid, Urinary Tract Infections, Wounds and Sores

Usage: Echinacea was the most popular herbal remedy of the Eclectic physicians, who practiced in the late 1800s and early 1900s. It has the ability to stimulate the immune system to fight infection and can be very effective against very serious infections, including strep throat, tonsillitis and even gangrene. It is also an effective remedy both topically and internally for snakebites, bee stings and insect bites.

For maximum effectiveness, echinacea needs to be taken in large, frequently repeated doses. A low dose for adults is 15-20 drops (1 ml), but most herbalists suggest 45-90 drops (3-5 ml) three times daily. When fighting infection take every two hours. A good dose for children is about 1/3-1/2 the adult dose.

Warnings: Do not use with auto-immune disorders where the immune system is over active (i.e. M.S. Lupus, Hodgkins, etc.). In excessive amounts, can cause excessive salivation and scratchy, tingling sensation in throat. No known toxic effects, but some herbalists feel that because it stimulates the immune response it is best to use it for short periods, then take a break and allow the immune system to rest before resuming ingestion. General recommendation is two weeks on, two weeks off. This may also be wise because the body tends to adapt to remedies taken over a long period of time so they become less effective.

Ultimate Echinacea (2 fl. oz.) - Liquid (Glycerite)

Available in: US & Canada - Stock #3181-2

Ingredients: Echinacea purpurea, Echinacea angustifolia, Echinacea pallida, glycerin.

Uña de Gato Combination

Other Names: Cat's Claw

Product Type: Formula

Properties: Anti-inflammatory, Antibacterial, Anticancer, Antiherpetic, **Antiviral**, Detoxifying, Immune Stimulant, Tonic, Vulneraries (for intestinal system)

Systems Affected: Female Reproductive, Immune System, Liver, Respiratory System, Structural System

Conditions: Acquired Immune Deficiency Syndrome (AIDS/HIV), Allergies (food), Arthritis, Bursitis, Cancer (natural therapy for), Cartilage Damage, Chemotherapy (reducing side effects), Crohn's Disease, Diabetes, Diverticulitis (Diverticuli), Epstein Barr Virus (Chronic Fatigue Syndrome, CFS), Herpes, **Leaky Gut Syndrome**, Pregnancy (herbs and supplements to avoid during), Radiation Sickness, Reye's Syndrome

Usage: Modern research has verified a number of herbs that have the capacity to enhance the function of the immune system. Una de Gato Combination contains three of these herbs with proven immune-enhancing benefits.

Una de Gato Combination has several potential uses. First, it helps maintain intestinal health and may be useful for Crohn's disease, inflammatory bowel disorders, gastritis, irritable bowel syndrome and leaky gut syndrome.

All three of the herbs in this blend have strong anti-viral action and may be helpful for herpes, chronic fatigue syndrome or other chronic viral infections. Una de Gato Combination may also be beneficial for other immune-related disorders such as cancer and AIDS. Another benefit of Una de Gato Combination is its anti-inflammatory properties. This makes it potentially useful for chronic inflammatory disease such as diverticulitis, bursitis or arthritis.

Warnings: Because of its immune stimulating properties, this formula may cause problems for people with autoimmune disorders. Because cat's claw has contraceptive effects, this blend should probably be avoided by women who are pregnant or who are trying to conceive.

Uña de Gato (Cat's Claw) - Capsule (100)

Available in: US - Stock #175-0
Ingredients: Uña de Gato, astragalus, echinacea.

Cat's Claw Combination - Capsule (100)

Available in: US & Canada - Stock #175-0
Ingredients: Same as Stock #175-0

Urinary Maintenance

Other Names: URY

Product Type: Formula

Properties: Diuretic, Lithotriptic

Systems Affected: Bladder (Urinary)

Conditions: Bedwetting, Bladder (irritable), Bladder Infection, Cystitis, Edema (Dropsy, Water Retention, Swelling), Incontinence (urinary), Urination (burning or painful), Urine (scant)

Usage: This formula has been improved over the old URY and is one of the gentlest diuretics in the NSP product line. It is the only formula that is not contraindicated with kidney inflammation. It strengthens the kidney's ability to flush toxins as well as increasing the flow of urine. It is a good choice for long term use as a general kidney tonic. Use 2 capsules three times daily with plenty of pure water during kidney or bladder infections, or as a diuretic. As a general aid to strengthen the kidneys, use 1 capsule three times a day with meals.

Warnings: No known warnings.

Urinary Maintenance - Capsule (120)

Available in: US - Stock #2884-4
Ingredients: Magnesium, potassium, asparagus, dandelion, parsley, cornsilk, watermelon seed, dong quai, horsetail, hydrangea, uva ursi, eleuthero root, schizandra

Uva Ursi

Latin Name: *Arctostaphylos uva ursa*

Product Type: Single Herb

Properties: Antiseptic, **Astringent**, Diuretic

Systems Affected: **Bladder (Urinary)**, Mucus Membranes, Pancreas, Reproductive Glands, Urinary System, Uro-genital Tract, Uterus

Conditions: Bedwetting, Bladder (irritable), **Bladder (ulcerated)**, **Bladder Infection**, Bleeding (external), Cuts, Cystitis, Diabetes, Edema (Dropsy, Water Retention, Swelling), Gonorrhea, Gout, **Incontinence (urinary)**, **Kidney Infection**, Kidney Stones, Nephritis, Pancreatitis, **Poison Ivy or Oak**, **Prostatitis**, Urethritis, **Urinary Tract Infections**, Urination (burning or painful), Urination (frequent), Urine (scant)

Usage: A reliable diuretic with strong disinfectant and infection fighting properties, making it useful for kidney and bladder infections, irritated female organs and other uro-genital problems. Works well with cranberry juice. Contains a substance that disinfects the urinary tract. Take 1-2 capsules one to two times daily between meals with large glass of water, or 1 cup of tea each day. Works better as a tea than as capsules. Use 2-3 capsules per cup. Add peppermint or chamomile to improve taste.

Warnings: Not for use in cases involving fluid deficiency, wasting or dryness. Not recommended for long term use because of strong astringency. Prolonged use may irritate the stomach and cause constipation. Since animal studies show that uva ursi stimulates the uterus to contract, pregnant women may wish to avoid it, although we don't believe it would do any harm in a low doses in case of a bladder infection. The herb contains a large amount of tannins, which could cause stomach upset in large quantities. As a result, uva ursi should not be given to children under the age of two. Begin with smaller amounts of uva ursi and increase if necessary. You may notice that the herb often turns urine a dark green. Do not become alarmed; this is normal.

Uva Ursi - Capsule (100)

Available in: US & Canada - Stock #710-9
Ingredients: Arctostaphylos uva ursa (uva ursi) leaf

V-X

Product Type: Formula

Properties: Uterine

Systems Affected: Female Reproductive, Liver, Rectum, Uterus

Conditions: **Fibroids (uterine)**, Menstruation (scant)

Usage: V-X is a combination of herbs that draw toxins out of the body and encourage tissue repair and healing. Its original purpose was to be used as a vaginal suppository or rectal bolus for problems such as uterine fibroids, vaginal irritation, cysts, tumors and hemorrhoids. So, it was sold in bulk.

V-X is made into suppositories by blending it with enough melted cocoa butter to make a paste the consistency of pie dough. These are rolled into pencil-like shapes approximately the size of the middle finger and about one inch long. They are then stored in the refrigerator until ready for use.

When inserted into the vagina or rectum, just prior to retiring for bed, the cocoa butter melts, leaving the herbs in contact with tissues to absorb toxins and promote healing. It is a good idea to use a sanitary napkin to help hold the suppository in place. The herbal powders can be washed out of the vagina with a douche the following morning. (This is not necessary with rectal insertion as the powders will simply pass with the next bowel movement.)

If heavy bleeding is a problem, the douche can be made using a decoction of an astringent herb (such as white oak bark or calendula) and retained for 5-10 minutes to tone up tissues. Bleeding can also be stopped by adding one ounce of colloidal minerals to each pint of water. The retention douche is performed on a slant board or with some pillows propped up under the hips. The fluid is taken into the vaginal area and retained for 5-10 minutes, then expelled.

This procedure can cause fibroids to start passing and bleeding problems to ease up in as little as a week. Unfortunately, although the procedure is very effective, many people simply will not take the time to make the suppositories or do the douching, so the formula was eventually encapsulated for internal use.

Internally, V-X soothes the irritated tissues of the digestive tract, reducing inflammation and absorbing toxins in a similar manner to Intestinal Soothe and Build. It also helps supply iron to build the blood and improves the flow of lymph in the body. It helps eliminate toxins and can be used for diarrhea, reducing cholesterol and easing dry, irritated cough.

Even though it is more effective as a suppository or douche, herbalists have found that taken internally it still helps to eliminate uterine fibroids. The process is simply slower. When taken internally, V-X should be used in large doses, 3-4 capsules three times daily. Where heavy bleeding is a problem, yarrow and Menstrual Reg should be taken along with V-X.

Another excellent way to use V-X is as a poultice. Mix the powders with enough water to form a paste and apply topically to wounds, sores, insect bites and other injuries to reduce swelling and inflammation, ease pain, and promote rapid healing.

Warnings: No known warnings.

V-X - Capsule (100)

Available in: US - Stock #1382-4

Ingredients: Plantain herb, squaw vine herb, golden seal root, yellow dock root, marshmallow root, chickweed herb, mullein leaves, slippery elm bark

Vaccine Detox

Product Type: Homeopathic

Properties: Detoxifying

Systems Affected: Immune System

Conditions: Fever, **Vaccines (detoxification from)**

Usage: Used for the relief of vaccination side effects, including restlessness, fever and headaches and to help the body detoxify from vaccinations. Adults and children 12 and older: Take 15-20 drops under the tongue every one-two hours until symptoms are relieved. Children and infants over 4 months: Take three-five drops under the tongue every 15-20 minutes until symptoms decrease, then every two-four hours until they are relieved. For detoxifying from vaccines take 15-20 drops three times daily along with Heavy Metal Detox and/or Enviro-Detox.

Warnings: Administer to children under 4 months only on the advice of a health care professional.

Vaccination Detox (1 fl. oz.) - Homeopathic (Liquid)

Available in: US - Stock #8980-0

Ingredients: Active ingredients: Chamomilla (Chamomile) 3x, Atropa belladonna (Belladonna) 6x, Gelsemium sempervirens (Yellow Jessamine) 6x, Thuja occidentalis (Tree of Life) 6x, Chelidonium majus (Celandine) 6x, Silicea (Silica) 12x. Other ingredients: Purified water, glycerin, citric acid, potassium benzoate, and alcohol.

Valerian Root

Latin Name: *Valeriana officinalis*

Product Type: Single Herb

Properties: Analgesic (Anodyne), Antispasmodic, Aromatic, Bitter, CNS Depressant, **Hypnotic**, Hypotensive, **Nervine**, Parasympatholytic (Anticholinergic), **Relaxant**, **Sedative**, **Sympatholytic**, Tranquilizer

Systems Affected: Adrenal Medulla, **Central Nerves**, Circulation, GABA, Heart, Muscles, Nerves, Parathyroid

Conditions: Addictions (alcohol), Addictions (drugs), Addictions (tobacco smoking or chewing), Afterbirth Pain, Anxiety (Panic Attack), Arthritis, Children's Remedy, Colds (general remedies for), Colic (children), Colon (spastic), Cramps (leg), Cramps and Spasms (general), Headache (tension), Heart Fibrillation or Palpitations, Hemorrhoids, Insomnia, Myasthenia Gravis, Neuralgia and Neuritis, Neurosis, Pain (general remedies for), Parkinson's Disease, Stress, Tooth Extraction, Twitching

Usage: This is a potent nervine with strong tranquilizing effects on the central nervous system but does not depress the nerves. It slows the heart and increases its force. It has been used to treat a wide variety of nervous system conditions. It is initially stimulating in its effects, then relaxing. Use 1-2 capsule one to three times daily with meals or 1 capsule every hour for pain. Time release: take 1-2 tablets approximately one hour before bedtime.

Warnings: Not recommended for persons with "hot" disorders, i.e., high strung, nervous and excitable. (Skullcap or passion flower are better for such individuals.) Not recommended for long-term use in large doses, although there is no risk of dependence or addiction. Does not generally cause drowsiness that could affect driving. Possible side effects from high doses are vaso-dilative headaches, nausea and lessened alertness. In persons with a low thyroid or in hyperactive children, valerian sometimes acts as a stimulant rather than a sedative.

Valerian Root - Capsule (100)

Available in: US - Stock #720-0

Ingredients: Valeriana officinalis (valerian) root

Valerian Root Extract T/R - Capsule (60)

Available in: US - Stock #721-1

Ingredients: Standardized extract of Valeriana officinalis (valerian) root (0.8% valerenic acid)

Vari-Gone

Product Type: Formula

Properties: Anti-inflammatory, **Antithrombolytic**, Tonic, Vascular Tonics

Systems Affected: **Blood Vessels**, **Capillaries**, Gall Bladder, Legs, Parathyroid, Peripheral Blood Vessels, Rectum, **Veins**

Conditions: Bronchitis, Bruises (healing), Bruises (prevention), Capillary Weakness, Circulation (poor), Dysmenorrhea, **Hemorrhoids**, **Phlebitis**, Spider Veins, Thrombosis, **Varicose Veins**

Usage: Helps tone veins, and improve circulation in the legs. Use 1-2 capsules two times daily with meals. Vari-Gone cream can be applied topically to affected areas.

Warnings: No known warnings.

Vari-Gone - Capsule (90)

Available in: US - Stock #999-9

Ingredients: Horse chestnut extract, butcher's broom extract, rutin, hesperidin, lemon bioflavinoids, ascorbic acid

Vari-Gone - Capsule (90)

Available in: Canada - Stock #985-3

Ingredients: Butcher's broom root powder and extract (8.5% ruscogenins), horse chestnut seed extract (20% aescin), fenugreek seed, hesperidin, lemon bioflavonoid, and rutin

Vari-Gone Cream (2 oz. tube) - Topical Preparation

Available in: US - Stock #4947-5

Ingredients: Water, Carthamus tinctorius (safflower) seed oil, isopropyl palmitate, glycerin, glyceryl stearate, cetyl alcohol, stearic acid, Melilotus officinalis extract, Ruscus aculeatus root extract, Aesculus hippocastanum (horse chestnut) seed extract, Aloe barbadensis leaf juice, menthol, dimethicone, inulin lauryl carbamate, aminomethyl propanol, carbomer, disodium EDTA, caprylyl glycol, hexylene glycol, and phenoxyethanol.

Vegetable Seasoning Broth

Product Type: Food

Properties: Condiment

Systems Affected: Whole Body

Usage: A natural, salt-free seasoning for food.

Warnings: No known warnings.

Vegetable Seasoning Broth (9 oz.) - Bulk Powder

Available in: US - Stock #1393-5

Ingredients: Same as Stock #1835-1

Viral Recovery

Product Type: Homeopathic

Properties: Antiviral

Systems Affected: Immune System

Conditions: Epstein Barr Virus (Chronic Fatigue Syndrome, CFS), Infection (viral), Warts

Usage: Assists the body in detoxifying and regaining strength and vitality following viral infections. Take 10-15 drops under the tongue two-six times daily, depending upon the severity of the symptoms.

Warnings: NSP does not recommend this product for children under the age of 12 except on the advice of a health professional. Other than the alcohol content, which is not good for infants, we consider this product safe for children.

Viral Recovery (1 fl. oz.) - Homeopathic (Liquid)

Available in: US - Stock #8850-8

Ingredients: Active ingredients: Alfalfa (Alfalfa) 4x, Arnica montana (Mountain Arnica) 4x, Gelsemium sempervirens (Yellow Jessamine) 4x, Asclepias tuberosa (Butterfly Weed) 6x, Pulsatilla (Wind Flower) 6x, Calcarea Carbonica (Calcium Carbonate) 8x, Selenium Metallicum (Selenium) 10x, Sulphur (Sulfur) 10x, Epstein-Barr Virus 12x, Cytomegalovirus 12x, Herpes Simplex and Herpes Zoster 12x, Coxsackie Virus 12x, Morbillinum (Measles) 12x, Parotitis (Mumps) 12x, Hepatitis 12x, Infectious Mononucleosis 12x. Other ingredients: Purified water and 20% USP alcohol.

Vita Lemon

Product Type: Formula

Properties: Nutritive

Systems Affected: Whole Body

Conditions: Energy (lack of)

Usage: Mix 1 scoop in 8 ounces of water. Stir briskly or blend in electric blender. In addition to your diet program, before eating in the morning, add 1/2 scoop of powder to 5 ounces of hot Nature's Spring water. Stir briskly or blend in electric blender. Drink immediately.

Warnings: No known warnings.

Vita Lemon (25 oz.) - Bulk Powder

Available in: US - Stock #2932-7

Ingredients: Crystalline fructose, dehydrated lemon juice, maltodextrin, citric acid, soy fiber, guar gum, microcrystralline cellulose, molasses, sodium chloride (salt), citrus pectin, xanthan gum, potassium citrate, natural yellow color, l-carnitine hydrochloride (25 mg), vitamin C (ascorbic acid), viatmin A (palmitante), niacin (niacinamide), pantothenic acid (d-calcium pantothenate), vitamin D, vitamin B6 (pyridoxine hydr9ochloride), vitamin B2 (riboflavin), vitamin B1 (thiamine mononitrate), vitamin B12 (cyanocobalamin), folic acid, biotin.

Vitamin A & D

Product Type: Nutrient

Properties: Antibacterial, Antioxidant, Nutritive, Nutritive

Systems Affected: Liver, Structural System, Thymus

Conditions: Acne (Pimples, Blackheads), Age Spots, Bronchitis, Colds (general remedies for), Conjunctivitis (Pink Eye), Cystic Breast Disease, Cysts, Disks (spinal - bulging or slipped), Eczema, Emphysema, Endometriosis, Eye Infections, Eye Problems (general), Floaters, Flu, Free Radical Damage, Glaucoma, Leucorrhea, Lyme Disease, Macular Degeneration, Poison Ivy or Oak, Polyps, Psoriasis, Radiation Sickness, Rashes and Hives, Rosacea, Seborrhea, Skin (infections), Skin Care (general), Sprains, Staph Infections, Stye, Wrinkles

Usage: Take 1 capsule daily. Increase during the cold/flu season, when exposed to inner-city pollution, when traveling abroad or during any cleanse (especially a liver cleanse). Up to a 3 month supply of vitamin A can be stored in the liver. Vitamin A&D can also be used externally. Open a capsule and apply it around the eyes for infection or macular degeneration. It can also be applied directly to areas afflicted with skin disorders.

Warnings: If headaches or nausea ensue from a mega-dose regime of Vitamins A & D simply cut back on the dosage until those symptoms abate (the liver is limited on the amount of Vitamins A & D it can process at one time and it is different for every individual). Large amounts of Vitamin A (over 25,000 daily) are contraindicated in liver disease and pregnancy. Massive doses (over 40,000 IU) are reported to be toxic, especially if taken for long periods of time. Antibiotics, laxatives, Fat Grabbers, and cholesterol-lowering drugs can interfere with vitamin A absorption.

Vitamin A & D (10,000/400 IU) - Softgel (100)

Available in: US - Stock #4065-3

Ingredients: Fish oils (10,000 IU Vitamin A and 400 IU Vitamin D)

Vitamin A & D - Softgel (100)

Available in: Canada - Stock #1724-2

Ingredients: Same as stock #4065-3

Vitamin B-12

Product Type: Formula

Properties: Nervine, Nutritive

Systems Affected: Blood, Brain, Ears, Nerves, **Red Blood Cells**, Taste Buds

Conditions: Acquired Immune Deficiency Syndrome (AIDS/HIV), Adrenals (exhaustion, weakness or burnout), Alzheimer's Disease, **Anemia**, Body Odor, Calcium Deficiency, Calcium Deposits (Calcification), Celiac Disease, Children's

Remedy, Cholesterol (low), Coordination, Cystic Fibrosis, Dehydration, Depression, Dermatitis, Diabetes, Dizziness (Vertigo), Eyes (spots before), Heart Fibrillation or Palpitations, **Hemochromatosis**, Memory and Brain Function, Nervousness, Neuralgia and Neuritis, **Pernicious Anemia**, Schizophrenia, Sore or Geographic Tongue, Tinnitus (Ringing in the Ears)

Usage: B-12 is essential for normal formation of red blood cells, metabolism and the nervous system. It acts as a cofactor or essential component in DNA synthesis. This makes B-12 important for the production of all cells in the body, the full development of red blood cells, normal myelination or covering of nerve cells and the production of neurotransmitters. B-12 promotes normal growth development and is responsible for binding with calcium for calcium uptake and utilization. It is not found in plant foods, so total vegetarians typically need supplementation and become anemic without this vitamin. Anyone who smokes, has digestive insufficiencies or is on an acid-blocking agent will need to supplement with this vitamin as well.

Warnings: No known toxicity, even in very large doses. Anti-gout drugs, anticoagulant drugs and potassium supplements may block absorption of vitamin B-12.

Vitamin B12 Complete, Liquid - Liquid

Available in: US - Stock #1588-7

Ingredients: B12 (1000 mcg), niacin, B6, B2, B1, distilled water, glycerin.

Liquid B12 - Liquid

Available in: Canada - Stock #1588-7

Ingredients: Same as Stock #1588-7

Vitamin B-6

Product Type: Nutrient

Properties: Diuretic, Nervine, Nutritive

Systems Affected: Brain, Female Reproductive, Liver, Mouth, Muscles, Nerves

Conditions: Acne (Pimples, Blackheads), Anemia, Anxiety (Panic Attack), Autism, Birth Control (countering side effects), **Breasts (swelling and tenderness)**, Burning Feet or Hands, Carpal Tunnel Syndrome, Cramps (leg), Depression, Diabetes, Edema (Dropsy, Water Retention, Swelling), Fatigue, Hemochromatosis, Morning Sickness, Motion Sickness, Nausea and Vomiting, PMS (general), PMS Type A, PMS Type C, PMS Type D, **PMS Type H**, Pregnancy (herbs and supplements for), Psoriasis, Sore or Geographic Tongue, Tooth Decay (prevention)

Usage: The primary role of vitamin B6 is as a cofactor for over 100 enzymes related to amino acid metabolism. It is involved in changing homocysteine to cysteine, the production of heme (a component of hemoglobin), the production of niacin from tryptophan, and the production of neurotransmitters, particularly serotonin. It also binds to steroid hormone receptors in cells, weakening their effects. B-6 is necessary for metabolism of nutrients. It aids in formation of antibodies and helps maintain balance of sodium and phosphorus in the body. It is also needed for the synthesis of RNA and DNA. Doses of 100 milligrams two or three times a day help relieve nerve compression injuries (like carpal tunnel syndrome), premenstrual syndrome (PMS), and some cases of depression and arthritis. It is often used to treat high homocysteine levels along with folic acid and vitamin B12. It is the only water-soluble vitamin that can be toxic. High doses over time may result in numbness and tingling in the extremities.

Warnings: Anti-depressants, oral contraceptives and estrogen may increase need.

B6, Vitamin (50 mg) (120) - Tablet (120)

Available in: US & Canada - Stock #1626-6

Ingredients: B-6, Calcium, phosphorus

Vitamin B-Complex

Product Type: Formula

Properties: Adrenal Tonic, Nervine, Nutritive

Systems Affected: Blood, Brain, Cuticle, DNA, Ears, Female Reproductive, Liver, Mouth, Muscles, Nerves, Pineal, Pituitary (anterior), Red Blood Cells, Spleen, Taste Buds, Thyroid

Conditions: Acne (Pimples, Blackheads), Addictions (alcohol), Alzheimer's Disease, Anemia, Anxiety (Panic Attack), Anxiety Disorders, Asthma, Autism, Bipolar Mood Disorder (Manic Depressive Disorder), **Birth Control (countering side effects)**, Body Odor, **Burning Feet or Hands**, Carpal Tunnel Syndrome, **Chemotherapy (reducing side effects)**, Cholesterol (low), Cold Sores (Fever Blisters), Coordination, Cramps (leg), Cushing's Disease, Cystic Fibrosis, Depression, Depression, Dizziness (Vertigo), Down Syndrome, Dyspepsia, Eczema, Edema (Dropsy, Water Retention, Swelling), Epilepsy, Eyes (spots before), Fatigue, Fingernail Biting, Grave's Disease, Gray Hair, Grief (excessive), Hair Care (general), Hangover, Heart Fibrillation or Palpitations, Inflammatory Bowel Disorders (Colitis, IBS), Insomnia, Lupus, **Memory and Brain Function**, Menopause, **Mental Illness**, **Morning Sickness**, **Muscular Dystrophy**, Myasthenia Gravis, Narcolepsy, Nausea and Vomiting, Nervous Exhaustion (Enervation), Nervousness, Neuralgia and Neuritis, **Neurosis**, Pancreatitis, Parkinson's Disease, Pernicious Anemia, Pregnancy (herbs and supplements for), Psoriasis, **Restless Leg Syndrome**, **Schizophrenia**, **Sore or Geographic Tongue**, **Stress**, **Tics**, Tinnitus (Ringing in the Ears), Tooth Decay (prevention), Vitiligo

Usage: B-complex vitamins are essential for the formation of red blood cells, metabolism, nervous system function, promotes normal growth and metabolism of nutrients and proteins. They aid in formation of antibodies and help maintain balance of sodium and phosphorus in the body. They are also needed for the synthesis of RNA and DNA, growth and division of cells, fetal development, especially neural tube development. B-complex vitamins are valuable for raising low cholesterol levels (below 170). They help with high estrogenic types of PMS. Individuals with low cardiac output or low zinc and potassium levels (sluggish metabolism) benefit from a higher amount of B-Complex. This group of vitamins also reduces blood levels of homocysteine, which is an amino acid that contributes to cardiovascular disease by damaging the endothelium (thin layer of cells that protect the artery walls.)

Deficiencies of B-vitamins in pregnant women may cause birth defects. NSP's B-Complex capsules provide "stress levels" of B-complex vitamins. Take 1 capsule with a meal, three times daily. Balanced B-Complex tablets are suitable for vegetarians both because of its high B12 level and because no animal by-products are used. Take 1 tablet daily with a meal.

Warnings: No known toxicity. Anti-gout drugs, anticoagulant drugs and potassium supplements may block absorption of vitamin B-12. Anti-depressants, oral contraceptives and estrogen may increase need. Avoid high doses in hormone-related cancer cases or convulsive disorders.

Balanced B-Complex - Tablet (120)

Available in: US - Stock #1625-4

Ingredients: Vitamin B1 (5 mg), vitamin B2 (6 mg), niacin (50 mg), vitamin B6 (9 mg), folic acid (400 mcg), vitamin B12 (50 mcg), biotin (100 mcg), panothenic acid (45 mg), calcium (120 mg), wheat germ, choline, inositol, PABA, cabbage, wild lettuce, watercress, rice polish, phosphorus

B-Complex - Capsule (100)

Available in: US - Stock #1778-9

Ingredients: Vitamin B1 (33 mg), vitamin B2 (33 mg), niacinamide (33 mg), vitamin B6 (33 mg), folic acid (133 mcg), vitamin B12 (33 mcg), biotin (100 mcg), panthothenic acid (33 mg), choline (33 mg), acerola, inositol, lemon bioflavinoids, PABA, rose hips, rutin, wheat germ

Vitamin B-Complex - Tablet (120)

Available in: Canada - Stock #1723-7

Ingredients: Vitamin B12 (50 mcg), folic acid (0.4 mg), biotin (0.1 mg), vitamin B1 (5 mg), vitamin B6 (9 mg), panothenic acid (45 mg), niacinaminde (50 mg), wheat germ, choline, inositol, para-aminobenzoic acid, rice polish, watercress, cabbage, lettuce

B Vitamin Complex, SynerPro - Tablet

Available in: Canada - Stock #1773-8

Ingredients: Vitamin B1, vitamin B2, vitamin B3, vitamin B6, folic acid, vitamin B12, biotin, pantothenic acid, SynerPro concentrate of broccoli, carrot, red beet, rosemary, tomato, turmeric, cabbage, grapefruit, orange bioflavonoids, hesperidin.

Vitamin C

Product Type: Nutrient

Properties: Acidifer, Adrenal Tonic, Anti-abortive, **Anti-allergenic**, Antibacterial, **Antihistamine**, Antioxidant, **Antiscorbutic**, Antiseptic, Hepatoprotective, **Mast Cell Stabilizer**, Nutritive, Vascular Tonics, Vulnerary

Systems Affected: Adrenal Glands, Brain, Circulation, **Collagen**, **Connective Tissue**, Immune System, Intestinal System, Liver, Mucus Membranes, Muscles, Peripheral Blood Vessels, Skin, Structural System, Veins

Conditions: Addictions (tobacco smoking or chewing), Adrenals (exhaustion, weakness or burnout), **Allergies (respiratory)**, Anemia, **Appendicitis**, Appetite (deficient), Backache (Back Pain, Lumbago), Bites and Stings, Bladder Infection, Bleeding (external), Bleeding (internal), Blood in Urine, Blood Poisoning, **Bruises (prevention)**, Burns and Scalds, Cancer (prevention), **Capillary Weakness**, **Cataracts**, Chemotherapy (reducing side effects), Cholesterol (high), Colds (general remedies for), Colds (prevention), Congestion (general), Cystic Breast Disease, Dermatitis, Fatigue, Floaters, Free Radical Damage, Gingivitis (Bleeding Gums, Gum Disease, Pyorrhea), Glaucoma, Infection (viral), Irritability, Jaundice (adults), Labor and Delivery, Ligaments (torn or injuried), Lou Gehrig's Disease, Lupus, Lyme Disease, Macular Degeneration, Meningitis, Mercury Poisoning, Myasthenia Gravis, **Nose Bleeds**, **Overalkalinity**, Parkinson's Disease, Poison Ivy or Oak, Poison Ivy or Oak, Poisoning (general), Post Partum Weakness, Pregnancy (herbs and supplements for), Prolapsed Colon, Psoriasis, **Rhinitis**, Schizophrenia, **Scurvy**, Skin Care (general), Snake Bite, **Spider Veins**, Staph Infections, Strokes, Sunburn, Surgery (healing from), **Teeth (loose)**, Toxemia, **Ulcerations (external)**, Vaccines (detoxification from), Varicose Veins, Warts, Wrinkles

Usage: Vitamin C is extremely important for tissue integrity (healthy gums, wound healing, etc.), adrenal function (stress, fatigue, etc.), the immune system (colds, aids interferon production, etc.), and much more. 250-1000 mg. daily or more as desired. RDA infants: 35 mg., RDA children: 45-50 mg., RDA adults: 60 mg., dietary amount: 60 mg., therapeutic range: 50-10,000 mg. Use with vitamin E to extend the antioxidant activity. Use the ascorbate forms for better absorption and tolerance. Vitamin C with Ascorbates (2,000 mg. per teaspoon), Chewable Vitamin C (250 mg. per tablet), Vitamin C with Bioflavonoids (500 mg. per tablet) and Vitamin C Timed Release (1000 mg. per tablet). Vitamin C may be taken to bowel tolerance, meaning you can increase the dose until it starts causing diarrhea. If diarrhea occurs with large doses of vitamin C reduce the dose until the diarrhea stops.

Warnings: Non-toxic, although massive doses can cause cankers and kidney problems in some people. Aspirin, analgesics, alcohol, anti-depressants, anticoagulants, steroids and oral contraceptives may reduce vitamin C levels in the body. Diabetic and sulfa drugs may not be effective when taken with vitamin C. In some people who are overacidic, vitamin C can cause mouth ulcers and related problems. If this occurs, reduce the dose or discontinue the vitamin C. Rose hips can be helpful for people who are intolerant of vitamin C.

C, Vitamin T/R (1000 mg) - Tablet (60)

Available in: US - Stock #1635-5

Ingredients: 1,000 mg. vitamin C, rose hips, acerola, rutin, lemon bioflavionids, hesperidin.

Vitamin C 1000 mg Time Release - Tablet (60)

Available in: Canada - Stock #1727-2

Ingredients: Same as Stock #1635-5

C, Vitamin T/R (1000 mg) - Tablet (180)

Available in: US - Stock #1636-0

Ingredients: Same as Stock #1635-5

C, Vitamin - Chewable (250 mg) - Tablet (120)

Available in: US - Stock #1633-8

Ingredients: 250 mg vitamin C, freeze-dried orange juice, rose hips.

Vitamin C - 250 mg Chewable - Tablet (120)

Available in: Canada - Stock #1726-1

Ingredients: 250 mg vitamin C, freeze dried orange juice, fructose

C, Vitamin - Citrus Bioflavinoids (500 mg) - Tablet (90)

Available in: US - Stock #1646-4

Ingredients: 500 mg vitamin C, lemon bioflavionids, orange bioflavionids, grapefruit bioflavionids, hesperidin, rutin, rose hips

Vitamin C 500 mg with Citrus Bioflavionids - Tablet (180)

Available in: Canada - Stock #1651-7

Ingredients: Same as Stock #1646-4

C, Vitamin Ascorbates (9 oz.) - Bulk Powder

Available in: US - Stock #1606-3

Ingredients: Vtiamin C, maltodextrin, acerola fruit extract (Malpighia glabra), xanthan gum, natural orange flavor, rutin and hesperidin (bioflavonoids), and beta carotene (contains soy)

Vitamin D3

Product Type: Nutrient

Properties: Anticancer, Nutritive

Systems Affected: Cardiovascular System, Immune System, Skeletal System

Conditions: Arthritis, Autoimmune Disorders, Broken Bones, Cancer (prevention), **Depression**, Diabetes, Gingivitis (Bleeding Gums, Gum Disease, Pyorrhea), Multiple Sclerosis (MS), **Osteoporosis**, Schizophrenia, Seasonal Affective Disorder, Tuberculosis (Consumption, Scrofula)

Usage: Do you get 10-15 minutes of exposure to sunlight everyday? If not, you may have a deficiency of vitamin D. Vitamin D is produced in your skin when you are exposed to the ultra violet (UV) frequencies found in natural sunlight. Unfortunately, many people get very little exposure to natural sunlight and, as a result, may wind up deficient in vitamin D.

Perhaps you work indoors. In the winter, many people in Northern climates drive to work in the dark, work all day under artificial lights and then drive home in the dark. Sunlight filtered through a window will not produce vitamin D. Cloud cover and smog reduce UV light and vitamin D production.

Vitamin D is primarily known for its role in promoting calcium absorption from the intestinal tract. It helps to maintain adequate levels of calcium and phosphorus in the blood to enable the mineralization of bone. Without sufficient vitamin D, bones become thin, brittle or misshapen. A severe deficiency produces the disease known as rickets in children and osteomalacia in adults. Vitamin D is needed to prevent osteoporosis in the elderly, too.

However, the benefits of vitamin D do not end with the role in plays in maintaining proper calcium and phosphorous levels for bone health. Vitamin D also affects the immune system. It promotes phagocytosis (anti-tumor activity) and helps modulate the immune system. Some evidence suggests it may have a role in protecting the body against cancer.

Insufficient levels of vitamin D may also be linked to an increased susceptibility to other chronic diseases. There is debate about whether vitamin D plays a protective role in preventing heart disease, but heart attacks are more frequent in winter and are lowest in summer in temperate climates. Also cholesterol levels were found to be lower in gardeners during the summer months.

One billion people in the world are currently Vitamin D deficient. Obese individuals often have lower levels of vitamin D and are at increased risk for deficiency. Cases of rickets are still occurring periodically in the United States, particularly among African American infants and children. This severe vitamin D deficiency usually occurs in cases of prolonged, exclusive breast feeding of dark-skinned infants whose mothers are vitamin D deficient. The use of sunscreen on children and the increase of indoor activities with limited exposure to sunlight is also casing vitamin D deficiencies in children.

NSP's Vitamin D3 contains 2,000 IU natural vitamin D3 derived from lanolin harvested from the wool fat of sheep from New Zealand and Australia. These animals are certified BSE-free. Take 1-2 tablets daily with a meal.

Warnings: Although there are no known warnings, we suggest you don't exceed the recommended dose of 1-2 tablets per day.

Vitamin D3 - Tablets (60)

Available in: US - Stock #1155-1

Ingredients: Vitamin D3, Dicalcium phosphate, cellulose (plant fiber), stearic acid, and magnesium stearate (vegetable).

Vitamin E

Product Type: Nutrient

Properties: Anticoagulant (Blood Thinner), Antioxidant, **Antithrombolytic, Cicatrisant**, Nutritive

Systems Affected: Capillaries, Circulation, Hair, Heart, Liver, Muscles, Red Blood Cells, Reproductive Glands, Skin

Conditions: Age Spots, Aging (prevention), Arteriosclerosis (Atherosclerosis, Hardening of the Arteries), Bites and Stings, **Blood Clots (prevention of)**, Breast Lumps, Breast Milk (increase or enrich), Breasts (swelling and tenderness), **Cardiac Arrest (Heart Attack)**, Cardiovascular Disease (Heart Disease), **Cataracts**, Chemotherapy (reducing side effects), Chicken Pox, Circulation (poor), **Corns, Cradle Cap**, Croup, Cuts, Cystic Breast Disease, Dandruff, Denture Sores, Dermatitis, Endometriosis, Erectile Dysfunction, Fibroids (uterine), Fibrosis, Free Radical Damage, Glaucoma, Hair Care (general), Hashimoto's Disease (Thyroiditis), Heart Fibrillation or Palpitations, Hemorrhoids, Infertility, Itching, Lou Gehrig's Disease, Menopause, Miscarriage (prevention), Muscle Tone (lack of), Muscular Dystrophy, Myasthenia Gravis, Narcolepsy, **Oxygen Deficiency**, Parkinson's Disease, **Phlebitis**, PMS (general), PMS Type H, Pregnancy (herbs and supplements for), Psoriasis, Restless Leg Syndrome, Rosacea, **Scars / Scar Tissue**, Sex Drive (low), Skin Care (general), **Stretch Marks**, Surgery (healing from), Surgery (preparation for), **Thrombosis**

Usage: Enables the body to utilize oxygen on a cellular level efficiently. It protects fat-soluble vitamins and red blood cells. Also inhibits coagulation to prevent blood clots. RDA infants: 4-6 IU, RDA children: 7-12 IU, RDA men: 15 IU, RDA women: 12 IU, dietary levels: 12-15 IU, therapeutic range: 100-1000 IU. 400 IU is a good maintenance dose. Works well in combination with: selenium, other antioxidant vitamins, hawthorn berries, ginkgo and capsicum. Zinc is needed to maintain levels of vitamin E.

Warnings: Essentially non-toxic, but avoid high doses with diabetes, rheumatic heart disease or overactive thyroid. Begin with low doses in cases of hypertension. Do not take iron at the same time as vitamin E.

E, Vit.-Complete w/Selenium (400 IU) - Softgel (60)

Available in: US - Stock #1509-8

Ingredients: 400 IU natural vitamin E (d-alpha, Beta, Delta, and Gamma tocopherol), 25 mcg selenium, glycerin.

E, Vit.-Complete w/Selenium (400 IU) - Softgel (200)

Available in: US - Stock #1508-5

Ingredients: Same as Stock #1509-8

E, Vitamin (100 IU) - Capsule (180)

Available in: US - Stock #1650-6

Ingredients: 100 IU natural vitamin E (d-alpha), cold-pressed soybean oil, glycerin.

Vitamin E with Selenium - Capsule (60)

Available in: US & Canada - Stock #1667-8

Ingredients: Same as Stock #1509-8

Vitamins & Minerals, Multiple

Product Type: Formula

Properties: Nutritive

Systems Affected: Whole Body

Usage: Multiple vitamin and mineral supplements are taken for general health purposes to supply nutrients that may be lacking in modern diets. Many people take them for basic health prevention.

See also *Super Supplemental and Children's Vitamins*

Warnings: No known warnings.

Super Supplemental Vit. & Min. - Tablet (120)

Available in: US - Stock #1792-7

Ingredients: Vitamin A, Vitamin C, Vitamin D, Vitamin E, Vitamin B1, Vitamin B2, Niacin, Vitamin B6, Folic Acid, Vitamin B12, Biotin, Pantothenic Acid, Calcium, Iron, Phosphorus, Iodine, Magnesium, Zinc, Selenium, Copper, Manganese, Chromium, Potassium, Inositol, PABA, Choline, Lutein, Lycopene

Solstic Nutrition - Packets

Available in: US - Stock #6505-4

Ingredients: Vitamin A (beta-carotene) 5,000 IU, Vitamin C (ascorbic acid) 60 mg, Vitamin D3 (cholecalciferol) 400 IU, Vitamin E (d-alpha tocopherol from soy) 30 IU, Vitamin K2 (MK-7, contains milk) 16 mcg, Vitamin B1 (thiamine) 4.5 mg, Vitamin B2 (riboflavin) 5.1 mg 300, Niacin (niacinamide) 60 mg, Vitamin B6 (pyridoxine HCl) 6 mg, Folic Acid 400 mcg, Vitamin B12 (cyanocobalamin) 18 mcg, Biotin 300 mcg, Pantothenic acid (d-calcium pantothenate) 30 mg, Iodine (potassium iodide) 200 mcg, Zinc (oxide) 9 mg, Selenium (amino acid chelate) 70 mcg, Copper (gluconate) 2 mg, Manganese (amino acid chelate) 2 mg, Chromium (amino acid chelate) 120 mcg, Molybdenum (amino acid chelate) 75 mcg, Sodium 2 mg, stevia, lutein, inositol, para amino benzoic acid, lycopene and choline bitartrate

Super Vitamins and Minerals - Tablets (120)

Available in: Canada - Stock #1791-2

Ingredients: Vitamin A (palmitate), Beta Carotene (a source of Vitamin A), Vitamin B1 (Thiamine mononitrate), Vitamin B2 (Riboflavin), Vitamin B6 (Pyridoxine hydrochloride), Vitamin B12 (Cyanocobalamin), Biotin, Vitamin C (Ascorbic acid), Vitamin D3, Vitamin E (d-alpha tocopherol succinate), Folic Acid,

Pantothenic Acid, Niacinamide, Calcium (dicalcium phosphate, bone meal), Chromium (amino acid chelate), Copper (gluconate), Iodine (potassium iodide), Iron (Ferrous gluconate), Magnesium (oxide), Manganese (amino acid chelate), Potassium (citrate), Selenium (amino acid chelate), Zinc (gluconate), barley grass juice, rosehips, lemon bioflavonoids, hesperidin, dulse plant, broccoli, asparagus and cabbage powders, and alfalfa herb.

Super Supplemental without Iron - Tablet (120)

Available in: US - Stock #1809-0

Ingredients: Vitamin A, Vitamin C, Vitamin D, Vitamin E, Vitamin B1, Vitamin B2, Niacin, Vitamin B6, Folic Acid, Vitamin B12, Biotin, Pantothenic Acid, Calcium, Phosphorus, Iodine, Magnesium, Zinc, Selenium, Copper, Manganese, Chromium, Potassium, Inositol, PABA, Choline, Lycopene

Super Supplemental Trial Pack - Packet (20)

Available in: US - Stock #2488-3

Ingredients: Same as Stock #1792-7

Multiple Vit. & Min., T/R - Capsule (60)

Available in: US - Stock #1619-5

Ingredients: Vitamin A, vitamin C, vitamin D, vitamin E, vitamin B1, vitamin B2, niacin, vitamin B6, folic acid, vitamin B12, biotin, pantothenic acid, calcium, iron, phosphorus, iodine, magnesium, zinc, selenium, copper, manganese

Multiple Vit. & Min., SynerPro - Capsule (60)

Available in: US - Stock #1644-1

Ingredients: Vitamin A, vitamin C, vitamin D, vitamin E, vitamin B1, vitamin B2, niacin, vitamin B6, folic acid, vitamin B12, biotin, pantothenic acid, calcium, iron phosphorus, iodine, magnesium, zinc, selenium, copper, maganese, chromium, molybdenum

SynerPro Super Vitamins & Minerals - Capsule (60)

Available in: Canada - Stock #4057-9

Ingredients: Vitamin(e) A (beta carotene), Vitamin(e) B1 (riboflavin), Calcium (calcium di-phosphate), Niacin, Pantothenic acid (d-calcium pantothenate), Vitamin(e) B6 (pyridoxine hydrochloride), Vitamin(e) B12 (cyanocobalamin), Vitamin(e) C (ascorbic acid), Vitamin(e) D3, Vitamin(e) E, Folic Acid, Biotin, Copper (gluconate), Iron (fumarate), Zinc (gluconate), Iodine (potassium iodide), Selenium, Magnesium (gluconate), Phosphorus (dicalcium phosphate), Chromium, Potassium, Maganese, Molybdenum

VitaWave Liq. Vit./Min. (32 fl. oz.) - Liquid

Available in: US - Stock #3332-3

Ingredients: Vitamin A, Vitamin D, Vitamin E, Vitamin K, Thiamin, Riboflavin, Niacin, Vitamin B6, Vitamin B12, Biotin, Pantothenic Acid, Calcium, Iodine, Magnesium, Zinc, Selenium, Copper, Manganese, Chromium, Molybdenum, Folic Acid, Korean ginseng, damiana, oat straw, saw palmetto, nettle, green tea extract (decaffeinated), bilberry extract, grape seed extract, taurine, alanine, arginine, aspartic acid, cysteine, glutamic acid, glycine, histadine, isoleucine, leucine, lysine, methionine, phenylalanine, proline, serine, threonine, tyrosine, valine, alpha-lipoic acid, citrus bioflavinoids, inositol, p-amino benzoic acid, choline bitartrate, lutein, lycopene, dead sea salt, boron, and plant-derived trace minerals.

Vitamins, Children's Multi

Other Names: Herbasaurs

Product Type: Formula

Properties: Nutritive

Systems Affected: Whole Body

Conditions: Children's Remedy

Usage: Children, like adults, can benefit from nutritional supplementation. These vitamin and mineral products can be helpful for children.

Warnings: No known warnings.

Multiple Vit. & Min., Chewable - Capsule (90)

Available in: US - Stock #1593-0

Ingredients: Vitamin A, vitamin C, vitamin E, vitamin B1, vitamin B2, niacin, vitamin B6, folic acid, vitaminB12, biotin, pantothenic acid, iron, iodine, zinc, copper, manganese, chromium, selenium

Multiple Vitamins + Iron (4 fl. oz.) - Liquid

Available in: US - Stock #3330-6

Ingredients: Vitamin A, C, D, E, B1, B2, Niacin, B6, Folic Acid, B12, Biotin, Pantothenic Acid, Iron, Zinc, Water, Brown Rice Syrup, Fructose, glycerin, citric acid, natural lemon, lime, orange, mango, vanilla, and sodium benzoate.

Herbasaurs Multiple Vitamin & Mineral Chewable - Tablet (90)

Available in: Canada - Stock #1722-9

Ingredients: Vitamin A (Palmitate), Vitamin D3, Vitamin E, C, B1, B2, B6, B12, Niacinamide, Folic Acid, Biotin, Pantothenic Acid, Iron

VitaWave

See *Vitamins & Minerals, Multiple*

VS-C

Product Type: Formula

Properties: **Alterative (Blood Purifier)**, Anti-inflammatory, **Antiherpetic, Antiviral**

Systems Affected: Immune System, Liver, Mouth, Mucus Membranes, **Tongue**

Conditions: Acquired Immune Deficiency Syndrome (AIDS/HIV), Adenitis, **Bell's Palsy**, Bleeding (internal), **Canker Sores (Mouth Ulcers), Chicken Pox, Cold Sores (Fever Blisters)**, Contagious Diseases, Dizziness (Vertigo), **Epstein Barr Virus (Chronic Fatigue Syndrome, CFS), Herpes**, Infection (bacterial), **Infection (viral), Measles, Mononucleosis, Reye's Syndrome, Shingles**, Thrush, Vaccines (detoxification from), **Warts**

Usage: This formula was developed by Dr. Wenwei Xie of Beijing, China, for use against the herpes simplex virus. It was formerly sold under the trade name HRP-C. Testing in the United States confirmed its effectiveness against the virus, but no useful drugs were found in any of the herbs and research was abandoned. The formula is used for cold sores, canker sores and herpes infections. Its Chinese indications are "weak chi (or vital energy) with external heat." Hence, it is useful for viral disorders in which the immune system is weak, but external heat (sores, fever, etc.) is present. Dr. Xie thought it might also benefit AIDS patients. Use 4 capsules two times a day for mild cases, 4 capsules three times daily for moderate cases and 4 capsules four times daily for severe cases.

Warnings: No known warnings.

VS-C Capsules - Capsule (100)

Available in: US - Stock #937-7

Ingredients: Dandelion, purslane, indigo, thlaspi, bupleurum, typhonium rhizome, scute, cinnamon twig, licorice, Panax ginseng

VS-C TCM Conc. - Capsule (30)

Available in: US - Stock #949-2

Ingredients: Same as Stock #937-7

HRP-C - Capsule (100)

Available in: US & Canada - Stock #937-7

Ingredients: Same as Stock #937-7

VS-C Extract (2 fl. oz.) - Liquid (Glycerite)

Available in: US - Stock #3167-6

Ingredients: Same as Stock #3167-6

HRP-C Extract (2 fl. oz.) - Liquid (Glycerite)

Available in: Canada - Stock #3167-6

Ingredients: Same as Stock #3167-6

~W~

White Oak Bark

Latin Name: *Quercus alba*

Product Type: Single Herb

Properties: Anti-emetic (Antinauseous), **Anticarious, Antidiarrheal**, Antiseptic, Antivenomous, **Astringent, Dentifrice, Hemostatic**, Styptic, Vermifuge

Systems Affected: Blood Vessels, Bones, Eyes, **Gums**, Intestinal System, Large Intestine (Colon), Liver, Mucus Membranes, Parathyroid, Parotids, **Rectum**, Skeletal System, Skin, **Spleen**, Stomach, Structural System, **Teeth**, Uterus

Conditions: **Anal Fistula or Fissure**, Bedwetting, Bites and Stings, Bladder (ulcerated), Bladder Infection, Bleeding (external), Bleeding (internal), Blood in Stool, Blood in Urine, Burns and Scalds, Canker Sores (Mouth Ulcers), Cuts, Diarrhea, Dysentery, Edema (Dropsy, Water Retention, Swelling), **Gingivitis (Bleeding Gums, Gum Disease, Pyorrhea), Hemorrhoids**, Injuries, Itching (rectal), Menorrhagia (Heavy Menstrual Bleeding), Oral Surgery, **Phlebitis, Poison Ivy or Oak**, Sprains, **Teeth (loose)**, Tonsillitis (Adenoids), **Tooth Decay (prevention)**, Tooth Extraction, **Varicose Veins**, Wounds and Sores

Usage: Use internally for hemorrhoids and varicose veins. Use 2 capsules between meals one to three times daily. Make a decoction by simmering 2-4 capsules per cup of water for 20 minutes. Decoction can be used as rectal injection for hemorrhoids, douche to stop bleeding, fomentation for swellings, varicose veins and other injuries. Gargle with decoction for sore throat or use as mouthwash for bleeding gums. Use white oak bark powder with black walnut powder as tooth powder for bleeding gums and loose teeth. Sprinkle powder in cuts to stop bleeding.

Warnings: Can be constipating when taken internally. Can interfere with digestion (take between meals). Contains large amounts of tannin, which may be associated with mouth and stomach cancer with consistent, long-term use. Use only for short periods internally. No warnings for external use.

White Oak Bark - Capsule (100)

Available in: US & Canada - Stock #730-7

Ingredients: Quercus alba (white oak) bark

Wild American Ginseng

See *Ginseng (Wild American)*

Wild Yam

Latin Name: *Dioscorea villosa*

Product Type: Single Herb

Properties: Anti-emetic (Antinauseous), **Anti-inflammatory**, Antirheumatic, **Antispasmodic**, Cholagogue, Contraceptive, Diaphoretic, Diuretic

Systems Affected: Adrenal Glands, Cortisol, Digestive System, Female Reproductive, **Gall Bladder**, Intestinal System, Nerves, Ovaries, Uterus

Conditions: Addictions (alcohol), Arthritis, Autoimmune Disorders, Birth Control (natural), **Colic (adults)**, **Cramps (menstrual)**, Cramps and Spasms (general), **Crohn's Disease**, Dermatitis, Diarrhea, Diverticulitis (Diverticuli), Dysmenorrhea, Gall Bladder (sluggish), Inflammatory Bowel Disorders (Colitis, IBS), Lupus, Neuralgia and Neuritis, Ovarian Pain, Progesterone (low)

Usage: Common use is 1-2 capsules of wild yam with a meal twice daily. Women who are using the plant as a contraceptive are taking 4 capsules of wild yam daily. It is reported that it takes two months of continuous use before the contraceptive effect takes hold. If the contraceptive effect is to be maintained, the plant must be taken daily, i.e., you cannot skip a day or the contraceptive effect is lost and you have to start over.

Wild yam does not contain progesterone and there is no evidence that taking wild yam affects DHEA or progesterone levels. Wild yam contains diosgenin that is used in the synthetic manufacture of progesterone.

Warnings: Excessive doses may cause nausea, vomiting and diarrhea.

Wild Yam - Capsule (100)

Available in: US - Stock #745-2

Ingredients: Dioscorea villosa (wild yam) root

Wild Yam & Chaste Tree

Product Type: Formula

Properties: **Anaphrodisiac**, Antispasmodic, Contraceptive, Female Tonic

Systems Affected: Digestive System, Female Reproductive, Nerves, Ovaries, Pituitary (anterior)

Conditions: **Acne (Pimples, Blackheads)**, Arthritis, Birth Control (natural), Colic (adults), Cramps (menstrual), Cramps and Spasms (general), Diarrhea, Diverticulitis (Diverticuli), Dysmenorrhea, Lupus, Menopause, **Menstrual Irregularity**, Nocturnal Emission (Wet Dreams), **Ovarian Pain**, Progesterone (low), **Puberty (hormone balancer)**, **Sex Drive (excessive)**

Usage: This combination helps regulate excessive estrogen levels in the body. It may ease severe menstrual cramping or reduce "raging" hormones in teenagers. It can be helpful for teenage acne caused by hormones. The formula may also help women who have been on birth control pills and are experiencing irregular cycles after discontinuing those pills. Take 1-2 capsules with a meal twice daily.

Warnings: Chaste tree may reduce sexual desire, but this effect is probably directed at men, not at women. This formula is not recommended for women who are trying to get pregnant (see *Wild Yam*). It is probably not a good idea to take this formula while taking birth control pills or other female hormone replacement drugs. Very high doses could cause nausea, vomiting and mild headache.

Wild Yam & Chaste Tree - Capsule (100)

Available in: US & Canada - Stock #1108-7

Ingredients: Wild yam, chaste tree.

Women's Formula

See *Female Comfort*

Wood Betony

Latin Name: *Stachys officinalis*

Product Type: Single Herb

Properties: Anticephalalgic, Antispasmodic, Nervine, Parasympathomimetic, Sedative

Systems Affected: Central Nerves, Circulation, Digestive System, Liver, Nerves, Pineal

Conditions: Attention Deficit Disorder (ADD, ADHD), Backache (Back Pain, Lumbago), **Bell's Palsy**, Cramps and Spasms (general), Headache (sinus), Headache (tension), Indigestion, Nerve Damage, **Neuralgia and Neuritis**, Nose Bleeds, Parkinson's Disease, Pregnancy (herbs and supplements to avoid during), Senility, Tics, Tinnitus (Ringing in the Ears), Tremors

Usage: A sedative that helps calm tension headaches and hyperactive children. Take 1 capsule with a meal twice daily.

Warnings: Not recommended for use during pregnancy.

Wood Betony - Capsule (100)

Available in: US - Stock #740-6

Ingredients: Betonica officinalis (wood betony) herb

X-A

Product Type: Formula

Properties: **Aphrodisiac**, Glandular, Stimulant

Systems Affected: Adrenal Glands, Circulation, Prostate, Reproductive Glands

Conditions: Aging (prevention), Endurance (lack of), Energy (lack of), Erectile Dysfunction, Fatigue, Hot Flashes, Infertility, Menopause, Night Sweating, Nocturnal Emission (Wet Dreams), Senility, **Sex Drive (low)**

Usage: This formula helps to stimulate the sex drive. As a tonic to the male/female glands for frigidity, impotency and loss of sex drive take 1-2 capsules two to three times daily. It can also be used for hormonal imbalances in the reproductive system and for infertility. As a glandular tonic use 1 capsule three times daily.

Warnings: No known warnings.

X-A - Capsule (100)

Available in: US - Stock #1130-2

Ingredients: Eleuthero, parthenium, saw palmetto berries, gotu kola, damiana, sarsaparilla, horsetail, garlic bulb, capsicum fruit, chickweed

X-A - Capsule (100)

Available in: Canada - Stock #1112-5

Ingredients: Siberian ginseng, parthenium, saw palmetto berries, alfalfa, damiana, sarsaparilla, cornsilk, garlic bulb, capsicum fruit, chickweed

X-Action (Men's)

Product Type: Formula

Properties: Androgenic, **Aphrodisiac**, Stimulant

Systems Affected: Adrenal Glands, Circulation, Male Reproductive, Testosterone, Urinary System

Conditions: Circulation (poor), Debility, Depression, **Erectile Dysfunction**, Infertility, **Sex Drive (low)**, Testosterone (low)

Usage: Men's X-Action is a combination of herbs that enhances male energy, activity and vitality. These herbs also support the male reproductive organs including prostate function. The formula also increases blood flow to the pelvic region. It may be helpful for impotence or lack of desire in men. Men's X-Action helps restore male vitality. Take 1 capsule with a meal three times daily.

Warnings: Not recommended for "hot" constitutions or teenagers. Excessive use can over-stimulate sexual organs.

X-Action (Men's) - Capsule (100)

Available in: US - Stock #1113-7

Ingredients: Muira puama, yohimbe, l-arginine, damiana, oatstraw, saw palmetto, DHEA, horny goat weed

X-Action (Women's)

Product Type: Formula

Properties: **Aphrodisiac**, Nervine, Uterine Tonic

Systems Affected: Adrenal Glands, Digestive System, Female Reproductive, Nerves

Conditions: Amenorrhea, Anxiety (Panic Attack), Cystitis, Depression, Infertility, Menopause, Pregnancy (herbs and supplements to avoid during), **Sex Drive (low)**, Vaginitis

Usage: Used to support the female reproductive organs and to increase sexual desire in women. Take 2 capsules with a meal three times daily.

Warnings: No known warnings.

X-Action (Women's) - Capsule (100)

Available in: US - Stock #1121-6

Ingredients: Maca Root (Lepidium meyenii), L-Arginine, Eleuthero root (Eleutherococcus senticosus), Oatstraw (Avena Sativa), Red Raspberry leaves (Rubus ideaus), Damiana leaves (Turnera diffusa), Licorice Root (Glycyrhiza glabra), Sarsaparilla Root (Smilax app.).

X-Action Gel

Product Type: Formula

Properties: Aphrodisiac, Stimulant

Systems Affected: Female Reproductive

Conditions: Pregnancy (herbs and supplements to avoid during), Sex Drive (low)

Usage: Apply topically for vaginal dryness or as needed to enhance the sensuality of intimate pleasure.

Warnings: Do NOT use during pregnancy. Avoid exposure to eyes. Discontinue use if irritation occurs.

X-Action Gel (15 ml tube) - Topical

Available in: US - Stock #4937-6

Ingredients: Water, glycerin, butylene glycol, pentylene glycol, cellulose gum, carbomer, methylparaben, menthol, arginine PCA,sodium hydroxide, dipotassium glycyrrhizinate, red clover flower extract, panax ginseng root extract, dulse plant extract, propylparaben, green tea extract, black cohosh root extract, saw palmetto fruit extract, red raspberry fruit extract, dong quai root extract.

Xylitol

Product Type: Nutrient

Properties: Anticarious, Antidiabetic, Low Glycemic, Nutritive, Sweetener

Systems Affected: Ears, Gums, Intestinal System, Pancreas, Sinuses, Teeth, **Weight Loss**

Conditions: Diabetes, Ear Infection or Earache, Gingivitis (Bleeding Gums, Gum Disease, Pyorrhea), **Gingivitis (Bleeding Gums, Gum Disease, Pyorrhea),** Halitosis (Bad Breath), Hyperinsulinemia (Syndrome X), **Hypoglycemia, Osteoporosis,** Sinus Infection, **Sugar Cravings, Tooth Decay (prevention), Weight Loss (aids for)**

Usage: Too much of a good thing can be harmful, and that's especially true when it comes to sugar. Although sugar is a major energy source for the body, the excess consumption of one sugar has lead to an epidemic of obesity and diabetes, which is affecting not only older people, but young children as well. Furthermore, excess sugar consumption triggers the release of excess insulin, which has been linked with increased inflammation and the development of degenerative diseases like heart disease and cancer.

Most of us know we should eat less sugar, but it's hard to give up our sweets, even if we know they're ruining our health. To satisfy their sweet tooth, without the calories, many people turn to artificial sweeteners like aspartame and sucralose to satisfy their cravings. Unfortunately, these artificial sweeteners have a dubious safety record and often create health problems of their own.

Fortunately, there is a natural alternative to refined sugar that has recently been introduced into the American marketplace. This product is a form of sugar, but one with both a long safety record and some great health benefits.

We're talking about xylitol. Xylitol is a sugar alcohol. (Don't worry, this has no relationship to the alcohol you drink--it's just a chemical structure.) This sugar is usually made from the fibers of corn husks or birch tree bark, but can also be found in beets, oats, certain mushrooms and some fruits and vegetables.

Xylitol has been used in Europe and China for over twenty years and has an excellent track record for safety. It has about the same sweetness as table sugar (sucrose) so it can be substituted for equal amounts of sugar in most cases. It does have caloric value, but it has 40% fewer calories than sugar. So, eating xylitol instead of sugar immediately cuts down your caloric intake without sacrificing sweetness.

Even better, xylitol has been shown to actually have some health benefits. For starters, xylitol has a glycemic index of 7, which means it does not trigger insulin production. (By reference, refined sugar has a glycemic index of 100.) This means that xylitol can be used safely by both hypoglycemics and diabetics.

But, the benefits of xylitol don't stop there. The bacteria which cause cavities and gum disease can't live on xylitol. This means that using xylitol regularly will actually reduce cavities and prevent gum disease.

Even better, xylitol helps the body bind calcium, which actually aids remineralization of the teeth and bones. So, it is helpful for creating strong teeth and preventing osteoporosis.

Xylitol also inhibits the bacteria that cause middle ear infections and sinus problems. It also helps reduce unfriendly bacteria in the intestines.

Yeast can't live on xylitol, either, which means it isn't good for making leavened (yeast) breads. (That's one of the few applications you can't use xylitol in place of sugar.) However, that's not bad news, because it also means that xylitol doesn't feed Candida and other yeast infections. Instead, it actually helps the body get rid of yeast-and that's good news!.

Xylitol increases saliva production. So, it is helpful for people who experience dry mouth. It can also improve bad breath problems.

Best of all, consumed regularly, xylitol actually balances blood sugar and reduces carbohydrate cravings. That means, it can help you break your addiction to refined sugar and simple carbohydrates without sacrificing your need for something sweet occasionally. This makes it very helpful for weight loss programs.

Xylitol is available in bulk for cooking and baking. Use it just like sugar in most recipes. It is also available in gum, chocolate bars and breath mints which can be used as alternatives to treats made with refined sugar. Xylitol is also available in a xylitol mouthwash and toothpaste to help prevent cavities and gum disease.

Warnings: Xylitol alters the balance of friendly flora in the digestive tract and can cause diarrhea in some people. To avoid this, start using xylitol in small amounts and gradually build up a tolerance for it.

Xylitol Bulk - Bulk Powder

Available in: US - Stock #5435-3

Ingredients: 100% granulated xylitol.

Xylitol Mouthwash - Liquid

Available in: US - Stock #5425-6

Ingredients: Water, xylitol, vegetable glycerin, peppermint (Mentha piperita), citric acid, PEG-40 hydrogenated castor oil, calcium glycerophosphate, Aloe barbadensis leaf juice,sodium benzoate, potassium sorbate, grapefruit seed extract (Citrus grandis).

Xylibrite Toothpaste - Paste

Available in: US - Stock #5420-2

Ingredients: Water (aqua), xylitol, sorbitol (vegetable source), glycerin (vegetable source), silica (natural mineral), hydrated silica (natural mineral), sodium lauroyl sarcosinate (from coconut), sodium bicarbonate (baking soda), natural flavors (spearmint, cinnamon, anise), cocoamidopropyl betaine (foaming agent from coconut), stevia (Stevia rebaudiana), sodium hydroxide (mineral source), cellulose gum (from cellulose), titanium dioxide (natural mineral), carrageenan (seaweed extract)

Xylitol Gum (Cinnamon)

Available in: US & Canada - Stock #5400-8

Ingredients: Xylitol, gum base, natural flavor (cinnamon oil), vegetable glycerin, gum arabic, soy lecithin, and beeswax.

Xylitol Gum (Green Tea)

Available in: US - Stock #5403-3

Ingredients: Xylitol, gum base, green tea (natural flavor), vegetable glycerin, gum arabic, non-GMO soy lecithin, and beeswax.

Xylitol Gum (Peppermint)

Available in: US - Stock #5408-4

Ingredients: Xylitol, gum base, natural flavor (peppermint oil), vegetable glycerin, gum arabic, soy lecithin, and beeswax

Xylitol Gum (Spearmint)

Available in: US & Canada - Stock #5405-0

Ingredients: Xylitol, gum base, natural flavor (spearmint oil), vegetable glycerin, gum arabic, soy lecithin, and beeswax.

Xylitol Mints (Berry)

Available in: US & Canada - Stock #5412-2

Ingredients: Xylitol, calcium lactate, magnesium stearate, gum arabic, natural flavor (berry), glazing agent (beeswax)

Xylitol Mints (Lemon)

Available in: US & Canada - Stock #5415-1

Ingredients: Xylitol, calcium lactate, mag- nesium stearate, gum arabic, natural flavor (lemon oil), glazing agent (beeswax).

Xylitol Mints (Peppermint)

Available in: US - Stock #5410-7

Ingredients: Xylitol, calcium lactate, mag- nesium stearate, gum arabic, natural flavor (peppermint oil), glazing agent (beeswax)

Dark Chocolate - Bar

Available in: US - Stock #5453-7

Ingredients: Sugar Free Dark Chocolate Coating (Maltitol*, Chocolate Liquor, Cocoa Butter, Milk Fat, Soy Lecithin Emulsifiers, Vanilla), Xylitol

Dark Chocolate (Orange Memory) - Bar

Available in: US - Stock #5452-5

Ingredients: Sugar Free Dark Chocolate Coating (Maltitol, Chocolate Liquor, Cocoa Butter, Milk Fat, Soy Lecithin Emulsifiers, Vanilla), Xylitol, Soy Lecithin, Orange Peel, Cocoa Butter, Dha From Algae, Omega-3 Powder (Glucose Syrup Solids, Algal Oil, Sodium Caseinate (Milk), Sodium Ascorbate, Dipotassium Phosphate, Glyceryl Mono- And Distearate, Soy Lecithin, Tricalcium Phosphate And Mixed Natural Tocopherols And Ascorbyl Palmitate As Antioxidants), Natural Flavors, Ginkgo

Yarrow

Latin Name: *Achillea millefolium*

Product Type: Single Herb

Properties: Alterative (Blood Purifier), Anti-inflammatory, Antibacterial, Antifungal, **Antiphlogistic**, Antiseptic, Antiviral, Aromatic, **Astringent**, Bitter, **Coagulant**, **Diaphoretic**, Diuretic, Emmenagogue, Febrifuge, Hypotensive, Lymphatic, Nervine, **Parasympatholytic (Anticholinergic)**, Stimulant, Stomachic, **Styptic**, Tonic, Uterine, Vulnerary

Systems Affected: Bladder (Urinary), Capillaries, Eyes, Female Reproductive, Hair, Immune System, Intestinal System, Large Intestine (Colon), Lymph Nodes, Lymphatic System, Mucus Membranes, Progesterone, Skin, Stomach, **Sweat Glands**, Thymus, Urinary System, Uro-genital Tract, Uterus

Conditions: Amenorrhea, **Aneurysm**, Bites and Stings, **Bleeding (external)**, Bleeding (internal), Bruises (healing), Bruises (prevention), **Capillary Weakness**, **Chicken Pox**, Colds (antiviral), Colds (general remedies for), Colds (with fever), **Cuts**, Diarrhea, Dysmenorrhea, **Fever**, **Fibroids (uterine)**, Flu, Fungal Infections (Yeast Infections, Candida albicans), Glands (swollen lymph), Infection (viral), Inflammation, Injuries, Measles, **Meningitis**, **Menorrhagia (Heavy Menstrual Bleeding)**, Mumps, **Nose Bleeds**, **Perspiration (deficient)**, Pleurisy, Pneumonia, Scars / Scar Tissue, Scratches and Abrasions, Shingles, Skin (infections), Sweat Baths (herbs for), Vaccines (detoxification from), Varicose Veins, Wounds and Sores

Usage: For colds, fever, or infection use 1-3 capsules two to four times per day with warm water. Can be used internally for bleeding, internal injuries, kidney infections and inflammation. Must be used for long periods for anti-inflammatory action (combine with chamomile and other anti-inflammatories). Use as a decoction for stomach and intestinal problems. Make into decoction by simmering 3-4 capsules per pint of water for 20 minutes. Apply powder directly to injuries to stop bleeding and aid healing. Indians chewed this herb for relief of toothache. Make into a tea with equal parts peppermint and drink warm for fevers. Warm tea helps induce perspiration.

Warnings: Probably a very safe remedy, but some herbalists feel it should not be used long term.

Yarrow - Capsule (100)

Available in: US - Stock #750-2

Ingredients: Achillea millefolium (yarrow)

Yeast/Fungal Detox

Product Type: Formula

Properties: **Antifungal**, Immune Stimulant

Systems Affected: Digestive System, Immune System, Intestinal System, Large Intestine (Colon), Liver, Urinary System, **Vagina**

Conditions: Addictions (sugar or food), Aging (prevention), **Antibiotics (side effects of)**, **Athlete's Foot**, Belching, Birth Control (countering side effects), Body Odor, Cellulite, Cholesterol (high), Ear Infection or Earache, **Fungal Infections (Yeast Infections, Candida albicans)**, Inflammatory Bowel Disorders (Colitis, IBS), **Itching Ears**, **Jock Itch**, Leaky Gut Syndrome, **Leucorrhea**, Mental Illness, **Sugar Cravings**

Usage: This is a powerful formula for eliminating yeast from the body. Take 1 capsule with a meal twice daily. For even more effective yeast control add 1 tablet of L. Reuteri twice daily and 1 capsule of Paw Paw Cell-Reg once daily.

Warnings: Do not exceed recommended dose. If symptoms of headache, nausea or diarrhea develop, reduce dosage or discontinue. This is a sign of a too rapid die-off of yeast and a resultant "cleansing crisis." Use some of the detoxification methods discussed in this book to rapidly flush the toxins from the system.

Yeast Fungal Detox - Capsule (90)

Available in: US & Canada - Stock #508-9

Ingredients: Caprylic acid, Sodium Propionate, sorbic acid, Echinacea Pupurea root, oregano, garlic, pau d'arco, selenium, zinc

Yellow Dock

Latin Name: *Rumex crispus*

Product Type: Single Herb

Properties: Alterative (Blood Purifier), **Antipruritic**, **Aperient**, Astringent, Bitter, **Cholagogue**, Hepatic, Laxative (general), Mineralizer, Parasiticide

Systems Affected: Blood, **Gall Bladder**, Liver, Lymphatic System, Nerves, Red Blood Cells, Skin, Spleen, Tongue

Conditions: Acne (Pimples, Blackheads), **Anemia**, Cancer (natural therapy for), Chicken Pox, Constipation (children), Depression, Hepatitis, Inflammatory Bowel Disorders (Colitis, IBS), **Itching**, **Jaundice (adults)**, Leprosy, Myasthenia Gravis, **Oxygen Deficiency**, Pernicious Anemia, Poison Ivy or Oak, Pregnancy (herbs and supplements for), Rashes and Hives, **Sore or Geographic Tongue**, Wounds and Sores

Usage: Yellow dock is high in organic iron compounds and liberates iron stored in the liver. This herb is useful for anemia, especially when combined with alfalfa, beets and other iron-rich herbs. It is also used as a blood purifier for skin disorders (acne, boils, etc.) and general liver problems. Yellow dock combines well with sarsaparilla, dandelion and burdock. Also, this herb stimulates the flow of bile and acts as a mild laxative. Use 2 capsules one to three times daily.

Warnings: Large doses can cause nausea and diarrhea.

Yellow Dock - Capsule (100)

Available in: US - Stock #760-1

Ingredients: Rumex crispus (yellow dock) root

Ylang Ylang

Product Type: Essential Oil

Properties: Antidepressant, Antiseptic, Antispasmodic, Aphrodisiac, Cardiac, Hypertensive, Nervine, **Perfume**, Sedative, Sympatholytic

Systems Affected: Breasts, Hypothalamus, Pineal, Reproductive Glands

Conditions: Aging (prevention), Anger (excessive), Anxiety (Panic Attack), Asthma, Blood Pressure (high), Breasts (enhance size), Cramps and Spasms (general), Depression, Diarrhea, Fatigue, Fear (excessive), Gas and Bloating, Guilt, Hair Care (general), Menopause, Mood Swings, Nervous Exhaustion (Enervation), Nervousness, PMS Type C, Sex Drive (low), Skin (oily), Stress

Usage: Ylang Ylang is a wonderfully exotic and enticing flora oil. The name Ylang Ylang means "flower of flowers." In Indonesia they spread the delicate and sweet-smelling flower petals on the bed of a newly married couple. Ylang Ylang helps to balance the nervous system, inhibiting excess adrenal activity and symptoms of anxiety like rapid breathing and heart rate. It can be used in a massage oil for muscle cramps and spasms, high blood pressure, intestinal gas and diarrhea, PMS and stress. It can be used in a shampoo to stimulate hair growth. It also stimulates breast growth. Emotionally, ylang ylang arouses the senses and draws out sensuality and passion. It eases fear and anxiety and inspires confidence and joy. It replaces jealousy, doubt, frustration and guilt with a feeling of caring and contentment.

Warnings: Do not take internally. Excessive use may lead to headaches and nausea.

Ylang Ylang Complete BIO (5ML) - Essential Oil

Available in: US - Stock #3917-7

Ingredients: Ylang Ylang Complete Bio (Cananga odorata) Essential Oil

Yucca

Latin Name: *Yucca glauca*

Product Type: Single Herb

Properties: Alterative (Blood Purifier), Analgesic (Anodyne), **Anti-inflammatory**, Antirheumatic, Bitter

Systems Affected: Cortisol, Intestinal System, Parotids, Structural System

Conditions: Addictions (alcohol), Adenitis, Adrenals (exhaustion, weakness or burnout), Arthritis, **Autoimmune Disorders**, Crohn's Disease, Dermatitis, Eczema, **Fibromyalgia Syndrome (FMS)**, **Inflammation**, **Lupus**, Myasthenia Gravis, Neuralgia and Neuritis, Overacidity, Rheumatoid Arthritis (Rheumatism), Uric Acid Retention

Usage: Yucca is a natural alternative to corticosteroid drugs. It has an anti-inflammatory action and a blood cleansing effect. Take 1 capsule two to three times daily as a blood purifier. For chronic inflammation take 2 capsules three times daily.

Warnings: Occasionally produces laxative effect with intestinal cramping.

Yucca - Capsule (100)

Available in: US & Canada - Stock #770-3

Ingredients: Yucca Root (Yucca baccata)

Zambroza

See *Thai-Go*

Zinc

Product Type: Nutrient

Properties: Appetite Stimulant, Nutritive

Systems Affected: Blood, Brain, Cuticle, Hair, Immune System, Liver, Nails, Pancreas, Pancreas Tail, Parotids, Pineal, **Prostate**, Reproductive Glands, Skin, Stomach, Taste Buds, Testes, Thymus

Conditions: Aging (prevention), Anemia, Appetite (deficient), Arteriosclerosis (Atherosclerosis, Hardening of the Arteries), Benign Prostate Hyperplasia (BPH), Body Odor, Burns and Scalds, Cancer (prevention), Cataracts, Cold Sores (Fever Blisters), Colds (prevention), **Copper Toxicity**, Croup, Cystic Breast Disease, Cystitis, Dermatitis, Diabetes, Eczema, Erectile Dysfunction, Floaters, Free Radical Damage, Gray Hair, Hair (loss or thinning), Hair Care (general), Hashimoto's Disease (Thyroiditis), **Hemochromatosis**, Inflammation, Lead Poisoning, Lupus, Lyme Disease, Macular Degeneration, PMS Type C, Psoriasis, Rosacea, **Schizophrenia**, **Smell (loss of sense of)**, **Stretch Marks**, Tinnitus (Ringing in the Ears), **Ulcerations (external)**, Vitiligo, Wounds and Sores

Usage: Zinc is important for immune system function, male reproductive function and tissue healing. It is an antioxidant and free radical scavenger. It is a component of insulin (along with chromium) and male reproductive fluid. It aids in the digestion and metabolism of phosphorus. RDA infants: 3-5 mg., RDA children: 10-15 mg., RDA adults: 15 mg., dietary levels: 15 mg., therapeutic range: 20-100 mg. NSP Zinc contains 25 mg. of zinc in the form of zinc gluconate in a base of kelp, thyme and alfalfa.

Warnings: Relatively non-toxic. Daily doses over 100 mg. can depress the immune system. Copper and zinc levels should be balanced.

Zinc (25 mg) - Tablet (150)

Available in: US & Canada - Stock #1657-9

Ingredients: 25 mg zinc (zinc gluconate), kelp, thyme, alfalfa, Calcium Phosphorus.

Zinc Lozenges

Product Type: Formula

Properties: Antiseptic, Antiviral, Immune Stimulant

Systems Affected: Immune System, Mouth, Mucus Membranes, Tongue

Conditions: Colds (general remedies for), Free Radical Damage, **Laryngitis (Hoarseness)**, **Sore Throat**

Usage: Take 1 lozenge each hour or as needed. Allow lozenge to dissolve slowly in the mouth. Do not use more than 6 lozenges in a 24-hour period.

Warnings: No known warnings.

Zinc Lozenges - Lozenge

Available in: US - Stock #1596-8

Ingredients: Vitamin C (ascorbic acid), sodium ascorbate, echinacea root, slippery elm bark, zinc citrate, licorice root, eucalyptus oil and menthol, natural sweeteners (sorbitol and xylitol) and natural orange, peach and spearmint flavors.

Zinc and Vitamin C Lozenges - Lozenge

Available in: Canada - Stock #1596-8

Ingredients: Same as Stock #1596-8

Section Three

Conditions

Section Three

Index to Conditions (Ailments and Health Problems)

This section allows you to look up specific conditions, ailments and health problems. In most cases, a definition of the problem is included along with a brief description of natural approaches to working with that problem. Our best product picks for each condition are highlighted in bold. A list of products that have been used to help each condition is also included. Be sure to read Section One before trying to put a program together, paying particular attention to the basic therapies listed for that conditon.

Abdominal Pain and Inflammation

See Also *Acid Indigestion (Heartburn, Acid Reflux), Fungal Infections (Yeast Infections, Candida albicans), Hiatal Hernia, Inflammatory Bowel Disorders (Colitis, IBS), Parasites (general)*

Pain or cramping in the abdominal area is a sign of irritation and inflammation in the digestive tract. It may be accompanied by gas and bloating or intestinal cramps. Intestinal Soothe and Build and CLT-X both help to reduce inflammation in the intestinal tract. CLT-X also contains wild yam, which is anti-spasmodic, relieving intestinal cramping. When using essential oil products for this problem, apply them topically over the abdominal area, rather than taking them internally. If problems persist, see a medical doctor for a proper diagnosis. See related conditions for more detailed information on possible causes of this problem.

Therapies: Gall Bladder Flush (pg. 23), Hiatal Hernia Correction (pg. 12), Colon Cleansing (pg. 18), Enzymes (pg. 15), Probiotics (pg. 16), Fiber (pg. 17)

Remedies: Catnip & Fennel, CLT-X, Gall Bladder Formula, Intestinal Soothe & Build, ALJ, Anti-Gas Formula, Capsicum, Chamomile, Chamomile (Roman), Clove Bud, Deep Relief Oil, Grapefruit (Pink), IF Relief, Influenza Remedy, Kidney Activator (Chinese), Liver Balance, Magnesium, Myrrh, Parasites, Spleen Activator, Stress Relief, Sunshine Heroes Probiotic Power, Super ORAC

Abrasions

See Also *Wounds and Sores*

An injury caused by a scraping away of a portion of skin or a mucus membrane. Aloe Vera Gel, Golden Salve, Silver Shield Gel and Nature's Fresh are all remedies that can be applied topically to help abrasions heal more quickly. Healing AC Cream should only be applied if there are no cuts that have punctured the skin.

Remedies: Golden Salve, Healing AC Cream, Silver Shield, Aloe Vera, Black Ointment, IF-C, Lavender, Lymphatic Drainage, Nature's Fresh, Tea Tree Oil

Abscesses

An abscess is either an open sore, usually surrounded by inflamed tissue, from which there is an oozing of pus, or a cavity formed by a collection of pus-like material in solid tissue. Most remedies should be applied topically for the treatment of abscesses. However, alteratives or blood purifiers can be taken internally to help drain abscesses.

Therapies: Colon Cleansing (pg. 18)

Remedies: BP-X, Golden Salve, Silver Shield, Black Ointment, Chamomile, Chamomile (Roman), Chickweed, Flax Seed Oil, Ho Shou Wu, Intestinal Soothe & Build, Lemon Oil, Lobelia, Lymph Gland Cleanse, Lymphatic Drainage, Marshmallow, Slippery Elm, Tea Tree Oil

Aches

See *Pain (general remedies for)*

Acid Indigestion (Heartburn, Acid Reflux)

See Also *Hiatal Hernia*

Just about everyone has experienced heartburn at least once, but many people suffer from it on a regular basis. In fact, the American Gastroenterological Association estimates that about 60 million Americans have a case of heartburn each month. Of course, heartburn really has nothing to do with the heart. It's a form of acid indigestion in which acid leaves the stomach and enters the esophagus causing burning and pain.

When we eat, food passes down the esophagus and into the stomach. There, the stomach secretes hydrochloric acid and pepsin to break down proteins and begin the process of digestion. The stomach has a mucous lining to help protect it against this acid. The esophagus does not. A muscular valve at the top of the stomach holds the stomach's contents, including the stomach acid, in the stomach. This valve opens to permit belching, then closes again.

Heartburn occurs when the valve at the top of the stomach allows acid to seep back (reflux) into the esophagus. This acid burns and inflames the esophageal lining, which does not have the same degree of mucous protection as the stomach. This creates the burning sensation in the center of the chest we call heartburn. The more technically accurate term is acid reflux.

Although uncomfortable, occasional heartburn or acid reflux is not a serious condition. However, if it happens frequently and persistently, then the repeated burning and inflammation of the esophagus can result in more serious damage. This more serious condition is called gastro esophageal reflux disease or GERD. GERD is surprisingly common, affecting an estimated 5-7% of the American population.

The problem with GERD is that it can cause an erosion of the esophagus due to the continual acid damage. Scar tissue can form which will narrow the passageway. It also increases the risk of esophageal cancer.

Several things cause acid indigestion. Poor digestion, irritating foods and mechanical pressure all contribute to indigestion and acid reflux.

First, let's look at the problem of poor digestion. The more protein foods are cooked, the more their proteins are denatured. Denatured proteins become leathery making it more difficult for digestive juices to penetrate the tissues and break down the food particles. Greasy fried foods create the same problem because water and oil do not mix. The grease coats the food particles, preventing penetration by the hydrochloric acid.

When food is not thoroughly chewed, it is harder for digestive juices to penetrate the food; so food needs to be chewed thoroughly to help digestive secretions blend with food particles. Overeating also causes acid indigestion, as the stomach becomes overburdened with more food than it can handle. Food winds up not being digested properly, which causes the body to signal for more production of acid.

Certain foods may trigger acid indigestion because they trigger a type of allergic response. Common foods that cause excess acid production in some people include onions, peppermint, chocolate, coffee, citrus fruits, tomatoes, garlic and spicy foods. While these foods may cause acid indigestion, other factors cause the acid to reflux into the esophagus. These are generally due to mechanical pressures. Anything that puts pressure on the value at the top of the stomach will allow acid to enter the esophagus, even if acid production isn't excessive.

Anything that pushes the stomach upward can put pressure on this valve including intestinal gas and bloating, excess body weight, tight fitting clothes, pregnancy, and lying down after eating. Stress also tenses the solar plexus area and draws the stomach upward.

When a portion of the upper stomach passes through the opening in the diaphragm where the esophagus enters (known as the hiatus), this is called a hiatal hernia. A hiatal hernia creates chronic digestive problems and eventually chronic digestive weakness. Pressure on the stomach from below (as described previously) can cause a hiatal hernia, but it can also be created by repeated stress and swallowing of emotions.

A hiatal hernia causes a kink in the valve at the top of the stomach, creating chronic heartburn or GERD. It can also put pressure on the heart, weaken the thyroid, and contribute to chronic gas and bloating in the intestines. This is mechanical problem which must be corrected mechanically before any permanent relief can be found. Herbs and supplements alone will not correct this problem.

Since acid reflux usually begins with poor digestion brought on by lack of digestive enzymes, food allergies, or processed foods that are difficult for the body to digest, taking enzymes to help foods break down better is the first step. Believe it or not, acid reflux can actually be a sign that the body isn't producing enough hydrochloric acid to properly digest proteins. The undigested food ferments, producing waste acids and fermentation. In these cases, the acid indigestion is usually accompanied by poor digestion of proteins, intestinal gas and bloating, and a heavy feeling in the stomach after eating. The acid indigestion usually occurs about an hour after eating, and the tongue is pale, often with a heavy coating.

For immediate relief from this type of acid indigestion take any bitter herb in liquid or powder form (golden seal works well), place it on the tongue and suck on it. This will relieve the burning very quickly. For long term relief start using enzymes, digestive bitters and/or hydrochloric acid supplements.

When there really is too much acid production in the stomach, the symptoms will be different from those described above. Protein digestion is good, and food digests rapidly. The tongue is red and the person is usually younger. This condition is usually due to overeating and stress. It may also be due to eating too many grains and refined carbohydrates and not enough protein and vegetables. An antiacid like Stomach Comfort can be used for this kind of occasional acid indigestion.

If bloating and gas are problems, they may also be contributing to the acid reflux. Intestinal gas and bloating put pressure on the stomach, pressing it upwards against the diaphragm. Use carminatives to relieve gas pressure.

To heal damage to the esophagus and digestive tract due to acid reflux, soothing mucilaginous remedies are needed. One of the best is Aloe Vera Juice, but there are many others. By sipping small amounts of aloe diluted in water, the burning or inflammation in the esophagus due to the acid reflux can be cooled and soothed. Aloe will also help to normalize the pH of the body, reducing overacidity in general.

These remedies, along with appropriate lifestyle changes, can eliminate acid reflux and the need for antacids.

Therapies: Hiatal Hernia Correction (pg. 12), Hydration (pg. 10), Enzymes (pg. 15),Fiber (pg. 17)

Remedies: Catnip & Fennel, Golden Seal, Papaya Mint, Stomach Comfort, Aloe Vera, Anti-Gas Formula, Blessed Thistle, Caffeine Detox, Catnip, Coral Calcium, Devil's Claw, Digestive Bitters Tonic, Food Enzymes, Gastro Health, Intestinal Soothe & Build, Lemon Oil, MSM, Oregon Grape, PDA, Red Raspberry, Safflowers, Small Intestine Detox, Sunshine Heros Whole Foods Papayazyme

Acid pH

See *Overacidity*

Acid Reflux

See *Acid Indigestion (Heartburn, Acid Reflux), Acid Indigestion (Heartburn, Acid Reflux)*

Acne (Pimples, Blackheads)

See Also *Hyperthyroid, Leaky Gut Syndrome*

Acne is an inflammatory condition of the skin. The small glands that excrete oil to lubricate the skin become irritated and inflamed. This is due to fat-soluble toxins in the body that are being eliminated through the skin pores. Microbes may also start feeding on this waste, which exacerbates the condition. As the body fights the infection, the skin pores fill with pus.

Several imbalances are at work here. First of all, acne is a sign of a general toxic condition of the blood and lymph. So, the first step is to cleanse the colon and strengthen the detoxification capacity of the liver and kidneys. Alteratives or blood purifiers have cleared up some cases of acne. Burdock has been very effective in some cases. Other possibilities include E-Tea, EnviroDetox, All Cell Detox or BP-X.

Since the irritation is affecting glands that secrete oil, the toxins involved here are fat-soluble and may be resulting from a problem with the breakdown and utilization of fats in the body. Hydrogenated oils are irritating to the body and should be replaced with healthier fats such as flax seed oil, Super Omega-3 EPA and olive oil.

Both burdock and chickweed aid the body in properly metabolizing fats. Fat-soluble vitamins help protect against the oxidation of fats. Oxidation causes fats to become rancid and irritating to the system. This may explain why large doses of vitamin A have helped clear up some cases of acne.

One of the reasons teenagers are so prone to acne is that their hormones are out of balance. For this reason, chaste tree berries have also been beneficial in clearing up teenage acne. Other glandular herbs that might be helpful include sarsaparilla (which is also a blood purifier) for teenage boys and dong quai (also a blood tonic) for teenage girls.

In some cases, low thyroid can contribute to this problem. The thyroid hormones are needed to properly combust fats in the body. Remedies that help the thyroid include Thyroid Activator and Thyroid Support.

Cleansing the skin thoroughly to remove excess oil from the glands and to get rid of unwanted microorganisms is also helpful. However, the problem with internal toxicity and combustion of fats must still be addressed.

Therapies: Colon Cleansing (pg. 18), Essential Fatty Acids (pg. 17), Fiber (pg. 17)

Remedies: Acne Treatment Gel, All Cell Detox, BP-X, Silver Shield, Skin Detox, Wild Yam & Chaste Tree, Acne, Aloe Vera, Bentonite (Hydrated), Bergamot, Burdock, C-X, Chamomile (Roman), Dandelion, E-Tea, Echinacea Purpurea, Enviro-Detox, FCS II, Female Comfort, Herbal Trace Minerals, Hi Lipase, Kelp, Mandarin (Red), Master Gland, MSM, Nature's Fresh, NF-X, PS II, Red Beet Formula, Red Clover, Red Clover Blend, Sandalwood, SF, Sunshine Heroes Probiotic Power, Ultimate Echinacea, Vitamin A & D, Vitamin B-6, Vitamin B-Complex, Yellow Dock

Acquired Immune Deficiency Syndrome (AIDS/HIV)

Acquired Immune Deficiency Syndrome, more commonly known as AIDS, is a disease in which the body's immune system is depressed. It is believed to be the result of an infection by the HIV virus, although there are some researchers who dispute this explanation. There are many drugs that can also cause the immune system to be depleted. Malnutrition causes immune deficiency. Late stages of the disease include severe susceptibility to infections of all kind and a general weakening of the body. Supporting normal immune function is the key, although preventing secondary infections is also important. Professional assistance and medical supervision should be sought in dealing with this very serious condition. The following products have been used to help those suffering with AIDS.

Therapies: SuperFood (pg. 15), Minerals (pg. 15), Enzymes (pg. 15), Essential Fatty Acids (pg. 17)

Remedies: Astragalus, Garlic, Lymphomax, Silver Shield, 7-Keto, AdaptaMax, Aloe Vera, Barley Grass, BP-X, Cellular Energy, Collatrim, Defense Maintenance, DHEA-F, DHEA-M, E-Tea, Geranium, GreenZone, Guardian, IF-C, Immune Stimulator, Lymphatic Drainage, Milk Thistle Combination, N-acetyl Cysteine, Pau D Arco, Protease, RG-Max, Silver Shield, Spirulina, St. John's Wort, SUMA Combination, Super Algae, Thai-Go, THIM-J, Uña de Gato Combination, Vitamin B-12, VS-C

ADD/ADHD

See *Attention Deficit Disorder (ADD)*

Addictions (alcohol)

See Also *Addictions (General), Hypoglycemia*

Alcohol is, essentially, a pure, refined carbohydrate, which contributes to hypoglycemia. While a moderate amount of alcohol (such as a glass of wine with a meal) does not seem to cause any serious health problems, excessive alcohol consumption does. Over-consumption of alcohol damages the liver and brain, destroys personal relationships, and is the number one cause of traffic accidents.

As is stressed under *Addictions (General)* alcoholics need to seek outside assistance to obtain the social and emotional support they need to overcome the habit. Good nutrition can help the process. Cravings for alcohol increase with poor nutrition, so following the Zone Diet plan and dealing with the problem of hypoglycemia will be helpful.

B-Complex or Balanced B-Complex vitamins, Chromium GTF and licorice root can all help stabilize sugar (glucose) levels in the blood and reduce alcoholic cravings.

Some research indicates that evening primrose oil helps reduce the craving for alcohol. Also try 1,000 mg. of vitamin C a day. Alcohol robs the body of large amounts of magnesium, so Magnesium Complex may also be helpful.

To soothe the nerves and resolve headaches try hops, chamomile or a nervine combination like Stress-J, Nerve Eight or Nutri-Calm.

An excellent formula to aid the recovering alcoholic is Kudzu/St. John's Wort. Kudzu is a vine common to the southern states with the remarkable properties of reducing high blood pressure, relieving pain (analgesic), and relieving cramps (antispasmodic). Studies have demonstrated it also possesses the ability to help control cravings for alcohol.

In laboratory studies, alcoholic golden hamsters voluntarily and significantly reduced their alcohol consumption when given a water extract of kudzu. In clinical practice, some people have reduced (not stopped) their alcohol consumption. The flowers of kudzu have been used to treat alcohol poisoning (hangover). St. John's wort adds the benefits of bringing the nervous system back into balance and helping to dispel negative emotions associated with alcoholism.

The most addictive thing about beer drinking isn't the alcohol; it's an herb used in making the beverage hops. This herb contains a sedative substance with a mild addictive effect. Hops can be taken to reduce cravings for beer, providing a similar relaxing effect without the negative problems associated with alcohol.

Since the liver works overtime to neutralize alcohol when levels are too high in the blood, the liver itself breaks down after long abuse. The liver can repair itself; all it needs is rest. Milk Thistle can be especially helpful for protecting the liver from the effects of alcohol, or in helping the liver to heal. When the liver has suffered severe damage due to alcoholism, Milk Thistle and SAM-e, along with topical application of helichrysum essential oil over the liver area will promote healing.

Therapies: Hydration (pg. 10), Colon Cleansing (pg. 18)

Remedies: Adrenal Support, Blood Build, Kudzu/St.John's Wort, Milk Thistle Combination, Nutri-Calm, SAM-e, Bergamot, Black Currant Oil, Chamomile, Clary Sage, Eucalyptus, Evening Primrose Oil, FCS II, Gastro Health, Helichrysum, Herbal Sleep, Hops, HY-A, L-Glutamine, Lemon Oil, Licorice Root, LIV-J, Liver Cleanse Formula, Magnesium, Marjoram (Sweet), Mood Elevator, Nerve Control, Nerve Eight, Niacin, Passion Flower, Rose Bulgaria, Spirulina, St. John's Wort, Thai-Go, Thyme, Valerian Root, Vitamin B-Complex, Wild Yam, Yucca

Addictions (coffee, caffeine)

See Also *Addictions (General)*

Caffeine may be the most widely used of all drugs, because it'simply isn't regarded as a drug. Adults use it freely, and parents allow children to drink caffeinated sodas as casually as if it were lemonade. In spite of its innocuous reputation, excessive use of caffeine may cause serious damage to the brain and central nervous system. High blood pressure cases are particularly at risk.

The continued, habitual use of caffeinated beverages (coffee, tea, kola drinks) to sustain energy causes a person to become addicted to these substances. This addiction continues until the person is unable to sustain normal energy levels without the use of caffeine. Chocolate also contains some caffeine.

Caffeine stimulates epinephrine production, upping the function of the central nervous system. Habitual use of caffeine causes adrenal fatigue, anxiety, nervousness, insomnia and other nervous symptoms. It depletes the adrenals, causing fatigue. The longer caffeine is used, the more depleted the adrenals become and the more the person craves caffeine. Herbs that strengthen the adrenals such as licorice root, Adrenal Support and Nervous Fatigue Formula are helpful in caffeine withdrawal. These same products can also help with sugar addiction, which is often related to caffeine addiction.

NSP's Caffeine Detox homeopathic may also help reduce caffeine (and chocolate) addiction. When a pick-me-up is needed, Energ-V, Cellular Energy or Target Endurance are safer ways to increase energy levels. Liquid Chlorophyll added to a glass of pure water provides a natural energy boost.

Herbal Beverage can be used as a coffee substitute. Many people start by blending coffee and Herbal Beverage, then gradually increase the amount of Herbal Beverage while reducing the amount of coffee.

Therapies: Hydration (pg. 10), Low Glycemic Diet (pg. 11), Stress Management (pg. 8), Essential Fatty Acids (pg. 17)

Remedies: Adrenal Support, Caffeine Detox, Herbal Beverage, Target Endurance, Cellular Energy, Eleuthero, Energ-V, Juniper Berries, Licorice Root, Nervous Fatigue Formula

Addictions (drugs)

See Also *Addictions (General)*

Drug addictions are a compulsive psychological and often painful physiological craving for a drug. Withdrawal from drug addiction requires both detoxification and nervous and glandular system support.

When it comes to drugs, you have to define your terms. Drugs may refer to the socially acceptable drugs such as caffeine, tobacco or alcohol, or hard drugs such as meth, rave, crystal, ice and crack cocaine. Drugs could also mean medicinal drugs available only by prescription, or over-the-counter drugs, the ones that are so heavily promoted on TV.

Like any other addiction, social support and general good nutrition will be helpful in withdrawal from any drug. Professional assistance should be sought for help with drug withdrawal, but good nutrition and supplements can be helpful.

Whatever the drug, one thing is certain. It will become a burden on the liver. So, herbs which help to detoxify the blood and strengthen the liver are probably central to any nutritional program for drug withdrawal. Consider formulas like Milk Thistle Combination, All Cell Detox or Enviro-Detox. Although the amounts will vary from individual to individual, a person will generally need three to four capsules of these herbs three times per day. It is also important to drink lots of pure water to help flush drugs from the system.

Anyone withdrawing from drugs is going to experience mental and emotional stress. Hence, a nervine formula such as Stress-J, Stress Relief, or Nervous Fatigue Formula would be another important component of a drug withdrawal program.

Overcoming an addiction to tranquilizers can be aided by herbs that also provide a relaxing effect, such as Herbal Sleep, valerian or kava kava. In other words, if you can find an herb that has a similar, but milder, more natural effect, use it as a substitute for the drug to ease withdrawal.

As with alcohol and tobacco addiction, adrenal support may be necessary because blood sugar imbalances are often part of addictive behavior. Licorice root and Nutri-Calm can be very helpful.

Therapies: Hydration (pg. 10), Affirmation and Visualization (pg. 8), Colon Cleansing (pg. 18), Stress Management (pg. 8), Essential Fatty Acids (pg. 17)

Remedies: Adrenal Support, Milk Thistle Combination, Nutri-Calm, All Cell Detox, Bayberry, BP-X, Chamomile, Enviro-Detox, GABA Plus, Ginseng (Korean), Herbal Sleep, Kava Kava, L-Glutamine, Licorice Root, Nature's Chi, Nerve Control, Nerve Eight, Nervous Fatigue Formula, Niacin, St. John's Wort, Stress-J, THIM-J, Valerian Root

Addictions (general remedies for)

See Also *Addictions (Alcohol), Addictions (Coffee, Caffeine), Addictions (Drugs), Addictions (Sugar or Food), Addictions (Tobacco)*

Addictions are all around us, but they are often difficult for people to acknowledge or talk about. An addiction is a dependency that creates a compulsive or habitual need to repeat an experience. The addiction may be mild or severe, socially acceptable or socially unacceptable.

When we think of addictions, we commonly think of socially unacceptable addictions, specifically addiction to drugs, but many addictions are socially acceptable. Two of the most common addictive substances in North America are tobacco and alcohol, and both are widely accepted socially, in spite of their potential dangers. A car is a lethal weapon when the driver is under the influence of alcohol, while smoking damages the lungs not only of smokers, but of people who are inhaling the secondhand smoke.

Alcoholism and cigarette smoking are obvious addictions, but an even more common addiction is caffeine. How many people just can't get through the day without a cup of coffee or a caffeinated soda? And what about food? With 70-80% of the people in this country being overweight, how many people are suffering from food addictions?

Behaviors can be addictive, too. People can become compulsive about sex, gambling, or shopping. What makes a behavior addictive is its compulsive and repetitive nature. The Greeks had a saying about this: Everything in moderation; nothing in excess. When we can't stop a behavior that is damaging to ourselves and to others, it is addictive.

What causes addiction? It's the desire to feel good! One reason people may become addicted to certain substances or behaviors is because they stimulate the release of neurotransmitters in their brain and nervous system, such as endorphins, dopamine, or epinephrine. Most addictive substances either mimic or trigger the release of these chemicals, all of which elevate mood and help us feel good. The body becomes accustomed to this outside stimulus, which becomes necessary to trigger the release of these chemicals, and the person becomes addicted.

Moderation is the key, and it's easier said than done. It's a dilemma that is deeply rooted in human nature, our need for ritual, our love of pleasure, and our cravings. Here are some basic suggestions.

First, it is critical to understand that one does not overcome addictions by willpower. It'simply doesn't happen. The instinctive drive for pleasure and self-satisfaction is too strong for us to resist. It easily subverts people's good intentions.

That's why the first thing anyone needs to do to overcome an addiction is to seek outside help. One of the reasons for the remarkable success of organizations like Alcoholics

Anonymous (AA) is they provide a support system that gives people accountability to a power outside of themselves.

Forming groups of people that meet together to support one another in overcoming addictions to drugs, alcohol or even overeating (weight loss) has proved to be one of the most successful models for helping people become free of addictive behaviors. So first, seek assistance from other people who have overcome the addiction you wish to overcome. Allow yourself to be accountable to outside influences, including spiritual sources, so you are not relying on your own willpower.

Secondly, since addictions are motivated by the inner desire we all share to feel good, improving overall health and nutrition will make it much easier to overcome addictions. When our diet contains a proper balance of nutrients, and we are otherwise taking care of the body, it produces the chemicals that make us feel good. This is the healthy way to feel good.

Finally, cleansing the body is very helpful in overcoming addictions, especially in helping a person going through withdrawal. A good general cleanse with a general blood and liver-cleansing formula can help the body flush the toxins created by the addictive substance out of the system.

Therapies: Stress Management (pg. 8), SuperFood (pg. 15)

Addictions (sugar or food)

The compulsive desire to eat sugar usually comes from a glandular imbalance. Adrenal and pancreatic imbalances are common. Often there is a deficiency of protein (or a lack of proper protein digestion and metabolism), good fats and trace minerals in the diet. Yeast infections may also be a culprit. Food addictions in general signify glandular and biochemical imbalances.

Excessive cravings for sweets or food in general can also come from a lack of sweetness (joy) in one's life which causes one to excessively seek pleasure through food. Learning to find other ways to have joy and pleasure in one's life can be helpful.

Therapies: Low Glycemic Diet (pg. 11), Hiatal Hernia Correction (pg. 12), Stress Management (pg. 8), Minerals (pg. 15), Enzymes (pg. 15), Probiotics (pg. 16), Essential Fatty Acids (pg. 17),Fiber (pg. 17)

Remedies: Adrenal Support, Target Endurance, 5-HTP, Bee Pollen, Candida Clear, Cellular Energy, L-Glutamine, Love and Peas, Lymphatic Drainage, Spirulina, Stevia, Sunshine Heroes Multiple Vitamin & Mineral, Sunshine Heroes Omega 3 with DHA, Sunshine Heroes Probiotic Power, Super Algae, Yeast/Fungal Detox

Addictions (tobacco smoking or chewing)

See Also *Addictions (General)*

Withdrawal from tobacco products requires support to the respiratory and nervous system. Nicotine is an alkaloid present in the tobacco plant that attaches to receptor sites in the sympathetic nervous system. Lobelia contains lobeline, an alkaloid with a similar structure that attaches to these sites and blocks them. The difference is that lobeline relaxes the nerves, while nicotine stimulates them. Using lobelia can help reduce the craving for nicotine while lessening withdrawal symptoms.

Nervines like chamomile and Nutri-Calm help to calm the nerves and ease withdrawal symptoms. Since smoking depletes vitamin C in the body, a vitamin C supplement is often helpful.

Where smoking has damaged the lungs, Lung Support can help promote healing. This formula is good for dry and weak lungs. Cigarettes contain heavy metals, so consider doing a colon cleanse with the addition of Heavy Metal Detox and Tobacco Detox homeopathic.

Therapies: Antioxidants (pg. 17), Heavy Metal Detoxification (pg. 22), Affirmation and Visualization (pg. 8), Colon Cleansing (pg. 18), Stress Management (pg. 8)

Remedies: Lobelia, Nutri-Calm, Tobacco Detox, APS II, Catnip, Chamomile, Enviro-Detox, Herbal Sleep, Lemon Oil, Lung Support, Nerve Control, Nerve Eight, Pine Needle, St. John's Wort, Stress-J, Valerian Root, Vitamin C

Addison's Disease

See Also *Adrenals (Exhaustion, Weakness or Burnout)*

Addison's disease is a severe depletion of the adrenal cortex. It results in extreme weakness, loss of weight, low blood pressure, gastrointestinal disturbances, and brown pigmentation of the skin and mucous membranes.

Therapies: Low Glycemic Diet (pg. 11), Stress Management (pg. 8)

Remedies: Adrenal Support, Licorice Root, AdaptaMax, Eleuthero, SUMA Combination

Adenitis

See Also *Congestion (lymphatic)*

Inflammation of a lymph node or gland. The following remedies may be helpful.

Remedies: Lymph Gland Cleanse, Lymph Gland Cleanse-HY, Lymphatic Drainage, BP-X, CBG Extract, Golden Seal, IN-X, Kidney Drainage, Nature's Fresh, SF, Thai-Go, Ultimate Echinacea, VS-C, Yucca

Adrenals (exhaustion, weakness or burnout)

The adrenals are one of the endocrine glands. They sit on top of the kidneys and produce a wide variety of hormones that regulate energy, fluid balance and other important body processes. They also help the body cope with or adapt to stress. Long-term stress, excessive use of sugar and caffeine and loss of sleep can weaken the adrenal glands. Symptoms of adrenal exhaustion include fatigue with restless sleep, difficulty concentrating, quivering tongue, dark circles under the eyes and pulsing pupils. Post traumatic stress disorder and Addison's disease involve exhausted adrenals. Adaptagens, such as AdaptaMax, eleuthero or Nervous Fatigue Formula are the most important remedies for normalizing adrenal function.

Two of the best remedies for rebuilding the adrenals are Adrenal Support and Nervous Fatigue Formula. When the adrenals are severely depleted Adrenal Support is the best choice. Where sugar cravings are a problem licorice root is helpful. Licorice has a sparing effect on adrenal hormones and strengthens adrenal function.

It is essential to eliminate caffeine from the diet when treating adrenal exhaustion. It is also helpful to eliminate refined sugar.

Therapies: Antioxidants (pg. 17), Stress Management (pg. 8)

Remedies: AdaptaMax, Eleuthero, Licorice Root, Mineral-Chi Tonic, Nervous Fatigue Formula, Nutri-Calm, Pantothenic Acid, Adrenal Support, Astragalus, Devil's Claw, Energ-V, Geranium, Herbal Beverage, Kelp, Maca, Magnesium, Master Gland, Parsley, Sea Salt, Spirulina, SUMA Combination, Sunshine Heroes Whole Foods Antioxidant, Target P-14, Thai-Go, Vitamin B-12, Vitamin C, Yucca

Afterbirth Pain

See Also *Pain (general remedies for)*

After childbirth there is often discomfort in the abdominal and pelvic regions from the exertion, contractions and stretching that occurred in labor. Baths (particularly sitz baths) using the essential oils of lavender and rose bulgaria may also be helpful. Using red raspberry during pregnancy helps to prevent this problem by toning the uterine muscles. Other remedies to consider include nervines and pain relievers.

Remedies: Lavender, Red Raspberry, APS II, IF Relief, Jasmine Absolute, Magnesium, Rose Bulgaria, Valerian Root

Age Spots

See Also *Free Radical Damage, Sunburn*

Also called liver spots or solar lentigo, age spots are pigments on the skin that are usually caused by overexposure to the sun. Age spots are typically treated with methods that cause superficial destruction of the skin which can leave white spots and occasional scars. They may be reduced or eliminated by using antioxidants and other remedies that heal and protect the skin against free radical damage, such as the following.

Therapies: Antioxidants (pg. 17)

Remedies: Nature's Fresh, Triple Effect Age Relief, Ginkgo & Hawthorn, Master Gland, Pau D Arco, Rose Bulgaria, Thai-Go, Vitamin A & D, Vitamin E

Aging (prevention)

See Also *Free Radical Damage*

Many people spend a lot of time and energy saving, investing and planning for retirement. Unfortunately, most of these people will develop chronic and degenerative health problems as they age, which diminish the quality of life they experience in their senior years. It won't do much good to have a fat bank account if they die of a heart attack or cancer, or suffer from crippling arthritis or other conditions that prevent them from enjoying life as they grow older.

When we are planning and preparing for our senior years, we ought to invest some time, effort and money into improving and maintaining our health at the same time. And remember that government health care programs like Medicare and private health insurance policies don't really insure good health. All they cover is the cost of disease care the use of drugs and surgery to treat symptoms of disease after they develop. This is not the same as investing in creating good health.

An investment in good health isn't all that complex. It doesn't even require that much self-discipline. Self-discipline suggests some type of self-deprivation, but caring for your health is the exact opposite of self-deprivation. Instead, it is self-nurturing.

What it takes to invest in one's health is to form positive health habits. Once you start making these investments and notice how much better you feel, you ll want to keep investing because the physical, mental and emotional dividends will become obvious. Here is a checklist of places to invest time, energy and money in your health.

Eat Quality Food

Obviously, good nutrition is the place to start. The basic investment rule here is simple; avoid putting your money into refined and processed foods and purchase whole, natural and organically grown foods instead.

The most important dietary habit you can establish is to eat five to seven 1/2 cup servings of fresh fruits and vegetables every day. It's also important to select whole grain products and natural sugars (such as raw honey, xylitol, organic natural brown sugar, etc.) over their refined and processed counterparts. If you crave sugary snacks and white flour products, you need to eat more healthy fats and protein.

When selecting fats, avoid margarine, shortening and partially hydrogenated vegetable oils. Instead, use olive oil, organic butter and cream from grass-fed cows, coconut oil, avocados and nuts as good sources of quality fats. Starting your day with a tablespoon of coconut oil will greatly reduce sugar cravings, too.

Food is both the fuel that energizes your body functions and the source of raw materials to produce healthy structures. Eating cheap junk food is no way to save money. It will reduce your energy, weaken your tissues and you ll wind up spending far more in doctor and hospital bills than you saved by eating low quality food.

Drink Water and Breathe Deeply

Two very simple things one can do to insure one's health are to drink plenty of pure water and practice deep breathing. For optimal health one needs to drink about 1/2 ounce of water per pound of body weight every day. Practicing any form of deep, abdominal breathing will also greatly benefit one's health. These two practices alone can reduce pain and inflammation throughout the body while increasing energy levels and overall health.

Balance Rest and Exercise

It is very important to stay physically active as one grows older. A rigorous exercise program isn't necessary, but some form of moderate physical activity such as walking, swimming or gentle bouncing on a mini-trampoline is essential to good health. The lymphatic system stagnates when we aren't breathing deeply and moving around. This causes toxins to accumulate in the system and contributes to aging and degenerative disease. A daily stretching routine is also beneficial.

Of course, it's also important to balance activity with rest. As we grow older, it's natural to slow down a little. Taking short naps or otherwise resting when tired is good for our health. It's also important to get a sound night's sleep, something that often becomes a challenge as we get older.

If you re having trouble sleeping, there are numerous herbs and supplements that can help such as Herbal Sleep or Nervous Fatigue Formula.

Use Appropriate Supplements

Even if you re eating a healthy diet, you can take out some additional health insurance by selecting a few well-chosen supplements. These can be general supplements, designed to support overall nutrition, or specific supplements to address common health concerns associated with aging. Here are a few ideas for some basic supplements.

Super Trio is a great basic program containing three important supplements: Super Supplemental, a high quality multiple vitamin and mineral; Super Omega-3, a source for the essential omega-3 fatty acids; and Super ORAC, a powerful antioxidant and anti-inflammatory blend. Super Trio can provide basic nutrients that may be lacking in the diet, reduce the inflammation and oxidative damage associated with aging and degenerative disease and support general health.

Three other basic supplements to consider are Mineral Chi Tonic, Thai Go and Ultimate GreenZone. Various tonic herbs, like ginseng, ginkgo, cordyceps and ho shou wu are used in other cultures to help people stay healthier as they age. Other possibilities follow; pick the remedies that fit your needs the best.

Therapies: Hydration (pg. 10), Stress Management (pg. 8), SuperFood (pg. 15), Minerals (pg. 15), Enzymes (pg. 15), Essential Fatty Acids (pg. 17), Antioxidants (pg. 17), Low Glycemic Diet (pg. 11)

Remedies: **AdaptaMax, Cordyceps, Ginkgo/Gotu Kola, Ginseng (Korean), Ginseng (Wild American), Ho Shou Wu, Mineral-Chi Tonic, Thai-Go,** Alpha Lipoic Acid, Barley Grass, Bee Pollen, Brain-Protex, Carotenoid Blend, Co-Q10, Colostrum, DHEA-F, DHEA-M, Eleuthero, Ginkgo Biloba, Gotu Kola, Grapine, Green Tea Extract, GreenZone, IGF-1, L-Glutamine, Lemon Oil, Master Gland, Mega-Chel, Melatonin Extra, Nature's Gold, Noni (Morinda), Pantothenic Acid, Patchouli, Pregnenolone, Protector Pak, SUMA Combination, Super Antioxidant, Super ORAC, Super Trio, Trace Mineral Maintenance, Trigger Immune, Triple Effect Age Relief, Vitamin E, X-A, Yeast/Fungal Detox, Ylang Ylang, Zinc

AIDS

See *Acquired Immune Deficiency Syndrome (AIDS/HIV)*

Alcoholism

See *Addictions (Alcohol)*

Alkalosis

See *Overalkalinity*

Allergies (food)

See Also *Leaky Gut Syndrome*

A food allergy is an excessive immune reaction to a particular food. This is often due to the fact that the food is either improperly digested, or incompatible with a person's basic blood chemistry (that is, their blood type). Symptoms of food allergies are many and varied, but may include gastrointestinal disturbances (inflammatory bowel disorders and leaky gut syndrome), respiratory and lymphatic congestion (excess mucus, swollen lymph nodes), rashes, fatigue, headaches, autoimmune disorders, and even emotional disturbances. In fact, many common health problems trace back to food allergies. They may also be responsible for hyperactivity, restlessness, irritability, anxiety and depression.

There are several ways to test for possible food allergies. Using muscle testing may be helpful, but is not 100% reliable. A better method is to take the pulse, eat a small amount of a food, and test the pulse again. Any food that raises the pulse more than six beats per minute after ingestion is probably an allergen.

The best method, however, is to eliminate suspect foods for a few days. You can even fast for a few days consuming only water or juice. If symptoms improve, then food allergies may be involved. When a food is reintroduced and symptoms reoccur, that confirms food allergies are a problem.

In the search for possible food allergies, start with the foods that are incompatible with your blood type (O, A, B or AB). Tree of Light Publishing has Blood Type, pH and Nutrition charts that can help you in addition to Peter D Adamo's Eat Right for Your Type books.

The most important supplements to help with food allergies are enzymes. The liver may also need to be detoxified to overcome food allergies. A lack of trace minerals necessary for enzyme functions may also be involved. Colloidal minerals may help. Homeopathics may also help to desensitize a person's allergies. By taking a highly diluted supplement of the foods a person is allergic to, they can be desensitized.

Many food allergies involve Leaky Gut Syndrome. When a person suffers from allergic responses to raw, unprocessed foods like strawberries, pineapple or papaya, it is probably not an allergic response to the food at all. Typically, there is some toxin or irritant present in the person's body that this particular food stirs up. By using the above procedure (i.e., taking very tiny amounts of the food and gradually increasing them), the person's body will be able to gradually discharge the irritant and the food will no longer cause a reaction. This also works for many so-called allergic reactions to herbs.

Therapies: Antioxidants (pg. 17), Blood Type Diet (pg. 13), Hiatal Hernia Correction (pg. 12), Colon Cleansing (pg. 18), Enzymes (pg. 15), Probiotics (pg. 16)

Remedies: HistaBlock, Lactase Plus, Proactazyme Plus, Adrenal Support, Alfalfa, ALJ, Aloe Vera, Anti-Gas Formula, Bentonite (Hydrated), Black Currant Oil, Blood Build, Burdock, Chickweed, CLA, CleanStart, Co-Q10, Colostrum with Immune Factors, Devil's Claw, Energ-V, Food Enzymes, LIV-J, Liver Balance, MSM, Nature's Gold, Omega-3, Para-Cleanse, Protease, Red Beet Formula, Small Intestine Detox, Sunshine Heroes Probiotic Power, Sunshine Heros Whole Foods Papayazyme, Super GLA, Tofu Moo, Uña de Gato Combination

Allergies (respiratory)

See Also *Fungal Infections (Yeast Infections, Candida albicans), Leaky Gut Syndrome, Rhinitis, Allergic*

Respiratory allergies are caused by an overly sensitive immune response reacting to environmental substances. As the immune system reacts to neutralize irritants it causes mast cells to burst which releases histamine into the respiratory tissues causing symptoms such as respiratory congestion, sinus discharge, sneezing, watery eyes, itchy eyes, sinus pain and/or pressure, coughing and/or sore throat.

General detoxification of the body, especially the colon and liver, will desensitize and stabilize the immune reactions in the respiratory membranes as there is a strong connection between the health of the digestive membranes and the respiratory membranes. Yeast infections may also be a root cause.

HistaBlock is a formula that counteracts histamine reactions and can help ease allergic reactions. The herbal combination Four is also good at easing allergic reactions. Eyebright is a specific remedy for upper respiratory allergies involving thin, runny mucus with itchy nose and eyes and redness. The herbal formula EW, which contains eyebright, is helpful for this as well.

Remedies that stabilize mast cells will also help counteract allergies. Vitamin C, feverfew and burdock all help with mast cell stabilization. ALJ also has anti-allergic effects as well as helping to expel excess mucus from the respiratory tract. Other expectorant and decongestant formulas that may benefit respiratory allergies are Bronchial Formula and Sinus Support.

Homeopathics can help to desensitize the body to allergens. By taking a highly diluted remedy containing the frequency pattern of the allergy-producing substance, the body becomes desensitized. Where pollen allergy is a problem, taking small amounts of bee pollen may help to desensitize the body to allergens, however locally gathered pollen works best for this purpose.

Therapies: Antioxidants (pg. 17), Blood Type Diet (pg. 13), Hiatal Hernia Correction (pg. 12), Enzymes (pg. 15), Essential Fatty Acids (pg. 17), Colon Cleansing (pg. 18), Fiber (pg. 17)

Remedies: ALJ, Allergies-Hayfever/Pollen, Allergies-Mold/Yeast/Dust, Allergy, EW, Four, HistaBlock, IF Relief, Vitamin C, Bee Pollen, Black Currant Oil, Bronchial Formula, Burdock, CLA, CleanStart, Co-Q10, Colostrum with Immune Factors, Cough Syrup-DH, Devil's Claw, Elderberry Plus, Eyebright, Fenugreek & Thyme, Helichrysum, Licorice Root, Lobelia, Marshmallow & Fenugreek, MSM, Mullein, Nature's Gold, Omega-3, Para-Cleanse, Seasonal Defense, Small Intestine Detox, Sunshine Heroes Omega 3 with DHA, Sunshine Heroes Whole Foods Antioxidant, Sunshine Heros Whole Foods Papayazyme, Super GLA, Tei Fu Oil

Alzheimer's Disease

See Also *Dementia, Free Radical Damage, Heavy Metal Poisoning*

Alzheimer's is a degenerative disease of the central nervous system characterized by mental deterioration. Neurons in the brain that produce a neurotransmitter called acetylcholine are destroyed by free radical damage. Most of the remedies listed here are more for prevention than they are for a cure. Most herbal and nutritional remedies will slow the progress of Alzheimer s, but may not completely halt or reverse the progress of the disease.

High levels of aluminum have been found in the brains of Alzheimer's patients, but most experts now agree this is not a cause of the disease. However, it is still wise to protect the body from this heavy metal by avoiding aluminum cookware, especially when preparing acidic foods like tomatoes. Also avoid antiperspirant deodorants that contain aluminum compounds and baking powder with aluminum.

To prevent Alzheimer's eat good fats and avoid bad fats. Minimize exposure to alcohol, tobacco and environmental toxins. Eat generous servings of fresh fruits and vegetables every day and stay mentally active.

Once Alzheimer's has started, antioxidants can slow deterioration. Ginkgo/Gotu Kola may be helpful in enhancing cognitive function in Alzheimer's patients. It may also be helpful to improve circulation to the brain and chelate toxic metals by doing an Oral Chelation program.

Therapies: Antioxidants (pg. 17), Heavy Metal Detoxification (pg. 22), Oral Chelation (pg. 21), Essential Fatty Acids (pg. 17)

Remedies: Brain-Protex, Ginkgo/Gotu Kola, Heavy Metal Detox, Mega-Chel, 7-Keto, Alpha Lipoic Acid, Clary Sage, Co-Q10, DHEA-F, DHEA-M, Focus Attention, Ginkgo Biloba, Green Tea Extract, Lecithin, Lemon Oil, MSM, N-acetyl Cysteine, Super Algae, Super Antioxidant, Thai-Go, Vitamin B-12, Vitamin B-Complex

Amenorrhea

Amenorrhea is a lack of periods or menstrual flow. When periods are scant, infrequent, or absent it is important to try to identify the cause. The most common causes are extreme diets (especially diets low in fats and protein), excessive exercise, stress, general poor health, thyroid imbalances, and medications. Seek appropriate assistance in identifying the cause. Depending on the underlying cause, the following remedies may be helpful.

Therapies: Stress Management (pg. 8), Essential Fatty Acids (pg. 17), Minerals (pg. 15)

Remedies: Aloe Vera, Anamu, Artemisia Combination, Blue Cohosh, Clary Sage, Colloidal Minerals, Damiana, Dong Quai, False Unicorn, FCS II, Female Comfort, Feverfew, Flax Seed Oil, Geranium, Ginger, Myrrh, Rose Bulgaria, Safflowers, Stress-J, Super Algae, Thyroid Support, X-Action (Women s), Yarrow

Amyotrophic Lateral Sclerosis

See *Lou Gehrig's Disease*

Anal Fistula or Fissure

See Also *Hemorrhoids*

An anal fistula is an abnormal passageway in the anal area. It is also known as an anal fissure. Take fiber and lots of water to keep the stool soft and easy to pass. Apply astringent herbs like white oak bark mixed with Golden Salve topically to help close the fissure and promote healing.

Therapies: Fiber (pg. 17)

Remedies: White Oak Bark, Every Body's Fiber, Golden Salve, Horsetail, IF-C, Intestinal Soothe & Build, Magnesium

Anemia

Anemia indicates that there is a deficiency in red blood cells, hemoglobin (the component in red blood cells that carries oxygen) or in total blood volume. This results in fatigue, reduced resistance to illness and other symptoms. Iron alone is not necessarily the solution, since there are many other nutrients necessary for the utilization of iron and the production of hemoglobin.

To supply the body with iron take chelated iron or alfalfa, I-X, and/ or yellow dock. A good dose is 8-12 alfalfa or I-X capsules per day along with 4 capsules of yellow dock. This works better than most iron supplements. The Blood Build formula is also beneficial for overcoming anemia, especially in women who lose blood through menses.

Although it does not contain iron, liquid chlorophyll may be helpful in building up iron poor blood. Supplements that help the utilization of iron in the body include l-glutamine, vitamin B-12, vitamin B-6 and B-Complex.

Remedies: Alfalfa, Blood Build, Chlorophyll, Dong Quai, I-X, Vitamin B-12, Yellow Dock, All Cell Detox, Blue Vervain, Chickweed, Cordyceps, FCS II, Ginseng (Korean), Ginseng (Wild American), GreenZone, Herbal Trace Minerals, Iron, L-Glutamine, Marshmallow, Nervous Fatigue Formula, SF, Small Intestine Detox, Super Algae, Super GLA, Thai-Go, Vitamin B-6, Vitamin B-Complex, Vitamin C, Zinc

Aneurysm

An aneurysm is a blood-filled dilation of a blood vessel wall as a result of a blood vessel disease or weakness. Appropriate medical attention should be sought for this condition. The following may assist in recovery.

Remedies: Butcher's Broom, Capsicum, Yarrow, Alfalfa, Golden Seal, Magnesium, Rose Hips

Anger (excessive)

See Also *Irritability*

Excessive anger is often a sign of a hot or overactive liver function. It may also be due to stress and blood sugar issues. It can also be a sign of unresolved hurts and disappointments from the past that need to be worked through, forgiven and released. Liver Balance is often helpful in reducing excess anger and irritability. Distress Remedy can calm feelings and restore composure during acute episodes of anger.

Therapies: Gall Bladder Flush (pg. 23), Low Glycemic Diet (pg. 11), Colon Cleansing (pg. 18), Stress Management (pg. 8)

Remedies: Liver Balance, Bergamot, Caffeine Detox, Chamomile (Roman), Guardian, Helichrysum, Lavender, LIV-J, Patchouli, Rose Bulgaria, Sunshine Heroes Whole Foods Antioxidant, Ylang Ylang

Angina

See Also *Anxiety Disorders, Cardiovascular Disease (Heart Disease)*

Angina is chest pain that can range from mild to severe. It may be associated with heart disease, anxiety, a gallbladder problems or digestive problems. Severe abdominal bloating with a hiatal hernia can cause this problem. Seek appropriate medical attention for a proper diagnosis and then select appropriate remedies based on the cause.

Therapies: Antioxidants (pg. 17), Hiatal Hernia Correction (pg. 12), Stress Management (pg. 8), Essential Fatty Acids (pg. 17)

Remedies: Co-Q10, Magnesium, Nervous Fatigue Formula, RG-Max, RG-Max, APS II, Black Cohosh, Black Currant Oil, Blessed Thistle, Cardio Assurance, Folic Acid Plus, Gastro Health, Ginkgo & Hawthorn, Hawthorn Berries, HS II, Kidney Drainage, L-Carnitine, Lobelia, Marshmallow & Fenugreek, Neroli, Nutri-Calm, Olive Leaf, Sarsaparilla, Stress Relief

Anorexia

See Also *Anxiety Disorders, Appetite (Deficient)*

A prolonged loss of appetite creates the condition called anorexia. Without appetite a person does not eat and hence loses weight. Anorexia nervosa is a psychological disorder common among teenage women and includes a fear of weight gain. Seek appropriate medical attention when dealing with this condition.

To stimulate appetite take bitter herbs such as gentian or golden seal. It is absolutely essential that you taste these herbs as it is the bitter taste that has the effect. Use a liquid product or open the capsules and put the powders directly into the mouth. Digestive Bitters is OK, but is too sweet. Mix a little goldenseal powder with it to make it more effective. Carminatives and stomachics may also stimulate appetite when taken as teas prior to mealtime. Chamomile tea is a good choice.

For anorexia nervosa, support the nervous system. It may also be necessary to support the heart with HS II and stabilize blood sugar levels. Trigger Immune and Spleen Activator can also be used to increase appetite and overcome wasting and deficiency.

Therapies: Low Glycemic Diet (pg. 11), Stress Management (pg. 8)

Remedies: Digestive Bitters Tonic, HY-A, Nutri-Calm, Trigger Immune, Alfalfa, Anti-Gas Formula, Astragalus, Bee Pollen, Bergamot, Blessed Thistle, Catnip, Chamomile, Cinnamon, Geranium, Golden Seal, Grapefruit (Pink), Green Tea Extract, HS II, Licorice Root, Probiotics, Spleen Activator, Stress-J, Thai-Go

Antibiotics (alternatives to)

See Also *Infection (bacterial)*

There is no question about it. Antibiotics are one of the wonders of modern medicine and they have saved countless lives through their appropriate use.

Unfortunately, these valuable drugs are also commonly prescribed for conditions where they have little or no effect. For starters, antibiotics only work on bacterial infections, so they are worthless on viral or fungal infections. This means that there is absolutely no reason to take an antibiotic for the common cold or flu. Antibiotics are also ineffective in many, if not most, cases of sore throats, sinus infections, bronchitis, respiratory congestion and earaches (otitis media).

In spite of these facts, many people run to their doctor and practically insist on getting a prescription for an antibiotic for these types of health problems. What these people don't realize is that using antibiotics in this inappropriate manner will actually harm their health in the long run.

This is partly because antibiotics kill friendly bacteria in the intestinal tract. When these friendly bacteria are destroyed, yeast and infectious bacteria proliferate causing intestinal inflammation, Leaky Gut Syndrome and a weakened immune system. This makes the person even more susceptible to future infections.

An even more serious problem created by antibiotic overuse is the development of antibiotic resistant bacteria, sometimes called superbugs. Here's how this happens. Antibiotics never kill all the bacteria and the few that survive are the ones that are most resistant to the drug. Over time, the process of natural selection gradually creates strains of bacteria which can't be killed by that particular antibiotic.

Antibiotic resistance can develop very quickly. For instance, penicillin became widely used after World War II; it only took four years for microbes to start becoming resistant to penicillin. This is why new antibiotics have to be introduced regularly, but it's becoming a losing battle. Antibiotic resistance is now a worldwide problem, especially in hospitals and medical clinics. It has made diseases such as tuberculosis, gonorrhea, malaria, and childhood ear infections more difficult to treat than they were a few decades ago.

Prescribing antibiotics for colds and other viral infections and feeding livestock antibiotics for prevention has hastened the development of antibiotic resistance, which is why we need to put a halt to this abuse of antibiotics, especially when there are natural ways to treat most infections. Many of these natural remedies are not only effective against bacterial infections, they also work on viral and fungal infections. More importantly, bacteria do not seem to develop resistance to natural substances. Here are a few of the best remedies:

Remedies: Garlic, Silver Shield, Tea Tree Oil, Echinacea/Golden Seal, Fizz Active-Immune, Golden Seal, Guardian, Immune Stimulator, Thyme, Ultimate Echinacea

Antibiotics (side effects of)

See Also *Fungal Infections (Yeast Infections, Candida albicans)*

Antibiotics are antibacterial agents. They have no effect against viral conditions for which they are frequently prescribed. Unfortunately, antibiotics also destroy beneficial bacteria in the gastrointestinal and female reproductive tract. This allows yeast to multiply out of balance with other organisms in the body. Yeast produce a toxin that weakens the immune system. Long-term use of antibiotics results in reduced immunity.

After completing a round of antibiotics it is important to take probiotics in order to replace the friendly bacteria that have been destroyed. There are many species of friendly bacteria or lactobacillius besides the commonly recommended Lactobacillus acidophilus. See *Probiotics* in the *Product Section.*

It may also be helpful to take antifungal herbs after taking antibiotics to knock down yeast.

Therapies: Probiotics (pg. 16)

Remedies: Probiotics, Yeast/Fungal Detox, Candida Clear, Garlic, Immune Stimulator, Lymphatic Drainage, Milk Thistle Combination, Pantothenic Acid, Pau D Arco, Silver Shield, Sunshine Heroes Probiotic Power, Tea Tree Oil

Anxiety (Panic Attack)

See Also *Anxiety Disorders*

When a person experiences mild to severe apprehension or uneasiness over a present or impending event, with symptoms such as a cold sweat, heart palpitations, trembling, faintness, a sense of pressure in the chest over the heart area, and/or dry mouth, he or she is experiencing anxiety. A panic attack occurs when a state of anxiety is so severe that the person begins to breathe in a rapid, shallow manner and become tense to the point of cramping and the inability to act. A person who is prone to anxiety and panic attack usually needs adrenal, thyroid and nervous system support.

People with blood type A are more prone to anxiety and panic because they have a harder time breaking down the stress hormones produced by the adrenal glands. Adaptagens can help overcome the tendency to anxiety by reducing the production of stress hormones. Avoid caffeine and other stimulants. Taking B-complex vitamins, particularly B6, and eliminating refined sugar from the diet will also be helpful.

To relieve acute anxiety or panic attacks, it is necessary to calm down the function of the sympathetic nervous system with sympatholytic agents and to increase the activity of the parasympathetic nervous system with parasympathomemetic agents. To do this, try taking 10-15 drops of lobelia every 2-3 minutes or one capsule of kava kava every ten minutes while concentrating on breathing slowly and deeply. You can also take Distress Remedy.

Therapies: Hydration (pg. 10), Low Glycemic Diet (pg. 11), Hiatal Hernia Correction (pg. 12), Stress Management (pg. 8)

Remedies: AdaptaMax, Adrenal Support, Distress Remedy, Kava Kava, Lobelia, Magnesium, Nervous Fatigue Formula, 5-HTP, Bergamot, Blue Vervain, Chamomile (Roman), Damiana, FCS II, GABA Plus, Ginseng (Korean), Ginseng (Wild American), Gotu Kola, Herbal Beverage, Herbal Sleep, Hops, HSN-W, Kelp, Lavender, Master Gland, Neroli, Nerve Control, Nerve Eight, Nutri-Calm, Pantothenic Acid, Passion Flower, Patchouli, Rose Bulgaria, Rose Bulgaria, St. John's Wort, St. John's Wort with Passion Flower, Stress Relief, Stress-J, Thyroid Support, Valerian Root, Vitamin B-6, Vitamin B-Complex, X-Action (Women s), Ylang Ylang

Anxiety Disorders

See Also *Adrenals (Exhaustion, Weakness or Burnout), Insomnia, Stress*

Anxiety is a complex combination of feeling apprehension, dread, fear, nervousness and worry, in anticipation of problems or misfortune. So, when facing unknown or scary situations, it's perfectly normal to experience the sensation of butterflies or knots in your stomach that signal a touch of anxiety. However, for a large number of people, anxiety is something far more serious and persistent.

When anxiety is severe enough to interfere with family relations, socializing and work, it can be debilitating. It can manifest as shortness of breath, rapid heartbeat or heart palpitations, muscle tension, trembling, insomnia, irritability, chest pain, cold sweats, feeling faint and general feelings of stress. These symptoms are bad enough, but to make matters worse, anxiety contributes to the development of other health problems, including heart disease, high blood pressure, cancer, diabetes, and pain-related disorders such as arthritis and fibromyalgia. There is also a high correlation between anxiety and addiction to alcohol, smoking and drug use.

Clinicians recognize about 12 relatively distinct subtypes of anxiety disorder, but the major ones are: panic disorder, phobias, obsessive-compulsive disorder, post-traumatic stress syndrome and generalized anxiety disorder. These anxiety-related problems have reached epidemic proportions in the United States. According to the Surgeon General, 16 percent of adults ages 18 to 24 experience one subtype of an anxiety disorder that lasts at least a year. That's a lot of anxiety!

The large majority of those suffering with these disorders are holding full-time jobs, many at executive and managerial levels, and are experiencing a relatively high degree of workplace stress. Most are just masking the symptoms of their problem by taking some sort of medication, such as tranquilizers, antidepressants or sleeping pills. Others are self-medicating through alcohol, cigarettes and drugs. There are better ways to deal with anxiety, however.

To understand how to reduce anxiety in a healthy, natural way, we first need to understand the nature of anxiety. Anxiety is not a bad thing. It is a natural reaction to real or perceived dangers. The perception of possible danger triggers the release of hormones and neurotransmitters that prime the body and mind for action. These physiological changes can help us push beyond our normal limits and may actually help us perform better.

For instance, if a person feels anxious at the thought of giving a speech or performing in front of a group, the perceived danger of humiliating oneself in front of a large group of people creates a release of hormones that actually prime the person to perform better. So, the key is not to eliminate anxiety, but to keep our anxiety at manageable levels so we can perform well in the tasks before us, and not be paralyzed from action by excessive anxiety.

Fortunately, there are many simple, natural therapies that can regulate the production of stress-related hormones and ease anxiety-related problems. These natural approaches don't just temporarily relieve the symptoms, either. They can actually resolve anxiety problems and help a person get rid of that crippling anxiety for good.

Start by reducing the output of stress hormones by taking adaptagenic herbs like Eleuthero Root, Adaptamax, Suma Combination and Nervous Fatigue Formula. Also, start making more time for rest and relaxation. One doesn't have to try to avoid stress to reduce the effects of stress in one's life. Pleasurable experiences trigger the release of hormones and neurotransmitters that counteract the effects of stress and a pleasurable experience creates more positive benefits than a stressful experience causes harm. So, instead of reducing stress, we should be deliberately creating pleasure and enjoyment in our lives.

It's likely that a major part of the reason anxiety-related disorders are epidemic in our society is because we are just too busy. We are constantly on the go, and take very little time for pleasure and recreation. Making sure we plan time to do enjoyable things is very important to our emotional and physical health. Consider activities like a warm bath, a soak in a hot tub, a massage, listening to relaxing music or taking a walk in the park.

As part of this process, make sure you re getting enough sleep. If you have trouble getting to sleep there are many remedies you can try. What works for you will depend on the specific problems inhibiting your sleep. See Insomnia for more information.

People who suffer from anxiety usually have a lot of tension in their muscles. Anything that helps to stretch muscles and get them to relax will reduce feelings of stress and anxiety. Good choices include stretching exercises, yoga, tai chi or massage therapy.

You can also use supplements to help muscles relax. Lobelia and kava kava are great antispasmodic herbs that can relax muscles and reduce anxiety. For long-term use, kava kava is the best choice. It can help you stay relaxed, improve your mood, but keep your mind alert and focused.

Finally, learn to quiet your mind. All of us have a constant flow of verbal monkey chatter going on in our minds. In some people, these thoughts can become so obsessive that they lead to constant states of worry, fear, anxiety and obsession. Learning to calm the mind through prayer and meditation can be very helpful. In many cases, counseling may be needed to help a person learn to step back from these negative thoughts and replace them with more positive ones.

One way of calming the mind is to center one's attention more on directly experiencing the world around us. While breathing slowly and deeply, look around you and notice colors, sounds, textures and smells. Touch things and feel their texture. Coming to your senses in this manner counteracts the effects of stress and reduces anxiety. And it doesn't cost anything!

These techniques, along with basic good health practices, can help you recover from anxiety disorders. If the problem is severe, be sure to seek appropriate professional help.

Therapies: Hydration (pg. 10), Low Glycemic Diet (pg. 11), Stress Management (pg. 8), Hiatal Hernia Correction (pg. 12), Affirmation and Visualization (pg. 8), Colon Cleansing (pg. 18), Stress Management (pg. 8), SuperFood (pg. 15), Essential Fatty Acids (pg. 17)

Remedies: Adrenal Support, Eleuthero, Kava Kava, Nutri-Calm, Stress Relief, AdaptaMax, Lobelia, Nervous Fatigue Formula, Stress-J, SUMA Combination, Vitamin B-Complex

Apathy

See Also *Depression*

Apathy is an I don't care attitude that may have its roots in wounds to the self-esteem or a lack of energy or drive. Aromatherapy or flower essence therapy can be helpful.

Therapies: Affirmation and Visualization (pg. 8)

Remedies: Bergamot, Depressaquel, Geranium, Jasmine Absolute, Lemon Oil, Patchouli, Peppermint, Rose Bulgaria, Rosemary

Appendicitis

Inflammation of the appendix is called appendicitis. It is characterized by inflammation (heat, swelling and pain) in the lower abdomen, just inside the right hip. This is a serious condition because the appendix can rupture resulting in a life-threatening infection of the abdominal cavity. For this reason, stimulant laxatives and enemas should be avoided when having an acute episode of appendicitis.

Dr. Henry Bieler, in his book *Food is Your Best Medicine*, speaks of treating appendicitis without surgery by putting the patient on a diet of clear liquids (no solid food) and giving doses of antibiotics. It is highly possible that the same results could be achieved with herbs using large doses of garlic and goldenseal (two cloves of garlic every two hours or one High Potency Garlic every four hours along with four capsules of goldenseal every two hours). Silver Shield in high doses (1/4 to 1/2 bottle per day) would be another option. Five hundred milligrams of vitamin C every 2-4 hours would also be helpful.

Due to the seriousness of this condition, however, we recommend that appropriate medical attention be sought before undertaking any course of treatment.

Remedies: Garlic, Golden Seal, Silver Shield, Vitamin C, Cellular Energy, IF Relief, Lymphatic Drainage, Target Endurance

Appetite (deficient)

See Also *Anorexia, Wasting*

A deficient appetite is a lack of desire for food. A loss of appetite is often due to illness, nervous or glandular system problems, stress, or a congested digestive tract and liver. A loss of appetite can result in a loss of weight and health. The appetite can be stimulated by tasting bitter herbs such as Digestive Bitters, goldenseal or gentian about 20 minutes prior to eating. Digestive Bitters is quite sweet, so mixing a little goldenseal powder into it will make it more effective. Chamomile tea before meals may also be helpful.

Therapies: Colon Cleansing (pg. 18), Enzymes (pg. 15)

Remedies: Digestive Bitters Tonic, 5-HTP, Alfalfa, Anti-Gas (Chinese), Anti-Gas Formula, Bee Pollen, Blessed Thistle, Cellular Energy, Chamomile, Dandelion, Golden Seal, Lemon Oil, Licorice Root, Liver Cleanse Formula, Lymphatic Drainage, Myrrh, Oregon Grape, PDA, Pine Needle, Proactazyme Plus, Vitamin C, Zinc

Appetite (excessive)

See Also *Weight Loss (aids for)*

When a person craves food beyond what is nutritionally necessary to sustain life, health and optimum body weight, there may be a need for substances to curb appetite. Often the problem is related to eating too many empty calories, that is, foods that have a very low content of vitamins, minerals and other micronutrients such as refined sugar and white flour. By supplying the body with micronutrients, the body will crave less food, as its nutritional requirements will be satisfied. There may also be hormonal or nervous system imbalances.

Supplements that help to curb appetite in this manner include mineral rich supplements like alfalfa, Colloidal Minerals or Mineral Chi Tonic, protein-rich foods like Super Algae, spirulina and barley juice powder, and essential fatty acids like Super Omega 3 EPA and flax seed oil. Coconut oil is also helpful.

Excessive appetite can also be a sign of hormonal imbalances. Serotonin helps control both mood and appetite, so 5-HTP Power may help when excessive appetite is linked to feelings of depression and/or insomnia. Other supplements that may help to curb appetite are listed below.

Therapies: Hydration (pg. 10), Low Glycemic Diet (pg. 11), Stress Management (pg. 8), SuperFood (pg. 15), Minerals (pg. 15), Enzymes (pg. 15), Essential Fatty Acids (pg. 17), Fiber (pg. 17)

Remedies: Appetite Control, Mineral-Chi Tonic, 5-HTP, 7-Keto, Barley Grass, Black Walnut, Blood Build, Catnip & Fennel, Chickweed, Colloidal Minerals, Flax Seed Oil, Garcinia Combination, Grapefruit (Pink), MetaboMax, Nature's Hoodia, Nerve Eight, Nutri-Burn, Omega-3, Psyllium, Red Beet Formula, SF, Spirulina, Spleen Activator, Sunshine Heroes Multiple Vitamin & Mineral, Sunshine Heroes Omega 3 with DHA, Super Algae, Super GLA

Arrhythmia

Arrhythmia refers to an irregular heartbeat. This problem can be difficult to treat with herbs and nutrition. It is sometime congenital (inherited) and in such case cannot be effectively eliminated. It can be due to a lack of minerals or nervous stress. A hiatal hernia may also cause arrhythmia. Seek appropriate medical assistance to help determine the severity of the problem and, if possible, the cause.

Arrhythmia will sometimes respond to cardiac tonics such as hawthorn berries, CardioAssurance or HS II. In some cases it will respond to nervines such as passion flower, lobelia or valerian. Magnesium or trace minerals like Colloidal Minerals are often needed. Adaptagens like wild American ginseng or eluthero Root may also help to regulate the cardiac rhythm in some cases. Mineral Chi Tonic may help in some cases because it'supplies minerals and has adaptagenic properties.

Some of the best herbs for arrhythmia are toxic botanicals not available from Nature's Sunshine. You may be able to find them through a professional herbalist. These include lily of the valley, night blooming cerus, arnica and mistletoe. In the absence of being able to find appropriate herbal remedies, drug medications may be needed to stabilize the condition. When suffering from arrhythmia avoid stimulants like caffeine, tobacco, alcohol, ephedra, licorice root and peppermint essential oil.

Therapies: Minerals (pg. 15), Hiatal Hernia Correction (pg. 12)

Remedies: Hawthorn Berries, Magnesium, Passion Flower, Black Cohosh, Cardio Assurance, Eleuthero, Ginseng (Korean), Ginseng (Wild American), HS II, Lobelia, Mineral-Chi Tonic, Rose Bulgaria

Arteriosclerosis (Atherosclerosis, Hardening of the Arteries)

See Also *Cardiovascular Disease (Heart Disease), Free Radical Damage, Hyperinsulinemia (Syndrome X)*

The number one cause of death in the United States is cardiovascular disease. Through improper diet and toxic damage to the body, people develop a thickening of interior vessel walls, usually by fatty or mineral deposits that reduce the opening size and obstruct blood flow. This process is probably precipitated by free radical damage.

Restricted blood flow prevents tissues from getting oxygen and nutrients, which prevents wounds from healing, causes coldness in the extremities and affects memory and cognitive function. When a blood clot forms in the circulatory system it may get lodged in the heart, causing cardiac arrest, or in the brain, causing a stroke.

Although cholesterol has been implicated as the primary culprit in hardening of the arteries, more recent research is showing this is simply not the case. High insulin levels are a much better indicator of cardiovascular risk than is high cholesterol. Free radical damage or oxidative stress is at the root of this problem, so reducing free radical damage with antioxidants like Co-Q10 and Super ORAC can be very helpful in preventing this problem.

A little known factor that contributes significantly to this problem is the chlorination of water supplies. Chlorine is a free radical that turns high-density lipoproteins into low-density lipoproteins. Put into layman's terms, chlorine makes cholesterol gummy so that it'sticks to the lining of the arteries. Purifying both your drinking water, and the water in which you bathe or shower, can help to prevent this condition. Supplementation with iodine will counteract chlorine as well.

Once arteriosclerois is present, an oral chelation program can help reverse it. Garlic, capsicum and herbs to support liver detoxification will also be helpful.

Therapies: Oral Chelation (pg. 21), Essential Fatty Acids (pg. 17),Fiber (pg. 17), Antioxidants (pg. 17), Hydration (pg. 10), Low Glycemic Diet (pg. 11)

Remedies: Garlic, Mega-Chel, Black Currant Oil, Butcher's Broom, Capsicum & Garlic with Parsley, Cardio Assurance, Carotenoid Blend, Chromium GTF, Flax Seed Oil, Folic Acid Plus, GC-X, Ginkgo Biloba, Ginseng (Korean), Ginseng (Wild American), Grapefruit (Pink), Green Tea Extract, Hawthorn Berries, Hydrangea, Kelp, Lecithin, Olive Leaf, Protease, Rose Hips, SF, Thai-Go, Vitamin E, Zinc

Arthritis

See Also *Rheumatoid Arthritis (Rheumatism)*

Arthritis is quite simply inflammation of a joint. But, behind this deceptively simple definition, there are actually many different kinds of arthritis, as well as a welter of misconceptions and misinformation about its causes, treatment and prognosis. When dealing with this disturbingly common disease (more than 20 million Americans are thought to suffer from it) it is necessary to look at its underlying causes of and deal with them to achieve any kind of real results.

Most people are at least somewhat familiar with two of the most common forms of arthritis. Osteoarthritis affects more than 15 million Americans and occurs most frequently in older people. In osteoarthritis, the cartilage which coats the ends of the bones in our joints begins to break down, starting a vicious cycle of damage, reduced function and health, leading to more damage. It is not a systemic disease, meaning it is the result of damage from local wear and tear, trauma, surgery or infection to a specific joint. It can also result from the effects of other diseases.

Symptoms of osteoarthritis include:

Pain in the affected joint(s) after repeated use, especially later in the day.

Swelling, pain and stiffness after long periods of inactivity, like sleep, that subside with movement and activity.

Continuous pain, even at rest, with advanced osteoarthritis.

By contrast, rheumatoid arthritis is much more rare, occurring in less than one percent of the population. Unlike osteoarthritis, rheumatoid arthritis is an autoimmune disease, meaning the tissues that surround and cushion the joints are attacked by the body's own immune system. This happens throughout the whole body, not just in joints that have been subjected to wear and tear. It usually occurs between the ages of 25 and 50, but it can develop at any age, and generally strikes women three times as often as men.

Symptoms of rheumatoid arthritis include:

Swollen, warm, painful joints, especially after long periods of inactivity.

Fatigue and occasional fever.

Symmetrical pattern of inflammation if one wrist is involved, the other will be also.

The small joints of the body (hands, fingers, feet, toes, wrists, elbows and ankles) are usually affected first.

As the disease progresses, the joints often will become deformed and may freeze in one position, making it difficult to move them.

In addition to these most common forms of arthritis, there are some other recognized forms.

Allergic arthritis, as the name implies, is triggered by an allergic reaction. This type of arthritis has actually been produced in the laboratory by injecting research subjects with allergenic substances. The connection between allergies and arthritis is easier to understand if we consider that both of these conditions are caused by inflammation in the body, and that the inflammation itself can become chronic when an individual is repeatedly exposed to allergenic substances over time.

Gonorrheal arthritis is inflammation of the joints resulting from gonorrheal infection.

Gouty arthritis was the most widely known variety up until the 20th century. Caused by an imbalance of uric acid in the blood, it also causes joint inflammation and usually affects one joint at a time.

Hemophilic arthritis results from bleeding into a joint in someone who is a hemophiliac. This often results in joint stiffness and inflammation.

Menopausal arthritis can occur because of hormonal imbalances experienced during menopause.

Tuberculous arthritis is joint inflammation found in people infected by tuberculosis where the infection has spread into the joints.

What all these forms of arthritis have in common is that the joints have been subjected to some kind of stress: mechanical, biochemical or infectious. The stress leads to irritation of the tissues that causes inflammation. The irritation and inflammation lead to physical and chemical changes that send the joint(s) into a downward spiral of breakdown, then toxification, which creates more irritation, more inflammation, and further breakdown.

In advanced cases of arthritis, irreversible deformation and degeneration may have occurred. But even in these situations, the foundation for healing lies in breaking the vicious cycle by cleansing the body of the toxins that have caused or resulted from the tissue damage. Once this is done, then nutrition can be used to aid the body's ability to heal itself.

One of the misperceptions people have about bones is they associate them with the bleached-white and dried-out bones from creatures which have died. Living bones aren't static, dead objects, they are composed of living tissue. This makes them capable of growth, change and repair. In fact, if joints weren't alive they couldn't become inflamed in the first place. Recognizing that bones and joints are living tissue helps us realize that they are capable of self-repair, if we remove the sources of irritation and supply them with the tools they need to repair themselves.

Think of an engine that hasn't had its oil changed in a long time as it gets dirtier and dirtier, it will start to function less and less effectively until it finally it'stops working. To extend the analogy, an engine that has been stressed by contaminants or poor quality fuel or by being driven at excessive speeds or under excessive loads will eventually malfunction.

Joints are meant to endure a certain amount of wear and tear, but when toxins and inflammation are present, it creates more friction in the joints (just like the dirty oil in the car). Furthermore, when nutrients needed for joint health aren't there then repairs can't be made, which makes the joint more easily damaged and inflamed.

There are three things that need to be done to help arthritis to heal. First we need to identify and remove sources of stress, whether they are mechanical, biochemical or infectious. Secondly, we need to reduce inflammation and tissue toxicity, which is like changing the dirty oil in the engine and replacing it with fresh oil. Lastly, we need to supply nutrients necessary for joint health to aid the body in effecting repairs.

1. Remove Sources of Irritation

In osteoarthritis, reducing mechanical stress to the joints is an important key. This mechanical stress is often the result of repetitive habits of movement and posture that were not properly balanced. Correcting structural alignment through stretching, yoga, massage, or other forms of bodywork will help to take mechanical stress off joints and allow better blood flow and alignment. If excess weight is putting stress on joints, then obviously losing weight is going to help. Mild exercise that doesn't put stress on the joints will improve blood and lymph flow to bring healing energy to the joints. Self-massage will also improve blood and lymph flow.

Where there is an infectious cause, obviously the infection will need to be dealt with using whatever remedies are appropriate to that type of infection. Where the cause of stress is biochemical, as in rheumatoid, allergic or gouty arthritis, improving the diet is a crucial step. Here are some things to consider.

High levels of acidity in the body have been correlated with overactive inflammatory response. Studies have shown that we can alkalize our body chemistry by replacing acid-forming meats, dairy, grain, nuts and beans with more alkaline foods like fruits and vegetables. These also have higher levels of the naturally occurring antioxidants necessary to fight inflammation.

Wheat, dairy and corn in particular have been implicated in triggering arthritis through what is called chemical onset toxicity. Acidifying citrus fruits should be avoided (aside from highly-alkalizing lemon juice), as should nightshade vegetables like eggplant, tomatoes, potatoes and green peppers. Coffee and tobacco have both been linked to increased risk of arthritis.

Gentle fasting, fruit and vegetable juice diets and mild food diets can all help to strengthen the body's cleansing abilities and to remove the toxins that are causing inflammation. Inflammation can be combated nutritionally by cooking with herbs like ginger and turmeric which have been shown to have anti-inflammatory properties. Oily fish such as salmon, as well as walnuts and freshly ground flax seed are excellent sources of omega-3 fatty acids which also combat inflammation.

2. Reduce Inflammation

Since arthritis is an inflammatory condition, remedies that reduce inflammatory reactions are an obvious place to start. Consider Thai Go or Super ORAC.

Herbs containing salicylates have been used for thousands of years to ease arthritic pain. Salycilates, the forerunners of modern aspirin, reduce joint swelling and inflammation and ease pain. The most famous of these salycilate-bearing herbs is white willow bark, which has been used since the time of Hippocrates for arthritis. Other plants containing salicylates include black cohosh and wintergreen. APS II with White Willow Bark, Triple Relief and IF Relief are herbal blends which utilize these natural anti-inflammatory and pain-reducing agents. They can be used as effective natural replacement for NSAIDs, without the side-effects. It goes without saying that short-term pain relief also needs to be accompanied by working on the underlying causes of the pain.

Certain plant seed oils containing the fatty acid GLA (gamma-linolenic acid) can help alleviate the pain and discomfort of arthritis. Found in evening primrose, borage, black currant, and flax seed oils, GLA is important because the body converts it to compounds with strong anti-inflammatory and immune regulating effects. Super GLA is a great way to supplement the diet with GLA and may also be helpful for arthritis. Omega 3 Essential Fatty acids are also beneficial.

There are encapsulated blends of anti-inflammatory herbs that not only help reduce inflammation and pain, they also aid in detoxification of tissues. As toxins are reduced, irritation and inflammation diminish, helping to slow or even reverse the downward spiral of damage. Joint Support is an anti-arthritic blend created by the famous nutritionist Paavo Airola that reduces inflammation, alkalizes the system and supports detoxification.

Joint Health is a traditional combination of Ayurvedic herbs that have been used for arthritis. It has similar actions to Joint Support. Other remedies that can reduce inflammation in arthritis include Whole Leaf Aloe Vera, Devil's claw, and yucca.

3. Provide Nutrition to the Joints

Even in osteoarthritis, where the original stress is mechanical, nutrition plays a critical role. Healthy joints need good nutrition. When joints have the nutrients they need, they have a greater capacity to resist damage or to heal from damage when it occurs. So, appropriate supplements should be considered with all forms of arthritis.

Minerals are extremely critical to aiding joint repair. Silica adds resiliency to joints so they are less susceptible to damage. It is found in horsetail, dulse and HSN-W.

Calcium is important for joints, but taking calcium supplements doesn't help unless other elements are present for assimilation and utilization, including vitamin D, silica, boron and magnesium. Skeletal Strength contains all of these nutrients and is the best calcium supplement for helping with joint repair. Still, many people actually do better with the herbal calcium found in Herbal CA, which also promotes bone and joint healing.

A very helpful formula is EverFlex, which combines MSM, glucosamine and chondroitin. MSM (MethylSulfonylMethane) is a sulfur compound. Sulfur, the eighth most abundant element in the human body, has a long history as a healing agent. For centuries, mankind has soaked in sulfur-rich mineral hot springs to help heal a variety of ailments. MSM supplies biologically active sulfur. It helps with liver detoxification and studies show it helps ease arthritis pain in many individuals.

Glucosamine is an amino sugar normally found in the human body and is the base material for making up mucous membranes, ligaments, tendons and synovial fluid in the joints. It helps joints to heal and can help them become more fluid and well lubricated.

Chondroitin is a long chain of repeating sugars found naturally in the joints and connective tissues. It helps to produce new cartilage and protects existing cartilage. Chondroitin helps by interfering with enzymes that destroy cartilage molecules and enzymes that prevent nutrients from reaching the cartilage.

Collagen is another major supportive tissue in the human body. Cartilage, ligaments and tendons are primarily made of collagen. This means the collagen found in Collatrim can be useful in preventing damaged cartilage from hardening and in promoting healing.

Looking closely at the causes and effects of arthritis shows both the complexity and the simplicity that underlie its symptoms complexity in just how many factors there are to consider. Simplicity because all the factors and symptoms ultimately boil down to some very straightforward concepts. If we remove the stresses causing the disease, detoxify our bodies from the effects of the disease and build our systems up to be able to more effectively combat the disease, we ll have gone a long way, not only in eliminating the symptoms, but also in creating health for ourselves at every level.

Therapies: Hydration (pg. 10), Antioxidants (pg. 17), Hiatal Hernia Correction (pg. 12), Affirmation and Visualization (pg. 8), Colon Cleansing (pg. 18), SuperFood (pg. 15), Minerals (pg. 15), Enzymes (pg. 15), Essential Fatty Acids (pg. 17)

Remedies: APS II, Devil's Claw, HSN-W, Joint Support, Tei Fu Oil, 7-Keto, Alfalfa, All Cell Detox, Aloe Vera, Anamu, Arthritis, Black Currant Oil, Blue Cohosh, Bone/Skin Poultice, Burdock, Capsicum, Chondroitin, Collatrim, Colloidal Minerals, Dandelion, Deep Relief Oil, DHA, Evening Primrose Oil, EverFlex, Everflex Pain Cream, Feverfew, Flax Seed Oil, Garlic, Germanium Combination, Glucosamine, Gotu Kola, Grapine, Helichrysum, Herbal CA, Horsetail, HSN Complex, Hydrangea, IF Relief, IF Relief, IF-C, IGF-1, Ionic Minerals, Joint Health, KB-C, Kidney Drainage, Lemon Oil, Licorice Root, Lymphatic Drainage, Magnesium, MSM, Nature's Fresh, Nature's

Gold, Nature's Phenyltol with NEM, Nerve Eight, Noni (Morinda), Omega-3, Oregano (Wild), Pau D Arco, Phyto-Soy, PLS II, Red Beet Formula, Red Clover, Red Clover Blend, Rosemary, S-O-D With Gliadin, Safflowers, Sarsaparilla, SC Formula, Skeletal Strength, Super GLA, Super ORAC, Super Trio, Thai-Go, Uña de Gato Combination, Valerian Root, Vitamin D3, Wild Yam, Wild Yam & Chaste Tree, Yucca

Asthma

See Also *Adrenals (Exhaustion, Weakness or Burnout), Allergies (Respiratory)*

Helplessness. Panic. Fear. These are the feelings invoked by an asthma attack, both for the sufferer and for parents or other loved ones. During an asthma attack the muscles surrounding the bronchial passages in the lungs constrict. This interferes with the outflow of stale air and causes a feeling of suffocation in the victim. Typical symptoms of asthma include a feeling of tightness in the chest and difficulty breathing often coupled with wheezing and coughing.

Asthma is on the rise, affecting about seventeen million Americans (five million children and twelve million adults). The dramatic increase in asthma appears to be linked to the increase in air pollution and other lung irritants. For instance, in Mexico City, the most heavily air-polluted city in the world, as high as 50% of the children may have asthma.

This respiratory disorder is often unpredictable; and this is what makes it'so intimidating. Those who suffer from it experience bouts of breathlessness which can come on suddenly during periods of stress, anxiety, exercise, low blood sugar, laughing, changes in temperature, extremes of dryness or dampness or exposure to allergens such as dust, animal dander, smoke, mold or food additives. Asthma attacks can last from minutes to hours and can come daily or annually.

Everyone's lungs will react to irritants by the process of inflammation, swelling, mucus production and coughing. Yet, for the person with asthma these reactions appear to be exaggerated or hyperactive. Swelling and inflammation in the lung tissue triggers spastic reactions in the lungs which further constrict airways. As air is trapped in the lungs, excess carbon dioxide builds up in the blood creating the suffocating feeling.

Asthma is commonly treated with antihistamines (substances which reduce allergic reactions), anti-inflammatories (substances that reduce swelling and inflammation) and bronchial dilators (substances that relax the bronchial passages, allowing air to escape). These therapies are effective for symptomatic relief and can ease attacks and even save lives. However, they do not help to relieve any of the underlying causes of this disease. Here are some of the underlying causes to consider:

Food and Respiratory Allergies

Most individuals who experience asthma notice that it is prompted by substances such as pollen, dander, smoke, cold air or excessive exercise. Along with the dust, mites, molds and pet dander that tend to cause an onset of allergies or asthma, diet has also become a part of the picture.

Dairy and wheat have come to be known as contributing factors. Most asthma sufferers notice tremendous relief once these allergens have been eliminated. Dairy, in particular, contains the protein casein that generates mucus. When a victim experiences an asthma attack, wouldn't it be better if his or her body were not insulted with mucus forming foods?

A number of remedies can also be used to reduce allergic reactions. HistaBlock is very helpful. Other possible remedies include burdock, which stabilizes mast cells, and cordyceps, which will particularly benefit the asthmatic athlete.

Excess Estrogen

Adult asthma tends to be most prevalent in women. Many physicians believe this to be hormonally related. With the rising number of asthma cases, this could be due to the influence of xenoestrogens (estrogen-like chemicals such as pesticides and plastics) present in our environment. Women who are estrogen dominant (with estrogen too high relative to progesterone) display signs of asthma more often than those who are more hormonally balanced.

The balancer for estrogen is progesterone, which can be applied topically in a progesterone cream such as the Pro-G-Yam cream. This has helped to ease symptoms in some women. Synthetic progestins, conversely, have shown no such benefit.

Stress and Adrenal Fatigue

The adrenal hormone ephinephrine acts as a bronchial dilator. Forms of it can be injected or are used in bronchial inhalers in order to halt asthma attacks. Ephinephrine is an adrenal hormone and a sympathetic neurotransmitter.

Corticosteroid drugs are also used to treat asthma. These drugs are mimics of another adrenal hormone, cortisol, which reduces inflammation in the body.

This suggests a connection between stress and adrenal fatigue in asthma cases. For any of us who have ever been involved in any type of organized athletics, we are bound to have met individuals with exercise-induced asthma. These individuals, as well as others under chronic stress, generally have exhausted their adrenals and are lacking the naturally produced steroids necessary to prevent these attacks.

So, rebuilding the adrenals is critical to overcoming asthma. The difficulty in this is that the drugs usually used to treat asthma contribute to adrenal fatigue. The best approach, therefore, is to start rebuilding the adrenals while slowly backing off the medications.

A number of nutrients can help to rebuild the adrenals. The combination of B-complex and vitamin C in Nutri-Calm has a rebuilding effect on the adrenals. It also contains herbs to relax tension and ease stress. Pantothenic acid is very important for adrenal function and can also help to rebuild exhausted adrenal glands. Nervous Fatigue Formula is another great adrenal rebuilding agent.

Licorice root helps preserve cortisol levels (the adrenal hormone that reduces inflammation) and can help to rebuild exhausted adrenals. Licorice is also anti-inflammatory. This herb can be taken regularly by many asthmatic children to reduce frequency and severity of attacks.

When rebuilding the adrenals, it is very important to avoid refined sugar, and foods and beverages containing caffeine, as these substances tax the adrenals. Licorice root or Mineral Chi Tonic can be taken to reduce caffeine and sugar cravings and maintain energy levels without these addictive and harmful substances.

Digestive Problems

Most asthmatics have a hiatal hernia which inhibits free movement of the diaphragm and contributes to poor digestion of proteins which causes mucus congestion. Food allergies, leaky gut and lymphatic congestion are all common in asthmatics.

Start by correcting the hiatal hernia, if present. Taking digestive enzymes, such as Protease Plus or Proactazyme Plus, can help reduce allergic reactions to foods that contribute to asthma.

Cleansing the liver and colon also has tremendous benefits for healing the lungs. Therefore the Tiao He Cleanse or CleanStart may also be beneficial. Blood purifiers like Enviro-Detox or All Cell Detox have also proved helpful in some cases.

It can also be helpful to take remedies that clear mucus from the lungs. Remedies to consider include ALJ and CC-A with Yerba Santa. As an expectorant and nervine, yerba santa is beneficial in easing most allergic or asthmatic conditions.

Natural Bronchial Dilators

To stop asthma attacks it is necessary to dilate the bronchials to let in more air. Lobelia has been long used to relieve asthma attacks. It can be rubbed onto the chest in tincture form or taken orally to relieve feelings of tightness and to relieve coughing while maintaining expectorant properties. Lobelia is a very effective bronchial dilator and antispasmodic that can be used as a natural alternative to inhalers. The extract of lobelia can be administered in doses of about 10-20 drops at one to two minute intervals starting at the beginning of the attack until it'subsides.

Occasionally, this therapy will cause the person to vomit. However, the attack nearly always subsides as soon as the person expels the contents of the stomach (which interestingly enough often contains a large quantity of mucus).

Another herb that can help to stop asthma attacks is black cohosh. Although not commonly used for this purpose by modern herbalists, eclectic physicians at the turn of the century used both lobelia and black cohosh for asthma. Distress Remedy taken internally and Tei Fu oil rubbed on the chest have also been used to open the bronchials during an asthma attack.

Asthma is a serious condition, yet one that can be dealt with effectively by natural means. It will however take some determination, study and a commitment to a generally healthier lifestyle.

Therapies: Hydration (pg. 10), Antioxidants (pg. 17), Hiatal Hernia Correction (pg. 12), Affirmation and Visualization (pg. 8), Colon Cleansing (pg. 18), Stress Management (pg. 8), Enzymes (pg. 15), Essential Fatty Acids (pg. 17), Avoid Xenoestrogens (pg. 23)

Remedies: Adrenal Support, ALJ, Breathe EZ, Licorice Root, Lobelia, Alfalfa, Aloe Vera, Anti-Gas Formula, Asthma, Black Currant Oil, Blue Cohosh, Blue Vervain, Breathe Free, Bronchial Formula, C-X, Chamomile (Roman), Chickweed, Clary Sage, Clove Bud, Co-Q10, Cordyceps, Deep Relief Oil, Fenugreek & Thyme, Four, Frankincense, Garlic, GC-X, Ginkgo Biloba, Ginseng (Korean), Ginseng (Wild American), Helichrysum, IF Relief, IF-C, Kava Kava, Lemon Oil, Lung Support, Mandarin (Red), Marshmallow & Fenugreek, Men's Formula, MSM, Mullein, Nerve Eight, Nutri-Calm, Oregano (Wild), Pantothenic Acid, Passion Flower, Pine Needle, Rose Hips, Sage, Sea Salt, Sinus Support, St. John's Wort, Stress Relief, Stress-J, Sunshine Heroes Omega 3 with DHA, Sunshine Heroes Whole Foods Antioxidant, Sunshine Heros Whole Foods Papayazyme, Super GLA, Super ORAC, Tei Fu Oil, Vitamin B-Complex, Ylang Ylang

Atherosclerosis

See *Arteriosclerosis (Atherosclerosis, Hardening of the Arteries)*

Athlete's Foot

See Also *Fungal Infections (Yeast Infections, Candida albicans)*

Athlete's foot is a fungal infection of the foot most often between the toes and on the soles of the feet. Symptoms include pain, burning and itching. Antifungal remedies that may be applied topically for athlete's food include garlic oil, Pau d Arco Lotion and Paw Paw Cell Reg (empty the capsules and mix with Black Ointment or Golden Salve). Silver Shield Gel is also beneficial, especially when mixed with antifungal essential oils like tea tree and lavender. When there is a topical fungal infection like this, it is wise to treat for yeast internally as well.

Remedies: Candida Clear, Silver Shield, Tea Tree Oil, Yeast/Fungal Detox, Black Walnut, Garlic, Guardian, Myrrh, Neroli, Pau D Arco, Paw Paw

Athletic Performance or Exercise (aids to)

These are products that can help to enhance athletic performance, muscle building and exercise.

Therapies: Hydration (pg. 10), Affirmation and Visualization (pg. 8), Minerals (pg. 15), Low Glycemic Diet (pg. 11)

Remedies: Cordyceps, Target Endurance, Black Walnut, Blood Build, Free Amino Acids, IGF-1, Recovery, Safflowers, Super Algae, Thai-Go

Attention Deficit Disorder (ADD, ADHD)

See Also *Hypoglycemia, Leaky Gut Syndrome*

ADD (Attention Deficient Disorder) and ADHD (Attention-Deficit Hyperactive Disorder) are characterized by inappropriate inattention, impulsivity and hyperactivity. There is a tendency to haphazard, poorly organized activity. In true ADD, there is often a weakness of the sympathetic nervous system and a corresponding overactivity of the parasympathetic nervous system. It is easy to tell whether a child is suffering from anxiety or ADD by looking at their pupils. Enlarged pupils signal an excess of sympathetic nervous system activity, so standard nervines such as Stress-J, lavender essential oil or lobelia will calm them down. If the pupils are small and contracted, then there is an excess of parasympathetic nervous system activity and stimulants like caffeine will have a calming effect.

In true ADD, nutrients that support sympathetic nervous system activity and adrenal function are usually critical. Products that stimulate sympathetic nerves include Adrenal Support, ENERG-V, licorice root, HistaBlock (which helps leaky gut) and small amounts of caffeine (preferably green tea). Note: NSP's green tea extract won't work because it is caffeine-free.

Essential fatty acids are absolutely essential to brain function since the brain structure is mostly composed of fat. Protein is also very important, both for balancing blood sugar and creating neurotransmitters. The amino acid l-tyrosine is very important because it is the precursor to epinephrine and norepinephrine. Its richest source is in red meat. Feeding children with ADD a hearty breakfast that includes eggs and red meat will help them become more focused. Simple carbohydrates (such as sugar sweetened breakfast cereals) should not be eaten for breakfast. Super Algae or Free Amino Acids can provide essential amino acids (including l-tyrosine) for vegetarians.

Heavy metal poisoning, particularly with lead or mercury, can be another root problem in learning disabilities such as ADHD. These metals may have been introduced into the nervous system through vaccines.

There are specific neurotransmitters that calm down excess nervous system reactions. Focus Attention and GABA Plus help to elevate levels of these neurotransmitters and calm down excess activity in the brain, making it very helpful in some cases of ADHD.

Hypoglycemia and leaky gut syndrome may also play a role in these disorders. Licorice root may be helpful in both. Don't substitute artificial sweeteners for sugar. Food additives, including aspartame, can be linked with hyperactivity and other behavioral disorders.

Therapies: Blood Type Diet (pg. 13), Low Glycemic Diet (pg. 11), Stress Management (pg. 8), Essential Fatty Acids (pg. 17), Probiotics (pg. 16)

Remedies: Adrenal Support, Energ-V, Focus Attention, GABA Plus, Sunshine Heroes Omega 3 with DHA, AdaptaMax, Astragalus, Black Currant Oil, Chamomile, Chamomile (Roman), Clary Sage, Flax Seed Oil, Free Amino Acids, Heavy Metal Detox, Herbal Sleep, HistaBlock, Licorice Root, Lobelia, Magnesium, Master Gland, Nerve Eight, Nutri-Calm, Spirulina, Stress-J, SUMA Combination, Super Algae, SynerProTein, TS II, Wood Betony

Autism

Autism is a mental condition of self-centered subjective mental activity (daydreams, fantasies, delusions, hallucinations, etc.) often accompanied by a marked withdrawal from reality. Many natural healers believe this condition is caused by damage to the central nervous system from vaccines or other sources of chemical solvents and heavy metals. Autistic children have differences in the shape and structure of their brain. Professional assistance and support should be sought for this condition, but the following may be helpful.

Therapies: Antioxidants (pg. 17), Heavy Metal Detoxification (pg. 22), Essential Fatty Acids (pg. 17), Probiotics (pg. 16)

Remedies: Flax Seed Oil, Frankincense, Germanium Combination, Ginkgo Biloba, Green Tea Extract, Heavy Metal Detox, Heavy Metal Detox, Lavender, Magnesium, Myrrh, Protease, St. John's Wort, Sunshine Heroes Omega 3 with DHA, Sunshine Heros Whole Foods Papayazyme, Target P-14, Vitamin B-6, Vitamin B-Complex

Autoimmune Disorders

See Also *Fibromyalgia, Grave's Disease, Hashimoto's Disease (Thyroiditis), Lou Gehrig's Disease, Lupus, Multiple Sclerosis (MS), Myasthenia Gravis*

The immune system's job is beautifully orchestrated to determine what is self and what is not self and to get rid of anything that is not self by destroying or eliminating that not self substance. But sometimes things go awry and the immune system becomes confused enough to begin to attack the body's own tissues. This is what causes autoimmune disorders. In essence, the immune system goes renegade.

Think of the renegade immune system as an army that has received confusing messages on who and where the enemy is. Acting on this critical misinformation, the commanding officers order an attack on what they think is the enemy, while in reality they are attacking their own troops and destroying them with what is called friendly fire .

Like the confused commanding officers, instead of being the protector of health, the confused immune system has become the destroyer of health. These attacks are directed at different tissues, systems and parts of the body in different people, creating symptom clusters that have been named as various diseases.

Before our modern society started monkeying around with our health and environment, immune disorders were relatively uncommon. In the last few decades, the incidences of immune disorders have risen at alarming rates. In order to turn the tide and stop the onslaught, we need to understand what we are doing to ourselves and to our environment that may be major contributors to these increases in autoimmune disorders. We also need to know what we can do to prevent further health damage, and how to recapture our health through diet, life style changes, and natural health supplements.

There are many possible causes for autoimmune disorders. Here are the major ones and what to do about them.

Viruses

Certain viruses are known to trigger immune problems. A good example is Epstein-Barr virus which is often at the root of many fatigue and pain related autoimmune disorders. Antiviral remedies may be helpful as long as they are not immune stimulants. VS-C and Silver Shield would be good choices.

Vaccinations

Immunizations bypass our body's first line of immune defense the skin and mucus membranes. This can confuse the body's immune system and may cause the viral material in the vaccine to get mixed in with the body's own DNA. The body then associates some of its own tissues with disease organisms. The polio vaccine is particularly suspect in multiple sclerosis. Vaccines have been associated with type I diabetes and autism. They may also be involved in multiple sclerosis (MS).

Mercury Fillings

One of the most toxic of all substances on earth is the liquid metal mercury. Among other things, it damages our immune system. Just touching it is extremely dangerous, and yet we mix it with silver and put it in our mouths where it'stays for the rest of our lives, leaching tiny amounts into our bodies. Consider having a properly trained dentist remove the mercury/silver fillings and replace them with composite fillings, then do a heavy metal cleanse.

Air Pollution

If the air we breathe is loaded with toxins and chemicals, it will negatively affect the immune system. If you live where the air has been damaged by pollutants, use filters and ionizers in your home and make sure you take frequent breaks to clean air places. We have to do whatever we can to protect our air. Plant trees and other greenery and have an active voice in environmental issues.

Water Pollution

Our bodies are made of nearly 70% water, and we need to constantly replenish our water supply. When we put bad water and chemically loaded liquids into our bodies, we suffer immediate loss of energy. Some of the most dangerous viruses are waterborne, as are chemicals and toxins, so filter and purify your water.

Chemicals in Food

Commercially grown and processed foods typically contain numerous chemicals. These chemicals also contribute to immune problems. Non-organic foods are grown with artificial chemical fertilizers and toxic chemical pesticides in dead soil. This greatly reduces the nutritional value of the foods, as well as being questionably safe for human consumption. Non-organic meats are raised with steroids, antibiotics, and hormones. These are passed through to us when we eat them. So, wherever possible, insist on certified organic foods.

Avoid eating saturated fats, hydrogenated oils, fried foods, prepackaged foods, artificial dyes, flavorings and additives. Hydrogenated fats are a major problem for people with autoimmune problems.

Other Negative Environmental Influences

Many cleaning fluids are made of highly toxic chemicals. Toxic paint fumes, formaldehyde in new carpets and chemicals in particle board and the electromagnetic frequencies from high tension electrical wires that scramble our energy can be invisible enemies. Microwave ovens and cellular phones are also suspect in our quest to find and eliminate factors that can be negatively affecting our immune systems. Some medications can also negatively affect the immune system.

Aids to Balanced Immunity

The key to overcoming and healing autoimmune disorders lies in calming down the immune system response, detoxifying the body gently, and building health from within. Natural therapy for autoimmune disorders includes undoing the damage we have done to our bodies and our environment. Massage, accupressure, acupuncture, chiropractic treatment, spas, gentle exercise, regular stress-breaks, calm and quiet, and rest are valuable possibilities. The only way to re-cover from autoimmune disorders is to pamper and take care of yourself.

Stress

Avoid or reduce stress wherever possible. Stress can push one's health to the breaking point and exacerbates autoimmune problems. Practice meditation, prayer, yoga or tai chi, take walks, do gentle stretching, garden, journal, enjoy life, take a nice bath or listen to quiet music. Nearly everyone with an autoimmune disorder has exhausted adrenals. The adrenals make cortisol which helps regulate the immune system by dampening excessive inflammation. Adrenal tonics like Adrenal Support and adaptagens like Nervous Fatigue Formula can be helpful.

Inflammation

Autoimmune disorders involve chronic, low grade inflammation. Antioxidants and anti-inflammatory remedies can be particularly helpful. Licorice root, yucca and wild yam all have a cortisol-like anti-inflammatory action and can be used as alternatives to the corticosteroid drugs often used to treat these conditions. Antioxidants like Thai-Go and Super ORAC and anti-inflammatories like IF Relief and APS II with White Willow Bark may also be helpful.

Emotional Factors

Find the attitude or feeling that may be presenting itself as a health condition in your life. (i.e. Multiple Sclerosis (MS) eats at the nervous system, so take a look at what in your life could be eating at your nerves.) If the inner body is attacking itself in any way because of self-guilt, shame, anger, etc., then self-care and self-love, the antithesis of the attack, can be a powerful tool for turning the autoimmune condition around.

Exercise

Exercise has to be the right kind for the autoimmune profile. It's very important not to overdo on exercise and stress the body! Hard aerobics and intense team sports are out.

What is needed most is oxygen in the system and staying flexible. Yoga, tai chi and other flowing martial arts and dance disciplines, hiking, walking, swimming, light bicycle riding and other enjoyable activities that stretch the body and gently oxygenate the blood are appropriate choices. Keep the exercise regime on the gentle side, but be regular about it.

Diet

Stick to white meats (fish and organic poultry), cut off extra fats and avoid fried foods. It may take total abstinence from meats for awhile for the body to heal, then careful reintroduction. Hydrogenated oils are very aggravating to autoimmune conditions. Use olive oil, coconut oil, flax seed oil and other good fats. Essential fatty acid supplements are often helpful for autoimmune disorders.

Make sure your diet is two-thirds fresh veggies and fruits, organic when possible. Watch out for dairy, corn and wheat/gluten sensitivities. You may also need to abstain from these completely for a time while the body heals. Stay off of sugars and sweets.

Supplements

The autoimmune disorder profile includes digestive weakness and malabsorption, so enzyme supplementation is essential. Consider taking Proactazyme Plus or Food Enzymes. Extra doses of protease enzymes like High Potency Protease may also be helpful.

As mentioned under inflammation above, essential fatty acids can be very helpful for reducing pain and inflammation in arthritis. Consider flax seed oil, Super Omega-3 EPA or Super GLA.

Antioxidants are usually very helpful for autoimmune conditions. Green Tea Extract and Grapine are particularly helpful. Also consider Thai-Go, Super ORAC and Carotenoid Blend.

Remedies that balance the immune system include astragalus, Mineral Chi Tonic, black walnut and colostrum.

Gentle detoxification with one capsule per day of Enviro-Detox is helpful. Other helpful supplements can be found in the list below or under the headings for specific autoimmune disorders.

Herbs and Supplements to Avoid

Avoid all herbs and supplements that stimulate the immune system, such as Colostrum with Immune Factors. Immune stimulants can be devastating to autoimmune disorders because the immune system is already in a state of hyperactivity. (Small amounts in formulas may be moderated enough, but use with care.) Herbs and supplements to avoid include echinacea, golden seal, yarrow, dandelion root, Nature's Immune Stimulator and Trigger Immune. Caffeine containing foods and beverages may also overstimulate the immune system. Avoid ylang ylang, geranium and thyme essential oils, as they are too stimulating to a hyperactive immune system.

Therapies: Hydration (pg. 10), Antioxidants (pg. 17), Blood Type Diet (pg. 13), Heavy Metal Detoxification (pg. 22), Hiatal Hernia Correction (pg. 12), Affirmation and Visualization (pg. 8), Stress Management (pg. 8), SuperFood (pg. 15), Enzymes (pg. 15), Probiotics (pg. 16), Essential Fatty Acids (pg. 17), Fiber (pg. 17)

Remedies: Adrenal Support, Fibralgia, GreenZone, Mineral-Chi Tonic, Omega-3, Proactazyme Plus, Yucca, AdaptaMax, Colostrum, Cordyceps, Glyco Essentials, Heavy Metal Detox, HSN-W, IF-C, Licorice Root, Protease, Super Antioxidant, Super GLA, Super ORAC, Vitamin D3, Wild Yam

Backache (Back Pain, Lumbago)

Pain in the back, often relates to a lack of minerals, acid pH and/or poor kidney function. Some of these remedies are for topical use for immediate pain relief and others work internally for more long-term relief.

When a person has severe pain in the back, several things may be to blame. An overacid condition of the tissues may be resulting in mineral depletion and muscle spasms. Strengthening the kidneys in their ability to flush acid waste using KB-C and drinking more pure water is often helpful. Alkalizing the diet by consuming more fresh fruits and vegetables and less meat and grains will also help.

Some people may find that they are mineral deficient and need more minerals to create greater structural strength in the spine. Colloidal minerals, Skeletal Strength, Calcium w/ Magnesium, Collatrim, PLS II and Herbal CA have all been effective in some cases.

For immediate pain relief apply lobelia extract along both sides of the spine and massage the muscles to help them relax. Follow this with an application of Tei Fu Essential Oils and massage some more. This helps the spine relax and draws oxygen and nutrients to the area. Triple Relief or APS II w/White Willow may also help relieve inflammation that is associated with the pain. Spraying Nature's Fresh along the spine can also relieve inflammation, especially when the cause of the back pain is problems with the spinal disks.

Bodywork can also be helpful for backache. Try stretching exercises, yoga, massage therapy and/or chiropractic care. Maintaining good posture is essential to a healthy back, as well.

Therapies: Hydration (pg. 10), Minerals (pg. 15)

Remedies: Kava Kava, KB-C, Lobelia, Magnesium, Nature's Fresh, APS II, Bone/Skin Poultice, Collatrim, Cordyceps, Deep Relief Oil, Devil's Claw, EverFlex, Everflex Pain Cream, Germanium Combination, Herbal CA, Hydrangea, IF Relief, JP-X, Kidney Activator, Kidney Activator (Chinese), Liver Balance, Nervous Fatigue Formula, Niacin, Pain, PLS II, Sciatic, Skeletal Strength, Tei Fu Oil, Thai-Go, Trigger Immune, Triple Relief, Vitamin C, Wood Betony

Bacterial Infection

See *Infection (bacterial)*

Bad Breath

See *Halitosis (Bad Breath)*

Baldness

See *Hair Loss*

Bedwetting

See Also *Incontinence (urinary)*

When a child loses bladder control while sleeping and wets the bed, it is often related to the same reasons why some adults have to wake up two or three times a night to go to the bathroom adrenal fatigue, because the adrenals help to regulate fluid balance in the body. It can also be due to fear, hypoglycemia, mineral imbalances or a weakness of the kidneys and bladder. If the child also tends to wet themselves during the day, they may have a urinary tract infection.

Licorice root or HY-A can help build the adrenals and regulate blood sugar. A protein snack before bedtime (such as a couple of ounces of peanut butter, cottage cheese or beef jerky) may also help. Collatrim taken before bed will have the same effect.

Mineral imbalances may also contribute to this problem. Magnesium deficiency in particular may be a root cause.

Kidney herbs (cornsilk, JP-X, Kidney Activator, marshmallow, parsley) are often used to encourage urination during the day. If the problem is a lack of tone in the sphincter of the bladder then uva ursi taken as a tea may be helpful.

Therapies: Hydration (pg. 10), Low Glycemic Diet (pg. 11)

Remedies: Adrenal Support, Licorice Root, Magnesium, Black Cohosh, Chromium GTF, Collatrim, Cornsilk, Cranberry & Buchu, Eleuthero, Free Amino Acids, Green Tea Extract, Herbal CA, HY-A, JP-X, Juniper Berries, Kava Kava, Kidney Activator, Kidney Drainage, Marshmallow, Master Gland, Parsley, Pro-Pancreas, Red Raspberry, St. John's Wort, Super Algae, Urinary Maintenance, Uva Ursi, White Oak Bark

Belching

See Also *Gas and Bloating*

Belching, gas bubbles from the stomach expelled through the esophagus and mouth, usually signals a problem with poor digestion. Enzymes (Proactazyme Plus, Food Enzymes, Papaya Mint) and remedies that improve digestive function (Anti-Gas formulas) will be helpful.

When a person belches and a foul, rotten egg odor and taste accompany the belch, this is a sign that proteins are not being properly broken down in the stomach. In fact, they are decaying. A lack of hydrochloric acid and digestive enzymes causes this problem.

Therapies: Enzymes (pg. 15), Hiatal Hernia Correction (pg. 12)

Remedies: Anti-Gas (Chinese), Food Enzymes, Papaya Mint, Proactazyme Plus, Anti-Gas Formula, Catnip & Fennel, Charcoal (Activated), Clove Bud, Gastro Health, Kudzu/St.John's Wort, Mandarin (Red), PDA, Peppermint, Protease, Yeast/Fungal Detox

Bell's Palsy

Paralysis of the facial nerve producing distortion on one side of the face is called Bell's Palsy. Wood betony has proven beneficial in many cases. Other remedies that may be helpful are listed below.

Therapies: Essential Fatty Acids (pg. 17)

Remedies: Nerve Eight, VS-C, Wood Betony, Alfalfa, Anti-Gas Formula, Herbal Sleep, Horsetail, HSN-W, N-acetyl Cysteine, Nutri-Calm, Omega-3, St. John's Wort, Stress Relief, Super Algae

Benign Prostate Hyperplasia (BPH)

See Also *Prostatitis*

This is a nonmalignant, abnormal growth of the prostate tissue. The severity is measured in stages I-IV (mild-serious), which refers to the size of the growth (walnut-sized to grapefruit-sized) and the impact this enlargement has on the quality of life.

Research suggests this condition is due to an excess of a special form of testosterone called dihydrotestosterone (DHT). DHT is necessary for the development of the male reproductive organs in the womb. It is also needed to complete the development of male sexual characteristics during puberty. Testosterone is converted to DHT by the enzyme 5a-reductase. When too much of this conversion occurs, usually later in life, DHT stimulates enlargement of the prostate, baldness, unusual hair growth and acne.

A key to preventing prostate problems, then, is to interfere with the body's production of DHT. The mineral zinc, long known for its beneficial effects on the male prostate, has been found to be a potent inhibitor of 5a-reductase. Supplementing with zinc helps prevent testosterone from being converted to DHT.

Herbs can also be used to inhibit this conversion. One of the most popular herbs for this problem is saw palmetto. Saw palmetto is an endocrine agent that has a normalizing and tonic action on the prostate. It reduces the abnormal growth of glandular tissue. It'suppresses the expression of estrogen, progesterone and androgen receptors in the prostate and blocks testosterone and DHT from binding to androgen receptors. High in essential fatty acids, saw palmetto provides nutritional support for the prostate gland and urinary system, relaxes smooth muscles and provides powerful anti-inflammatory action.

Other herbs that have beneficial effects on BPH include nettle root and pygeum. These herbs are combined with saw palmetto and zinc in Men's Formula. This blend is a good supplement for most men who are concerned about preventing prostate enlargement or reducing an already enlarged prostate. Breast Assured will also help here.

Xenoestrogens are a major factor in BPH. Reduce or eliminate dairy products and minimize exposure to pesticides, soft plastics and other sources of xenoestrogens. Take Indole-3-Carynol and eat cruciferous vegetables to help break down excess estrogens. Avoid grapefruit as this inhibits estrogen breakdown. Beer should also be avoided as the hops in beer is very estrogenic.

Therapies: Avoid Xenoestrogens (pg. 23), Colon Cleansing (pg. 18), Essential Fatty Acids (pg. 17)

Remedies: Herbal Pumpkin, Men's Formula, Omega-3, PS II, Breast Assured, Damiana, DHEA-M, Eleuthero, False Unicorn, Ginseng (Korean), Indole 3 Carbinol, KB-C, P-X, Saw Palmetto, Zinc

Bipolar Mood Disorder (Manic Depressive Disorder)

See Also *Allergies (Food)*

Bipolar mood disorder, also known as manic depressive disorder, is a psychological condition where a person has dramatic swings between depression and mania. In the depressed phase the person may sleep a lot, have very little motivation and feel discouraged. During the manic phase the person feels they can do anything and will launch unreasonable projects. People with the O blood type are more prone to this disorder because they can develop high levels of dopamine (the neurotransmitter that creates the manic feelings) and then have these neurotransmitters rapidly converted to epinephrine and used up, leaving the person depressed.

Adequate intake of the amino acid l-tyrosine is essential to maintaining adequate levels of dopamine in the brain. This amino acid is found primarily in red meat. Wheat is also high in this amino acid, but people with the O blood type tend to be allergic to wheat. Excessive consumption of wheat also contributes to the development of hypoglycemia, which further contributes to this problem. Eating red meat for breakfast, or taking a protein supplement, such as Super Algae, can be helpful.

Supporting the nervous system with another amino acid, l-taurine, B Complex vitamins or Nutri-Calm and essential fatty acids from flax seed oil can also be helpful. Lithium is used to treat this disorder medically. Lithium is one of the trace elements found in Colloidal Minerals.

This disorder may involve environmental sensitivities and food allergies. Avoid food additives and chemicals and screen for food allergies.

Therapies: Blood Type Diet (pg. 13), Enzymes (pg. 15), Low Glycemic Diet (pg. 11), Affirmation and Visualization (pg. 8), Colon Cleansing (pg. 18), Stress Management (pg. 8), Essential Fatty Acids (pg. 17), Minerals (pg. 15), Heavy Metal Detoxification (pg. 22)

Remedies: Mineral-Chi Tonic, Mood Elevator, Colloidal Minerals, Flax Seed Oil, Ionic Minerals, Nature's Gold, Nutri-Calm, Super Algae, Super GLA, Vitamin B-Complex

Birth Control (countering side effects)

Birth control pills contain hormones which can disrupt the body's normal hormonal balance. They may also contribute to yeast overgrowth. These remedies help counter side effects from using birth control pills.

Remedies: Vitamin B-Complex, FCS II, Magnesium, NF-X, Vitamin B-6, Yeast/Fungal Detox

Birth Control (natural)

These remedies have been used as natural birth control. There is some controversy over the effectiveness of using wild yam as a natural form of birth control. Many report it to be very effective, but it must be used precisely according to directions.

Four capsules of wild yam must be taken every day for a period of thirty days before it will take effect. You must continue to take four capsules each day as long as you don't wish to become pregnant. You can never miss a day. If you miss one day, you must take the wild yam for another thirty days before having unprotected sex.

If you wish to try this method, do so at your own risk. It may not work.

If a woman has a regular cycle, you can chart her cycle. On a normal 28 day menstrual cycle, a woman is fertile about 14 days after the beginning of her period, and this period of fertility lasts from 5-7 days. Generally speaking there is little risk of pregnancy if unprotected intercourse is limited to 7 days after menstrual flow ceases and the 7 days prior to the beginning of the period. This approach, coupled with the use of the wild yam, would probably be more effective.

Another natural approach to birth control is natural lambskin condoms. These do not interfere with pleasure as much as latex condoms. They are effective at preventing conception, but not sexually transmitted diseases.

Remedies: Wild Yam, Wild Yam & Chaste Tree

Birth Defects (prevention)

See Also *Pregnancy (herbs and supplements for), Pregnancy (herbs and supplements to avoid during)*

When a woman is pregnant, she needs to get extra nutrition to ensure she has adequate nutrients for the developing child. Many birth defects are caused by toxins and nutritional deficiencies. Women planning to get pregnant should do a cleanse before getting pregnant. They should also minimize their exposure to tobacco, alcohol, drugs (including over-the-counter and prescription medications), food additives and chemicals of all kinds. These are remedies that may help provide extra nutrition to prevent birth defects.

Therapies: SuperFood (pg. 15), Essential Fatty Acids (pg. 17), Stress Management (pg. 8), Minerals (pg. 15), Antioxidants (pg. 17)

Remedies: Colloidal Minerals, Nature's Prenatal, Red Raspberry, Folic Acid Plus, GreenZone, Ionic Minerals, Kelp

Bites and Stings

There are many herbs that can be applied topically to help heal bites and stings from ants, chiggers, mosquitos and other insects. These natural remedies reduce swelling and inflammation, relieve itching and ease pain, often very rapidly. Tei Fu oil, tea tree oil and Nature's Fresh are good choices, but there are many others.

For black widow bites apply a poultice of crushed fresh plantain leaves if available or moisten some activated charcoal and apply it as a poultice. Take one capsule of black cohosh internally, along with 1,000 milligrams of vitamin C every two hours. Brown recluse spider bites can be particularly difficult to treat, but have been dealt with successfully by applying charcoal poultices and changing them every hour or two. Medical assistance should be sought for in treating poisonous spider bites.

Bites from poisonous snakes have been eased by the use of poultices of herbs like plantain, echinacea and black cohosh. However, these treatments should be used enroute to the emergency room for medical treatment.

For allergic reactions to bee stings HistaBlock may be helpful along with plantain, white oak bark or any other astringent herb applied topically as a poultice, but again medical assistance should be sought immediately.

Remedies: Black Cohosh, Charcoal (Activated), Lobelia, Nature's Fresh, Tei Fu Oil, Allergy, Aloe Vera, Bayberry, Dong Quai, Echinacea Purpurea, Feverfew, Golden Salve, HistaBlock, IN-X, Intestinal Soothe & Build, Lemon Oil, Lymph Gland Cleanse-HY, Patchouli, PLS II, Sage, St. John's Wort, Ultimate Echinacea, Vitamin C, Vitamin E, White Oak Bark, Yarrow

Blackheads

See *Acne (Pimples, Blackheads)*

Bladder (irritable)

See Also *Urethritis, Urinary Tract Infections, Urination (burning or painful), Urination (frequent)*

With an irritable bladder there is a constant urge to urinate, even when there is only a small amount of urine in the bladder. This can be due to dehydration (lack of water) or inflammation of the bladder.

Most people who have this problem drink less water trying to avoid having to urinate so frequently, but this is the wrong approach as it concentrates toxins even more, causing greater irritation. Drinking more water dilutes the toxins and helps the body flush them more effectively.

Herbs that soothe the urinary passages, like cornsilk, marshmallow or kava kava, will be helpful. If these don't work try finding pippsessiwa (an herb that is not available from NSP).

You should also consider the possibility of a urinary tract infection. Uva ursi, goldenseal and Silver Shield are possible remedies if this is the cause.

Emotionally, an irritable bladder may be a symptom of unresolved angry feelings, in other words, being p**sed off. Dealing with whatever is making you angry will ease the irritation.

Therapies: Hydration (pg. 10), Stress Management (pg. 8)

Remedies: Cornsilk, Marshmallow, False Unicorn, Golden Seal, Kava Kava, Kidney Drainage, Lymphatic Drainage, Silver Shield, Urinary Maintenance, Uva Ursi

Bladder (ulcerated)

When bladder tissues becomes damaged through continuous inflammation, they may become chronically inflamed and sore. Ulcers may develop in the bladder just as they can in the stomach or intestines. The following remedies may be helpful for an ulcerated bladder.

Remedies: Golden Seal, Uva Ursi, Marshmallow, White Oak Bark

Bladder Infection

See Also *Urinary Tract Infections*

The bladder can easily become infected, especially in women, from bacteria in the feces. This is especially true when the microflora of the intestinal tract has been upset due to the use of antibiotics, birth control pills and other medications. An imbalance in the pH of the body will also make one more prone to bladder infections. Invasion of the bladder by bacteria can result in pain, itching, burning urination, and frequent urgency to void.

To prevent bladder infections unsweetened cranberry juice is helpful. The Cranberry/Buchu combination will also help prevent, but not treat bladder infections. When there is an active infection, use remedies such as goldenseal, Silver Shield, uva ursi or Ultimate Echinacea to fight the infection.

Therapies: Hydration (pg. 10)

Remedies: Cranberry & Buchu, Golden Seal, Silver Shield, Uva Ursi, Golden Seal/Parthenium, JP-X, KB-C, Olive Leaf, Parthenium, Ultimate Echinacea, Urinary Maintenance, Vitamin C, White Oak Bark

Bleeding (external)

See Also *Cuts, Nose Bleeds*

To stop external bleeding, styptic herbs are used. To stop bleeding, styptic herbs should be applied directly into the bleeding wound. Two of the best styptics are yarrow and capsicum. Just about any astringent herb will also be helpful.

For arterial bleeding, pressure should be applied as instructed in any good first aid manual. For any serious bleeding or blood loss, seek appropriate medical attention.

Remedies: Bayberry, Capsicum, Yarrow, Bone/Skin Poultice, Gastro Health, Golden Seal, Horsetail, IF-C, Menstrual-Reg, Red Raspberry, Sage, Skin Detox, St. John's Wort, Uva Ursi, Vitamin C, White Oak Bark

Bleeding (internal)

To stop internal bleeding, homeostatic herbs are used. These herbs should be taken internally at frequent intervals (anywhere from every few minutes to every couple of hours, depending on the situation). Vitamin C with Citrus Bioflavonoids is helpful for strengthening blood vessels and preventing easy bleeding. Horsetail is very helpful for strengthening mucus membranes in the kidneys and lungs for minor bleeding in these organs. For serious bleeding, especially internally, seek appropriate medical attention.

Remedies: Bayberry, Capsicum, Horsetail, HSN-W, Menstrual-Reg, Aloe Vera, Blessed Thistle, Bone/Skin Poultice, Chlorophyll, Cordyceps, Golden Seal, IF-C, Red Raspberry, Vitamin C, VS-C, White Oak Bark, Yarrow

Bleeding Gums

See *Gingivitis (Bleeding Gums, Gum Disease, Pyorrhea)*

Blisters

Blisters are formed by an irritation that causes a bump filled with lymphatic fluid to be raised on the skin. These remedies are applied topically to speed healing.

Remedies: Silver Shield, Tea Tree Oil, Eucalyptus, Golden Salve, Gotu Kola, Silver Shield

Bloating

See *Gas and Bloating*

Blood Clots (prevention of)

See Also *Thrombosis*

Our blood contains a protein called fibrinogen. When we cut ourselves, fibrinogen helps form blood clots, an important defense mechanism that protects us from bleeding to death. Unfortunately, this same substance can cause our death when a blood clot forms inside the circulatory system and lodges in our heart or brain, resulting in a heart attack or stroke.

When a person's blood is too thick with fibrin, doctors prescribe blood thinners. There are some natural options, however, to prevent the formation of blood clots in the circulatory system. One of these options is vitamin E, which acts as an antioxidant and helps to naturally thin the blood. Super Omega 3 EPA will also do this.

Butcher's broom is an herb that is most commonly used to treat varicose veins. However, it also appears to inhibit clot formation in blood vessels without thinning the blood, especially when taken with vitamin E.

The latest, and most promising, addition to the arsenal of natural tools to prevent blood clots naturally is nattokinase. This enzyme, found in the fermented soy product natto, breaks down the fibrin mesh that forms blood clots. Research has demonstrated that taking nattokinase may prevent blood clots from forming in the circulatory system and may even dissolve blood clots that have already formed.

Nattozimes Plus is a formula that contains fermented soy enzymes that have the benefits of nattokinase. It contains the natto enzymes, hawthorn berries, capsicum, dandelion leaf and resveratol. Taking 1 capsule of Nattozimes Plus between meals twice daily on an empty stomach may help reduce one's risk of forming blood clots while enhancing overall cardiac health.

Therapies: Hydration (pg. 10), Enzymes (pg. 15), Essential Fatty Acids (pg. 17)

Remedies: Butcher's Broom, Nattozimes Plus, Nattozimes Plus, Omega-3, Vitamin E, Alfalfa, Anamu, Astragalus, Blood Pressurex, Cardio Assurance, Cholester-Reg II, Flax Seed Oil, Green Tea Extract, Mega-Chel, Protease

Blood in Stool

See Also *Inflammatory Bowel Disorders (Colitis, IBS)*

Blood in the stool can be a sign of severe intestinal inflammation or other injuries in the colon or rectum. It can also be a sign of cancer or other serious problems. Appropriate medical diagnosis should be sought to determine the exact nature of the problem before determining what remedies to use. The following may be helpful in healing.

Remedies: Bayberry, Intestinal Soothe & Build, Menstrual-Reg, Aloe Vera, Green Tea Extract, Red Raspberry, White Oak Bark

Blood in Urine

See Also *Bleeding (Internal)*

Blood in the urine can be caused by severe irritation and inflammation in the kidneys or bladder. It can also be caused by tumors or other serious problems. Appropriate medical diagnosis should be sought to determine the exact nature of the problem before determining what remedies to use. The following may be helpful.

Remedies: Horsetail, HSN-W, Bayberry, Kidney Activator, Kidney Drainage, Marshmallow, Vitamin C, White Oak Bark

Blood Poisoning

See Also *Infection (bacterial)*

Blood poisoning involves an invasion of the bloodstream by microorganisms. This is usually accompanied by chills, fever and weakness, which may result in secondary abscess in other organs. Seek appropriate professional assistance for this serious problem.

Remedies that can be applied topically for blood poisoning include Silver Shield Gel, charcoal poultices, Lobelia Essence and Ultimate Echinacea. Internally, Silver Shield, Ultimate Echinacea and High Potency Garlic are good options.

Remedies: Charcoal (Activated), Echinacea Purpurea, Garlic, IN-X, Lobelia, Silver Shield, Ultimate Echinacea, Black Cohosh, BP-X, Burdock, Chickweed, Dandelion, Enviro-Detox, IF-C, LIV-J, Lymph Gland Cleanse, Lymph Gland Cleanse-HY, MSM, Pau D Arco, PLS II, Red Clover, Red Clover Blend, SAM-e, Sarsaparilla, Vitamin C

Blood Pressure (high)

See Also *Arteriosclerosis (Atherosclerosis, Hardening of the Arteries), Cardiovascular Disease (Heart Disease), Hyperinsulinemia (Syndrome X), Weight Loss (aids for)*

When there is excessive arterial tension, the heart has to pump harder in order for blood to reach the extremities of the body. This results in high blood pressure. When this happens, most people start taking high blood pressure medication prescribed by their doctor.

How many people do you know who have cured high blood pressure by taking drug medications? Probably none. That's because high blood pressure medicine isn't caused by a deficiency of high blood pressure medications, and the medicine prescribed isn't designed to cure hypertension. In fact, the whole issue of hypertension is a perfect example of the major shortcoming of modern Western medicine. Modern medicine focuses on symptom management, but does little to address the causes of most diseases, including high blood pressure.

High blood pressure is a problem associated with lifestyle. It is virtually unknown in undeveloped areas of the world where people are living on their traditional diets. For example, high blood pressure is not found in Africa among natives living a traditional lifestyle, even among the elderly. In contrast, a high percentage of Americans have this problem about 60 million.

Hypertension greatly increases the risk of other diseases. It increases the risk of heart attack, stroke, and kidney failure. Research done by insurance companies has shown that even slight increases in blood pressure can result in decreased survival rates. Clearly, solving the problem of high blood pressure is important, but 90% of all cases are treated with drugs that only address the symptoms without examining the causes.

About 80% of all hypertension cases involve mild to moderate symptoms and can be effectively managed with dietary and lifestyle changes accompanied by herbs and dietary supplements. For example, all of the following have been scientifically demonstrated to reduce blood pressure: consuming sufficient amounts of pure water, increasing dietary fiber, loosing weight, reducing salt consumption, eliminating caffeine, alcohol and tobacco, exercising, consuming omega-3 essential fatty acids, and balancing intake of calcium and magnesium. These are all options individuals can utilize. In addition, they can examine the specific causes of hypertension and choose other supplements and diet and lifestyle changes that address their individual situation.

High blood pressure is a symptom of other imbalances, not a disease in itself. The job of the heart is to pump blood throughout the body. The blood has to reach every part of the body, including the fingers and toes. When the heart is beating harder, raising the blood pressure, it is usually a signal that something is restricting the flow of blood to the extremities of the body. Using a medication that decreases the pressure does not remove the obstructions interfering with blood flow. The result is decreased blood pressure with poor circulation to the extremities. We have simply traded one health problem for another.

If we can determine why the heart is pumping harder and increasing the blood pressure, then fix the cause, the heart will stop pumping harder. If we can make it easier for the blood to flow to the extremities, the blood pressure will automatically drop. So, let's examine some of the factors that impair circulation and cause the blood pressure to rise.

Hardening of the Arteries

Arterial plaque creates obstructions in the blood vessels. Like hard water deposits in a water pipe, these deposits reduce the size of blood vessels and restrict blood flow. As a result, the heart is forced to pump harder to get the blood through the narrower pipes.

When hardening of the arteries is the cause of high blood pressure, there is a very effective solution oral chelation using Mega-Chel. The instructions for this effective chelation program are found in the *Introduction* to this book under *Basic Cleansing Therapies.*

Vasoconstriction

Blood vessels have muscular walls that can either tense or relax. When they tense, there is vasoconstriction. It's very similar to the problem asthmatics have when the bronchial pipes constrict, reducing the flow of air into the lungs. They begin gasping for oxygen. Vasoconstriction does the same thing to the arteries. The constricted artery walls limit the flow of blood, and the heart pumps harder trying to force the life-giving blood through the constricted vessels. There are several root causes of vasoconstriction, each with its own remedies.

Stress In response to stress, real or perceived, the sympathetic nervous system becomes more active and the body tenses. We ve all felt the results of this fight-or-flight response when someone suddenly startled us and adrenaline started pumping. The heart started beating harder and blood pressure rose as the body went on red alert. This sensation is due to the release of a hormone and neurotransmitter called epinephrine (or adrenaline).

Nerve receptors that react to epinephrine are called adrenergic receptors. There are two types, alpha and beta. When the beta adrenergic receptors are stimulated, they cause blood vessels to contract and the heart to beat harder. Perhaps you ve heard of beta blockers. These are drugs that help to lower blood pressure, and they work by blocking these beta adrenergic receptor sites.

Caffeine and other stimulants. Caffeine stimulates the sympathetic nervous system and causes the release of more epinephrine. So excessive caffeine consumption increases stress responses and raises blood pressure. Other substances that trigger a sympathetic nervous reaction and stress response include alcohol, tobacco, chocolate, cheese, sugar, alcoholic beverages, and cured pork products such as ham and sausages. All of these substances can contribute to hypertension.

Magnesium Deficiency. Contrary to what most people believe, the number one mineral deficiency in most Americans is not calcium it is magnesium. When muscles contract, calcium ions flow into the muscle cells; as the muscle relaxes, there is an exchange of magnesium for calcium. In other words, calcium helps muscles contract and have tone, while magnesium helps muscles relax.

This is why calcium channel blockers are sometimes used to lower blood pressure. These drugs block calcium from entering the muscle tissues, causing them to be more relaxed. Taking extra magnesium usually creates the same results. It helps blood vessels relax and increases blood flow.

Hyperinsulinemia or Syndrome X. High insulin levels in the blood due to the consumption of refined carbohydrates cause inflammation in the blood vessels which constrict blood flow. Simple sugars also react with proteins to reduce elasticity, causing blood vessels to lose flexibility. Eliminating simple carbohydrates from the diet can be very helpful for preventing and reversing high blood pressure.

Magnesium Complex can be used in conjunction with vasodilative herbs to dilate blood vessels and reduce blood pressure. Hawthorn and ginkgo have both been found to dilate peripheral blood vessels and improve blood flow to the extremities, thus reducing hypertension. Numerous studies have also shown that garlic can reduce blood pressure. Taken regularly, garlic will usually reduce blood pressure by 10-15 points. Besides having a vasodilative effect, it also decreases blood cholesterol and triglycerides. Onions also have this effect, as do many pungent spices and herbs, which can all be safely consumed as part of the regular diet.

RG-Max contains l-arginine, an amino acid that acts on a chemical messenger called nitric oxide which dilates blood vessels. It has proven helpful in many cases of hypertension.

Lobelia contains a compound called lobeline which acts as a natural beta blocker. It combines well with capsicum and a small amount of black cohosh to reduce cardiac stress and angina, improve circulation to the heart, and lower blood pressure caused by tension. Other nervines like kava kava and black cohosh may also help.

Adaptagens can also help to reduce blood pressure. These include eleuthero root, American and Korean ginseng, schizandra (found in Nutri-Calm and Nervous Fatigue Formula) and SUMA Combination.

Water Retention and Kidney Function

When the tissues of the body are filled with fluid, this will put pressure on the blood vessels, again constricting blood flow. The kidneys also have an influence on the heart, so problems with the kidneys can also cause the blood pressure to rise. This is why diuretics are sometimes be used to bring down blood pressure. Where fluid retention is a problem, diuretics like Chinese Kidney Activator can be used as part of a blood pressure lowering program. Reducing table salt consumption and replacing it with an all-natural sea salt like the one available from NSP can also help reduce fluid retention. Kidney issues are often an undiagnosed issue behind blood pressure problems.

Excess Weight

Excess weight alone can increase blood pressure simply because there are blood vessels the heart has to pump blood through. There is also a link between excess insulin production, which contributes to excess weight, and imbalances in messenger chemicals that cause arterial constriction. If you have excess weight start following the suggestions in this book found under weight loss.

Therapies: Antioxidants (pg. 17), Low Glycemic Diet (pg. 11), Affirmation and Visualization (pg. 8), Oral Chelation (pg. 21), Stress Management (pg. 8), Essential Fatty Acids (pg. 17)

Remedies: Blood Build, Blood Pressurex, Capsicum & Garlic with Parsley, Eleuthero, Lobelia, Magnesium, Mega-Chel, RG-Max, RG-Max, 7-Keto, AdaptaMax, All Cell Detox, Aloe Vera, Alpha Lipoic Acid, Astragalus, Black Cohosh, Capsicum, Cellular Energy, Co-Q10, Cornsilk, Evening Primrose Oil, Garlic, GC-X, Ginseng (Korean), Ginseng (Wild American), Grapefruit (Pink), Guggul, Hawthorn Berries, Herbal Sleep, HS II, IGF-1, Kidney Activator (Chinese), Kidney Drainage, Lemon Oil, Love and Peas, Lymphatic Drainage, Marjoram (Sweet), N-acetyl Cysteine, Nature's Gold, Olive Leaf, Passion Flower, Pro-Pancreas, Rose Bulgaria, Sea Salt, Stevia, Stress Relief, Super GLA, Thai-Go, Ylang Ylang

Blood Pressure (low)

See Also *Adrenals (Exhaustion, Weakness or Burnout)*

While not as readily recognized as high blood pressure, low blood pressure can also be a serious problem. Low blood pressure can result in fatigue, fainting or dizziness. Low blood pressure may be the result of blood loss, but can also be due to glandular problems, particularly adrenal fatigue. An adrenal glandular like Adrenal Support or licorice root will often help this problem.

Shepherd's purse is one of the best vasoconstrictive herbs for tightening blood vessels. It is not available as a single from NSP, but can be found in the Menstrual-Reg formula.

Capsicum and garlic tend to normalize blood pressure, reducing high blood pressure and increasing low blood pressure. Adaptagens can do the same thing.

Therapies: Stress Management (pg. 8)

Remedies: Adrenal Support, Capsicum, Mineral-Chi Tonic, Capsicum & Garlic with Parsley, Clove Bud, Dandelion, Garlic, Ginkgo & Hawthorn, Ginseng (Korean), Ginseng (Wild American), GreenZone, HS II, Jasmine Absolute, Kidney Activator, Omega-3, Oregon Grape, Thyroid Support, TS II

Bloodshot Eyes

Blood shot eyes are eyes with many red blood vessels showing in the whites of the eyes. This may be caused by fatigue, stress or irritation. The herbs on this list are used as eyewashes for bloodshot eyes.

Remedies: Chamomile, EW, Eyebright, Golden Seal

Body Building

See Also *Exercise*

These products can help with body building.

Therapies: SuperFood (pg. 15), Minerals (pg. 15), Hydration (pg. 10), Avoid Xenoestrogens (pg. 23)

Remedies: Collatrim, Cordyceps, Free Amino Acids, Adrenal Support, EveryBody's Formula, GreenZone, IGF-1, Love and Peas, Nutri-Burn, Safflowers

Body Odor

When there is an excessively offensive odor associated with sweat, it is an indication of a need for cleansing the body. The following products can work either internally or externally to help reduce or eliminate unpleasant body odor.

Liquid chlorophyll is a natural deodorizer when taken internally. Pau d Arco Lotion and Nature's Fresh have been used as natural underarm deodorants. They work better if some essential oils are added.

Therapies: Hydration (pg. 10), Colon Cleansing (pg. 18)

Remedies: Chlorophyll, Nature's Fresh, All Cell Detox, Cellular Energy, Enviro-Detox, GreenZone, Heavy Metal Detox, HSN-W, Lymphatic Drainage, Neroli, Parsley, Pau D Arco, Proactazyme Plus, Sage, Vitamin B-12, Vitamin B-Complex, Yeast/Fungal Detox, Zinc

Boils

A boil is an infection of a skin gland resulting in localized swelling and inflammation, having a hard central core often filled with pus and/or watery fluid. Boils are best treated by using remedies that cleanse the blood and help drain the lymphatics. Topical, as well as internal, remedies can be used.

Remedies: All Cell Detox, Burdock, Skin Detox, Ultimate Echinacea, Aloe Vera, Black Walnut, BP-X, Cellular Energy, Chickweed, Devil's Claw, Echinacea Purpurea, Flax Seed Oil, Ho Shou Wu, Lemon Oil, LIV-J, Lymphatic Drainage, Marshmallow, Nature's Fresh, Red Clover, Red Clover Blend, Rose Bulgaria, SF, Tea Tree Oil

Bone Spur

See *Calcium Deposits (Calcification)*

BPH

See *Prostate Problems (Benign Prostate Hyperplasia, Prostatitis)*

Breast (infection)

See *Mastitis*

Breast Lumps

The following remedies may be helpful for breast lumps that are not cancerous in nature. Be sure to obtain an appropriate medical diagnosis. Many of these remedies also help to protect the breasts against lumps and breast cancer.

Therapies: Antioxidants (pg. 17), Avoid Xenoestrogens (pg. 23), Colon Cleansing (pg. 18)

Remedies: All Cell Detox, Breast Assured, Nature's Fresh, Black Currant Oil, Evening Primrose Oil, False Unicorn, Hydrangea, Liver Balance, Lymphatic Drainage, Master Gland, Mood Elevator, Pro-G-Yam Cream, Super GLA, Vitamin E

Breast Milk (dry up)

The following remedies help to dry up breast milk. Placing cold cabbage leaves on the breast will also help to dry up breast milk.

Remedies: Parsley, Sage, Clary Sage

Breast Milk (increase or enrich)

The following herbs have been used to enrich or increase the flow of breast milk.

Therapies: Essential Fatty Acids (pg. 17)

Remedies: Blessed Thistle, Marshmallow, Milk Thistle, Alfalfa, Chlorophyll, Herbal CA, Marshmallow & Fenugreek, Vitamin E

Breasts (enhance size)

Certain phytoestrogens in plants appear to encourage breast development. Saw palmetto, in particular, has been reported helpful for this purpose. Breast Enhance contains saw palmetto and other herbs which have been reported to enhance breast size. The effect is not likely to be dramatic in most women.

Remedies: Saw Palmetto, Breast Enhance, Ylang Ylang

Breasts (swelling and tenderness)

See Also *PMS Type H*

Tenderness or pain in the breast tissue may be due to lymphatic congestion. Hormonal imbalances may also cause breast swelling and tenderness.

Formulas that increase lymphatic drainage may help, and include Lymphomax and Lymphatic Drainage. Poultices with herbs like mullein and slippery elm have also eased this problem.

When associated with PMS this may be due to elevated aldosterone. Vitamin B6, magnesium, vitamin E, omega-3 fatty acids, evening primrose oil and the essential oils of frankincense and lemon are all remedies that may help here. Monthly Maintenance is a possible remedy.

Estrogenic herbs and hormonal balancers such as black cohosh and dong quai, both of which are found in Flash Ease may also help ease breast swelling when hormones are involved. Female Comfort or FCS-II are other options.

Therapies: Avoid Xenoestrogens (pg. 23), Essential Fatty Acids (pg. 17)

Remedies: Magnesium, Monthly Maintenance, Vitamin B-6, Black Cohosh, Black Currant Oil, Black Walnut, Breast Assured, Dong Quai, Eleuthero, FCS II, Female Comfort, Flash-Ease T/R, Frankincense, Geranium, Kidney Activator, Kidney Activator (Chinese), Lemon Oil, Lymphatic Drainage, Mullein, NF-X, Omega-3, Parsley, Pro-G-Yam Cream, Slippery Elm, Vitamin E

Broken Bones

The following may been taken internally to speed the healing of broken bones.

Therapies: SuperFood (pg. 15), Minerals (pg. 15)

Remedies: Calcium, Herbal CA, HSN-W, Skeletal Strength, Bone/Skin Poultice, C-X, Collatrim, Colloidal Minerals, Colostrum with Immune Factors, Free Amino Acids, Horsetail, HSN Complex, IGF-1, Ionic Minerals, Mineral-Chi Tonic, Nature's Fresh, Phyto-Soy, PLS II, Pro-G-Yam Cream, Proactazyme Plus, Silver Shield, Slippery Elm, Sunshine Heroes Calcium Plus D3, Thai-Go, Vitamin D3

Bronchitis

See Also *Congestion (bronchial)*

Bronchitis is an inflammation of the small tubes of the lungs, usually as a result of a respiratory infection. Expectorants, decongestants, and anti-inflammatory remedies can help bronchitis. One of the best remedies providing these actions in one formula is Bronchial Formula.

Therapies: Antioxidants (pg. 17)

Remedies: ALJ, Bronchial Formula, Cordyceps, Lobelia, Astragalus, Bergamot, Black Cohosh, Blue Vervain, Breathe EZ, Breathe Free, Capsicum, Capsicum & Garlic with Parsley, Catnip, CC-A, Clove Bud, Deep Relief Oil, Echinacea/Golden Seal, Elderberry Plus, Four, Garlic, GC-X, Ginger, Golden Seal, Heavy Metal Detox, Helichrysum, HistaBlock, IF Relief, IN-X, Influenza Remedy, Kava Kava, Lemon Oil, Licorice Root, Lung Support, Lymph Gland Cleanse-HY, Marjoram (Sweet), Marshmallow, Marshmallow & Fenugreek, Milk Thistle Combination, MSM, Mullein, Myrrh, Neroli, Oregano (Wild), Pine Needle, Sandalwood, Thyme, Ultimate Echinacea, Vari-Gone, Vitamin A & D

Bruises (healing)

Bruises are caused by an injury to the tissues where stagnation sets in causing the area to turn a purplish black color. Some of the best remedies for helping bruises heal quickly are yarrow, Silver Shield Gel, Nature's Fresh or Tei Fu Oil applied topically.

Remedies: Healing AC Cream, Helichrysum, Rose Hips, Aloe Vera, Burdock, Butcher's Broom, L-Glutamine, Lobelia, Marjoram (Sweet), Nature's Fresh, Rose Bulgaria, St. John's Wort, Tei Fu Oil, Thai-Go, Vari-Gone, Yarrow

Bruises (prevention)

To prevent easy bruising take rose hips, Vitamin C with Citrus Bioflavonoids, VariGone or butcher's broom internally. To prevent bruising apply Healing AC Cream or Tei Fu Oil topically to injuries, where the skin isn't broken, right after the injury occurs.

Therapies: Antioxidants (pg. 17)

Remedies: Butcher's Broom, Rose Hips, Vitamin C, Bilberry Fruit, Bone/Skin Poultice, Cardio Assurance, Ginkgo Biloba, Grapine, Hawthorn Berries, Healing AC Cream, Noni (Morinda), Sunshine Heroes Whole Foods Antioxidant, Tei Fu Oil, Thai-Go, Vari-Gone, Yarrow

Bulimia

See Also *Anorexia, Anxiety Disorders*

Bulimia is a disorder characterized by binge eating followed by self-induced vomiting, and the use of laxatives, fasting or diuretics to prevent weight gain. Herbs and supplements may help with appetite or nerves, but this disorder is psychologically based and requires emotional healing work. Seek appropriate professional assistance.

Therapies: Affirmation and Visualization (pg. 8), Stress Management (pg. 8), Low Glycemic Diet (pg. 11), Minerals (pg. 15), Enzymes (pg. 15), Essential Fatty Acids (pg. 17)

Remedies: Digestive Bitters Tonic, Spleen Activator, Distress Remedy

Bunions

A localized swelling at a joint in the foot caused by an inflammation of the bursa is called a bunion. Remedies that may help bunions heal include the following.

Remedies: Chlorophyll, Ho Shou Wu, Lobelia, Safflowers, St. John's Wort, Triple Relief

Burning Feet or Hands

See Also *Circulation (poor), Diabetes (Type II), Hyperinsulinemia (Syndrome X), Inflammation*

Burning hands and feet are caused by an abnormal nervous system signal or by a lack of circulation. This causes a tingling sensation in the extremities.

To help the nerves to function better, B-Complex Vitamins and HSN-W can be very helpful. Where the cause is circulatory in nature, the oral chelation program may be helpful. This circulatory imbalance may be due to blood sugar problems (hyperinsulinemia or diabetes), in which case HY-C is helpful. Where general inflammation is the cause, IF-C may be of benefit.

Therapies: Low Glycemic Diet (pg. 11), Oral Chelation (pg. 21), Essential Fatty Acids (pg. 17), Low Glycemic Diet (pg. 11), Hydration (pg. 10), Antioxidants (pg. 17)

Remedies: HY-C, IF-C, Mega-Chel, Vitamin B-Complex, Collatrim, HSN-W, HSN-W, Lobelia, Nerve Eight, Nutri-Calm, Safflowers, Super GLA, Vitamin B-6

Burnout

See *Adrenals (Exhaustion, Weakness or Burnout)*

Burns and Scalds

Burns and scalds can be very painful injuries. Minor burns (1st degree) involve normal symptoms of inflammation redness, pain and swelling. More severe burns (2nd degree) can result in blisters and the most severe burns (3rd degree) can have permanent skin damage that prevents skin regeneration and can threaten life due to infection or fluid loss if the area is extensive. Most of the following can be applied topically for relief of pain and irritation from 1st and 2nd degree burns. A few are taken internally for pain.

Topical remedies include: aloe vera gel, tea tree essential oil, lavender essential oil, Silver Shield Gel, pure vanilla extract, honey, slippery elm or chickweed. Vitamin E can be applied once the pain is gone from the burn to prevent scarring and speed tissue repair. Zinc and vitamin C, taken internally, help to speed the healing of burns.

Remedies: Aloe Vera, Herbal Trim Skin Treatment, Lavender, Nature's Fresh, Silver Shield, Bergamot, Burdock, Chickweed, Distress Remedy, Flax Seed Oil, Golden Salve, Gotu Kola, Healing AC Cream, Intestinal Soothe & Build, Marshmallow, Peppermint, Phyto-Soy, Rose Bulgaria, Sarsaparilla, Slippery Elm, Tea Tree Oil, Triple Relief, Vitamin C, White Oak Bark, Zinc

Bursitis

See Also *Arthritis*

Bursitis is an inflammation of the connective tissue capsule of joints resulting in pain and inflammation. It is treated naturally in a similar manner to arthritis.

Remedies: Deep Relief Oil, Joint Support, Tei Fu Oil, Alfalfa, Aloe Vera, Anti-Gas Formula, Bone/Skin Poultice, Burdock, Collatrim, Devil's Claw, Herbal CA, IF Relief, IF-C, IGF-1, Licorice Root, Magnesium, Nature's Phenyltol with NEM, PLS II, Sarsaparilla, Uña de Gato Combination

Calcium Deficiency

A deficiency of calcium in the body may be helped by some of the following remedies.

Therapies: Minerals (pg. 15), Essential Fatty Acids (pg. 17)

Remedies: Herbal CA, HSN-W, Skeletal Strength, Calcium, Damiana, Food Enzymes, Kelp, Magnesium, Mineral-Chi Tonic, Omega-3, PDA, Super Algae, Vitamin B-12

Calcium Deposits (Calcification)

In the presence of mineral imbalances, calcium can come out of solution and form gravel, stones, hardened tissue, or bone spurs. Calcium deposits often signal a lack of magnesium or other nutrients used in conjunction with calcium. Herbs that have lithotriptic properties can help bring calcium back into solution in the body. These include hydrangea, lemon juice and gravel root.

If you have been taking calcium supplements, discontinue their use and take 500-1,000 milligrams of magnesium each day. If you feel you still need a calcium supplement try Herbal CA.

Remedies that reduce inflammation can be helpful when bone spurs are involved. These include Joint Support and Joint Health. For hardened tissues, use the oral chelation program with lithotriptic herbs.

Therapies: Oral Chelation (pg. 21), Minerals (pg. 15)

Remedies: HSN-W, Hydrangea, Magnesium, Mega-Chel, Blessed Thistle, Joint Support, KB-C, Kidney Drainage, Omega-3, SF, Vitamin B-12

Cancer (natural therapy for)

See Also *Hodgkin's Disease, Leukemia*

Anyone who has ever had cancer, or had a loved one with cancer, knows the feelings of fear, anxiety, worry and often hopelessness that this very serious illness can bring. This is understandable considering cancer is the second leading cause of death in civilized nations. Furthermore, conventional treatments such as chemotherapy, radiation and surgery are often dangerous in and of themselves. So, it's little wonder that cancer usually causes intense emotional distress in everyone involved. However, one must never believe that there is no hope, even when orthodox medicine doesn't offer any. As long as the body has life, there is hope.

Of course, the subject of cancer is far too involved to adequately address in this book. We can acquaint you with some important information about cancer from a natural healing perspective and give you some ideas about options you may not be familiar with. However, we strongly encourage you to seek professional help when dealing with cancer. You need competent health care professionals helping you with your program and monitoring your progress, but you should also do some study on your own and learn about things you can do for yourself.

Cancer is a disease involving cells that have undergone a genetic mutation so they are no longer responsive to messages from the body that regulate cell metabolism and growth. These mutations are believed to be the result of free radical damage that causes the cells to develop an anaerobic metabolism and turn cancerous. Normal cells have an aerobic metabolism, which means they produce energy by means of oxygen and oxidation. Anaerobic cells produce energy without oxygen via a process of fermentation.

This is important to know because if the body is highly oxygenated, the environment for cancer does not exist. In fact, in 1931, Dr. Otto Warburg won a Nobel Prize for proving that whenever any cell is denied 60% of its oxygen requirements, it can become cancerous. So, conditions that deprive cells of oxygen (such as chronic inflammation, buildup of toxins or problems with red blood cells or circulation) increase the risk of cancer. An overly acidic environment in the body is also a breeding ground for cancer.

Another important thing you should know is that cancer cells are forming in the body on a regular basis. Very likely, you have a few inside you right now. Don't worry, the immune system normally recognizes these deviant cells and destroys them.

Therefore, two factors must exist for you to develop cancer. First, your body has to have a toxic, low oxygen environment that encourages the development of anaerobic cancer cells, and second, your immune system must be weakened so that it is not able to recognize and destroy these cells.

So, while killing cancer cells (the goal of conventional cancer therapy) is an important part of treating cancer, it does not fix the underlying problems that created the cancer in the first place. This is the weakness of the standard medical approach to cancer. An effective protocol for cancer should do more than just destroy cancer cells it'should try to restore a normal, healthy environment in the body and rebuild the immune system.

So, even if one chooses to use orthodox cancer therapies to destroy the cancer cells, they would be wise to consider doing natural therapy both to restore the body's state of health and prevent the cancer from reoccurring.

Principle Number One: Increase Oxygen Levels

Cancer cells are anaerobic, which means they live and thrive in a low oxygen environment. They are able to get their energy by metabolizing nutrients, notably sugars and carbohydrates, without oxygen, in a fermentative process. Cancer cells cannot survive in a high oxygen environment, so keeping the body well oxygenated inhibits cancer. Do this by getting plenty of fresh air and exercise. Breathe deeply. If you smoke, quit.

There are several supplements that can enhance oxygenation in the body. For starters, Chinese Lung Support helps with oxygen delivery and uptake and is very useful for anyone with respiratory weakness. Another Chinese remedy that strengthens the lungs and enhances oxygen transport is Cordyceps. Cordyceps also helps the immune system.

Liquid chlorophyll is another great way to enhance oxygen transport. It prevents clumping of red blood cells and helps them carry more oxygen to the tissues. Research has shown that it can reduce the risk of cancer.

Principle Number Two: Balance pH

In order to have optimal health, the body must maintain a proper internal pH. Cancer cells prefer an acid pH environment. Their fermentative metabolism also releases large amounts of lactic acid that inhibit transportation of oxygen to neighboring cells. This spreads the environment for cancer. A buildup of acid waste also inhibits toxins from being released by cells.

The best way to counteract this acidic environment is to eat large quantities of fresh, preferably organic, fruits and vegetables every day. This is well-recognized as one of the best ways of decreasing your risk of cancer, but unfortunately, its benefit in helping people recover from cancer is often ignored. Fresh fruits and vegetables not only help alkalize the body, they also strengthen immunity, aid detoxification and provide antioxidants.

Besides eating fresh fruits and vegetables, there are several other supplements that can help alkalize the body. The first is Green Zone, a whole food supplement that alkalizes the environment, aids in detoxification and builds the immune system to help it recognize and destroy cancer cells.

Digestive enzymes are very important to helping balance pH. Food Enzymes and PDA are good choices for people with cancer. A lack of hydrochloric acid (an ingredient found in both Food Enzymes and PDA) leads to excess lactic acid in the body, which sets the stage for cancer. They are taken with meals to help the body metabolize food correctly so as to avoid a buildup of acid waste and toxins. They are also taken between meals to help rid the body of excess proteins that may be causing acid in the blood.

Principle Number Three: Strengthen the Immune System

The body normally and regularly produces cancer cells due to free radical damage, environmental factors or other causes. A healthy immune system recognizes and destroys these defective cells. When the immune system is unable to recognize these deviant cells or is too weak to destroy them, the disease we call cancer develops.

There are many reasons why the immune system becomes weakened. Poor nutritional intake is a major factor. The loss or destruction of friendly bacteria in the intestinal tract is another. Excessive sugar consumption supplies cancer cells with the energy they need to proliferate quickly, while contributing to chronic inflammation that distresses the immune system.

Improving nutritional intake, especially eating those 7-9 servings of fruits and vegetables daily, will help the immune system, as will eliminating chemically-laden processed foods. Reducing intake of sugar, white flour and other simple carbohydrates will also help.

Trigger Immune helps to rebuild a weakened immune system, enhance vital energy and immune function and increase white blood cell production and platelet count. It is very helpful in recovering from chemotherapy and can also be taken during chemotherapy to strengthen the system to resist toxic effects from chemotherapy.

Immune Stimulator is another supplement that can boost the immune system to fight cancer. It not only enhances natural killer cells (lymphocytes), which kill cancer cells, it also increases T-cell production and antibody production. An important benefit of Immune Stimulator is its ability to enhance cellular communication, which helps the immune system identify and target cells it needs to destroy.

Una de Gato Combination is another herbal formula that strengthens a weakened immune system. The principle ingredient, Una de Gato, has been used in South America for cancer treatment and contains compounds that inhibit tumor growth in animals.

Principle Number Four: Detoxify

The human body is bombarded with toxins, heavy metals, chlorine and thousands of chemicals that we breathe in, consume in our diet or absorb through our skin. These all cause free radicals in the body and contribute to the environment of cancer.

Avoiding these toxins is part of both cancer prevention and holistic cancer therapy. In particular, avoid or eliminate refined and processed foods (especially foods raised with pesticides, antibiotics or steroids), toxic cleaning products (such as laundry detergents, skin care items, fluoridated toothpaste, etc.) and chlorinated and fluoridated water. Also, avoid microwaved or irradiated food and protect yourself from electrical equipment (as electromagnetic pollution may be a major causal factor in cancer).

It is also helpful to assist the body in detoxifying from these substances using any of the following supplements.

For starters, E-Tea or Essiac Tea is a famous anticancer formula that helps the body eliminate toxins. It improves lymphatic drainage and stimulates the immune system.

All Cell Detox is another good, general cleansing formula that is based on a Native American medicine man's cancer remedy. It reduces inflammation, supports the liver and digestive system and helps neutralize acid.

Blood Build nourishes the blood and helps increase its volume. It is an excellent formula for helping the body rebuild after chemotherapy.

Principle Number Five: Use Antioxidants

Our need for oxygen exceeds the demand for any nutrient, even water, because we need oxygen for normal energy production in the cell. However, oxygen can also produce free radicals that can damage normal cells and cause cancer. Antioxidant nutrients protect the body from this free radical damage, thereby reducing cancer risk. Antioxidants can also be used in a treatment program for cancer, because they help protect the body from harmful side effects of radiation and chemotherapy.

Some antioxidants to consider here include High Potency Grapine, an extract of grape seed that helps reduce inflammation and prevent cell damage, and Green Tea Extract, which contains polyphenols called catechins, powerful antioxidants that protect cells from cancer and kill cancer cells. One of these catechins is epigallocatechin gallate (EGCG), which was shown in several lab studies to kill cancer cells without harming healthy tissue.

Thai-Go contains xanthones, powerful antioxidants that have been shown in numerous studies to inhibit cancer cells and aid in tumor reduction. These compounds cause apoptosis (or preprogrammed cell death) in cancer cells. Xanthones exert cytotoxic (cancer cell killing) effects against human hepatocellular carcinoma cells, and have been shown to inhibit the growth of human leukemia HL60 cells. Xanthones have also been shown to be effective against human breast cancer SKBR3 cells. IF Relief also contains xanthones and is helpful for reducing chronic inflammation and pain.

Principle Number Six: Kill Cancer Cells

For those diagnosed with cancer, it is important to kill the cancerous cells. The problem is that chemotherapy and radiation also cause damage to healthy cells. Killing cancer cells also produces toxins that the body must eliminate.

There are some natural compounds that can help kill cancer cells, too. Paw Paw Cell Reg(c) is a standardized extract of acetogenins from the American paw paw tree. These compounds have been shown in scientific research to cause apoptosis (preprogrammed cell death) in cancer cells by inhibiting their energy production.

Paw Paw Cell Reg(c) is not appropriate for all cancers and should be used as part of a comprehensive cancer program. However, even if one chooses to use chemotherapy drugs, Paw Paw may still be beneficial. It has also been shown to reduce chemotherapy drug resistance. By itself, paw paw doesn't have the side effects of hair loss, weight loss, extreme nausea and compromised immunity. It may cause nausea in larger doses, however.

Paw paw is not recommended as a supplement for preventing cancer. It is only appropriate when a person has cancer. Paw Paw is also ineffective against leukemia, lymphoma and other cancers that do not involve tumors.

Another supplement that can help the body destroy cancer cells is High Potency Protease. It breaks down the protein coat on cancer cells and also helps prevent cancer cells that are breaking down from creating a toxic load on nearby healthy cells. It is taken between meals for this purpose.

There are other natural cancer killing herbs that are not sold by NSP that a person may wish to research if Paw Paw doesn't work for them. These include European mistletoe (Vibasum alba), Venus fly trap (Dionae muscipula), bloodroot (Sanguinarea canadensis), chaparral (Larrea tridentada), poke root (Phytolacca sp.) and black nightshade (Solanum nigrum). Many of these plants are toxic and should only be used under the guidance of a skilled professional herbalist.

Principle Number Seven: Increase Joy and Pleasure

One German study shows a commonality that all cancer patients experienced a trauma and an unresolved psychological issue shortly before the cancer developed. Stress is a big component of cancer because psychological stress creates physical stress that dramatically reduces immune function.

Reducing stress should be a stress-less task. That's why the goal here is not to reduce stress, but rather to deliberately seek out joy and pleasure. A pleasurable, happy experience has a more positive effect on the immune system and healing, than the stressful effects of an experience.

So, seek out pleasurable experiences. Find things that make you laugh. Spend time with family, friends or pets. Take a walk in the fresh air and sunshine. Surround yourself with pleasing colors, smells and sounds. Listen to your favorite music; get up and dance. Listen to calming, meditative music. Take a hot bath in Epsom salts and lavender oil. Treat yourself to a massage, take a mini-vacation or go to a spa for the day.

You can also reduce stress by taking products like Nutri-Calm, which replenishes the nervous system and adrenals. Adrenal Support will build stamina and promote healing, as well as giving the body more ability to cope with stress. Chinese Stress Relief is another option that can improve the body's ability to relax, counteracting stress.

Cancer is a difficult disease to work with, but many people have successfully recovered from cancer using both natural therapies and conventional therapies or a combination of the two. Seek out professional assistance in designing the holistic program that's right for you. Remember, there is hope!

Therapies: Hydration (pg. 10), Antioxidants (pg. 17), Avoid Xenoestrogens (pg. 23), Heavy Metal Detoxification (pg. 22), Hiatal Hernia Correction (pg. 12), Affirmation and Visualization (pg. 8), Colon Cleansing (pg. 18), Stress Management (pg. 8), Enzymes (pg. 15)

Remedies: E-Tea, Immune Stimulator, Paw Paw, Protease, Red Clover, Red Clover Blend, Trigger Immune, Alfalfa, All Cell Detox, Aloe Vera, Anamu, Astragalus, Black Currant Oil, Blood Build, BP-X, Carotenoid Blend, Chickweed, Chlorophyll, Co-Q10, Colostrum, Colostrum with Immune Factors, Defense Maintenance, Echinacea Purpurea, Eleuthero, Folic Acid Plus, Food Enzymes, Frankincense, Garlic, Gastro Health, Germanium Combination, Ginseng (Korean), Ginseng (Wild American), Glyco Essentials, Grapine, Green Tea Extract, GreenZone, IF Relief, Indole 3 Carbinol, Kelp, Lung Support, Lutein, N-acetyl Cysteine, Noni (Morinda), Papaya Mint, Pau D Arco, PDA, Phyto-Soy, Pro-G-Yam Cream, Probiotics, Red Beet Formula, S-O-D With Gliadin, SC Formula, Super Algae, Thai-Go, THIM-J, Ultimate Echinacea, Uña de Gato Combination, Yellow Dock

Cancer (prevention)

Since cancer is caused primarily by environmental toxins and electromagnetic pollution, the best way to prevent it is to minimize exposure to these influences. Eating organic food, using natural personal care and household cleaning products, and otherwise avoiding chemicals will help. For those toxins that can't be avoided, using antioxidant and hepatoprotective substances will minimize damage from chemicals.

Therapies: Antioxidants (pg. 17), Avoid Xenoestrogens (pg. 23), Heavy Metal Detoxification (pg. 22), Stress Management (pg. 8), SuperFood (pg. 15), Essential Fatty Acids (pg. 17),Fiber (pg. 17)

Remedies: Breast Assured, Chlorophyll, Thai-Go, All Cell Detox, Astragalus, Black Cohosh, Burdock, Carotenoid Blend, Enviro-Detox, Flax Seed Oil, Grapine, IF Relief, Indole 3 Carbinol, Milk Thistle, Milk Thistle Combination, Noni (Morinda), Omega-3, Parsley, Phyto-Soy, Protector Pak, Red Clover, Small Intestine Detox, Sunshine Heroes Whole Foods Antioxidant, Super Antioxidant, Super GLA, Super ORAC, Super Trio, Vitamin C, Vitamin D3, Zinc

Candida Albicans or Candidiasis

See *Fungal Infections (Yeast Infections, Candida albicans)*

Canker Sores (Mouth Ulcers)

A small painful ulcer usually in the mouth is called a canker sore. It has a grayish-white base surrounded by a red inflamed area.

Golden seal, taken internally and applied topically, is very effective at healing canker sores. A drop of Tei Fu Essential Oils or tea tree oil, applied topically to the sore, will relieve the pain within minutes. L-lysine is an amino acid that helps prevent canker sores. Where canker sores are frequent and severe, VS-C taken regularly will help to prevent them.

Remedies: Golden Seal, Tei Fu Oil, VS-C, Aloe Vera, Black Walnut, Blood Build, Burdock, Echinacea Purpurea, Free Amino Acids, HSN-W, L-Lysine, Lemon Oil, Lobelia, Myrrh, Sage, Stress Relief, Tea Tree Oil, Ultimate Echinacea, White Oak Bark

Capillary Weakness

Capillaries are the smallest of blood vessels that allow the passage of nutrients and oxygen from the bloodstream to the cells of the body. Some are so small that red blood cells must pass through them in single file. Nutritional deficiencies can cause these thin walls to become fragile and prone to rupture causing bleeding and bruising.

Therapies: Antioxidants (pg. 17), Essential Fatty Acids (pg. 17)

Remedies: Bilberry Fruit, Rose Hips, Vitamin C, Yarrow, Bee Pollen, Blood Build, Butcher's Broom, HSN-W, Lemon Oil, Thai-Go, Vari-Gone

Carbuncles

A carbuncle is a deep-seated infection of the skin, usually arising from several hair follicles that are close together. The following may be helpful.

Remedies: Silver Shield, Skin Detox, Aloe Vera, Ho Shou Wu, Ultimate Echinacea

Cardiac Arrest (Heart Attack)

A cardiac arrest or heart attack is caused by an acute episode of insufficient blood supply to the heart muscle often resulting in damage to the heart and possibly even death. This lack of blood supply to the heart may be triggered by a blood clot or a muscle spasm constricting already partially blocked blood vessels.

When a person is having a heart attack, capsicum and lobelia extracts or powders placed under the tongue can help to support the heart and may save the person's life. Massive doses of vitamin E (400 IU every 10-20 minutes) can also be helpful.

After a heart attack, high doses of Co-Q10 (100 milligrams or more) will aid repair of the damage. Other supplements that may aid recovery from a heart attack include CardioAssurance, HS II, hawthorn and magnesium. These nutrients can also be taken to help prevent heart disease.

Therapies: Stress Management (pg. 8), Affirmation and Visualization (pg. 8)

Remedies: Capsicum, Co-Q10, Hawthorn Berries, Lobelia, Magnesium, Vitamin E, Cardio Assurance, HS II

Cardiovascular Disease (Heart Disease)

See Also *Arteriosclerosis (Atherosclerosis, Hardening of the Arteries), Blood Clots (Prevention of), Blood Pressure (High), Cholesterol (high)*

Cardiovascular disease is still the leading cause of death in Western civilization. One out of two people die from it. So, it makes sense to do what we can to reduce our risk of becoming one of the one in two statistics. Unfortunately, much of the information in the popular media about reducing one's risk of heart disease is based on outdated research.

For instance, most people believe that high cholesterol causes heart disease and that the lower your cholesterol level, the less risk you have of dying of heart disease. This simply isn't true. More recent research shows that chronic inflammation (not cholesterol) is the cause of heart disease and that having your cholesterol get too low is more dangerous to your health than having high cholesterol.

Most people also believe that fats cause heart disease and that low fat diets will prevent heart disease. This is partially true because the wrong kinds of fats (such as margarine and partially hydrogenated vegetable oils) do contribute to the development of heart disease. However, it's also true that good fats (such as olive oil, omega-3 essential fatty acids and the medium chain saturated fats found in organic butter from grass fed cows) actually protect your heart and reduce your risk of heart disease. Foods marketed as fat free or low fat often contain high amounts of refined sugars that actually increase inflammation and heart disease risk.

Furthermore, eating refined carbohydrates is far worse for your heart than eating fats. This is because sugar, white flour and other products spike insulin levels. High insulin levels are a bigger risk factor for heart disease than high cholesterol or high triglycerides. So, if this information comes as a surprise to you, it's time to update your knowledge a little by reading this newsletter. But first, let's look at some tools for evaluating your risk of heart disease.

Evaluating Your Risk of Heart Disease

Most people feel that heart disease strikes without warning, but the truth is that there are many subtle clues that demonstrate the heart needs help long before a person has a heart attack. Besides high blood pressure and high cholesterol, here are some things to consider.

Gum Disease There is a high correlation between inflammation of the gums and the risk of dying of a heart attack. If your gums are inflamed, so are your arteries.

Varicose Veins and Hemorrhoids These problems are reflections of sluggish circulation and poor blood vessel tone.

Fatigue and Shortness of Breath Feeling no desire for physical activity, getting winded with minor exertion and feelings of pressure or pain in your chest are early warning signs your heart may need some help.

Facial Clues A red, bulbous tip on the nose, spider veins on nose and vertical crease in the left earlobe are all early warning signs that your heart may need help. A bright red tip and pointed tongue is also an indicator of heart stress.

Iridology If you know an iridologist or are familiar with iridology, markings in the heart area of the iris, having a spleen heart transversal and/or having a lipemic diathesis (lipid ring) are all indicators of a genetic tendency to heart disease.

Blood Tests Besides cholesterol and triglycerides, consider tests for homocysteine, fibrinogen, C-reactive protein, hemoglobin A1C, Lp(a) and ferritin (iron) checked. These tests can be more revealing of heart disease risk. If you are concerned about your heart and circulation, consider getting these blood tests done.

If you show signs of needing help with your heart, take action now. Here are some steps to take.

Step One: Reduce Inflammation and Free Radical Damage with Antioxidants

Oxidative stress and the inflammation that accompanies it is what allows cholesterol and minerals to stick to our arteries, forming arterial plaque. This lessens blood flow to the heart, brain and other parts of the body, increasing the risk of heart attack, stroke and other arterial blockages.

That's why the single most important thing you can do to reduce your risk of heart disease is to obtain adequate amounts of antioxidant and anti-inflammatory nutrients. If you re one of the millions of Americans who aren't eating enough fresh fruits and vegetables, supplementing your diet with extra antioxidants is one of the best things you can do to reduce your risk of heart disease, cancer, dementia and other degenerative diseases associated with aging.

When it comes to protecting your heart, one of the best antioxidants is Co-Q10. It reduces blood pressure, aids recovery from heart attacks, keeps LDL cholesterol from oxidizing and improves energy production in the heart muscle. Statin drugs deplete Q-10, so this supplement should always be taken by people using statin drugs to lower cholesterol. Other options include Thai-Go, Super ORAC and IF Relief.

Step Two: Get an Oil Change

For a long time we ve heard the dogma preached to us that high fat diets contribute to heart disease, and that margarine and vegetable oils are healthier for us than butter, coconut oil or animal fats. In response to this propaganda many people have adopted low fat diets, avoiding eggs, whole milk and red meat in an effort to stay healthier. Unfortunately, this hasn't reduced deaths from heart disease.

The fact is that fatty acids are the preferred fuel of the heart. In other words, the heart needs fats to be healthy, but not just any kind of fats; it needs good fats.

Margarine, shortening, processed vegetable oils and most deep, fat-fried foods are examples of bad fats. These fats have been molecularly altered and do contribute to chronic inflammation, heart disease and other health problems. But, the natural fats found in high quality foods actually have the opposite effect. So, if you want a healthy heart, keep it well-oiled with the right kinds of fats. So don't eliminate them from your diet, just make an oil change and change the kinds of fats you eat.

What has blown the whole high fat equals heart disease myth is the discovery of cultures (such as Mediterranean and Eskimo) that have both high fat diets and low incidence of heart disease. Part of the secret is an essential fatty acid called Omega-3, an essential fatty acid in short supply in most Western diets. Taking omega-3 fatty acids actually reduces the risk of heart disease.

Besides omega-3 supplements, use other good fats. Butter from organically raised, grass-fed cows is a very healthy fat. So is organic, virgin coconut oil. The medium chain saturated fats in these oils are the preferred fuel of the heart and are also important for your immune system. For a particularly healthy spread, try blending one pound of softened butter with 1 cup of flax seed oil to make a tasty and nutritious soft spread butter. The flax seed oil will add additional omega-3s.

Step Three: Get Physically Active

The benefits of exercise on the heart are well-known. You don't have to go to the gym, just get out and walk, or hike, swim, play golf, garden, or do anything else pleasurable that gets your body moving. This helps keep your blood flowing properly and helps maintain a healthy cardiovascular system.

Step Four: Control Your Temper

Anger damages the heart. It is well documented that angry people are more prone to heart disease. If you have a problem with your temper, learn how to manage your anger and develop closer relationships. Having loving relationships reduces your risk of heart disease.

Step Five: Use Appropriate Supplements to Support Heart Health.

There are numerous herbs and nutritional supplements that can help to both prevent and reverse heart disease. We can't cover them all, but here are a few of the most important ones besides Co-Q 10 and Omega 3 (which we ve already discussed).

L-Carnitine for Heart Energy: This important amino acid, found primarily in red meat, transports fatty acids to be metabolized for energy in the mitochondria. It improves energy production and oxygen utilization in the heart and can be very helpful for improving heart health.

Magnesium to Prevent Spasms: About half of all Americans are deficient in magnesium, a critical mineral for heart health. Magnesium helps the heart and blood vessels to relax properly, which reduces stress on the heart, helps protect the heart against spasms and helps lower blood pressure. Magnesium is also essential for energy production in the heart.

Cardio-Assurance for Cardiac Health: This blend contains nutrients that help to metabolize homocysteine and acts as an antioxidant to protect the heart. It can help maintain normal cholesterol and inhibit the formation of clots in the circulatory system. Use it as a preventative supplement to strengthen and support the heart.

Mega Chel to Remove Arterial Plaque: This powerful oral chelation product can be used to help reverse arterial plaque buildup and improve circulation to heart, brain and peripheral areas of the body. It can also be used in place of a daily multivitamin and mineral for people who wish to maintain healthy circulation. See the instructions on how to do the oral chelation program in the Introduction.

RG-Max to Reduce Blood Pressure: This blend of amino acids contains 5 grams of L-arginine. L-arginine helps nitric oxide, which dilates the blood vessels to reduce blood pressure. It also contains citruline, L-carnitine and other amino acids that support healthy circulation. RG-Max can even help with erectile dysfunction in some men.

HS II for Improved Circulation: HS II is a basic herbal remedy for the heart. It reduces inflammation and helps normalize blood pressure.

Ginkgo/Hawthorn for Brain and Peripheral Circulation: Ginkgo enhances peripheral circulation and improves blood flow to the brain. It also helps prevent blood clots from forming. It is combined with hawthorn, one of the best general herbs for heart health.

Don't be one of the statistics. Alter your lifestyle and start using some of the many supplements that can keep your heart healthy.

Therapies: Antioxidants (pg. 17), Low Glycemic Diet (pg. 11), Affirmation and Visualization (pg. 8), Oral Chelation (pg. 21), Stress Management (pg. 8), SuperFood (pg. 15), Essential Fatty Acids (pg. 17), Hydration (pg. 10)

Remedies: Co-Q10, HS II, Magnesium, Mega-Chel, Omega-3, RG-Max, Alpha Lipoic Acid, Blood Pressurex, Cardio Assurance, Chromium GTF, Collatrim, Evening Primrose Oil, Flax Seed Oil, Garcinia Combination, Germanium Combination, Ginkgo & Hawthorn, Ginseng (Korean), Ginseng (Wild American), Hawthorn Berries, HY-C, IF Relief, IGF-1, Kelp, Lecithin, Nattozimes Plus, Pro-G-Yam Cream, Protector Pak, S-O-D With Gliadin, SF, Super ORAC, Thai-Go, Vitamin E

Carpal Tunnel Syndrome

Carpal tunnel syndrome is a narrowing of the bony passage in the wrist that constricts blood vessels and nerves passing to and from the hand causing pain and disturbances of sensation in the hand. Repetitive movements such as constant typing cause carpal tunnel syndrome.

Vitamin B-6 has helped many people with carpal tunnel syndrome. Anti-inflammatories such as IF-C may also be helpful along with massage and stretching exercises that keep the wrists flexible. Chiropractors and other body workers can also make adjustments to the wrists to aid in healing. Topical remedies like Deep Relief oil or Healing AC Cream may also be used to relieve inflammation and pain.

Therapies: Antioxidants (pg. 17)

Remedies: Collatrim, Deep Relief Oil, Healing AC Cream, IGF-1, MSM, Tei Fu Oil, Thai-Go, Vitamin B-6, Vitamin B-Complex

Cartilage Damage

See Also *Arthritis*

Cartilage is a spongy material that cushions the ends of bones at the joints. Nutritional deficiencies may prevent its proper production or immune disorders may cause enzymes to destroy healthy cartilage. Cartilage may also be torn during injuries such as those that occur in sports. Cartilage generally heals slower than bone since it relies on passive fluid movement for nutrients rather than blood vessels.

Collatrim has proven very beneficial in helping to repair damaged cartilage. Nature's Fresh, sprayed topically over damaged areas, can also speed healing.

Therapies: Minerals (pg. 15), Antioxidants (pg. 17)

Remedies: Collatrim, EverFlex, Bilberry Fruit, Bone/Skin Poultice, Cellu-Smooth, Glucosamine, Herbal CA, IGF-1, Joint Health, Joint Support, Mineral-Chi Tonic, MSM, MSM/Glucosamine Cream, Nature's Fresh, SAM-e, Spirulina, Super Algae, Thai-Go, Uña de Gato Combination

Cataracts

See Also *Free Radical Damage*

A clouding of the lens of the eye is called a cataract. Cataracts usually develop from age-associated free radical damage but they may also be caused by injury. Symptoms include blurred vision and glare, the latter especially at night.

For protection against cataract development antioxidants such as Thai-Go, Super Antioxidant or Super ORAC may be helpful.

The herbal eyewash formula EW has been used to dissolve cataracts. It is made as a tea using 2-4 capsules per cup of boiling water. The tea is then used as an eyewash several times per day. It'should be refrigerated and a fresh batch made every couple of days to the tea from becoming contaminated and causing an eye infection. This process often results in the eyes becoming weepy and red as toxins are eliminated. If the eyes start becoming severely irritated discontinue the use of the eyewash.

Therapies: Antioxidants (pg. 17), Essential Fatty Acids (pg. 17)

Remedies: EW, Super Antioxidant, Vitamin C, Vitamin E, Blood Build, Carotenoid Blend, Chickweed, Eyebright, Folic Acid Plus, Free Amino Acids, Germanium Combination, Ginseng (Korean), Heavy Metal Detox, Ho Shou Wu, L-Carnitine, Lactase Plus, Mega-Chel, Melatonin Extra, MSM, N-acetyl Cysteine, Perfect Eyes, SF, Spirulina, Thai-Go, Zinc

Cavities

See *Tooth Decay (prevention)*

Celiac Disease

See Also *Inflammatory Bowel Disorders (Colitis, IBS)*

Celiac disease is an inflammatory condition of the colon, a chronic condition that causes breakdown of the intestines due to a gluten allergy. Gluten is a protein found in wheat and other grains such as oats, barley and rye. This disease requires avoiding gluten foods and taking extra enzymes such as Protease to aid breakdown of foods. Basic therapies for inflammatory bowel disorders should also be used.

Therapies: Probiotics (pg. 16), Enzymes (pg. 15), Antioxidants (pg. 17)

Remedies: Intestinal Soothe & Build, Probiotics, Every Body's Fiber, Magnesium, Protease, Vitamin B-12

Cellulite

Cellulite is made of fatty deposits trapped by collagen that give the skin a dimpled, orange peel look. It tends to develop on the thighs, hips and buttocks of women.

Herbs that help the body metabolize fats and release toxins will help to burn cellulite. One of these herbs is chickweed, which is incorporated into the formula Cellu-Smooth, an herbal product for burning cellulite. Applying the essential oil blend Cellu-Tone topically will increase the effectiveness of this program.

Therapies: Essential Fatty Acids (pg. 17)

Remedies: Cellu-Smooth, Cellu-Tone, All Cell Detox, Chickweed, Grapefruit (Pink), L-Carnitine, Lemon Oil, MetaboMax, Patchouli, SF, Yeast/Fungal Detox

Chemical Poisoning

See Also *Heavy Metal Poisoning*

In the modern world we are exposed to thousands of chemicals we are exposed to on a regular basis. Many of them are toxic. These chemicals come from several sources. Pesticides, herbicides and fungicides are used on our crops. Antibiotics and hormones are routinely fed to animals. Artificial colorings, flavoring agents, preservatives and other food additives are also added to processed foods. In fact, the average person eats two to three pounds of chemical food additives each year.

We are also exposed to chemicals in household cleaning products and personal care products. Chemicals are found in building materials, paint and solvents, carpets and fabrics and our water and air due to pollution.

The liver bears the primary burden of having to process all of these chemicals. Avoiding as many of these chemicals as possible takes stress off of the liver. Consider products like Enviro-Detox and the Tiao He Cleanse to detoxify from chemical exposure. Milk thistle helps protect the liver against environmental poisons. SAM-e can also help with liver detoxification. Antioxidants also help protect the body against chemical poisons. The body needs essential fatty acids to bind certain types of toxins for elimination, so good fats help, too.

To avoid these chemicals, buy organic food as much as possible and wash commercial produce in Sunshine Concentrate to remove pesticide residues. Use natural cleaning and personal care products and drink purified water.

For acute chemical poisoning it is best to consult a poison control center for advice. Activated charcoal can be taken to absorb many kinds of poisons after they have been ingested. Ask the poison control center for specific instructions.

Therapies: Hydration (pg. 10), Antioxidants (pg. 17), Probiotics (pg. 16), Essential Fatty Acids (pg. 17), Fiber (pg. 17)

Remedies: Charcoal (Activated), Milk Thistle Combination, N-acetyl Cysteine, Omega-3, Tiao He Cleanse, All Cell Detox, Enviro-Detox, Flax Seed Oil, Heavy Metal Detox, Intestinal Soothe & Build, Liver Balance, LOCLO, Milk Thistle, Probiotics, Red Clover, Red Clover Blend, Sunshine Concentrate Cleaner, Super GLA

Chemotherapy (reducing side effects)

See Also *Cancer (Natural Therapy for), Chemical Poisoning, Radiation Sickness*

Chemotherapy is the administration of poisons in an attempt to kill cancerous cells. Cancer cells are generally weaker than normal body cells, so the trick is to give enough poison to kill the cancer but not the patient. This overall poisoning of the body is very destructive and the symptoms and side effects can be widespread, severe and varied. Many chemotherapy agents target rapidly growing cells. Since the cells that line the digestive tract, white blood cells, and skin and hair cells all grow rapidly, these cells often take the brunt of chemotherapy. This is why it is common to have digestive upset, loss of hair and a lowering of the immune response associated with standard chemotherapy.

Herbs and supplements can be used in conjunction with standard cancer therapies like chemotherapy, but people who choose to go this route will need to find a doctor who will cooperate with their wishes. Many cancer doctors forbid their patients from using herbs or supplements because it creates reactions in the body that are different than what they expect.

The digestive tract of patients on chemotherapy can be soothed using Mucilant herbs to absorb irritants and reduce inflammation. Whole leaf Aloe Vera juice is very effective. Eating bulk slippery elm as a food is also very effective. Other products that might help include Noni Juice and Intestinal Sooth and Build.

Antioxidants can be used to help protect the immune system from the oxidative damage caused by chemotherapy and radiation. Grapine, vitamin C, vitamin E, Super Antioxidants or Grapine with Protectors are all products that could be used. Using herbs that stimulate and rejuvenate the immune system may also be helpful, such as Immune Stimulator, Pau d Arco tea, Eleuthero root and Uña de Gato. These products will help raise white blood cell counts.

Products that aid the detoxification process of normal cells will also help protect the body in the process of standard chemotherapy. B-complex vitamins, in particular, will help to ease many side effects of chemotherapy.

Paw Paw Cell Reg(c) has anticancer activity of its own, but it also has the ability to inhibit a pumping mechanism in the membranes of the cancer cells that enable them to eliminate toxic drugs and become drug resistant. Cancer patients who have a relapse after a few years often have drug resistant cancer cells. Paw Paw Cell Reg(c) can be used to restore the effectiveness of chemotherapy.

There are other options besides the standard medical treatments used for cancer. The route a person chooses to take is a matter of his or her free will and choice, but people should carefully consider their options.

Therapies: Antioxidants (pg. 17), Affirmation and Visualization (pg. 8), Probiotics (pg. 16), Essential Fatty Acids (pg. 17),Fiber (pg. 17)

Remedies: Aloe Vera, Astragalus, Slippery Elm, Super Antioxidant, Vitamin B-Complex, Co-Q10, Eleuthero, Germanium Combination, Immune Stimulator, Intestinal Soothe & Build, Noni (Morinda), Pau D Arco, Paw Paw, Red Raspberry, S-O-D With Gliadin, Ultimate Echinacea, Uña de Gato Combination, Vitamin C, Vitamin E

Chest Pain

See Also *Acid Indigestion (Heartburn, Acid Reflux), Acid Indigestion (Heartburn, Acid Reflux), Angina, Gall Bladder (sluggish), Pain (general remedies for), Pleurisy*

Chest pain can have several causes, including a gallbladder attack, heartburn or acid reflux, inflammation in the pleura, muscle tension or angina. Get a proper diagnosis and see appropriate related conditions.

Chicken Pox

See Also *Itching (Topical Remedies for), Shingles*

Chicken pox is an acute, contagious infection of the Herpes Zoster virus. The immune response to this infection creates a low-grade fever and oozing, itching sores almost anywhere on the surface of the body. They may even occur on mucous membrane surfaces such as the throat. If sores are scratched or picked they can leave scars.

Chicken pox is aided naturally by enhancing the immune system's ability to expel the virus from the body. This is accomplished by taking frequent doses (every 2-4 hours) of alterative or blood purifying herbs such as VS-C, BP-X, Enviro-Detox, burdock, golden seal, Oregon grape, safflowers or yellow dock along with plenty of water.

IF-C or Nerve Eight can be used to help reduce fever and inflammation. Aspirin is not recommended because it'suppresses the body's ability to eliminate the virus, which causes problems with shingles later in life.

Remedies used to ease topical itching may be helpful as well. These include baths, compresses or fomentations using burdock, comfrey, goldenseal, Hydrated Bentonite, chickweed or Oregon grape. Aloe Vera Gel, Black Walnut extract and Silver Shield are also useful topical remedies.

Remedies: Blue Vervain, Olive Leaf, Silver Shield, VS-C, Yarrow, Aloe Vera, Bergamot, Black Walnut, BP-X, Burdock, Catnip, Chickweed, Cordyceps, E-Tea, Enviro-Detox, Golden Seal, Hydrated Bentonite, IF-C, Lobelia, Nerve Eight, Oregon Grape, Safflowers, St. John's Wort, Vitamin E, Yellow Dock

Childbirth

See *Labor and Delivery, Pregnancy (herbs and supplements for)*

Children's Remedy

The following are good remedies and products for young children.

Therapies: SuperFood (pg. 15), Probiotics (pg. 16), Essential Fatty Acids (pg. 17), Antioxidants (pg. 17), Hydration (pg. 10), Low Glycemic Diet (pg. 11), Affirmation and Visualization (pg. 8), Minerals (pg. 15), Blood Type Diet (pg. 13)

Remedies: Sunshine Heroes Omega 3 with DHA, ALJ, Aloe Vera, Bedwetting, Bedwetting, Bee Pollen, Berry Healthy, Blue Vervain, Calming, Catnip, Catnip & Fennel, CBG Extract, CC-A, Chamomile, Charcoal (Activated), Colloidal Minerals, Colostrum, Cough Syrup (Childrens), Distress Remedy, Elderberry Plus, Flax Seed Oil, Focus Attention, Healing AC Cream, Herbal Punch, Herbal Trim Skin Treatment, Ionic Minerals, Lemon Oil, Love and Peas, Mandarin (Red), Mineral-Chi Tonic, Oregon Grape, Papaya Mint, Red Clover Blend, Sunshine Heroes Multiple Vitamin & Mineral, Sunshine Heroes Probiotic Power, Sunshine Heroes Whole Foods Antioxidant, Sunshine Heros Whole Foods Papayazyme, Super Algae, Super GLA, Tea Tree Oil, Teething, Thai-Go, Tofu Moo, Ultimate Echinacea, Valerian Root, Vitamin B-12, Vitamins, Children's Multi

Chills

Chills are usually associated with an increased internal body temperature during a fever but can also be present in severe blood loss or shock. Mild, chronic feelings of cold, especially in the extremities, is usually caused by poor circulation, or hormonal imbalances, particularly low thyroid. The hypothalamus regulates the body temperature and opens and closes the vents or pores in the skin to help regulate the body temperature.

For chills associated with fever, IF-C is a helpful formula. Where chills are associated with night sweats sage may be helpful. Chills associated with fatigue and lowered resistance to disease may benefit from Trigger Immune or licorice root. Where cold limbs are associated with arteriosclerosis, the oral chelation program is very effective.

Therapies: Oral Chelation (pg. 21)

Remedies: Capsicum, IF-C, Licorice Root, Alpha Lipoic Acid, Bayberry, Capsicum & Garlic with Parsley, Cinnamon, Eleuthero, HY-C, Influenza Remedy, Mega-Chel, Sage, Trigger Immune

Cholera

See Also *Diarrhea*

An acute infection caused by the bacterium Vibrio cholerae. Generally causes diarrhea and dehydration. Remedies to consider include the following:

Remedies: Garlic, Golden Seal, Silver Shield, Charcoal (Activated), Marshmallow, Slippery Elm

Cholesterol (high)

See Also *Cholesterol (low)*

High cholesterol is not really a disease. It is a symptom of metabolic imbalance, not a root cause of any health problem. The lab ranges for cholesterol have been artificially reduced due to pressure from the pharmaceutical industry in order to sell more highly profitable statin drugs. Normal cholesterol ranges should be 175 to 275 with Blood Type O people running at the higher end of this spectrum because of the way their body utilizes protein.

These are the pathological ranges, meaning that if you are above or below these values then your body is becoming seriously imbalanced. For optimal health, you should be in the middle third of this range. So healthy cholesterol should be between 200 and 250. Cholesterol below 175 is too low and can cause serious health risks.

Cholesterol plays a very important role in our body. The primary use of cholesterol (60-80%) is to make bile for the digestion of fats. Cholesterol is also used to make adrenal and reproductive hormones and to sequester toxins in the body.

The following can help to lower cholesterol when it is too high.

Fiber in the diet helps to reduce cholesterol levels for a couple of reasons. First, it binds toxins in the gut and secondly it binds to cholesterol being released in the bile to prevent it from being reabsorbed. All of the following have been shown to have this effect: activated charcoal, Fat Grabbers, Nature's Three, LOCLO and Psyllium Hulls Combination.

Cholesterol levels can also be lowered by obtaining adequate quantities of high quality fats. Eating a lot of olive oil will actually help to lower cholesterol because more bile has to be produced to break down the fats. Essential fatty acids in flax seed oil and Super Omega-3 EPA will also help lower cholesterol.

Fiber and good fats are the first, and most important, approaches to regulating high cholesterol. Other effective aids to lowering cholesterol include Red Yeast Rice, ho shu wu, niacin, lecithin, garlic and guggul. Red Yeast Rice should be taken with Co-Q 10-75. Anyone on statin drugs should also be taking Co-Q 10. A complete list of products that may help lower cholesterol follows.

Therapies: Gall Bladder Flush (pg. 23), Antioxidants (pg. 17), Colon Cleansing (pg. 18), Oral Chelation (pg. 21), Essential Fatty Acids (pg. 17),Fiber (pg. 17)

Remedies: Charcoal (Activated), Fat Grabbers, Ho Shou Wu, Nature's Three, Niacin, Psyllium, Psyllium Hulls Combination, Alfalfa, Algin, Aloe Vera, Bee Pollen, Black Currant Oil, Blood Build, Bowel Detox, Cardio Assurance, Cellular Energy, Chickweed, Cholester-Reg II, Chromium GTF, CLA, Cordyceps, Devil's Claw, DHA, Flax Seed Oil, Gall Bladder Formula, Garcinia Combination, Green Tea Extract, Guggul, Herbal Trace Minerals, HS II, Lecithin, Lemon Oil, LOCLO, LOCLO, Mega-Chel, Milk Thistle, Myrrh, N-acetyl Cysteine, Omega-3, Phyto-Soy, Red Yeast Rice, Safflowers, Sea Salt, SF, Stress Relief, SugarReg, Super Algae, Vitamin C, Yeast/Fungal Detox

Cholesterol (low)

See Also *Fat Metabolism (poor), Gall Bladder (sluggish), Hyperthyroid*

A cholesterol level below 175 is too low and increases risk of death from cardiovascular diseases and cancer. The lower the cholesterol, the higher the risk of cancer. Low cholesterol also interferes with glandular function, especially the adrenal and reproductive hormones. It may be a cause of infertility and depression.

Extremely low cholesterol levels are sometimes the result of taking statin drugs. Low thyroid or an iodine deficiency may be the cause of low cholesterol. A lack of bile production or congestion in the liver is another possible cause. The following can help correct the problems causing cholesterol to be too low.

Therapies: Essential Fatty Acids (pg. 17)

Remedies: Adrenal Support, L-Carnitine, Thyroid Support, Blood Build, Catnip & Fennel, Chlorophyll, Digestive Bitters Tonic, Dulse, Folic Acid Plus, Gall Bladder Formula, Gastro Health, Heavy Metal Detox, Helichrysum, Lecithin, LIV-J, Master Gland, MSM, Omega-3, Oregon Grape, SAM-e, SF, Super Algae, Vitamin B-12, Vitamin B-Complex

Chron's Disease

See *Inflammatory Bowel Disorders (Colitis, IBS)*

Chronic Fatigue Syndrome (CFS)

See *Epstein Barr Virus (Chronic Fatigue Syndrome, CFS)*

Circulation (poor)

Poor circulation is characterized by a variety of symptoms. It may result in cold hands and feet, fatigue, loss of memory, wounds or sores that won't heal in the extremities, loss of eyesight, swelling in the legs and tingling sensations in the arms, hands, legs and feet. These remedies help improve circulation.

Therapies: Oral Chelation (pg. 21)

Remedies: Capsicum, Capsicum & Garlic with Parsley, Ginkgo & Hawthorn, Mega-Chel, AdaptaMax, Anti-Gas (Chinese), Astragalus, Bilberry Fruit, Blood Pressurex, Blue Vervain, Cellu-Smooth, Chamomile, Cinnamon, DHA, Dong Quai, Eleuthero, Energ-V, Flax Seed Oil, Garlic, Ginger, Ginkgo Biloba, Ginkgo/Gotu Kola, Guggul, HS II, Maca, Men's Formula, Neroli, Olive Leaf, Omega-3, Pine Needle, Sea Salt, Thai-Go, Vari-Gone, Vitamin E, X-Action (Men s)

Circulation (to the brain)

See Also *Alzheimer's Disease, Dementia, Memory and Brain Function*

These remedies improve circulation to the brain, which can help with senility and cognitive functions as a person ages.

Therapies: Oral Chelation (pg. 21), Antioxidants (pg. 17), Essential Fatty Acids (pg. 17)

Remedies: Ginkgo/Gotu Kola, Mega-Chel, Blessed Thistle, Brain-Protex, Ginkgo Biloba, Gotu Kola

Cirrhosis of the Liver

See Also *Hepatitis*

Cirrhosis is an end stage liver disease associated with functional failure of liver tissues and eventual liver failure. The liver tissue becomes scarred from viral infections, alcohol abuse, reactions to drugs and exposure to environmental toxins. This disease usually results in the need for a liver transplant. The best natural remedy is to go on a completely mild food diet (nothing but fresh fruits and vegetables, preferably in juice form) to give the liver a rest. Juices made from greens (chard, celery, etc.) are extremely beneficial. Salads are also beneficial. Of course, all alcohol, drugs and chemicals must be avoided. Also avoid all spicy food and heavy fats and oils.

Seek appropriate medical assistance with this very severe disease. Helichrysum essential oil can be applied topically over the liver area. Other remedies in the following list can be taken internally.

Remedies: Helichrysum, Milk Thistle Combination, SAM-e, Blood Build, Dandelion, Germanium Combination, Licorice Root, Milk Thistle, MSM, Thyroid Support

Cold Hands and Feet

See Also *Circulation (Poor), Hypothyroid*

Mild, chronic feelings of cold, especially in the extremities, can be a symptom of several different health problems. Low thyroid is a common cause and taking remedies to boost the thyroid often resolves the problem.

Another common cause is poor circulation to the extremities. If there are symptoms of arteriosclerosis, then the oral chelation program may be helpful. For less serious circulatory issues, try taking Capsicum or Capsicum, Garlic and Parsley.

Cold hands and feet may also be a symptom of a lowered immune response. If the person is tired and catches colds and flu easily, then try Trigger Immune. If the lungs are weak and the person has problems with wheezing or shortness of breath, then Lung Support may be helpful. If the person is pale and thin or emaciated, then poor digestion may be a root cause. Spleen Activator is helpful in that case.

Therapies: Oral Chelation (pg. 21)

Remedies: Capsicum, Mega-Chel, Niacin, Thyroid Support, Anti-Gas (Chinese), Blood Build, Capsicum & Garlic with Parsley, Flax Seed Oil, Ginkgo & Hawthorn, Guggul, IN-X, KC-X, Lung Support, Omega-3, Spleen Activator, Trigger Immune, TS II

Cold Sores (Fever Blisters)

A cold sore is an infection of the Herpes simplex virus that causes a painful, oozing group of blisters usually located around the lips. They turn into a scabby sore. Cold sores are also known as fever blisters.

VS-C is very effective at eliminating cold sores and preventing their recurrence. It was originally formulated for herpes. L-lysine has also been used to both treat and prevent the reoccurrence of cold sores. Black walnut extract, Tei Fu Essential Oils, tea tree oil, and peppermint oil can be applied topically to help heal the sores. Try applying Tei Fu or tea tree oil to the sore at the first sign of tingling. This will usually clear them in less than a day, and often in just a few hours. Where the cold sore has scabbed over, try applying golden salve.

Remedies: Black Walnut, Silver Shield, VS-C, Bergamot, Chlorophyll, Garlic, Golden Salve, GreenZone, Intestinal Soothe & Build, L-Lysine, Olive Leaf, Omega-3, Peppermint, Probiotics, Red Clover, St. John's Wort, Tea Tree Oil, Tei Fu Oil, Vitamin B-Complex, Zinc

Colds (antiviral)

Remedies that are helpful against cold and flu viruses include the following.

Remedies: Elderberry Defense, Elderberry Plus, Immune Stimulator, Astragalus, CC-A, Echinacea Purpurea, Garlic, Olive Leaf, Thyme, Ultimate Echinacea, Yarrow

Colds (decongestant)

Remedies to break up respiratory congestion associated with colds.

Remedies: ALJ, HCP-X, Tei Fu Oil, Eucalyptus, Eyebright, Fenugreek & Thyme, Garlic, HistaBlock, HS II, Lemon Oil, Lobelia, Rosemary, Seasonal Defense, Sinus Support, Thyme

Colds (general remedies for)

See Also *Cough (general), Infection (viral)*

Colds are a group of acute, contagious infections characterized by malaise, fever, chills (thus the name) and respiratory congestion. The goal in working with colds should be to help the body eliminate the toxins it is trying to expel from the body. The following are general remedies that have been used to fight colds. Other headings contain more specific remedies for specific types of cold symptoms.

There are many herbs that can be used to aid this process. Ultimate Echinacea or Echinacea purpurea, taken at the onset of a cold often prevents the cold from taking place, but this may be a slightly suppressive effect.

The best approach we have found is to take herbs that help flush the irritants in the early stages of the cold. Drink plenty of fluids and take CC-A or Elderberry Plus every 1-2 hours. Two teaspoons of Silver Shield taken 2-3 times per day will help kill the virus and relieve symptoms.

Spicy herbs taken in large doses with a lot of water are very good for eliminating colds with thin, watery mucus drainage. Take ginger, garlic or capsicum every 1-2 hours with a large glass of water. Avoid eating any heavy foods and, if possible, only consume juices, soups and fruits until the cold symptoms subside.

For aches and pains associated with colds use APS II w/ White Willow Bark or Nerve Eight. To break up and expel mucus in the respiratory tract, use expectorants like ALJ, garlic and lobelia. Garlic is probably the very best herb for dealing with respiratory infections. Golden seal is good for the later stages of the cold when the mucus begins to thicken and discolor.

Tei Fu Essential Oils or Breath Free Essential Oil Blend can be inhaled to help relieve sinus drainage and pressure. Where the cold involves a fever, boneset, IF-C or yarrow are good remedies. Where there are aches and pains in the muscles and bones, boneset is an excellent cold remedy. ALJ or lobelia can be helpful for colds with coughs.

Therapies: Colon Cleansing (pg. 18)

Remedies: ALJ, Capsicum, Elderberry Defense, Elderberry Plus, HCP-X, IF-C, Silver Shield, Ultimate Echinacea, Zinc Lozenges, Aloe Vera, Anamu, APS II, Astragalus, Bayberry, Berry Healthy, Blue Vervain, Breathe EZ, Breathe Free, Bronchial Formula, Capsicum & Garlic with Parsley, Catnip, CBG Extract, CC-A, Cinnamon, Clove Bud, Cold, Colostrum with Immune Factors, Echinacea Purpurea, Echinacea/Golden Seal, Eleuthero, Eucalyptus, EW, Eyebright, Fenugreek & Thyme, Fizz Active-Immune, FV, Garlic, Ginger, Golden Seal, Helichrysum, HistaBlock, Immune Stimulator, IN-X, Jasmine Absolute, Lemon Oil, Lobelia, Lymph Gland Cleanse-HY, Marjoram (Sweet), Marshmallow, Myrrh, Olive Leaf, Pine Needle, Prevention, Red Raspberry, Red Raspberry Liquid, Rose Hips, Rosemary, Seasonal Defense, Sinus Support, Thai-Go, THIM-J, Thyme, Valerian Root, Vitamin A & D, Vitamin C, Yarrow

Colds (prevention)

Silver Shield gel can be used as a natural hand disinfectant. It kills pathogens that come in contact with the skin for a period of four hours or until the hands are washed. The following are remedies one can take when colds are going around to keep from catching them.

Therapies: Probiotics (pg. 16)

Remedies: Trigger Immune, Ultimate Echinacea, Berry Healthy, Echinacea Purpurea, Elderberry Defense, Eleuthero, Fizz Active-Immune, Garlic, Immune Stimulator, Lung Support, Prevention, Silver Shield, Sunshine Heroes Probiotic Power, Sunshine Heroes Whole Foods Antioxidant, Thai-Go, Vitamin C, Zinc

Colds (with fever)

See Also *Fever (general remedies for)*

The following remedies are for colds accompanied by fever.

Therapies: Hydration (pg. 10)

Remedies: IF-C, Garlic, Lymph Gland Cleanse, Yarrow

Colic (adults)

See Also *Gas and Bloating*

Colic involves acute abdominal pain, characterized by cramping and gas. It is common in infants, but adults can have this problem, too. These remedies are for colic in adults.

Remedies: CLT-X, Gall Bladder Formula, Lobelia, Wild Yam, Bergamot, Chamomile, Digestive Bitters Tonic, Hops, Passion Flower, Sandalwood, Stress-J, Wild Yam & Chaste Tree

Colic (children)

See Also *Fungal Infections (Yeast Infections, Candida albicans)*

Colic involves acute abdominal pain, characterized by cramping and gas. It is common in infants, but adults can have this problem, too. These remedies are for colic in children.

Catnip and Fennel is an old fashioned remedy for colic. Place several drops in the child's mouth or apply it to the bottom of the feet.

If the Catnip and Fennel extract doesn't work, try making some catnip and fennel tea, as this is often more effective. Chamomile tea is also helpful. Lobelia and magnesium can help colic in infants because the colic is often due to muscle cramps. You can also rub peppermint essential oil and lobelia on the infant's stomach.

Yeast infections in infants can also cause colic. Use a little lavender and tea tree essential oil topically to help with the yeast infections. A garlic enema, or a plain water enema with a drop or two of lavender oil essential in it, can help knock down yeast and relieve severe colic. If yeast is a problem, then giving the infant probiotics, especially bifidophilus, will help. This can be administered orally by mixing with food or rectally in an enema.

Therapies: Probiotics (pg. 16)

Remedies: Catnip & Fennel, Chamomile, Peppermint, Aloe Vera, Bergamot, Catnip, Chamomile (Roman), Lavender, Lobelia, Mandarin (Red), Marjoram (Sweet), Neroli, Oregon Grape, Safflowers, St. John's Wort, Sunshine Heroes Whole Foods Antioxidant, Thyme, Valerian Root

Colitis

See Also *Inflammatory Bowel Disorders (Colitis, IBS)*

Colitis is inflammation of the colon and small intestine. Besides the general remedies for Inflammatory Bowel Disorders, the following may be specifically helpful for colitis.

Therapies: Antioxidants (pg. 17), Fiber (pg. 17), Probiotics (pg. 16), Essential Fatty Acids (pg. 17), Stress Management (pg. 8)

Remedies: Aloe Vera, CLT-X, Every Body's Fiber, Gentle Move, Intestinal Soothe & Build, Slippery Elm, Bowel Detox, Chamomile, Cinnamon, Gastro Health, Golden Seal, IF Relief, IF-C, L-Glutamine, Marshmallow, Omega-3, PLS II, Psyllium, Sarsaparilla, Small Intestine Detox, Stress-J

Colon (atonic)

An atonic colon is one that has lost muscular tone and balloons, lacking sufficient peristaltic strength to push material forward. This is the less common form of constipation and results in very slow, but regular, bowel eliminations. The following remedies help tone an atonic colon.

Therapies: Enzymes (pg. 15),Fiber (pg. 17)

Remedies: LBS II, Psyllium Hulls Combination, Cascara Sagrada, LB Extract, LB-X, Liquid Cleanse, Senna Combination

Colon (spastic)

See Also *Stress*

A spastic colon is caused by muscle spasms in the colon that inhibit peristalsis. It is characterized by irregular bowel movements and constipation that is aggravated by stress. Most adults and children who are constipated have a spastic bowel condition.

Therapies: Stress Management (pg. 8)

Remedies: Gentle Move, Lobelia, Magnesium, Black Cohosh, Blue Cohosh, Cellular Energy, Cramp Relief, Every Body's Fiber, Gentle Move, Hops, Intestinal Soothe & Build, Psyllium, Small Intestine Detox, Stress-J, Valerian Root

Concentration (poor)

See Also *Adrenals (Exhaustion, Weakness or Burnout), Circulation (To the Brain)*

Poor concentration is often due to a lack of circulation to the brain or imbalances in glandular function. The following remedies may help to improve concentration.

Therapies: Low Glycemic Diet (pg. 11), Oral Chelation (pg. 21)

Remedies: Focus Attention, Ginkgo/Gotu Kola, Nervous Fatigue Formula, Bergamot, Blessed Thistle, Breathe Free, Eucalyptus, Eucalyptus, GABA Plus, Ginkgo & Hawthorn, Ginkgo Biloba, Gotu Kola, Lemon Oil, Mega-Chel, Mineral-Chi Tonic, Mood Elevator, Peppermint, Pine Needle, Pine Needle, Rosemary, Sunshine Heroes Omega 3 with DHA, Tei Fu Oil

Concussions

A concussion is an injury to the soft tissue of the brain. It usually results from being shaken violently, or receiving a blow to the head. The following may assist healing. Take St. John's wort internally, use the other listed remedies topically.

Remedies: St. John's Wort, Healing AC Cream, Nature's Fresh

Confusion

See Also *Adrenals (Exhaustion, Weakness or Burnout), Hypoglycemia*

This is often a sign of low blood sugar or adrenal exhaustion. Remedies that may help when a person is easily confused include the following.

Therapies: Hydration (pg. 10), Low Glycemic Diet (pg. 11), Stress Management (pg. 8)

Remedies: Adrenal Support, Nervous Fatigue Formula, Bergamot, Focus Attention, Frankincense, Geranium, Grapefruit (Pink), Helichrysum, HY-A, Lavender, Lemon Oil, Licorice Root, Magnesium, MSM, Patchouli, Peppermint, Pine Needle, Rosemary, Super Algae, Thyme

Congestion (bronchial)

See Also *Bronchitis*

The following remedies may be helpful when the bronchial passages are swollen or congested with mucus. Decongestant remedies and remedies that reduce inflammation are particularly helpful here.

Therapies: Hydration (pg. 10), Colon Cleansing (pg. 18)

Remedies: ALJ, Breathe EZ, Bronchial Formula, Adrenal Support, Chickweed, Cordyceps, Cough Syrup-LP, EW, Gastro Health, HistaBlock, IF-C, MSM, Mullein, Oregon Grape, Pantothenic Acid

Congestion (general)

Excessive mucous collected in the nasal and/or lung passages causes congestion, which can interfere with breathing. The lymphatics can also become congested, resulting in swollen lymph nodes, earaches, sore throats and frequent respiratory infections.

Decongestants and expectorants are used to break up mucus and help to expel it from the body. It is important to understand that expectorant and decongestant remedies won't necessarily dry up the sinuses. Since they promote breakup of congestion and the expulsion of mucus, they may cause a temporary increase in drainage as the lungs and sinuses clear out irritants.

The long-term solution to congestion involves normalizing the intestinal tract. Enzymes such as Proactazyme Plus or Protease will help the body break down food better, which results in less congestion. A colon cleanse such as the Tiao He Cleanse can also be beneficial.

The following are general remedies for congestion.

Therapies: Hydration (pg. 10), Colon Cleansing (pg. 18), Enzymes (pg. 15), Probiotics (pg. 16),Fiber (pg. 17)

Remedies: Breathe Free, Fenugreek & Thyme, Tei Fu Oil, ALJ, All Cell Detox, Allergies-Mold/Yeast/Dust, Alpha Lipoic Acid, Anamu, Anti-Gas (Chinese), Blue Vervain, Capsicum, Capsicum & Garlic with Parsley, Cascara Sagrada, CC-A, Cellular Energy, Cinnamon, Elderberry Defense, Elderberry Plus, Eucalyptus, Four, Geranium, HCP-X, Lobelia, Lymphatic Drainage, Marjoram (Sweet), Marshmallow & Fenugreek, Mullein, N-acetyl Cysteine, Protease, Rosemary, Small Intestine Detox, Thyme, Ultimate Echinacea, Vitamin C

Congestion (lungs)

See Also *Cough (damp)*

The following are specific remedies for congestion in the lungs.

Therapies: Hydration (pg. 10), Colon Cleansing (pg. 18)

Remedies: ALJ, Bayberry, Breathe EZ, Bronchial Formula, Cordyceps, Cough Syrup-LP, Garlic, Seasonal Defense

Congestion (lymphatic)

Isn't indoor plumbing a wonderful thing? We can all be thankful that we can turn on a tap and have fresh, running hot and cold water for drinking, bathing, and washing.

On the other hand, we ve all experienced the frustration of a clogged drain. When the dirty water we ve washed with won't go down the drain, we ve got a stagnant mess on our hands. Because we all understand that stagnant water isn't healthy, we do whatever is necessary to unclog that drain, so the wastewater isn't standing around polluting the internal environment of our home.

Most people aren't aware that the body has a drainage system, too. It's called the lymphatic system, and it works hand in hand with the circulatory system to keep the various tissues and organs alive and healthy. It's the lymphatic system's job to make certain the fluid around the cells in the body doesn't become stagnant. Sometimes, however, the lymphatic system becomes congested, and like a clogged or sluggish drain an unhealthy stagnation of fluids occur. Without the lymphatic drainage working properly, the tissues in the body become like a clogged kitchen sink, a trash-laden back alley, or a stagnant swamp none of which can be considered healthy conditions.

Lymphatic congestion contributes to swollen lymph nodes, earaches, sore throats, chronic sinus and respiratory congestion, tonsillitis, appendicitis, breast swelling, lumps and tumors, lymphatic cancers and other health problems.

Here are some ways to clear the lymphatic congestion. First, deep breathing, combined with muscular movements, is the key to pumping our lymph and keeping it moving. Exercise, for example, increases lymph flow as much as five to fifteen times. One of the best forms of lymphatic exercise is gentle bouncing on a mini trampoline. If a person is unable to stand on the mini trampoline, he or she can still obtain benefit by sitting in a chair next to the trampoline with his or her feet on the trampoline. Another person stands on the trampoline and gently bounces up and down. This passively moves the lymphatics as the seated person's legs move up and down. If you don't own a mini trampoline, don't worry. Just walking and breathing deeply will greatly enhance lymphatic circulation, as will any other form of moderate exercise.

The second key to reducing lymphatic sluggishness is to drink an adequate amount of water. Even moderate dehydration will contribute to poor lymphatic drainage.

Dietary therapy may also be helpful. Certain foods seem to clog up the lymphatic system more than others. For many people, dairy products are major culprits. Wheat is another lymphatic clogger for many people. However, any food that creates allergic reactions for a person may contribute to lymphatic stagnation. Avoiding these foods is the third key to improving lymph drainage.

The fourth, and final key, is using herbs that improve lymphatic function such as the following.

Therapies: Hydration (pg. 10),Fiber (pg. 17), Colon Cleansing (pg. 18)

Remedies: Lymphatic Drainage, Lymphomax, Ultimate Echinacea, ALJ, Burdock, Capsicum & Garlic with Parsley, Echinacea Purpurea, Geranium, IN-X, Lobelia, Lymphostim, Mullein, Oregon Grape, Red Clover, Red Clover Blend

Congestion (sinus)

See Also *Headache (sinus), Polyps, Sinus Infection*

Inhaling Tei Fu essential oil is a great way to relieve sinus congestion. Using Silver Shield in a nasal aspirator and spraying it into the sinuses is another good remedy. Silver Shield Gel can also be applied inside the nose. Bayberry can be snuffed into the sinuses to shrink nasal polyps and improve sinus drainage. Fenugreek and Thyme is helpful for easing sinus pressure and headaches.

Therapies: Fiber (pg. 17), Colon Cleansing (pg. 18), Hydration (pg. 10), Enzymes (pg. 15), Blood Type Diet (pg. 13)

Remedies: **Bayberry, Fenugreek & Thyme, Sinus Support, Tei Fu Oil,** ALJ, Allergies-Hayfever/Pollen, Cold, Pine Needle, Seasonal Defense, Silver Shield, Sinus, SnorEase

Congestive Heart Failure

See Also *Cardiovascular Disease (Heart Disease), Heart (to strengthen)*

Congestive heart failure is a decline in the function of the heart due to fluid retention around the heart. Many of the best botanical remedies for congestive heart failure are toxic botanicals like lily of the valley that require the assistance of a professional herbalist. Appropriate medical assistance should be sought for this condition. The following remedies may be helpful.

Therapies: Antioxidants (pg. 17), Essential Fatty Acids (pg. 17)

Remedies: **Co-Q10, Hawthorn Berries, L-Carnitine, Magnesium, RG-Max,** Kidney Drainage, Lymphatic Drainage, Oregon Grape, Sarsaparilla

Conjunctivitis (Pink Eye)

Conjunctivitis is also known as pink eye because of the redness that occurs in the whites of the eyes due to the inflammation in the eye. The inflammation occurs in the clear covering over the white of the eyes and the lining of the inner eyelids. It may be caused by an allergic reaction or by a bacteria or virus, the latter two forms being contagious.

The fastest way to help this problem is to make eye drops or a compress out of any of the following products: Eyebright, EW, chamomile or golden seal. Make one of these products into a tea and put drops of the tea into the eyes after the tea has been carefully strained through a fine cloth and cooled. You can also saturate a cotton ball with the tea and place it over the closed eyes, allowing it to rest there for about 15 minutes. Repeat this process at least four times per day.

Silver Shield can be used as eye drops to kill pathogens. The Silver Shield Gel can also be rubbed around the eyes and over the eyelid. Another remedy that helps conjunctivitis is to rub vitamin A&D around, but not in, the eyes. Vitamin A&D can also be taken internally to help the problem.

Remedies: **Chamomile, EW, Silver Shield,** Eyebright, Golden Seal, Lemon Oil, Nature's Fresh, Rose Bulgaria, Vitamin A & D

Constipation (adults)

See Also *Colon (spastic), Fungal Infections (Yeast Infections, Candida albicans), Gall Bladder (sluggish), Inflammatory Bowel Disorders (Colitis, IBS), Parasites (general)*

Dehydration is the main cause of constipation in adults. So start by drinking plenty of pure water.

Temporary constipation can be relieved by the use of stimulant laxatives. In order of strength, stimulant laxative products include cascara sagrada, LB Extract, LB-X, LBS-II, Liquid Cleanse and Senna Combination. Gentle Move is a good product for encouraging normal bowel eliminations. It tonifies the colon while hydrating it and can improve bowel function when taken over time. Chlorophyll capsules are also a gentle laxative that can rebuild colon function over time.

Bulk laxatives are better as a solution for long-term constipation. These include Nature's Three, Psyllium Hulls Combination, LOCLO and Everybody's Fiber. These products must be taken with large glasses of water in order to work properly. They can actually cause constipation when taken with an insufficient amount of water.

Some people have problems with constipation due to a spastic colon which clamps down under stress. Intestinal inflammation, yeast infections, a sluggish liver and gallbladder, poor digestion and parasites are other possible cause of constipation.

Therapies: Hydration (pg. 10), Enzymes (pg. 15), Probiotics (pg. 16),Fiber (pg. 17), Antioxidants (pg. 17), Colon Cleansing (pg. 18), Hiatal Hernia Correction (pg. 12), Gall Bladder Flush (pg. 23)

Remedies: **All Cell Detox, Every Body's Fiber, Gentle Move, LB-X, LBS II, Magnesium, Nature's Three, Psyllium Hulls Combination,** Aloe Vera, Anti-Gas (Chinese), Artemisia Combination, Bentonite (Hydrated), Bowel Detox, Cascara Sagrada, Cellular Energy, Chlorophyll, CleanStart, CLT-X, Flax Seed Oil, Gall Bladder Formula, HY-C, LB Extract, Liquid Cleanse, LIV-J, Liver Cleanse Formula, LOCLO, Mandarin (Red), Milk Thistle, Mineral-Chi Tonic, Nervous Fatigue Formula, Para-Cleanse, Peppermint, Probiotics, Psyllium, Red Beet Formula, Senna Combination, SF, Stress Relief, Tiao He Cleanse

Constipation (children)

These are milder laxatives, suitable for use with children.

Therapies: Hydration (pg. 10), Enzymes (pg. 15), Probiotics (pg. 16)

Remedies: **Gentle Move, Magnesium,** Flax Seed Oil, LB Extract, Licorice Root, Slippery Elm, Sunshine Heroes Probiotic Power, Sunshine Heros Whole Foods Papayazyme, Yellow Dock

Contagious Diseases

See Also *Chicken Pox, Colds (general remedies for), Flu, Infection (bacterial), Infection (viral), Measles, Mumps*

These are remedies that are generally good for fighting contagious diseases. They can also be taken to help prevent contagious disease.

Therapies: Probiotics (pg. 16), Enzymes (pg. 15)

Remedies: Garlic, Immune Stimulator, Silver Shield, Ultimate Echinacea, Echinacea Purpurea, Elderberry Defense, Golden Seal, Lobelia, Lymph Gland Cleanse-HY, Lymphomax, Olive Leaf, Probiotics, Seasonal Defense, Sunshine Heroes Probiotic Power, Trigger Immune, VS-C

Convalescence

After a prolonged illness, the body is often in a debilitated state and needs special nutritional support to aid in rebuilding and recovering good health. This period of recovery from debility is called convalescence. The following remedies can aid this healing process.

Therapies: Antioxidants (pg. 17), SuperFood (pg. 15), Enzymes (pg. 15), Essential Fatty Acids (pg. 17)

Remedies: Cordyceps, GreenZone, Mineral-Chi Tonic, Thai-Go, Trigger Immune, Alfalfa, Aloe Vera, Astragalus, Barley Grass, Bee Pollen, Blood Build, Carotenoid Blend, Cellular Energy, CLT-X, Colostrum with Immune Factors, Digestive Bitters Tonic, Green Tea Extract, Lung Support, Marshmallow, Nature's Gold, PLS II, Proactazyme Plus, Slippery Elm, Spleen Activator, SUMA Combination, Super Algae, Super Trio, THIM-J

Convulsions

See Also *Epilepsy*

Convulsions are abnormal, violent and involuntary contractions or a series of contractions of the muscles are called convulsions. The following remedies may help convulsions.

Therapies: Antioxidants (pg. 17), Essential Fatty Acids (pg. 17)

Remedies: Blue Vervain, Kava Kava, Lobelia, AdaptaMax, Aloe Vera, Catnip, Geranium, Passion Flower, Stress Relief, Stress-J

Coordination

The following remedies may help improve coordination.

Remedies: Free Amino Acids, Mineral-Chi Tonic, Nervous Fatigue Formula, Super Algae, SynerProTein, Vitamin B-12, Vitamin B-Complex

Copper Toxicity

The following can help with copper toxicity (excessive copper in the system).

Remedies: Zinc, Heavy Metal Detox, HSN-W, Intestinal Soothe & Build, LIV-J, MSM, Potassium

Corns

A lesion of the skin formed between two toes as a result of pressure between them. The surface of the skin is macerated and yellowish in color. Most of these remedies are applied topically for relief.

Remedies: Nature's Fresh, Vitamin E, Chamomile, Garlic, Germanium Combination, GreenZone, Herbal Trim Skin Treatment, Tea Tree Oil

Cough (damp)

A damp cough is a cough that produces a lot of phlegm. There is excess mucus production and fluid in the lungs. These cough remedies help to dry up the excess dampness.

Remedies: ALJ, Breathe EZ, Garlic, HCP-X, Anamu, Bayberry, Breathe Free, Bronchial Formula, Capsicum & Garlic with Parsley, Clove Bud, Cordyceps, Cough Syrup-LP, Eucalyptus, EW, Ginger, Jasmine Absolute, Lymphatic Drainage, Sinus Support, Sunshine Heros Whole Foods Papayazyme, Tei Fu Oil, Thyme

Cough (dry)

When a cough is unproductive (dry and hacking) so that there is little mucus production, the lungs are dehydrated. Moistening expectorants and decongestants, like those that follow, are needed.

Therapies: Hydration (pg. 10)

Remedies: Astragalus, Cordyceps, Licorice Root, Lung Support, Marshmallow, Adrenal Support, Aloe Vera, Cough Syrup-DH, Flax Seed Oil, HY-C, Lemon Oil, Lung Support, Marshmallow & Fenugreek, Mullein, Pine Needle, PLS II, Sandalwood, Sarsaparilla, St. John's Wort

Cough (general)

Under normal conditions, the lungs and sinuses secrete a thin, protective layer of mucus that traps dust and other particles in the air. Thin, hair-like projections called cilia sweep this mucus to the back of the throat (from the sinuses) or to the top of the throat (from the lungs). When the mucus gets trapped in the lungs and the cilia are unable to move it out of the lungs, this creates an involuntary, explosive expulsion of air from the lung in an attempt to expel the mucus and irritants from the lungs.

All over-the-counter cough medicines contain cough suppressants, which suppress this cough reflex. This does not help the body eliminate the irritants. Decongestants help thin the mucus so that it can move more freely and expectorants stimulate the cilia to move it out of the system. The following are all herbs that act as expectorants and decongestants to help make coughs productive:

Therapies: Hydration (pg. 10)

Remedies: ALJ, Garlic, Lobelia, Pine Needle, Rosemary, Thyme, Asthma, Cordyceps, Cough Syrup (Childrens), Cough Syrup-DH, Cough Syrup-LP, Cough Syrup-NT, Eucalyptus, Four, HCP-X, Marshmallow, Marshmallow & Fenugreek, Mullein, Myrrh, Tei Fu Oil

Cough (spastic)

See Also *Croup, Pertussis (Whooping Cough)*

A spastic cough is one in which there is muscle constriction in the bronchials. This can come as a result of muscle exhaustion after an extended period of coughing. Coughs that have a whooping sound or constricted quality to them need these remedies.

Therapies: Stress Management (pg. 8)

Remedies: Blue Vervain, Lobelia, Black Cohosh, Cough Syrup-NT, Rosemary, Sandalwood, Thyme

Cradle Cap

Cradle cap is a skin condition affecting the scalp of babies. It is characterized by large oily, yellow flaking from the babies scalp. It is a form of dermatitis and is related to eczema. It may involve yeast infections. Often babies who have this are severely deficient in essential fatty acids, most likely due to a deficiency in the mother's diet during pregnancy. The following remedies may be applied topically.

Therapies: Essential Fatty Acids (pg. 17)

Remedies: Herbal Trim Skin Treatment, Vitamin E, Aloe Vera, Black Walnut, Healing AC Cream, Omega-3, Pau D Arco, Probiotics, Super GLA

Cramps (leg)

Cramps in the legs are usually a sign of magnesium and/or potassium deficiencies. Stress can be a contributing factor. The following are possible remedies for cramps in the legs.

Remedies: Magnesium, Potassium, Hops, HSN-W, Mineral-Chi Tonic, Nature's Fresh, Skeletal Strength, Valerian Root, Vitamin B-6, Vitamin B-Complex

Cramps (menstrual)

See Also *Dysmenorrhea, PMS (general)*

Cramp Relief was specifically formulated to help relieve menstrual cramps. Magnesium Complex is also very helpful for menstrual cramping. Other possible remedies are listed below.

Therapies: Stress Management (pg. 8)

Remedies: Black Cohosh, Blue Cohosh, Cramp Relief, Magnesium, Wild Yam, C-X, Clary Sage, Deep Relief Oil, DHEA-F, FCS II, Female Comfort, Jasmine Absolute, Mandarin (Red), Mineral-Chi Tonic, Nature's Fresh, Nerve Eight, NF-X, Potassium, Skeletal Strength, Spleen Activator, Super Algae, Thyme, Wild Yam & Chaste Tree

Cramps and Spasms (general)

See Also *Tension*

A cramp is a painful, involuntary spasmodic contraction of a muscle. A spasm is an involuntary and abnormal contraction of a muscle that can be extremely painful.

Cramps can also be signs of magnesium and potassium deficiencies. Magnesium Complex is a good choice. Fibralgia contains magnesium and malic acid and can be very helpful for cramps.

Antispasmodic herbs, such as lobelia, black cohosh and kava kava, will help to relax cramps and muscle spasms both taken internally and applied topically. Cramp Relief is a combination of antispasmodics that can relax menstrual cramps, intestinal cramps, leg cramps and other muscle cramps and spasms.

Muscles expend energy to contract and must regenerate energy in order to relax again. A muscle that camps is low in energy. Because of this Cellular Energy can help to relax cramps.

Liquid chlorophyll, peppermint essential oil and Spleen Activator can be beneficial for stomach cramps. Wild yam and CLT-X are good for intestinal cramps.

Therapies: Affirmation and Visualization (pg. 8), Stress Management (pg. 8)

Remedies: Cramp Relief, Kava Kava, Lobelia, Magnesium, Black Cohosh, Blood Build, Cellu-Smooth, CLT-X, Deep Relief Oil, Dong Quai, Fibralgia, FV, Ginger, Herbal CA, Hops, Kelp, Mineral-Chi Tonic, Mood Elevator, Nature's Fresh, Pantothenic Acid, Potassium, Sea Salt, Skeletal Strength, Spleen Activator, Super Algae, Valerian Root, Wild Yam, Wild Yam & Chaste Tree, Wood Betony, Ylang Ylang

Crohn's Disease

See Also *Inflammatory Bowel Disorders (Colitis, IBS)*

Crohn's disease is an inflammatory condition of the small intestine and colon. It is a severe form of colitis that causes fistulas to form in the colon. There is atrophy and ulceration of the intestines. Symptoms include diarrhea, cramping, loss of appetite, loss of weight and intestinal abscesses and scarring. Remedies to reduce inflammation, absorb irritants and restore normal intestinal flora are needed.

Therapies: Antioxidants (pg. 17), Blood Type Diet (pg. 13), Probiotics (pg. 16), Essential Fatty Acids (pg. 17), Fiber (pg. 17), Stress Management (pg. 8)

Remedies: Every Body's Fiber, Intestinal Soothe & Build, Probiotics, Slippery Elm, Wild Yam, Aloe Vera, Blue Vervain, Bowel Detox, Devil's Claw, Gastro Health, IF-C, Magnesium, Mineral-Chi Tonic, Nutri-Calm, Pau D Arco, Protease, Psyllium, Skeletal Strength, Stress-J, SUMA Combination, Super Algae, Thai-Go, Uña de Gato Combination, Yucca

Croup

See Also *Cough (spastic)*

A spasmodic laryngitis, especially in infants, is called croup. Symptoms include difficulty breathing and a hoarse metallic cough. Antispasmodic herbal remedies are helpful.

Remedies: Blue Vervain, Lobelia, Rosemary, Carotenoid Blend, Catnip, Eucalyptus, Fenugreek & Thyme, Mullein, Omega-3, Tei Fu Oil, Thyme, Ultimate Echinacea, Vitamin E, Zinc

Cushing's Disease

Cushing's disease is characterized by elevated adrenal glands, meaning that the adrenal glands are overactive and are overproducing adrenal hormones. Symptoms include excess cortisol, low blood sugar, poor wound healing, lowered immune response, thinning hair, muscle wasting, abdominal fat and high blood pressure.

In this disease licorice root and other supplements that enhance adrenal function should be avoided. Adaptagens like eleuthero root, schizandra berries and SUMA Combination can be used to calm down elevated adrenal function.

Therapies: Stress Management (pg. 8), Affirmation and Visualization (pg. 8)

Remedies: Eleuthero, Nature's Cortisol, AdaptaMax, Blue Vervain, Chamomile, Chamomile (Roman), Kava Kava, Lavender, Nutri-Calm, Passion Flower, Rose Bulgaria, Stress Relief, SUMA Combination, Vitamin B-Complex

Cuts

See Also *Bleeding (external)*

Styptics are herbs, usually astringents, that have the power to stop bleeding, close cuts and speed the healing of cuts. Some of the herbs that have this property include bayberry, capsicum and yarrow, which can be poured directly into cuts and followed by applying pressure.

To prevent infection in cuts, you can apply Silver Shield Gel or essential oils like tea tree oil or lavender. To prevent scarring apply vitamin E and helicrysum essential oil.

Remedies: Capsicum, Silver Shield, Yarrow, Bayberry, Bone/Skin Poultice, Eucalyptus, Geranium, Golden Salve, Golden Seal, Helichrysum, Lavender, Nature's Fresh, PLS II, St. John's Wort, Tea Tree Oil, Tei Fu Oil, Uva Ursi, Vitamin E, White Oak Bark

Cystic Breast Disease

See Also *Breast Lumps*

The formation of benign fluid-filled sacs in the breast tissue is called cystic breast disease. Caffeine contributes to this problem and all forms should be avoided, including coffee, tea, kola drinks and chocolate. There is a need to improve lymphatic drainage from the breast tissue. A colon and liver cleanse is appropriate. There may also be a problem with low thyroid and a lack of iodine.

Therapies: Antioxidants (pg. 17), Avoid Xenoestrogens (pg. 23), Stress Management (pg. 8), Enzymes (pg. 15), Essential Fatty Acids (pg. 17), Fiber (pg. 17)

Remedies: Caffeine Detox, Nature's Fresh, Pro-G-Yam Cream, BP-X, Breast Assured, Chamomile, Evening Primrose Oil, False Unicorn, Herbal Trim Skin Treatment, Hydrangea, Indole 3 Carbinol, Lymphatic Drainage, Magnesium, Master Gland, Mood Elevator, Super GLA, Thai-Go, Vitamin A & D, Vitamin C, Vitamin E, Zinc

Cystic Fibrosis

Cystic fibrosis is a hereditary disease that appears in early childhood and involves a generalized disorder of exocrine glands. Symptoms include faulty digestion due to a deficiency of pancreatic enzymes, difficulty breathing and excessive loss of salt in the sweat.

Professional assistance should be sought, but attention must also be paid to overall diet and nutrition. High quality, nutrient-packed foods need to be eaten. The following supplements may be helpful.

Therapies: Enzymes (pg. 15), Minerals (pg. 15), SuperFood (pg. 15)

Remedies: Alfalfa, Black Currant Oil, Blue Vervain, Cordyceps, Evening Primrose Oil, Fenugreek & Thyme, Flax Seed Oil, Food Enzymes, Gastro Health, Juniper Berries, Marshmallow, N-acetyl Cysteine, Papaya Mint, PDA, Pro-Pancreas, Probiotics, Protease, Super GLA, Vitamin B-12, Vitamin B-Complex

Cystitis

Cystitis is inflammation of the urinary bladder. Remedies to reduce inflammation and soothe irritation in the urinary passages are helpful.

Therapies: Hydration (pg. 10), Antioxidants (pg. 17), Essential Fatty Acids (pg. 17)

Remedies: Cornsilk, Kidney Drainage, Marshmallow, Bergamot, Collatrim, Colostrum, Cranberry & Buchu, Damiana, Garlic, Horsetail, HSN-W, Hydrangea, IF Relief, IF-C, Intestinal Soothe & Build, JP-X, KB-C, Kidney Activator, Parsley, Pro-Pancreas, Probiotics, Red Raspberry, RG-Max, Sandalwood, Silver Shield, Skeletal Strength, Slippery Elm, Thai-Go, Urinary Maintenance, Uva Ursi, X-Action (Women s), Zinc

Cysts

Cysts are closed sacs with distinct membranes that are usually fluid filled. They form abnormally in a cavity or other structure of the body. Detoxification is essential for eliminating cysts. General cleansers like All Cell Detox and Enviro-Detox may be helpful. Plantain is a very good herb for breaking up cysts. Coconut oil is beneficial when applied to sebaceous cysts on the scalp. Other possible remedies are listed below.

Therapies: Colon Cleansing (pg. 18)

Remedies: All Cell Detox, Enviro-Detox, Nature's Fresh, Barley Grass, Chickweed, Cornsilk, Helichrysum, Kelp, Kidney Activator, Lymphatic Drainage, Peppermint, Protease, Red Beet Formula, Red Clover, SF, Vitamin A & D

Dandruff

Dandruff is caused by extreme dryness of the scalp, which results in white flakes. Jojoba oil rubbed into the scalp may relieve dandruff problems. Coconut oil may also be applied to the scalp. Adding tea tree, rosemary or other essential oils to one's shampoo to stimulate circulation to the scalp can also help dandruff. Dandruff can be related to stress, so NutriCalm or vitamin B-complex may be needed. It can also be linked with Candida. Increase intake of essential fatty acids.

Therapies: Essential Fatty Acids (pg. 17), Stress Management (pg. 8)

Remedies: Jojoba Oil, Tea Tree Oil, Aloe Vera, Black Walnut, Evening Primrose Oil, Flax Seed Oil, Herbal Trim Skin Treatment, HSN Complex, HSN-W, Kelp, Mineral-Chi Tonic, Nature's Fresh, Vitamin E

Deafness

Deafness is partial or complete loss of hearing. These remedies can help to prevent or slow the progress of deafness, or correct problems with the ears causing a temporary loss of hearing. They will not reverse complete deafness.

Remedies: CBG Extract, Helichrysum, St. John's Wort

Debility

Debility is a general weakness of the body brought on by prolonged illness or general poor health. The following remedies help strengthen the body overall, helping to overcome debility.

Therapies: SuperFood (pg. 15), Affirmation and Visualization (pg. 8), Essential Fatty Acids (pg. 17), Minerals (pg. 15), Enzymes (pg. 15)

Remedies: GreenZone, Mineral-Chi Tonic, Thai-Go, Trigger Immune, CLT-X, Co-Q10, Colostrum, Digestive Bitters Tonic, Food Enzymes, Ginseng (Korean), Ginseng (Wild American), IGF-1, Nature's Chi, Slippery Elm, Spleen Activator, X-Action (Men s)

Defensiveness

See Also *Anger (excessive), Irritability*

These are remedies for people who feel defensive, who respond like they are being attacked even when they are not.

Remedies: Distress Remedy, Liver Balance, Oregon Grape, 5-HTP, Chamomile, Dandelion, Guardian, Milk Thistle, MSM, Nerve Eight, Rose Bulgaria, SAM-e

Dehydration

Dehydration is a lack of water in the tissues of the body. Although water needs to be consumed to overcome dehydration, sometimes water is not being properly taken up and utilized by the tissues. The following are remedies that help to rehydrate the tissues of the body when they are not holding enough moisture.

Therapies: Hydration (pg. 10)

Remedies: HY-C, Licorice Root, Recovery, Adrenal Support, Blood Build, Cellular Energy, Nervous Fatigue Formula, Nopal, Potassium, Super Algae, Vitamin B-12

Dementia

Dementia is a loss of cognitive and intellectual function, without the loss of perception. Symptoms include disorientation, impaired memory and judgment, and a loss of intellectual capacity. It may be caused by toxins or diseases of the brain. The following remedies may be helpful in preventing, slowing the progress of, or helping the body reverse the process of dementia.

Therapies: Heavy Metal Detoxification (pg. 22), Oral Chelation (pg. 21), Essential Fatty Acids (pg. 17)

Remedies: Ginkgo/Gotu Kola, Mega-Chel, Omega-3, Brain-Protex, Ginkgo Biloba, Gotu Kola

Denture Sores

Dental sores are abrasions or ulcerations on the gums or gingiva as a result of improperly fitted dental appliances. The following remedies may help. They should be applied topically.

Remedies: Aloe Vera, Nature's Fresh, Chamomile, Clove Bud, Intestinal Soothe & Build, Silver Shield, Tea Tree Oil, Vitamin E

Depression

See Also *Hypothyroid, Inflammatory Bowel Disorders (Colitis, IBS), Liver (toxic), PMS Type D, Post Partum Depression, Seasonal Affective Disorder*

Are you singing the blue s? If you are, you are not alone. Doctors are handing out prescriptions for antidepressants like candy. Unfortunately, the number one noted side effects for these popular drugs is, believe it or not, depression.

Depression is a misunderstood, misdiagnosed and often mistreated condition. So, let's unravel some of the causes of depression and provide some natural alternatives to the prescription drug scene.

Once named melancholia , shamans and sages felt that bad thoughts or demons caused the disorder. Our modern understanding of depression comes from research done on chemicals called neurotransmitters. Here's how these chemicals work.

The area where two nerves meet and exchange information is referred to as a synapse . This synaptic region is filled with fluid in which neurotransmitters travel and transmit information from one nerve to another. As one nerve releases a neurotransmitter, the other nerve has its receptor that will then accept and process the information.

Once the receptor has been stimulated by this chemical message, the neurotransmitter may be returned to the originating nerve for recycling or it may be destroyed. Destruction of neurotransmitters involves converting them into substances that can be absorbed into the bloodstream and eliminated in the urine. Various enzymes, such as monamine oxidase (MAO) accomplish this task.

The nerves must replenish their supply of neurotransmitters in order to send the next message. This can be accomplished by reabsorbing the previously released neurotransmitters (a process called reuptake) or by manufacturing a new supply. Neurotransmitters are created from amino acids using vitamins and minerals.

Modern scientists believe that during depression there may be an imbalance of neurotransmitters, either a depleted supply or excess. Some of the neurotransmitters believed to be involved in depression include serotonin, norepinephrine and dopamine. When the supply of one of these chemicals is depleted, the nerve is unable to send signals properly. When the chemical is in excess, the message is over-sent.

Today's antidepressant drugs are all designed to alter neurotransmitters in some fashion. In the 1940 s, tricyclic drugs, better known as MAOIs (monoamine oxidase inhibitors) began to hit the market. These drugs inhibited the enzyme we mentioned earlier that breaks down some of these neurotransmitters. Although this approach worked for some, it had serious side effects for others, including death. In some cases individuals who were depressed, ended up with anxiety and vice versa.

In 1988, Prozac, a SSRI (Selective Serotonin Reuptake Inhibitor) was introduced into the scene. Soon 22 million Americans were taking the drug. With the fad popularity of Prozac, soon more antidepressant drugs appeared on the scene.

Antidepressants such as Wellbutrin and Zyban work by inhibiting the reuptake of serotonin and dopamine. Effexor inhibits serotonin, norepinephrine and dopamine. SSRIs such as Prozac, Paxil and Zoloft cause more of the serotonin to remain in the synapses for extended periods of time.

These drugs all have numerous side effects. For starters, they mask the real causes of depression, which relate to diet, exercise, stress and unresolved emotional issues. Instead of making appropriate life-style adjustments or dealing with life's issues, patients are given a false sense of reality and often feel detached and indifferent to family, friends and environment.

These drugs are also noted in many cases of joint pain, abnormal blood cell counts, sexual dysfunction and altered appetite/weight. An overdose or faulty interaction with other drugs can also cause a SSRI to induce hallucinations, irregular heartbeat and seizures. As an odd irony, depression is one of the common side effects of these drugs.

A common cause of depression is prescription medication. Your busy family physician may overlook the fact that the antibiotic or antihistamine you are taking could be the cause of your being down and out. Diuretics, beta-blockers, hormone replacement therapies, painkillers, sleep aids, Tagamet and Zantac can all create the imbalance of neurotransmitters that may color your days gray. You can see the importance of educating yourself about the medications and supplements that you use.

Diet plays an important role in relieving depression. A deficiency of amino acids or B-complex vitamins will result in a deficiency of neurotransmitters. Tyrosine and phenylalanine will both elevate levels of norepinephrine, seen to improve mood in depression and anxiety case studies. Tryptophan (and its derivative 5-HTP) will increase serotonin levels.

Amino acids are derived from vitalized protein. Overcooked meat is not a good source of amino acids. Furthermore, many people have problems with protein digestion. Super Algae or Ultimate Green Zone can be beneficial here, along with protease and high quality vegetable or animal protein sources.

B-complex has remarkable results for many sufferers, because the B-vitamins are critical to the synthesis of neurotransmitters. Dr. Peter D Adamo has successfully treated depression, hyperactivity and Attention Deficit Disorder in many O Blood Types with high doses of folic acid and B12. One should always make sure to incorporate a B-complex when taking a single B vitamin.

Essential fatty acids (flax seed oil, Super Omega 3 EPA, etc.) are also important to nerve function and may help depression and other nervous system disorders.

Where depression is associated with anxiety, kava kava or St. John's wort may help to eliminate nervous tension. Nutri-Calm is an excellent source of B-vitamins and also contains herbs and nutrients to help anxiety.

Since 1976 s-adenosyl l-methionin (SAM) has elevated mood, eliminated suicidal tendencies and improved intellectual performance. This improvement was noted in approximately 80% of the cases. This is as effective as clomipramine and amitriptyline.

Getting off Antidepressants

For individuals who have already been subject to synthetic antidepressants, it is essential that you wean yourself gradually from these drugs while transitioning to other therapies. Sometimes the sudden loss of said chemicals can contribute to a deeper and more disturbing condition for the depression sufferer.

Balancing the adrenals with products such as Adrenal Support and Licorice may prove beneficial. While transitioning, it is vital to keep the bowel and liver in good condition in order to aid in elimination of wastes and to support the other organs in their responsibilities as well.

We have been led to believe that if we aren't giddy and full of laughter at all times, then we must be depressed. It is up to you to know the difference between being a little blue and having depression. It is normal and natural to feel blue when we have experienced grief, sadness or setbacks. The key here is to find a constructive solution to the life situation so that we can move forward again.

In many cases there is no need for medication at all. Sometimes our down time is just that, a time to rest and reflect. Keeping a journal, exercise, dietary changes, daily exposure to sunshine, fresh air, talking to a trusted friend or advisor and taking a few supplements can work wonders.

The supplements that will work best for you will depend on what is causing your depression. Here are some of the common causes and remedies that can help.

As a precursor to serotonin, 5-HTP (5-HTP Power) aids in balancing mood and regulating sleep patterns.

In recent years, St. John's wort has been widely touted as a natural product for mild to moderate depression. St. John's wort does have antidepressant action, but works best on depression associated with anxiety.

Kava Kava targets the limbic system in the brain, as well as relaxing muscles. It helps promote a calm, mentally alert state coupled with a feeling of well-being.

High cortisol levels from chronic stress may contribute to feeling anxious and blue. Adaptagens such as eleuthero root or SUMA Combination can help to reduce stress levels and calm distressed feelings.

Licorice root strengthens the adrenal function and may also help balance blood sugar levels. This can help to stabilize a person's mood.

Black cohosh can be an effective antidepressant for depression associated with PMS, menopause or childbirth. It is also a good antidepressant for women who feel trapped. Damiana is another mood elevator, which can treat depression in both men and women caused by low reproductive hormones.

Taking 2000 IU of Vitamin D3 daily can cause significant improvement in seasonal depression. St. John's wort is also helpful for seasonal depression.

Folk medicine has long recognized a connection between the liver and digestive function and depression. Constipation was thought to produce melancholy in traditional Western medicine. Recent research shows there is scientific validity to this point of view as the bowel produces more serotonin than the brain. Toxic irritants can also lead to depression.

A particularly effective formula for depression here is the Chinese formula Mood Elevator. This formula helps sagging energy, feelings of sadness and sagging spirits. It decongests the digestive organs and liver, and helps calm anxiety.

The thyroid plays an elemental function in this physiology. Hypothyroidism is often overlooked and the depression, a mere symptom, diagnosed. Depression is noted and is sometimes the first diagnosed in 40% of hypothyroid patients. If a group of depression patients were tested, approximately 10-15% of them would be found to have thyroid disorder.

In many cases feeding the thyroid with iodine (found in Liquid Dulse, concentrated black walnut, Thyroid Activator and TS II with Hops) can alleviate the depression and turn hypothyroidism around. Thyroid Support may also help.

A complete list of possible remedies for depression follows.

Therapies: Hydration (pg. 10), Blood Type Diet (pg. 13), Low Glycemic Diet (pg. 11), Stress Management (pg. 8), Probiotics (pg. 16), Essential Fatty Acids (pg. 17), Affirmation and Visualization (pg. 8), Heavy Metal Detoxification (pg. 22)

Remedies: **5-HTP, Black Cohosh, Kava Kava, Mood Elevator, Thyroid Support, Vitamin D3**, AdaptaMax, Adrenal Support, Bergamot, Black Walnut, Blood Build, Bowel Detox, Cinnamon, CLA, Clary Sage, Damiana, Depressaquel, Dulse, Eleuthero, Energ-V, Flax Seed Oil, Folic Acid Plus, Frankincense, Geranium, Ginkgo/Gotu Kola, Ginseng (Korean), Ginseng (Wild American), Gotu Kola, Grapefruit (Pink), GreenZone, Helichrysum, Jasmine Absolute, Kudzu/St.John's Wort, L-Glutamine, Magnesium, Mandarin (Red), Master Gland, Mega-Chel, Melatonin Extra, Mineral-Chi Tonic, Nature's Gold, Neroli, Nervous Fatigue Formula, Niacin, Nutri-Calm, Omega-3, Passion Flower, Pine Needle, Pro-G-Yam Cream, Rose Bulgaria, SAM-e, St. John's Wort, Stress Relief, Stress-J, SUMA Combination, Super Algae, Thyroid Activator, Trigger Immune, TS II, Vitamin B-12, Vitamin B-6, Vitamin B-Complex, Vitamin B-Complex, X-Action (Men s), X-Action (Women s), Yellow Dock, Ylang Ylang

Dermatitis

See Also *Eczema, Rashes and Hives*

Inflammation of the skin is known as dermatitis. The following remedies may be helpful used topically and/or taken internally.

Therapies: Probiotics (pg. 16), Essential Fatty Acids (pg. 17), Antioxidants (pg. 17), Stress Management (pg. 8), Colon Cleansing (pg. 18)

Remedies: Dulse, HistaBlock, Omega-3, Skin Detox, Aloe Vera, BP-X, Chamomile (Roman), Flax Seed Oil, Geranium, Gotu Kola, Herbal Trace Minerals, IF Relief, IF-C, Jasmine Absolute, Lemon Oil, Milk Thistle Combination, MSM, Nature's Fresh, Nutri-Calm, Oregon Grape, Pantothenic Acid, Pau D Arco, Pro-G-Yam Cream, SC Formula, St. John's Wort, Super GLA, Thai-Go, Vitamin B-12, Vitamin C, Vitamin E, Wild Yam, Yucca, Zinc

Diabetes

See Also *Cardiovascular Disease (Heart Disease), Hyperinsulinemia (Syndrome X)*

Diabetes is a problem with insulin, a hormone produced by the pancreas. Insulin allows the cells of the body to utilize glucose or blood sugar, their major source of fuel. There are two types of diabetes.

In Type I (or insulin dependent) diabetes the pancreas has been damaged or destroyed and doesn't produce the necessary insulin. It is believed that this damage is the result of an autoimmune response. Insulin production may be as low as 4 units (versus the normal 31) in this type of diabetes. People with Type I diabetes have to have regular insulin shots and constant monitoring of insulin levels to keep their levels of insulin as close to the normal 31 as possible. Supplements that reduce insulin resistance can reduce the need for insulin in this type of diabetes, but they will not restore normal pancreatic function.

In Type II (or insulin resistant) diabetes, the pancreas has no trouble producing insulin. The problem is that the cells of the body develop resistance to insulin, so the pancreas produces more insulin in order to try to overcome the cells resistance. Insulin levels may exceed 100 units (versus the normal 31). Type II develops slowly, over the course of years. It is really a long-term and severe case of Syndrome X.

Insulin or oral hypoglycemic medications may be necessary to stabilize a Type II diabetic, but changes in life-style and nutrition can cure this condition.

Both types of diabetics should avoid high glycemic carbohydrates and eat a diet with a balanced intake of fats, proteins and carbohydrates. Exercise and weight loss are also very helpful in overcoming insulin resistance in cells. Supplements which help reduce insulin resistance include: SugarReg, Target P-14, ProPancreas, PBS, Nopal, and Goldenseal.

Since diabetes also involves a deterioration of circulation, herbs that help cardiovascular disease are also important in diabetes. These include Co-Q10, garlic, hawthorn and Mega-Chel.

Diabetes is a serious disorder. Seek appropriate medical assistance.

Therapies: Antioxidants (pg. 17), Low Glycemic Diet (pg. 11), Affirmation and Visualization (pg. 8), Oral Chelation (pg. 21), Stress Management (pg. 8), Essential Fatty Acids (pg. 17),Fiber (pg. 17), Oral Chelation (pg. 21)

Remedies: Alpha Lipoic Acid, Chromium GTF, Golden Seal, SugarReg, Target P-14, 7-Keto, AdaptaMax, Adrenal Support, Aloe Vera, Anamu, Astragalus, Black Currant Oil, Blood Sugar Formula, C-X, Carbo-Grabbers, Chickweed, Co-Q10, Collatrim, Dandelion, Devil's Claw, DHA, DHEA-F, DHEA-M, Eucalyptus, Fenugreek & Thyme, Free Amino Acids, Garlic, Ginseng (Korean), Ginseng (Wild American), Grapine, Hawthorn Berries, HY-C, IGF-1, Juniper Berries, Kidney Drainage, L-Carnitine, Lemon Oil, Lymphatic Drainage, Mega-Chel, Mineral-Chi Tonic, MSM, Nervous Fatigue Formula, Noni (Morinda), Nopal, Nutri-Burn, Nutri-Burn, Olive Leaf, Omega-3, P-X, PBS, Pro-Pancreas, Proactazyme Plus, Safflowers, Saw Palmetto, Sea Salt, SF, Spirulina, Stevia, SUMA Combination, Super GLA, Super Trio, Uña de Gato Combination, Uva Ursi, Vitamin B-12, Vitamin B-6, Vitamin D3, Xylitol, Zinc

Diaper Rash

Diaper rash is a form of contact dermatitis (inflammation of the skin) that is common in infants where there is prolonged exposure to urine-soaked or soiled diapers or the chemicals found in disposable diapers.

Apply Golden Salve when changing diapers to help prevent diaper rash. To soothe diaper rash apply Aloe Vera Gel or sprinkle slippery elm powder into the diapers when changing. Be sure to change diapers promptly after a child wets or soils them.

A fiery-looking diaper rash is typically caused by yeast infections. The infant may also have thrush. Probiotics are beneficial in this case. Open the capsules and put the sweet tasting powders into the baby's mouth. Probiotic powders can also be sprinkled in the diaper to help prevent diaper rash.

All of the remedies in the following list can be applied topically.

Therapies: Probiotics (pg. 16)

Remedies: Aloe Vera, Golden Salve, Nature's Fresh, Chickweed, Chlorophyll, Cornsilk, Probiotics, Probiotics, Slippery Elm

Diarrhea

See Also *Giardia, Infection (bacterial), Inflammatory Bowel Disorders (Colitis, IBS), Parasites (general)*

Excessively loose, watery, and frequent bowel movements are a sign that the bowels are trying to eliminate toxic irritants.

The best approach to treating this condition naturally is to use agents that absorb toxins in the digestive tract. These include Psyllium Hulls Combination, LOCLO and Nature's Three. For severe diarrhea, activated charcoal is the best choice. Be careful not to use too much of this product, however, as it is so effective it can actually cause constipation. For diarrhea in infants, slippery elm or marshmallow are good choices.

When diarrhea is extremely watery, astringents may be needed to tone bowel tissue and halt fluid loss. Some of the herbs that can be used for this purpose include white oak bark or yarrow although almost any astringent may help.

For diarrhea caused by infectious organisms, Silver Shield and High Potency Garlic may be helpful. Golden seal is useful for diarrhea caused by giardia. Use two capsules every two hours until relief is obtained. Probiotics can be helpful for both treating and preventing diarrhea from infectious microbes. When traveling, taking PDA and probiotics can help prevent diarrhea.

Constant diarrhea is a symptom of inflammation in the bowels. Diarrhea alternating with constipation can be a sign of poor digestion and/or a spastic colon. Magnesium may be helpful in this case. Spleen Activator or slippery elm may also help.

After a bout with diarrhea it is always good to take probiotics and minerals to replace lost mineral electrolytes. Combination Potassium, Recovery or Mineral Chi tonic are good sources for minerals in such cases.

Therapies: Probiotics (pg. 16),Fiber (pg. 17)

Remedies: Bayberry, Charcoal (Activated), Gastro Health, Golden Seal, Nature's Three, Probiotics, Psyllium Hulls Combination, Silver Shield, Slippery Elm, ALJ, Anamu, Anti-Gas (Chinese), Bentonite (Hydrated), Bilberry Fruit, Catnip & Fennel, Cinnamon, Clove Bud, CLT-X, Every Body's Fiber, EW, FV, Garlic, Ginger, Green Tea Extract, Intestinal Soothe & Build, Lemon Oil, Liver Balance, LOCLO, Magnesium, Marshmallow, Mineral-Chi Tonic, Myrrh, Neroli, Niacin, Parasites, Pau D Arco, PDA, Peppermint, Potassium, Psyllium, Recovery, Red Raspberry, Sage, Small Intestine Detox, Spleen Activator, St. John's Wort, Stress Relief, Sunshine Heroes Probiotic Power, White Oak Bark, Wild Yam, Wild Yam & Chaste Tree, Yarrow, Ylang Ylang

Dieting

See *Weight Loss (aids for)*

Digestion (poor)

See Also *Hiatal Hernia, Wasting*

Poor digestion leads to an inability to properly break down food into usable nutrients. Symptoms include pain, bloating, gas, cramping and/or heartburn. This is usually due to a deficiency of stomach acid or digestive enzymes. When poor digestion becomes chronic and severe it can lead to weight loss, the inability to develop muscle mass, wasting, pallor and fatigue.

Taking digestive enzymes is the first step towards correcting this problem. Proactazyme Plus is the place to start. Protease Plus, Food Enzymes or PDA are other options. It is also a good idea to try to stimulate the body's natural digestive secretions as well. Products like Chinese Anti-Gas, Anti-Gas Formula with Lobelia, Digestive Bitters Tonic, safflowers, or Papaya Mint Tablets can help.

Weak digestion, coupled with lack of muscle mass, pale skin color and fatigue are indications for the Chinese formula, Spleen Activator. In Chinese medicine, the spleen refers to the body's ability to transform food into muscle mass. A weakness of the spleen chi, therefore, results in an inability to gain muscle mass. Trigger Immune, American ginseng and saw palmetto may also help with protein metabolism and spleen chi deficiency.

Therapies: Hiatal Hernia Correction (pg. 12), Enzymes (pg. 15), Blood Type Diet (pg. 13)

Remedies: Digestive Bitters Tonic, Food Enzymes, Proactazyme Plus, Spleen Activator, All Cell Detox, Anti-Gas (Chinese), Anti-Gas Formula, Artemisia Combination, Astragalus, Bergamot, Blue Vervain, Bowel Detox, Cellular Energy, Chamomile, CLT-X, Colloidal Minerals, Colostrum with Immune Factors, Dandelion, Free Amino Acids, Gall Bladder Formula, Gastro Health, Ginseng (Wild American), Golden Seal, GreenZone, Helichrysum, Herbal Trace Minerals, Lactase Plus, LB Extract, Lemon Oil, Lymphatic Drainage, Magnesium, Mandarin (Red), Oregon Grape, Papaya Mint, PDA, Probiotics, Protease, Rose Bulgaria, Rosemary, Safflowers, Saw Palmetto, Small Intestine Detox, Stress-J, Thyme

Diphtheria

See Also *Contagious Diseases*

An infectious disease caused by a bacteria that affects mucus membrane linings. The microorganism produces a toxin that causes degeneration of the nerves, heart and other tissues. The following may be helpful remedies.

Remedies: Garlic, Silver Shield, Ultimate Echinacea, Black Cohosh, Blue Vervain, Cordyceps, Echinacea Purpurea, Guardian, Helichrysum, Lobelia, Marjoram (Sweet), Trigger Immune

Disks (spinal - bulging or slipped)

Nature's Fresh has been sprayed topically over the spine to help heal problems with disks. Goldenseal can also be applied topically to help disks heal. Triple Relief or IF Relief may be helpful for pain. Seek appropriate professional assistance. Chiropractic care, improved posture and stretching exercises may be helpful. The following remedies can help repair bulging or slipped spinal disks. They may also help to repair damaged or degenerating disks.

Remedies: Golden Seal, KB-C, Nature's Fresh, Bone/Skin Poultice, C-X, Chlorophyll, Collatrim, Deep Relief Oil, GreenZone, Heavy Metal Detox, Helichrysum, HistaBlock, IF Relief, IF-C, IGF-1, Joint Health, Lymphatic Drainage, Master Gland, Mineral-Chi Tonic, MSM, PLS II, Pro-G-Yam Cream, Red Beet Formula, Skeletal Strength, Super Algae, Thai-Go, Thyme, Triple Relief, Vitamin A & D

Dislocation

A dislocation is the displacement of a body part, such as an internal organ or the bones in a joint, from its proper position. Dislocations should be mechanically corrected by a professional therapist, but the remedies listed here may speed healing and ease pain.

Remedies: Healing AC Cream, Herbal CA, HSN-W, Bone/Skin Poultice, Deep Relief Oil, Lobelia, MSM, Mullein, Nature's Phenyltol with NEM, PLS II, Tei Fu Oil, Triple Relief

Diverticulitis (Diverticuli)

See Also *Inflammatory Bowel Disorders (Colitis, IBS)*

Diverticulii are abnormally formed pockets in the bowel. The diverticulii require a colon cleanse to clean them out and tone the bowel. When they become inflamed, the afflicted person has diverticulitis. These products may help clean the bowel and/or reduce inflammation.

Therapies: Fiber (pg. 17)

Remedies: Bentonite (Hydrated), Intestinal Soothe & Build, Slippery Elm, Black Walnut, Bowel Detox, Chamomile, IF-C, LB Extract, LBS II, Marshmallow, Nature's Gold, PLS II, Small Intestine Detox, Stress Relief, Stress-J, Uña de Gato Combination, Wild Yam, Wild Yam & Chaste Tree

Dizziness (Vertigo)

A loss of the sense of balance, vertigo can be caused by an ear infection or damage to the inner ear. Try using CBC Extract in the ear. Other possible causes include poor circulation to the brain, low blood sugar, and glandular or nervous system imbalances. The remedies that help will depend on the cause.

Remedies: CBG Extract, Ginkgo Biloba, Licorice Root, Black Cohosh, Blood Build, CLA, Clary Sage, Eleuthero, EW, Ginger, Ginkgo & Hawthorn, Ginkgo/Gotu Kola, Ho Shou Wu, HSN-W, HY-A, IF-C, KB-C, Kidney Activator (Chinese), L-Glutamine, Lung Support, Mood Elevator, Omega-3, Peppermint, Phyto-Soy, Stress Relief, Trigger Immune, Vitamin B-12, Vitamin B-Complex, VS-C

Down Syndrome

Down Syndrome is a genetic defect characterized by slow physical development and moderate to severe mental retardation. The following are not cures for Down's syndrome, but may be helpful in promoting general health in someone suffering from the disease.

Therapies: Minerals (pg. 15), Enzymes (pg. 15), Probiotics (pg. 16), SuperFood (pg. 15), Affirmation and Visualization (pg. 8), Antioxidants (pg. 17)

Remedies: Black Currant Oil, Co-Q10, Colostrum, Evening Primrose Oil, Free Amino Acids, Kelp, Lecithin, Omega-3, Super GLA, Vitamin B-Complex

Dropsy

See *Edema (Dropsy, Water Retention, Swelling)*

Drug Detox or Withdrawal

See *Addictions (Drugs)*

Duodenal ulcers

See Also *Ulcers*

An ulceration of the first part of the small intestine. The following remedies may be helpful.

Therapies: Fiber (pg. 17)

Remedies: Aloe Vera, Gastro Health, Intestinal Soothe & Build, Licorice Root

Dysentery

See Also *Diarrhea*

A severe form of diarrhea characterized by frequent watery stools, often with blood and mucus. Symptoms include pain, fever and dehydration. The following remedies may help.

Therapies: Probiotics (pg. 16), Fiber (pg. 17)

Remedies: Bayberry, Charcoal (Activated), Golden Seal, Silver Shield, Artemisia Combination, Chamomile, Every Body's Fiber, Garlic, Intestinal Soothe & Build, Lemon Oil, Para-Cleanse, Peppermint, Probiotics, Psyllium Hulls Combination, Slippery Elm, St. John's Wort, White Oak Bark

Dysmenorrhea

See Also *Cramps (menstrual), PMS (general)*

Painful menstruation or pain at the time of the passage of the menses is called dysmenorrhea. Native people felt that cramping is nature's way of exercising muscles for childbirth, so minor cramping was considered normal. For excessive pain, remedies that decongest pelvic circulation or relax muscle spasms can be helpful.

When pains are dull, the pain is probably congestive pain, caused by a lack of good blood flow in the pelvic region. Pelvic decongestants will help. Two of the best herbs for this purpose are ginger and yarrow. Formulas like Female Comfort or FCS II may also be helpful. Caster oil packs applied topically over the lower abdomen will also ease congestive pain.

Liver congestion is often involved in congestive menstrual pain. Liver Balance, Milk Thistle Combination and other products that encourage liver detoxification may be helpful. If varicose veins are a problem, too, then VariGone may be helpful in relieving both the varicose veins and the dysmenorrhea.

When the pains are sharp and cramping they are probably due to muscle spasms. Antispasmodics such as Cramp Relief, black cohosh, kava kava, lobelia and wild yam can be helpful. Magnesium is also helpful for menstrual cramping. Peppermint, clary sage and lavender essential oils applied topically will also help spasmodic pain.

Therapies: Colon Cleansing (pg. 18), Antioxidants (pg. 17), Hydration (pg. 10), Avoid Xenoestrogens (pg. 23)

Remedies: Cramp Relief, Magnesium, Monthly Maintenance, APS II, Black Cohosh, Blood Build, Blue Cohosh, Capsicum, Chamomile (Roman), Clary Sage, Deep Relief Oil, EverFlex, False Unicorn, FCS II, Female Comfort, Ginger, Ginseng (Korean), Kava Kava, Lavender, Liver Balance, Lobelia, Menstrual, Menstrual-Reg, Milk Thistle Combination, Myrrh, Passion Flower, Peppermint, PMS, SAM-e, Stress Relief, Triple Relief, Vari-Gone, Wild Yam, Wild Yam & Chaste Tree, Yarrow

Dyspepsia

See Also *Gastritis, Indigestion*

Impaired gastric function due to some disorder of the stomach. Symptoms include pain and sometimes nausea or burning. The following remedies may be helpful.

Therapies: Hiatal Hernia Correction (pg. 12), Enzymes (pg. 15)

Remedies: Anti-Gas (Chinese), Digestive Bitters Tonic, Ginger, Alfalfa, ALJ, All Cell Detox, Aloe Vera, Anti-Gas Formula, Catnip, Catnip & Fennel, Chamomile, Clary Sage, CLT-X, Fenugreek & Thyme, Food Enzymes, Garlic, GC-X, Herbal Trace Minerals, Liver Balance, Mandarin (Red), Neroli, Nerve Eight, Omega-3, Papaya Mint, PDA, Proactazyme Plus, Probiotics, Small Intestine Detox, Spleen Activator, Stomach Comfort, Stress-J, Thyme, Vitamin B-Complex

Ear Infection or Earache

See Also *Allergies (food), Fungal Infections (Yeast Infections, Candida albicans)*

An earache is the result of inflammation or infection in the ear. The irritation or infection can happen in the outer ear canal (otitis externa or swimmer's ear), in the middle ear behind the eardrum (otitis media) or in the inner ear (otitis interna). An inner ear infection causes the sudden onset of vomiting, vertigo, and loss of balance.

Children are more prone to earaches than adults. Part of the reason that children are prone to ear infections is that their ear passages are small, so an amount of fluid/mucous that would not cause a problem in an adult can cause severe problems in a child.

There is evidence that ear infections are often allergy induced. The allergy is usually toward cow's milk and/or formula, but wheat is another common allergen. The allergy causes irritation to the mucous membranes, which results in swelling. This causes constriction of the eustachian tube, a tube between the middle ear and the throat. This tube allows for equalization of pressure on both sides of the eardrum and for drainage of fluid from the inner ear. When this tube is swollen, pressure builds up in the inner ear causing pain.

Antibiotics are typically used to treat ear infections, but they are very ineffective and contribute to problems in the intestinal tract that create more lymphatic stagnation and perpetuate the tendency to earaches. Parents are often afraid to treat earaches naturally for fear that the ear infection will cause permanent hearing loss. This is simply not the case. There is no scientific evidence that antibiotics do anything but shorten the duration of pain and infection. The trade off, however, is more frequent earaches.

There are many very effective natural remedies for earaches and ear infections. Garlic oil is very effective for ear infections. Warm the oil to body temperature and place a few drops in the ear. You can also rub some of the oil down the side of the neck to improve lymphatic drainage. For adults or older children, cut a clove of raw garlic, put a little olive oil on it, and place it on the outside of the ear (sort of like a hearing aid). This will rapidly eliminate infection, stimulate lymphatic drainage and relieve the pain. High Potency Garlic or raw garlic can also be taken internally.

Another highly effective therapy is to bake or steam an onion and place a few drops of the onion juice into the ear. (Make certain it has cooled to body temperature first.) You can also cut the cooked onion in half and place it over the ear while it is still warm. In most cases, the onion or onion juice almost instantly relieves the pain and eases the inflammation and infection.

Other good ear drops include lobelia, Silver Shield and essential oils like lavender, tea tree or thyme diluted in olive oil (1 part essential oil to 10 parts olive oil). Make certain to warm all of these substances to body temperature before putting them in the ear. An excellent ear oil can be made using garlic, mullein and/or St. John's wort in combination.

Where children have frequent ear infections, eliminate the possibility of food allergies and yeast infections. Eyebright is very effective at preventing the eustachian tubes from swelling, or opening them after they are swollen. It is the herbal equivalent of tubes in the ears. The herb works best as a tincture of the fresh plant, but large doses of the dried powder or the EW formula can also be effective. Taken regularly, ALJ can also help reduce allergic reactions that create earaches.

CBG Extract is also widely used for relieving congestion in the ears in both children and adults. It is used as ear drops.

Internally, echinacea can help fight infection and reduce inflammation in earaches. Herbs that promote lymphatic drainage such as Lymphatic Drainage can also be helpful.

Therapies: Blood Type Diet (pg. 13), Enzymes (pg. 15), Colon Cleansing (pg. 18)

Remedies: CBG Extract, Eyebright, Garlic, IF-C, Lobelia, Silver Shield, Tea Tree Oil, ALJ, Echinacea Purpurea, Echinacea/Golden Seal, EW, Golden Seal, Helichrysum, IF Relief, IN-X, L-Glutamine, Lavender, Lymph Gland Cleanse, Lymph Gland Cleanse-HY, Lymphatic Drainage, Mullein, St. John's Wort, Thyme, Ultimate Echinacea, Ultimate Echinacea, Xylitol, Yeast/Fungal Detox

Eczema

See Also *Dermatitis, Itching, Rashes and Hives*

Eczema is a very common skin condition that usually involves a rash, itching, redness and flaking of the skin. Acute inflammation of the skin is known as dermatitis, and usually occurs when the body comes in contact with an irritating substance (such as poison ivy). Eczema is a chronic form of dermatitis, or in other words, a chronic inflammation of the skin.

All inflammatory diseases happen when damaged cells release histamine and bradykinin into surrounding fluids. This causes capillary pores to dilate allowing fluid and protein to enter the tissue spaces. Tissues become red, swollen, hot and painful. In both eczema and dermatitis, the skin becomes irritated, resulting in red and itchy skin.

In eczema, the skin is repeatedly irritated and inflamed which causes the upper layer of the skin (epidermis) to thicken as skin cells multiply rapidly. This creates a scaly effect on the surface of the skin. Oil glands become obstructed and the skin becomes dry. The scaly skin inhibits elimination through the skin, causing toxins to become trapped under the skin. This causes itching, which leads the person to scratch. Scratching breaks the upper dermis layer so the skin develops a broken and cracked appearance.

Eczema is caused by the body being hypersensitive to certain irritants, so it is closely related to allergic asthma, hayfever and food allergies. These conditions are actually caused by a healthy immune system that is overburdened with toxins. Simply put, the body is being overwhelmed by more irritants than it can effectively handle.

This may explain why children are extremely prone to eczema. Almost thirty percent of all newborn babies may develop this condition, which affects about one in eight young children. It often occurs on the scalp or the cheeks, but can spread over other parts of the body, making children itchy and miserable. Although 75% will outgrow this condition by their mid-teens, we wonder if this isn't occurring just because their immune system's become too depressed to manifest it. Adults who had eczema as children will remain prone to dry skin in later years and to occasional flare-ups of skin inflammation.

Diet can be very important. A good place to start is to avoid foods that are incompatible with your blood type. Some of the common allergenic foods that may contribute to skin irritation include wheat, dairy, corn, orange juice, coffee, black tea, soda pop and sugar.

Eczema is often treated medically with corticosteroid drugs that mimic the anti-inflammatory action of the adrenal hormone cortisol. People with eczema often suffer from adrenal exhaustion (with a corresponding deficiency in the production of cortisol). This helps explain why the excessive inflammation is present and why eczema can flare up under stress. Stress depletes the adrenals.

So, along with learning good stress management techniques, herbs that support adrenal function and have a cortisol-like action may be helpful. Both licorice root and yucca have this effect. Licorice, however, should be avoided with weeping eczema. Other adrenal enhancing remedies such as Adrenal Support and Energ-V may also be of benefit. HistaBlock can also help because it blocks the histamine reactions that cause the inflammation.

The following remedies may also be helpful.

Therapies: Antioxidants (pg. 17), Blood Type Diet (pg. 13), Enzymes (pg. 15), Probiotics (pg. 16), Essential Fatty Acids (pg. 17),Fiber (pg. 17)

Remedies: Adrenal Support, Bergamot, Burdock, Flax Seed Oil, HistaBlock, Omega-3, Skin Detox, Sunshine Heroes Omega 3 with DHA, Aloe Vera, Black Currant Oil, Black Walnut, BP-X, Chickweed, Clary Sage, Colloidal Minerals, Dandelion, Echinacea Purpurea, Eczema/Psoriasis, Energ-V, Enviro-Detox, Evening Primrose Oil, Geranium, Gotu Kola, Herbal CA, Herbal Trace Minerals, Hi Lipase, HSN Complex, HSN-W, IF Relief, Jasmine Absolute, Kelp, Lemon Oil, Licorice Root, Liver Balance, Myrrh, Nature's Fresh, Nature's Gold, Nutri-Calm, Pantothenic Acid, Pau D Arco, Red Clover, Rose Bulgaria, Sandalwood, Sunshine Heroes Probiotic Power, Sunshine Heroes Whole Foods Antioxidant, Super GLA, Thyroid Support, Ultimate Echinacea, Vitamin A & D, Vitamin B-Complex, Yucca, Zinc

Edema (Dropsy, Water Retention, Swelling)

See Also *Congestive Heart Failure*

Tissues hold water when they are damaged through inflammation. Using anti-inflammatory and vulnerary herbs on injuries can prevent or reduce swelling and aid repair. When water retention is general, it is usually a problem with lymphatic drainage and kidney function. Occasionally, it is also a cardiac problem, which is why some cardiac remedies are also listed.

There are two major classes of diuretic herbs. The first are those that stimulate the kidneys to work harder. These could also be called irritating diuretics. Juniper berry is one of the stronger herbs in the class. Other stimulating diuretics include buchu and uva ursi. Formulas with this action include Kidney Activator, Chinese Kidney Activator, Kidney Drainage and JP-X. These formulas are contraindicated when the kidneys are inflamed.

Another class of diuretics are herbs that nourish and improve kidney function without stimulating kidney activity. These are known as non-irritating diuretics. They include nettles, goldenrod, horsetail and cornsilk. These remedies are useful for kidney inflammation and problems where the kidneys are not filtering wastes from the body very effectively. Potassium Combination, Urinary Maintenance and KB-C are combinations that have more of this quality.

Another cause of edema is problems with the heart. Hawthorn, Co-Q10 and other supplements that aid the heart can be helpful here.

Believe it or not, drinking more water and using all natural sea salt can actually help to relieve water retention. Detoxification of the liver and bowel can also help.

Therapies: Hydration (pg. 10), Minerals (pg. 15), Colon Cleansing (pg. 18)

Remedies: Kidney Activator, Kidney Activator (Chinese), Kidney Drainage, Sea Salt, Blue Vervain, Co-Q10, Cornsilk, Cranberry & Buchu, Dandelion, FCS II, Geranium, Golden Seal, Grapefruit (Pink), Hawthorn Berries, Horsetail, Hydrangea, JP-X, Juniper Berries, Lavender, Lymphatic Drainage, NF-X, Parsley, Patchouli, Potassium, Sarsaparilla, SF, Thyme, Urinary Maintenance, Uva Ursi, Vitamin B-6, Vitamin B-Complex, White Oak Bark

Electromagnetic Pollution

In modern society, we are constantly exposed to electromagnetic radiation from computers, microwave ovens, radar, TV sets, digital clocks and other electrical items. Cell phones are another growing source of electromagnetic pollution. There is a link between electromagnetic pollution and cancer. The following remedies may be helpful for countering the negative influences of electromagnetic pollution in the body.

There are a number of devices available that can help protect the body against electromagnetic pollution Wayne Cook's Diodes, for example. Horse chestnuts can be carried as natural diodes. Barley grass, Mineral Maintenance and Trigger Immune can be helpful for countering the negative influences of electromagnetic pollution in the body.

Remedies: Trigger Immune, Barley Grass, Mineral-Chi Tonic, Monthly Maintenance, Thai-Go

Emotional Sensitivity

See Also *Adrenals (Exhaustion, Weakness or Burnout), Anxiety (Panic Attack)*

The following remedies may be helpful when people are feeling excessively sensitive emotionally.

Therapies: Stress Management (pg. 8), Affirmation and Visualization (pg. 8), Low Glycemic Diet (pg. 11)

Remedies: Adrenal Support, Nervous Fatigue Formula, Nutri-Calm, Chamomile, Chamomile (Roman), Jasmine Absolute, Mood Elevator, Nature's Phenyltol with NEM, Stress Relief, Trigger Immune

Emphysema

Emphysema is a chronic lung condition involving a loss of elasticity of the alveoli, the tiny air sacs in the lungs. It is usually caused by smoking. The alveoli eventually collapse, causing drowning in lung fluid. Symptoms include labored breathing, wheezing and a husky cough. This is a serious condition requiring appropriate medical assistance.

In addition to following a general program for improving health, a number of specific supplements may be helpful in emphysema. Lung Support is a Chinese herb formula designed to help weak lungs and is indicated for emphysema. Mullein is very helpful for this condition when taken in doses of about six capsules per day for at least six months. It has a slow cumulative effect in helping the lungs become more moist and supple. Horsetail can also be used to help restore elasticity to lung tissue.

A complete list of possible remedies follows.

Therapies: Antioxidants (pg. 17), SuperFood (pg. 15), Essential Fatty Acids (pg. 17), Affirmation and Visualization (pg. 8), Stress Management (pg. 8), Heavy Metal Detoxification (pg. 22)

Remedies: Cordyceps, Horsetail, Lung Support, Mullein, Capsicum & Garlic with Parsley, Carotenoid Blend, Chlorophyll, Co-Q10, Echinacea Purpurea, Fenugreek & Thyme, IN-X, Licorice Root, Lobelia, Lymph Gland Cleanse-HY, Marshmallow, Marshmallow & Fenugreek, Thai-Go, Ultimate Echinacea, Vitamin A & D

Endometriosis

The presence of functioning tissue from the uterine lining in places where it is not normally found, such as outside of the uterus in the uterine tubes or abdominal cavity is called endometriosis. Symptoms include severe pain and cramping at the time of menstruation.

Endometriosis tissue is affected by estrogens, so avoiding xenoestrogens and helping the body detoxify from excess estrogen helps. Paw Paw Cell Reg and IF-C have proven helpful in many cases. A complete list of possible remedies follows.

Therapies: Avoid Xenoestrogens (pg. 23), Essential Fatty Acids (pg. 17), Colon Cleansing (pg. 18)

Remedies: False Unicorn, IF-C, Paw Paw, Alfalfa, Bayberry, Breathe Free, Female Comfort, Pau D Arco, Super GLA, Vitamin A & D, Vitamin E

Endurance (lack of)

See Also *Fatigue*

The following have been reported to help increase endurance and stamina.

Therapies: Hydration (pg. 10), SuperFood (pg. 15), Affirmation and Visualization (pg. 8), Low Glycemic Diet (pg. 11), Antioxidants (pg. 17), Minerals (pg. 15), Enzymes (pg. 15)

Remedies: Target Endurance, Adrenal Support, Bee Pollen, Eleuthero, Energ-V, Fatigue/Exhaustion, Germanium Combination, Ginseng (Korean), Ginseng (Wild American), Herbal Trace Minerals, Licorice Root, SUMA Combination, Trigger Immune, X-A

Energy (lack of)

See Also *Adrenals (Exhaustion, Weakness or Burnout), Anemia, Ennervation / Nervous Exhaustion, Fatigue, Hypothyroid*

A lack of energy can have many different causes, but will usually involve glandular imbalances or poor oxygenation of the blood. The remedies will vary depending on the cause.

Therapies: Hydration (pg. 10), Low Glycemic Diet (pg. 11), Affirmation and Visualization (pg. 8), Stress Management (pg. 8), Antioxidants (pg. 17), Low Glycemic Diet (pg. 11), Hiatal Hernia Correction (pg. 12)

Remedies: Adrenal Support, Mineral-Chi Tonic, Nature's Chi, Nervous Fatigue Formula, Target Endurance, Thyroid Support, Trigger Immune, AdaptaMax, Barley Grass, Bee Pollen, Blood Build, Chlorophyll, Cinnamon, Co-Q10, Energ-V, Fatigue/Exhaustion, Fibralgia, GC-X, Ginseng (Korean), Ginseng (Wild American), Herbal Punch, Maca, Master Gland, MetaboMax, Mood Elevator, Nature's Gold, Peppermint, RG-Max, SAM-e, Spirulina, Spleen Activator, SUMA Combination, Super Algae, Super Trio, Tei Fu Oil, Thyroid Activator, TS II, Vita Lemon, X-A

Enteritis

See Also *Inflammatory Bowel Disorders (Colitis, IBS)*

Enteritis is an inflammation of the intestines, often the ileum, marked by diarrhea. The following are possible remedies.

Remedies: Intestinal Soothe & Build, Slippery Elm, Every Body's Fiber, IF Relief, Probiotics

Enuresis

See *Incontinence (urinary)*

Environmental Pollution (protection from)

See Also *Chemical Poisoning, Heavy Metal Poisoning*

Pollutants from the environment are a growing health concern. People who work in polluted areas or with toxic chemicals of any kind (dry cleaners, beauty salons, labs, etc.) should consider taking remedies to protect their body from these pollutants. The following may help.

Therapies: Hydration (pg. 10), Antioxidants (pg. 17), Colon Cleansing (pg. 18), Probiotics (pg. 16), Essential Fatty Acids (pg. 17),Fiber (pg. 17)

Remedies: Enviro-Detox, Milk Thistle Combination, Alpha Lipoic Acid, Grapine, Milk Thistle, N-acetyl Cysteine, Omega-3, Sunshine Heroes Omega 3 with DHA, Sunshine Heroes Whole Foods Antioxidant, Super Antioxidant, Super GLA, Super ORAC, Super Trio, Thai-Go

Epilepsy

Epilepsy refers to disorders marked by disturbed electrical rhythms of the central nervous system. Episodes of excess firing in the neurons may create convulsions or a clouding of consciousness. Appropriate professional help should be sought for this problem. If there has been any blow or head trauma in the person's life, then a cranial adjustment may be helpful. This can relieve pressure on the brain.

GABA is an inhibitory neurotransmitter that calms brain activity. It may be helpful in some cases of epilepsy. Focus Attention also contains substances that affect inhibitory neurotransmitters.

The following remedies may also be helpful.

Therapies: Essential Fatty Acids (pg. 17)

Remedies: Focus Attention, GABA Plus, Lobelia, Black Currant Oil, Blue Cohosh, Evening Primrose Oil, Flax Seed Oil, Gotu Kola, Hops, Magnesium, Passion Flower, Sunshine Heroes Omega 3 with DHA, Super GLA, Thyroid Activator, TS II, Vitamin B-Complex

Epstein Barr Virus (Chronic Fatigue Syndrome, CFS)

Chronic infection with the Epstein-Barr virus creates severe lack of energy, lymphatic swelling, muscle aches and sleep disturbances. The following remedies may be helpful.

VS-C is helpful for low-grade viral infections, but is indicated primarily for conditions where there is excess heat in the body. Trigger Immune is an alternative choice here because it is for weakened immunity coupled with cold, weakness and fatigue. Germanium Combination helps natural interferon levels which also helps viral conditions. L-lysine also helps some low-grade viral conditions. Other possible remedies include the following.

Therapies: Stress Management (pg. 8), Enzymes (pg. 15), Essential Fatty Acids (pg. 17), Affirmation and Visualization (pg. 8)

Remedies: Immune Stimulator, Trigger Immune, VS-C, Aloe Vera, Bee Pollen, Black Walnut, Colostrum, Cordyceps, Echinacea Purpurea, Eleuthero, Energ-V, Fibralgia, Germanium Combination, Heavy Metal Detox, IN-X, L-Lysine, Licorice Root, Lymph Gland Cleanse, Lymphatic Drainage, Maca, Magnesium, MSM, Mullein, Nerve Eight, Noni (Morinda), Olive Leaf, Pau D Arco, PLS II, Pro-G-Yam Cream, Proactazyme Plus, Probiotics, SUMA Combination, Super Algae, Thai-Go, THIM-J, Ultimate Echinacea, Uña de Gato Combination, Viral Recovery

Erectile Dysfunction

Erectile dysfunction is when a man is unable to achieve or maintain an erection for normal sexual activity. Formerly known as impotency, this problem can be physical or psychological. Where the problem is physical, it often involves a lack of good blood circulation to the penis or low levels of testosterone. It can also be a side effect of certain drugs. The following may be helpful in cases of impotency that have a physical cause.

Therapies: Avoid Xenoestrogens (pg. 23), Affirmation and Visualization (pg. 8), Stress Management (pg. 8), Essential Fatty Acids (pg. 17)

Remedies: Damiana, DHEA-M, Ginseng (Korean), RG-Max, X-Action (Men s), C-X, Clary Sage, Cordyceps, Eleuthero, False Unicorn, Ginkgo Biloba, Ginseng (Wild American), KB-C, Maca, Mega-Chel, Men's Formula, Nervous Fatigue Formula, Noni (Morinda), Sarsaparilla, Trigger Immune, Vitamin E, X-A, Zinc

Estrogen (low)

See Also *PMS Type D*

The following remedies contain phytoestrogens (estrogen-like compounds) that can help with low estrogen levels.

Remedies: Black Cohosh, Flash-Ease T/R, Phyto-Soy, Breast Assured, C-X, Hops, Licorice Root, Red Clover

Exercise

See Also *Body Building*

The following remedies may be helpful for exercise programs. They can improve muscle development or energy, or aid in preventing or reducing muscle soreness.

Therapies: Hydration (pg. 10), Essential Fatty Acids (pg. 17), SuperFood (pg. 15), Minerals (pg. 15), Low Glycemic Diet (pg. 11)

Remedies: Eleuthero, Energ-V, Safflowers, Target Endurance, AdaptaMax, Barley Grass, Bee Pollen, Cellular Energy, CLA, Deep Relief Oil, EverFlex, Everflex Pain Cream, EveryBody's Formula, Free Amino Acids, IGF-1, Nature's Chi, Nutri-Burn, Recovery, Sea Salt, Super Algae, Super Trio, SynerProTein, Tei Fu Oil, Thai-Go

Eye Infections

The following can be helpful for treating eye infections. They can be used both internally and topically. For topical application make herbal powders into a tea and apply as a compress, eyedrops or eyewash. Be sure to strain the tea thoroughly and allow it to cool before using it in the eyes.

Remedies: EW, Golden Seal, Silver Shield, Chamomile, Eyebright, Liver Balance, Sarsaparilla, Vitamin A & D

Eye Problems (general)

See Also *Conjunctivitis (Pink Eye), Eye Infections, Macular Degeneration*

The following can be useful remedies for the eyes. For bloodshot or irritated eyes there are several herbs that can be used as eyewashes, eye drops or compresses over the eyes. These include goldenseal, chamomile, chickweed, and eyebright. The herbal eyewash formula, EW is especially effective.

When eyes are dry, HY-C may be of benefit because it hydrates tissues. It is indicated for dry mouth, dry skin and dry eyes where there is constant thirst. Blood Build may also help dry eyes. It is a liver formula, but there is a strong connection in Chinese medicine between the liver and the eyes.

Night vision can be aided by bilberry. The formula Perfect Eyes may also help here.

Therapies: Antioxidants (pg. 17), Essential Fatty Acids (pg. 17), Oral Chelation (pg. 21)

Remedies: Bilberry Fruit, EW, Perfect Eyes, Blood Build, Carotenoid Blend, Chamomile, Chickweed, DHA, Eyebright, Germanium Combination, Golden Seal, HistaBlock, HY-C, Mega-Chel, Silver Shield, Super Antioxidant, Thai-Go, Vitamin A & D

Eyes (red or itching)

See Also *Allergies (respiratory), Rhinitis*

The following may be helpful when eyes are red and/or itching.

Therapies: Antioxidants (pg. 17), Hydration (pg. 10), Stress Management (pg. 8)

Remedies: ALJ, EW, HistaBlock, Allergies-Hayfever/Pollen, Allergies-Mold/Yeast/Dust, Allergy, Allergy, Chickweed, Eyebright, Golden Seal, IF Relief, Lymph Gland Cleanse, Lymphatic Drainage, Marshmallow, Sinus Support, Thai-Go

Eyes (spots before)

See Also *Floaters*

The following may be helpful when seeing spots before the eyes.

Remedies: Vitamin B-12, Vitamin B-Complex

Eyesight (to improve)

The following have been helpful in improving eyesight when taken regularly. If nothing else, these remedies can help slow deterioration of eyesight.

Dr. John Christopher used the herbs in EW as an eyewash to help with near and far sightedness. The eyewash must be used daily for several months before results will be seen. Perfect Eyes w/Lutein may also help eyesight when taken regularly.

Remedies: Mega-Chel, Perfect Eyes, Bilberry Fruit, EW, Eyebright, MSM, Thai-Go

Failure to Thrive

Failure to thrive is a problem when infants do not gain weight and develop properly. This is often a digestive problem and mixing a small amount of Proactazyme Plus enzymes with food is helpful. A little bit of black strap molasses and some gruel made with slippery elm can also be helpful.

Therapies: Enzymes (pg. 15)

Remedies: Flax Seed Oil, Proactazyme Plus, Tofu Moo, Digestive Bitters Tonic, Distress Remedy, Free Amino Acids, GreenZone, Lactase Plus, Marshmallow, PDA, Slippery Elm, Sunshine Heroes Multiple Vitamin & Mineral, Sunshine Heroes Omega 3 with DHA, Sunshine Heros Whole Foods Papayazyme, Super Algae, Super GLA

Fainting

Fainting happens when a person suddenly feels weak or as if they are about to lose consciousness. This may be caused by a variety of factors including poor circulation, low blood pressure, low blood sugar or shock. Persistent fainting should be checked by a medical doctor to determine the cause. The following remedies may be helpful, depending on the cause.

Remedies: Distress Remedy, HY-A, Licorice Root, Capsicum, Ginkgo & Hawthorn, Ginkgo/Gotu Kola, Lavender

Fat Cravings

See Also *Gall Bladder (sluggish)*

When a person craves fats constantly, he or she is probably not getting enough essential fatty acids in their diet, or not digesting or metabolizing fats correctly. The following may help.

Start by giving Super GLA, flax seed oil or omega-3 essential fatty acids as a supplement. If this does not correct the problem add Hi-Lipase to improve fat digestion. A gall bladder flush may also be needed.

Therapies: Gall Bladder Flush (pg. 23), Enzymes (pg. 15), Essential Fatty Acids (pg. 17)

Remedies: Fat Grabbers, Hi Lipase, Omega-3, Black Currant Oil, Blood Build, Chickweed, Evening Primrose Oil, Flax Seed Oil, Gall Bladder Formula, Liver Balance, SF, Super GLA

Fat Metabolism (poor)

The inability to properly or adequately digest and assimilate lipid or fat containing foods is common in individuals who have had their gallbladders removed. This is because the body no longer stores adequate amounts of bile, which is necessary to emulsify or break down fats.

There may also be a problem utilizing fats in the body. This is often associated with the liver, but can also be due to the thyroid, spleen, prostate or uterus. Having the uterus or prostate removed can create problems with fat metabolism, since these organs are involved in the combustion of fats. Supplementation with Hi-Lipase or Food Enzymes is recommended to improve fat digestion. Liver Balance or other herbs to cleanse the liver may be helpful. Burdock and chickweed are very helpful herbs in helping the body properly metabolize fats. Other remedies that may help include the following.

Therapies: Gall Bladder Flush (pg. 23), Enzymes (pg. 15)

Remedies: Chickweed, Hi Lipase, SF, Burdock, Digestive Bitters Tonic, Food Enzymes, Gall Bladder Formula, L-Glutamine, LIV-J, Liver Balance, MetaboMax, Thyroid Support

Fatigue

See Also *Adrenals (Exhaustion, Weakness or Burnout), Anemia, Hyperthyroid, Hypoglycemia*

Fatigue is a symptom that something else is wrong in the body. The cause can be a disease, a glandular problem, lack of sleep, low blood sugar, low blood pressure or a host of other problems. The products that will help fatigue vary greatly depending on the cause, but for starters, Target Endurance or Cellular Energy can improve overall energy reserves, increasing stamina and endurance.

One of the major causes of fatigue is glandular problems involving reduced function of the thyroid, adrenals, pancreas or pituitary. When low thyroid is the problem, the person may be cold, have difficulty losing weight, and have problems with dry skin. Thyroid Activator or Thyroid Support may be of help when the thyroid is the cause.

When the adrenals are the problem, the person may have symptoms like a quivering tongue, dark circles under their eyes and restless sleep. Licorice root, Nervous Fatigue Formula and Adrenal Support may be of help in this case.

Low blood sugar will cause fatigue in the afternoons or sudden fatigue a few hours after eating. The nose or limbs will suddenly go cold and the person may have problems concentrating. Licorice root or Super Algae may be of benefit here.

When the pituitary is out of balance, it will upset the general balance of the glandular system. Herbs that help the pituitary as a cause of fatigue include Super Algae and bee pollen.

Anemia may also be a cause of fatigue. I-X may be of help in this case.

When fatigue is accompanied by a run-down immune system and general weakness, Trigger Immune may be of help. When accompanied by stress, adaptagens such as Suma Combination or AdaptaMax may be the answer.

It goes without saying that loss of sleep will cause fatigue. If insomnia is a problem, then Nervous Fatigue Formula during the day or Herbal Sleep in the evenings may help.

Fatigue associated with depression is an indication for Mood Elevator. Fatigue after eating is an indication for Chinese Anti-Gas and digestive enzymes like Proactazyme.

Therapies: Hydration (pg. 10), Antioxidants (pg. 17), Enzymes (pg. 15), SuperFood (pg. 15), Low Glycemic Diet (pg. 11)

Remedies: **AdaptaMax, Adrenal Support, Energ-V, Ginseng (Korean), Ginseng (Wild American), Licorice Root, Nature's Chi, Target Endurance, Trigger Immune,** Anti-Gas (Chinese), Astragalus, Bee Pollen, Blood Build, Blue Vervain, C-X, Capsicum, Capsicum & Garlic with Parsley, Cellular Energy, Chlorophyll, Co-Q10, Damiana, Dulse, Eleuthero, Fatigue/Exhaustion, Fibralgia, Food Enzymes, Germanium Combination, Ginger, Gotu Kola, HS II, HY-A, I-X, IGF-1, Iron, Jasmine Absolute, Kava Kava, KB-C, Kelp, Kidney Activator (Chinese), L-Carnitine, L-Glutamine, Lemon Oil, Maca, Magnesium, Melatonin Extra, MetaboMax, Milk Thistle Combination, Mineral-Chi Tonic, Mood Elevator, Nature's Chi, Nervous Fatigue Formula, Niacin, Pantothenic Acid, Pine Needle, Proactazyme Plus, Red Beet Formula, Sea Salt, SF, Spirulina, St. John's Wort, SUMA Combination, Super Algae, Super Trio, Target TS-II, Thai-Go, Thyroid Activator, Thyroid Support, TS II, Vitamin B-6, Vitamin B-Complex, Vitamin C, X-A, Ylang Ylang

Fatty Liver Disease

The accumulation of fat in the liver of people who drink little or no alcohol is called fatty liver disease. It is common and in most people causes no signs or symptoms. In some people, however, it can lead to inflammation, scarring of the liver and eventually liver failure. It is probably related to our diet of refined carbohydrates and processed oils. Essential fatty acid supplements should be avoided by people with fatty liver disease. Some supplements that may be helpful in getting rid of the fat in the liver follow:

Therapies: Enzymes (pg. 15), Gall Bladder Flush (pg. 23), Fiber (pg. 17), Low Glycemic Diet (pg. 11)

Remedies: **Chickweed, Hi Lipase, SF,** Burdock, Milk Thistle Combination

Fatty Tumors or Deposits

These are abnormal growths that are mostly composed of lipid or fatty material. The following remedies have been used to help dissolve them.

Therapies: Gall Bladder Flush (pg. 23)

Remedies: **Burdock, Chickweed, SF,** All Cell Detox, E-Tea, Garcinia Combination, Hi Lipase

Fear (excessive)

See Also *Anxiety Disorders*

Excessive fear can be a symptom of adrenal or kidney weakness. Excessive fear is one of the indications for the Chinese formula KB-C, which strengthens the kidney chi or energy. Supporting the adrenals with licorice root, Korean ginseng or Wild American ginseng may also be helpful. Essential oils may also be helpful in combatting excessive fears. Consider any of the following remedies.

Therapies: Affirmation and Visualization (pg. 8), Stress Management (pg. 8)

Remedies: **Adrenal Support, KB-C,** Bergamot, Black Currant Oil, Chamomile (Roman), Clary Sage, Distress Remedy, Eucalyptus, Frankincense, Ginseng (Wild American), Jasmine Absolute, Kidney Activator (Chinese), Lavender, Lemon Oil, Licorice Root, Neroli, Sandalwood, St. John's Wort, Stress Relief, Ylang Ylang

Fever

There are many herbs that can help when fevers are present.

Fevers will usually go down once the bowel has been opened with an enema, especially in young children. Catnip or garlic tea can be used as an enema solution to help bring down the fever.

Herbs that promote perspiration can also be used to bring down fevers when taken as warm teas, or as extracts taken with warm water. Some of the herbs that have proven useful here include yarrow, peppermint, chamomile, catnip and capsicum. Yarrow is particularly helpful for serious fevers when taken in the form of a hot tea.

Boneset is a good herb for fever associated with the flu where the bones ache. (It is an ingredient in ALJ.) IF-C is a good formula for fevers, or for fevers that alternate with chills. High Potency Garlic is another product that can help with fevers, especially when caused by bacterial infection or respiratory infection.

Herbs that promote perspiration can also be used to bring down fevers when taken as warm teas, or as extracts taken with warm water. Essential oils can be used topically. Dilute them in water, then moisten a wash cloth with the aromatic water and gently sponge off the fevered person.

Therapies: Hydration (pg. 10), Colon Cleansing (pg. 18)

Remedies: IF-C, Silver Shield, Thai-Go, Yarrow, APS II, Bergamot, Blue Vervain, Capsicum, Catnip, Chamomile, Chamomile (Roman), Devil's Claw, Echinacea Purpurea, Elderberry Defense, Feverfew, FV, Garlic, Ginger, HCP-X, HY-C, Immune Stimulator, IN-X, Influenza Remedy, Jasmine Absolute, Lavender, Lemon Oil, Lobelia, Lymph Gland Cleanse, Lymph Gland Cleanse-HY, Pau D Arco, Peppermint, Rose Bulgaria, Rose Hips, Rosemary, Sunshine Heroes Whole Foods Antioxidant, Thyme, Ultimate Echinacea, Vaccine Detox

Fever Blisters

See *Cold Sores (Fever Blisters)*

Fibroids (uterine)

See Also *Fibrosis*

Fibroids are abnormal growths of connective tissue, usually benign, that are found in the uterus or breast. Fibroids may form as large elevated scars after large infected wounds. Excess levels of estrogen in the body usually cause fibroids. High levels of estrogen are often the result of xenoestrogens found in pesticides and plastics.

Symptoms of uterine fibroids include heavy menstrual bleeding, abdominal bloating, menstrual cramps, and spotting between periods. Anemia may result from loss of blood, and there may be abdominal pain, pain during intercourse, painful urination, or constipation. Since symptoms are often obscure, it is important to get an accurate medical diagnosis.

Factors that increase the risk of uterine fibroids are too much caffeine and too much fat in the diet, deficiencies of essential fatty acids, hormone imbalances, underactive thyroid, birth control pills and X-rays. Obesity and family history of fibroids further increase the tendency for them to develop.

Commercial meat and dairy products typically contain hormones that aggravate fibroids. Soft plastic containers (like plastic milk bottles) leech xenoestrogens into food that contribute to fibroids. Pesticide residues on commercially grown produce also stimulate estrogen receptor sites and aggravate the condition. Coffee contributes heavily to fibroid growth. All of these should be avoided or eliminated.

Progesterone cream massaged into the abdomen may help by shifting the balance between estrogen and progesterone.

Yarrow has been helpful in breaking up fibroids in many cases. Matthew Wood, a professional herbalist, has cleared up many cases of uterine fibroids with this herb alone.

Fibroids can be related to congestion in the liver, so a liver cleanse may be of help. All Cell Detox is another formula that may be of help here. The liver is responsible for breaking down and eliminating excess estrogen in the system.

V-X is an excellent formula for breaking up uterine fibroids. It can be taken internally, but works even better when used as a vaginal suppository. It can be blended with cocoa butter to form finger-sized suppositories that can be stored in the refrigerator until ready for use.

The uterus helps the body metabolize fats and having it removed can cause permanent problems with fat metabolism, which can lead to weight and heart problems. A woman's uterus serves many important functions and she should do her best to keep it her entire life. Using some of the remedies described here, it is possible to get rid of uterine fibroids without having a hysterectomy.

Therapies: Avoid Xenoestrogens (pg. 23), Colon Cleansing (pg. 18), Enzymes (pg. 15),Fiber (pg. 17)

Remedies: All Cell Detox, V-X, Yarrow, Dong Quai, Enviro-Detox, FCS II, Female Comfort, Indole 3 Carbinol, Lecithin, Menstrual-Reg, NF-X, Omega-3, PDA, Pro-G-Yam Cream, Super GLA, Vitamin E

Fibromyalgia Syndrome (FMS)

See Also *Autoimmune Disorders, Epstein Barr Virus (Chronic Fatigue Syndrome, CFS)*

Fibromyalgia Syndrome (FMS) can cause severe pain and impair deep sleep. It is a stress-related autoimmune disorder and many of the symptoms mimic those of chronic fatigue syndrome (CFS) and arthritis.

Symptoms of fibromyalgia include painful, tender and recurrent aches in various points all over the body. There is a persistent, but diffused pain in the structural system (bones and muscles), accompanied by fatigue, headaches, general weakness, irritable bowel, poor sleep patterns, digestive problems, and nervous system problems (depression and anxiety).

Modern medicine hasn't identified the cause of FMS, but it is likely caused by diet and lifestyle factors, as evidenced by the fact that many people have experienced relief from FMS by making diet and lifestyle changes. While there is no magic bullet formula or secret recipe of things to do to cure FMS, working on the suspected causes will probably bring relief, if not a complete recovery.

Nutrition should be a person's first concern in overcoming FMS. A mild food diet consisting primarily of fresh vegetables with some fruits has led to a reduction in joint stiffness and pain in many FMS suffers. In the beginning, about 80% of the diet should be fresh vegetables with some fruits. These are alkalizing foods, which aid in reducing acid waste in the tissues. They also contain antioxidants which reduce inflammation and tissue damage.

The remaining 20% of the diet can consist of high quality proteins and whole grains. Sugar, refined carbohydrates, processed vegetable oils, shortening, margarine and processed, packaged foods should be avoided. In addition, FMS suffers usually benefit from some specific supplements, including (in order of importance) magnesium, iodine, omega-3 essential fatty acids and antioxidants.

Magnesium is essential for the synthesis of adenosine triphosphate (ATP), the energy powerhouse for cells. Deficiencies of magnesium cause cells to have to use anaerobic pathways to produce energy, which leads to lactic acid formation in muscles and to pain. Magnesium ions are also essential to helping muscles relax, so deficiencies lead to muscle stiffness.

Fibralgia is a supplement containing magnesium with malic acid. It aids in the production of ATP in the muscles and helps reduce muscle pain and stiffness. It also helps reduce fatigue in people suffering from FMS.

Iodine is an essential nutrient for the thyroid gland and many FMS sufferers also have a dysfunctional thyroid. Dr. David Brownstein, author of *Iodine: Why You Need It, Why You Can't Live Without It*, claims that iodine supplements alone have cured fibromyalgia. This helps explain why several natural healers independently discovered that black walnut was helpful for FMS. Black walnut, especially ATC Concentrated Black Walnut, is a good source of natural iodine. It is also antimicrobial and antiparasitic and mildly detoxifying. Liquid Dulse is another good iodine supplement.

Omega-3 essential fatty acids are essential for the health of nerve fibers and aid in the production of chemical messengers resulting in reduced inflammation and pain. Most modern diets are deficient in omega-3 fatty acids, so a supplement like Super Omega-3 EPA may also be helpful.

FMS suffers who are having a hard time eating as many vegetables and fruits as they should, will find supplementing their intake of antioxidant nutrients helpful. Thai-Go is a tasty way to improve antioxidant intake. It has the highest oxygen radical absorbing capacity (ORAC) rating of any liquid antioxidant supplement, and this has been certified by the company that developed the ORAC testing. It contains mangosteen, which contains xanthones proven to have antioxidant and anti-inflammatory activity.

People who suffer from FMS tend to be low in enzymes. This is due to the high level of enzyme inhibitors and the lack of raw and enzyme-rich foods in most modern diets. Supplementing enzymes with Proactazyme Plus or Food Enzymes is important for people suffering from just about any chronic health problems. In most cases, these people also have a hiatal hernia that needs to be corrected.

One of the biggest contributing factors to FMS and its related conditions may be chronic intestinal inflammation and leaky gut syndrome. Intestinal inflammation can be caused by enzyme deficiency, poor digestion, environmental toxins, yeast or bacterial infections, parasites or drug medications. The inflammation causes the intestines to become excessively porous, which allows toxins from the digestive tract to easily enter the bloodstream, causing irritation and inflammation to tissues.

Regular use of a fiber supplement like Psyllium Hulls Combination or Everybody's Fiber can reduce intestinal inflammation and aid in the repair of leaky gut. Uña d gato (also known as Cat's Claw) is one of best herbal remedies available for toning up intestinal membranes and reducing gut leakage. Kudzu/St. John's wort is another great herbal supplement that helps here.

A yeast or parasite cleanse may also be needed. It is very probable that environmental toxins play a role in FMS. However, harsh cleansing is usually contraindicated in these situations because it will aggravate symptoms. What is needed is gentle detoxification, starting with fiber to help bind toxins in the gut.

In addition to the fiber, a small amount of a supplement to help cleanse the liver and aid its detoxification process is also helpful. Taking just one capsule of Enviro-Detox or one to two capsules of Liver Balance daily is all that is needed. These formulas will encourage the liver to break down chemicals in the system. The fiber will help bind them and remove them. As your body gets stronger, you can take more.

Heavy metal toxicity may play a role in some of the problems associated with FMS. If this is a problem, try adding one capsule of Heavy Metal Detox per day along with six to eight capsules of Algin, a mucilaginous fiber from kelp that binds heavy metals in the intestines for elimination.

One other factor to consider in detoxifying the body is helping the cells to remove toxins that may be lodged inside the tissues. Cellular Energy aids the production of energy in the cell, which helps cells and tissues eliminate toxins. It can also help reduce fatigue and weakness and increase stamina in FMS.

FMS is most common in hardworking perfectionists. K.P. Khalsa, RH(AHG), has noted that FMS sufferers are usually high achieving women that are burning the candle at both ends and in the middle, too. From this point of view, FMS can be considered a health collapse related to chronic stress.

The adrenal glands are responsible for helping the body cope with stress. They synthesize cortisol to control inflammation and balance blood sugar in times of crisis. They also produce hormones like epinephrine (adrenaline) to give us the energy we need to face challenges.

Chronic stress depletes the adrenal glands, which makes it difficult for the body to control pain and inflammation. Adrenal exhaustion also leads to a collapse of a person's drive and energy and produces fatigue coupled with restless and disturbed sleep patterns.

To overcome adrenal exhaustion, it is necessary to avoid refined carbohydrates (particularly sugar), alcohol and caffeinated beverages. An adrenal glandular like Adrenal Support will help to rebuild the adrenal glands and restore energy levels and sleep. Rebuilding the adrenals will also help to reduce pain and inflammation throughout the body. If the thyroid is also low, then adding Thyroid Support will also be necessary for restoring energy to the system.

Another formula to consider is Nervous Fatigue Formula. This Chinese blend helps with mental confusion and muddled thinking (brain fog) and fatigue coupled with disturbed and restless sleep. It gently aids liver detoxification, too.

Of course, it also helps to learn to deal better with stress. People suffering from FMS should learn to pace themselves by setting realistic goals and balancing work with rest and recreation. Taking breaks when one is tired, getting a good night's sleep and allowing time in one's life for relaxing activities is a must. Consider stretching, meditation, yoga, tai chi, relaxing baths or long walks in nature.

Therapies: Hydration (pg. 10), Antioxidants (pg. 17), Heavy Metal Detoxification (pg. 22), Hiatal Hernia Correction (pg. 12), Affirmation and Visualization (pg. 8), Stress Management (pg. 8), Enzymes (pg. 15), Essential Fatty Acids (pg. 17),Fiber (pg. 17), Blood Type Diet (pg. 13)

Remedies: Black Walnut, Fibralgia, Yucca, Adrenal Support, APS II, Cellular Energy, Enviro-Detox, Food Enzymes, IGF-1, Kava Kava, Licorice Root, Lobelia, Magnesium, Mineral-Chi Tonic, MSM, Proactazyme Plus, Probiotics, Protease, Thai-Go

Fibrosis

See Also *Fibroids (uterine)*

Fibrosis is the abnormal formation of fibrous tissue in a part of the body, such as the lungs, breast or uterus. The following may be helpful.

Therapies: Enzymes (pg. 15)

Remedies: Joint Support, APS II, Black Currant Oil, Devil's Claw, Evening Primrose Oil, Licorice Root, Mullein, Red Clover, Thyroid Activator, Thyroid Support, Vitamin E

Fingernail Biting

Nail biting may be due to nervousness, parasites or mineral deficiencies. It may also be a sign of parasites. Nutri-Calm, Distress Remedy or vitamin B-complex may help calm the nerves. HSN-W, Herbal CA or Colloidal Minerals may supply missing minerals. A parasite cleanse like the ParaCleanse will help with parasites. All of the following are possible remedies, depending on the cause.

Therapies: Minerals (pg. 15), Stress Management (pg. 8)

Remedies: HSN-W, Mineral-Chi Tonic, Nutri-Calm, Artemisia Combination, Colloidal Minerals, Distress Remedy, Herbal CA, Herbal Pumpkin, Horsetail, HSN Complex, Ionic Minerals, Para-Cleanse, Stress-J, Vitamin B-Complex

Fingernails (weak or brittle)

Weak or brittle fingernails are a sign of a lack of silica and other trace minerals. The following may help.

Therapies: Minerals (pg. 15), Enzymes (pg. 15)

Remedies: Horsetail, HSN-W, Herbal CA, Herbal Trace Minerals, HSN Complex, Ionic Minerals, Kelp, Mineral-Chi Tonic

Flatulence

See *Gas and Bloating*

Fleas

The following may be helpful for fleas on pets. A few drops of any of the listed essential oils may be blended with Nature's Fresh and applied topically as a repellant. Paw Paw can be added to a shampoo and used to give the animal a bath.

Remedies: Nature's Fresh, Paw Paw, Eucalyptus, Lemon Oil, Pine Needle, Tea Tree Oil

Floaters

Floaters are small black specks that appear to dance or float across the visual field. Possible causes include an unhealthy diet, excessive eye strain and inadequate rest. A generally good diet is helpful, along with extra doses of vitamin A, zinc, vitamin C and magnesium. An eyewash or compress over the eyes using a tea made from EW or chamomile may also be helpful.

Therapies: Antioxidants (pg. 17), Low Glycemic Diet (pg. 11)

Remedies: Chamomile, EW, Heavy Metal Detox, Magnesium, Mega-Chel, Perfect Eyes, Vitamin A & D, Vitamin C, Zinc

Flu

Also known as influenza, the flu refers to viral infections accompanied by nausea, vomiting, diarrhea, fever, malaise, body pain, and respiratory symptoms. The media is constantly warning people about flu epidemics to scare people into getting vaccines, but the truth is that very few people die of the flu. Most of the people who die of the flu are elderly or have compromised immune systems.

Besides, there are numerous effective natural remedies for the flu. For prevention, simply wash hands frequently and use Silver Shield Gel as a natural hand disinfectant. Take something to boost the immune system, such as Immune Stimulator, Fizz-Active Immune or Ultimate Echinacea when the flu is going around.

If you do get the flu try taking a good antiviral remedy such as Elderberry Plus or Una de Gato Combination. Other great flu remedies include Silver Shield, HCP-X, FV and IF-C. For settling the stomach flu, two of the best remedies are ginger and peppermint essential oil. For flu with aches and fever try APS II with White Willow or yarrow. Other remedies for the flu include the following.

Therapies: Hydration (pg. 10)

Remedies: Elderberry Plus, Fizz Active-Immune, FV, Ginger, HCP-X, IF-C, Silver Shield, Aloe Vera, Anamu, APS II, Bayberry, Blood Build, Breathe Free, Capsicum & Garlic with Parsley, Catnip, CC-A, Cinnamon, Echinacea/ Golden Seal, Elderberry Defense, Garlic, HS II, IN-X, Influenza Remedy, Jasmine Absolute, Lemon Oil, Lung Support, Lymph Gland Cleanse, Lymph Gland Cleanse-HY, Peppermint, Pine Needle, Prevention, Red Raspberry, Red Raspberry Liquid, Sunshine Heroes Whole Foods Antioxidant, THIM-J, Vitamin A & D, Yarrow

Foot Odor

The following remedies may be helpful for severe foot odor.

Remedies: Chlorophyll, Nature's Fresh, Pine Needle

Fractures

See *Broken Bones*

Free Radical Damage

Experts suggest that about 50-80% of all chronic and degenerative diseases, including heart disease, cancer, diabetes, arthritis, macular degeneration, Alzheimer's and dementia are caused by oxidative stress, also known as free radical damage. Free radical damage also causes the cosmetic problems we associate with aging, dry skin, wrinkles, age spots and so forth. Read about the basic supplement Antioxidants for more information. The following remedies are antioxidants that help prevent free radical damage.

Therapies: Antioxidants (pg. 17)

Remedies: Super Antioxidant, Super ORAC, Thai-Go, AdaptaMax, Barley Grass, Brain-Protex, Cardio Assurance, Carotenoid Blend, Chlorophyll, Co-Q10, Grapine, Green Tea Extract, Immune Stimulator, Mega-Chel, Nature's Gold, Protector Pak, S-O-D With Gliadin, Sunshine Heroes Whole Foods Antioxidant, Super ORAC, Super ORAC, Vitamin A & D, Vitamin C, Vitamin E, Zinc, Zinc Lozenges

Frigidity

See *Sex Drive (low)*

Frostbite (prevention)

Sprinkle tiny amounts of capsicum in socks or gloves to prevent frostbite.

Remedies: Capsicum

Fungal Infections (Yeast Infections, Candida albicans)

See Also *Athlete's Foot, Jock Itch, Vaginal Discharge*

There are many positive benefits of yeast. Yeast helps bread to rise. Yeast creates the fermentation process that allows brewers to make beer and wine. Yeast microorganisms are a type of fungus, like mushrooms. They are present in the soil and part of the mix of microbes needed for soil health. Yeasts are also included in the dozens of species of microorganisms that inhabit our intestines. This blend of microbes are known collectively as the intestinal microflora, and are critical to health. So, yeast can be very beneficial under the right conditions.

Under the wrong conditions, however, yeast can create problems with our health. Yeast or fungal infections have become a serious problem for most people living in modern Western society. In particular, one species of yeast, Candida albicans, has been shown to be at the root of a wide variety of health problems including chronic sinus problems, vaginal yeast infections, frequent colds and flu, earaches, swollen lymph nodes, fatigue, reduced immunity, brain fog, leaky gut syndrome, athlete's foot, jock itch and more.

It is becoming widely recognized that antibiotics contribute to yeast overgrowth because they upset the balance of the intestinal microflora by killing the friendly bacteria, but antibiotics aren't the only reason yeast gets out of control. Many other chemicals in our environment also have negative effects on our intestinal flora. These include alcohol, chlorinated drinking water, MSG, nitrates, and sulfates. Since yeast feed on sugar, excess consumption of sugar also plays a role in yeast overgrowth.

Once the yeast is out of control, it'secretes substances that weaken the integrity of the intestines (resulting in intestinal inflammation and leaky gut syndrome) and are absorbed into the blood stream, weakening the immune system and causing us to crave more sugar. It's almost like the yeast hijack the body and cause us to want to perpetuate the environment that sustains their existence.

Determining If You Have a Yeast Infection

Candida often becomes one of those catch-all diagnosis. Many people think they have yeast overgrowth even though they have done extensive yeast cleanses. Here is a little quiz to help you determine if yeast overgrowth may be contributing to your health problems. If your answer yes to any of the following questions check the box on the left. If you check five or more boxes, you may have a problem with yeast overgrowth.

- Do you generally feel fatigued or have low energy?
- Do you experience food sensitivities or food allergies?
- Do you have nail fungus, athlete's foot or jock itch?
- Do you have recurrent vaginal yeast infections?
- Have you taken broad spectrum antibiotics?
- Do you crave sugar or sweets (candy, soda pop, etc.)?
- Do you often have gas, bloating or indigestion?
- Do you crave refined white flour (bread, pasta, baked goods)?
- Have you been on birth control pills for 6 months or more?
- Do you experience brain fog, mental confusion or mental fatigue?

If you do appear to have a problem with yeast overgrowth, here are some steps to getting it under control.

Step One: Modify the Diet to Stop Feeding the Yeast

The first, and most important, step in eliminating yeast overgrowth is to stop feeding the yeast. Yeast love carbohydrates, especially simple sugars. So, you need to get all simple carbohydrates out of the diet for a period of time. For two to four weeks eliminate all simple sugars and refined grain products from your diet. Simple sugars include table sugar (or sucrose), glucose, fructose, corn syrup and even natural sugars like honey, brown sugar and fruit juices. Refined grain products include white flour, white rice, corn chips and breakfast cereals. You re going to have to read labels carefully to do this because sugars and refined grains are added to most prepackaged foods.

It is also important to avoid alcohol because it is also converted to sugar in the body. In fact, if your problem is severe, you may wish to avoid even whole grains, most fruit and starchy foods like potatoes for at least the first two weeks.

It is also a good idea to avoid foods that contain yeast or mold, such as bread, beer, aged cheeses and so forth. Many experts also recommend avoiding pickled and fermented foods and vinegar. These foods don't cause yeast overgrowth, but eliminating them for a period of time seems to help get yeast under control.

Eat a low glycemic diet and include some good fats in your diet. A particularly good fat for fighting candida is coconut oil because it contains a medium chain saturated fatty acid called caprylic acid that helps control yeast.

Step Two: Improve General Digestive and Intestinal Health

Yeast get out of control when the environment becomes conducive to their growth. So, if we want to get them back under control, we need to change the environment of the digestive tract. Normally, the hydrochloric acid and enzymes found in our stomach help keep these microbes in check. These can be stimulated by taking Digestive Bitters 15-20 minutes before meals. It will also help to relieve the gas and bloating common in people with yeast overgrowth. Also consider taking Proactazyme or Food Enzymes with meals. Taking High Potency Protease between meals will also help to regulate digestive microbes.

Yeast overgrowth is often accompanied by intestinal inflammation and leaky gut syndrome. You may want to check out remedies to bring these problems under control, too.

Step Three: Use Anti-Fungal Agents to Reduce Yeast Overgrowth

After cutting off the yeast's food supply and altering the digestive environment to make it unfriendly for yeast growth, we can knock it down using antifungal herbs and supplements. A convenient way to do this is with the Candida Clear Pack. This is a great pre-packaged program for controlling yeast overgrowth. Other options include taking Pau D Arco in capsules or tea form, taking Yeast/Fungal Detox by itself and/or using Silver Shield.

Essential oils are also powerful allies in dealing with yeast infections. Antifungal essential oils include tea tree, lavender, thyme, clove, and oregano. These can be used in baths, diffused into the room or taken internally in one drop doses per day for no more than one to two weeks.

Step Four: Repopulate the Body with Friendly Bacteria (Probiotics)

The final step in taming the yeast is to repopulate the intestines with friendly bacteria or probiotics. Naturally fermented foods such as yoghurt, raw sauerkraut and miso are good dietary sources of these friendly microbes to take after your cleanse, but you will probably want to also take probiotic supplements.

Therapies: Colon Cleansing (pg. 18), Enzymes (pg. 15), Probiotics (pg. 16),Fiber (pg. 17)

Remedies: Candida Clear, Pau D Arco, Silver Shield, Tea Tree Oil, Yeast/Fungal Detox, Candida, Caprylic Acid Combination, Caprylimune, Clove Bud, Garlic, Guardian, MSM, Neroli, Olive Leaf, Oregano (Wild), Para-Cleanse, Patchouli, Paw Paw, Peppermint, Probiotics, Sunshine Heroes Probiotic Power, THIM-J, Yarrow

Gall Bladder (sluggish)

Sluggish activity of the gall bladder will result in poor digestion of fats. It will also make it difficult for the body to eliminate excess cholesterol. Clay-colored stools or greasy stools that float and are difficult to flush are symptoms of sluggish gall bladder function. Herbs that stimulate gall bladder function are called cholagogues. Here are some remedies to aid gall bladder function.

Therapies: Gall Bladder Flush (pg. 23)

Remedies: **Hi Lipase, Liver Balance, Milk Thistle,** Blood Build, Burdock, Cascara Sagrada, Dandelion, Fat Grabbers, Gall Bladder Formula, LIV-J, Liver Cleanse Formula, Milk Thistle Combination, Wild Yam

Gall Stones

Deposits, resembling small rocks, that form in the gallbladder are called gall stones. They are usually composed primarily of cholesterol. If they are large or numerous enough, they may cause severe abdominal pain. Besides doing a gall bladder flush, drinking more water and taking fiber is very helpful. The following can help to prevent or eliminate gallstones.

Therapies: Gall Bladder Flush (pg. 23), Hydration (pg. 10),Fiber (pg. 17), Enzymes (pg. 15)

Remedies: Gall Bladder Formula, Hi Lipase, Lemon Oil, Liver Balance, Liver Cleanse Formula, Magnesium, Milk Thistle, Milk Thistle Combination, Nature's Three, Psyllium Hulls Combination

Gangrene

Gangrene is localized death of the skin and underlying soft tissue due to lack of blood supply. This is a serious illness and medical attention should be sought.

Silver Shield liquid and gel have been proven in scientific studies to halt the progression of gangrene and often prevent the need for amputation. Apply the gel topically and take the liquid orally.

Oral chelation with Mega Chel has cleared up cases of gangrene brought on by diabetes. Other remedies that may be helpful include Ultimate Echinacea internally and tea tree essential oil (applied topically). Alternating soaks with ice water and warm water to stimulate circulation to the afflicted areas can also be helpful.

The following remedies may be helpful.

Therapies: Oral Chelation (pg. 21), Low Glycemic Diet (pg. 11), Antioxidants (pg. 17)

Remedies: **Mega-Chel, Silver Shield, Tea Tree Oil,** BP-X, Capsicum, Echinacea Purpurea, Echinacea/Golden Seal, Garlic, Golden Seal, Ultimate Echinacea

Gas and Bloating

See Also *Belching, Hiatal Hernia, Ileocecal Valve*

Intestinal gas is created by the action of friendly flora on foods we eat. Some amount of gas production is normal. Excessive gas is usually due to imbalances in the friendly flora or problems with digesting certain foods. Digestive enzymes can help reduce intestinal gas. Taking fiber or probiotics sometimes temporarily increases intestinal gas by altering intestinal microbes.

Bloating is an abnormal feeling of fullness in the gastrointestinal tract caused by excessive gas that builds up pressure in the abdomen. Bloating puts pressure on the stomach and contributes to the development of a hiatal hernia. It may even cause stress on the heart. Where bloating is frequent, the ileocecal valve is probably swollen and inflamed and needs to be closed.

Herbs used to relieve gas and bloating are called carminatives. These herbs relieve intestinal gas by increasing blood flow to the abdominal cavity. Most of the gas produced in the intestines is actually absorbed into the bloodstream and excreted through the lungs. Carminative herbs aid this process. They also aid digestion to slow gas production.

Where bloating is severe and accompanied by foul belching, there is petrification taking place in the digestive tract. Activated charcoal can be very helpful for relieving this condition. Chinese Anti-Gas is also helpful.

Here is a complete list of remedies helpful for gas and bloating.

Therapies: Enzymes (pg. 15), Probiotics (pg. 16), Fiber (pg. 17), Colon Cleansing (pg. 18), Hiatal Hernia Correction (pg. 12)

Remedies: **Anti-Gas (Chinese), Anti-Gas Formula, Catnip & Fennel, Charcoal (Activated), Food Enzymes, Peppermint, Proactazyme Plus,** Bergamot, Blessed Thistle, Bowel Detox, Catnip, Chamomile, Chamomile (Roman), Clove Bud, Fenugreek & Thyme, Garlic, Ginger, JP-X, Kava Kava, Lactase Plus, Lemon Oil, Liquid Cleanse, Mandarin (Red), Myrrh, Neroli, Papaya Mint, Parasites, PDA, Probiotics, Red Raspberry Liquid, Rosemary, Safflowers, Tei Fu Oil, Thyme, Ylang Ylang

Gastritis

See Also *Indigestion*

Gastritis is an inflammation of the stomach. The following may be helpful.

Therapies: Antioxidants (pg. 17)

Remedies: **Aloe Vera, Gastro Health, Licorice Root, Slippery Elm,** Anamu, IF Relief, Lymph Gland Cleanse, Marshmallow, Red Raspberry, Red Raspberry Liquid, Sandalwood, St. John's Wort

Gastroesophageal Reflux Disease (GERD)

See *Acid Indigestion (Heartburn, Acid Reflux), Acid Indigestion (Heartburn, Acid Reflux)*

Generalized Anxiety Disorder

See *Anxiety Disorders*

Giardia

See Also *Parasites (general)*

A single celled organism, giardia (Giardia lambila) is the most common cause of waterborne disease in the United States. It is picked up by drinking contaminated water. Because giardia form cysts that are not destroyed by chlorination they must be filtered from water.

Symptoms of giardia infection include diarrhea, gas, upset stomach or stomach cramps, nausea and/or greasy stools that tend to float. Symptoms usually appear 7-14 days after exposure. The following are helpful remedies for giardia infections.

Remedies: Golden Seal, Silver Shield, Charcoal (Activated), Echinacea/Golden Seal

Gingivitis (Bleeding Gums, Gum Disease, Pyorrhea)

See Also *Cardiovascular Disease (Heart Disease)*

When bacteria get into the space between the teeth and the gums, they can cause inflammation (gingivitis) and a breakdown of the gum tissue. This leads to bleeding of the gums. Pyorrhea is a discharge of pus or an advanced form of periodontal disease associated with a discharge of pus and loose teeth.

A thorough cleaning by a dental hygienist is highly recommended. The following may also help.

It is absolutely essential that the teeth be brushed and flossed several times daily. After cleaning the teeth, brush with a mixture of black walnut, white oak bark and goldenseal powders. Leave some of the powder on your gums for 5-10 minutes (or even overnight) and then rinse. These herbs fight infection, strengthen tooth enamel and gums, and contract tissues to arrest bleeding. Other herbs that can be applied topically to the gums to reduce bleeding and inflammation include bayberry root bark and calendula.

Brushing the teeth with XyliBrite toothpaste and using the Xylitol Mouthwash will inhibit the growth of the bacteria involved in gingivitis. You can also apply Silver Shield Gel directly to the gums to kill the bacteria and promote healing.

Using Colloidal Minerals as a mouth wash will also help to stop bleeding. Swallow the Colloidal Minerals after swishing them around in your mouth so that you increase mineral levels in the body, too.

Internally, Co-Q 10 helps reduce the gum inflammation. It also helps with reducing inflammation in the cardiovascular system. In fact, gingivitis is an early warning sign of developing heart disease. A complete list of helpful remedies follows.

Therapies: Antioxidants (pg. 17), Minerals (pg. 15), Low Glycemic Diet (pg. 11)

Remedies: Black Walnut, Co-Q10, Silver Shield, White Oak Bark, Xylitol, Aloe Vera, Bayberry, Colloidal Minerals, Echinacea Purpurea, Golden Seal, IF Relief, IF-C, Lemon Oil, Lymph Gland Cleanse, Tea Tree Oil, Ultimate Echinacea, Vitamin C, Vitamin D3, Xylitol

Glands (swollen lymph)

See Also *Congestion (lymphatic), Tonsillitis (Adnoids)*

The lymphatic system contains lymph nodes that filter lymphatic fluid to remove infectious microbes and toxins. When these glands become irritated, they swell. This happens in the neck, throat, armpits, groin and chest where these nodes are located. Swollen lymph nodes are usually present in respiratory congestion, earaches, sore throats and breast swelling. Lymphomax is an excellent formula for lymphatic swelling. Another excellent formula is Lymphatic Drainage, which can be added to water and sipped throughout the day.

The following can also help reduce swollen lymph glands.

Therapies: Hydration (pg. 10)

Remedies: Lobelia, Lymph Gland Cleanse-HY, Lymphatic Drainage, Lymphomax, Mullein, ALJ, Burdock, Ho Shou Wu, IN-X, Lymph Gland Cleanse, Oregon Grape, SUMA Combination, Thyroid Activator, TS II, Ultimate Echinacea, Yarrow

Glaucoma

See Also *Free Radical Damage*

Glaucoma is a serious eye disease characterized by abnormally elevated fluid pressure within the eye. This is caused by a tiny mesh in the eye that allows fluid to drain becoming clogged. The clogging is usually due to free radical damage causing debris in the lymph. Untreated, this pressure can damage the retina and destroy the optic nerve, resulting in vision loss or blindness. In fact, glaucoma is the most common cause of blindness. Risk factors for this disease include people of African ancestry, people with diabetes, high blood pressure, severe myopia (nearsightedness), or a family history of glaucoma, and those taking corticosteroid preparations.

Other contributing factors include diabetes, food allergies, excess caffeine and sugar consumption, allergies, adrenal exhaustion, liver and thyroid problems and arteriosclerosis.

Coffee should be eliminated from the diet along with other sources of caffeine, including chocolate and soft drinks.

Taking remedies that can relax the eye, allowing the drain to open may be helpful. One such remedy that is highly effective, but illegal is cannabis. It's possible remedies like kava kava or lobelia could have a similar effect.

Therapies: Antioxidants (pg. 17)

Remedies: Magnesium, N-acetyl Cysteine, Super ORAC, Bilberry Fruit, Blood Pressurex, EW, Eyebright, Germanium Combination, Ginkgo Biloba, Kava Kava, Lobelia, Mega-Chel, Perfect Eyes, Spirulina, Vitamin A & D, Vitamin C, Vitamin E

Goiter

See Also *Hyperthyroid*

An enlargement of the thyroid gland that can be seen as a swelling in the neck is a goiter. This results from insufficient intake of iodine. The following supplements supply iodine.

Remedies: Black Walnut, Dulse, Thyroid Support, Kelp, Thyroid Activator, TS II

Gonorrhea

Gonorrhea is a sexually transmitted bacterial infection that causes inflammation of the genital mucous membranes. Symptoms of gonorrhea in men include painful urination and a thick discharge from the penis. Women may have no initial symptoms at all, but painful urination and vaginal discharge, along with abnormal menstrual bleeding, are typical symptoms. In advanced states the disease can cause fever, muscle aches and inflamed joints. It can also cause infertility.

Seek medical attention for this condition. The following may be supportive of medical therapy.

Remedies: Silver Shield, Black Cohosh, Garlic, Golden Seal, Kava Kava, Kidney Activator, Lemon Oil, Marshmallow, Ultimate Echinacea, Uva Ursi

Gout

See Also *Arthritis*

Gout is a metabolic disease characterized by excessive amounts of uric acid in the blood and deposits of uric acid crystals in the joints. Uric acid is a byproduct of protein metabolism and is filtered from the blood by the kidneys.

It is important, therefore, to alkalize the diet. Animal protein consumption should be reduced and when it is consumed, PDA should be taken with it. Soda pop should be avoided completely. Make sure you are drinking sufficient amounts of pure water to flush metabolic waste from the body. Complex carbohydrates such as fresh fruits, dark green vegetables and other alkaline-forming foods should be increased in the diet. The following may be helpful.

Therapies: Hydration (pg. 10), Blood Type Diet (pg. 13), Antioxidants (pg. 17), SuperFood (pg. 15)

Remedies: Alfalfa, Devil's Claw, Joint Health, Joint Support, KB-C, PDA, Burdock, Chickweed, Chlorophyll, Dandelion, DHA, Herbal CA, Kava Kava, Kidney Drainage, L-Glutamine, Lemon Oil, Pine Needle, Red Beet Formula, Red Clover, Red Clover Blend, Safflowers, Sarsaparilla, Uva Ursi

Grave's Disease

See Also *Hashimoto's Disease (Thyroiditis)*

Graves disease is an overactive or hyper-thyroid condition, where the thyroid is overproducing hormones. This means the thyroid is over stimulating metabolism. You can think of this as having the thermostat is set too high. As a result, fuel burns too quickly, which results in weight loss, intolerance to heat and hyperactivity and restlessness. For example, some of the specific symptoms associated with Grave's disease (the most common hyperactive thyroid condition) include bulging eyes, rapid pulse rate (90-160), heart palpitations, tremors, restlessness and anxiety, lack of periods, muscle weakness and impaired sleep.

The reason why heart rate is linked with thyroid function is because the heart prefers fatty acids over carbohydrates for fuel. So, when fat burning is hot, the heart is over stimulated. When fat burning is slow, the heart tends to beat more slowly, too.

This is a serious medical condition and needs proper medical attention. The rapid heart beat can over-stress the heart and circulation resulting in life-threatening effects. It is essential that a physician monitor someone with a hyperthyroid condition, even if the patient is opting to try a natural approach. Medication may be needed to lower the heart rate and inhibit the thyroid while you work on removing the underlying health problems.

While it is important to have proper medical monitoring of a hyperthyroid situation, medical treatments for hyperactive thyroid conditions leave much to be desired. While drugs can be used to inhibit thyroid function, physicians usually convince the patient to destroy the thyroid gland with radioactive iodine.

This therapy is literally designed to fry the thyroid gland. The radioactive iodine is taken up by the thyroid gland, causing it to be destroyed. Thereafter, the person will have to take medications for low thyroid, as their thyroid gland will no longer function properly. Obviously, there has to be a better way.

There are herbs which inhibit thyroid function. Several plant species contain substances known to bind to TSH receptor sites in the thyroid, inhibiting them and reducing thyroid output. These include bugleweed and lemon balm. Motherwort and mistletoe can also be used to calm the heartbeat. (Unfortunately, none of these herbs are available from NSP.)

However, simply inhibiting the thyroid (even with herbs) isn't correcting the underlying problem or cause. According to Dr. Henry Bieler in *Food is Your Best Medicine*, the glands act as a third line of immune defense. When toxins get past the intestinal membranes and the liver and enter the blood stream, the glandular system becomes overexcited in an effort to increase metabolic rate in order to drive the toxins out of the body. So, according to this theory, a hyperactive thyroid would signal a need to cleanse the blood of toxins. Heavy metals, in particular, may be at the root of hyperthyroid function.

NSP's formula IF-C is a heat-reducing formula, indicated for rapid heartbeat, a red tongue, and conditions involving excess heat, such as fever or inflammation. It clears toxins from the blood and can be very helpful for hyperthyroid.

The adrenals tend to work with and balance the thyroid. People with hyperactive thyroid function also tend to have adrenal problems. The stress hormone, cortisol, is an anti-inflammatory, so hyperthyroidism may be a sign of excess stress, accompanied by adrenal weakness. So the cooling effect of the adrenal hormone, cortisol, is reduced. In this case, licorice root, Adrenal Support or Nervous Fatigue Formula have also proved helpful.

Other supplements that may be helpful for hyperthyroid include Magnesium Complex, Stress-J, Chinese Stress Relief, Intestinal Soothe and Build, Co Q10 and hawthorn.

Diet can also play a role in helping to balance an overactive thyroid. High carbohydrate diets, coupled with low protein and/or fat intake, tend to elevate thyroid function. So, a properly balanced diet with correct proportions of fats, proteins, and low glycemic carbohydrates is helpful.

Cruciferous vegetables, such as cabbage, broccoli and cauliflower tend to have an inhibiting effect on the production of thyroid hormones. Soy also has a a strong thyroid inhibiting effect. Millet has a mild effect. These foods should be consumed freely.

Hyperthyroid patients may actually be deficient in iodine. If your body is saturated with iodine, it would not take up radioactive iodine, so the fact that the medical profession can use radioactive iodine to kill the thyroid suggests the tissues are not properly saturated with iodine. The best formula for supplementing iodine is TS II with Hops because the hops also has a calming affect on the thyroid.

Again, hyperthyroid conditions can be serious and life-threatening, so the situation should be monitored by a physician to make certain the therapy is working, even when the person chooses to go the natural route.

Therapies: Heavy Metal Detoxification (pg. 22), Hiatal Hernia Correction (pg. 12), Stress Management (pg. 8), Antioxidants (pg. 17), Hydration (pg. 10)

Remedies: IF-C, Nervous Fatigue Formula, SUMA Combination, TS II, 7-Keto, Adrenal Support, Co-Q10, Hawthorn Berries, Heavy Metal Detox, Hops, Intestinal Soothe & Build, L-Carnitine, Licorice Root, Magnesium, Mineral-Chi Tonic, Phyto-Soy, Stress Relief, Stress-J, THIM-J, Vitamin B-Complex

Gray Hair

Gray hair is associated with aging and is typically the result of mineral deficiencies, as minerals help to add color to the hair. It can also be a sign of declining hydrochloric acid production (resulting in poor protein digestion and mineral absorption). A lack of B-Complex vitamins may also contribute to this problem. The following may help.

Therapies: Minerals (pg. 15)

Remedies: Colloidal Minerals, Ho Shou Wu, KB-C, PDA, Aloe Vera, Ionic Minerals, Pantothenic Acid, Sage, Vitamin B-Complex, Zinc

Grief (excessive)

Abnormally severe and/or prolonged sadness and grief is believed in traditional medicine to damage the lungs. The loss of a dear loved one frequently causes pneumonia, chronic coughing and other respiratory ailments. Various essential oils can help with excessive grief and sadness, but one of the best is rose, which has a wonderful healing effect on the heart. Distress Remedy can also help someone who is in shock from grief. The Chinese formula Lung Support is helpful for protecting and healing the lungs in someone suffering from severe grief. Extra B-Complex vitamins or Nutri-Calm can help keep grief from turning into depression.

Therapies: Affirmation and Visualization (pg. 8)

Remedies: Breathe EZ, Rose Bulgaria, ALJ, Bergamot, Breathe Free, Chamomile (Roman), Frankincense, Helichrysum, Lung Support, Lung Support, Marjoram (Sweet), Myrrh, Nutri-Calm, Pine Needle, Vitamin B-Complex

Guilt

The following essential oils may be helpful for working through feelings of guilt.

Therapies: Affirmation and Visualization (pg. 8)

Remedies: Distress Remedy, Pine Needle, Clary Sage, Jasmine Absolute, Rose Bulgaria, Ylang Ylang

Gum Disease

See *Gingivitis (Bleeding Gums, Gum Disease, Pyorrhea)*

Hair (loss or thinning)

See Also *Hypothyroid*

Loss of hair or baldness may be caused by hormonal imbalances, oxidative damage, poor nutrition or poor circulation to the scalp. One of the most common causes is low thyroid. Lack of adequate protein in the diet or difficulty in digesting and assimilating protein is another common cause. Each case must be evaluated individually to determine the underlying cause. The following remedies may help slow or prevent hair loss.

Therapies: Hiatal Hernia Correction (pg. 12), Minerals (pg. 15), Enzymes (pg. 15)

Remedies: **HSN-W, Jojoba Oil, Mineral-Chi Tonic, Thyroid Support**, Aloe Vera, Clary Sage, Co-Q10, Dulse, Eleuthero, Ho Shou Wu, HSN Complex, Iron, Kelp, Maca, Noni (Morinda), Pro-G-Yam Cream, Protease, PS II, Rosemary, Sage, Sarsaparilla, SF, Super GLA, Thai-Go, Trigger Immune, TS II, Zinc

Hair Care (general)

The following products may be helpful with creating healthy hair.

Therapies: Minerals (pg. 15), Essential Fatty Acids (pg. 17), Enzymes (pg. 15)

Remedies: **HSN-W, Jojoba Oil**, Chamomile, Clary Sage, Dulse, Herbal Trace Minerals, Horsetail, HSN Complex, Rosemary, Sage, Tea Tree Oil, Vitamin B-Complex, Vitamin E, Ylang Ylang, Zinc

Halitosis (Bad Breath)

Halitosis is a bad odor from the mouth, also called bad breath. Cleansing the colon can be very helpful. The following remedies are also helpful.

Therapies: Enzymes (pg. 15), Colon Cleansing (pg. 18)

Remedies: **Anti-Gas (Chinese), Chlorophyll, Food Enzymes, Parsley, Proactazyme Plus**, Anti-Gas Formula, Clove Bud, Dieter's Cleanse, Lavender, Licorice Root, Niacin, Peppermint, Tiao He Cleanse, Xylitol

Hangover

See Also *Addictions (alcohol)*

Acute symptoms such as headache, nausea and/or vomiting as a result of recent, excessive consumption of alcohol. Milk Thistle Combination can aid the metabolism of alcohol. Alcohol causes low blood sugar (hypoglycemia) and dehydration. Eating protein and taking remedies to balance blood sugar, along with drinking plenty of pure water is helpful. The following may also be helpful for hangovers.

Therapies: Low Glycemic Diet (pg. 11), Essential Fatty Acids (pg. 17), Hydration (pg. 10)

Remedies: **Kudzu/St.John's Wort, Licorice Root, Liver Balance, Milk Thistle Combination**, Black Currant Oil, Capsicum, Evening Primrose Oil, Liver Cleanse Formula, Milk Thistle, Vitamin B-Complex

Hardening of the Arteries

See *Arteriosclerosis (Atherosclerosis, Hardening of the Arteries)*

Hashimoto's Disease (Thyroiditis)

See Also *Hyperthyroid*

Hashimoto thyroiditis is the most common cause of hypothyroidism. It is an autoimmune condition where the thyroid becomes inflamed and is eventually destroyed. It is often associated with Type I diabetes and celiac disease. This suggests it may be due to chemical toxicity and/or heavy metal poisoning. Here are some remedies that may be helpful.

Therapies: Affirmation and Visualization (pg. 8), Stress Management (pg. 8), Heavy Metal Detoxification (pg. 22)

Remedies: **Astragalus, Eleuthero, Ho Shou Wu, IF-C, Thyroid Support**, AdaptaMax, APS II, Black Cohosh, Carotenoid Blend, Co-Q10, Dulse, Flax Seed Oil, Gotu Kola, IF Relief, Lecithin, Lobelia, Milk Thistle, Mullein, Myrrh, Olive Leaf, Sage, Saw Palmetto, TS II, Vitamin E, Zinc

Hayfever

See *Allergies (respiratory), Rhinitis, Allergic*

Headache (general)

See Also *Pain (general remedies for)*

Headaches have a variety of causes, including dehydration, digestive upset, toxicity of the liver and colon, tension in the neck and shoulders causing misalignment of vertebrae and stress. Rubbing Tei Fu oil and lobelia into the neck and shoulders can relax muscles and relieve many headaches. APS II with White Willow Bark, Nerve Eight and IF Relief can take the place of aspirin and other OTC pain relievers. The following are general remedies for headaches.

Therapies: Hydration (pg. 10), Affirmation and Visualization (pg. 8), Colon Cleansing (pg. 18), Stress Management (pg. 8)

Remedies: **APS II, IF Relief, Nerve Eight, Tei Fu Oil**, Caffeine Detox, Helichrysum, IF-C, Lemon Oil, Neroli, Rose Bulgaria, Thyme, Triple Relief

Headache (Migraine)

See *Migraine*

Headache (sinus)

See Also *Congestion (sinus)*

The following may be helpful for sinus headaches.

Therapies: Enzymes (pg. 15), Colon Cleansing (pg. 18)

Remedies: **ALJ, Fenugreek & Thyme, Tei Fu Oil**, Four, Sinus, SnorEase, St. John's Wort, Wood Betony

Headache (tension)

The following are good for relieving tension (vasoconstrictive) headaches). These are headaches where there is a sensation of pressure on the head, as if the head were in a vise or being squeezed by a belt.

Therapies: Stress Management (pg. 8), Hydration (pg. 10)

Remedies: Black Cohosh, Lavender, Lobelia, Nerve Eight, Tei Fu Oil, APS II, Eleuthero, Herbal Sleep, Hops, Jasmine Absolute, Kava Kava, Nerve Control, Passion Flower, Stress Relief, Valerian Root, Wood Betony

Hearing Loss

See *Deafness*

Heart (weakness)

The following are general remedies to strengthen the heart.

Therapies: Antioxidants (pg. 17), Essential Fatty Acids (pg. 17)

Remedies: Co-Q10, Hawthorn Berries, HS II, L-Carnitine, Magnesium, Cardio Assurance, Eleuthero, Ginkgo & Hawthorn, Nervous Fatigue Formula, Passion Flower

Heart Attack

See *Cardiac Arrest (Heart Attack)*

Heart Disease

See *Cardiovascular Disease (Heart Disease)*

Heart Fibrillation or Palpitations

Rapid, irregular contractions of the heart muscle are called heart palpitations. Symptoms can include an irregular or rapid heart rate, skipped heartbeats, shortness of breath, and chest discomfort. They may be caused by anxiety, lack of exercise, high blood pressure, diabetes or other problems. Many of the best herbal remedies for this problem are toxic botanicals that are not readily available, such as lily of the valley, so professional assistance should be sought. However, one of the best natural solutions is to make sure the person has adequate amounts of minerals, especially magnesium, potassium and calcium. This can be a very difficult condition to treat naturally, but the following may help.

Therapies: Hiatal Hernia Correction (pg. 12), Stress Management (pg. 8), Affirmation and Visualization (pg. 8)

Remedies: Co-Q10, Hawthorn Berries, Magnesium, Passion Flower, Calcium, Capsicum, Chlorophyll, Eleuthero, Iron, L-Carnitine, Neroli, Potassium, Stress Relief, Valerian Root, Vitamin B-12, Vitamin B-Complex, Vitamin E

Heart Rate (irregular)

See *Arrhythmia*

Heart Rate (rapid)

See *Tachycardia*

Heart Valves

The following remedies may be helpful for problems with heart valves.

Remedies: Ginkgo & Hawthorn, Magnesium, Hawthorn Berries

Heartburn

See *Acid Indigestion (Heartburn, Acid Reflux), Acid Indigestion (Heartburn, Acid Reflux)*

Heavy Metal Poisoning

See Also *Lead Poisoning, Mercury Poisoning*

Rome may not have been built in a day, but it was destroyed by heavy metal poisoning in its water supply! The Roman aqueduct system and the plumbing in it's famous public baths and in the residences of Rome's ruling class were incredible for their time. However, the lead pipes in the civic water system caused neurological disorders that led to the decadent behavior that caused the Roman Empire to collapse.

Today, heavy metals and other toxic substances in our environment are bringing about a similar decline in the mental (and physical) health of society. Learning disabilities and behavioral problems are rampant, and a new category of diseases, autoimmune disorders, have been increasing at an alarming rate. These include rheumatoid arthritis, chronic fatigue, type I diabetes, fibromyalgia, lupus, Lou Gehrig's disease, myasthenia gravis and multiple sclerosis. Besides being a major factor in autoimmune diseases, heavy metals play a role in cancer, heart disease, dementia, Alzheimer's, reduced immunity, mental illness and parasitic infections.

Ideally, we should do all we can to avoid them, so here are some important tips for reducing your exposure to these health-destroying elements. Purify your water! Make sure your water pipes have no lead; avoid lead-based painted objects; don't store liquids in lead crystal containers. Be especially cautious of imports from China as they often have lead contamination.

Buy and prepare fresh, organic food as much as possible. Keep the chemicals in your life, especially cleaning chemicals, to a minimum. Avoid cooking with aluminum pans or using anything that is aluminum with your food, especially acidic foods like citrus. Read labels many processed foods, including iodized salt and baking powder, may contain aluminum as an anti-caking agent. Antiperspirant deodorants always contain aluminum; use a natural deodorant (or Nature's Fresh) instead.

Insist on composite fillings from your dentist, not mercury/silver amalgams. Avoid vaccinations as they contain heavy metals.

You can also periodically do a cleanse to pull heavy metals from the body. This is especially important for people who work around a lot of chemicals or are starting to develop signs of neurological problems. A good heavy metal cleanse would consist of Heavy Metal Detox, N-acetyl Cysteine, algin and some type of essential fatty acid supplement. Doing foot spa baths and drawing baths using clay can also be helpful. Oral chelation is helpful for getting rid of heavy metals, too. Here is a complete list of products that can help rid the body of heavy metal poisoning.

Therapies: Antioxidants (pg. 17), Heavy Metal Detoxification (pg. 22), Oral Chelation (pg. 21), Essential Fatty Acids (pg. 17),Fiber (pg. 17),Fiber (pg. 17)

Remedies: Algin, Heavy Metal Detox, Mega-Chel, N-acetyl Cysteine, Omega-3, Alpha Lipoic Acid, Bentonite (Hydrated), Bowel Detox, Detoxification, Flax Seed Oil, Milk Thistle Combination, Nature's Three, Psyllium Hulls Combination

Hemochromatosis

When a person has too much iron in the blood the condition is known as hemochromatosis. Since zinc and iron are antagonists, this problem is most often due to zinc deficiency. The following may be helpful for this condition. The form of B12 that is beneficial is the liquid B12.

Remedies: Vitamin B-12, Zinc, Chlorophyll, Folic Acid Plus, HSN-W, Liver Balance, Vitamin B-6

Hemorrhage

See *Bleeding (external), Bleeding (internal)*

Hemorrhoids

See Also *Varicose Veins*

A hemorrhoid is a mass of dilated veins in the rectum that cause painful bowel eliminations. It is important to keep the stool soft with fiber and a gentle bowel cleanser like Gentle Move. A good treatment is to mix Golden Salve and white oak bark and apply them topically. Hemorrhoids are in indication that blood vessels, in general, need toning. It is common for an individual with hemorrhoids to have other varicosities like varicose veins. When this is the case, Vari-Gone capsules or cream can be helpful. Here is a complete list of possible remedies.

Therapies: Enzymes (pg. 15),Fiber (pg. 17), Antioxidants (pg. 17), Hydration (pg. 10)

Remedies: Gentle Move, Golden Salve, Intestinal Soothe & Build, Vari-Gone, White Oak Bark, Aloe Vera, Butcher's Broom, Chickweed, Every Body's Fiber, Geranium, Golden Seal, Horsetail, LOCLO, Marshmallow, Menstrual-Reg, Milk Thistle, Nature's Three, Noni (Morinda), Psyllium, Psyllium Hulls Combination, Slippery Elm, Spleen Activator, St. John's Wort, Valerian Root, Vitamin E

Hepatitis

Hepatitis is inflammation of the liver. Hepatitis A is an infectious hepatitis that can be transmitted through poor sanitary conditions such as food handlers or child care workers not washing their hands. Hepatitis B or serum hepatitis is passed through a blood transfusion or other contact with blood. Hepatitis can also be caused by chemicals, poor diet and other lifestyle factors that damage the liver. Seek appropriate medical assistance. SAM-e and Milk Thistle Combination taken internally with helichrysum essential oil applied topically over the liver area is generally an effective program. Eat mild foods and drink plenty of water when recovering from hepatitis. Any of the following remedies may be helpful.

Therapies: Antioxidants (pg. 17)

Remedies: Helichrysum, Liver Balance, Milk Thistle Combination, SAM-e, Aloe Vera, Astragalus, Blood Build, Dandelion, Germanium Combination, IF Relief, Liver Cleanse Formula, Lobelia, Milk Thistle, Oregon Grape, Red Beet Formula, Spirulina, THIM-J, Yellow Dock

Hernias

Protrusion of an organ through connective tissue or the wall of a cavity by which it is normally enclosed is called a hernia. Medical attention should be sought. The following can help hernias to heal.

Remedies: PLS II, Red Raspberry, Bone/Skin Poultice, Herbal CA, Spleen Activator

Herniated Disks

See *Disks (bulging or slipped spinal disks)*

Herpes

See Also *Chicken Pox, Cold Sores (Fever Blisters)*

Herpes is a name for any of several inflammatory viral diseases characterized by blister like sores. Herpes simplex can cause cold sores or fever blisters around the mouth or genital herpes. Herpes zoster is responsible for chicken pox and shingles. The following may be helpful for these disorders.

Therapies: Probiotics (pg. 16), Stress Management (pg. 8)

Remedies: Black Walnut, Paw Paw, Silver Shield, VS-C, Aloe Vera, Bergamot, Burdock, Grapefruit (Pink), L-Lysine, Lemon Oil, Olive Leaf, Pau D Arco, Rose Bulgaria, Uña de Gato Combination

Hiatal Hernia

See Also *Ileocecal Valve*

In a hiatal hernia the stomach pushes upward through the opening in the diaphragm for the esophagus. This can cause acid reflux, put stress and pressure on the heart, create poor digestive function and generally weaken the body. See the instructions for correcting this in the Introduction to this book under *General Building Remedies*. Mechanical correction is needed, but the following remedies can help.

Therapies: Stress Management (pg. 8), Enzymes (pg. 15), Hiatal Hernia Correction (pg. 12)

Remedies: Anti-Gas Formula, Blessed Thistle, Dandelion, Intestinal Soothe & Build, Lobelia, PLS II, Slippery Elm, Spleen Activator

Hiccups

Hiccups are involuntary, spasmodic contractions of the diaphragm causing a quick inhalation of air that makes a strange sound. The following may be helpful for hiccups.

Remedies: Lobelia, Blessed Thistle, Clove Bud, Colloidal Minerals, Mandarin (Red)

High Blood Pressure

See *Blood Pressure (high)*

High Cholesterol

See *Cholesterol (high)*

HIV (AIDS)

See *Acquired Immune Deficiency Syndrome (AIDS)/HIV*

Hives

See *Rashes and Hives*

Hoarseness

See *Laryngitis (Hoarseness)*

Hodgkin's Disease

This is a disease that causes enlargement of the lymphatic tissue, spleen and liver. The following may be helpful.

Therapies: Enzymes (pg. 15), SuperFood (pg. 15), Colon Cleansing (pg. 18)

Remedies: Lymphomax, Lymphatic Drainage, Mullein, Red Clover

Hormone Replacement

See *Estrogen (low)*

Hot Flashes

See Also *Menopause*

Hot flashes are a complex symptom associated with menopause. There is an initial feeling of discomfort followed by a sensation of heat moving towards the head. The face becomes red, which is followed by sweating and fatigue. Hot flashes usually involve a congested liver and weak adrenal glands.

Essential oils can stimulate the hypothalamus and help to regulate hormonal levels to control hot flashes. One remedy for hot flashes is to mix 10 drops of clary sage oil, 10 drops of germanium oil and 5 drops of lemon oil into 5 teaspoons of flax seed oil and massage this blend on the abdomen.

A hydrosol spray can also be made for cooling hot flashes by putting some rose and peppermint essential oils in a glass spray bottle with a little purified or distilled water. Shake well and mist the area around the face (with the eyes closed) to help cool hot flashes.

Any of the following remedies may help with hot flashes.

Therapies: Antioxidants (pg. 17), Essential Fatty Acids (pg. 17), Stress Management (pg. 8), Low Glycemic Diet (pg. 11), Avoid Xenoestrogens (pg. 23)

Remedies: Adrenal Support, Clary Sage, Flash-Ease T/R, Grapefruit (Pink), Pantothenic Acid, Pro-G-Yam Cream, C-X, Damiana, FCS II, Female Comfort, Geranium, HY-C, Lemon Oil, Licorice Root, Master Gland, Menopause, Natural Changes, Nervous Fatigue Formula, NF-X, Peppermint, PMS, Rose Bulgaria, Rose Bulgaria, Sage, X-A

Hydrophobia

See *Rabies*

Hyperactivity

See *Attention Deficit Disorder (ADD)*

Hyperinsulinemia (Syndrome X)

See Also *Diabetes (Type II), Weight Loss (aids for)*

Research has brought to light a previously hidden cause of many modern illnesses. Dubbed metabolic syndrome X, this condition involves cellular resistance to a hormone called insulin, which causes insulin levels in the blood to rise, a condition known as hyperinsulinemia.

We ve all heard that a deficiency of insulin produces a condition known as diabetes, but that's only true in type I diabetes (10-15% of cases). In most cases of diabetes (type II) there is an excess of this pancreatic hormone. In type II diabetes, insulin is being produced in excess, but isn't working due to cellular resistance. If this sounds like type II diabetes and syndrome X have the same cause, you got it. Type II diabetes is one of the serious health problems syndrome X can cause, but it isn't the only one.

Too much insulin is linked to high blood pressure and arteriosclerosis. In fact, excess insulin is a bigger risk factor for cardiovascular disease than excess cholesterol.

Hyperinsulinemia is also a major cause of obesity, because insulin causes our body to store more fat. It disrupts sodium metabolism, increasing water retention. By depressing neurotransmitters in the brain syndrome X contributes to depression. In women, 75% of all cases of pilocystic ovarian syndrome are also related to too much insulin.

In the initial stages, producing too much insulin causes a rapid lowering of blood sugar levels, which causes hypoglycemia or low blood sugar. This increases the craving for sweets and stresses other hormone systems. It interferes with the conversion of thyroid hormone T-4 to T-3 which can result in functional hypothyroidism. Another negative effect is a rise in cortisol production from the adrenals. This reduces our ability to cope with stress, lowers our immune response and eventually exhausts our adrenals. Excess cortisol also contributes to aging.

If you want to know if you have Syndrome X, you could have lab tests run to check your insulin levels. (Fasting levels of insulin should be below 10 units.) However, there is an easier way measure your waist and hips.

Abdominal obesity is a major indicator of excess insulin production. So, grab a tape measure and check your circumference at the navel and at the widest part of your hips. In men, if your waist measurement is larger than your hips, you ve got Syndrome X. In women, the waist should be less than 80% of the hip measurement.

Another indicator of Syndrome X is your triglyceride and HDL levels in routine blood tests. If your triglyceride level is greater than 200 or your HDL level is less than 35, you re having problems with excess insulin production and insulin resistance.

Here are some contributing factors to the development of syndrome X:

Low fat diets or diets high in saturated fat.

Too many omega-6 essential fatty acids in the diet, with insufficient omega-3.

Sedentary life-styles and lack of exercise.

Deficiencies of dietary chromium and magnesium. Deficiencies of zinc, manganese, vanadium, B-vitamins and vitamin A may also be involved.

Eating foods that trigger excess insulin production especially refined carbohydrates.

High carbohydrate, low protein diets.

The secret to correcting metabolic syndrome X lies in some basic life-style changes. The first is resistance exercise. Resistance exercise trains muscles to take up glucose without the need for insulin, thereby decreasing insulin requirements. After just five days of no exercise insulin resistance increases. When we do exercise like weight lifting sufficient to make our muscles burn a little, we are causing our muscle tissue to take up sugar without insulin. Thus, a program of muscle building exercise (at least three times per week) will help Syndrome X and reduce our risk of heart disease, diabetes and obesity.

A second secret is changing the kinds of fats we consume. Transfatty acids, found in margarine and vegetable oils, and saturated fats increase cellular resistance to insulin. Most vegetable oils are high in omega-6 fatty acids, but deficient in omega-3 fatty acids, which decrease insulin resistance. Switch from vegetable oils and hydrogenated fats (fries, chips, pastries, bagels, etc.) to high quality fats like olive oil, butter and flax seed oil.

A third secret is to avoid refined carbohydrates. Get on a low glycemic diet, cutting out grain, sugars and high starch foods from your diet. Eat lots of non-starchy fruits and vegetables instead.

An appropriate program of dietary supplements will also help to overcome hyperinsulinemia (but not without appropriate dietary changes). Here are some suggestions.

Therapies: Essential Fatty Acids (pg. 17),Fiber (pg. 17), Low Glycemic Diet (pg. 11)

Remedies: **HY-C**, **Pro-Pancreas**, Chromium GTF, CLA, Licorice Root, LIV-J, Omega-3, Rosemary, Stevia, SugarReg, Super Algae, Xylitol

Hypertension

See *Blood Pressure (high)*

Hyperthyroid

See *Grave's Disease*

Hypochondria

A hypochondriac is a person with abnormal or excessive interest in diseases, who fears they have conditions that they do not have. This is often due to liver problems. The following may be helpful for people who are always feeling sick and may feel like they are becoming a hypochondriac. These formulas may also be helpful for people who are healthy, but live in constant fear of disease.

Therapies: Affirmation and Visualization (pg. 8), SuperFood (pg. 15)

Remedies: **Adrenal Support**, **Blood Build**, **Liver Balance**, Jasmine Absolute, Kidney Activator, Marjoram (Sweet), Oregano (Wild), Peppermint, Rosemary, Tea Tree Oil, Thyme

Hypoglycemia

See Also *Hyperinsulinemia (Syndrome X)*

Hypoglycemia is low blood sugar. This results in dizziness, weakness, inability to concentrate, irritability, mood swings, fatigue and more. There tends to be a constant craving for sugar and simple carbohydrates. Eating small meals with some protein in them throughout the day is very beneficial. The following supplements can also be helpful.

Therapies: Low Glycemic Diet (pg. 11), Essential Fatty Acids (pg. 17), SuperFood (pg. 15), Fiber (pg. 17)

Remedies: Chromium GTF, HY-A, Licorice Root, Love and Peas, Stevia, Super Algae, Xylitol, Alpha Lipoic Acid, Bee Pollen, Blood Build, Burdock, Eleuthero, EveryBody's Formula, Food Enzymes, Ho Shou Wu, Juniper Berries, L-Glutamine, Liver Balance, Nutri-Burn, Safflowers, Spirulina, SynerProTein, Target P-14

Hypotension

See *Blood Pressure (low)*

Hypothyroid

See Also *Grave's Disease, Hashimoto's Disease (Thyroiditis)*

Sitting at the base of the neck is a butterfly shaped gland known as the thyroid. This important endocrine gland helps regulate metabolism, the rate at which the body burns fuel. It can be likened to the gas pedal on your car. When the thyroid is hyperactive, the body's engine races, burning hot and fast. When the thyroid activity is low, the body engine sputters, runs slowly and stalls.

A malfunctioning thyroid gland can be the cause of many health problems. For hypothyroid, the most important symptoms are: feeling cold and fatigue. If you are tired and get cold easily, even when others feel hot, it is very likely you have low thyroid function. Other important symptoms of low thyroid are excess weight and difficulty losing weight, dry skin and thinning hair (hair loss). An easy way to check for thyroid problems is to take your temperature every morning before you get out of bed for 3-7 days. If your body temperature is consistently below normal (98 degrees), you probably have a low thyroid.

Thyroid problems are extremely common. About 5 million people suffer from low thyroid (hypothyroidism) in the United States. Worldwide, it has been estimated that as many as one and a half billion people are at risk for thyroid disorders. Thyroid problems are much more common in women than they are in men. About 90% of the people with thyroid disorders are women.

So, why are thyroid disorders so prevalent? While the exact reasons aren't clear, there are a number of causal factors to consider.

First, iodine is essential to the production of thyroid hormones. This nutrient, while found in abundance in sea foods, is not found in high concentrations in plants or animals raised inland. Furthermore, fluoride, chlorine and bromide are all found in the same group as iodine on the periodic table of elements. This means they can displace iodine in the body; so, the chlorination of water supplies and the use of fluorides may be a contributing factor. Drugs, corticosteroids, aspirin (salicylates) and anticoagulants can depress thyroid activity.

To understand how to deal effectively with low thyroid using natural substances, it is necessary to know a little bit about how the body produces thyroid hormones. The hypothalamus, a stalk of the brain, is the master regulator of most of the body's major endocrine hormones. When the hypothalamus detects the need for thyroid hormones, it produces the thyroid releasing hormone (TRH). The TRH travels to the pituitary gland where it'stimulates the release of the thyroid stimulating hormone, TSH or thyrotrophin.

TSH travels through the blood stream and binds to receptor sites in the thyroid gland. It'stimulates the thyroid to produce two hormones thyroxin (T4) and tri-iodotyrosine (T3). Target TS II stimulates the pituitary to produce more TSH.

In response to TSH, T4 and T3 are released in a ratio of about a 4:1 (4 times more T4 than T3). T3 is the more active form. T4 is a storage form of the hormone. T4 is converted to T3 in peripheral tissues, particularly the liver. Cortisol, a stress hormone, tends to stimulate the conversion of T4 to T3, while insulin tends to suppress the production of T4 to T3.

The thyroid can also produce relatively inactive reverse T3 (RT3). During times of grief, trauma and illness, the body produces more RT3 and less T3, presumably to conserve energy and force us to slow down.

The primary job of these thyroid hormones is to regulate metabolism and to help burn fuel, especially fats. The thyroid acts sort of like a metabolic thermostat. When the thyroid output is low, the fats tend to be stored instead of burned, resulting in weight gain. Since the body burns fat primarily to keep warm, the body temperature tends to be low. The skin is usually dry, again due to a lack of proper fat metabolism, because fats are what keep the skin moist and supple. Reproductive hormones may also be thrown out of balance (since they are made of fat) and energy levels tend to be low because the metabolism is slow.

When low thyroid is a problem, the first thing to try is increasing one's intake of dietary iodine. Adding foods rich in natural iodine to the diet will often improve thyroid function. While the primary use of iodine is in the thyroid gland, it may have other functions. For example, iodine is also concentrated around the nipples in female breast tissue and is critical to breast health. Iodine is also important for the immune system and helps the body fight infection.

Iodine is a very rare nutrient in land plants but is common in fish and sea vegetables like kelp, dulse, bladderwrack, and Irish moss. Sea vegetables, like kelp, can be sprinkled on food or added to soups, stews, etc. They add a pleasant salty taste to foods. Two formulas are available which contain these sea vegetables and are designed to feed the thyroid gland and aid its function. They are TS II with Hops and Thyroid Activator. These formulas can be very helpful in cases of moderately low thyroid. Liquid Dulse is another great source of natural iodine.

Another powerful supplement that can help solve low thyroid problems is Thyroid Support. This formula contains thyroid glandular substance as well as pituitary and hypothalamus substance. Thyroid Support helps rebuild the thyroid gland, not just improve its function. If one is already on thyroid medication (and still has a thyroid gland that is at least partially functional), they may wish to try rebuilding their thyroid and reducing or weaning off of their thyroid medication.

Eating coconut oil stimulates the thyroid. Natural Sea Salt is a good way to get extra iodine in the diet to stimulate the thyroid. Cruciferous vegetables and soy have a thyroid inhibiting effect.

Even if levels of thyroid hormones are low, one can still have thyroid problems if the liver and other tissues are not converting T4 into T3 properly. Weak adrenals may contribute to this problem, so Adrenal Support or licorice root may have indirect benefits to the thyroid by supporting the adrenal glands. 7-Keto increases T4 to T3 conversion and is sometimes used to stimulate the burning of fat for weight reduction.

Eating a properly balanced diet (especially reducing simple carbohydrates) will also aid this conversion. Since most of the T4 to T3 conversion takes place in the liver, the liver is often involved in thyroid problems. Some liver supplements that can indirectly help the thyroid by aiding the liver include SF, the Tiao He Cleanse, and SAM-e taken with MSM.

All of the following remedies may be helpful for the thyroid.

Therapies: SuperFood (pg. 15), Low Glycemic Diet (pg. 11), Essential Fatty Acids (pg. 17)

Remedies: Black Walnut, Dulse, Ho Shou Wu, Kelp, Target TS-II, Thyroid Support, 7-Keto, Adrenal Support, Licorice Root, Mood Elevator, MSM, SAM-e, Saw Palmetto, Sea Salt, SF, SF, Thyroid Activator, Tiao He Cleanse, TS II

Hysteria

Hysteria is a neurotic condition where there is no recognizable organic disease, but there can be symptoms mimicking various diseases. The person is calm, but aloof, but may become very emotional (laughing or crying) for no apparent reason. This emotional state is almost like a second personality and there may be a forgetting of what happened in this other state when the normal personality reasserts itself. There is emotional instability with a marked craving for sympathy.

The following may be helpful in working with hysteria.

Therapies: Affirmation and Visualization (pg. 8), Stress Management (pg. 8)

Remedies: Black Cohosh, Chamomile, Lavender, Nerve Eight, Nervous Fatigue Formula, Aloe Vera, APS II, Bergamot, Blue Cohosh, Chamomile (Roman), Neroli, Passion Flower, Peppermint, Rosemary, Thyme

IBS

See *Inflammatory Bowel Disorders (Colitis, IBS)*

Ileocecal Valve

See Also *Hiatal Hernia*

The ileocecal valve is the valve between the small intestine and the large intestine. This valve may become irritated and inflamed and not shut properly. This causes a leakage of material from the colon back into the small intestine which weakens the body. It often causes serious gas and bloating.

Massage to the area helps. To locate the ileocecal valve draw an imaginary line from your navel (belly button) to the right hip. The ileocecal valve is located about halfway along that line. Massage the area in a clockwise motion to close the value. The following can also help.

Therapies: Enzymes (pg. 15),Fiber (pg. 17)

Remedies: Intestinal Soothe & Build, Aloe Vera, Proactazyme Plus, Slippery Elm, Spleen Activator

Impetigo

Impetigo is an inflammatory skin disease caused by a contagious staph or strep infection. It is characterized by isolated pustules, usually around the nose and mouth. These pustules become crusted and rupture. The following may be helpful. Most of these remedies would be applied topically.

Remedies: Black Walnut, Echinacea/Golden Seal, Silver Shield, BP-X, IN-X, Peppermint, Tea Tree Oil

Impotency

See *Erectile Dysfunction*

Incontinence (urinary)

Incontinence is the inability to retain urine through the loss of sphincter control in the bladder. Uva ursi tea is a great remedy for this problem. The following may be also be helpful.

Therapies: Hydration (pg. 10), Minerals (pg. 15)

Remedies: KB-C, Uva Ursi, Co-Q10, Juniper Berries, Kidney Activator, Noni (Morinda), St. John's Wort, Urinary Maintenance

Indigestion

See Also *Acid Indigestion (Heartburn, Acid Reflux), Acid Indigestion (Heartburn, Acid Reflux), Gastritis*

The following remedies may be helpful in easing ordinary indigestion.

Therapies: Enzymes (pg. 15), Hydration (pg. 10)

Remedies: Anti-Gas (Chinese), Anti-Gas Formula, Catnip & Fennel, Chamomile, Gall Bladder Formula, Papaya Mint, Peppermint, Proactazyme Plus, Alfalfa, ALJ, Aloe Vera, Blessed Thistle, Blue Vervain, Capsicum, Clove Bud, Devil's Claw, Fenugreek & Thyme, Food Enzymes, FV, Ginger, Kidney Activator (Chinese), Lavender, LIV-J, PDA, Rose Bulgaria, Rosemary, Safflowers, Sage, Slippery Elm, Small Intestine Detox, Spleen Activator, Stress Relief, Thyme, Wood Betony

Infection (bacterial)

See Also *Antibiotics (alternatives to)*

The following remedies may be helpful in combating bacterial infections.

Therapies: Enzymes (pg. 15), Probiotics (pg. 16)

Remedies: Echinacea/Golden Seal, Garlic, Golden Seal, Immune Stimulator, IN-X, Silver Shield, Tea Tree Oil, Ultimate Echinacea, Black Ointment, Echinacea Purpurea, Enviro-Detox, Eucalyptus, Fizz Active-Immune, Guardian, Lymph Gland Cleanse, Lymph Gland Cleanse-HY, Myrrh, Oregano (Wild), Oregon Grape, Parthenium, Rosemary, Tei Fu Oil, Thyme, VS-C

Infection (fungal)

See *Fungal Infections (Yeast Infections, Candida albicans)*

Infection (viral)

The following may be helpful in combating viral infections.

Remedies: Echinacea Purpurea, Elderberry Defense, Elderberry Plus, Immune Stimulator, Silver Shield, Ultimate Echinacea, VS-C, Aloe Vera, Astragalus, Capsicum, Capsicum & Garlic with Parsley, CC-A, Fizz Active-Immune, Garlic, Germanium Combination, Glyco Essentials, Guardian, Lemon Oil, Olive Leaf, Oregano (Wild), Paw Paw, Prevention, St. John's Wort, Thyme, Viral Recovery, Vitamin C, Yarrow

Infertility

The inability to conceive a baby may be helped by some of the following remedies. Lack of certain nutrients, particularly minerals, and low cholesterol are often root causes. Improving overall health is important, too.

Therapies: Hiatal Hernia Correction (pg. 12), Low Glycemic Diet (pg. 11), Avoid Xenoestrogens (pg. 23), Essential Fatty Acids (pg. 17), Stress Management (pg. 8)

Remedies: Colloidal Minerals, FCS II, GreenZone, Master Gland, Mineral-Chi Tonic, C-X, Damiana, Dong Quai, Eleuthero, False Unicorn, Female Comfort, Geranium, Ginseng (Korean), Ginseng (Wild American), Ho Shou Wu, KB-C, NF-X, Super Supplemental, Vitamin E, X-A, X-Action (Men s), X-Action (Women s)

Inflammation

See Also *Free Radical Damage*

Inflammation is the body's normal response to any kind of tissue damage. When you cut, bruise, burn, bump, scrape or break some part of your body, inflammation sets in. Inflammation also occurs from chemical and microbial damage (toxins and infection). So, no matter how the body gets injured, inflammation is going to be the body's primary response to the damage.

Itis is the Latin term for inflammation, which is characterized by heat, swelling, redness and pain at the site of injury. Many traditional names for diseases are simply naming the location of the heat, swelling, redness and pain. Thus, appendicitis is inflammation of the appendix; bronchitis is inflammation of the bronchials; tonsillitis, inflammation of the tonsils, and so forth. When you consider all the itises there are arthritis, tendonitis, bursitis, colitis, dermatitis, gingivitis, conjunctivitis, diverticulitis, sinusitis, etc. it's already clear that inflammation is involved in a lot of health problems.

Normally, inflammation resolves itself naturally, and the injuries heal. However, when healing isn't completed, tissues become chronically inflamed. A slow process of deterioration ensues, resulting in the development of chronic and degenerative diseases, including cardiovascular disease, arthritis, obesity and mental deterioration, to name just a few. The fact is that just about any chronic disease probably involves inflammation.

We ve already established that inflammation starts with tissue damage. To understand why inflamed tissues don't heal, we need to understand the normal process of inflammation, which works like this:

When the tissues are initially damaged, there is a release of histamine, which is followed by a release of bradykinin, serotonin and other chemical mediators. These dilate capillary pores and initiate inflammation by allowing fluid and protein to enter the tissue spaces (creating swelling). This is the first phase of inflammation.

In the second phase, chemical messengers are released to further open blood vessels so white blood cells can reach the damaged area. This causes further swelling. At this stage, if the inflammation is in the respiratory tract, histamine and leukotrienes will cause bronchial constriction and increased mucus production to flush toxins from mucus membranes. Pain receptors are also activated at this stage.

During the third phase, white blood cells use free radicals to destroy microbes and cellular debris. Healthy cells need adequate levels of antioxidants in order to protect themselves from these free radicals. If antioxidant levels are too low, healthy tissues will get damaged causing inflammation to spread.

Once white blood cells have completed their cleanup of the area, a healing phase is initiated. Cortisol from the adrenals is secreted to shut down production of the chemical messengers that mediate the inflammatory process. Macrophages clean up the remaining debris and a regenerative cycle begins as chemical messengers are released which stimulate tissue repair.

In chronic inflammation, the body is never able to complete the healing phase of the inflammatory process. It gets stuck in the earlier phases. Meanwhile, the free radical activity in phase three causes more and more cells to get damaged, causing the inflammatory fires to spread. It's like a forest fire, which starts when dry, dead plant material catches fire, but can get hot enough that even the green trees get burned up in the process.

So, here are the factors that cause inflammation to become chronic. First, inflammation can't heal if there is a lack of nutrients needed for the healing and repair phase. Second, chronic tissue irritation from environmental toxins and poor diet continually re-irritates tissues, preventing healing. Third, adrenal fatigue from chronic stress (which shuts down cortisol production), prevents initiation of the healing phase. Finally, lack of adequate lymphatic drainage prevents the removal of excess fluid from the tissues.

Now that we understand what the inflammatory process is, we can understand what we can do to put out the fires of chronic inflammation. Here are seven keys to locking up the inflammation arsonist in your body.

Key #1 is to detoxify. Your body can't heal if it is constantly being re-inflamed by environmental toxins and microbial parasites. So, start by avoiding chemicals as much as possible (food additives, pesticide residues, cleaning solutions, etc.). Buy organic food wherever possible and use natural household cleaning products and personal care products. A good cleanse will help eliminate toxins already in the body.

Key #2 is to eat the right kinds of fats. The chemical messengers that mediate the inflammatory process are made from omega-6 and omega-3 essential fatty acids (EFA). A ratio of 4 parts omega-6 EFA to 1 part omega-3 EFA is important in order to keep inflammation in check. That's because many of the chemical messengers that promote the healing process are made from omega-3 EFA. If there are too many omega-6 EFAs and not enough omega-3 EFAs, then the body will be unable to heal properly and chronic inflammation will ensue.

Key #3 is to avoid simple carbohydrates. Refined sugars and grains cause spikes in insulin production. High levels of insulin inhibit the conversion of essential fatty acids to anti-inflammatory chemical messengers. The result is chronic inflammation.

Key #4 is to improve lymphatic drainage. One of the major effects of inflammation is the pooling of lymphatic fluid in the spaces around the cells. The only way this fluid can be removed is via the lymphatic system. This is one of the little known secrets to reducing chronic inflammation. The lymph system has no pump, so moderate exercise (walking, swimming, bouncing up and down on a mini-trampoline, etc.) and deep breathing are needed to encourage lymphatic drainage and reduce inflammation. When lymph glands are congested, Lymphatic Drainage formula or Lymphomax will also improve lymph drainage.

Key #5: is to use antioxidant nutrients. Inflammation and oxidative stress go hand in hand. An adequate level of antioxidant nutrients will help reduce both oxidative stress (free radical damage) and control inflammation.

Key #6 is to support the adrenal glands. The adrenals produce the hormone cortisol, which keeps inflammation in check. Corticosteroid drugs mimic this hormone. Chronic stress, caffeine and sugar use exhaust these important glands and reduce their ability to control inflammation. Nervous Fatigue Formula or Adrenal Support can help rebuild the adrenal glands and keep chronic inflammation in check. Also, Yucca and Licorice Root are two herbs which have a cortisol-like action.

Key #7 is to use natural anti-inflammatory remedies. Nature has supplied us with many natural remedies that reduce chronic inflammation and promote tissue healing. IF Relief, Nerve Control and IF-C are all excellent formulas for reducing inflammation. IF-C works best for more acute inflammation. Nerve Control is helpful for inflammation and pain.

Using these seven keys, we can keep inflammation from damaging our health. Here are some major anti-inflammatory herbs and supplements to help with these keys.

Therapies: Antioxidants (pg. 17), Enzymes (pg. 15), Essential Fatty Acids (pg. 17), Low Glycemic Diet (pg. 11)

Remedies: Co-Q10, Devil's Claw, Healing AC Cream, IF Relief, IF-C, Nature's Fresh, Nerve Eight, Omega-3, Super ORAC, Thai-Go, Yucca, Aloe Vera, APS II, Berry Healthy, Blue Cohosh, Bone/Skin Poultice, Butcher's Broom, Chamomile, Chamomile (Roman), Chickweed, CLA, CLT-X, Deep Relief Oil, Distress Remedy, Elderberry Defense, Feverfew, Frankincense, Gastro Health, Glucosamine, Golden Salve, Golden Seal, Gotu Kola, Grapine, Green Tea Extract, Guardian, Helichrysum, Herbal Trim Skin Treatment, HY-C, Inflammation, Joint Health, Kudzu/St.John's Wort, Licorice Root, Lymph Gland Cleanse, Lymph Gland Cleanse-HY, Marshmallow, MSM, MSM/Glucosamine Cream, Myrrh, N-acetyl Cysteine, Nature's Phenyltol with NEM, Noni (Morinda), Pau D Arco, PLS II, Prevention, Rose Hips, Skin Detox, St. John's Wort, Sunshine Heroes Whole Foods Antioxidant, Super Trio, Yarrow, Zinc

Inflammatory Bowel Disorders (Colitis, IBS)

See Also *Celiac Disease, Colitis, Leaky Gut Syndrome*

Inflammatory bowel disorders, including colitis, Crohn's disease, celiac disease, and ulcerative colitis involve a breakdown of the intestinal mucous due to inflammatory processes. This can be caused by antibiotics, food allergies, drugs, chemicals or microorganisms, including parasites. The process involved in healing these problems involves: 1) removing the source of irritation, 2) using digestive enzymes to enhance food breakdown, 3) rebuilding intestinal flora with probiotics, and 4) repairing and toning damaged intestinal tissues. There is often a stress component that must also be identified and addressed.

One of the best formulas for inflammatory bowel disorders is Intestinal Soothe and Build. For best results it'should be taken with Stress-J. Aloe vera juice and slippery elm are also excellent remedies for reducing intestinal inflammation. The following is a complete remedies may be used in the four step process of healing inflammatory bowel disorders.

Therapies: Antioxidants (pg. 17), Blood Type Diet (pg. 13), Hiatal Hernia Correction (pg. 12), Affirmation and Visualization (pg. 8), Stress Management (pg. 8), Enzymes (pg. 15), Probiotics (pg. 16), Essential Fatty Acids (pg. 17),Fiber (pg. 17)

Remedies: Aloe Vera, CLT-X, Intestinal Soothe & Build, Licorice Root, Probiotics, Slippery Elm, Stress-J, Alfalfa, Artemisia Combination, Bilberry Fruit, Black Walnut, Bowel Detox, Catnip, Chamomile, Every Body's Fiber, Fat Grabbers, Food Enzymes, Gentle Move, Herbal Pumpkin, Horsetail, IF Relief, IF-C, Kudzu/St.John's Wort, Liver Balance, LOCLO, Magnesium, Marshmallow, Nature's Gold, Nature's Three, Nutri-Calm, Para-Cleanse, Proactazyme Plus, Protease, Psyllium, Psyllium Hulls Combination, Small Intestine Detox, Sunshine Heroes Omega 3 with DHA, Sunshine Heroes Probiotic Power, Sunshine Heroes Whole Foods Antioxidant, Super Trio, Thyme, Vitamin B-Complex, Wild Yam, Yeast/Fungal Detox, Yellow Dock

Injuries

The following can help injuries heal more quickly.

Therapies: Minerals (pg. 15), Affirmation and Visualization (pg. 8)

Remedies: Bone/Skin Poultice, Distress Remedy, Healing AC Cream, Helichrysum, IF Relief, Nature's Fresh, PLS II, Tei Fu Oil, Bayberry, Black Ointment, Collatrim, Deep Relief Oil, EverFlex, Golden Salve, Golden Seal, HSN-W, IF-C, Marshmallow, MSM, MSM/Glucosamine Cream, Mullein, Nerve Eight, Sprains and Pulls, Tea Tree Oil, Thai-Go, White Oak Bark, Yarrow

Insect Bites

See *Bites and Stings*

Insects

The following can help to kill or repel insects.

Remedies: Eucalyptus, Paw Paw, Rosemary, Tea Tree Oil, Black Cohosh, Garlic, Geranium, Lemon Oil, Patchouli, Peppermint, Pine Needle

Insomnia

See Also *Adrenals (Exhaustion, Weakness or Burnout), Hypoglycemia, Stress, Tension*

Insomnia, the inability to fall asleep or obtain a good night's rest, can be caused by anxiety and stress, depression, adrenal exhaustion or liver congestion. Sometimes, just taking something that helps relax the nervous system shortly before bedtime is enough. Herbal Sleep formula may be helpful here. Take 3-6 capsules about one hour before bedtime to help you relax. If you have a lot of muscle tension, add 1-2 capsules of Kava Kava.

Melatonin, a hormone produced by the pineal gland to induce sleep, can be helpful for resetting your biological clock when crossing time zones due to travel or when situations have disrupted your normal sleep pattern.

If sleep problems are accompanied by depression and carbohydrate cravings, you may be low in serotonin. Serotonin is the building block for melatonin in the brain. 5-HTP is a precursor to serotonin and helps increase serotonin levels in the brain. Try taking 5-HTP Power about one hour before bedtime. Since the pineal gland converts serotonin to melatonin when it's dark, simply darkening the room helps improve sleep patterns. Light pollution (excess light at night) is a common cause of sleep disturbances. (Watching TV or playing on the computer doesn't help you relax and get to sleep because you are staring at a source of light, which inhibits melatonin production. Try reading with a reading lamp or listening to music in a darkened room instead.)

If you have trouble staying asleep or have disturbed sleep with restless dreams, this can be a sign of adrenal fatigue or blood sugar problems. Nervous Fatigue Formula or Adrenal Support can help.

If you wake up suddenly in the middle of the night, wide awake and unable to go back to sleep, then your adrenals are kicking in gear with a shot of adrenaline to boost very low blood sugar at night. Get off of refined sugar and simple carbohydrates and try eating a snack containing some fat and protein about one hour before bedtime. Good choices would be a few spoonfuls of cottage cheese or nut butter, nuts or a spoonful of coconut oil.

If you re a coffee and/or sugar addict, then it's time to start breaking these addictions. Replace coffee with Herbal Beverage or other natural beverages and replace sugar with complex carbohydrates such as whole grains. Xylitol is a natural alternative to refined sugar that can also help reduce sugar cravings and balance blood sugar levels.

If you have a difficult time getting to sleep, then wake up groggy in the morning, you have a toxic liver. Liver formulas like Liver Balance can be helpful in this case.

Therapies: Hydration (pg. 10), Low Glycemic Diet (pg. 11), Stress Management (pg. 8)

Remedies: **5-HTP, Herbal Sleep, Kava Kava, Liver Balance, Magnesium, Melatonin Extra, Nervous Fatigue Formula,** AdaptaMax, Adrenal Support, Blue Vervain, Caffeine Detox, Calcium, Calming, Catnip, Catnip & Fennel, Chamomile, Eleuthero, Enviro-Detox, Fatigue/Exhaustion, Frankincense, Ho Shou Wu, Hops, Jasmine Absolute, Lobelia, Mandarin (Red), Marjoram (Sweet), Mood Elevator, Neroli, Nerve Control, Nutri-Calm, Passion Flower, Rose Bulgaria, Sea Salt, Sea Salt, St. John's Wort, St. John's Wort with Passion Flower, Stress Relief, Sunshine Heroes Calcium Plus D3, Valerian Root, Vitamin B-Complex

Irregular Heart Rate

See *Arrhythmia*

Irritability

See Also *Anger (excessive)*

Irritability is often a sign of liver congestion, blood sugar problems and/or hormonal imbalances. Depending on the cause, the following may help with irritability.

Therapies: Hydration (pg. 10), Affirmation and Visualization (pg. 8), Colon Cleansing (pg. 18)

Remedies: **Caffeine Detox, Chamomile, Chamomile (Roman), Liver Balance, Magnesium,** Aloe Vera, Blue Vervain, Calming, Catnip & Fennel, Clary Sage, Enviro-Detox, Frankincense, Ginkgo Biloba, Grapefruit (Pink), Guardian, Iron, Kudzu/St.John's Wort, Lavender, Marjoram (Sweet), Nervousness, SAM-e, St. John's Wort, Stress Relief, Sunshine Heroes Whole Foods Antioxidant, Vitamin C

Irritable Bowel Syndrome (IBS)

See *Inflammatory Bowel Disorders (Colitis, IBS)*

Itching

See Also *Chicken Pox (internal), Poison Ivy or Oak, Rashes and Hives*

Itching is an irritating sensation on the surface of the skin that compels one to scratch the area affected. It is common in allergic reactions and is a sign of irritants affecting the skin.

To ease itching and prevent scarring when a person is itching due to rashes, exposure to poison ivy or oak, chicken pox or other afflictions of the skin, the afflicted person can put any of the following into a bath or soak or apply them topically using fomentations or compresses: burdock, comfrey, golden seal, yellow dock, chickweed, hydrated bentonite, and Oregon grape. You can also use oatmeal in a bath. Just put a handful of uncooked oatmeal into a hot bath.

Mixing tea tree oil with vitamin E, one can also make a topical application for pox and other irritations. Aloe vera gel and black walnut tincture can also be applied topically to soothe itching. Silver Shield applied topically will help prevent open sores from becoming infected and may decrease the itching and speed healing time.

Blood purifiers or alteratives can be taken internally to ease itching. These include BP-X, burdock, chickweed, yellow dock and Oregon grape. HistaBlock may be used internally for itching caused by allergic reactions.

A complete list of potential remedies for itching follows.

Therapies: Essential Fatty Acids (pg. 17), Antioxidants (pg. 17)

Remedies: **Bentonite (Hydrated), Burdock, Chickweed, HistaBlock, Nature's Fresh, Oregon Grape, Pau D Arco, Yellow Dock,** Allergies-Mold/Yeast/Dust, Allergy, Aloe Vera, Blood Build, BP-X, Golden Seal, Herbal Trim Skin Treatment, Intestinal Soothe & Build, Jojoba Oil, MSM, Pantothenic Acid, Tea Tree Oil, Vitamin E

Itching (rectal)

See Also *Hemorrhoids, Parasites (general)*

Rectal itching may be a sign of parasites. It can also be a sign of hemorrhoids. Try doing a parasite cleanse or applying white oak bark mixed with Golden Salve topically.

Remedies: **Black Walnut, Para-Cleanse,** Golden Salve, White Oak Bark

Itching Ears

See Also *Fungal Infections (Yeast Infections, Candida albicans)*

An irritating sensation of the ears that compels one to scratch is often a sign of fungal (yeast) infection. A little Herbal Trim or tea tree oil swabbed in the ear helps relieve the itching. Remedies for fungal infection may be taken internally.

Therapies: Low Glycemic Diet (pg. 11)

Remedies: **Candida Clear, Herbal Trim Skin Treatment, Yeast/Fungal Detox,** Garlic, Pau D Arco, Probiotics, Tea Tree Oil

Jaundice (adults)

Jaundice is caused by a buildup of bilirubin in the blood. This causes a yellowing of the skin. Several blood or liver disorders can cause jaundice. Seek medical attention for an accurate diagnosis and treatment. The following remedies may be helpful.

Therapies: Fiber (pg. 17), Antioxidants (pg. 17)

Remedies: **Charcoal (Activated), Milk Thistle Combination, Yellow Dock,** Alfalfa, Blood Build, BP-X, Butcher's Broom, Dandelion, Gall Bladder Formula, Gotu Kola, Lemon Oil, Liver Cleanse Formula, Milk Thistle, Oregon Grape, Red Beet Formula, Safflowers, St. John's Wort, Vitamin C

Jaundice (infants)

It is common for newborn infants to have a small amount of jaundice. Exposure to 5-10 minutes of sunlight per day is helpful. Safflower tea or activated charcoal mixed with water can also be given to infants to help clear up jaundice.

Remedies: Charcoal (Activated), Safflowers

Jet Lag

Fatigue and irritability after a long flight on an airplane is called jet lag. It is especially a problem when a person crosses several time zones creating a disruption of the circadian rhythms of the body. The following remedies may help decrease problems with jet lag.

Remedies: Adrenal Support, Melatonin Extra, Licorice Root, Nervous Fatigue Formula, Thyroid Support

Jock Itch

See Also *Fungal Infections (Yeast Infections, Candida albicans)*

Jock itch is a fungal infection that affects the folds of skin in the thigh area. Signs are persistent itching and eruptions of small red bumps or flaking skin. It is more common in men. Pau d Arco lotion, tea tree oil (diluted in a carrier oil) or Herbal Trim can be applied topically over affected areas. Remedies for yeast should also be taken internally.

Therapies: Probiotics (pg. 16)

Remedies: Herbal Trim Skin Treatment, Silver Shield, Yeast/Fungal Detox, Candida Clear, Candida Clear, Garlic, Pau D Arco, Probiotics, Tea Tree Oil

Kidney Infection

See Also *Infection (bacterial)*

When there is an infection in the kidneys, the following may be helpful.

Therapies: Hydration (pg. 10)

Remedies: Cranberry & Buchu, Golden Seal, Silver Shield, Uva Ursi, Capsicum & Garlic with Parsley, CBG Extract, IN-X, JP-X, Juniper Berries, Kava Kava, KB-C, Parthenium, PS II

Kidney Stones

Deposits resembling small rocks that form in the kidneys are called kidney stones. If they are large or numerous enough, they may cause severe back pain, blood in the urine or interfere with the elimination of urine.

Most (80%) of kidney stones are made of calcium oxalate and are the result of minerals solidifying out of overly concentrated urine. They obstruct the flow of urine and are very painful to pass.

People in primitive societies rarely develop kidney stones. Why should we? Drinking plenty of clean water will help to keep minerals in solution. Avoid foods that increase urinary oxalate significantly including nuts, chocolate, tea, and peanuts. Caffeine, carbonated beverages, table salt and animal protein all increase the risk of forming kidney stones.

A good-quality calcium supplement will actually help bind oxalate in the gut. Magnesium and vitamin B6 help the body to convert oxalate into other substances. People who consume plenty of fiber and potassium have a lower risk of forming kidney stones. Fruits and vegetables are high in fiber and potassium.

Hydrangea and lemon water help to dissolve the rough edges from stones when they are being passed. Marshmallow soothes urinary membranes and may also be helpful. Kava kava and lobelia can be used to relax urinary passages to help stones pass.

Therapies: Hydration (pg. 10)

Remedies: Hydrangea, Lobelia, Magnesium, Juniper Berries, Kava Kava, Kidney Activator, Kidney Activator (Chinese), Lemon Oil, Marshmallow, Parsley, Uva Ursi

Knees (weak)

The following may help weakness in the knees.

Therapies: Minerals (pg. 15)

Remedies: KB-C, Collatrim, EverFlex

Labor (to induce)

The following can help to induce labor in some women.

Remedies: Blue Cohosh, Master Gland

Labor and Delivery

There are a number of remedies that can ease labor and delivery. Taking 5-W for the last five weeks of pregnancy has helped many women. Blue Cohosh can be used to strengthen uterine contractions. Other remedies in this list can provide energy, relax muscle cramps, stop bleeding and ease pain. Consult with a midwife or professional herbalist for assistance.

Remedies: 5-W, Bayberry, Blue Cohosh, Capsicum, Red Raspberry, APS II, Black Cohosh, Chamomile, Chlorophyll, Distress Remedy, Dong Quai, Jasmine Absolute, Lobelia, Magnesium, Vitamin C

Lactose Intolerance

Lactose intolerance results in bloating and gas after eating dairy products due to the inability to break down the lactose or milk sugar in dairy products that have not been cultured. Lactase is the enzyme that helps break down this sugar.

Therapies: Probiotics (pg. 16)

Remedies: Lactase Plus

Laryngitis (Hoarseness)

The following remedies can help with laryngitis, an inflammation of the larynx or voice box that causes a complete or partial loss of voice. The same remedies will help with hoarseness. Apply essential oils like Tei Fu topically and following with the topical application of lobelia. Use sage tea as a gargle. Other possible remedies include the following.

Remedies: Cellu-Tone, Sage, Zinc Lozenges, Capsicum, Clove Bud, Cough Syrup-DH, Frankincense, IF Relief, Jasmine Absolute, Kava Kava, Lobelia, Marshmallow & Fenugreek, Pine Needle, Sore Throat/Laryngitis, Tea Tree Oil

Lead Poisoning

See Also *Heavy Metal Poisoning*

Lead is one of the most toxic metals known. It's been many years since our society was made aware of the damage that exposure to lead-based paints was doing to our health, especially to young children who suck on and chew anything they can get their hands on.

When the lead reaches toxic levels in the body, it can damage the kidneys, liver, heart and nervous system. The body can't tell the difference between lead and calcium, so pregnant women, children and other people who are deficient in calcium absorb lead more easily, with infants and children affected most severely. Possible symptoms of lead poisoning include anxiety, arthritis, confusion, chronic fatigue, behavioral problems, juvenile delinquency, hyperactivity, learning disabilities, metallic taste in the mouth, tremors, mental disturbances, loss of memory, mental retardation, impotence, reproductive disorders, infertility, liver failure and death.

Exposure to lead can come from food that is grown near roads or factories, lead-based paint, hair products, food from lead-soldered cans, imported ceramic products (especially from Mexico and China), lead crystal glassware, ink on bread bags, batteries in cars, bone meal, insecticides, tobacco, lead pipes, and lead solder in the water pipes. If you suspect you could have lead pipes or lead solder in your water system, have the water tested.

Seek medical assistance if you think you have lead poisoning. The following may be helpful for removing lead from the body.

Therapies: Heavy Metal Detoxification (pg. 22), Oral Chelation (pg. 21)

Remedies: Algin, Bentonite (Hydrated), Heavy Metal Detox, Lobelia, Mega-Chel, Garlic, Kelp, N-acetyl Cysteine, Nature's Three, Zinc

Leaky Gut Syndrome

See Also *Fungal Infections (Yeast Infections, Candida albicans), Inflammatory Bowel Disorders (Colitis, IBS)*

Leaky gut is when the intestinal membranes have been damaged by inflammation so they do not filter out toxins and irritants as effectively. This allows toxins to be absorbed into the blood stream, irritating the rest of the body.

Just imagine for a minute that your sewer or septic system started backing up into your kitchen. It's not a pleasant thought, is it? Yet, many people have a similar problem happening right inside their own bodies they have Leaky Gut Syndrome.

Leaky gut syndrome occurs when the intestinal membranes have lost structural integrity. This causes them to become excessively porous, which allows toxic material and large molecules of unprocessed foodstuffs to pass into the blood and lymph. Numerous physical and mental health problems have been linked with this leakage in the intestines, including ADHD, autism, depression, allergies, asthma and skin diseases like eczema and psoriasis. Leaky gut syndrome may also be a factor in autoimmune diseases like arthritis, chronic fatigue and fibromyalgia.

Leaky gut starts with intestinal inflammation, which is brought on by a combination of factors, including drugs, infections and parasites, food allergies and chemicals. Enzyme deficiencies and a high carbohydrate diet are culprits as well.

The intestines do not absorb nutrients correctly when they are inflamed, which can cause fatigue and bloating. When large, undigested food particles are absorbed because of the excessive porousness in the membranes, this contributes to allergic and autoimmune responses like asthma, hay fever, arthritis and fibromyalgia.

The inflammation damages carrier proteins that help nutrients to be assimilated. This can cause nutritional deficiencies. Leaking toxins also burden the liver and immune system. Finally, the damaged intestinal membranes also allow bacteria, viruses and yeast to pass more readily into the system to damage other organs and systems.

Reducing intestinal inflammation and rebuilding damaged intestinal membranes to stop gut leakage can help numerous health problems. Here are seven steps you can take to reduce intestinal inflammation, promote healing and stop intestinal leakage.

Step one is to avoid intestinal irritants, such as allergens, food additives, drugs and chemicals.

Step Two is to bind intestinal toxins with fiber. Fiber can protect your colon from toxic or irritating substances you are unable to avoid.

Step three is to improve colon transit time. Colon transit time is the length of time it takes for material to travel from one end of the alimentary canal to the other. In a healthy colon, this should be about 18-24 hours. To test your own colon transit time, eat a food that dyes the stool (like beets) and see how long it takes for the color to show up in the stool and be eliminated. If it takes more than a day, then you have a sluggish colon transit time.

Just drinking plenty of water and taking some dietary fiber will usually improve colon transit time, but you may need enzymes or an herbal laxative like LBS II. Laxative products should not be used on a long term basis. If a person finds it difficult to have a bowel movement without stimulant laxatives, Gentle Move is a good product to try.

Step four is to eliminate harmful organisms. Yeast, bacteria and parasites can all contribute to intestinal inflammation. Yeast, in particular, secretes a toxin that damages the intestinal lining and increases membrane permeability. Candida Clear is a good product for getting rid of yeast.

Another good combination for removing harmful microbes from the digestive tract is Una de Gato Combination. Una de gato, or cat's claw, is both antimicrobial and anti-inflammatory.

H. pylori is a bacteria associated with ulcers and intestinal inflammation. Gastro Health is an herbal formula that helps knock down H. pylori and other bacterial infections in the gastrointestinal tract.

Parasites are a more common problem than most people think. People can easily pick up parasites from pets, contaminated food and water, or foreign travel, which in turn, contribute to intestinal inflammation and leaky gut. A periodic parasite cleanse using the ParaCleanse Packets can help eliminate these unfriendly organisms.

Step five is to reduce intestinal inflammation. Up to this point, we ve focused on helping to clean out the intestines. It's also important to reduce the intestinal inflammation to promote tissue regeneration and repair. For problems like colitis, irritable bowel or Crohn's disease, Intestinal Soothe and Build has proven to be a very dependable formula for reducing inflammation and promoting healing.

Another great product that can help reduce inflammation in the intestinal lining (and everywhere else in the body) is Thai-Go. For serious inflammation, also consider IF Relief.

Step six is to plug the leaks. Of course, the primary goal in everything we re trying to do here is to restore tone and integrity to the intestinal membranes. There are a number of products which can be helpful here.

The first is Intestinal Soothe and Build, a good formula for promoting healing to the intestinal membranes and reducing gut leakage. Una de Gato Combination, mentioned earlier, is also helpful here.

Another formula people have had great success with is Kudzu/St. John's Wort. Kudzu is very helpful for repairing gut leakiness and St. John's wort aids the nerves that regulate the digestive process.

Finally, the amino acid glutamine is a major nutrient that intestinal cells need for maintenance and repair. It reinforces this first line of the immune system. There is considerable evidence that taking l-glutamine can aid the gut in its role of protecting against viral, bacterial, and food antigen invaders.

Step seven is to repopulate the colon with friendly bacteria or probiotics. One way to repopulate the colon with friendly bacteria is to eat fermented foods with live cultures, such as yoghurt or raw sauerkraut. Another way is to take probiotic supplements like Acidophilus or Bifidophilus Flora Force.

People are often amazed at how many health problems disappear (and how much better their overall health and energy is) when they heal their intestinal tract by reducing inflammation and putting a halt to gut leakage.

The following remedies help repair leaky gut.

Therapies: Enzymes (pg. 15), Probiotics (pg. 16), Essential Fatty Acids (pg. 17),Fiber (pg. 17), Antioxidants (pg. 17), Hydration (pg. 10), Colon Cleansing (pg. 18)

Remedies: Gentle Move, Kudzu/St.John's Wort, L-Glutamine, Proactazyme Plus, Probiotics, Uña de Gato Combination, Candida Clear, Caprylic Acid Combination, Chamomile, Every Body's Fiber, Gastro Health, IF Relief, Intestinal Soothe & Build, LBS II, Licorice Root, Liver Balance, Para-Cleanse, Sarsaparilla, Super ORAC, Thai-Go, Yeast/Fungal Detox

Leg Cramps

See *Cramps (leg)*

Leprosy

Leprosy is a chronic infection caused by Mycobacterium leprae. It usually affects the skin, peripheral nerves and testes. The following herbs have been reported in historical literature to help with leprosy, however this is a condition requiring medical attention.

Remedies: Silver Shield, Gotu Kola, Sarsaparilla, Slippery Elm, Yellow Dock

Lesions

See Also *Acne (Pimples, Blackheads), Boils, Wounds and Sores*

A lesion is an area of pathologically altered tissue such as an injury, abscess, boil, mole, pimple, rash, or wound. Look up the specific type of problem for suggested remedies.

Leucorrhea

See Also *Vaginitis*

Leucorrhoea is a whitish discharge from the vaginal area and uterus, usually the result of a chronic bacterial or fungal infection. It is associated with inflammation of the vagina (vaginitis). Possible remedies include the following.

Remedies: Candida Clear, Pau D Arco, Probiotics, Silver Shield, Yeast/Fungal Detox, Bergamot, Clove Bud, False Unicorn, Garlic, Kava Kava, Super GLA, Tea Tree Oil, Vitamin A & D

Leukemia

See Also *Cancer (natural therapy for)*

A cancer involving a proliferation of abnormal white blood cells (leukocytes). In addition to the general protocols listed under cancer, the following have been reported helpful in cases of leukemia. Seek medical attention for this serious health problem.

Therapies: Antioxidants (pg. 17), Heavy Metal Detoxification (pg. 22), Enzymes (pg. 15)

Remedies: E-Tea, Immune Stimulator, Lymphomax, Aloe Vera, Blood Build, Germanium Combination, Pau D Arco, Protease, Red Clover

Lice

The following have been used to control head lice, an insect that can infest hair. Mix essential oils or the contents of Paw Paw Cell Reg capsules with shampoo and wash the hair, leaving the shampoo in the hair for about 5-10 minutes before rinsing.

Remedies: Tea Tree Oil, Black Walnut, Cinnamon, False Unicorn, Oregano (Wild), Paw Paw

Ligaments (torn or injuried)

See Also *Sprains*

A torn ligament is similar to a sprain, but more serious. Torn ligaments cause severe swelling, bruising and pain, and may require surgical intervention. The following may help to speed the healing of ligaments that have been injured or torn.

Therapies: Minerals (pg. 15), Enzymes (pg. 15), SuperFood (pg. 15), Affirmation and Visualization (pg. 8)

Remedies: Bone/Skin Poultice, Collatrim, Healing AC Cream, Herbal CA, PLS II, Deep Relief Oil, Nature's Phenyltol with NEM, Super GLA, Tei Fu Oil, Vitamin C

Liver (fatty)

See *Fatty Liver Disease*

Liver Spots

See *Age Spots*

Lockjaw

See *Tetanus*

Lou Gehrig's Disease

See Also *Autoimmune Disorders*

Lou Gehrig's disease is a rare, fatal, progressive degenerative condition that usually begins in middle age and is characterized by increasing and spreading muscular weakness leading to paralysis; also called amyotrophic lateral sclerosis. It is an autoimmune disorder and a very difficult condition to treat. It is very similar to MS, but there are high levels of iron in the tissue causing highly rapid oxidative damage. Professional assistance should be sought, but the following might be helpful.

Therapies: Antioxidants (pg. 17), Essential Fatty Acids (pg. 17), SuperFood (pg. 15), Enzymes (pg. 15), Affirmation and Visualization (pg. 8)

Remedies: All Cell Detox, Colloidal Minerals, Damiana, LBS II, Pau D Arco, Sarsaparilla, Silver Shield, Small Intestine Detox, Trace Mineral Maintenance, Vitamin C, Vitamin E

Lumbago

See *Backache (Back Pain, Lumbago)*

Lungs (congestion)

See *Congestion (lungs)*

Lungs (fluid in)

See Also *Pneumonia*

The following are helpful for relieving fluid in the lungs.

Remedies: ALJ, Garlic, Breathe EZ, Capsicum, Four

Lupus

See Also *Autoimmune Disorders*

Lupus is a chronic inflammatory and autoimmune disease that attacks multiple organs. It affects the skin in many people creating a butterfly rash over the face. Immune stimulates should be avoided and general therapies for autoimmune diseases should be applied. This is a serious illness and professional assistance should be sought. The following may be helpful.

Therapies: Antioxidants (pg. 17), Enzymes (pg. 15), Hydration (pg. 10), Low Glycemic Diet (pg. 11), Hiatal Hernia Correction (pg. 12), Blood Type Diet (pg. 13)

Remedies: Adrenal Support, DHEA-F, DHEA-M, **Licorice Root, Omega-3, Yucca**, 7-Keto, Aloe Vera, APS II, Astragalus, Barley Grass, Black Currant Oil, Black Walnut, Blood Build, EverFlex, Flax Seed Oil, Garlic, Grapine, Joint Support, Nerve Eight, Probiotics, Protease, Red Beet Formula, Super GLA, Thai-Go, Vitamin B-Complex, Vitamin C, Wild Yam, Wild Yam & Chaste Tree, Zinc

Lyme Disease

Lyme disease is most commonly a tick-borne illness. It is a bacterial infection that can be difficult to eradicate and become very debilitating. Medical attention should be sought. It is easily cured with antibiotics in the early stages. Silver Shield taken in very high doses (1/4 to 1/2 bottle daily) can be very effective. Teasel root (an herb not available through NSP) is very effective in treating this condition. The following may also help with recovery from Lyme disease along with improving general health and nutrition.

Remedies: Silver Shield, Food Enzymes, Garlic, Golden Seal, Super GLA, THIM-J, Ultimate Echinacea, Vitamin A & D, Vitamin C, Zinc

Lymph Nodes or Glands (swollen)

See Also *Congestion (lymphatic)*

The following remedies will help reduce swelling in lymph nodes or glands.

Therapies: Hydration (pg. 10), Colon Cleansing (pg. 18)

Remedies: Improved Lymphomax, Lobelia, Lymph Gland Cleanse, Lymph Gland Cleanse-HY, Lymphatic Drainage, Mullein, Ultimate Echinacea

Macular Degeneration

See Also *Free Radical Damage*

Macular degeneration involves a loss of central vision in both eyes produced by pathological changes in the center of the retina, the region of most visual acuity. It is believed to be an inflammatory condition and the result of free radical damage. High blood pressure and hardening of the arteries increase the risk of this condition. The following remedies may be helpful in preventing, slowing or even partially reversing this condition.

Therapies: Antioxidants (pg. 17), Oral Chelation (pg. 21), Essential Fatty Acids (pg. 17)

Remedies: Mega-Chel, Perfect Eyes, Super ORAC, Bilberry Fruit, Carotenoid Blend, Ginkgo Biloba, Grapine, Lutein, Thai-Go, Vitamin A & D, Vitamin C, Zinc

Malaria

Malaria is an acute or chronic disease caused by parasites that invade the red blood cells. It is transmitted from an infected person to an uninfected person by the bite of a mosquito. Symptoms include chills, fever, mass destruction of red blood cells and the parasitic release of toxic substances. The following may be helpful for malaria.

Remedies: Artemisia Combination, Silver Shield, Aloe Vera, Black Walnut, Echinacea Purpurea, Garlic, Ho Shou Wu, Lemon Oil, Noni (Morinda), Pau D Arco, Probiotics, Ultimate Echinacea

Manic Depressive Disorder

See *Bipolar Mood Disorder (Manic Depressive Disorder)*

Mastitis

Mastitis is an infection in the breast that can occur during breastfeeding causing inflammation and tenderness. A poultice of slippery elm, plantain, mullein and/or lobelia is often helpful. Any of the following remedies may help reduce swelling and tenderness.

Remedies: Lymph Gland Cleanse-HY, Lymphatic Drainage, Mullein, IF Relief, Lobelia, Lymph Gland Cleanse, Red Clover, Slippery Elm, Ultimate Echinacea

Measles

Measles is an acute, contagious viral disease that begins with inflammation of mucus membranes, conjunctivitis and cough. This is followed on the third or fourth day by an eruption of distinct circular red spots. The following remedies may be helpful for measles.

Remedies: Oregon Grape, Silver Shield, VS-C, Black Cohosh, Burdock, Catnip, E-Tea, Golden Seal, Yarrow

Melanoma (Skin Cancer)

See Also *Cancer (natural therapy)*

Herbalists have used drawing salves and ointments successfully on skin cancers to pull them out. A simple recipe for one is to mix Paw Paw Cell Reg(c) into some Black Ointment and apply a little bit of this mixture directly over the skin cancer. Cover with a bandage and change twice daily. Use this in conjunction with other natural therapies for cancer and seek appropriate medical help.

Therapies: Antioxidants (pg. 17), Enzymes (pg. 15), SuperFood (pg. 15), Colon Cleansing (pg. 18), Affirmation and Visualization (pg. 8), Stress Management (pg. 8)

Remedies: Black Ointment, E-Tea, Paw Paw, Red Clover, Red Clover Blend

Memory and Brain Function

See Also *Alzheimer's Disease, Dementia*

Understanding the biochemical make-up of the mind is helpful in understanding how to nourish it properly. For starters, the brain is 70% water, so it is very sensitive to dehydration. So, if you want to think more clearly and protect your gray stuff, start by drinking more pure water.

Next, 50-60% of the dry weight of the brain is fat, with 35% of that fat being omega-3 fatty acids. So, if you want to keep your brain healthy you need to be a fathead by eating the right kinds of fats. The most abundant omega-3 fatty acid in the brain is DHA. DHA is essential for proper brain function and is found primarily in fish oil. Adding cold-water fish, such as halibut, mackerel, salmon, herring, anchovies, trout and tuna, to the diet is very beneficial for your brain. DHA is available as a single supplement and is found in Super Omega-3 EPA Vegetarians, in particular, are often very low in this fatty acid and should consider supplementation.

At the same time, avoid hydrogenated oils (shortening and margarine) and transfatty acids from processed and deep fried foods. These fats aren't good for your brain.

Besides fat, the brain also needs amino acids from protein. Amino acids are necessary for the production of neurotransmitters. Meals containing adequate protein tend to increase levels of dopamine and norepinephrine (which results in a more alert mind and a better mood). Protein foods are especially important at breakfast. Having protein for breakfast helps to increase mental alertness and stabilize blood sugar levels. A good supplement for neurotransmitter production is Super Algae. It contains significant amounts of amino acids and has been used to improve alertness, energy and pituitary function.

Foods high in the B-vitamins are essential for lowering the risk of Alzheimer's and dementia, as well as cardiovascular disease. B-vitamins are involved in helping the formation of neurotransmitters such as dopamine, epinephrine, and serotonin. Recent studies have shown a link between declining memory and Alzheimer's disease and inadequate levels of folic acid, vitamin B-12 and vitamin B-6. Nutri-Calm is a good choice for getting B-vitamins and other nutrients to support healthy brain function.

An increasing body of evidence implicates most problems associated with aging to inflammation and oxidative stress, including deteriorating mental conditions such as dementia, Parkinson's and Alzheimer's disease. The brain is particularly sensitive to damage from inflammation and free radicals, so another part of feeding your brain properly is to eat plenty of antioxidant-rich fresh fruits and vegetables, organically grown where possible.

Thai-Go is a convenient, easy and tasty way to get more antioxidants into the diet. Another way to combat inflammation and reduce free radical damage is to take Super Trio. A third option for protecting the brain is Brain Protex. Brain Protex contains an extract of Chinese club moss which contains a substance called huperzine A. Huperzine A has been shown to help prevent and slow the progress of both Alzheimer's and dementia.

The brain consumes more blood sugar (glucose) than any other organ and takes this sugar up directly from the blood stream without the need for insulin. The amount of sugar in your blood directly controls the amount of sugar reaching your brain. Too much sugar and the brain is over stimulated, which results in agitation, irritability and nervousness. Too little sugar in the blood (hypoglycemia) can cause mental confusion, irritability, shakiness and difficulty concentrating.

The pancreas regulates the blood sugar level to keep the brain stable. Blood sugar problems are linked to age-related cognitive decline, dementia and Alzheimer's disease. These blood sugar problems also increase inflammation, which can damage brain cells.

So, although the brain needs sugar as an energy source, simple sugars actually contribute to brain fog and mental decline. Select complex carbohydrates, fruits, vegetables and whole grains, over products containing refined sugar, high fructose corn syrup, white flour and other processed grains.

Because the brain has such a high demand for oxygen and sugar, it requires a constant supply of oxygen-rich blood. Because blood flowing to the head must fight gravity, exercise is extremely important for maintaining good blood flow to the brain. This is why something as simple as taking a walk can help to clear your mind. Exercise regularly for a healthy brain and a keen mind.

If blood flow to the brain starts to become impaired due to arterial constriction or hardening the arteries, brain function will suffer. Fortunately, there are herbs and supplements that can improve blood flow to the brain and enhance cognitive processes at the same time.

One of these is Ginkgo/Gotu Kola with Bacopa. Ginkgo biloba keeps blood vessels and capillaries flexible, aiding circulation and increasing blood flow to the brain improving the neurons uptake of glucose and oxygen. It is also a powerful antioxidant that prevents cell damage and has been shown to increase neurotransmission by normalizing levels of acetylcholine and norepinephrine. Gotu kola revitalizes the nerves and brain cells, strengthens the adrenal glands and purifies the blood. It has been traditionally used as a memory-enhancing herb, but is also a powerful rejuvenating tonic.

Hardening of the arteries will impair blood flow to the brain and reduce cognitive function. People with reduced blood flow to the brain often get sleepy when they sit for long periods and have problems with being absent minded. Oral chelation with Mega-Chel is helpful in these cases.

Toxins can seriously damage the brain, especially fat-soluble toxins (such as petrochemical solvents) and heavy metals like mercury, aluminum and lead. In addition, drugs and alcohol do serious damage the brain.

To keep your mind clear and active avoid as many chemicals as possible, don't use drugs and minimize the consumption of alcohol. Specifically, to avoid chemicals, don't use aluminum cookware, purify your drinking water, use non-toxic household cleaning products and personal care items and read labels carefully.

Two great supplements that help to rid the body of unwanted chemicals are Heavy Metal Detox and Enviro-Detox. Be sure to drink plenty of water when doing any kind of detoxification program. It may also help to take a fiber supplement. Detoxification will reduce inflammation, improve sleep quality, enhance brain function and support general health.

Here are some general remedies to help the brain.

Therapies: Oral Chelation (pg. 21), Minerals (pg. 15), Essential Fatty Acids (pg. 17), Antioxidants (pg. 17), Hydration (pg. 10), Low Glycemic Diet (pg. 11), Colon Cleansing (pg. 18), Heavy Metal Detoxification (pg. 22)

Remedies: Brain-Protex, DHA, Ginkgo/Gotu Kola, HSN-W, Mega-Chel, Nervous Fatigue Formula, Omega-3, Sunshine Heroes Omega 3 with DHA, Super Algae, Vitamin B-Complex, Blessed Thistle, Breathe Free, Enviro-Detox, Focus Attention, Ginkgo & Hawthorn, Ginkgo Biloba, Gotu Kola, Hawthorn Berries, Heavy Metal Detox, HY-C, Lecithin, MSM, Nutri-Calm, Pregnenolone, Rosemary, Sage, Super Antioxidant, Super GLA, Super ORAC, Super Trio, Thai-Go, Trigger Immune, Vitamin B-12

Meningitis

Inflammation of the membranes that surround the brain and spinal cord is called meningitis. This is often caused by a bacteria or virus, but can also be caused by adverse reactions to vaccines. It is a serious condition for which medical attention should be sought. The following may be of benefit.

Remedies: Immune Stimulator, St. John's Wort, Ultimate Echinacea, Yarrow, Echinacea Purpurea, Germanium Combination, Golden Seal, Gotu Kola, Grapine, IF Relief, Lobelia, Pau D Arco, Thai-Go, Vitamin C

Menopause

See Also *Aging (prevention), Hot Flashes, Osteoporosis*

Anyone who watches TV has seen the commercials pitching hormone replacement therapy (HRT). These commercials promise to do away with the symptoms of menopause, as if it were some kind of unnatural disease. But menopause is not a disease. In her work, A Woman's Book of Life, Joan Borysenko notes that many cultures don't view menopause as a time of loss and diminishment. Instead, they view this time as a wonderful life transition and a time of increased value and self-esteem. She also documents that women living in cultures where post-menopausal women are revered as wise women have fewer negative symptoms of menopause than are experienced by women in the United States.

Menopause is a normal transition where a woman begins to ovulate less frequently. The ovaries also cut back on their production of estrogen. This triggers the release of larger quantities of another hormone from the pituitary called the Follicle Stimulating Hormone (FSH for short). When FSH levels rise, they can disturb the hypothalamus gland in the brain. Since the hypothalamus influences body temperature, sleep, mood, weight, energy levels and the autonomic nervous system (which is responsible for breathing, circulation, digestion and blood pressure, among other things) all these functions can be disrupted by the elevated levels of FSH. In greatly oversimplified terms, this is what gives rise to many of the disturbances during menopause.

So, on a purely biochemical level, menopause is about the body adjusting to fluctuating levels of hormones caused by the aging process. But why is it that women in some cultures experience less negative effects of that process? Well, to start with, the hypothalamus is not just sensitive to levels of FSH. It's also sensitive to stress. Joan Borysenko cites a report published in the Journal of Health Physiology in which 21 post-menopausal women who were being monitored in the laboratory reported significantly more hot flashes when they were subjected to stress than in stressless sessions. And in a culture that places such high value on youthfulness (and often views menopause as a tragedy), it's understandable that many women feel sadness, grief or stress at the onset of a transition that's often seen as the end of that youthfulness. This is coupled with other stresses of middle-aged life children growing up and becoming more independent, health challenges, and difficulties in finances and career. All of these stresses serve to exacerbate these hormonal imbalances and aggravate the symptoms of menopause.

Stress has an impact on menopause because the adrenal glands can also manufacture sex hormones like estrogen and progesterone. However, long-term stress (especially coupled with caffeine and sugar in the diet) can exhaust the adrenals. So, many women have discovered that taking herbs or supplements to rebuild the adrenal glands can reduce problems associated with menopause.

Adrenal Support is an excellent formula for rebuilding the adrenals. It contains nutrients that support adrenal function. Nutri-Calm is another good formula which can help reduce stress and support the adrenals to make the transition of menopause easier. Pantothenic acid is also helpful for rebuilding exhausted adrenal glands.

There are other lifestyle-related sources of hormonal imbalance during menopause. Poor diet and lack of exercise also cause problems. Studies done at Harvard Medical School, as well as in Scandinavia, showed that women who engaged in regular exercise experienced fewer and less intense hot flashes. Other research has indicated that women in Japan who eat a less-processed diet and more soy-based food generally experience fewer menopausal symptoms than women in the U.S.

Soy is one of the many foods which contain phytoestrogens (plant chemicals that have an estrogen-like effect). However, soy is not the only source of these phytoestrogens. They are also found in other beans and legumes, dark green vegetables and whole grains. Consuming these foods regularly prior to and during menopause will decrease the likelihood of severe hormonal imbalances.

Black cohosh has been used as a natural estrogen replacement during menopause. Not only does black cohosh contain phytoestrogens, it is also anti-inflammatory (so it reduces heat in the body) and anti-spasmodic (which means it helps to relax muscle spasms and ease stress). It is also very helpful for depression associated with menopause or other hormonal imbalances in women (PMS or postpartum related depression).

In large quantities, black cohosh can dilate the blood vessels enough to create vasodilative headaches or dizziness in some women. The formula Flash-Ease helps remedy this problem by providing a time-released dose of black cohosh and dong quai so that there is a more sustained effect, and a reduced risk of headache and dizziness.

Many women have found that progesterone helps them more during menopause than estrogen. Progesterone can help prevent bone loss and may even help affix calcium into the bones. Pro-G-Yam cream has helped many post-menopausal women to feel better.

HRT with estrogen has been commonly promoted to help reduce bone loss after menopause. However, progesterone has a better effect in preventing bone loss than estrogen. Pro-G-Yam Cream and Skeletal Strength in combination with a more alkaline diet can help prevent bone loss and may even help to build the bones. Calcium supplements alone are not very effective in preventing bone loss because calcium is needed in combination with many other nutrients in order to lay down bone. Skeletal Strength provides these other nutrients.

In addition to these herbs and supplements, here are some self-help measures you can take to support yourself in menopause:

Drink plenty of water.
Smoking, alcohol and caffeine all upset hormone balance, so eliminate them as much as possible.
Avoid excessive hot drinks and spicy or fatty foods, since they can trigger heat in the body.
Many women report that they feel better if they cut down on meat and dairy consumption;
Wear natural fibers as much as possible to allow your skin to breathe and to enhance skin temperature regulation.
Stress reduction exercises, such as yoga, tai chi and abdominal breathing can help to elicit relaxation and to diminish hot flashes.

As much as possible, allow yourself to seek out and accept the support of others, especially women who have had similar experiences and who can validate your own.

As one can see, there are many natural aids that can ease this important transition in a woman's life, without resorting to HRT. The following is a complete list of possible menopausal aids to consider.

Therapies: SuperFood (pg. 15), Minerals (pg. 15), Enzymes (pg. 15), Essential Fatty Acids (pg. 17), Antioxidants (pg. 17), Hydration (pg. 10), Low Glycemic Diet (pg. 11), Avoid Xenoestrogens (pg. 23), Stress Management (pg. 8), Affirmation and Visualization (pg. 8)

Remedies: Adrenal Support, Black Cohosh, Flash-Ease T/R, NF-X, Pro-G-Yam Cream, Super GLA, Aloe Vera, Bergamot, Breast Enhance, C-X, Chamomile (Roman), Clary Sage, Cordyceps, DHEA-F, Dong Quai, Evening Primrose Oil, False Unicorn, FCS II, Female Comfort, Geranium, Herbal CA, HSN-W, L-Glutamine, Menopause, Mood Elevator, Natural Changes, Peppermint, Phyto-Soy, Red Clover, Rose Bulgaria, Skeletal Strength, St. John's Wort, Vitamin B-Complex, Vitamin E, Wild Yam & Chaste Tree, X-A, X-Action (Women s), Ylang Ylang

Menorrhagia (Heavy Menstrual Bleeding)

Excessive menstrual bleeding or menorrhagia is often caused by excess estrogens or xenoestrogens. It may be due to uterine fibroids, too. Get an appropriate medical diagnosis if the problem is persistent and doesn't respond to natural remedies.

Menstrual-Reg was specifically formulated to help with this problem. If it doesn't work on its own, take it with extra yarrow. Blood Build and I-X can be used to help rebuild the blood with heavy bleeding. You can also try douching with diluted colloidal minerals or a tea made of an astringent herb like bayberry root bark. The following are possible remedies for menorrhagia.

Therapies: Avoid Xenoestrogens (pg. 23)

Remedies: Bayberry, Blood Build, Menstrual-Reg, Yarrow, Blue Cohosh, Capsicum, Colloidal Minerals, Frankincense, Geranium, I-X, Sarsaparilla, White Oak Bark

Menstrual Cramps

See *Cramps (menstrual)*

Menstrual Irregularity

See Also *PMS (general)*

When a woman's menstrual cycle does not follow the normal 28-day pattern (too long, too short, irregular, etc.) the following remedies can be used to help bring hormones into balance to regulate the periods.

Therapies: Stress Management (pg. 8), Essential Fatty Acids (pg. 17)

Remedies: Blood Build, False Unicorn, Pro-G-Yam Cream, Wild Yam & Chaste Tree, Black Cohosh, Blessed Thistle, Clary Sage, DHEA-F, Dong Quai, Evening Primrose Oil, Feminine Tonic, Menstrual, Milk Thistle Combination, Super GLA, Thyroid Support

Menstruation (heavy bleeding)

See *Menorrhagia (Heavy Menstrual Bleeding)*

Menstruation (painful)

See *Dysmenorrhea*

Menstruation (scant)

The following can help when menstruation is abnormally light.

Remedies: Artemisia Combination, Clary Sage, Marjoram (Sweet), Red Raspberry, Trigger Immune, V-X

Mental Illness

See Also *Fungal Infections (Yeast Infections, Candida albicans), Heavy Metal Poisoning, Hypoglycemia*

Mental health problems can be a touchy subject because most of us find the idea of mental illness disturbing or frightening. These attitudes about mental health problems have been around since ancient times, when the mentally ill were considered to be possessed by devils. In fact, the phrase casting out devils used in the New Testament is an Aramaic idiom which would be accurately translated today as, curing the mentally ill.

Even though our understanding of mental illness has improved since ancient times, the social stigma attached to these conditions is still negative. Furthermore, even modern treatments for mental conditions leave much to be desired. Does drugging the brain or locking up the severely disturbed really solve the problem?

Just like any other disease human beings suffer from, the so-called mental illnesses have causes. While the causes of mental illness aren't fully understood, it's very unlikely that a drug deficiency is one of them. So, while drugs might be of use in stabilizing some situations, they aren't the ultimate cure for mental health problems.

Furthermore, it's not like a person wakes up one morning and thinks, I ve decided I m going insane today, so people with mental health problems should be treated with the same compassion and respect we would treat anyone suffering from a physical disease. Using criticism, rewards and punishments to try to correct the behavior of one suffering from mental health problems isn't going to work, either.

Likewise, the labels given to mental conditions are often confusing and don't really help. Labels are just names attached to certain patterns of symptoms, and are really of little use in helping to cure anything.

We must get past the stigmas, misunderstandings and symptomatic treatments and start understanding the root causes of mental health issues. Only when we understand the cause can we formulate effective treatment.

To begin our journey, we need to recognize the difference between three different classes of mental health issues. First, there are mood disorders, such as depression, anxiety, learning disabilities, etc. These are not debilitating issues. People who have them can still function in society. In fact, many are problems from which just about everyone suffers from time to time. Who hasn't felt depressed or anxious once in a while?

Unfortunately, in order to push their wares, drug companies appear to be turning these ordinary issues of mood and mental attitude into diseases. The child who is restless, bored and fidgety at school is suddenly suffering from a disease called ADHD. The person who is socially shy or a little bit anxious is a victim of social anxiety disorder. A person who is exhausted and run down from long-term stress has post traumatic stress disorder. The person who is rightfully down because of difficult experiences in life needs medication.

This is not to downplay the idea that mood disorders aren't real; it's just that if a person is basically able to function in their life, they probably don't need drugs to get over these mood problems. Some counseling (or even a listening ear from a friend), good parenting (for kids), and attention to basic health needs like nutrition, rest and exercise will usually clear up mood disorders.

A second category of mental health problems is those that are caused by actual destruction of the brain. Alzheimer's disease and dementia are examples of these problems. These are caused by either physical or chemical damage to the brain tissues, which result in the destruction of brain cells. Much of this damage occurs from oxidative stress and inflammation in the brain. These diseases can be prevented by good nutrition and health practices (antioxidants, detoxification, etc.), but once they occur natural therapy will slow deterioration, but may be insufficient to turn them around.

The third kind of mental health problem we want to consider are those problems where there is no actual damage to the brain, but the person has mental problems which are keeping them from being a functioning human being. These are the people who we usually consider insane, people who suffer from schizophrenia, severe depression or whatever other labels get concocted to describe the person's symptoms.

The following information may help people in this third category, but they will probably need some professional help, too. At a minimum, counseling or some kind of psychotherapy are necessary. But, people with these disorders are not doomed to a life of insanity. If root causes are addressed, and dealt with, they can become functioning people again. It takes dedication and effort, but it can happen.

Like any other organ, the brain needs nutrients to function properly. If the brain isn't nourished properly problems will occur. The brain needs water, fats, protein, B-complex vitamins and other nutrients to function properly. It also needs balanced blood sugar. Hypoglycemia or low blood sugar is common in people who have been labeled mentally ill, and stabilizing their blood sugar with a low glycemic diet and appropriate supplements can create big improvements.

For example, Barbara Reed, a parole officer from New York, found that diet contributed to juvenile delinquency. As a parole officer for young offenders, she noticed that most of them were living on a junk food diet. When she was able to get these kids to go on a hypoglycemic diet, along with a B-complex plus C vitamin supplement they never got in trouble with the law again. This is not the only case where behavior has been linked to diet, studies have also been done showing that prisoners have had remarkable improvements in behavior when put on a diet for hypoglycemia.

In the documentary Super Size Me, a school for troubled teens is shown, where the teens are calmer because they are being fed a high quality diet. Research on the link between nutrition and mental ability has been around for a long time. Michael Lesser testified before the Senate Select Committee on Nutrition and Human Needs in the 1970s that 70% of all previously uncontrollable schizophrenics showed improvement when put on a diet to counter hypoglycemia. They also showed improvement with vitamin supplements, particularly certain B vitamins.

The brain is probably the most chemically sensitive organ in the body, so it is highly susceptible to damage from environmental toxins. So, chemical toxins are another major root cause of mental illness.

For example, intestinal inflammation and leaky gut syndrome have been linked with depression, hyperactivity, ADHD and schizophrenia. So, a good cleanse can often help lighten a person's mood and clear their thinking.

Yeast infections can also mess up one's thought processes. Since yeast feed on sugar, yeast infections contribute to sugar cravings, too.

A major contributing factor to mental breakdown is heavy metal poisoning. Mercury, lead, cadmium and aluminum all contribute to the breakdown of brain and nervous system tissue. Obviously, detoxification of heavy metals will help protect the brain and may improve brain function.

Since toxins damage brain tissue by causing inflammation and free radical damage, antioxidant supplements, such as Grapine, Thai-Go and Brain Protex, may also protect the brain from deterioration due to chemical toxicity and free radical damage.

In addition to the issues of nutrition and detoxification, we also have the issue of unresolved mental and emotional stress. The simple fact is that many of the people who have been labeled insane have simply suffered trauma or abuse that has overwhelmed their ability to cope. Sexual and physical abuse as a child can set up patterns that contribute to poor mental (and physical) health.

As Peter R. Breggin, M.D. points out in his book *Toxic Psychiatry*, many people who are labeled as mentally ill are simply in emotional and spiritual crisis. Their language is metaphoric. It'sounds crazy because people can't hear what the person is really trying to communicate. Delusions of grandeur (I m God or Napoleon , for example) can indicate a person is struggling with their sense of importance.

Dr. Breggin stresses that locking these people up, drugging them and shocking them doesn't help them work through their inner crises or deal with their repressed emotional pain. He also suggests that drugs (which includes medications, alcohol, tobacco and illegal street drugs) only act to chemically lobotomize the brain, numbing a person to their inner pain.

What people in crisis really need is a good psychologist or counselor who can help them work through their unresolved issues. Flower essences (such as Distress Remedy) may also be helpful in bringing repressed emotions and issues to the surface. Aromatherapy oils can also be used to help improve mood. Space doesn't permit a full discussion of these tools, but there are plenty of resources for learning more about them.

Many people who do bodywork from massage therapists to Rolfers have observed that people seem to store stress and pain in the body. When people experience intense feelings they don't know how to deal with, they hold their breath and contract around the emotions to avoid feeling them. This creates tension in the body and stores the memory of the trauma in the tissues.

Bodywork can release these tense areas and help a person reconnect with pain and emotions they have repressed, promoting release and healing. So, bodywork techniques like chiropractic care, rolfing, deep tissue massage, yoga and other bodywork techniques can be very helpful in improving mental health. A number of innovative psychologists have used movement and bodywork to improve mental health.

Supplements that will help a person suffering from mental illness vary,depending on the underlying health problems. The following are possibilities to consider.

Therapies: Heavy Metal Detoxification (pg. 22), Stress Management (pg. 8), Essential Fatty Acids (pg. 17), SuperFood (pg. 15), Affirmation and Visualization (pg. 8), Minerals (pg. 15), Antioxidants (pg. 17), Low Glycemic Diet (pg. 11), Blood Type Diet (pg. 13), Heavy Metal Detoxification (pg. 22)

Remedies: Adrenal Support, Brain-Protex, DHA, Distress Remedy, Heavy Metal Detox, Licorice Root, Mood Elevator, Nervous Fatigue Formula, Nutri-Calm, Vitamin B-Complex, Calcium, Candida Clear, Chromium GTF, Grapine, HY-A, Magnesium, Mega-Chel, Omega-3, Pro-Pancreas, Stress Relief, Stress-J, Super Trio, Thai-Go, Thyroid Support, Yeast/Fungal Detox

Mercury Poisoning

See Also *Autism, Heavy Metal Poisoning, Multiple Sclerosis (MS)*

Mercury is one of the most toxic substances we can be exposed to and we are exposed to it on a fairly regular basis. It is found in fungicides, pesticides, dental fillings, contaminated seafood, thermometers and a host of products, including various cosmetics, fabric softeners, inks, tattoo ink, latex, medications, paints, plastics, polishes, solvents and wood preservatives.

Mercury can be absorbed through the skin or inhaled. It passes through the blood-brain barrier and is attracted to and absorbed by nerve endings. This neurotoxin lodges inside neuron cells disrupting cellular communication. It can cause autoimmune disorders, arthritis, blindness, candidiasis, depression, dizziness, fatigue, gum disease, hair loss, insomnia, memory loss, muscle weakness, multiple sclerosis, lateral sclerosis (ALS), Alzheimer s, Parkinson s, paralysis, lupus, food and environmental allergies, menstrual disorders, miscarriages, behavioral changes, depression, irritability, hyperactivity, allergic reactions, asthma, metallic taste in the mouth, loose teeth and more.

Remember those silver fillings that dentists put into your mouth? Well, they are really 50% mercury! Although the American Dentist's Association will do everything they can to deny that amalgam fillings can affect health, many European countries now completely outlaw the use of silver/mercury amalgam fillings! According to the World Health Organization, amalgam dental fillings are a major source of mercury exposure. Mercury fillings may be a contributing factor to many chronic, degenerative and autoimmune conditions, because many people with these conditions have had significant improvement to their health when they have been removed. It is important to note that a mercury detox program is usually needed after fillings have been removed.

Mercury damages the nervous system, the immune system and can be a contributing factor in a wide variety of ailments, including autoimmune disorders. The following may be helpful in removing mercury from the body.

Therapies: Heavy Metal Detoxification (pg. 22), Essential Fatty Acids (pg. 17),Fiber (pg. 17)

Remedies: Algin, Bentonite (Hydrated), Heavy Metal Detox, N-acetyl Cysteine, Alpha Lipoic Acid, Bee Pollen, Detoxification, Echinacea Purpurea, Eleuthero, Garlic, Germanium Combination, Lobelia, Mega-Chel, Milk Thistle Combination, Omega-3, Red Clover, Red Clover Blend, Sarsaparilla, Sunshine Heroes Omega 3 with DHA, Super Algae, Ultimate Echinacea, Vitamin C

Migraine

A migraine is a recurrent, severe headache often accompanied by visual symptoms, nausea and vomiting caused by dilated blood vessels in the brain. There may be food allergy triggers, including chocolate, citrus, alcohol or aged or cured foods. There are two kinds of migraines, vasoconstrictive and vasodilative. In vasoconstrictive migraines, the head feels like it is in a vice or being squeezed. There is too little blood flow to the brain so remedies that enhance circulation to the brain and relax muscles such as ginkgo, lobelia or ginger may help. With vasodilative headaches there is too much blood flow going to the brain, so blood needs to be drawn to the digestive tract with bitter herbs like Digestive Bitters. Feverfew acts as a preventive remedy for vasodilative migraines. There is usually a liver problem in migraine headaches, so liver remedies can help eliminate them. Applying Tei Fu oil and lobelia to the neck and shoulders regularly to keep these muscles relaxed will also help.

Therapies: Hydration (pg. 10)

Remedies: Feverfew, Liver Balance, Tei Fu Oil, Digestive Bitters Tonic, Dong Quai, Ginger, Ginkgo Biloba, Helichrysum, HistaBlock, Lobelia, Magnesium, Marjoram (Sweet), Migraquel, Nature's Gold, Niacin, Proactazyme Plus, RG-Max, Tiao He Cleanse

Miscarriage (prevention)

The body has to maintain a higher level of progesterone than estrogen during pregnancy. If this balance is upset, bleeding can occur and eventually result in miscarriage. This is not the only cause of miscarriage, but one of the more common ones. The following remedies have been used to help prevent miscarriage. They may also help to avert a threatened miscarriage.

Therapies: Avoid Xenoestrogens (pg. 23)

Remedies: False Unicorn, Lobelia, Pro-G-Yam Cream, Red Raspberry, Capsicum, Vitamin E

Mononucleosis

Mononucleosis (mono) is an infectious disease that causes an increase in a specific type of white blood cell. The following may be of help in fighting mononucleosis.

Remedies: Immune Stimulator, Ultimate Echinacea, VS-C, Astragalus, BP-X, L-Lysine, Liver Cleanse Formula, Probiotics, Red Clover, Red Clover Blend

Mood Swings

Mood swings are abnormal, often rapid, changes in one's state of mind or predominant emotions. They may be caused by blood sugar, liver or hormonal imbalances. The following remedies may help.

Therapies: Blood Type Diet (pg. 13), Stress Management (pg. 8), Low Glycemic Diet (pg. 11)

Remedies: HY-A, Licorice Root, Liver Balance, Bergamot, Blood Build, C-X, Geranium, Grapefruit (Pink), Grapefruit (Pink), Helichrysum, Jasmine Absolute, Lemon Oil, Nature's Phenyltol with NEM, Neroli, Patchouli, SAM-e, Sunshine Heroes Omega 3 with DHA, Super Algae, Ylang Ylang

Morning Sickness

See Also *Pregnancy (herbs and supplements for)*

Morning sickness is mild to severe nausea, often accompanied by vomiting, that occurs in pregnancy, usually only in the early stages, but in some cases throughout pregnancy. Although common upon rising in the morning, hence the name, it can occur during any part of the day or night.

Remedies: Ginger, Liver Balance, Red Raspberry, Red Raspberry Liquid, Vitamin B-Complex, Alfalfa, Anti-Gas Formula, Blood Build, False Unicorn, Golden Seal, Liver Cleanse Formula, Peppermint, Vitamin B-6

Motion Sickness

See Also *Nausea and Vomiting, Vertigo*

Nausea usually caused by travel in a vehicle such as a car, ship or plane is called motion sickness. Studies have suggested that ginger can be very effective in preventing motion sickness when taken prior to travel. A drop of Tei Fu or peppermint oil on the back of the tongue may also be helpful.

Remedies: Ginger, Peppermint, Charcoal (Activated), FV, L-Glutamine, Tei Fu Oil, Vitamin B-6

Mouth Ulcers or Sores

See *Canker Sores (Mouth Ulcers)*

Mucus

See *Congestion (general)*

Multiple Sclerosis (MS)

See Also *Autoimmune Disorders*

Multiple Sclerosis is an autoimmune disease that attacks the insulating coverings (myelin sheath) of the nerves and brain cells. Symptoms often include weakness, fatigue, debility and numbness and occur in varied regions of the body depending on the nerves affected. Fats are very important for the myelin sheath. One possible cause of MS is mercury poisoning. Some MS patients have improved after having all metal fillings removed from their mouth, followed by a mercury detoxification program. MS may also be due to the polio virus from vaccines. General therapy for autoimmune disorders can be helpful.

Therapies: Affirmation and Visualization (pg. 8), Minerals (pg. 15), Enzymes (pg. 15), Essential Fatty Acids (pg. 17), Antioxidants (pg. 17), Heavy Metal Detoxification (pg. 22), Blood Type Diet (pg. 13)

Remedies: Heavy Metal Detox, HSN-W, N-acetyl Cysteine, Omega-3, Black Currant Oil, DHA, Evening Primrose Oil, Flax Seed Oil, Ginkgo Biloba, Herbal CA, IGF-1, LOCLO, Nature's Three, Nerve Eight, Psyllium, Psyllium Hulls Combination, Super GLA, Trace Mineral Maintenance, Vitamin D3

Mumps

Mumps is an acute contagious viral disease marked by fever and the swelling of lymph nodes causing a swollen, puffy, full appearance to the cheeks. The following have been used to aid recovery from mumps.

Remedies: Lobelia, Lymphomax, Mullein, Silver Shield, Ultimate Echinacea, Burdock, Catnip, Clove Bud, Garlic, Lymphatic Drainage, Peppermint, Yarrow

Muscle Spasms or Cramps

See *Cramps and Spasms (general)*

Muscle Tone (lack of)

The following remedies can help improve muscle tone.

Therapies: Minerals (pg. 15)

Remedies: Spleen Activator, Co-Q10, Germanium Combination, Iron, Lung Support, Red Clover, Red Clover Blend, Vitamin E

Muscle Twitch

See *Twitching*

Muscular Dystrophy

Muscular dystrophy is a disease that progressively destroys muscle tissue. Since it is a genetic disorder, it does not have a cure. However, the following may help ease some of the symptoms and improve quality of life.

Therapies: Minerals (pg. 15), SuperFood (pg. 15), Antioxidants (pg. 17)

Remedies: Target Endurance, Vitamin B-Complex, Barley Grass, Co-Q10, Flax Seed Oil, Free Amino Acids, Licorice Root, Saw Palmetto, Vitamin E

Myasthenia Gravis

See Also *Autoimmune Disorders*

Myasthenia gravis is a disease of progressive weakness and exhaustion of the voluntary muscles of the body without any wasting; caused by a defect at nerve and muscle junctions. It is considered an autoimmune disease. The following may be helpful.

Therapies: Antioxidants (pg. 17), Heavy Metal Detoxification (pg. 22), Minerals (pg. 15), Enzymes (pg. 15), Essential Fatty Acids (pg. 17)

Remedies: **Magnesium**, Blood Build, BP-X, Gotu Kola, Lecithin, Licorice Root, Master Gland, Red Raspberry, St. John's Wort, Thai-Go, THIM-J, Valerian Root, Vitamin B-Complex, Vitamin C, Vitamin E, Yellow Dock, Yucca

Narcolepsy

Narcolepsy is a sleeping disorder that causes an overwhelming need to sleep during the day. This can happen very suddenly and attempts to stay awake will fail. It is caused by damage to the nerves that control sleep and wakefulness. Strengthening the nerves may help. Eliminating food allergens has helped in some cases. Possible nutritional aids include the following, but there is no simple natural therapy for this problem, so professional help should be sought.

Therapies: Blood Type Diet (pg. 13), Low Glycemic Diet (pg. 11), SuperFood (pg. 15)

Remedies: Calcium, Chromium GTF, Co-Q10, Free Amino Acids, Ginkgo Biloba, Gotu Kola, L-Glutamine, Lecithin, Magnesium, Nutri-Calm, Omega-3, St. John's Wort, Vitamin B-Complex, Vitamin E

Nausea and Vomiting

Nausea is a queasiness in the stomach that makes one resist food and have the urge to vomit. Vomiting is disgorging the contents of the stomach through the mouth. The following remedies may help to settle the stomach in cases of nausea and vomiting.

Remedies: **FV, Ginger, Peppermint**, Aloe Vera, Blessed Thistle, Blood Build, Cinnamon, Clove Bud, Grapefruit (Pink), Lavender, Lemon Oil, Red Raspberry Liquid, Vitamin B-6, Vitamin B-Complex

Nephritis

See Also *Inflammation*

Nephritis is inflammation of the kidney. Juniper berry and other kidney stimulates should be avoided. Only cooling and soothing urinary remedies, like the ones below, should be used.

Therapies: Antioxidants (pg. 17), Essential Fatty Acids (pg. 17)

Remedies: **Cornsilk, Dandelion, Marshmallow, Thai-Go**, Astragalus, Echinacea Purpurea, FCS II, Golden Seal, Horsetail, IF Relief, IF-C, NF-X, Uva Ursi

Nerve Damage

The following may help heal damage to the nerves.

Therapies: Essential Fatty Acids (pg. 17)

Remedies: **DHA, Helichrysum, HSN-W, Nutri-Calm, Omega-3, St. John's Wort**, Flax Seed Oil, Super GLA, Wood Betony

Nervous Disorders

See *Anxiety Disorders*

Nervous Exhaustion (Enervation)

See Also *Adrenals (Exhaustion, Weakness or Burnout)*

Closely related to adrenal exhaustion, enervation occurs after long periods of stress where the nervous system becomes depleted. The person feels shaky, tired and on edge. They have a hard time holding their hands steady. The following remedies act as nerve tonics to replenish the nervous system. Epsom salt baths with the addition of some of the essential oils in this list can be very helpful.

Therapies: Hydration (pg. 10), Stress Management (pg. 8), Essential Fatty Acids (pg. 17), Affirmation and Visualization (pg. 8)

Remedies: **Blue Vervain, Distress Remedy, Flax Seed Oil, Nervous Fatigue Formula, Nutri-Calm**, Adrenal Support, Bergamot, Cellular Energy, Clary Sage, Cramp Relief, Fatigue/Exhaustion, Free Amino Acids, Grapefruit (Pink), Kava Kava, Lavender, Nervousness, Noni (Morinda), Pantothenic Acid, Patchouli, Pine Needle, Pine Needle, Rose Bulgaria, Rosemary, Stress-J, Thai-Go, Vitamin B-Complex, Ylang Ylang

Nervousness

See Also *Anxiety Disorders, Stress*

The following remedies may help reduce nervousness.

Therapies: Hydration (pg. 10), Stress Management (pg. 8), Essential Fatty Acids (pg. 17)

Remedies: **Adrenal Support, Chamomile, Kava Kava, Lavender, Nervous Fatigue Formula, Nutri-Calm, Stress-J**, Bergamot, Chamomile (Roman), Clary Sage, Patchouli, Rose Bulgaria, Sunshine Heroes Calcium Plus D3, Vitamin B-12, Vitamin B-Complex, Ylang Ylang

Neuralgia and Neuritis

Neuralgia is a pain that radiates along the course of one or more nerves. In neuralgia, irritated nerves cause severe pain in the body. Neuritis is inflammation of nerves, similar to neuralgia, except there may not be any pain. There may also be degeneration of the nerves. Remedies like the following that reduce inflammation and/or nourish nerves may be helpful for both conditions.

Therapies: Heavy Metal Detoxification (pg. 22), Essential Fatty Acids (pg. 17), Antioxidants (pg. 17), Heavy Metal Detoxification (pg. 22)

Remedies: **Devil's Claw, DHA, IF Relief, Joint Support, Nerve Eight, Omega-3, St. John's Wort, Wood Betony**, Black Cohosh, Blue Vervain, Deep Relief Oil, Eucalyptus, Evening Primrose Oil, Lavender, Lecithin, Neroli, Pain, Pantothenic Acid, Passion Flower, Super GLA, Super ORAC, Thai-Go, Thyme, Valerian Root, Vitamin B-12, Vitamin B-Complex, Wild Yam, Yucca

Neurosis

See Also *Mental Illness*

Neurosis is a mental and emotional disorder that affects only part of the personality. It usually involves a less distorted image of reality than a psychosis and does not result in a disturbance of language. It is accompanied by various physical, physiological and mental disturbances such as anxieties and phobias. The following are remedies that may be helpful.

Therapies: Low Glycemic Diet (pg. 11), Heavy Metal Detoxification (pg. 22), SuperFood (pg. 15)

Remedies: Distress Remedy, Magnesium, Nutri-Calm, Vitamin B-Complex, Calcium, Ginseng (Korean), Ginseng (Wild American), Kava Kava, Mood Elevator, St. John's Wort, Stress Relief, Super GLA, Valerian Root

Night Blindness

Night blindness is a difficulty seeing at night. The following may be helpful.

Remedies: Bilberry Fruit, Perfect Eyes, Thai-Go

Night Sweating

See Also *Perspiration (excessive)*

The following may help when there is profuse perspiration during sleep. This is often caused by problems with the hypothalamus or adrenals.

Remedies: HY-C, Nervous Fatigue Formula, Sage, Astragalus, Clary Sage, X-A

Nightmares

See Also *Restless Dreams*

The following may be helpful where nightmares are a problem. Nightmares are often a sign of a toxic liver, blood sugar problems or adrenal fatigue.

Therapies: Essential Fatty Acids (pg. 17), SuperFood (pg. 15)

Remedies: Adrenal Support, Nervous Fatigue Formula, St. John's Wort, Frankincense, Liver Balance, Milk Thistle, Milk Thistle Combination, Sandalwood, Thyme

Nocturnal Emission (Wet Dreams)

The following may help reduce nocturnal emissions.

Remedies: KB-C, False Unicorn, Herbal Sleep, Ho Shou Wu, Wild Yam & Chaste Tree, X-A

Nose Bleeds

The following may help stop or reduce nose bleeds. Yarrow and bayberry may be snuffed directly into the sinuses to arrest bleeding. Bayberry and capsicum should be removed from the capsules and the powders placed directly on the tongue. Other remedies may help strengthen capillaries to prevent nose bleeds. Proper hydration is important.

Therapies: Hydration (pg. 10)

Remedies: Bayberry, Capsicum, Menstrual-Reg, Rose Hips, Vitamin C, Yarrow, Bilberry Fruit, Eyebright, Geranium, Horsetail, Wood Betony

Numbness

The following may help ease sensations of numbness and tingling in the extremities by enhancing circulation. Numbness and tingling may also involve irritation or pressure on nerves which may be corrected by chiropractic care or bodywork. St. John's wort stimulates nerve regeneration and repair where the problem is caused by bruised or damaged nerves.

Remedies: Ginkgo & Hawthorn, Mega-Chel, St. John's Wort, Capsicum, Ho Shou Wu, HY-C

Nursing

See *Breast Milk (increases and/or enriches)*

Obesity

See *Weight Loss (aids for)*

Obsessive Compulsive Disorder

See Also *Anxiety Disorders, Mental Illness*

Obsessive-compulsive disorder (OCD) is one of a number of Anxiety Disorders. It is characterized by involuntary thoughts that cause the sufferer to develop a dread that something bad will happen. This makes them feel compelled to perform involuntary, irrational and time-consuming behaviors. Symptoms range from repetitive hand-washing to preoccupation with sexual, religious, or aggressive impulses.

This condition requires counseling and emotional healing work, but progress can be aided by using the general principles for working with anxiety disorders.

Therapies: Affirmation and Visualization (pg. 8), Low Glycemic Diet (pg. 11), Essential Fatty Acids (pg. 17)

Remedies: Adrenal Support, Distress Remedy, Nervous Fatigue Formula

Oral Surgery

The following may help the mouth to heal after oral surgery.

Remedies: Golden Seal, IF Relief, Colloidal Minerals, Deep Relief Oil, Lymphatic Drainage, Myrrh, Super ORAC, White Oak Bark

Osteoporosis

See Also *Menopause*

A condition of decreased bone mass and density accompanied by increase in bone spaces creating bone fragility and increased risk of fracture. This is caused by a lifetime of poor dietary choices, particularly consumption of acid forming foods. Soda pop, excessive consumption of meat and grains without fruits and vegetables, and sugar consumption all contribute to an overacid condition in the body. This causes the body to withdraw calcium from the bones to maintain pH balance. The following may be helpful in preventing and even reversing osteoporosis.

Therapies: Minerals (pg. 15), Essential Fatty Acids (pg. 17)

Remedies: Black Cohosh, HSN-W, KB-C, Pro-G-Yam Cream, Skeletal Strength, Vitamin D3, Xylitol, Alfalfa, C-X, Germanium Combination, Horsetail, PDA, Phyto-Soy, Sea Salt

Ovarian Cysts

See *Cysts*

Ovarian Pain

See Also *Pain (general remedies for)*

The following may be helpful for easing pains arising from the ovaries.

Remedies: False Unicorn, Wild Yam & Chaste Tree, Frankincense, Wild Yam

Overacidity

The following remedies help to alkalize the body, raising pH levels in the urine and saliva.

Therapies: Hydration (pg. 10), Enzymes (pg. 15), Low Glycemic Diet (pg. 11), SuperFood (pg. 15), Stress Management (pg. 8)

Remedies: Calcium, Chlorophyll, GreenZone, KB-C, Magnesium, Noni (Morinda), Alfalfa, Coral Calcium, Food Enzymes, Joint Support, Lymphatic Drainage, Potassium, Proactazyme Plus, Protease, Skeletal Strength, Thai-Go, Yucca

Overalkalinity

The following remedies may help to balance an overalkaline pH in the urine and saliva.

Therapies: Enzymes (pg. 15)

Remedies: Food Enzymes, Vitamin C, Cellular Energy, Love and Peas, Proactazyme Plus, Probiotics, Protease, SynerProTein

Oxygen Deficiency

The following help the body to better oxygenate the tissues.

Remedies: Chlorophyll, Vitamin E, Yellow Dock, Alfalfa, Germanium Combination, I-X

Pain (general remedies for)

See Also *Arthritis, Headache (general), Migraine*

Pain is the primary and nearly universal symptom of all our human afflictions, whether they are physical or emotional. Pain is why we seek help when we are sick. If illness didn't cause disease (lack of ease or pain), we would not be motivated to avoid doing things that damage the body.

So, no matter how much we dislike it and want to make it go away, pain is not an enemy. Pain is a form of communication. It is how the body tells us it is having a problem. It is the 911 system that the cells of our body use to call for help. Pain is also the teacher that motivates us, if we let it, to not abuse our body and pursue a healthy lifestyle.

Modern pain-relieving drugs are wonderful things. They enable us to undergo necessary surgery or dental work without pain. They can also be of great relief to the person who is suffering because of a serious accident or illness.

Unfortunately, pain-relieving medications allow people to disconnect from taking responsibility for their health. Instead of asking why they are getting headaches, upset stomachs, muscle aches or other pain, they simply pop a painkiller. Thus, they never make the connection that their pain is originating from their diet, lifestyle and stress and usually don't make the necessary changes they need to make to be healthy.

We need to stop seeing pain as an enemy that needs to be killed. We need to see pain as a teacher that is giving us a wake up call that we need to change the way we are treating our body.

Let's take headaches for example. Headaches have causes. The cause may be as simple as dehydration. Drinking more water has greatly reduced the number of headaches some people have. A headache can also be a sign of an overacid system, of poor bowel elimination or of excess stress and tension.

Taking pain killers may relieve today's headache, but it won't stop us from having another one tomorrow. When we start listening to the headache's message, we can actually learn to stop having headaches altogether. Headaches are a normal part of most people's lives only because the average person has bad health habits.

When we just keep taking painkillers without changing our bad health habits, we keep doing small things that damage our health. After twenty years of failing to heed these little warnings that something is wrong, we get a bigger wake up call in the form of a heart attack, cancer, diabetes or some other serious illness. Then we wonder, How could this happen to me?

Besides, painkillers themselves are toxic drugs. Sure, they re okay for occasional relief of pain, but when used frequently, they can damage the liver, kidneys, nerves and other organs. They can even increase the risk of heart disease and other serious health problems.

Since pain has a cause, we need to try to identify the source of our pain. When we suffer an acute injury such as a burn, cut or bruise, it's usually pretty easy to see what caused it. When we experience more long-lasting, chronic pain, it's often more difficult to see what is causing the damage. In fact, you may have to search and experiment a little to discover what is causing your pain, but the reward will be worth it. Here are some places to start.

At a cellular level, the primary cause of pain is lack of oxygenation to the tissues. The sharp pain of acute injuries is caused by the inflammatory process depriving cells of oxygen. When cells are chronically deprived of oxygen, you ll get chronic dull pain. Many people have experienced a great deal of relief from chronic pain just by practicing deep breathing. So, if you are in chronic pain, start practicing deep abdominal breathing for 5-10 minutes twice daily. The results may amaze you.

Accumulation of toxins in the tissues is another underlying cause of pain. Dehydration inhibits oxygen transport and allows toxins to accumulate. Another simple, but highly effective strategy for reducing chronic pain is to increase your water intake. This simple practice can greatly reduce the frequency of headaches, constipation, indigestion and muscle aches.

When tissues become congested due to poor lymphatic drainage, cells experience accumulation of toxic acid waste and low oxygenation. Ear aches, sore throats, menstrual pain and headaches can all occur because of swelling in the lymph nodes and congestion of the lymphatic system.

Massage can relieve stagnant lymph, especially if you use a topical analgesic essential oil when you do the massage. Tei Fu oil and Deep Relief oil massaged into painful areas of the body have worked minor miracles in easing pain. The secret is to massage the painful areas 6-8 times per day or more to keep the lymph flowing.

Moderate physical activity such as walking or swimming helps increase lymphatic drainage. Gentle bouncing on a mini-trampoline while doing deep breathing is an excellent way to move lymph and has proven very effective in reducing many kinds of chronic pain.

Another tip for reducing lymphatic congestion is to put one teaspoon of Lymphatic Drainage Formula and one teaspoon of Kidney Drainage Formula into a quart of water. Sip this throughout the day. This will aid hydration, too.

Cells produce acid waste in the process of metabolism. If this acid waste isn't flushed properly from the system it leads to an over acid condition in the tissues. This reduces oxygenation and causes muscle tension and pain. Drinking enough water and breathing deeply will help reduce overacidity, but you can also alkalize the body through other means.

Try drinking alkaline water or going on a short fast using lemon water. Use the juice of four lemons in 1/2 gallon of water, sweetened with a little real maple syrup. Don't eat anything for a couple of days and just drink this mixture whenever you are hungry. You should also drink lots of pure water. Often after two or three days of fasting like this, the body becomes more alkaline and many aches and pains simply disappear.

To maintain a more alkaline system, eat a diet that is 70% fruits and vegetables and only 30% proteins and grains. Drinking water with Liquid Chlorophyll can also help. Chlorophyll helps alkalize the system and helps the blood carry more oxygen to the tissues.

Muscle tension is a frequent cause of pain. This may be due to stress, but it can also be due to repetitive movements or bad posture that stress the structural alignment of the body. For example, sitting at a computer and typing all day can cause chronic tension in the neck, shoulders and upper back. This can lead to sore throats, headaches and back pain.

Periodic stretching and better posture will often correct this type of pain. Chiropractic adjustments, massage therapy or other forms of body work can help a person have better structural alignment, thereby easing pain.

Antispasmodic herbs can also help. Lobelia and kava kava are two great herbs for easing muscle tension to relieve pain. Muscle tension can also be eased by using Magnesium Complex.

All of these causes of pain are linked with chronic inflammation. Antioxidants and anti-inflammatories will help to reduce this inflammation and relieve chronic pain. Eating a diet that is 70% fresh fruits and vegetables (which are loaded with antioxidants) will greatly reduce chronic inflammation.

Besides these changes in diet, consider taking an antioxidant supplement. Thai-Go and Super ORAC are two excellent antioxidant supplements that have helped reduce chronic pain in many people.

Omega-3 essential fatty acids have also proven helpful in easing chronic inflammation and pain. They help with the production of anti-inflammatory and pain-relieving prostaglandins in the tissues. It also helps to avoid bad fats (margarine, shortening and processed vegetable oils) and refined carbohydrates.

Therapies: Hydration (pg. 10), Colon Cleansing (pg. 18), Enzymes (pg. 15), Essential Fatty Acids (pg. 17)

Remedies: APS II, Deep Relief Oil, IF Relief, Kava Kava, Lobelia, Magnesium, Omega-3, Tei Fu Oil, Triple Relief, Aloe Vera, Anamu, Catnip, Chlorophyll, Detoxification, Everflex Pain Cream, Flax Seed Oil, Geranium, Germanium Combination, Grapine, Herbal Sleep, Hops, IF-C, Inflammation, Kidney Activator (Chinese), Kidney Drainage, Lung Support, Lymphatic Drainage, Nature's Gold, Nature's Phenyltol with NEM, Noni (Morinda), Pain, Safflowers, St. John's Wort, Stress-J, Super ORAC, Thai-Go, Valerian Root

Palpitations

See *Heart Fibrillation or Palpitations*

Pancreatitis

Pancreatitis is an inflammation of the pancreas. This can be a serious, life-threatening condition requiring medical attention. The following may be helpful in recovery.

Therapies: Enzymes (pg. 15)

Remedies: Golden Seal, Licorice Root, Black Walnut, Food Enzymes, Horsetail, IF Relief, Lecithin, Proactazyme Plus, Probiotics, Uva Ursi, Vitamin B-Complex

Panic Attack

See *Anxiety Disorder (Panic Attack)*

Pap Smear (abnormal)

The following may be helpful in correcting an abnormal pap smear. Mix lemon and clove oils in a carrier oil (such as olive oil) and apply topically to the abdomen. Decongest the liver with liver-cleansing herbs. The following may help.

Remedies: All Cell Detox, Paw Paw, Tiao He Cleanse, Clove Bud, Enviro-Detox, Lemon Oil, Liver Balance

Paralysis

Paralysis is a complete or partial loss of function usually involving loss of motion or loss of sensation in any part of the body. This is usually due to nerve damage. The following may be helpful in aiding nerve repair, especially when used in the early stages of paralysis.

Therapies: Essential Fatty Acids (pg. 17), Antioxidants (pg. 17)

Remedies: Nutri-Calm, St. John's Wort, Capsicum, Lavender, Lobelia, Potassium

Parasites (general)

See Also *Giardia*

Imagine you have an unwanted house guest who is taking advantage of your hospitality. They eat your food, watch your TV, leave their dirty clothes all over your home and otherwise make a pest of themselves. Not only do they fail to help with any household chores, they make your work harder because you have to cook and clean for them, too. I m certain you would want to kick them out of your house at the earliest opportunity.

Well, there are tiny creatures that do similar things when they take up residence inside the body. These creatures feed off your nutrients, leave their waste products in your body and zap your energy. We call these unwanted creatures parasites.

There are many types of parasites that can make your body their unsuspecting host. Besides various worms (like tapeworms and ringworms), there are microscopic, single-celled organisms like giardia, H. pylori, candida and amoebas, which can all be considered parasites. In fact, there are over 1,000 organisms that are known to be parasites in human beings.

Many of us think that parasites are something that only affect people living in third-world countries. However, as the quotes on page four show, this simply isn't the case. Parasites are far more widespread in North America than most people are willing to believe. In fact, if single celled organisms are included, probably 80-95% of us have at least one form of parasite inside our body, which may, or may not, be affecting our health.

Parasites can be very hard to diagnose. A stool analysis may or may not reveal their existence, because even if parasites are present, they may or may not be present in any particular stool sample. Fortunately, they do leave tell-tale signs of their presence chronic fatigue or illness that just doesn't seem to get better, nervousness, teeth grinding, diarrhea, ulcers or digestive pain, extremes of appetite, weight loss or gain, anemia, itching (especially in the rectal area) and lowered immune response. These are some of the more common symptoms. If you have pets or animals and they have parasites, there is a high probability that you may, too. It's also easy to pick parasites up when traveling. So, if you ve experienced a change in health after traveling, it could be parasites.

Even when diagnosed correctly, parasites can be hard to eliminate. It takes powerful medications to kill parasites, and typically these medications are toxic to healthy tissue as well. Even some of the stronger antiparasitic herbs have some toxicity. However, our approach to parasites is going to be only partially effective if we just focus on trying to kill them.

One way to coax an unwanted guest to leave is to stop being a good host. In the body, this translates into altering the biological terrain to make it unfriendly for parasites. When one does this, smaller doses of less toxic antiparasitic remedies can be very effective in getting rid of them. This is where herbs can excel over chemical drugs. While they may not be as strong as drugs at directly killing the parasites, they are exceptionally good at helping to alter the host environment to make it unfriendly to parasites.

If you ve been diagnosed as having parasites, then you certainly want to serve up an eviction notice for the critters. However, it may be helpful to do a periodic parasite cleanse even if you haven't been diagnosed with parasites. For instance, if you have pets or animals, it's probably a good practice to do the Para-Cleanse program at least once a year, just for prevention. It's also a good idea to do a parasite cleanse after traveling (especially to third world countries). Another time to try a parasite cleanse is if you have chronic health problems that you can't seem to get rid of no matter what you do.

All of the following have been used to help destroy parasites in general. See listings for specific parasites, too.

Therapies: Colon Cleansing (pg. 18), Enzymes (pg. 15), Probiotics (pg. 16)

Remedies: Artemisia Combination, Black Walnut, Garlic, Herbal Pumpkin, Para-Cleanse, Paw Paw, All Cell Detox, Aloe Vera, Anamu, Cascara Sagrada, Cinnamon, Gastro Health, Golden Seal, Horsetail, LBS II, Lemon Oil, Neroli, Oregano (Wild), Parasites, Pau D Arco, Pine Needle, Sunshine Heros Whole Foods Papayazyme

Parasites (nematodes, worms)

See Also *Pain (general remedies for)*

Nematodes are tiny worms and include pinworms (Enterobius vermicularis), whipworm (Trichuris trichiura) and hookworms (Ancylostoma duodenale and Necator americanus). The most prevalent of these are pinworms, which are common in school children. They are highly contagious and easily passed around the family. Pinworm eggs can contaminate clothing, bed linens, toilet seats and even be transported through the air. When the eggs are ingested, the worms hatch in the intestines and their eggs are passed from the rectum. Pinworms can also infect the vulva, uterus and fallopian tubes in women.

Fortunately, they are one of the easier parasites to detect because they cause rectal itching. If one examines the rectal area at night with a flashlight, the worms appear as white threads at the anal opening. Other symptoms include: nervousness, inability to concentrate, lack of appetite and unusual dark circles around the eyes.

Precautions to prevent pinworms from spreading include washing bed linens, bed clothes and underwear of the entire family, having the infected child take daily morning showers to remove eggs deposited in the rectal region during the night, using disinfectants daily on the toilet seat and bathtub, sinks, handles, and being sure everybody washes their hands (and fingernails) before meals. Clean cat litter boxes daily. Also, avoid a diet high in sugar and other junk food, which encourages parasites.

Whipworms and hookworms are less common, but poise more serious health issues. Whipworms inject a fluid which liquefies colon tissue so the worms can ingest it. This creates severe nutritional deficiencies and infections. Whipworms affect about 6 million people worldwide.

Hookworms bite and suck on the intestinal wall causing bleeding and destruction of tissue. This can be severe enough to cause death. Since they consume iron, they cause severe anemia, which can help in detecting them.

Natural therapies for nematodes include garlic enemas or suppositories, Para-Cleanse, Artemesia Combination, High Potency Protease and Herbal Pumpkin.

Remedies: Artemisia Combination, Black Walnut, Blue Vervain, Cascara Sagrada, False Unicorn, Garlic, Herbal Pumpkin, Horsetail, LB Extract, LBS II, Papaya Mint, Para-Cleanse, Paw Paw, Sage, SF, Thyme

Parasites (tapeworm)

See Also *Pain (general remedies for)*

Other parasites may actually be more dangerous to one's health, but tapeworms (Taenia saginata-beef tapeworm and T. solium-pork tapeworm) are one of the most emotionally disturbing parasites because of their size. Tapeworms require an intermediate host, so they are usually ingested by eating improperly cooked beef, pork or fish.

Tapeworms are composed of numerous segments, about 3,000 to 4,000 per worm. New segments are formed near the head and the ones on the end are cast off with egg packets. When passed, these segments look like grains of uncooked rice or cucumber seeds. This is one of the ways they can be diagnosed. Other symptoms of tapeworms include diarrhea or (more commonly) constipation. They may also cause alternating diarrhea and constipation. Some people lose weight with tapeworms, but it is more common to gain weight and retain water. They raise blood sugar levels, cause anemia and interfere with vitamin B12 uptake.

Natural therapies for tapeworms include Para-Cleanse, Herbal Pumpkin, Artemesia Combination and High Potency Protease. Fasting on raw pineapple has helped in destroying them. Raw fig juice and pumpkin seeds have also been used.

Therapies: Colon Cleansing (pg. 18)

Remedies: Artemisia Combination, Para-Cleanse, Paw Paw, Protease, Aloe Vera, Black Walnut, Herbal Pumpkin, Silver Shield

Parkinson's Disease

Parkinson's disease is a chronic, progressive disease of the nervous system, usually occurring later in life. It involves the destruction of neurotransmitters that produce acetylcholine and dopamine and is marked by tremor and weakness in resting muscles and a gradual loss of muscle control. This disease is not easy to treat naturally, but natural remedies may slow the progress of the disease. Seek appropriate medical assistance.

Brain-Protex contains ingredients that help prevent the oxidative damage to neurons believed to be the cause of this disease. A complete list of other possible remedies follows.

Therapies: Antioxidants (pg. 17), Essential Fatty Acids (pg. 17), Oral Chelation (pg. 21), Heavy Metal Detoxification (pg. 22)

Remedies: Brain-Protex, 7-Keto, Alpha Lipoic Acid, Co-Q10, Evening Primrose Oil, GABA Plus, Ginkgo Biloba, Grapine, Herbal Sleep, Hops, Horsetail, HSN-W, Lecithin, Licorice Root, Mega-Chel, N-acetyl Cysteine, Nutri-Calm, Passion Flower, Super GLA, Valerian Root, Vitamin B-Complex, Vitamin C, Vitamin E, Wood Betony

Peptic Ulcer

See *Ulcers*

Periods (lack of)

See *Amenorrhea*

Pernicious Anemia

See Also *Anemia*

Pernicious anemia is a severe blood deficiency marked by a progressive decrease in number and increase in size and hemoglobin content of red blood cells. It is characterized by paleness, weakness, and gastrointestinal and nervous disturbances and a reduced ability to absorb Vitamin B-12 due to the absence of intrinsic factor. The Liquid B-12 is very important as it can be absorbed under the tongue.

Remedies: Vitamin B-12, Alfalfa, I-X, Vitamin B-Complex, Yellow Dock

Perspiration (excessive)

See Also *Night Sweating*

Excessive perspiration may be reduced with the help of some of the following.

Remedies: Sage, Trigger Immune, Lung Support, Nervous Fatigue Formula, Phyto-Soy, Stress Relief

Perspiration (deficient)

When a person has a difficult time sweating, the skin does not detoxify the body properly. The following can help promote normal perspiration. They can also be used to encourage sweating to help throw off acute illness.

Remedies: Capsicum, Yarrow, HCP-X

Pertussis (Whooping Cough)

Pertussis is a contagious bacterial infection, usually seen in children, marked by a spasmodic cough. It is also called whooping cough. The following have been used historically to treat whooping cough.

Remedies: Blue Vervain, Garlic, Lobelia, Rosemary, Silver Shield, Black Cohosh, Helichrysum, Licorice Root, Marshmallow, Oregano (Wild), Red Clover, Red Clover Blend, Thyme

Pets (supplements for)

Pets can benefit from herbs and nutritional supplements, too. Here are some basic suggestions.

Herbal Trace Minerals is a combination of kelp, alfalfa and dandelion. Rich in minerals, this blend was originally designed as a supplement for pets. It contributes to general pet health and healthy fur and skin. Just mix the powders from the capsules with the pet's food.

Many pet owners add Liquid Chlorophyll to the drinking water of their pets. Chlorophyll is found in green plants and is nature's blood-cleanser for animals. It helps red blood cells take up oxygen and supports the immune system. It also aids digestion and deodorizes the body. Taken daily it will help prevent halitosis, reduce body odors and dispel gas.

Animals also need good fats in their diet, so Super Omega-3 EPA can be a good supplement for your pet. It contains EPA and DHA fatty acids which benefit the cardiovascular system, the liver and the nervous system. These EFAs will help maintain healthy skin and bones and protects cell membranes form oxidative damage. They also support the immune system, helping your pet resist inflammation and arthritis.

In general, cats like the taste of omega-3, but should your cat or dog turn their nose up at it then try flax seed oil.

A good supplement for your pet's immune system is a mixture of goldenseal, echinacea and garlic powders in equal parts. Just mix the powders with their food. This strengthens the immune system, promotes gastrointestinal health and prevents parasites and infections. Silver Shield can also be used for pets for infections.

One of the big problems pets often have is parasites. Pets can have both external and internal parasites, the most common being fleas, lice, ear mites, fly larvae, ticks and Giardia.

Prevention is the best treatment. To avoid topical parasites, treat injuries as described above. Regular grooming will reveal the occasional hitchhiker, especially ticks. Keeping a clean environment for your pet and providing fresh drinking water everyday and good food will help prevent internal parasites.

Artemesia Combination is a good formula for helping to get rid of internal parasites in dogs and cats. Artemesia or wormwood has been used for centuries for the purpose of riding the body of amoebas, tapeworm and other parasites of the respiratory, digestive and intestinal system. It can be mixed with the animal's food. High Potency Garlic is another supplement that can help get rid of internal parasites.

If you have pets in your household, it is probably a good idea not only to give antiparasitic remedies to your pets, but to everyone in the family. Use the ParaCleanse with Paw Paw at least once per year for all members of the family. The ideal way to use this program is to take one package of ParaCleanse, wait ten days and then to a second package of the cleanse.

Generally speaking most dogs and cats will respond to the same remedies you use for human beings.

Nature's Fresh is a must-have product for pet owners. It eliminates odors and stains caused by having pets in the home.

All of the following are useful products and supplements for pets.

Therapies: Minerals (pg. 15), Essential Fatty Acids (pg. 17)

Remedies: Chlorophyll, Flax Seed Oil, Herbal Trace Minerals, Nature's Fresh, Omega-3, Silver Shield, Artemisia Combination, Colloidal Minerals, Echinacea Purpurea, Garlic, Glyco Essentials, Golden Seal, Herbal Pumpkin, Para-Cleanse, Super GLA, Ultimate Echinacea

Phlebitis

See Also *Varicose Veins*

Phlebitis is inflammation of a vein, usually in the legs. The following may be helpful for phlebitis.

Remedies: Butcher's Broom, Vari-Gone, Vitamin E, White Oak Bark, Capsicum, IF Relief, Lecithin, Lemon Oil, Thai-Go

Phobias

See Also *Anxiety Disorders*

A phobia is an excessive, unreasonable desire to avoid something because of fear. When this fear is beyond control and interferes with daily life, the phobia becomes an anxiety disorder. Counseling and emotional healing work will be necessary, but supplements that help anxiety disorders may also be useful.

Remedies: Adrenal Support, Distress Remedy, HY-A, Licorice Root, Nervous Fatigue Formula, Super Algae

Piles

See *Hemorrhoids*

Pimples

See *Acne (Pimples, Blackheads)*

Pin Worms

See *Parasites (nematodes, worms)*

Pink Eye

See *Conjunctivitis (Pink Eye)*

Pleurisy

Pleurisy is inflammation of the tissues that cover the lungs and line the thoracic cavity, creating painful and difficult breathing, cough, and collection of fluid or fibrous tissue in the thoracic cavity. The herb pleurisy root is a specific for this problem. Other possible remedies include the following.

Remedies: Four, Lobelia, Marshmallow & Fenugreek, Sarsaparilla, Thai-Go, Yarrow

PMS (general)

See Also *Dysmenorrhea, Menorrhagia (Heavy Menstrual Bleeding), Menstrual Cramps*

PMS is an abbreviation for premenstrual syndrome. PMS is not a specific ailment. A syndrome is a collection of symptoms with multiple causes. Pre-Menstrual Syndrome includes over 150 signs and symptoms which have been classified into four major types. Therapy for PMS will depend largely on which type of PMS you have. Please check out the four PMS types listed next to see which type or types match your symptoms best. The following are general remedies for PMS, which can be helpful for most, if not all, types.

Therapies: Avoid Xenoestrogens (pg. 23), Colon Cleansing (pg. 18), Essential Fatty Acids (pg. 17)

Remedies: Black Currant Oil, Blood Build, Evening Primrose Oil, FCS II, Female Comfort, Magnesium, Monthly Maintenance, PMS, Super GLA, Vitamin B-6, Vitamin E

PMS Type A

See Also *Anger (excessive), PMS (general)*

A stands for anxiety. This PMS type is characterized by high levels of estrogen and low levels of progesterone, and is the most common type of PMS. In fact, 80% of the cases of PMS usually involve too much estrogen and a deficiency of progesterone. Since symptoms include a tendency to anxiety, irritability, mood swings, moodiness and nervous tension, this is the type that is most commonly the source of PMS jokes.

The main therapy is to decrease estrogen levels and increase progesterone. Too much estrogen increases levels of adrenaline, noradrenaline, and serotonin, while the levels of dopamine and phenylethlamine drop. Estrogen also seems to block vitamin B6, which is instrumental in many important functions in the body, including maintaining normal blood sugar levels and stabilizing ones moods.

Pro-G-Yam cream and herbs that enhance progesterone are helpful, of course, since estrogen and progesterone compete for receptor sites. Higher levels of progesterone, therefore, neutralize the excess estrogens.

It is also important, however, to minimize exposure to xenoestrogens and reduce estrogen levels in the body by improving the liver's ability to detoxify excess estrogens. Indole 3 Carbinol and liver formulas are good possibilities for reducing estrogens.

Poor nutrition really aggravates this condition. Generally, women suffering from PMS A have a high consumption of animal fats, which often contain xenoestrogens and interfere with progesterone production. All these symptoms of type A PMS are further heightened by stress, too much sugar, refined carbohydrates, alcohol and caffeine.

To treat PMS A, reduce fats in the diet, especially animal fats of dairy and meat. Use organic meat and dairy products. Liver cleansing, especially in the first half of the cycle is also helpful, as is magnesium and B6. Monthly Maintenance, a Chinese herbal formula with magnesium, vitamin B6 and other nutrients is very helpful for this type of PMS.

Therapies: Avoid Xenoestrogens (pg. 23), Stress Management (pg. 8)

Remedies: Indole 3 Carbinol, Liver Balance, Pro-G-Yam Cream, Blood Build, BP-X, Chamomile (Roman), False Unicorn, Geranium, Lavender, Magnesium, Monthly Maintenance, Vitamin B-6

PMS Type C

See Also *Hypoglycemia, PMS (general)*

C stands for cravings. These cravings are generated by a hypoglycemic type of reaction. Blood sugar levels actually do drop significantly, but only between ovulation and the onset of menses. At this time there are strong desires for refined carbohydrates, chocolate, and pretty much anything that is sweet. Unfortunately, because the body's insulin balance is even more sensitive than it would normally be, the drop in blood sugar is faster and more dramatic, causing greater fatigue. It's the fatigue that prompts one to crave those sugars, thus the cycle starts again.

If magnesium levels are low, the pancreas will produce more insulin. Too much salt will also increase insulin production upon ingestion of sugar. In PMS C it'seems that prostaglandin 1 (PGE1) is deficient and this is also known to give rise to insulin production.

To treat PMS C, following a diet similar to that of a hypoglycemic diet can minimize the discomforts of PMS C. Stay off of refined sugars! Supplements of magnesium and good fats will help to keep the hormonal balance in check. Vitamin B6, and zinc help, also chromium to help serotonin levels and balance blood sugars and insulin. Reduce dairy, eat nuts, sesame seeds, millet and cashews. Licorice root helps to balance blood sugars. Essential oil of ylang ylang has been known to help.

Therapies: Low Glycemic Diet (pg. 11)

Remedies: Chromium GTF, Licorice Root, Magnesium, Super Algae, Evening Primrose Oil, HY-C, Omega-3, Vitamin B-6, Ylang Ylang, Zinc

PMS Type D

See Also *Depression, Lead Poisoning, PMS (general)*

D is for Depression. This type of PMS is associated with depression, anxiety, and rage coupled with confusion, forgetfulness, crying easily and being accident-prone. Often these women feel suicidal.

PMS D is due to too much progesterone and not enough estrogen. It is exactly the opposite problem of PMS Type A. In balance, progesterone has a calming effect. When progesterone is in excess, it becomes a depressant to the brain.

An interesting feature to this type of PMS is the high levels of lead found in hair samples. When a deficiency of magnesium occurs, the body seems to be more susceptible to taking in lead. High levels of lead are known to be the cause of some types of chronic depression. Heavy metal cleansing and/or oral chelation may be helpful here.

A diet of fresh fruits, raw or steamed vegetables of all types, leafy veggies, whole grains, seeds and nuts and olive oil, add essential fatty acids, sea salt, and herbal salts along with the right supplements will work wonders in alleviating PMS depression. St. John's wort or Mood Elevator are good for depression in general, but black cohosh is one of the best remedies for this type of depression. Flash-Ease also helps.

Vitamin B6 and magnesium are absolutely required for this type. Essential oils of clary sage, rose, lavender and bergamot can help to bring relief.

Therapies: Oral Chelation (pg. 21), Heavy Metal Detoxification (pg. 22)

Remedies: Bergamot, Black Cohosh, Clary Sage, Flash-Ease T/R, Heavy Metal Detox, Lavender, Magnesium, Mega-Chel, Mood Elevator, Rose Bulgaria, Sea Salt, St. John's Wort, Vitamin B-6

PMS Type H

See Also *Edema (Dropsy, Water Retention, Swelling), PMS (general)*

H is for Hyperhydration. This means that uncomfortable feeling you get caused by water retention. Water retention during times of PMS can be due to too much aldosterone, a hormone made by the adrenal glands, high levels of estrogen, or low levels of magnesium.

Again, when the levels of magnesium are low, it'seems to upset the balance of the hormones in so many ways. Vitamin B6 depends upon magnesium in order to convert into its active form, so a B6 deficiency is also indicated here with PMS H.

Diuretics are needed in PMS Type H. Other supplements that can help include: magnesium, vitamin B6, vitamin E and evening primrose oil. Adding beans, whole grains and leafy vegetables to the diet can be helpful, too. The essential oils of frankincense, juniper and lemon help.

Remedies: Kidney Activator (Chinese), Magnesium, Vitamin B-6, Evening Primrose Oil, Frankincense, Juniper Berries, Kidney Activator, Kidney Drainage, Lemon Oil, Lymphatic Drainage, Vitamin E

PMS Type P

See Also *Dysmenorrhea*

P is for Pain. Pain associated with menstruation is called dysmenorrhea.

Therapies: Hydration (pg. 10)

Remedies: Cramp Relief, Magnesium, Ginger, Kava Kava, Lobelia

PMS Type S

See Also *Acne (Pimples, Blackheads), Adrenals (exhaustion, weakness or burnout)*

S is for Skin. Some women get outbreaks of acne due to high levels of androgens which are a side affect of stress. Chronic stress will eventually fatigue the adrenal glands. Exhausted adrenal glands are aggravated by animal fats and dairy products. Eat more green leafy vegetables, vegetable proteins, and fruit. A colon and liver cleanse would add vital energy to the body and clear the skin. Liver formulas are good for helping clear skin. Decrease nicotine, caffeine, sugar and salt consumption.

Therapies: Colon Cleansing (pg. 18)

Remedies: BP-X, Liver Balance, Adrenal Support, Burdock, Dandelion, Nervous Fatigue Formula, Red Clover, Skin Detox

Pneumonia

Pneumonia is a disease of the lungs characterized by inflammation and fluid accumulation; usually caused by infection or irritation. The following have been used to combat pneumonia. Seek appropriate medical assistance.

Remedies: ALJ, Garlic, Immune Stimulator, Silver Shield, Breathe EZ, Bronchial Formula, Capsicum & Garlic with Parsley, CBG Extract, Four, Lemon Oil, Lobelia, Marshmallow & Fenugreek, Yarrow

Poison Ivy or Oak

Certain plants cause a mild to severe contact allergic reaction when touched, such as poison ivy or poison oak. Symptoms of this allergic reaction may include mild to severe redness, rash, itching, burning and/or oozing blisters. Wash the skin immediately with plenty of soap and water after contact with these plants. Then apply herbs topically to reduce swelling and inflammation such as aloe vera, black walnut, burdock, uva ursi or manzanita, jewelweed, etc. Make these herbs into a liquid form (infusion or decoction) for application.

Internally, vitamin C and blood purifiers like BP-X or yellow dock may be helpful. If you are hypersensitive to these plants, try using homeopathic poison ivy (Rhus tox) to desensitize the body. (Rhus tox is not an NSP product.) NSP products that may help include all of the following:

Remedies: Aloe Vera, Burdock, Uva Ursi, White Oak Bark, Allergy, Black Walnut, BP-X, Ginseng (Korean), Ginseng (Wild American), Lobelia, Sea Salt, Tea Tree Oil, Vitamin A & D, Vitamin C, Vitamin C, Yellow Dock

Poisoning (general)

There are numerous toxic substances that can accidentally be inhaled, ingested or absorbed through the skin. Call a poison control center near you for help with any kind of acute poisoning. Activated charcoal absorbs many toxins and milk thistle helps protect the liver against many toxins. For some toxins, lobelia or ipecac can be taken to induce vomiting. These and other remedies that may help recover from poisoning follow, but always contact a poison control center for advice.

Remedies: Charcoal (Activated), Lobelia, Milk Thistle, Milk Thistle Combination, Chlorophyll, Detoxification, Enviro-Detox, Vitamin C

Poisoning (food)

Consuming contaminated food can create the often severe symptoms of food poisoning, which include nausea, vomiting and often severe diarrhea. In severe cases medical attention should be sought.

Lobelia can be used to induce vomiting to expel the toxic material more quickly from the body. If lobelia isn't available, Ipecac, available at most drug stores, is an alternative. Other herbs that may help induce vomiting include boneset, blue vervain and mustard.

To absorb irritants in the digestive tract and ease diarrhea activated charcoal is the very best. Alternatives include Hydrated Bentonite or slippery elm. Taking probiotics and PDA can help prevent food poisoning.

Therapies: Probiotics (pg. 16), Fiber (pg. 17)

Remedies: Bentonite (Hydrated), Charcoal (Activated), Lobelia, Silver Shield, Ultimate Echinacea, Aloe Vera, Blood Build, Blue Vervain, Garlic, Golden Seal, Immune Stimulator, Liver Balance, Milk Thistle, Milk Thistle Combination, PDA, Probiotics, Psyllium

Polyps

A polyp is a projecting mass of swollen, overgrown or tumorous tissue, usually found in the nasal cavity or intestine. They are benign (non-cancerous) growths. The following have been used for polyps.

Remedies: All Cell Detox, Bayberry, HCP-X, Aloe Vera, Barley Grass, Blood Build, BP-X, Pau D Arco, Vitamin A & D

Post Partum Depression

Depression after having a baby is caused by low levels of hormones. Black cohosh can be very helpful. The following may also be helpful.

Remedies: Black Cohosh, Mood Elevator, FCS II, Jasmine Absolute, NF-X

Post Partum Weakness

The following may be safely used to help a woman gain strength after giving birth.

Therapies: Minerals (pg. 15), SuperFood (pg. 15), Essential Fatty Acids (pg. 17)

Remedies: Cellular Energy, GreenZone, Mineral-Chi Tonic, Chlorophyll, Colloidal Minerals, Energ-V, I-X, Super GLA, Target Endurance, Thai-Go, Vitamin C

Post Traumatic Stress Disorder

See Also *Adrenals (Exhaustion, Weakness or Burnout), Anxiety Disorders, Stress*

This condition used to be called shell shock or battle fatigue. It was identified as a condition affecting soldiers who had undergone so much stress that they simply couldn't cope anymore. You don't have to have gone to war to suffer post traumatic stress disorder. Anytime you ve had a series of extremely stressful situations that have overwhelmed your ability to cope and have left you in a chronic state of stress and anxiety you have post traumatic stress disorder to one degree or another.

Post Traumatic Stress Disorder is considered an Anxiety Disorder, so see that heading for more ideas on what to do for it. It typically involves extreme adrenal burnout and possible remedies include the following.

Therapies: Hydration (pg. 10), Stress Management (pg. 8), Low Glycemic Diet (pg. 11), Hiatal Hernia Correction (pg. 12)

Remedies: Adrenal Support, Kava Kava, Nervous Fatigue Formula, Pantothenic Acid, Distress Remedy, Nutri-Calm

Pregnancy (herbs and supplements for)

See Also *Labor and Delivery*

During pregnancy, a woman needs the nutrients necessary to form two extra pounds of uterine muscle, several pounds of amniotic fluid and the placenta. She also experiences a 50% increase in blood volume, and her liver and kidney cells need to process the waste from two living beings. And this is all in addition to forming the bones, muscles, skin, glands, nervous system and other vital organs of her developing child.

This means her body will require larger than normal amounts of proteins, good fats, vitamins and minerals-nutrients she isn't going to get from eating a diet of refined and processed foods. A pregnant woman needs to take extra good care of her body by consuming fresh fruits and vegetables, whole grains and organically raised meat and dairy products. She should also avoid alcohol, coffee, tobacco, refined sugar, white flour, shortening, margarine, commercially fried foods and hydrogenated oils. Ideally, all these dietary changes should take place several months before conception in order to prepare the body for a healthy pregnancy.

Chemicals are a major cause of birth defects and health problems in infants, so a pregnant woman should be very careful to minimize her exposure to toxic chemicals of all kinds, particularly pesticides. Many pesticides will cause a miscarriage because of their xenoestrogenic nature. A pregnant woman should also be careful to use only natural, non-toxic household cleaning products (such as Nature's Fresh and Sunshine Concentrate) and personal care products. It is also a good idea to do a cleanse prior to conception. This will minimize a woman's risk of morning sickness, toxemia and other health problems during pregnancy.

Traditional cultures used special foods for pregnant women to ensure healthy babies. Supplements can do the same for modern pregnant women. Mega-doses of vitamins and minerals aren't wise, but a good prenatal vitamin like Nature's Prenatal can be beneficial. Nature's Prenatal contains essential nutrients for energy and basic health during pregnancy, such as 800 mg. of folic acid, which is essential in the prevention of neural tube defects.

Many women, however, find that using whole foods and herbs are even more important than taking a prenatal in maintaining good health during pregnancy. Ultimate GreenZone, for instance, is a great whole food supplement for pregnant women.

Good fats are a must for pregnancy, as a developing child's brain and nervous system need good fats. Supplementing with omega-3 fatty acids has been shown to reduce the risk of developing pre-eclampsia, postpartum depression and pre-term labor. Deep ocean fish (especially sardines), walnuts, flax seeds and flax seed oil, hemp seed or hemp seed oil, avocados, coconut oil and organic butter from grass-fed cows are also great sources of good fats for pregnancy.

Due to modern agricultural practices, the population as a whole is deficient in minerals. Even when consuming a wholesome organic diet, pregnant women usually need extra amounts of trace minerals. Many problems in pregnancy have been attributed to trace mineral deficiencies because the developing infant pulls minerals from the mother's bloodstream.

Drinking a pregnancy tea of equal parts red raspberry leaf, alfalfa and peppermint is an easy way to get the needed trace minerals. Expectant mothers should steep 3-4 heaping tablespoons of this mixture in a quart of boiling water and drink at least a quart of this tea every day. Other mineral-rich herbs like nettles, oatstraw and horsetail can be added to the tea for an even better effect. This tea helps prevent morning sickness and strengthens a woman's body during pregnancy and delivery.

Red raspberry leaf is especially valuable to pregnant women as it tones the uterus and prepares the body for childbirth, making labor and delivery easier. It also reduces morning sickness, lowers the risk of miscarriage and decreases the risk of postpartum hemorrhage. Women who don't have time to make the pregnancy tea described above, can take Red Raspberry in capsules or use the liquid Red Raspberry Blend.

Other great ways to get minerals include taking Liquid Dulse, Colloidal Minerals, Mineral Chi Tonic and/or HSN-W. Generous intake of mineral-rich herbs like these has been found to eliminate most of the problems women experience in pregnancy, create healthier babies and make delivery easier.

Pregnant women especially need more iron. Eating dark, green leafy vegetables, organic red meat and iron-rich herbs can keep iron levels normal during pregnancy. Consider supplementing with yellow dock or I-X in addition to adding nettles to your pregnancy tea. Drinking Liquid Chlorophyll will increase utilization of iron. Taken after childbirth, chlorophyll also increases the quality of the mother's breast milk

Therapies: Antioxidants (pg. 17), SuperFood (pg. 15), Minerals (pg. 15), Enzymes (pg. 15), Essential Fatty Acids (pg. 17),Fiber (pg. 17)

Remedies: Alfalfa, Chlorophyll, Colloidal Minerals, Folic Acid Plus, Nature's Prenatal, Omega-3, Red Raspberry, Thai-Go, Bee Pollen, Blessed Thistle, Calcium, Dulse, Flax Seed Oil, Herbal Trace Minerals, HSN Complex, HSN-W, I-X, Ionic Minerals, Iron, Kelp, Love and Peas, Magnesium, Mineral-Chi Tonic, Red Raspberry Liquid, Super Trio, Vitamin B-6, Vitamin B-Complex, Vitamin C, Vitamin E, Yellow Dock

Pregnancy (herbs and supplements to avoid during)

The following supplements and herbs should be avoided during pregnancy. 5-W should only be used during the last five weeks of pregnancy. Black and blue cohosh can be used after the due date. Jasmine should not be used in the first trimester. Women should not do cleanses during pregnancy or use anti-parasitic herbs such as Artemesia Combination and Paw Paw Cell Reg(c) should definitely be avoided. Herbs and supplements that increase estrogen levels, such as black cohosh, may contribute to miscarriage. Glandular herbs that increase progesterone are fine, as progesterone helps a woman stay pregnant. Blue cohosh and anamu are potentially abortive and should be avoided if a woman is trying to get pregnant. They are especially dangerous during the first trimester. The supplements that are most likely to be a problem are highlighted in the following list. Supplements that are not highlighted may be safe during pregnancy, but should be used with caution.

Remedies: 7-Keto, Anamu, Artemisia Combination, Black Cohosh, Blue Cohosh, DHEA-F, DHEA-M, Flash-Ease T/R, Paw Paw, 5-HTP, 5-W, AdaptaMax, C-X, Cascara Sagrada, Cellu-Smooth, Cordyceps, Dong Quai, FCS II, Female Comfort, GABA Plus, Golden Seal, Green Tea Extract, Guggul, Heavy Metal Detox, IGF-1, Jasmine Absolute, JP-X, Marjoram (Sweet), Melatonin Extra, MetaboMax, MetaboStart, Myrrh, NF-X, Para-Cleanse, Pregnenolone, Rosemary, SAM-e, Senna Combination, SnorEase, Uña de Gato Combination, Wood Betony, X-Action (Women s), X-Action Gel

Premature Ejaculation

See Also *Erectile Dysfunction*

The following may be helpful with premature ejaculation in men.

Therapies: Avoid Xenoestrogens (pg. 23)

Remedies: Damiana, Ginseng (Korean), KB-C, Noni (Morinda)

Progesterone (low)

With exposure to xenoestrogens and an excess burden on liver function many women have too much estrogen and not enough progesterone. Progesterone and estrogen compete for the same receptor sites and good reproductive health requires a balance between these two hormones. Since the scale in many women is tipped towards estrogen, a natural progesterone supplement, such as Pro-G-Yam Cream is helpful for many women.

Don't overdo it with progesterone creams, however. You can also get too much progesterone. Symptoms of progesterone overdose include headache, weight gain, fatigue, water retention and depression. Consider getting a test to see where your current hormone balance lies.

Herbs such as sarsaparilla and false unicorn can also be used to counteract excess estrogen by enhancing progesterone. These herbs have been used to help sustain pregnancy and prevent miscarriage and to relieve heavy menstrual bleeding and cramps. Wild Yam and Chaste Tree, taken regularly for several months, can also balance out estrogen and progesterone.

Therapies: Avoid Xenoestrogens (pg. 23)

Remedies: False Unicorn, Pro-G-Yam Cream, Blue Cohosh, C-X, Magnesium, Monthly Maintenance, Sarsaparilla, Wild Yam, Wild Yam & Chaste Tree

Prolapsed Colon

A falling down or sagging of the colon from its usual position is called a prolapsed colon. This condition often involves the transverse or horizontal portion of the colon. Laying on a slant board with one's feet elevated and massaging the colon can help. The following remedies may also be beneficial.

Remedies: **Bowel Detox, Mood Elevator,** Calcium, Vitamin C

Prolapsed Uterus

A falling down or sagging of the uterus from its usual position is called a prolapsed uterus. It is more common after pregnancies. It may prevent conception and often puts pressure on the bladder, which may lead to incontinence.

Remedies: **Mood Elevator, Red Raspberry,** Astragalus, Bayberry, Black Walnut, Calcium, Cranberry & Buchu, Dong Quai, Magnesium, Spleen Activator

Prostate Problems

See *Benign Prostate Hyperplasia (BPH), Prostatitis*

Prostatitis

See Also *Benign Prostate Hyperplasia (BPH)*

Prostatitis is inflammation of the prostate gland that causes painful urination, frequent trips to the men's room, and miss-aim and dribbling because the weak stream of urine is insufficient to fully open the flaps at tip of penis. It is sometimes due to infection, but now is more often due to other, unknown causes.

One reason why the prostate may become inflamed involves its proximity to both the bladder and the rectum. If the body is toxic, the irritants being eliminated from the colon and urinary passages may be irritating the prostate gland, causing it to swell.

If the problem is due to an acute or chronic infection, consider some herbs with natural antibacterial action, such as golden seal or uva ursi. To reduce inflammation, consider IF Relief or Thai-Go, two excellent, anti-inflammatory formulas. PS II, a general herbal formula for the prostate, may also be helpful. KB-C is another good remedy.

Another good remedy for reducing prostate inflammation is Omega-3 essential fatty acids. Eskimo men who have a fish-rich diet have significantly lower rates of prostatitis and prostate cancer than other men. Omega-3 fatty acids have also been shown to inhibit prostate cell growth and reduce prostate enlargement. They help decrease pain and fatigue, reduce nighttime urination, increase elimination (stream) and increase libido.

Therapies: Colon Cleansing (pg. 18), Avoid Xenoestrogens (pg. 23), Probiotics (pg. 16), Essential Fatty Acids (pg. 17), Antioxidants (pg. 17), Hydration (pg. 10)

Remedies: **KB-C, PS II, Uva Ursi,** Golden Seal, Herbal Pumpkin, IF Relief, Kidney Activator (Chinese), Men's Formula, Omega-3, P-X, Silver Shield, Thai-Go

Protein Digestion (poor)

The following help to improve protein digestion and metabolism.

Therapies: Enzymes (pg. 15)

Remedies: **Digestive Bitters Tonic, Food Enzymes, PDA, Protease, Spleen Activator,** Ginseng (Wild American)

Psoriasis

Psoriasis differs from eczema because it involves rapid skin growth and appears to be an autoimmune disorder, like multiple sclerosis or lupus. Psoriasis primarily affects the skin, but in about 10% of the cases the joints are also affected. Research suggests that psoriasis is triggered when certain T-cells reproduce very rapidly, which starts an inflammatory reaction that causes skin cells to multiply seven to twelve times faster than normal. In natural medicine this may be taking place because the skin is malnourished and weak or because of allergic reactions to food.

Because this hyperactivity of the immune system also creates a form of inflammation, psoriasis has symptoms similar to eczema. The skin is often itchy and dry, and frequently cracking or blistering. Oils are needed to keep the skin moist. In particular, the omega-3 essential fatty acids may be helpful.

Diet is important in the effective treatment of psoriasis. Fasting and vegetarian diets have been shown to reduce symptoms. Eliminating gluten-bearing grains like wheat has also benefited some sufferers. Again, it would be a good idea to consider foods compatible with one's blood type, as food allergies are a contributing factor to psoriasis.

Incomplete protein digestion and bowel toxemia may be underlying factors in psoriasis. Proactazyme Plus or Protease Plus, taken between meals, will help break down undigested protein and detoxify the colon.

Detoxifying the liver is also important. Two good products for doing this are Milk Thistle Combination and All Cell Detox. Combined with a source of dietary fiber such as Nature's Three or Psyllium Hulls Combination, these formulas will also help to gently cleanse the bowel to get rid of the gut-derived toxins that may be involved in this disease.

Nutrients that have been reported helpful for psoriasis include vitamin A (in large doses of about 50,000 to 75,000 IU per day), vitamin E (400 to 800 IU per day), B-complex vitamins and vitamin B6 in particular, vitamin C, zinc and chromium. Feverfew can be used both internally and topically to ease psoriasis. Finally, the polyphenols in green tea can also help to reduce irritation of the skin and ease psoriasis.

Therapies: Enzymes (pg. 15), Fiber (pg. 17)

Remedies: Burdock, Feverfew, Green Tea Extract, Omega-3, Paw Paw, All Cell Detox, Aloe Vera, BP-X, Chamomile, Chickweed, Chromium GTF, Eczema/Psoriasis, Enviro-Detox, Feverfew, Flax Seed Oil, Gotu Kola, GreenZone, Hi Lipase, Licorice Root, LIV-J, Lymph Gland Cleanse-HY, Milk Thistle Combination, MSM, Nature's Three, Pau D Arco, PDA, Protease, Psyllium Hulls Combination, Red Clover, Red Clover Blend, Sandalwood, Sarsaparilla, Skin Detox, Super GLA, Vitamin A & D, Vitamin B-6, Vitamin B-Complex, Vitamin C, Vitamin E, Zinc

Puberty (hormone balancer)

It's no great secret that teenagers are undergoing major hormonal changes. These changes affect not only a teen's body, but also their thoughts and emotions, so it is important to talk with kids about these changes and help them through this critical time in their lives. Appropriate herbs and supplements can also help.

For young women who are just starting their periods, a female hormone balancing formula like NF-X can help. It cleanses the genito-urinary system and can make periods easier, reduce the risk of urinary tract infections and counteract environmental estrogens. It can also help reduce androgens in teenage boys to lessen acne.

Another good hormone balancer is Wild Yam and Chaste Tree. Chaste tree, also known as vitex, regulates the pituitary gland to balance hormone levels. It is a useful remedy for teen-age acne but can also help with excessive aggression in boys and menstrual cramps in girls.

If young women are having problems with painful periods, irregular periods and other menstrual problems, Monthly Maintenance can help. This blend contains vitamins and minerals and a blend of Chinese herbs that counteract PMS and balance the reproductive cycle.

Therapies: Minerals (pg. 15), Essential Fatty Acids (pg. 17), SuperFood (pg. 15), Avoid Xenoestrogens (pg. 23), Stress Management (pg. 8)

Remedies: Monthly Maintenance, Wild Yam & Chaste Tree, BP-X, FCS II, NF-X, NF-X, Red Raspberry

Puncture Wounds

See *Tetanus, Wounds and Sores*

Pyorrhea

See *Gingivitis (Bleeding Gums, Gum Disease, Pyorrhea)*

Radiation Sickness

When the body is exposed to radiation, cellular DNA is damaged and a toxic condition is created in the body. X-rays, radon, microwave ovens, radar and radiation treatments for cancer are among the ways the body can be exposed to radiation. One of the most important supplements to take when one has been or will be exposed to radiation is iodine (kelp, dulse, Thyroid Activator, etc.) as this protects the thyroid against radioactive iodine. Radiation also causes free radical damage,so antioxidants are also helpful. These remedies can also be used to counteract the damage caused by radiation treatments for cancer.

Therapies: Antioxidants (pg. 17)

Remedies: Algin, Aloe Vera, Dulse, Barley Grass, Blood Build, Carotenoid Blend, Eleuthero, Enviro-Detox, Ginseng (Korean), Ginseng (Wild American), Kelp, Nature's Gold, S-O-D With Gliadin, Super Algae, Super Antioxidant, Super ORAC, Thai-Go, Thyroid Activator, Uña de Gato Combination, Vitamin A & D

Rapid Heart Beat

See *Tachycardia*

Rashes and Hives

See Also *Dermatitis, Itching*

A rash is a skin eruption, which can be local or general. It is an inflammatory process and may involve allergic reactions or toxicity. Symptoms include redness, swelling, itching, burning and sometimes blisters.

The following herbs used in a bath may help relieve hives: chickweed, comfrey, ginger, marshmallow, mullein, and Oregon grape. Also consider topical applications of aloe vera gel or Herbal Trim. Internally, consider blood purifiers and/or HistaBlock to stop allergic reactions.

The following can be helpful for rashes and hives.

Remedies: BP-X, Burdock, Herbal Trim Skin Treatment, HistaBlock, Liver Balance, Nature's Fresh, Pau D Arco, Skin Detox, Aloe Vera, Black Walnut, Blood Build, Chamomile, Chickweed, Dandelion, E-Tea, Enviro-Detox, Helichrysum, IF-C, Lemon Oil, Lymph Gland Cleanse, Marshmallow, MSM, Mullein, Nature's Gold, Oregon Grape, Red Beet Formula, Rose Bulgaria, Tea Tree Oil, Vitamin A & D, Yellow Dock

Raynaud's Disease

Raynaud's disease is a vascular disorder marked by recurrent spasm of the capillaries especially those of the fingers and toes upon exposure to cold, skin changes from white, to blue, to red in succession and usually includes pain. The following may be helpful.

Remedies: Capsicum, Co-Q10, Ginkgo & Hawthorn, Butcher's Broom, Chlorophyll, Ginkgo Biloba, Lobelia

Recouperation

See *Convalescence*

Respiratory Congestion

See *Congestion (lungs)*

Respiratory Infections

See *Congestion (lungs), Infection (bacterial), Infection (viral), Pleurisy, Pneumonia*

Restless Dreams

See Also *Adrenals (Exhaustion, Weakness or Burnout)*

Restless and disturbing dreams are often one of the first symptoms that a person is under too much stress and is in danger of developing adrenal burnout. The following are indicated for restless and disturbing dreams.

Remedies: Adrenal Support, Nervous Fatigue Formula, Nutri-Calm

Restless Leg Syndrome

Restless leg syndrome is a condition where the legs itch, tickle or burn, often a night. Moving them brings temporary relief, but the urge to move them returns seconds or minutes later. It can hinder sleep. Food allergies, mineral deficiencies and stress could be underlying problems. The following remedies may help.

Therapies: Minerals (pg. 15), Stress Management (pg. 8)

Remedies: Kava Kava, Magnesium, Vitamin B-Complex, Folic Acid Plus, Herbal Sleep, Lobelia, St. John's Wort, Super GLA, Vitamin E

Reversed Polarity

Reversed polarity is a problem encountered by people who do muscle testing. The energy fields of the body reverse so the poles are incorrect. This causes a person to test incorrectly. People with reversed polarity are attracted to negative influences and have weakened immune systems. Exposure to electromagnetic pollution (computers, microwave ovens, cell phones, etc.) is often the cause. The following products help correct reversed polarity.

Therapies: Hydration (pg. 10)

Remedies: Bee Pollen, Mood Elevator, THIM-J, **Trigger Immune,** Damiana, Super Algae

Reye's Syndrome

Reye's syndrome is a serious disease. It affects internal organs such as the brain and liver and typically occurs after a viral infection. Taking aspirin with a viral disease dramatically increases the risk of this disorder. This serious disorder requires medical attention, but the following supplements may be helpful in recovery.

Therapies: Affirmation and Visualization (pg. 8)

Remedies: **VS-C**, Flax Seed Oil, Free Amino Acids, L-Carnitine, Lecithin, Uña de Gato Combination

Rheumatic Fever

Rheumatic fever is an acute, often recurrent, disease found mainly in children and young adults. It is characterized by fever, inflammation, pain and swelling in and around the joints. The inflammation also affects the surface and valves of the heart and may involve the formation of small nodules in the heart or other tissues. Appropriate medical assistance should be sought. The following may be helpful in recovery.

Remedies: Cardio Assurance, Silver Shield, Co-Q10, Garlic, Hawthorn Berries, L-Carnitine, Oregon Grape, Passion Flower, Peppermint

Rheumatoid Arthritis (Rheumatism)

See Also *Arthritis*

Rheumatoid arthritis is a type of arthritis that has an autoimmune component. It is characterized by inflammation or pain in muscles, joints or fibrous tissues. The following may prove helpful in cases of rheumatism.

Therapies: Hydration (pg. 10), Colon Cleansing (pg. 18), Essential Fatty Acids (pg. 17)

Remedies: Devil's Claw, Joint Health, Joint Support, Alfalfa, Anamu, Anti-Gas Formula, Blue Cohosh, Chickweed, Clove Bud, EverFlex, Glucosamine, Grapefruit (Pink), Kava Kava, Lemon Oil, Marjoram (Sweet), Noni (Morinda), Rosemary, Sarsaparilla, Super ORAC, Yucca

Rhinitis

See Also *Rhinitis, Allergic*

Rhinitis as an inflammatory condition that affects the sensitive membranes of the nasal and sinus passages, the eyes and the throat. In allergic rhinitis the inflammation is caused by allergic reactions. However, rhinitis can have other causes besides allergies. Whatever the cause, having congested nasal passages, a runny nose, itchy, watery eyes and an irritated throat can make life miserable.

Here's what's happening anytime the sensitive membranes in your upper respiratory tract are exposed to irritants, inflammation can occur. Tissues swell and mucus is secreted to try to flush the irritation away.

In most people, these symptoms include sneezing; wheezing; stuffiness; itchy, runny nose and throat; post-nasal drip; itchy, watery eyes; conjunctivitis; earaches and insomnia. Many feel a reduced sense of taste or smell and even difficulty hearing. Other suffers have a nasal voice, breathe noisily or snore and others complain of frequent headaches and feeling chronically tired. Some people are more sensitive and will experience nasal and respiratory congestion, pain and pressure in the face. In more severe cases, rhinitis can produce yellow or greenish discharge from the nose, a chronic cough that produces mucus, poor appetite, nausea and sometimes a fever.

To deal with this problem, you need to identify, if possible, the source of the irritation. In the case of non-allergic rhinitis, this is usually chemical in nature. So avoid household cleaning products or other chemicals that cause respiratory irritation. People have found permanent relief just by switching to non-toxic household cleaning products like Sunshine Concentrate and Nature's Fresh.

For symptomatic relief try using anti-inflammatories and respiratory decongestants. A complete list of potential remedies follows.

Therapies: Colon Cleansing (pg. 18), Enzymes (pg. 15)

Remedies: ALJ, EW, Nature's Fresh, Sunshine Concentrate Cleaner, Vitamin C, Eyebright, IF Relief, Sinus Support, Thai-Go

Ringing In Ears

See *Tinnitis (Ringing in the Ears)*

Ringworm

See *Parasites (nematodes, worms)*

Rosacea

Rosacea is a chronic inflammatory skin condition. It is very similar to facial acne, except that it typically appears after the age of thirty. Rosacea is usually restricted to the face, but occasionally spreads to other parts of the body. It is more commonly experienced by people with deficient amounts of hydrochloric acid and poor digestion. Red raspberry or feverfew can help when applied topically as a facial mask. The following may also be helpful.

Therapies: Essential Fatty Acids (pg. 17), Antioxidants (pg. 17)

Remedies: Red Raspberry, Alfalfa, Chlorophyll, Digestive Bitters Tonic, Evening Primrose Oil, Herbal Trim Skin Treatment, HSN Complex, HSN-W, IF-C, PDA, Proactazyme Plus, S-O-D With Gliadin, Tea Tree Oil, Thai-Go, Vitamin A & D, Vitamin E, Zinc

Runny Nose

See *Congestion (sinus)*

Scabies

Scabies is a parasitic mite that lives in the skin causing severe itching. The following may be helpful. Paw Paw Cell Reg(c) can be added to shampoo or soap and used as a wash. Silver Shield gel applied topically along with some essential oils could also be helpful.

Remedies: Garlic, Golden Seal, Lemon Oil, Paw Paw, Silver Shield, Tea Tree Oil, Thyme

Scars / Scar Tissue

A scar is a mark left in the skin by the healing of injured tissue. The following remedies help prevent scaring when applied to wounds.

Remedies: Helichrysum, Lavender, Vitamin E, Aloe Vera, Mandarin (Red), MSM, Patchouli, Tea Tree Oil, Yarrow

Schizophrenia

See Also *Mental Illness*

Schizophrenia is a psychotic disorder characterized by loss of contact with the environment, and by noticeable deterioration in the level of functioning in everyday life. There is a disintegration of personality expressed as disorder of feeling, thought and conduct.

This condition may be associated with high copper levels that depress levels of vitamin C and zinc. Prenatal zinc deficiency has been linked with development of this disorder later in life. Possible damage to the pineal gland, which contains high levels of zinc, is another possible cause.

Schizophrenia may be related to excess dopamine in the brain. Avoiding refined carbohydrates and consuming adequate protein often helps to stabilize this condition. Magnesium deficiencies and hypoglycemia are very common in schizophrenia. The following may be helpful in schizophrenia along with improved diet, counseling and emotional healing work.**Therapies:** Blood Type Diet (pg. 13), Heavy Metal Detoxification (pg. 22), Stress Management (pg. 8), SuperFood (pg. 15), Essential Fatty Acids (pg. 17), Low Glycemic Diet (pg. 11), Affirmation and Visualization (pg. 8)

Remedies: GABA Plus, Licorice Root, Magnesium, Nutri-Calm, Vitamin B-Complex, Zinc, Evening Primrose Oil, Flax Seed Oil, Folic Acid Plus, Free Amino Acids, HY-A, L-Glutamine, Nerve Control, Niacin, Omega-3, Stevia, Target TS-II, Vitamin B-12, Vitamin C, Vitamin D3

Sciatica

See Also *Backache (Back Pain, Lumbago)*

Sciatica is a pain along the course of the sciatic nerve, making pain common in the lower back, buttocks, hips and back of the thighs. This usually is isolated to one side of the body. It may involve pressure on the nerve from the hips being out of alignment. Chiropractic care can be very helpful. The following may also help.

Therapies: Hydration (pg. 10)

Remedies: KB-C, Black Cohosh, Herbal CA, Nature's Fresh, Nerve Eight, Pain, PLS II, Sciatic

Scoliosis

This is a disorder where the spine is abnormally curved like an s. It may also be rotated. It can be very painful. It is due to a number of underlying causes, but the following remedies may help to strengthen the spine along with appropriate physical therapy and body work.

Therapies: Minerals (pg. 15)

Remedies: HSN-W, Ionic Minerals, KB-C, Mineral-Chi Tonic, Nature's Fresh

Scratches and Abrasions

See Also *Wounds and Sores*

Wounds to the skin from sliding contact with sharp or rough objects or surfaces may benefit from the following.

Remedies: Healing AC Cream, Silver Shield, Tea Tree Oil, Aloe Vera, Distress Remedy, Golden Salve, Golden Seal, Herbal Trim Skin Treatment, Yarrow

Scrofula

See *Tuberculosis (Consumption, Scrofula)*

Scurvy

Scurvy is a deficiency of vitamin C.

Remedies: Vitamin C, Rose Hips

Seasonal Affective Disorder

See Also *Depression*

Seasonal affective disorder (SAD) is a form of depression that occurs in the dark and dreary fall and winter months. It is believed to be due to a lack of exposure to natural sunlight. Full spectrum lighting is very helpful in preventing this condition, as is taking the sunshine vitamin, vitamin D3. The following may also help.

Therapies: Affirmation and Visualization (pg. 8), Stress Management (pg. 8), SuperFood (pg. 15), Essential Fatty Acids (pg. 17), Low Glycemic Diet (pg. 11)

Remedies: Mood Elevator, St. John's Wort, Vitamin D3

Seborrhea

Seborrhea is characterized by scaly patches of skin. It is caused by a disorder of the oil producing glands. The following may be helpful.

Therapies: Essential Fatty Acids (pg. 17)

Remedies: Burdock, Hi Lipase, Jojoba Oil, Tea Tree Oil, Aloe Vera, Lemon Oil, Omega-3, SF, Super GLA, Vitamin A & D

Seizures

See Also *Epilepsy*

A seizure is a sudden convulsive attack. There is usually a clouding of consciousness involved. These are often a result of epilepsy. Helichrysum essential oil applied to the temples is very beneficial in preventing seizures.

Remedies: GABA Plus, Distress Remedy, Helichrysum, Lobelia

Senility

See Also *Alzheimer's Disease, Dementia, Memory and Brain Function*

One of the physical and mental infirmities of old age is senility, an increasing loss of brain function. It is often a sign of poor circulation to the brain or a lack of good nutrition for brain function. Staying physically and mentally active helps prevent senility. The following supplements may help.

Therapies: Heavy Metal Detoxification (pg. 22)

Remedies: Brain-Protex, Ginkgo Biloba, Ginkgo/Gotu Kola, Mega-Chel, Omega-3, Capsicum, Flax Seed Oil, GABA Plus, Gotu Kola, L-Glutamine, L-Glutamine, Lemon Oil, Rosemary, Sage, SUMA Combination, Super GLA, Thai-Go, Wood Betony, X-A

Sex Drive (excessive)

The following can help reduce an excessive sex drive.

Remedies: Hops, Wild Yam & Chaste Tree, Marjoram (Sweet)

Sex Drive (low)

See Also *Erectile Dysfunction*

Sex drive or libido is a very complex mechanism. It involves physical, psychological and emotional causes. Low libido may be due to low thyroid, low cholesterol, poor function of the adrenal and reproductive glands or just plain fatigue. Libido is also related to attraction, and a lack of sex drive in a relationship is often due to unresolved emotional issues in the relationship. The following can help increase a low sex drive when the cause is physical. Counseling or emotional healing work may be necessary, too.

Therapies: Stress Management (pg. 8), Essential Fatty Acids (pg. 17), Affirmation and Visualization (pg. 8), SuperFood (pg. 15), Avoid Xenoestrogens (pg. 23), Heavy Metal Detoxification (pg. 22), Blood Type Diet (pg. 13)

Remedies: Damiana, Ginseng (Korean), Ginseng (Wild American), Maca, Nervous Fatigue Formula, Thyroid Support, X-A, X-Action (Men s), X-Action (Women s), Black Cohosh, C-X, Clove Bud, Eleuthero, Energ-V, Feminine Tonic, Gotu Kola, IGF-1, Jasmine Absolute,

Kava Kava, Licorice Root, Noni (Morinda), Patchouli, Sandalwood, Sarsaparilla, SUMA Combination, Thyroid Activator, Vitamin E, X-Action Gel, Ylang Ylang

Shingles

See Also *Chicken Pox*

Shingles is an infection by the Herpes zoster virus that causes acute inflammation and severe pain along the path of a specific nerve or nerves. The following may be helpful for shingles.

Therapies: Stress Management (pg. 8)

Remedies: Paw Paw, VS-C, Bergamot, Herbal CA, L-Lysine, Licorice Root, Nerve Control, St. John's Wort, Yarrow

Shock

A sudden or violent disturbance in the mental or emotional faculties can put a person into a state of shock. Shock is characterized by paleness, rapid but weak pulse, rapid and shallow respiration, restlessness, anxiety or mental dullness, nausea or vomiting associated with reduced blood volume and low blood pressure and subnormal temperature. It usually results from severe injuries, hemorrhage, burns or major surgery.

Capsicum powder on the tongue can help relieve shock. Distress Remedy under the tongue is also very effective. Have the person lie down and cover them with a blanket. If there is not sign of a spinal injury, attempt to elevate their feet above the level of their heart. Other remedies that may be helpful are listed below.

Therapies: Hydration (pg. 10)

Remedies: Capsicum, Distress Remedy, Cinnamon, Ginger, Hawthorn Berries, Hops, Neroli, Peppermint

Shortness of Breath

See *Wheezing*

Sickle Cell Anemia

See Also *Anemia*

A hereditary form of anemia that causes abnormally shaped red blood cells. The following may be helpful.

Therapies: Minerals (pg. 15)

Remedies: Chlorophyll, I-X, BP-X, HY-A

Sinus Infection

See Also *Congestion (sinus)*

The following may be helpful for sinus infection. Make a solution of sea salt and water and use it as a nasal wash (neti pots are the perfect tool to use) or spray Silver Shield directly into the sinuses for rapid relief. Applying Silver Shield Gel topically over the sinus cavities can also bring relief. Snuffing a mixture of equal parts golden seal and bayberry into the sinuses may also bring rapid relief. A complete list of possible remedies follows.

Therapies: Colon Cleansing (pg. 18),Fiber (pg. 17), Enzymes (pg. 15), Probiotics (pg. 16)

Remedies: Bayberry, Echinacea/Golden Seal, Fenugreek & Thyme, Golden Seal, Sea Salt, Silver Shield, Sinus Support, Eucalyptus, EW, HCP-X, Pine Needle, Tei Fu Oil, Ultimate Echinacea, Xylitol

Sinusitis (Sinus Problems)

See *Congestion (sinus)*

Skin (acne)

See *Acne (Pimples, Blackheads)*

Skin (dry and/or flaky)

The following products specifically help dry or flaky skin.

Therapies: Essential Fatty Acids (pg. 17)

Remedies: Flax Seed Oil, Hi Lipase, HY-C, Omega-3, Thyroid Activator, Jasmine Absolute, Kelp, Lecithin, MSM, Neroli, Sandalwood, Sunshine Heroes Omega 3 with DHA, Super GLA, Thyroid Support

Skin (infections)

The following products can be helpful for infections in the skin.

Remedies: Echinacea/Golden Seal, Guardian, Silver Shield, Tea Tree Oil, Acne Treatment Gel, Burdock, Golden Seal, Gotu Kola, Helichrysum, Liver Balance, Pau D Arco, Rosemary, Skin Detox, Vitamin A & D, Yarrow

Skin (oily)

The following may help clear up oily skin.

Therapies: Colon Cleansing (pg. 18)

Remedies: Burdock, Bergamot, Geranium, Grapefruit (Pink), Rosemary, Ylang Ylang

Skin Care (general)

Applying Silver Shield Gel to the skin and following it with a spray of Nature's Fresh can promote healthier skin. HSN-W provides silica and other nutrients the promote healthier skin, fingernails and hair. Intake of good fats and a properly functioning liver and thyroid are also essential to healthy skin. The following are products that can be generally helpful for keeping the skin healthy and beautiful.

Therapies: Minerals (pg. 15), Antioxidants (pg. 17), Essential Fatty Acids (pg. 17)

Remedies: **Herbal Trim Skin Treatment, HSN-W, Nature's Fresh, Omega-3, Silver Shield, Skin Detox, Sunshine Concentrate Cleaner,** Aloe Vera, BP-X, Chamomile, Chamomile (Roman), Chickweed, Flax Seed Oil, Gotu Kola, Helichrysum, HSN Complex, Jojoba Oil, Liver Balance, Rose Bulgaria, Rosemary, Sandalwood, Super Trio, Thyroid Support, Vitamin A & D, Vitamin C, Vitamin E

Sleep (restless and disturbed)

See Also *Insomnia*

When a person tosses and turns throughout the night, wakes up frequently or had disturbing dreams, this can be a sign of adrenal fatigue or nervous depletion. The following remedies may help.

Therapies: Hydration (pg. 10), Low Glycemic Diet (pg. 11), Stress Management (pg. 8)

Remedies: **Adrenal Support, Nervous Fatigue Formula,** Licorice Root, Nutri-Calm

Slivers

Slivers are tiny pieces of wood or other material embedded in the skin. Black Ointment, pine gum or crushed leaves of lily of the valley are very good at drawing slivers. Apply the remedy and cover with a bandage. The following can be applied topically to help draw slivers or be used to help prevent infection.

Remedies: **Black Ointment,** Intestinal Soothe & Build, Silver Shield, Tea Tree Oil

Smell (loss of sense of)

See Also *Congestion (sinus)*

A loss of the sense of smell is sometimes due to a zinc deficiency. It can also be due to sinus congestion and nasal polyps, in which case a snuff made of bayberry and goldenseal may be helpful. Remedies that may help include the following.

Remedies: **Zinc,** Bayberry, Fenugreek & Thyme, Golden Seal, Silver Shield

Smoking

See *Addictions (Tobacco)*

Snake Bite

The following remedies have been applied topically to snake bites to promote healing and counteract the venom. They may also be taken internally. Seek medical attention if bitten by a poisonous snake and only use these remedies while in route to the hospital.

Remedies: **Black Cohosh, Echinacea Purpurea,** Pau D Arco, Ultimate Echinacea, Vitamin C

Sneezing

See Also *Colds (decongestant), Congestion (sinus)*

Sneezing is a sudden, violent, spasmodic, audible expiration of breath through the nose and mouth, usually as a reflex reaction to an irritant in the nasal mucous membrane. It is often associated with allergies. Possible remedies include the following.

Therapies: Hydration (pg. 10), Colon Cleansing (pg. 18)

Remedies: **Bayberry,** EW, Cold, Eyebright, Fenugreek & Thyme, HCP-X, Seasonal Defense

Snoring

See Also *Allergies (food), Allergies (respiratory), Congestion (sinus)*

Snoring is caused by blockage to the respiratory passages that narrows the airways when sleeping. Check for food allergies and clear respiratory congestion. The following may reduce snoring.

Therapies: Hydration (pg. 10), Colon Cleansing (pg. 18)

Remedies: **ALJ, SnorEase,** EW, HistaBlock, Seasonal Defense

Sore Gums

See *Gingivitis (Bleeding Gums, Gum Disease, Pyorrhea)*

Sore or Geographic Tongue

See Also *Inflammatory Bowel Disorders (Colitis, IBS)*

A tongue that is covered with bare red patches alternating with heavily coated areas is called a geographic tongue. Both a geographic tongue and a sore tongue are indications of cardiovascular inflammation. Bitter herbs, especially if taken as powders, teas or tinctures can be helpful for cooling down this digestive irritation. Mix some golden seal or yellow dock powder with Digestive Bitters and take 15-20 minutes prior to meals. Digestive enzymes may also be helpful.

Therapies: Fiber (pg. 17), Probiotics (pg. 16), Antioxidants (pg. 17)

Remedies: Digestive Bitters Tonic, Golden Seal, Vitamin B-Complex, Yellow Dock, Dandelion, Proactazyme Plus, Proactazyme Plus, Super ORAC, Vitamin B-12, Vitamin B-6

Sore Throat

Sore throat is a discomfort in the pharynx due to inflammation. Many of these remedies can be used as gargles for sore throat. Some of the liquids can be applied topically to the throat using a gentle massage to ease pain. Sucking on slippery elm or licorice powder may also help.

Therapies: Hydration (pg. 10)

Remedies: Bayberry, Capsicum, Echinacea/Golden Seal, IF-C, Licorice Root, Lobelia, Slippery Elm, Ultimate Echinacea, Zinc Lozenges, Allergies-Mold/Yeast/Dust, CBG Extract, Cold, Eucalyptus, Fizz Active-Immune, Garlic, Golden Seal, HY-C, IF Relief, Liver Balance, Myrrh, Sage, Sandalwood, Sore Throat/Laryngitis

Sores

See *Wounds and Sores*

Spasms

See *Cramps and Spasms (general)*

Spastic Colon

See *Colon (spastic)*

Spider Veins

See Also *Cardiovascular Disease (Heart Disease)*

The following may be helpful in getting rid of unsightly spider veins. Spider veins can be an early warning sign of cardiovascular inflammation that can lead to cardiovascular disease.

Therapies: Antioxidants (pg. 17), Oral Chelation (pg. 21), Essential Fatty Acids (pg. 17)

Remedies: Bilberry Fruit, Butcher's Broom, Co-Q10, Vitamin C, Mega-Chel, Nature's Fresh, Rose Hips, Vari-Gone

Spinal Meningitis

See *Meningitis*

Sprains

A sprain is caused by a sudden or violent twisting of a joint that causes stretching or tearing of ligaments, resulting in swelling, pain, inflammation, bruising and discoloration. Some of these products will work best when applied topically or used for making fomentations or soaks.

Remedies: Bone/Skin Poultice, Healing AC Cream, Nature's Fresh, Tei Fu Oil, Aloe Vera, Carotenoid Blend, Chickweed, Helichrysum, Jasmine Absolute, L-Lysine, Lobelia, Marjoram (Sweet), Marshmallow, Pau D Arco, Sprains and Pulls, St. John's Wort, Vitamin A & D, White Oak Bark

Staph Infections

See Also *Infection (bacterial)*

Staphlococcus are a particular type of bacteria that can invade the body. Serious infections may require medical attention, but the following herbs can help the body fight staph infections.

Therapies: Probiotics (pg. 16)

Remedies: Garlic, Golden Seal, Immune Stimulator, Silver Shield, Ultimate Echinacea, Aloe Vera, IN-X, Lymph Gland Cleanse, Lymph Gland Cleanse-HY, Oregon Grape, Tea Tree Oil, Vitamin A & D, Vitamin C

Stiff Neck

Pain, inflammation and lack of mobility between the head and shoulders can be aided by massaging Tei Fu Essential Oils, Tei Fu Massage Lotion and/or lobelia into the neck and shoulders. Chiropractic care can be helpful. Reduce stress and cleanse the body by working on the kidneys, colon and liver. Drink more water.

Therapies: Hydration (pg. 10), Colon Cleansing (pg. 18)

Remedies: Deep Relief Oil, Kava Kava, Lobelia, Tei Fu Oil, Kudzu/St.John's Wort, Nature's Fresh

Stomachache

See *Indigestion*

Strep Throat

See Also *Sore Throat*

The following may be helpful for strep throat, which is caused by a particular type of infectious bacteria. Serious bacterial infections may require medical attention.

Remedies: Echinacea/Golden Seal, Garlic, Immune Stimulator, Silver Shield, Echinacea Purpurea, IN-X, Tea Tree Oil, Ultimate Echinacea

Stress

See Also *Anxiety Disorders*

Stress is a physical, chemical or emotional factor that causes bodily or mental tension and is a contributing factor to disease.

The following can help improve a person's ability to cope with stress in their lives. Also check out the discussion of stress management under Activating Therapies in the Introduction.

Therapies: Hydration (pg. 10), Blood Type Diet (pg. 13), Stress Management (pg. 8)

Remedies: AdaptaMax, Eleuthero, Kava Kava, Magnesium, Nervous Fatigue Formula, Nutri-Calm, Pantothenic Acid, Stress-J, Vitamin B-Complex, 5-HTP, Fatigue/Exhaustion, Ginseng (Korean), Ginseng (Wild American), Gotu Kola, Herbal Sleep, Kudzu/St.John's Wort, Licorice Root, Liver Balance, Maca, Marjoram (Sweet), Nature's Cortisol, Nerve Control, Nerve Eight, Niacin, Pregnenolone, Sandalwood, Stress Relief, Tei Fu Oil, Valerian Root, Ylang Ylang

Stretch Marks

Stretch marks are lines of scarred tissue that form on the surface of the skin when the skin is stressed from rapid growth. For instance, stretch marks are common from pregnancy or rapid weight gain. In pregnancy, they are commonly found on the belly, hips and or thighs. Vitamin E and zinc can help prevent stretch marks. Massaging cocoa butter, coconut, olive or peanut oil into the skin also helps.

Remedies: Vitamin E, Zinc, Grapefruit (Pink), Helichrysum, Herbal Trim Skin Treatment, Mandarin (Red)

Strokes

See Also *Blood Clots (prevention of), Thrombosis*

A stroke causes temporary or permanent loss of blood flow to an artery of the brain and the part of the brain that artery feeds. It may be caused by an arterial rupture or a blood clot. The degree and type of damage depend on the size and location of the portion of the brain affected. Appropriate medical attention should be sought, but the following may help prevent strokes or aid in recovery from strokes.

Therapies: Enzymes (pg. 15), Antioxidants (pg. 17), Oral Chelation (pg. 21)

Remedies: Butcher's Broom, Mega-Chel, Capsicum, Colloidal Minerals, Flax Seed Oil, Ginkgo Biloba, Lecithin, Magnesium, Nattozimes Plus, Nutri-Calm, Omega-3, Vitamin C

Stye

See Also *Eye Infections*

An inflamed swelling of a sebaceous gland at the margin of the eyelid is called a stye. EW, made into a tea and applied to the eyes as a compress, can be very effective for treating styes.

Remedies: EW, Chamomile, Eyebright, Horsetail, Vitamin A & D

Sugar Cravings

See Also *Fungal Infections (Yeast Infections, Candida albicans), Hyperinsulinemia (Syndrome X), Hypoglycemia*

An excessive desire for sugar-containing foods is often a sign of yeast infections, mineral deficiencies or blood sugar problems. Xylitol and stevia can be used as natural alternatives to refined sugar, but the real cure is to start the day off with good quality fats and proteins and avoid refined carbohydrates. Follow the instructions for the *Low Glycemic Diet* in the *Introduction* and try some of the following supplements.

Therapies: Low Glycemic Diet (pg. 11), SuperFood (pg. 15), Minerals (pg. 15)

Remedies: Candida Clear, HY-A, Licorice Root, Stevia, Xylitol, Yeast/Fungal Detox, Anti-Gas (Chinese), Chromium GTF, Mineral-Chi Tonic, Super Algae

Sunburn

See Also *Burns and Scalds*

Varying degrees of damage to the skin due to overexposure to sunlight can be aided by applying Aloe Vera Gel, Silver Shield Gel or Herbal Trim topically. Other topical remedies include tea tree oil and lavender oil. Keeping the burn moist with Nature's Spring water in a spray bottle also speeds healing. Large doses of vitamin C taken internally also speed recovery.

Therapies: Hydration (pg. 10), Antioxidants (pg. 17)

Remedies: Aloe Vera, Herbal Trim Skin Treatment, Lavender, Nature's Fresh, Chamomile (Roman), Tea Tree Oil, Vitamin C

Surgery (healing from)

The following have been taken following surgery to speed healing or detoxify from drugs used during the surgery.

Therapies: Affirmation and Visualization (pg. 8), SuperFood (pg. 15), Minerals (pg. 15), Enzymes (pg. 15), Essential Fatty Acids (pg. 17)

Remedies: **Bone/Skin Poultice, Enviro-Detox, Herbal CA, IF Relief**, Blood Build, BP-X, Burdock, E-Tea, Grapine, HSN-W, Nature's Chi, Nature's Phenyltol with NEM, Pantothenic Acid, Super Algae, Vitamin C, Vitamin E

Surgery (preparation for)

The following have been taken prior to surgery to help the body prepare for it.

Remedies: **HSN-W, Mineral-Chi Tonic**, Bone/Skin Poultice, Butcher's Broom, Capsicum, Ionic Minerals, Vitamin E

Sweat Baths (herbs for)

The following can be helpful for sweat baths. Sweat baths are used to sweat out colds and flu.

Remedies: Capsicum, Ginger, HCP-X, Yarrow

Sweating

See *Perspiration (excessive)*

Swelling

See *Edema (Dropsy, Water Retention, Swelling)*

Syndrome X

See *Hyperinsulinemia (Syndrome X)*

Syphilis

The following have been used historically to treat syphilis. However, we recommend that medical attention be sought for this disease.

Remedies: **Silver Shield, Ultimate Echinacea**, Echinacea Purpurea, Golden Seal, Lemon Oil, Oregon Grape

Tachycardia

See Also *Grave's Disease*

Tachycardia is a rapid beating of the heart. This condition requires medical attention, but can be aided by natural remedies in some cases. Many of the herbs that work on rapid heart beat are for use by professional herbalists only or are not sold by Nature's Sunshine. These include motherwort and mistletoe. Some of the NSP remedies that might help include lobelia, which acts as a natural beta blocker, and magnesium, which can help calm and strengthen the heart beat. Grave's disease can cause tachycardia as can a lack of mineral electrolytes like magnesium, calcium and potassium.

Therapies: Hiatal Hernia Correction (pg. 12), Minerals (pg. 15)

Remedies: **Hawthorn Berries, Lobelia, Magnesium**, Calcium, Colloidal Minerals, IF-C, Ionic Minerals, Mineral-Chi Tonic, Potassium

Teeth (grinding)

The often unconscious habit of gritting the teeth together during sleep or periods of stress is often due to a lack of trace minerals, calcium deficiency, parasites or stress. Possible remedies include the following.

Therapies: Minerals (pg. 15), Stress Management (pg. 8)

Remedies: **HSN-W, Magnesium**, Calcium, Herbal CA, Nutri-Calm, Pantothenic Acid, Para-Cleanse, Sunshine Heroes Calcium Plus D3

Teeth (loose)

See Also *Gingivitis (Bleeding Gums, Gum Disease, Pyorrhea)*

Teeth which are not properly rooted or firmly attached into the gums of the mouth and thus wiggle back and forth in their sockets may be aided by remedies which strengthen tissue integrity and reduce inflammation. White oak bark tea can be used as a mouthwash. Brushing with a toothpowder made from equal parts white oak bark and black walnut can also be helpful.

Therapies: Minerals (pg. 15), SuperFood (pg. 15), Antioxidants (pg. 17)

Remedies: **Black Walnut, Co-Q10, Vitamin C, White Oak Bark**, Herbal CA, Lavender, Lemon Oil

Teething

In the process of developing the first set of teeth and having them push through the gums, infants often develop pain, irritability, fever and earache. Rubbing clove oil diluted in olive oil can ease gum pain. Lobelia can also be rubbed on the gums. Other remedies taken internally can also ease teething pain. Chamomile tea is a traditional remedy for this.

Remedies: **Chamomile, Chamomile (Roman), Clove Bud, Lobelia, Teething**, Calcium, Calming, Catnip & Fennel, Herbal CA, Marshmallow, Pain, Passion Flower

Tendonitis

Inflammation of a tendon or tendons of the body is often due to injury or overexertion. Anti-inflammatory and tissue healing (vulnerary) remedies may be helpful in speeding recovery. Topically, remedies like Tei Fu or Deep Relief can be applied to ease pain.

Remedies: Deep Relief Oil, Healing AC Cream, IF Relief, Safflowers, Tei Fu Oil, Aloe Vera, Bone/Skin Poultice, Herbal CA, Licorice Root, Sprains and Pulls

Tension

See Also *Anxiety Disorders, Stress*

Uneasiness or stress due to illness or emotion can cause muscles to become tense. Nervines and antispasmodics can help.

Therapies: Stress Management (pg. 8)

Remedies: Kava Kava, Lobelia, Magnesium, Nutri-Calm, APS II, Black Cohosh, Chamomile, Frankincense, Grapefruit (Pink), Mood Elevator, St. John's Wort, Stress-J

Testosterone (low)

Testosterone is the principle male hormone. It'stimulates sperm production, libido, muscular strength and the physical characteristics of the male. Low testosterone causes a number of symptoms including frustration and anxiety, mild depression, low self esteem, decreased sex drive, lack of muscle tone and muscle wasting. Men's testosterone levels have fallen sharply; most men have half the levels of testosterone men had 20 years ago. This is very bad for men's health. Here are some tips for increasing testosterone levels.

Excess estrogens upset the balance between testosterone and estrogen in men and xenoestrogens are a principle cause of this imbalance. While red meat stimulates testosterone levels in men, most commercially raised red meat contains xenoestrogens. If you can find organically raised, grass-fed beef, it not only helps you reduce your exposure to xenoestrogens, it will enhance your testosterone levels at the same time.

Pesticides and herbicides, such as DDT and organochlorines, are also xenoestrogens. Eating organically grown food wherever possible and making sure non-organically grown food is washed in a natural cleaning solution like Sunshine Concentrate will reduce your exposure to these chemicals.

PCP and thalates found in plastic materials are also xenoestrogens. They are accumulating in the world's soil, water and air. Because heat releases these chemicals, avoid microwaving in plastic containers, don't drink water from soft plastic bottles after they have been exposed to heat and avoiding placing hot food into plastic storage containers.

Another compound that reduces testosterone levels in men is fluoride. Don't use fluoridated toothpaste or other products containing fluoride. If your water is fluoridated, get a Nature's Spring Reverse Osmosis unit or some other water treatment appliance to remove the fluoride from your drinking water.

Men suffering from any kind of male reproductive problems should also be aware that too many phytoestrogens (estrogenic compounds found in plants) can also cause imbalances in testosterone and estrogen levels. One of the principle culprits here is soy. Widely touted as a beneficial health food, according to the Weston Price Foundation, Numerous animals studies show that soy foods cause infertility in animals. Japanese housewives feed tofu to their husbands frequently when they want to reduce his virility. So use soy sparingly.

Men should also avoid consuming large amounts of licorice, as it enhances cortisol levels and decreases testosterone levels. Hops is another highly estrogenic herb. It contains very potent estrogens that can reduce male sex drive. Since most beer is made with hops, men who are concerned about their fertility or are suffering from male reproductive health problems should avoid drinking beer made from hops. Finally, grapefruit interferes with estrogen breakdown and should also be consumed sparinglty by men who wish to enhance their testosterone levels.

Drugs can also effect testosterone levels. Classes of medications that may interfere with male reproductive function include anti-inflammatories, antibiotics, antifungals, statins (cholesterol-lowering medications), antidepressants, calcium channel blockers, sleeping pills and high blood pressure medications. Carefully read warning labels to discover if any medications you take may be affecting your reproductive health.

Diets that are high in refined carbohydrates and low in good fats and protein will also damage male reproductive health. High carbohydrate diets stress the adrenal glands and pancreas, resulting in increased levels of insulin and reduced levels of DHEA, a building block for male hormones. Also, the current drive to lower cholesterol levels is increasing depression and reproductive health problems in men. DHEA and all reproductive hormones are made from cholesterol, so driving cholesterol levels too low will actually cause reproductive problems in both sexes.

As suggested earlier, organic meat, eggs and dairy products are actually good for you, especially if they are from grass-fed animals. Get white bread and refined sugar out of your diet and eat fruits and vegetables instead, preferably organic. These foods also protect your body from heart disease and cancer.

Finally, exercise regularly. Regular exercise helps increase testosterone production. It also reduces your risk of cardiovascular disease, diabetes and other degenerative diseases. Resistance training with weights is especially important for men as they grow older.

The following supplements may also increase testosterone levels. Other remedies not listed here for low testosterone include pine tree pollen and puncture vine.

Therapies: SuperFood (pg. 15), Essential Fatty Acids (pg. 17), Blood Type Diet (pg. 13), Avoid Xenoestrogens (pg. 23)

Remedies: Cinnamon, DHEA-M, Ginseng (Korean), Damiana, Eleuthero, Ginseng (Wild American), Maca, Men's Formula, Sarsaparilla, X-Action (Men s)

Tetanus

See Also *Wounds and Sores*

Tetanus is an acute infectious disease characterized by tonic spasm of voluntary muscles, especially those of the jaw. It is caused by a toxin from a specific bacteria, which is usually introduced through a wound. Medical advice is to obtain a tetanus shot for any deep puncture wounds, however, if wounds are properly cleansed and topical antiseptics (such as Silver Shield and Tea Tree oil) are applied the risk of tetanus is very slight. If tetanus does develop, medical attention should be sought, but lobelia has been used in natural treatment.

Remedies: Silver Shield, Tea Tree Oil, Lobelia, Thyme

Thinking (cloudy)

See Also *Adrenals (Exhaustion, Weakness or Burnout), Hyperinsulinemia (Syndrome X), Hypoglycemia*

When a person's thinking is cloudy or they are having a difficult time concentrating, remembering or thinking clearly, they may be toxic and need cleansing or there may be a deficiency in blood flow to the brain. They may also be dehydrated, not getting enough sleep or lacking omega-3 fatty acids. Protein deficiencies and hypoglycemia may also be involved.

Therapies: Colon Cleansing (pg. 18), Enzymes (pg. 15), SuperFood (pg. 15), Heavy Metal Detoxification (pg. 22), Low Glycemic Diet (pg. 11), Blood Type Diet (pg. 13)

Remedies: Focus Attention, Mega-Chel, Nervous Fatigue Formula, Peppermint, Pine Needle, Rosemary, AdaptaMax, Blood Sugar Formula, Enviro-Detox, Eucalyptus, Free Amino Acids, Ginkgo Biloba, Ginkgo/Gotu Kola, Heavy Metal Detox, Mood Elevator, Nutri-Calm

Thrombosis

See Also *Blood Clots (prevention of)*

A blood clot within a blood vessel is called thrombosis. The formation of these clots is dangerous because if they break loose they can lodge in the heart or brain causing a heart attack or stroke. These remedies help prevent thrombosis.

Remedies: Butcher's Broom, Nattozimes Plus, Vitamin E, Alfalfa, HS II, Lemon Oil, Vari-Gone

Thrush

See Also *Fungal Infections (Yeast Infections, Candida albicans)*

Thrush is a Candida or yeast infection of the mouth marked by white patches in the oral cavity. It typically occurs in infants and children. For rapid relief mix equal amounts of tea tree oil, thyme oil, and lavender oil and dilute them 20-to-1 in olive oil (20 parts olive oil, 1 part essential oils). Give one drop of this mixture twice daily. You can also combine these essential oils about 40 to 1 with Silver Shield. Also give the child probiotics.

Therapies: Probiotics (pg. 16)

Remedies: Lavender, Silver Shield, Tea Tree Oil, Thyme, Licorice Root, Pau D Arco, Probiotics, VS-C

Thyroid (high)

See *Grave's Disease*

Thyroid (low)

See *Hypothyroid*

Tick

A tick is a small, blood-sucking insect whose bite can carry diseases such as Lyme's or Rocky Mountain spotted fever. Paw Paw may help to kill ticks if applied topically. Ticks should not be crushed when removed. Applying Silver Shield gel after removing the tick and taking Silver Shield internally may help prevent infection. Garlic is a natural tick repellant.

Remedies: Garlic, Paw Paw, Silver Shield

Tickle in Throat

See Also *Cough (dry)*

An annoying sensation in the pharynx that feels like a tickle can often be relieved by sucking on licorice or slippery elm powder.

Therapies: Hydration (pg. 10)

Remedies: Licorice Root, Slippery Elm, Marshmallow

Tics

See Also *Tremors, Twitching*

A tic is a spasmodic muscular contraction. The movement may appear voluntary or purposeful, but is involuntary. These are often due to deficiencies of mineral electrolytes like magnesium, potassium and calcium. Possible remedies include the following.

Remedies: Lobelia, Magnesium, Vitamin B-Complex, Calcium, Cellular Energy, Potassium, Wood Betony

Tinnitus (Ringing in the Ears)

See Also *Ear Infection or Earache*

A sensation of humming or ringing in the ears, without an external source for the sound, may be caused by ear infections, circulatory problems or nerve damage. CBG Extract mixed with Lobelia extract and warmed to body temperature to use as ear drops may be helpful. Appropriate remedies are based on cause.

Therapies: Oral Chelation (pg. 21)

Remedies: CBG Extract, Garlic, Ginkgo Biloba, Lobelia, Aloe Vera, Black Cohosh, Ginkgo/Gotu Kola, HY-C, IF-C, KB-C, Lemon Oil, Mega-Chel, St. John's Wort, Vitamin B-12, Vitamin B-Complex, Wood Betony, Zinc

TMJ

TMJ stands for temporomandibular joint or the point at which the lower jaw meets the temple region of the skull. Those who suffer from "TMJ" experience headaches and radiating pain from this region. Mechanical work by a massage therapist or chiropractor is often helpful.

Therapies: Stress Management (pg. 8)

Remedies: **Lobelia**, IF-C, Kava Kava, Lymph Gland Cleanse, Magnesium, Nature's Fresh

Tonsillitis (Adenoids)

When there is inflammation and swelling of the tonsils or lymph nodes located at the back of the throat (also known as adenoids), they can obstruct the nasal and ear passages resulting in mouth breathing, snoring and nasal discharge. If chronically inflamed, they can become a site of infection.

Remedies: **Echinacea Purpurea, Echinacea/Golden Seal, IN-X, Lymphatic Drainage, Lymphomax, Silver Shield, Ultimate Echinacea**, Bergamot, Blue Vervain, Capsicum & Garlic with Parsley, CBG Extract, Garlic, Golden Seal, IF-C, Lemon Oil, Liver Balance, Lymph Gland Cleanse, Lymph Gland Cleanse-HY, Slippery Elm, White Oak Bark

Tooth Decay (prevention)

Helping the teeth to have proper mineralization can help prevent tooth decay. Brushing with a toothpowder made of white oak bark and black walnut hulls can help strengthen enamel. The herbalist Dr. John Christopher claimed that alfalfa would help remineralize the teeth if the colon was clean. Xylitol powder, gum, toothpaste or mouthwash used after meals can also help prevent tooth decay.

Therapies: Minerals (pg. 15), Essential Fatty Acids (pg. 17)

Remedies: **Alfalfa, Colloidal Minerals, Sunshine Brite Toothpaste, White Oak Bark, Xylitol**, Black Walnut, Herbal CA, Ionic Minerals, Mineral-Chi Tonic, Sunshine Heroes Calcium Plus D3, Vitamin B-6, Vitamin B-Complex

Tooth Extraction

After a tooth has been removed by pulling or surgery these remedies may ease pain and promote more rapid healing.

Remedies: **Golden Seal, IF Relief**, APS II, Bone/Skin Poultice, Deep Relief Oil, Enviro-Detox, Intestinal Soothe & Build, Nature's Phenyltol with NEM, Valerian Root, White Oak Bark

Tooth Grinding

See *Teeth (grinding)*

Toothache

Pain in or around a tooth often as a result of a cavity or gum disease can be eased by anti-inflammatories and nervines. Clove oil, diluted in olive oil, can be applied topically to ease pain. You can also slice a piece of garlic, coat it with olive oil and put it in between the cheek and the gum to relieve pain and ease the swelling of dental abscesses until a dentist can repair the tooth. Silver Shield Gel can also be applied to inhibit the infection. These are temporary measures to use while seeking dental assistance.

Remedies: **Clove Bud, Garlic, Silver Shield**, Chamomile, Ginger, Kava Kava, Lobelia, Tea Tree Oil, Tei Fu Oil, Thyme

Toxemia

An abnormal condition of toxic substances in the blood is called toxemia. When this occurs in pregnancy it is marked by high blood pressure, protein in the urine, swelling, headache, visual disturbances and possibly even convulsions. Seek appropriate medical assistance. These remedies may help in recovery.

Therapies: Hydration (pg. 10), Colon Cleansing (pg. 18)

Remedies: **All Cell Detox, Enviro-Detox, Tiao He Cleanse**, Aloe Vera, BP-X, Chlorophyll, CleanStart, Germanium Combination, Liver Balance, MSM, Pau D Arco, Red Beet Formula, SF, Super Antioxidant, Thai-Go, Vitamin C

Toxic Blood

See *Blood Poisoning*

Tremors

See Also *Tics, Twitching*

A tremor is an involuntary quivering of a muscle. This can be due to severe depletion of muscle energy or nervous system problems. It often indicates low levels of potassium or magnesium and may also be a sign of exhausted adrenals.

Remedies: **Cellular Energy, Fibralgia, Magnesium, Potassium**, Adrenal Support, Ginkgo Biloba, Nervous Fatigue Formula, Wood Betony

Triglycerides (high)

Triglycerides are blood fats, composed of three fatty acids linked together. They travel with cholesterol in the blood stream and are used to produce energy. When triglycerides are high there may be problems with digestion, adrenal function or the hypothalamus.

Therapies: Fiber (pg. 17), Enzymes (pg. 15)

Remedies: Adrenal Support, Alpha Lipoic Acid, Barley Grass, Carbo-Grabbers, Cellular Energy, CLA, DHA, Food Enzymes, Free Amino Acids, Garcinia Combination, GreenZone, Guggul, Licorice Root, Melatonin Extra, Mineral-Chi Tonic, Nerve Control, Nutri-Calm, PBS, Potassium, Spirulina, Stress Relief, Stress-J, SUMA Combination, Super Algae, SynerProTein

Triglycerides (low)

Low triglycerides may be due to a lack of dietary fats, fatty congestion in the liver or digestive problems. The following may be helpful in raising low triglycerides.

Therapies: Essential Fatty Acids (pg. 17), Enzymes (pg. 15)

Remedies: L-Carnitine, Liver Balance, Omega-3, SF, Adrenal Support, Chickweed, Food Enzymes, Hi Lipase, Master Gland, Super GLA, Super Oil

Tuberculosis (Consumption, Scrofula)

An infectious disease caused by a bacterial infection, tuberculosis used to be called consumption. Scrofula is a particular variety of tuberculosis, involving lymph nodes, especially in the neck. Once very common, the discovery of antibiotics caused this condition to become very rare, but now that antibiotic resistant strains have developed, it is making a comeback. The following were historically used to treat tuberculosis or may be helpful in antibiotic resistant tuberculosis.

Remedies: Garlic, Silver Shield, Bergamot, Golden Seal, Ho Shou Wu, Lemon Oil, Lobelia, Lung Support, Mullein, Oregano (Wild), Red Clover, Red Clover Blend, St. John's Wort, Vitamin D3

Tumors

See *Cancer (natural therapy)*

Tumors (fatty)

See Also *Fatty Tumors or Deposits*

Twitching

See Also *Tics, Tremors*

A twitch is a quick spasmodic contraction of a muscle. When repeated in rapid succession this is twitching.

Remedies: Magnesium, Potassium, Calcium, Valerian Root

Typhoid

Typhoid is a severe, contagious disease marked by high fever, stupor alternating with delirium, intense headache, diarrhea, intestinal inflammation, and a dark red rash. The following may be helpful for typhoid.

Remedies: Echinacea Purpurea, Silver Shield, Golden Seal, Lemon Oil, Ultimate Echinacea

Ulcerations (external)

An open sore or break in the skin, often containing pus, can often be healed by topical application of herbal remedies such as Black Ointment, golden seal, or tea tree oil. Internally, remedies that enhance circulation and fight infection may be helpful.

Remedies: Golden Seal, Myrrh, Tea Tree Oil, Vitamin C, Zinc, Aloe Vera, Astragalus, Bayberry, Black Ointment, Food Enzymes, Gastro Health, Intestinal Soothe & Build, Marshmallow, Skin Detox, Slippery Elm, St. John's Wort

Ulcerative Colitus

See *Inflammatory Bowel Disorders (Colitis, IBS)*

Ulcers

An ulcer is an open sore or break in the mucous membrane lining of the stomach or first portion of the small intestine. It is exacerbated by stomach acid and other digestive juices but caused by H. pylori bacteria.

Therapies: Fiber (pg. 17), Antioxidants (pg. 17)

Remedies: Aloe Vera, Gastro Health, Golden Seal, Licorice Root, Slippery Elm, Alfalfa, Astragalus, Bayberry, Capsicum, Chamomile, CLT-X, Ginger, Gotu Kola, Intestinal Soothe & Build, L-Glutamine, Lobelia, Myrrh, Peppermint, PLS II

Underweight

See *Wasting, Weight Gain (aids for)*

Urethritis

See Also *Urinary Tract Infections, Urination (burning or painful)*

Inflammation of the urethra or tube that carries urine from the bladder to the outside of the body is called urethritis. Remedies that soothe inflammation in the urinary passages are helpful.

Therapies: Hydration (pg. 10)

Remedies: Cornsilk, Marshmallow, Damiana, Horsetail, Hydrangea, Kidney Activator (Chinese), Uva Ursi

Uric Acid Retention

See Also *Gout*

The inability of the body to eliminate uric acid results in painful uric acid crystal formation in the joints of the body. It can also weaken bones and joints and increase the risk of calcium deposits and kidney stones. When uric acid retention is a problem decrease protein consumption, eat more fresh fruits and vegetables (especially dark green, leafy vegetables), drink more water and use some of the following remedies to help neutralize and eliminate the acid.

Therapies: Hydration (pg. 10)

Remedies: Devil's Claw, Joint Support, KB-C, Alfalfa, Pine Needle, Safflowers, Yucca

Urinary Tract Infections

See Also *Bladder Infection*

The following products can help combat urinary tract infections. Cranberry & Buchu is helpful for preventing them. Some of the best remedies for curing them once they are present are Echinacea/Goldenseal, goldenseal, Silver Shield and uva ursi.

Therapies: Hydration (pg. 10)

Remedies: Cranberry & Buchu, Echinacea/Golden Seal, Golden Seal, Golden Seal/Parthenium, Silver Shield, Uva Ursi, Bergamot, Bilberry Fruit, Garlic, JP-X, Kava Kava, Kidney Activator (Chinese), Parthenium, Stress Relief, Ultimate Echinacea

Urination (burning or painful)

Painful urination is a sign of inflammation and/or infection. Remedies like the following can be used to reduce inflammation and combat infection.

Therapies: Hydration (pg. 10)

Remedies: Cornsilk, Kava Kava, Lobelia, Marshmallow, Golden Seal, Kidney Drainage, Nopal, Silver Shield, Sunshine Heroes Whole Foods Antioxidant, Super ORAC, Thai-Go, Urinary Maintenance, Uva Ursi

Urination (frequent)

The following may help with frequent urination or the frequent urge to urinate. Be sure to drink plenty of water to dilute toxins.

Therapies: Hydration (pg. 10)

Remedies: Cornsilk, HY-C, KB-C, Marshmallow, Nervous Fatigue Formula, Thai-Go, Uva Ursi

Urine (scant)

See Also *Edema (Dropsy, Water Retention, Swelling)*

When urine production is scant, the kidneys need stimulation and support. In men, this can be a sign of prostate swelling. These diuretic herbs and formulas can increase output of urine.

Remedies: JP-X, Juniper Berries, Kidney Activator, Kidney Activator (Chinese), Butcher's Broom, Cranberry & Buchu, Kidney Drainage, Lung Support, Lymphatic Drainage, Men's Formula, PS II, Urinary Maintenance, Uva Ursi

Uterine Fibroids

See *Fibroids*

Vaccines (detoxification from)

Vaccines contain chemicals like mercury, aluminum, formaldehyde and other toxic ingredients. These sometimes lodge in the tissues causing irritation, fever and chronic health problems. Vaccines may be an underlying cause of autoimmune disorders. The following can help the body detoxify from vaccinations.

Therapies: Heavy Metal Detoxification (pg. 22), Colon Cleansing (pg. 18), Essential Fatty Acids (pg. 17)

Remedies: Algin, CBG Extract, Enviro-Detox, Heavy Metal Detox, Vaccine Detox, BP-X, Chlorophyll, Elderberry Plus, Grapine, Lobelia, Lymphatic Drainage, Mullein, Nerve Control, Noni (Morinda), Omega-3, Oregon Grape, Thai-Go, Vitamin C, VS-C, Yarrow

Vaginal Discharge

See *Leucorrhea*

Vaginal Dryness

Vaginal dryness often occurs after menopause when hormone levels are reduced. Regular sexual activity can help to ease vaginal dryness. When vaginal dryness is a problem the following may be helpful.

Remedies: Geranium, Pro-G-Yam Cream, Aloe Vera, Black Cohosh, Flash-Ease T/R, Slippery Elm

Vaginitis

Inflammation of the vagina is known as vaginitis. It may come from a variety of causes including infection, parasites, or irritation from douches and sprays. In addition to using herbs internally douches with probiotics, red raspberry tea or antiseptic herbs may be helpful.

Therapies: Probiotics (pg. 16), Antioxidants (pg. 17), Stress Management (pg. 8)

Remedies: Golden Seal, Kava Kava, Pau D Arco, Silver Shield, Aloe Vera, Bayberry, Bergamot, Garlic, Oregon Grape, Probiotics, Red Raspberry, Tea Tree Oil, X-Action (Women s)

Varicose Veins

See Also *Hemorrhoids*

Externally visible, sometimes prominent veins are called varicose veins. They are common on the legs and are a sign of poor circulation and venous valve collapse. They are not just a cosmetic problem, they may be painful and indicate a lack of tone in the blood vessels.

Therapies: Oral Chelation (pg. 21),Fiber (pg. 17)

Remedies: Butcher's Broom, Healing AC Cream, Mega-Chel, Vari-Gone, White Oak Bark, 5-W, Aloe Vera, Bayberry, Bergamot, Bilberry Fruit, Capsicum, Geranium, Lemon Oil, Milk Thistle Combination, Neroli, Stress Relief, Vitamin C, Yarrow

Vertigo

See *Dizziness (Vertigo)*

Virus

See *Infection (viral)*

Vitiligo

Vitiligo is a skin condition where there are white patches of skin surrounded by a dark border. For some reason the cells that produce skin pigment have been damaged or destroyed. This problem can be difficult to treat, but the following may help.

Remedies: Evening Primrose Oil, HSN-W, Super GLA, Vitamin B-Complex, Zinc

Vomiting

See *Nausea and Vomiting*

Warts

A wart is a horny projection on the skin usually caused by a virus. Remedies may be taken internally and applied topically. A salve made of Paw Paw Cell-Reg(c) combined with Black Ointment can be applied topically to warts to remove them. Cover with a bandage and change twice daily.

Remedies: Black Ointment, Paw Paw, VS-C, Aloe Vera, Cinnamon, Clove Bud, Garlic, Germanium Combination, Nature's Gold, Viral Recovery, Vitamin C

Wasting

See Also *Weight Gain (aids for)*

Wasting is a condition where a person begins to lose muscle mass and general body weight, usually as a result of aging or a chronic disease. The following can help reverse wasting.

Therapies: Enzymes (pg. 15), SuperFood (pg. 15), Stress Management (pg. 8)

Remedies: Food Enzymes, Slippery Elm, Spleen Activator, Trigger Immune, Free Amino Acids, Ginseng (Wild American), Marshmallow, Proactazyme Plus, Saw Palmetto

Water Retention

See *Edema (Dropsy, Water Retention, Swelling)*

Weight Gain (aids for)

See Also *Hiatal Hernia, Wasting*

The following products may help people seeking to gain muscle weight. The inability to gain muscle weight is often due to a hiatal hernia. It can also be due to other digestive problems and glandular imbalances.

Therapies: Stress Management (pg. 8), Enzymes (pg. 15), Hiatal Hernia Correction (pg. 12)

Remedies: Licorice Root, Lung Support, Proactazyme Plus, Spleen Activator, EveryBody's Formula, Ginseng (Wild American), Nature's Gold, PDA, Protease, Saw Palmetto, Spirulina, Super Algae, Trigger Immune

Weight Loss (aids for)

See Also *Hyperinsulinemia (Syndrome X), Hyperthyroid, Sugar Cravings*

For years, the weight loss mantra has been, eat less, exercise more. Yet, the research shows that 90% of the people who lose weight in this manner simply gain it back. Why is this so?

Well, for starters, appetite, metabolism and mood are all controlled by messenger chemicals in the body. When you restrict calories, the cells of the body assume that there is a famine going on. In response, cells send chemical messengers that reduce metabolism (the rate at which you burn calories). This conserves the body's energy during the famine.

Because your metabolism is lower, your energy is reduced, so you become less physically active. Your mood also changes because you feel deprived, so your body is attracted to foods that enhance mood particularly carbohydrates. When food is available, other chemical messengers are released to stimulate the appetite and program the body to store energy (fat) in preparation for the next famine. Thus, a vicious cycle of feast and famine ensues.

In addition, fat itself acts like a gland, secreting its own chemical messengers. One of these is a hormone called leptin. Leptin is supposed to increase your metabolism and reduce your appetite. However, inflammation blocks the action of leptin. Weight can also cause leptin resistance, much like insulin resistance causes type II diabetes.

Inflammation is very common in most Americans because of the large quantity of chemicals in our food, water and air. Inflammation causes fluid retention, and since it is very common for Americans to have intestinal inflammation, it's very common to have excess fluid and fat in the abdominal area. Since a gallon of water weighs eight pounds, it is very easy to have 5-15 pounds of excess water stored in the tissues of the body.

One can see that unless one changes the type of chemical messages being sent by the cells, one is fighting a losing (or gaining) battle. Conversely, by getting our cells sending the right chemical messages, we can increase energy, reduce appetite, enhance mood and have a better functioning immune system. There are four basic keys to achieving this goal, as follows.

1. Start with an Attitude Adjustment

There is a simple, but powerful principle whatever we focus our mental energy on, we tend to create. If one is constantly thinking negative things about oneself, or one's body, these thoughts will both create and perpetuate health problems, including excess weight. Most people hate certain things about themselves and their bodies. For instance, they may not like their stomach, or their thighs, or their complexion. Even actresses and models who are considered the epitome of beauty have these kinds of issues, and often abuse their bodies trying to achieve some unrealistic image of perfection.

Mentally and emotionally, we associate being wrong with the need to be punished. So negative attitudes about the body cause us to want to punish the body for being wrong. This is why many people are driven to unhealthy diet or exercise regimes in trying to achieve the ideal image of weight and beauty. Real beauty comes from health and inner happiness. It radiates from within as a glowing complexion, sparkling clear eyes and a happy countenance.

If you stop and think about it, the biggest reason people pig out on junk food (or acquire any other bad habit, for that matter) is because they are unhappy. Being unhappy sends the wrong chemical messages to your cells. So, beating yourself up mentally and emotionally for being overweight is only going to perpetuate the problem.

Conversely, it has been scientifically documented that pleasurable experiences (such as loving relationships, laughter, enjoyable activities, time spent in nature, etc.) cause the body to send out chemical messages that reduce inflammation, enhance immunity, promote healing, improve metabolism and otherwise enhance health and well-being. In fact, the biggest single factor in having a long and healthy life isn't your weight, your diet or your exercise level it's your attitude. People who experience pleasure in life live longer, healthier lives.

So, instead of being hard on yourself, be gentle with yourself and find ways to experience pleasure in your life. Do things that help you find joy and fulfillment.

Furthermore, enjoy your food! Make a decision to enjoy whatever you decide to eat (even if it isn't the healthiest food). This means taking time to notice the color, aroma, texture and taste of each bite. It also means eating your food slowly and chewing it thoroughly. A good way to train yourself to do this is to put your fork or spoon down after each bite and take some nice deep breaths while chewing. If you just do this, you ll automatically eat less and feel more satisfied.

Remember that all healing, including losing weight, is inherently about love and nurturing, not fear and deprivation. So, as hard as it may be, start changing your self talk. Keep a journal and write down your feelings about yourself and your body. Then, start affirming that you love yourself, and that you are loving and caring for your body.

2. Instead of Focusing on Calories, Focus on Selecting Healthy Food

The single biggest reason why so many Americans are overweight and sick is because we are eating refined and processed foods. These foods are lacking in vitamins, minerals, enzymes and other phytonutrients the body is looking to obtain from food. When we eat these foods, we may be getting enough calories, but we still feel hungry because the body is still looking for the nutrition it needs.

Refined sugars (sucrose, high fructose corn syrup, fructose, etc.), white flour, polished rice, processed vegetable oils, margarine, shortening and just about all processed and packaged foods fall into this category. If you want to lose weight and be healthy, you must start eliminating these foods from your diet and replacing them with whole, natural, nutrient-rich, unprocessed foods. When you do so, your body will get the nutrition it needs and stop telling you it is hungry.

Of course, if you focus on the negative (what you shouldn't be eating) you ll never succeed. The way to succeed is to start incorporating more whole, natural foods into your diet. Eat plenty of fresh vegetables and fruits (with a greater emphasis on vegetables that are not starchy). Choose good sources of protein for your blood type and eat high quality protein and vegetables as your primary food source. Go easy on grains (bread, pasta, etc.) and only use whole grains when you do eat grains. Eat high quality fats, too. Eat the foods that are good for you first and your body will start craving the good foods while your desire for the junk foods will diminish.

3. Eat Small, but Balanced Meals on a Regular Basis

Here is a sure-fire recipe for gaining weight, even on good food. Skip breakfast and eat a big meal right before going to bed. This puts your blood sugar on a roller-coaster and results in a daily famine-feast cycle. You aren't hungry in the morning, so you don't eat. Your body, thinking it is starving sends messengers to lower your metabolism and energy level throughout the day. By night, your body is starving, and you eat too much. However, since you are inactive (going to sleep), the body stores the excess calories for tomorrow's famine.

To change this cycle, always break your fast by eating something for breakfast. Always start the day with some quality fat and protein. For example: avocado, eggs, meat, whole milk yogurt with fruit, nuts or oatmeal with butter or cream (not sugar). This sets your metabolism to start burning fat instead of storing it. When you eat carbohydrates for breakfast, you raise insulin levels, which prompts the body to store fat. Starting off with a meal that contains fat and protein, not just carbs, kicks a hormone called glucagon into action which mobilizes stored sugars.

Then, whenever you feel a little bit hungry during the day, eat a healthy snack such as nuts, fruit, organic cheese, fresh vegetables, tuna, a salad, etc. By eating small, regular meals your body realizes there's no more famine and will start adjusting your appetite accordingly. You won't be so hungry at night and won't overeat at bedtime. You will feel better and these meals will help you lose weight.

4. Get Physically Active

Forty-eight million Americans are considered sedentary, which contributes directly to obesity. Being sedentary means one doesn't get enough exercise to maintain health.

Almost every system of the body is affected by exercise. Exercise tones the heart, which is a muscle and needs exercise just as much as all other muscles do. It also increases the body's ability to use oxygen, raises the amount of blood pumping through the body, decreases blood pressure, lubricates the joints and makes the body function better in the use and storing of calories. That means better health and less excess body fat.

Don't worry. This doesn't mean you have to go to the gym. Chose physical activities that you enjoy such as swimming, riding a bike, playing a sport, gardening or hiking. Just getting out and taking a walk every day is helpful. You can also consider yoga or tai chi, as these exercises also help reduce stress. Resistance exercise (such as weight lifting) helps increase weight loss because muscle tissue burns more calories than fat. So consider doing at least a little bit of weight lifting.

5. Correct Underlying Health Problems

Part of the secret to weight loss is to identify some of the specific health issues that may be inhibiting your ability to lose weight. These can then be corrected with appropriate supplements or lifestyle changes. Consult a local natural health professional to help you create a program that is custom-tailored for you, but here are a few of the most important problems to consider.

Low Thyroid

The body burns fat in order to stay warm and the gland that sends the chemical messages to burn that fat is the thyroid. Low thyroid is extremely common, especially among women, and can result in weight gain, fatigue, depression, cold hands and feet and dry skin. If you have any of these symptoms, consider supporting your thyroid.

If lab tests show you have normal levels of thyroid hormones, but you still exhibit'symptoms of low thyroid, you may have a problem with conversion of T4 (the inactive form of the thyroid hormone) to T3, the active form. 7-Keto, a metabolite of DHEA can be very helpful in this case, as it increases the conversion of T4 to T3, which increases metabolic rate and increases fat metabolism.

Toxicity and Inflammation

Toxins contribute to weight gain in two ways. First, toxins cause inflammation and inflammation causes fluid retention in the tissues. The rapid weight loss most people experience at the beginning of any diet program or cleanse is typically due to a reduction of inflammation and fluid retention.

The second reason toxins contribute to weight gain has to do with the fact that many toxins we re exposed to are fat-soluble. So, if the body can't break them down, it'stores them in fat. It may also increase cholesterol levels to transport them. The body won't release the fat if it can't deal with the toxins.

This is why learning to eat natural foods is critical to weight loss. Not only are natural foods free of the chemical additives found in processed foods, they also contain more vitamins and minerals to break down toxins. It also explains why a cleanse can help a person lose weight.

Dieter's Cleanse is a convenient pre-packaged, cleansing program that can be helpful for anyone trying to lose weight. It not only contains herbs that promote detoxification and reduce fluid retention in the tissues, it also contains herbs that help balance hormones and metabolism. Be aware that most of the weight lost on a cleanse is not fat, it is simply water retained in the tissues from inflammation. It is not uncommon for people to lose 5-10 pounds on a cleanse, but the loss of fat proceeds more slowly (about one to two pounds per week).

Adding fiber to the diet is another way to increase weight loss. Fiber binds toxins for removal and results in a feeling of fullness that reduces appetite. Psyllium Hulls Combination or Fat Grabbers are good choices for helping with weight loss. Be sure to take fiber supplements with plenty of water.

Stress

Stress and unresolved emotional issues can also contribute to weight gain. Many people eat to feed emotional needs. Stress also releases hormones from the adrenals like cortisol, which causes a breakdown of muscle tissue and can contribute to creating fat deposits. Stress can also cause a release of other chemicals which create food cravings and otherwise upset the body's biochemistry.

Keeping an emotional journal where you can write down your feelings when you are having the desire to eat junk food or otherwise fail to take care of your body, can help you identify and fulfill your real emotional needs. Cravings for sugar and sweets often signals a lack of joy or sweetness in one's life. Finding ways to bring more joy into your life can help fulfill those emotional needs in more constructive ways.

Supplements that help you deal with stress more effectively can also be beneficial. These include Nature's Cortisol, which helps reduce stress levels and inhibits excess cortisol production. This can be particularly helpful where there is abdominal fat due to stress.

Along with Nature's Cortisol, other options are Nutri-Calm and Adrenal Support. These also help reduce stress levels. Nutri-Calm helps a person who is always running around like a chicken with it's head cut off, while Adrenal Support can rebuild the adrenals of those who are burned-out and addicted to stimulants like caffeine.

Depression may also be involved in weight problems and food cravings. 5-HTP can help increase serotonin levels, which can reduce appetite and improve mood.

Remember that cortisol is also released in response to chronic inflammation, so a supplement like Thai-Go that has antioxidant and anti-inflammatory properties can also help to reduce cortisol output. Reducing inflammation also increases metabolism and helps leptin, the fat burning hormone, work more efficiently.

Food Cravings

Sometimes, it seems that food cravings are our worst enemy when trying to change our diet. Therefore, supplements that reduce appetite or specific food cravings may also be helpful as part of a weight loss program. Nature's Hoodia can help reduce overall appetite. So can Garcinia Combination, which also boosts metabolism.

Cravings for sugar are signs of blood sugar problems and usually indicate that the diet is lacking fat and protein. Licorice Root and Super Algae are two supplements that can control blood sugar levels and reduce the cravings for sweets.

Addiction to sugar and other simple carbohydrates is a big problem for some people, so it is also helpful to substitute a healthier form of sweets,. For instance, xylitol-sweetened chocolate bars, gum or mints can satisfy the occasional cravings for sweets without causing the blood sugar spikes that lead to weight gain. Bulk xylitol may be substituted in equal amounts for refined sugar, but not in yeast breads.

Here is a complete list of products that may be helpful in a weight-loss program.

Therapies: Hydration (pg. 10), Blood Type Diet (pg. 13), Affirmation and Visualization (pg. 8), Colon Cleansing (pg. 18), Stress Management (pg. 8), SuperFood (pg. 15), Enzymes (pg. 15), Essential Fatty Acids (pg. 17), Fiber (pg. 17)

Remedies: Collatrim, Dieter's Cleanse, Garcinia Combination, Love and Peas, MetaboMax, MetaboStart, Nature's Hoodia, Nutri-Burn, Target TS-II, Thyroid Support, Xylitol, 5-HTP, 7-Keto, Adrenal Support, Appetite Control, Bee Pollen, Black Currant Oil, Carbo-Grabbers, Cellu-Smooth, Chickweed, CLA, Dulse, Evening Primrose Oil, EveryBody's Formula, Fat Grabbers, Herbal Trim Skin Treatment, IGF-1, Immune Stimulator, Kelp, Kidney Activator (Chinese), L-Carnitine, LB Extract, LBS II, Lemon Oil, Licorice Root, LOCLO, Nature's Chi, Nature's Cortisol, Nature's Three, Nutri-Calm, Psyllium, Psyllium Hulls Combination, SF, Stevia, Sunshine Heroes Omega 3 with DHA, Super Algae, Super GLA, Thyroid Activator, TS II

Wheezing

Wheezing can occur from constriction of the respiratory passages or by a loss of elasticity to the respiratory membranes. The following may help wheezing and shortness of breath.

Remedies: Cordyceps, Lung Support, Mullein, Asthma, Ginkgo Biloba, Horsetail, Lobelia, Sinus Support

Whiplash

An injury to the neck caused by auto accidents or any sudden distortion of the neck. These remedies may help with healing.

Remedies: Black Cohosh, Nature's Fresh

Whooping Cough

See *Pertussis (Whooping Cough)*

Worms

See *Parasites (worms)*

Worry

The following may help to ease excessive worry.

Remedies: Adrenal Support, Anti-Gas (Chinese), Chamomile, Chamomile (Roman), Frankincense, Guardian, Lavender, Myrrh, Spleen Activator

Wounds and Sores

These remedies are useful for helping wounds and sores to heal more quickly. To prevent tetanus, wounds, especially deep puncture wounds, should be thoroughly cleansed and a topical antiseptic such as Silver Shield and/or tea tree oil should be applied liberally. Black Ointment is good for drawing infection out of wounds and sores.

Remedies: Golden Salve, Healing AC Cream, Helichrysum, Herbal Trim Skin Treatment, Nature's Fresh, Tea Tree Oil, Aloe Vera, Bayberry, Bergamot, Black Ointment, Black Walnut, Bone/Skin Poultice, Capsicum, Chamomile (Roman), Chickweed, Devil's Claw, Echinacea Purpurea, Eucalyptus, Golden Seal, Guardian, Ho Shou Wu, Horsetail, Lemon Oil, Lymph Gland Cleanse-HY, Marshmallow, Pau D Arco, PLS II, Sage, Slippery Elm, Ultimate Echinacea, White Oak Bark, Yarrow, Yellow Dock, Zinc

Wrinkles

See Also *Skin Care (general)*

Avoid cigarette smoke and excessive sun exposure for more youthful, healthy skin. The following may be helpful in preventing or healing wrinkles.

Therapies: Antioxidants (pg. 17), Essential Fatty Acids (pg. 17), Minerals (pg. 15), Stress Management (pg. 8)

Remedies: Herbal Trim Skin Treatment, Nature's Fresh, Silver Shield, HSN Complex, HSN-W, Lemon Oil, Patchouli, Super GLA, Triple Effect Age Relief, Vitamin A & D, Vitamin C

Yeast Infections

See *Fungal Infections (Yeast Infections, Candida albicans)*

Section Four
Systems

Section Four

Index to Systems (Organs, Glands, Parts, etc.)

This section allows you to look up specific body systems, organs, and parts, including some neurotransmitters and hormones to learn what remedies can influence or aid these body parts and systems. Our best product selections for each system are highlighted in bold.

Abdomen

Remedies that aid the abdominal area.

Acetylcholine

The following may help increase levels of acetylcholine, or contain substances that bind to receptor sites for acetylcholine. This neurotransmitter that helps with memory and muscle tone. It is deficient in Alzhiemer's Disease.

Brain-Protex, HSN-W, Lecithin, Sage

Adrenal Cortex

Specific remedies for the adrenal cortex.

Geranium, Licorice Root, Maca, Pine Needle, Pregnenolone

Adrenal Glands

General remedies for the adrenal glands.

Adrenal Pack, **Adrenal Support**, C-X, Caffeine Detox, Chamomile (Roman), CLA, DHEA-F, DHEA-M, Eleuthero, Formula Chi with Ephedra, FV, Ginkgo & Hawthorn, Ginseng (Korean), Ginseng (Wild American), Hawthorn Berries, HY-A, Licorice Root, Mineral-Chi Tonic, Monthly Maintenance, Nature's Cortisol, Nervous Fatigue Formula, Nutri-Calm, Pantothenic Acid, PBS, Pro-Pancreas, PS II, Red Beet Formula, Saw Palmetto, Sea Salt, Stress Pack, SUMA Combination, Vitamin C, Wild Yam, X-A, X-Action (Men's), X-Action (Women's)

Adrenal Medulla

Specific remedies for the adrenal medulla.

Ginseng (Korean), Ginseng (Wild American), HY-C, Nervous Fatigue Formula, Pantothenic Acid, Pine Needle, Valerian Root

Aldosterone

Aldosterone is an adrenal hormone that regulates sodium and potassium balance. It reduces urine output and helps tissues hold onto fluid.

HY-A, HY-C, Licorice Root, Pine Needle, Rosemary, Thyme

Appendix

Remedies that aid the appendix.

Garlic, Golden Seal, **Intestinal Soothe & Build**, Lymphatic Drainage, Lymphomax, Oregon Grape

Arteries

Specific remedies to tonify or protect arteries.

Blood Pressurex, Cardio Assurance, Garlic, Ginkgo/Gotu Kola, Hawthorn Berries, Mega-Chel, Thai-Go

Bladder (Urinary)

Remedies for the urinary bladder.

Chickweed, **Cranberry & Buchu**, Horsetail, Hydrangea, JP-X, Juniper Berries, KB-C, Kidney Activator, Kidney Activator (Chinese), Licorice Root, Marshmallow, Oregon Grape, **Parthenium**, Red Clover, Red Clover Blend, St. John's Wort, Urinary Maintenance, **Uva Ursi**, Yarrow

Blood

Remedies that work on the blood.

Blessed Thistle, **Blood Build**, Burdock, Chickweed, Dong Quai, Herbal Trace Minerals, **I-X**, Iron, Liver Balance, Nattozimes Plus, Vitamin B-12, Vitamin B-Complex, Yellow Dock, Zinc

Blood Vessels

Remedies that work on blood vessels in general.

Bilberry Fruit, Butcher's Broom, **Mega-Chel**, Rose Hips, Thai-Go, **Vari-Gone**, White Oak Bark

Blood-Brain Barrier

Remedies that affect or can cross the blood-brain barrier.

Bones

Remedies that help the bones to heal.

Alfalfa, Bee Pollen, **Bone/Skin Poultice**, Calcium, Dulse, EverFlex, Grapine, Healing AC Cream, **Herbal CA**, Herbal Trace Minerals, **HSN-W**, Ionic Minerals, L-Lysine, Magnesium, Mullein, PLS II, **Skeletal Strength**, Sunshine Heroes Calcium Plus D3, White Oak Bark

Brain

Specific remedies for the brain.

5-HTP, Alpha Lipoic Acid, **Brain-Protex**, DHA, Focus Attention, Folic Acid Plus, GABA Plus, Germanium Combination, Ginkgo & Hawthorn, **Ginkgo Biloba**, **Ginkgo/Gotu Kola**, GLD-F, **Gotu Kola**, Herbal Sleep, HY-C, L-Glutamine, Lecithin, Magnesium, Mega-Chel, Mood Elevator, Nature's Phenyltol with NEM, Nervous Fatigue Formula, Niacin, Peppermint, Red Beet Formula, SUMA Combination, **Sunshine Heroes Omega 3 with DHA**, Super Algae, Vitamin B-12, Vitamin B-6, Vitamin B-Complex, Vitamin C, Zinc

Breasts

Specific remedies for the breasts.

Blessed Thistle, **Breast Assured**, **Breast Enhance**, Clary Sage, Frankincense, Helichrysum, Indole 3 Carbinol, Lutein, Lymphatic Drainage, Paw Paw, Ylang Ylang

Bronchials

Specific remedies for the bronchials.

ALJ, Aloe Vera, Asthma, Black Cohosh, Breathe EZ, Breathe Free, **Bronchial Formula**, Capsicum, Catnip, Chamomile, Cough Syrup-NT, HistaBlock, KB-C, Lobelia, Lung Support, Mullein, Oregon Grape, Thyme, Yerba Santa/Senega Combination

Capillaries

Specific remedies for the capillaries.

Cellu-Tone, Echinacea Purpurea, Ginkgo/Gotu Kola, Hawthorn Berries, HY-C, Mega-Chel, Neroli, Noni (Morinda), **Rose Hips**, Thai-Go, Ultimate Echinacea, **Vari-Gone**, Vitamin E, Yarrow

Cardiovascular System

General remedies for the cardiovascular system.

Alfalfa, Alpha Lipoic Acid, Bee Pollen, Black Currant Oil, Blood Pressurex, Caprylimune, Capsicum, Cardio Assurance, Chromium GTF, Eleuthero, Garcinia Combination, Ginkgo & Hawthorn, Ginseng (Korean), Ginseng (Wild American), Hawthorn Berries, HS II, IF Relief, IGF-1, Mega-Chel, Nature's Gold, Neroli, Red Yeast Rice, Super ORAC, Super Trio, Thai-Go, Vitamin D3

Central Nerves

Specific remedies for the central nerves. The central nerves are involved in muscle movement and sensory input. They also regulate pain.

Ginseng (Korean), Ginseng (Wild American), Gotu Kola, Hops, Horsetail, **HSN-W**, Kava Kava, Lavender, **St. John's Wort**, St. John's Wort with Passion Flower, Super GLA, Triple Relief, **Valerian Root**, Wood Betony

Cerebrospinal Fluid

Remedies affecting the cerebrospinal fluid.

Chest

Remedies that affect the chest area.

Circulation

Remedies that improve the circulation of blood.

Anamu, Blessed Thistle, Blood Pressurex, **Capsicum**, **Capsicum & Garlic with Parsley**, Cardio Assurance, Cellu-Smooth, Chamomile, **Chlorophyll**, Clove Bud, DHA, **Dong Quai**, Energ-V, Formula Chi with Ephedra, GC-X, Ginger, **Ginkgo & Hawthorn**, Ginkgo/Gotu Kola, Ginseng (Korean), Ginseng (Wild American), Gotu Kola, Grapefruit (Pink), Guggul, Hawthorn Berries, Helichrysum, **HS II**, Lecithin, Lemon Oil, Lobelia, Lymphatic Drainage, **Mega-Chel**, MetaboMax, Nattozimes Plus, Nature's Chi, **Niacin**, RG-Max, Rosemary, Sea Salt, Valerian Root, Vitamin C, Vitamin E, Wood Betony, X-A, X-Action (Men's)

Collagen

Remedies that help to build collagen.

Collatrim, Vitamin C

Connective Tissue

Remedies to help repair connective tissue.

EverFlex, Glucosamine, Gotu Kola, Horsetail, **HSN-W**, MSM, Rose Hips, Skeletal Strength, **Vitamin C**

Cortisol

Cortisol is an adrenal hormone which influences metabolism. It is a stress hormone, but it is also anti-inflammatory. Excessive cortisol leads to rapid aging, but deficient cortisol product can be a factor in excessive immune activity (auto-immune disorders) and inflammation.

Licorice Root, Trigger Immune, Wild Yam, Yucca

Cuticle

The cuticle is the outermost layer of the skin. It is also the hardened skin that grows around finger and toenails. These remedies help this layer of skin and may help with cuticles around finger and toenails.

Herbal Trim Skin Treatment, Horsetail, HSN-W, Mineral-Chi Tonic, **Nature's Fresh**, Omega-3, Skeletal Strength, Super Algae, Super GLA, Vitamin B-Complex, Zinc

Cyclic AMP (cAMP)

Cyclic AMP is a messenger chemical required for the synthesis of many different hormones. The following help promote cAMP.

Blood Pressurex, Feverfew, Ginseng (Korean), Hawthorn Berries, Liver Balance

Digestive System

General remedies for the digestive system.

5-HTP, Alfalfa, ALJ, All Cell Detox, Anamu, **Anti-Gas (Chinese)**, **Anti-Gas Formula**, Appetite Control, Artemisia Combination, AS with Gymnema, Barley Grass, Bergamot, Bilberry Fruit, Bronchial Formula, Burdock, Candida Clear, Carbo-Grabbers, Catnip, Catnip & Fennel, CBG Extract, Chamomile (Roman), Charcoal (Activated), Chlorophyll, Cinnamon, Clove Bud, Coral Calcium, Defense Maintenance, Digestive Bitters Tonic, Enviro-Detox, Every Body's Fiber, False Unicorn, Fat Grabbers, Fen-Chi, Food Enzymes, Gall Bladder Formula, Garcinia Combination, Garlic, Gastro Health, GC-X, Golden Salve, Golden Seal, Grapefruit (Pink), HCP-X, Healthy Blast, Heavy Metal Detox, Helichrysum, Herbal Beverage, Herbal Sleep, Hi Lipase, HistaBlock, Lactase Plus, Liquid Cleanse, LIV-J, Liver Balance, Mandarin (Red), Marjoram (Sweet), MetaboMax, MetaboStart, Myrrh, Nature's Gold, Nerve Eight, NF-X, Noni (Morinda), Oregano (Wild), Parsley, PBS, Peppermint, Potassium, **Proactazyme Plus**, Protease, Red Raspberry Liquid, Rose Bulgaria, Rosemary, Sandalwood, Sarsaparilla, Slippery Elm, SnorEase, Stress-J, Sunshine Heros Whole Foods Papayazyme, Super Algae, Tiao He Cleanse, Wild Yam, Wild Yam & Chaste Tree, Wood Betony, X-Action (Women's), Yeast/Fungal Detox

DNA

Remedies that may help build or repair the DNA inside the cells.

Cellular Build, Folic Acid Plus, **Super Algae**, Target Endurance, Vitamin B-Complex

Dopamine

The following remedies may help increase dopamine levels. Dopamine is a neurotransmitter involved in mood, muscle coordination and sexual drive.

5-HTP

Ears

Remedies to aid the ears.

Aloe Vera, **CBG Extract**, Echinacea Purpurea, **Garlic**, Ginkgo & Hawthorn, Ginkgo/Gotu Kola, Helichrysum, Herbal Trim Skin Treatment, HY-C, IN-X, KB-C, **Lavender**, **Lobelia**, Lymph Gland Cleanse, Lymph Gland Cleanse-HY, Mullein, Tea Tree Oil, Ultimate Echinacea, Vitamin B-12, Vitamin B-Complex, Xylitol

Epinephrine

The following remedies either enhance epinephrine or stimulate receptor sites for epinephrine.

Pine Needle, Rosemary, Sage, Thyme

Estrogen

There are three different estrogens produced in the body. They are estrais, estrone, and estradiol. These remedies have an estrogenic or estrogen-enhancing effect. (Wild Yam & Chaste Tree has an estrogen-inhibiting effect.)

Black Cohosh, Clary Sage, Geranium, Hops, Licorice Root

Eustachian Tubes

Remedies that help to open blocked Eustachian tubes.

EW, **Eyebright**, Garlic, HistaBlock, IF Relief, Lymphatic Drainage

Eyes

Remedies that aid the eyes.

Bilberry Fruit, Blood Build, Chamomile, Chickweed, DHA, **EW**, Eyebright, Germanium Combination, Golden Salve, Golden Seal, Gotu Kola, Horsetail, HY-C, IF-C, Lobelia, Lutein, Mega-Chel, Passion Flower, **Perfect Eyes**, Red Clover, Red Clover Blend, Red Raspberry, White Oak Bark, Yarrow

Female Reproductive

General remedies for the female reproductive system.

5-W, All Cell Detox, Alpha Lipoic Acid, Artemisia Combination, Bergamot, **Blood Build**, Blue Cohosh, C-X, Chamomile (Roman), CLA, Clary Sage, Deep Relief Oil, DHEA-F, False Unicorn, FCS II, Female Comfort, Flash-Ease T/R, Flax Seed Oil, Geranium, Golden Seal, Grapefruit (Pink), HY-C, Marjoram (Sweet), Menopause, Menstrual, Nerve Control, **NF-X**, Omega-3, Phyto-Soy, PMS, Pro-G-Yam Cream, **Red Raspberry**, Rose Bulgaria, Super GLA, Uña de Gato Combination, V-X, Vitamin B-6, Vitamin B-Complex, Wild Yam, Wild Yam & Chaste Tree, X-Action (Women's), X-Action Gel, Yarrow,

GABA

GABA is a neurotransmitter that calms down excessive nervous firing in the brain. It may be deficient in ADHD, epilepsy and schizophrenia. The following either enhance GABA or stimulate GABA receptors.

GABA Plus, Hops, Kava Kava, Valerian Root

Gall Bladder

Specific remedies for the gall bladder.

Aloe Vera, Blessed Thistle, Burdock, Cascara Sagrada, Chickweed, Dandelion, Devil's Claw, Digestive Bitters Tonic, Evening Primrose Oil, Food Enzymes, **Gall Bladder Formula**, Golden Seal, Horsetail, HSN-W, Hydrangea, Kelp, Lecithin, LIV-J, Liver Balance, Liver Cleanse Formula, LOCLO, Magnesium, Milk Thistle, Mullein, Oregon Grape, Thyme, Trigger Immune, Vari-Gone, **Wild Yam**, **Yellow Dock**

Glandular System

General remedies that help to build or balance the entire glandular system.

AdaptaMax, Breast Assured, Breast Enhance, Cellular Energy, Cinnamon, Energ-V, HY-C, Kelp, Maca, **Master Gland**, Menopause, Menstrual, MetaboStart, Monthly Maintenance, Natural Changes, PMS, Spirulina, SugarReg, SUMA Combination

Gums

Remedies that help to heal the gums.

Bayberry, **Black Walnut**, Capsicum, Clove Bud, **Co-Q10**, Geranium, Golden Seal, IF Relief, IF-C, Juniper Berries, Myrrh, **White Oak Bark**, Xylitol

Hair

Remedies that help the hair.

Clary Sage, Dulse, Ho Shou Wu, Horsetail, **HSN Complex**, **HSN-W**, KB-C, Kelp, MSM, Pantothenic Acid, Paw Paw Lice Remover Shampoo, Rosemary, Sage, Vitamin E, Yarrow, Zinc

Heart

Remedies to support the heart.

Astragalus, Black Cohosh, Blessed Thistle, Calcium, Capsicum, Capsicum & Garlic with Parsley, Cholester-Reg II, Clove Bud, **Co-Q10**, Garcinia Combination, GC-X, Germanium Combination, **Ginkgo & Hawthorn**, Ginseng (Korean), Ginseng (Wild American), Gotu Kola, **Hawthorn Berries**, HS II, Iron, **L-Carnitine**, Lecithin, **Magnesium**, **Mega-Chel**, Nervous Fatigue Formula, Olive Leaf, **Omega-3**, Passion Flower, Red Raspberry, RG-Max, Rose Bulgaria, Safflowers, Valerian Root, Vitamin E

Hypothalamus

Specific remedies for the hypothalamus.

5-HTP, Bee Pollen, Bee Pollen, Capsicum, Chlorophyll, Clary Sage, Eyebright, Garlic, Herbal CA, Hi Lipase, Jasmine Absolute, MSM, Patchouli, Potassium, Red Raspberry, Sage, Thyme, Ylang Ylang

Immune System

The immune system is not a single system, but rather a complex process in the body which involves multiple body systems, including the digestive system, lymphatic system, circulatory system, skin and more. It serves to protect the body from microbes and toxins that would interfere with health. Most natural remedies affect the immune system either directly or indirectly. Those listed here are some of the remedies with the most powerful and immediate immune system effects.

AdaptaMax, Algin, Allergies-Hayfever/Pollen, Allergies-Mold/Yeast/Dust, Alpha Lipoic Acid, Anamu, Antioxidant Arsenal, **Astragalus**, Barley Grass, Berry Healthy, Black Currant Oil, Black Walnut, Blue Vervain, Candida, Candida Clear, Caprylic Acid Combination, Cellular Build, Clove Bud, Cold, Colostrum, Colostrum with Immune Factors, **Cordyceps**, Defense Maintenance, Detoxification, Devil's Claw, E-Tea, Echinacea Purpurea, Echinacea/ Golden Seal, Elderberry Defense, Elderberry Plus, Eleuthero, Enviro-Detox, Eucalyptus, Evening Primrose Oil, Fizz Active-Immune, Frankincense, Frankincense, Garlic, Gastro Health, Germanium Combination, Ginseng (Korean), Glyco Essentials, Golden Seal, Golden Seal/Parthenium, Guardian, HCP-X, Heavy Metal Detox, HistaBlock, IF-C, **Immune Stimulator**, IN-X, Inflammation, Influenza Remedy, Ionic Minerals, L-Lysine, Lemon Oil, Lymph Gland Cleanse, Lymph Gland Cleanse-HY, Lymphatic Drainage, Lymphomax, Lymphostim, Mineral-Chi Tonic, MSM, Myrrh, Nature's Chi, Nature's Gold, Olive Leaf, Omega-3, Oregano (Wild), Oregon Grape, Para-Cleanse, Parasites, Pau D'Arco, Pau D'Arco Power Pack, Paw Paw, Prevention, Proactazyme Plus, Protease, Recovery, Red Clover, Red Clover Blend, Rosemary, SC Formula, Seasonal Defense, **Silver Shield**, Sunshine Heroes Omega 3 with DHA, Sunshine Heroes Probiotic Power, Sunshine Heroes Whole Foods Antioxidant, Sunshine Heros Whole Foods Papayazyme, Super Antioxidant, Super GLA, Super ORAC, Tea Tree Oil, Thai-Go, THIM-J, Thyme, Tiao He Cleanse, **Trigger Immune**, **Ultimate Echinacea**, Uña de Gato, Uña de Gato Combination, Vaccine Detox, Viral Recovery, Vitamin C, Vitamin D3, VS-C, Yarrow, Yeast/Fungal Detox, Zinc, Zinc Lozenges

Intestinal System

General remedies for the intestinal system.

Algin, **All Cell Detox**, Aloe Vera, Anti-Gas (Chinese), Anti-Gas Formula, Artemisia Combination, Bentonite (Hydrated), Black Walnut, Blessed Thistle, Bowel Detox, Burdock, Candida, Caprylimune, Cascara Sagrada, Catnip & Fennel, Chamomile, Charcoal (Activated), Chickweed, Chlorophyll, CleanStart, Clove Bud, CLT-X, Cramp Relief, Dandelion, Deep Relief Oil, Devil's Claw, Digestive Bitters Tonic, Enviro-Detox, Every Body's Fiber, Food Enzymes, Garcinia Combination, Garlic, GC-X, **Gentle Move**, Germanium Combination, Ginger, Ginseng (Korean), Ginseng (Wild American), Golden Seal, Grapefruit (Pink), Heavy Metal Detox, Herbal Pumpkin, Horsetail, HY-C, IF-C, LB Extract, LB-X, **LBS II**, Lemon Oil, Licorice Root, Magnesium, Mandarin (Red), Marjoram (Sweet), Marshmallow, Mullein, Neroli, Noni (Morinda), Oregon Grape, Para-Cleanse, Peppermint, Pro-Pancreas, Red Raspberry, Senna Combination, SF, Slippery Elm, **Small Intestine Detox**, Spleen Activator, St. John's Wort, Sunshine Heroes Probiotic Power, Sunshine Heroes Whole Foods Antioxidant, Super GLA, Target P-14, **Tiao He Cleanse**, Uña de Gato, Vitamin C, White Oak Bark, Wild Yam, Xylitol, Yarrow, Yeast/Fungal Detox, Yucca,

Kidneys

The kidneys have the job of filtering waste materials from the blood stream. They also play a role in maintaining pH and blood pressure.

Horsetail, Juniper Berries, Kava Kava, Kidney Activator, Kidney Activator (Chinese), Kidney Drainage, Love and Peas, Lymphatic Drainage, Parsley

Large Intestine (Colon)

Remedies for the large intestine or colon.

Aloe Vera, Bowel Detox, Candida Clear, Caprylic Acid Combination, Cascara Sagrada, Catnip & Fennel, Charcoal (Activated), Chlorophyll, CleanStart, CLT-X, Cramp Relief, Deep Relief Oil, Dieter's Cleanse, Enviro-Detox, Every Body's Fiber, Fat Grabbers, Flax Seed Oil, Gastro Health, Gentle Move, **Intestinal Soothe & Build**, Kudzu/St.John's Wort, LB Extract, LB-X, **LBS II**, Liquid Cleanse, LOCLO, Magnesium, Nature's Three, Para-Cleanse, Parasites, Pau D'Arco, Paw Paw, Peppermint, Probiotics, Psyllium, **Psyllium Hulls Combination**, Senna Combination, White Oak Bark, Yarrow, Yeast/Fungal Detox

Legs

Remedies that help the legs.

HSN-W, **KB-C**, Vari-Gone

Liver

Remedies that help the liver.

7-Keto, AdaptaMax, All Cell Detox, Alpha Lipoic Acid, Anamu, Anti-Gas Formula, Bee Pollen, Bergamot, Black Currant Oil, Blessed Thistle, **Blood Build**, Blue Vervain, BP-X, Brain-Protex, Burdock, C-X, Caprylimune, Carotenoid Blend, Cascara Sagrada,

Cellu-Smooth, Cellular Build, Chamomile, Chickweed, Chlorophyll, Cholester-Reg II, Chromium GTF, CLA, CleanStart, Co-Q10, Colloidal Minerals, Colostrum, Colostrum with Immune Factors, Dandelion, Digestive Bitters Tonic, E-Tea, Echinacea Purpurea, Echinacea/Golden Seal, Elderberry Defense, Elderberry Plus, Energ-V, **Enviro-Detox**, **Food Enzymes**, Gall Bladder Formula, Garcinia Combination, Garlic, Germanium Combination, Ginger, Ginseng (Korean), Ginseng (Wild American), Grapefruit (Pink), Grapine, Green Tea Extract, Heavy Metal Detox, Helichrysum, Herbal Beverage, Herbal Trace Minerals, Hi Lipase, HistaBlock, Ho Shou Wu, Hops, HY-A, I-X, IGF-1, IN-X, Indole 3 Carbinol, Iron, Jasmine Absolute, Joint Health, Joint Support, Lactase Plus, LB Extract, LB-X, LBS II, Lecithin, Lemon Oil, Licorice Root, LIV-J, **Liver Balance**, **Liver Cleanse Formula**, LOCLO, Lymph Gland Cleanse, Lymph Gland Cleanse-HY, Lymphatic Drainage, Lymphomax, Mandarin (Red), Master Gland, MetaboMax, **Milk Thistle**, **Milk Thistle Combination**, Mineral-Chi Tonic, Monthly Maintenance, Mood Elevator, **N-acetyl Cysteine**, Nature's Three, Nervous Fatigue Formula, Niacin, Noni (Morinda), Olive Leaf, Omega-3, Oregon Grape, P-X, Papaya Mint, Para-Cleanse, Pau D'Arco, Paw Paw, PBS, Perfect Eyes, Pro-Pancreas, Proactazyme Plus, Probiotics, Protease, Psyllium, Psyllium Hulls Combination, Red Clover, Red Clover Blend, Red Raspberry Liquid, Red Yeast Rice, Rose Bulgaria, Rose Hips, S-O-D With Gliadin, SAM-e, Sarsaparilla, SC Formula, Seasonal Defense, SF, Sinus Support, Skin Detox, Spirulina, Spleen Activator, St. John's Wort, Stevia, SUMA Combination, Super Algae, Super Antioxidant, Super GLA, Super ORAC, Thai-Go, THIM-J, Tiao He Cleanse, Trace Mineral Maintenance, Trigger Immune, Ultimate Echinacea, Uña de Gato Combination, V-X, Vitamin A & D, Vitamin B-6, Vitamin B-Complex, Vitamin C, Vitamin E, VS-C, White Oak Bark, Wood Betony, Yeast/Fungal Detox, Yellow Dock, Zinc

Lungs

Remedies that aid the lungs.

ALJ, Asthma, **Astragalus**, Black Cohosh, Bone/Skin Poultice, Breathe EZ, Bronchial Formula, CleanStart, **Cordyceps**, Cough Syrup (Childrens), Cough Syrup-DH, Cough Syrup-LP, Cough Syrup-NT, Cramp Relief, Devil's Claw, Devil's Claw, Enviro-Detox, Eucalyptus, Four, Frankincense, **Garlic**, Ginseng (Korean), Ginseng (Wild American), Golden Seal, Guardian, HCP-X, HistaBlock, Horsetail, Immune Stimulator, LH, Licorice Root, Lobelia, **Lung Support**, Lymphatic Drainage, Marshmallow, Marshmallow & Fenugreek, Mullein, Pine Needle, PLS II, Rosemary, Seasonal Defense, Sinus Support, St. John's Wort, Thyme, Tobacco Detox, Trigger Immune

Lymph Nodes

Specific remedies which affect the lymph nodes.

Blue Vervain, Echinacea Purpurea, IN-X, Lobelia, **Lymph Gland Cleanse**, **Lymph Gland Cleanse-HY**, **Lymphatic Drainage**, **Lymphomax**, Lymphostim, Mullein, Oregon Grape, Red Clover, Red Clover Blend, **Silver Shield**, **Ultimate Echinacea**, Yarrow

Lymphatic System

General remedies for the lymphatic system.

ALJ, Bergamot, Breast Assured, Breathe Free, Burdock, Echinacea Purpurea, Garlic, Geranium, Grapefruit (Pink), Helichrysum, I-X, Immune Stimulator, IN-X, Joint Support, Kelp, Kidney Activator (Chinese), Lobelia, Lymph Gland Cleanse, **Lymph Gland Cleanse-HY**, **Lymphatic Drainage**, Lymphomax, Lymphostim, Marshmallow, **Mullein**, Oregon Grape, Red Clover, Sage, **Ultimate Echinacea**, Yarrow, Yellow Dock

Male Reproductive

Specific remedies for the male reproductive system.

DHEA-M, KB-C, **Men's Formula**, PS II, Red Raspberry, RG-Max, Saw Palmetto, X-Action (Men's),

Mitochondria

The mitochondria are structures in the cell where energy is produced in the cell. These remedies affect the mitochondria, either slowing or enhancing cellular energy production.

All Cell Detox, Alpha Lipoic Acid, **Cellular Build**, Cellular Energy, Magnesium, Paw Paw, **Target Endurance**

Mouth

Remedies that may help various conditions in the mouth.

Barley Grass, **Black Walnut**, **Golden Seal**, L-Lysine, Peppermint, Vitamin B-6, Vitamin B-Complex, VS-C, Zinc Lozenges

Mucus Membranes

Remedies for the mucus membranes of the respiratory and digestive tracts.

ALJ, Aloe Vera, Barley Grass, Bayberry, Bergamot, Bone/Skin Poultice, Capsicum, Capsicum & Garlic with Parsley, CC-A, CLT-X, Eyebright, Fenugreek & Thyme, Feverfew, Four, FV, **Gastro Health**, GC-X, Golden Salve, **Golden Seal**, Golden Seal/Parthenium, **HCP-X**, **Intestinal Soothe & Build**, LH, Licorice Root, Lobelia, Marshmallow, Mullein, PLS II, Psyllium, Red Raspberry, Rose Hips, Sandalwood, Sinus Support, Slippery Elm, Small Intestine Detox, **Uña de Gato**, Uva Ursi, Vitamin C, VS-C, White Oak Bark, Yarrow, Zinc Lozenges

Muscles

General remedies for muscles.

APS II, Bee Pollen, Blood Build, Bone/Skin Poultice, Calcium, Chamomile, Co-Q10, Cramp Relief, Deep Relief Oil, Dong Quai, EverFlex, Everflex Pain Cream, **Fibralgia**, Germanium Combination, Hops, Joint Support, Kava Kava, KB-C, Lecithin, Liver Balance, Lobelia, Lung Support, **Magnesium**, Marjoram (Sweet), MSM, MSM/Glucosamine Cream, Neroli, Pantothenic Acid, Red Beet Formula, Safflowers, Spleen Activator, **Target Endurance**, Tei Fu Oil, Valerian Root, Vitamin B-6, Vitamin B-Complex, Vitamin C, Vitamin E

Nails

Specific remedies for fingernails and toenails.

Black Walnut, Dulse, **Horsetail**, **HSN Complex**, **HSN-W**, Kelp, MSM, Zinc

Nerves

General remedies for the nervous system.

5-HTP, AdaptaMax, Adrenal Pack, Alfalfa, Alpha Lipoic Acid, Appetite Control, APS II, Bee Pollen, Bergamot, Black Cohosh, Blue Cohosh, Blue Vervain, Brain-Protex, Caffeine Detox, Calcium, Calming, Capsicum, Catnip, Catnip & Fennel, CBG Extract, CC-A, Chamomile, Chamomile (Roman), Clary Sage, Cramp Relief, Damiana, Depressaquel, DHA, **Distress Remedy**, Dulse, Energ-V, Fenugreek & Thyme, Feverfew, Flax Seed Oil, Focus Attention, Frankincense, GABA Plus, Geranium, Geranium, Ginkgo & Hawthorn, Ginkgo Biloba, Ginkgo/Gotu Kola, Ginseng (Korean), Ginseng (Wild American), Gotu Kola, Grapefruit (Pink), Heavy Metal Detox, Herbal CA, Herbal Sleep, Hops, Horsetail, IF-C, Inflammation, Joint Support, Juniper Berries, Kava Kava, Kidney Activator (Chinese), Kudzu/St.John's Wort, L-Glutamine, Lavender, Lecithin, Lobelia, Magnesium, Mandarin (Red), Migraquel, Mineral-Chi Tonic, Monthly Maintenance, Mood Elevator, Mullein, Natural Changes, Nature's Phenyltol with NEM, Neroli, Nerve Control, Nerve Eight, **Nervous Fatigue Formula**, Nervousness, Niacin, Noni (Morinda), Nutri-Calm, Olive Leaf, Pain, Pantothenic Acid, Passion Flower, PS II, St. John's Wort, St. John's Wort with Passion Flower, Stress Pack, **Stress Relief**, **Stress-J**, SUMA Combination, Sunshine Heroes Calcium Plus D3, Sunshine Heroes Omega 3 with DHA, Tobacco Detox, Trigger Immune, Triple Relief, TS II, Valerian Root, Vitamin B-12, Vitamin B-6, Vitamin B-Complex, Wild Yam, Wild Yam & Chaste Tree, Wood Betony, X-Action (Women's), Yellow Dock

Nipples

These are remedies which can help heal the nipples.

Aloe Vera, Clove Bud, **Golden Salve**, Mullein, **Nature's Fresh**, **Silver Shield**, Slippery Elm

Ovaries

Remedies for the ovaries.

Aloe Vera, Black Cohosh, Blessed Thistle, **C-X**, Damiana, Dong Quai, **FCS II**, **Female Comfort**, Feminine Tonic, Frankincense, GLD-F, Liver Balance, Magnesium, Monthly Maintenance, Pro-G-Yam Cream, Spleen Activator, Wild Yam, Wild Yam & Chaste Tree

Pancreas

General remedies for the pancreas.

Anti-Gas (Chinese), Anti-Gas Formula, AS with Gymnema, Blood Sugar Formula, **Chromium GTF**, CLA, Cranberry & Buchu, Dandelion, Digestive Bitters Tonic, Eucalyptus, **Food Enzymes**, FV, GLD-F, Golden Salve, HY-A, Juniper Berries, Licorice Root, Liver Balance, Liver Cleanse Formula, Nopal, P-X, Patchouli, PBS, Pro-Pancreas, SF, **SugarReg**, **Target P-14**, Uva Ursi, Xylitol, Zinc

Pancreas Head

Specific remedies for the pancreatic head. The pancreatic head secretes digestive enzymes to help digest food in the small intestines.

Chromium GTF, **Food Enzymes**, Gastro Health, Marshmallow, PDA, Probiotics, Protease

Pancreas Tail

Specific remedies for the pancreatic tail. The pancreatic tail secretes insulin and glucagon to regulate blood sugar levels.

AS with Gymnema, Capsicum, Chickweed, Echinacea Purpurea, Fat Grabbers, Golden Seal, Grapine, HY-A, HY-C, Mineral-Chi Tonic, Noni (Morinda), Nopal, Nutri-Calm, PBS, Sage, Spirulina, **SugarReg**, **Target P-14**, Zinc

Parathyroid

Remedies for the parathyroid gland. The parathyroid gland regulates calcium and phosphorus.

7-Keto, Damiana, **Herbal CA**, Horsetail, HSN-W, Kelp, Pau D'Arco, Skeletal Strength, Thyroid Activator, Valerian Root, Vari-Gone, White Oak Bark

Parotids

Remedies that aid the parotid (salivary) glands.

Chlorophyll, Horsetail, HSN-W, Potassium, Sage, White Oak Bark, Yucca, Zinc

Peripheral Blood Vessels

Remedies that aid circulation and tone in the peripheral blood vessels

Butcher's Broom, **Capsicum**, Garlic, Ginkgo/Gotu Kola, Hawthorn Berries, **Rose Hips**, Vari-Gone, Vitamin C,

Peyer's Patches

Lymphatic nodules located in the ileum of the small intestines. These nodules are strategically placed in the intestines as part of the body's immune defenses.

Cellular Build, **Kudzu/St.John's Wort**, Ultimate Echinacea

Pineal

Remedies that aid the pineal gland. The pineal gland produces melatonin, which regulates sleep.

5-HTP, Alfalfa, Black Cohosh, Clary Sage, Dong Quai, Echinacea/Golden Seal, Enviro-Detox, Ginger, Ginger, GLD-F, Golden Seal, Gotu Kola, Horsetail, **HSN-W**, Jasmine Absolute, Kelp, Lavender, Lecithin, LIV-J, **Melatonin Extra**, Nutri-Calm, Patchouli, Sandalwood, Spirulina, SUMA Combination, THIM-J, Vitamin B-Complex, Wood Betony, Ylang Ylang, Zinc

Pituitary (anterior)

Specific remedies for the anterior pituitary. The anterior pituitary produces growth hormone, the thyroid stimulating hormone (TSH), the adreno-corticotrophic hormone (ACTH), the follicle stimulating hormone (FSH), the lutenizing hormone (LH), and prolactin. These hormones regulate the thyroid, adrenal and reproductive glands.

Alfalfa, APS II, Bee Pollen, Co-Q10, Devil's Claw, GreenZone, Kelp, Licorice Root, **Master Gland**, Nature's Prenatal, Spirulina, Super Algae, Vitamin B-Complex, Wild Yam & Chaste Tree

Pituitary (general)

General remedies for the pituitary gland. The pituitary is the master gland. It secretes hormones that regulate the other major endocrine glands.

Adrenal Support, **Alfalfa**, Bee Pollen, GLD-F, Kelp, Spirulina, Super Algae, **Target TS-II**, Thyroid Support

Pituitary (infundibulum)

Specific remedies for the pituitary infundibulum.

5-HTP, **Melatonin Extra**, Mood Elevator, SAM-e, St. John's Wort

Pituitary (posterior)

Specific remedies for the posterior pituitary. The posterior pituitary secretes oxytocin and the anti-diuretic hormone.

Alfalfa, Barley Grass, Blessed Thistle, Catnip, Gotu Kola, Hops, Kelp, Licorice Root, Parsley, **Potassium**, Rose Hips

Progesterone

Progesterone is a major female hormone which helps maintain pregnancy. The following enhance progesterone.

DHEA-F, Parsley, **Pro-G-Yam Cream**, Sarsaparilla, Yarrow

Prostaglandins

The following remedies affect prostaglandin production.

APS II, Black Cohosh, Black Currant Oil, CLA, Evening Primrose Oil, Flax Seed Oil, Nerve Eight, **Omega-3**, Super GLA, **Triple Relief**

Prostate

Specific remedies for the prostate gland.

Cranberry & Buchu, Echinacea Purpurea, Ginseng (Korean), Ginseng (Wild American), **Herbal Pumpkin**, Indole 3 Carbinol, **Men's Formula**, Omega-3, P-X, Phyto-Soy, **PS II**, Red Raspberry, Saw Palmetto, Ultimate Echinacea, X-A, **Zinc**

Rectum

Specific remedies for the rectum.

Every Body's Fiber, Geranium, Slippery Elm, V-X, Vari-Gone, **White Oak Bark**

Red Blood Cells

Specific remedies to build red blood cells.

Alfalfa, **Blood Build**, **Chlorophyll**, Folic Acid Plus, I-X, **Vitamin B-12**, Vitamin B-Complex, Vitamin E, Yellow Dock

Reproductive Glands

General remedies for the reproductive glands (male and female)

Clary Sage, **Damiana**, Eleuthero, Ginger, **Ginseng (Korean)**, **Ginseng (Wild American)**, Ho Shou Wu, IGF-1, Jasmine Absolute, KB-C, Kelp, Licorice Root, **Maca**, Pine Needle, Sandalwood, Sarsaparilla, Super GLA, **Trigger Immune**, Uva Ursi, Vitamin E, X-A, Ylang Ylang, Zinc

Respiratory System

General remedies for the overall respiratory system.

ALJ, Allergies-Hayfever/Pollen, Allergies-Mold/Yeast/Dust, Allergy, Artemisia Combination, Astragalus, Bayberry, Bee Pollen, Bergamot, Blue Vervain, **Breathe Free**, Catnip, Co-Q10, Cold, Echinacea Purpurea, Elderberry Defense, Elderberry Plus, Eucalyptus, **Fenugreek & Thyme**, Formula Chi with Ephedra, Four, Garlic, Ginseng (Korean), Ginseng (Wild American), Golden Seal, HCP-X, Helichrysum, **HistaBlock**, IF Relief, IN-X, Jasmine Absolute, LH, Lobelia, **Lung Support**, Marjoram (Sweet), Marshmallow, Marshmallow & Fenugreek, MetaboMax, Mullein, Myrrh, Nature's Chi, Noni (Morinda), Oregano (Wild), Pau D'Arco, Pine Needle, PLS II, Rosemary, Sandalwood, Saw Palmetto, Seasonal Defense, **Sinus Support**, Slippery Elm, SnorEase, Sunshine Heros Whole Foods Papayazyme, Tobacco Detox, Ultimate Echinacea, Uña de Gato Combination, Yerba Santa/Senega Combination

Scalp

Remedies that aid the scalp

Jojoba Oil, Patchouli, Paw Paw Lice Remover Shampoo, **Rosemary**, Sage

Serotonin

The following remedies may help increase serotinin levels in the body. Serotonin is a mood-elevating neurotransmitter.

5-HTP, Ginkgo Biloba, Passion Flower, St. John's Wort, St. John's Wort with Passion Flower

Sinuses

Specific remedies for the sinuses.

ALJ, Bayberry, Elderberry Plus, EW, Eyebright, **Fenugreek & Thyme**, Four, Garlic, LH, Pine Needle, Sinus, **Sinus Support**, **Tei Fu Oil**, Xylitol

Skeletal System

General remedies to support the skeletal system

Alfalfa, **Bone/Skin Poultice**, Calcium, Colloidal Minerals, EverFlex, Germanium Combination, **Herbal CA**, Herbal Trace Minerals, Horsetail, **HSN-W**, Hydrangea, Joint Health, Joint Support, KB-C, Nature's Gold, Noni (Morinda), PLS II, **Skeletal Strength**, Slippery Elm, Vitamin D3, White Oak Bark

Skin

General remedies to aid the skin.

Acne, Acne Treatment Gel, All Cell Detox, Allergy, Aloe Vera, Bayberry, Bentonite (Hydrated), Bergamot, Black Ointment, Black Walnut, Blood Build, Bone/Skin Poultice, BP-X, Burdock, Capsicum, Cellu-Tone, Chamomile, Charcoal (Activated), Chickweed, Clary Sage, CleanStart, Deep Relief Oil, DHA, Dulse, Echinacea Purpurea, Eczema/Psoriasis, Enviro-Detox, EverFlex, Feverfew, Germanium Combination, **Golden Salve**, Golden Seal, Gotu Kola, Grapefruit (Pink), Guardian, Healing AC Cream, Helichrysum, **Herbal Trim Skin Treatment**, HistaBlock, Horsetail, **HSN Complex**, **HSN-W**, Iron, Jasmine Absolute, Jojoba Oil, Juniper Berries, Kelp, Lavender, Mandarin (Red), Marjoram (Sweet), Marshmallow, MSM, MSM/Glucosamine Cream, Myrrh, **Nature's Fresh**, Nature's Gold, Neroli, Niacin, Oregon Grape, Passion Flower, Pau D'Arco, Paw Paw Lice Remover Shampoo, PLS II, Psyllium, Red Clover, Red Clover Blend, Rose Bulgaria, Rosemary, Safflowers, Sage, Sandalwood, Saw Palmetto, Skin Detox, Slippery Elm, Triple Effect Age Relief,

Ultimate Echinacea, Uña de Gato, Vitamin C, Vitamin E, White Oak Bark, Yarrow, Yellow Dock, Zinc

Small Intestines

Remedies to aid the small intestines.

Algin, ALJ, Aloe Vera, Bayberry, Blood Sugar Formula, Digestive Bitters Tonic, Fat Grabbers, Gastro Health, Ginger, HCP-X, Hi Lipase, **Intestinal Soothe & Build**, **Kudzu/St.John's Wort**, Marshmallow, Nature's Three, Papaya Mint, Red Beet Formula, **Slippery Elm**, **Small Intestine Detox**

Solar Plexus

The solar plexus is an area of the body just under the sternum (breastbone). There is an important nerve center here that connects with the digestive organs and provides us with "gut instinct."

Chamomile, Dandelion, **St. John's Wort**, St. John's Wort with Passion Flower

Spinal Disks

The following may be helpful in repairing slipped,bulging or damaged spinal disks.

Golden Seal, KB-C, **Nature's Fresh**

Spleen

Remedies to aid the spleen.

Cellular Build, Chamomile, Chickweed, Chlorophyll, Dandelion, Dandelion, Devil's Claw, E-Tea, Fat Grabbers, Folic Acid Plus, Ginseng (Korean), Ginseng (Wild American), GLD-F, Golden Salve, Hi Lipase, Immune Stimulator, Lecithin, Licorice Root, Milk Thistle, Nutri-Calm, Patchouli, Pau D'Arco, PS II, Red Clover, St. John's Wort, Trigger Immune, Ultimate Echinacea, Vitamin B-Complex, **White Oak Bark**, Yellow Dock

Stomach

General remedies for the stomach.

Aloe Vera, **Anti-Gas (Chinese)**, **Anti-Gas Formula**, Bayberry, Blessed Thistle, Bowel Detox, Capsicum, **Catnip**, Catnip & Fennel, Chamomile, Chickweed, CLT-X, Dandelion, Deep Relief Oil, **Food Enzymes**, FV, Garlic, Golden Salve, Hops, HSN-W, HY-A, I-X, Intestinal Soothe & Build, Juniper Berries, Kidney Activator (Chinese), Licorice Root, Lobelia, Marshmallow & Fenugreek, Papaya Mint, Passion Flower, PDA, **Peppermint**, **Proactazyme Plus**, Red Beet Formula, Safflowers, Saw Palmetto, Sinus Support, Slippery Elm, Small Intestine Detox, Spleen Activator, **Stomach Comfort**, White Oak Bark, Yarrow, Zinc

Structural System

General remedies for the structural system.

Alfalfa, Aloe Vera, Anamu, APS II, Arthritis, Barley Grass, Bayberry, Bergamot, Bilberry Fruit, Black Currant Oil, Black Ointment, Black Walnut, **Bone/ Skin Poultice**, BP-X, Burdock, Calcium, Cellu-Smooth, Cellu-Tone, Chickweed, Chondroitin, Collatrim, Colloidal Minerals, Coral Calcium, Cramp Relief, Devil's Claw, Dulse, Eczema/Psoriasis, Enviro-Detox, Evening Primrose Oil, EverFlex, Everflex Pain Cream, EveryBody's Formula, Fibralgia, Flax Seed Oil, Free Amino Acids, Glucosamine, Golden Salve, Golden Seal, Healing AC Cream, Helichrysum, **Herbal CA**, Herbal Trace Minerals, Herbal Trim Skin Treatment, Horsetail, HSN Complex, **HSN-W**, Hydrangea, Hydrangea, IF Relief, IF-C, IGF-1, Ionic Minerals, Joint Health, Marjoram (Sweet), **Mineral-Chi Tonic**, MSM, MSM/Glucosamine Cream, Mullein, Natural Changes, Nature's Gold, Nature's Phenyltol with NEM, Neroli, Nerve Eight, PLS II, Potassium, Protease, Red Raspberry, Rose Hips, S-O-D With Gliadin, Sage, Sciatic, Silver Shield, **Skeletal Strength**, Slippery Elm, Sprains and Pulls, St. John's Wort, Super Algae, Super GLA, SynerProTein, Target Endurance, Tea Tree Oil, Tei Fu Oil, Thai-Go, Trace Mineral Maintenance, Triple Relief, Uña de Gato Combination, Vitamin A & D, Vitamin C, White Oak Bark, Yucca

Sweat Glands

Remedies that can help to regulate the sweat glands.

Capsicum, Catnip, Chamomile, Cinnamon, Ginger, HCP-X, HSN-W, Lobelia, Sage, **Yarrow**

Taste Buds

Remedies that can help the taste buds.

Vitamin B-12, Vitamin B-Complex, Zinc

Teeth

Remedies that help problems with the teeth.

Black Walnut, Clove Bud, Colloidal Minerals, Herbal Trace Minerals, Sunshine Brite Toothpaste, Sunshine Heroes Calcium Plus D3, Teething, **White Oak Bark**, Xylitol

Testes

Specific remedies for the testes.

Black Walnut, C-X, **Ginseng (Korean)**, Ginseng (Wild American), **Herbal Pumpkin**, Kidney Activator (Chinese), Lecithin, Lung Support, **Maca**, Sarsaparilla, Zinc

Testosterone

The following remedies have a testosterone-enhancing effect. Testosterone is the male reproductive hormone.

Cinnamon, **DHEA-M**, Eleuthero, **Ginseng (Korean)**, Lemon Oil, Sarsaparilla, X-Action (Men's)

Throat

Remedies that can aid conditions involving the throat.

Aloe Vera, Barley Grass, Black Cohosh, Capsicum, Cough Syrup-DH, Garlic, HY-C, **Licorice Root**, Mullein, Myrrh, Pantothenic Acid, Sage, **Slippery Elm**, Sore Throat/Laryngitis

Thymus

Remedies which affect the thymus gland.

ALJ, Anamu, Barley Grass, Bergamot, Black Currant Oil, Defense Maintenance, E-Tea, Echinacea Purpurea, Echinacea/Golden Seal, Fenugreek & Thyme, Germanium Combination, GLD-F, HCP-X, Rose Hips, Super Algae, **THIM-J**, **Trigger Immune**, **Ultimate Echinacea**, Vitamin A & D, Yarrow, Zinc

Thyroid

Remedies that affect the thyroid.

7-Keto, **Black Walnut**, Cellular Build, Co-Q10, Dulse, Fat Grabbers, Garlic, Hawthorn Berries, Herbal Trace Minerals, Intestinal Soothe & Build, Kelp, L-Carnitine, Magnesium, MetaboStart, Mood Elevator, Oregon Grape, Potassium, Potassium, Sea Salt, Stress Relief, Stress-J, **Target TS-II**, THIM-J, **Thyroid Activator**, **Thyroid Support**, **TS II**, Vitamin B-Complex

Tongue

Specific remedies that may aid problems with the tongue.

Dandelion, **Golden Seal**, L-Lysine, Niacin, Tei Fu Oil, **VS-C**, Yellow Dock, Zinc Lozenges

Tonsils

Specific remedies that can aid the tonsils.

Echinacea Purpurea, Garlic, **IN-X**, Lobelia, Lymph Gland Cleanse, Lymph Gland Cleanse-HY, Lymphatic Drainage, **Lymphomax**, Mullein, Sage, **Silver Shield**, **Ultimate Echinacea**

Urinary System

General remedies that aid the urinary system.

Anti-Gas (Chinese), Astragalus, Bedwetting, Bergamot, Bilberry Fruit, Breathe EZ, Cellu-Smooth, Cornsilk, **Cranberry & Buchu**, Dandelion, **Golden Seal**, Golden Seal/Parthenium, Horsetail, Hydrangea, JP-X, Juniper Berries, Kava Kava, **KB-C**, **Kidney Activator**, **Kidney Activator (Chinese)**, **Kidney Drainage**, Lobelia, Lymphatic Drainage, Magnesium, Marshmallow, Master Gland, Men's Formula, NF-X, Noni (Morinda), Nopal, Parsley, Parthenium, Potassium, Red Raspberry, Sunshine Heroes Whole Foods Antioxidant, Uva Ursi, X-Action (Men's), Yarrow, Yeast/Fungal Detox

Uro-genital Tract

Remedies that aid the urinary and reproductive systems.

Chickweed, Feverfew, Golden Salve, Juniper Berries, **KB-C**, Licorice Root, Pine Needle, **Red Raspberry**, Saw Palmetto, Slippery Elm, Uva Ursi, Yarrow

Uterus

Specific remedies that help to strengthen and tone the uterus.

Astragalus, Barley Grass, Black Cohosh, Blessed Thistle, Breast Assured, Chamomile, Chickweed, Clove Bud, Dong Quai, FCS II, **Female Comfort**, Feminine Tonic, Frankincense, Ginkgo/Gotu Kola, Golden Salve, Horsetail, **Red Raspberry**, Red Raspberry Liquid, Uva Ursi, V-X, White Oak Bark, Wild Yam, Yarrow

Vagina

Specific remedies for the vaginal area.

Aloe Vera, Bayberry, Geranium, Golden Salve, Lavender, Myrrh, Sandalwood, **Silver Shield**, Slippery Elm, **Yeast/Fungal Detox**

Veins

Specific remedies which tone and strengthen the veins.

Bilberry Fruit, **Butcher's Broom**, Geranium, Ginkgo/Gotu Kola, **Mega-Chel**, Rose Hips, **Vari-Gone**, Vitamin C

Vocal Cords

Specific remedies for the vocal cords.

Cellu-Tone, **Lobelia**, **Sage**

Weight Loss

Remedies which aid the body in losing weight.

5-HTP, 7-Keto, Carbo-Grabbers, Cellu-Smooth, Cellu-Tone, Chromium GTF, CLA, CleanStart, Dieter's Cleanse, Every Body's Fiber, EveryBody's Formula, Fat Grabbers, Formula Chi with Ephedra, Herbal Trace Minerals, Hi Lipase, L-Carnitine, LOCLO, Master Gland, MetaboStart, Nature's Cortisol, Nature's Gold, Nature's Hoodia, Nature's Three, Nutri-Burn, Omega-3, Proactazyme Plus, Protease, Psyllium Hulls Combination, Red Beet Formula, Stevia, Super Algae, Super GLA, SynerProTein, Target TS-II, Target TS-II, Thai-Go, Thyroid Activator, **Thyroid Support**, **Tiao He Cleanse**, TS II, **Xylitol**

Whole Body

The following are remedies that act as tonics to strengthen the whole body.

Barley Grass, Cellular Energy, Colloidal Minerals, Exercise, Fatigue/Exhaustion, Flax Seed Oil, Free Amino Acids, **GreenZone**, Healthy Blast, Herbal Punch, Love and Peas, **Mineral-Chi Tonic**, Nature's Gold, Nature's Prenatal, Nutri-Burn, Protector Pak, Sea Salt, Sunshine Heroes Multiple Vitamin & Mineral, Sunshine Heroes Omega 3 with DHA, Super Algae, Super GLA, Super Supplemental, Super Trio, SynerProTein, Thai-Go, TNT, Tofu Moo, Trace Mineral Maintenance, **Trigger Immune**, Ultimate Build, Vegetable Seasoning Broth, Vita Lemon, Vitamins & Minerals, Multiple, Vitamins, Children's Multi, VitaWave

Section Five
Properties

Section Five

Index to Properties

This section allows you to look up the definition of the various properties used to describe the therapeutic actions of natural remedies. Indications and contraindications are included with many properties, and each property includes a list of products with that therapeutic action. Out best picks for products with each property are highlighted in bold.

Abortifacient

May cause abortion or miscarriage. Trying to abort a baby with herbs is not recommended. These herbs should be avoided during preganncy or if planning to get pregnant.

Anamu, **Blue Cohosh**

Absorbant

A substance used to absorb irritating toxins both internally and externally. May be applied topically as a poultice for bites, stings, or other irritations. Can also be taken internally to absorb toxins. Must be used with a large amounts of water to be effective.

Every Body's Fiber, Fat Grabbers, Intestinal Soothe & Build, LOCLO, Marshmallow, Marshmallow & Fenugreek, Mullein, **Nature's Three**, PLS II, **Psyllium**, **Psyllium Hulls Combination**, Slippery Elm

Acidifer

A substance used to increase gastric acidity; or a substance used to lower the pH of the body, especially urine and/or saliva.

Vitamin C

Acrid

A hot, biting, slightly irritating taste. Often found in herbs that are antispasmodic.

Black Cohosh, Kava Kava, Lobelia, Ultimate Echinacea

Adaptagen

Adaptagens help the body adapt to stressful situations and maintain normal function under mental or physical stress. Strengthen and support adrenal function by adjusting the hypothalamus/pituitary/ adrenal axis to reduce the output of stress hormones. Helps to build the immune system because stress hormones reduce the immune response.

AdaptaMax, Astragalus, **Eleuthero**, Energ-V, Ginkgo/Gotu Kola, Ginseng (Korean), Ginseng (Wild American), Gotu Kola, HY-A, Maca, Master Gland, **Mineral-Chi Tonic**, Nature's Chi, Nature's Cortisol, **Nervous Fatigue Formula**, Stress Pack, SUMA Combination, Trigger Immune

Adrenal Tonic

A substance that builds up and strengthens the function of the adrenal glands.

Adrenal Pack, **Adrenal Support**, Bee Pollen, Chamomile (Roman), Energ-V, HY-A, **Licorice Root**, **Mineral-Chi Tonic**, Nature's Chi, **Nervous Fatigue Formula**, Nutri-Calm, **Pantothenic Acid**, Stress Pack, Vitamin B-Complex, Vitamin C

Adrenergic

Stimulates the production of or mimics the action of epinephrine. See also Sympathomimetic. Useful for ADD where parasympathetic dominance is involved. Not recommended in anxiety and stress-related disorders.

Energ-V, **Licorice Root**

Adsorbant

An adsorbant causes other substances to stick to it. It is different from an absorbant, in that substances adhere to the surface of it, instead of being drawn into it. Used topically and internally to eliminate poisons and irritants. May be helpful to counteract chemical poisoning from caustic agents. Contact your poison control center for advice. Should not be taken with other nutritional supplements as they tend to inhibit nutrient absorption.

Bentonite (Hydrated), **Charcoal (Activated)**

Alexipharmic

Preventing the bad effects of poison inwardly; antidote.

Black Cohosh, **Lobelia**

Alkalinizer

An agent that neutralizes acid or causes alkalization; a substance used to raise the pH level of the body, especially urine and/or saliva.

Barley Grass, Calcium, **Chlorophyll**, Coral Calcium, **Magnesium**, Nature's Gold, Noni (Morinda), Potassium, Red Clover, Skeletal Strength, Stress Relief

Alterative (Blood Purifier)

Cleanses (or alters) the internal environment of the body without producing noticeable laxative or diuretic effects. Helps remove toxins from the blood, probably by strengthening liver and or lymphatic function. Purifies the blood, helping to combat impurity in the blood and organs. Used to treat torpid or stagnant conditions in the body. Traditionally used to clear up morbid conditions in the body, especially skin diseases, skin eruptions, cancer and wounds with pus.

Alfalfa, **All Cell Detox**, Black Cohosh, Blessed Thistle, Blood Build, Blue Vervain, BP-X, **Burdock**, Chickweed, Dandelion, E-Tea, Echinacea Purpurea, **Enviro-Detox**, Golden Seal, Gotu Kola, Grapefruit (Pink), Herbal Trace Minerals, I-X, IF-C, Joint Support, LIV-J, **Liver Balance**, Liver Cleanse Formula, Lymphatic Drainage, Milk Thistle, Monthly Maintenance, Pau D'Arco, Red Clover, Red Clover Blend, Safflowers, Sarsaparilla, SF, Sinus Support, **Skin Detox**, Tiao He Cleanse, Ultimate Echinacea, **VS-C**, Yarrow, Yellow Dock, Yucca

Analeptic

A central nervous system stimulant; a drug that acts as a restorative, such as caffeine, amphetamine, pentylenetetrazol.

Analgesic (Anodyne)

Helps to relieve pain without causing loss of sensation. An anodyne is a mild analgesic. Useful for minor pains. Don't expect herbal analgesics to have the strength of prescription pain-killing drugs. Often they simply take the "edge" off the pain so it is bearable.

Anamu, **APS II**, Arthritis, Black Cohosh, Black Currant Oil, Calming, Capsicum, Catnip, Chamomile, Clove Bud, Cramp Relief, **Deep Relief Oil**, Devil's Claw, Eucalyptus, Evening Primrose Oil, **Everflex Pain Cream**, Feverfew, Fibralgia, Geranium, Ginger, Grapefruit (Pink), Herbal Sleep, Hops, **IF Relief**, IF-C, Joint Support, **Kava Kava**, KB-C, Lavender, Lobelia, Marjoram (Sweet), Menstrual, Migraquel, MSM, MSM/Glucosamine Cream, Nature's Fresh, Nature's Phenyltol with NEM, Nerve Control, **Nerve Eight**, Noni (Morinda), Pain, Rosemary, Sciatic, Sore Throat/Laryngitis, Super GLA, Teething, **Triple Relief**, Valerian Root, Yucca

Anaphrodisiac

Decreases sex drive.

Hops, Marjoram (Sweet), **Wild Yam & Chaste Tree**

Androgenic

An agent that produces or enhances masculine characteristics.

Ginseng (Korean), Ginseng (Wild American), **Sarsaparilla**, X-Action (Men's)

Anesthetic

A drug or agent that is used to abolish the sensation of pain by numbing nerve endings.

Capsicum, **Clove Bud**, **Deep Relief Oil**, **Kava Kava**

Anhidrotic

Antiperspirant.

Anodyne

See *Analgesic*

IF Relief, Super ORAC

Antacid

Neutralizes excess stomach acid and/or relaxes the stomach and prevents or treats acid indigestion. Opposite action of digestive tonics. Antacids are used for excess stomach acid that usually occurs in younger people in response to overeating or stress. Indicated by sharp burning pains associated with eating. Dull, burning pains which occur about one to two hours after eating are not symptoms of acid indigestion, but rather a lack of hydrochloric acid and/or enzymes.

Calcium based antacids are generally contraindicated in this latter form of acid indigestion. However, herbal antacids do not appear to adversely affect acid indigestion caused by lack of hydrochloric acid.

Anti-Gas Formula, Catnip, **Catnip & Fennel**, Coral Calcium, Dandelion, Ginger, **Golden Seal**, Hops, Kelp, Marshmallow, Red Beet Formula, Red Raspberry, Red Raspberry Liquid, **Safflowers**, Sage, Sarsaparilla, Small Intestine Detox, **Stomach Comfort**

Anthelminthic

A vermifuge, destroying or expelling intestinal worms; an agent that is destructive to parasitic worms. See *vermifuge*.

Anti-abortive

Helps stop miscarriage or spontaneous abortion. Opposite of abortifacient. Used when bleeding or cramping starts in the early stages of pregnancy to help avoid a miscarriage.

False Unicorn, Lobelia, **Red Raspberry**, Vitamin C

Anti-adrenergic

Of or pertaining to a substance that opposes the physiological efects of epinephrine; an agent that reduces production or uptake of epinephrine (adrenaline) and other related compounds secreted by the adrenal glands. See *Sympatholytic*.

Anti-aging

Helps to counter the effects of aging

AdaptaMax, **Brain-Protex**, Co-Q10, DHEA-F, DHEA-M, **Ginkgo & Hawthorn**, **Ginkgo Biloba**, Ginkgo/Gotu Kola, Ginseng (Korean), Gotu Kola, Ho Shou Wu, IGF-1, Maca, **Mineral-Chi Tonic**, Nature's Chi, Super ORAC, **Thai-Go**

Anti-allergenic

Reduces allergic reactions. Includes herbs listed under antihistamine and mast cell stabilizers. These remedies are used to reduce the allergic responses in mucus membranes of the digestive and respiratory tract. They are useful for hayfever, allergy-induced asthma and earaches.

ALJ, **Allergies-Hayfever/Pollen**, **Allergies-Mold/Yeast/Dust**, Allergy, Burdock, Co-Q10, Echinacea Purpurea, EW, Eyebright, Ginkgo Biloba, **HistaBlock**, Seasonal Defense, **Vitamin C**

Anti-amebic

An agent that destroys or suppresses the growth of amebas.

Anti-arrhythmic

Combating an irregular heart beat; an agent that prevents or alleviates cardiac arrhythmia.

Calcium, Cardio Assurance, Ginkgo & Hawthorn, **Hawthorn Berries**, **Magnesium**, Mineral-Chi Tonic

Anti-arthritic

An agent that combats joint pain or arthritis (rheumatism).

Alfalfa, Aloe Vera, Anamu, Arthritis, Barley Grass, Black Cohosh, Capsicum, Chondroitin, Deep Relief Oil, **Devil's Claw**, Evening Primrose Oil, **EverFlex**, **Everflex Pain Cream**, Glucosamine, Herbal CA, **HSN-W**, Joint Health, **Joint Support**, **KB-C**, MSM, MSM/Glucosamine Cream, Nature's Gold, Nerve Eight, Nerve Eight, Omega-3, Super GLA, Triple Relief, Uña de Gato

Anti-emetic (Antinauseous)

Suppresses vomit reflex, relieves nausea. Used for flu, morning sickness, motion sickness, nausea, etc.

Chlorophyll, Clove Bud, **Ginger**, **Peppermint**, Red Raspberry, Red Raspberry Liquid, White Oak Bark, Wild Yam

Anti-epileptic

Opposed to epilepsy; relieving fits.

Blue Cohosh, Focus Attention, **GABA Plus**, Gotu Kola, Hops, Lobelia, **Magnesium**, Passion Flower

Anti-inflammatory

Reduces inflammation (heat, swelling and pain). Used to reduce inflammatory conditions in the body. Anti-inflammatories vary widely in their mode of action and hence in their specific applications. Almost any herb that heals tissue damage will also act as an anti-inflammatory agent.

Acne Treatment Gel, Adrenal Pack, Aloe Vera, Anamu, APS II, Arthritis, Bergamot, Berry Healthy, Blue Cohosh, Bone/Skin Poultice, Bronchial Formula, CC-A, Chamomile, Chamomile (Roman), CLA, Clove Bud, CLT-X, Co-Q10, **Devil's Claw**, DHA, Echinacea Purpurea, Eczema/Psoriasis, Evening Primrose Oil, EverFlex, Every Body's Fiber, Eyebright, Feverfew, Geranium, Ginkgo Biloba, Glucosamine, Gotu Kola, Grapine, **Healing AC Cream**, Helichrysum, **IF Relief**, **IF-C**, Inflammation, Joint Health, Joint Support, Kidney Activator (Chinese), LH, **Licorice Root**, Milk Thistle, MSM, MSM/Glucosamine Cream, Myrrh, **Nature's Fresh**, Nature's Gold, Nature's Phenyltol with NEM, Nerve Eight, Noni (Morinda), Omega-3, Patchouli, Pau D'Arco, Pine Needle, PLS II, Prevention, PS II, S-O-D With Gliadin, Sarsaparilla, Skin Detox, Sprains and Pulls, Sunshine Heroes Whole Foods Antioxidant, Super Antioxidant, Super GLA, **Super ORAC**, Tea Tree Oil, **Thai-Go**, Triple Relief, Uña de Gato, Uña de Gato Combination, Vari-Gone, VS-C, **Wild Yam**, Yarrow, **Yucca**

Anti-obesic

A remedy used in the treatment of weight reduction for persons suffering from obesity.

7-Keto, Carbo-Grabbers, Chickweed, Fat Grabbers, Fen-Chi, Garcinia Combination, Grapefruit (Pink), MetaboMax, MetaboStart, Nature's Cortisol, Nature's Gold, Nutri-Burn, **SF**, **Target TS-II**

Anti-smoking

An agent that helps a person quit smoking.

Chamomile, **Lobelia**, **Nutri-Calm**, St. John's Wort

Anti-urolithic

An agent that prevents the formation of urinary calculi (kidney stones)

Hydrangea, **Magnesium**

Antibacterial

Destroys or inhibits bacteria. Used to fight infection both internally and externally. Many of these remedies do not directly kill bacteria, but act in indirect ways to inhibit their growth or enhance the body's own ability to destroy them.

Caprylimune, Capsicum & Garlic with Parsley, Clove Bud, Cranberry & Buchu, Echinacea Purpurea, **Echinacea/Golden Seal**, Eucalyptus, Feverfew, **Garlic**, Gastro Health, GC-X, Golden Seal, Gotu Kola, Helichrysum, IN-X, Lavender, Lemon Oil, Lung Support, Lymph Gland Cleanse, Lymph Gland Cleanse-HY, Marjoram (Sweet), Myrrh, Neroli, Olive Leaf, Pau D'Arco, Seasonal Defense, **Silver Shield**, Sinus Support, **Tea Tree Oil**, Thyme, **Ultimate Echinacea**, Uña de Gato, Uña de Gato Combination, Vitamin A & D, Vitamin C, Yarrow

Anticancer

Remedies that help to fight cancer and shrink tumors.

Anamu, Burdock, E-Tea, **Immune Stimulator**, Indole 3 Carbinol, Pau D'Arco, **Paw Paw**, Protease, Red Clover, Red Clover Blend, SC Formula, Uña de Gato, Uña de Gato Combination, Vitamin D3

Anticarious

Preventing or suppressing the development of dental caries (cavities).

Alfalfa, **Black Walnut**, Thyme, **White Oak Bark**, **Xylitol**

Anticatarrhal

Help the body remove excess catarrh, whether in the sinus area or other parts of the body.

ALJ, EW, Eyebright, **Fenugreek & Thyme**, **Sinus Support**, Small Intestine Detox, SnorEase

Anticephalalgic

Curing or preventing headache.

APS II, **Feverfew**, Lobelia, **Nerve Eight**, Tei Fu Oil, Triple Relief, Wood Betony

Anticholesteremic

Agent that helps to reduce cholesterol.

Algin, Cholester-Reg II, Evening Primrose Oil, Fat Grabbers, Garcinia Combination, Garlic, Guggul, **Ho Shou Wu**, **LOCLO**, Milk Thistle, Nature's Three, **Niacin**, Psyllium, **Psyllium Hulls Combination**, **Red Yeast Rice**

Anticoagulant (Blood Thinner)

Inhibits coagulation of blood and blood clotting. Opposite of blood tonic. Used where there is a risk of thrombosis (blood clots forming in the circulatory system). These remedies also help blood stagnation (thick, heavy blood) which in natural medicine is believed to contribute to a wide variety of conditions including varicose veins, uterine fibroids and liver congestion. These remedies may be contraindicated in anemia or when the person is already using blood thinners.

Alfalfa, **Butcher's Broom**, Capsicum, Chlorophyll, Energ-V, Evening Primrose Oil, Flax Seed Oil, Garlic, **Ginkgo Biloba**, Ginkgo/Gotu Kola, **Guggul**, Helichrysum, Nattozimes Plus, Noni (Morinda), Pau D'Arco, **Vitamin E**

Antidepressant

Relieves depression. Helps alleviate depression. Used to help lift the spirits and relieve depression. Caution: do not take people "cold turkey" off antidepressant drugs.

5-HTP, Bergamot, **Black Cohosh**, Clary Sage, Clove Bud, Damiana, Depressaquel, Frankincense, Grapefruit (Pink), Jasmine Absolute, Kava Kava, Kudzu/St.John's Wort, **Mood Elevator**, Neroli, Pro-G-Yam Cream, Red Raspberry, Rose Bulgaria, St. John's Wort, St. John's Wort with Passion Flower, Ylang Ylang

Antidiabetic

An agent that prevents or alleviates diabetes.

AS with Gymnema, Bilberry Fruit, Blood Sugar Formula, Chromium GTF, Garlic, **Golden Seal**, Milk Thistle, Noni (Morinda), Nopal, Nutri-Burn, Olive Leaf, PBS, Pro-Pancreas, Stevia, **SugarReg**, **Target P-14**, **Xylitol**

Antidiarrheal

Arrests diarrhea.

Bentonite (Hydrated), **Charcoal (Activated)**, Psyllium Hulls Combination, Red Raspberry, Red Raspberry Liquid, Sage, Slippery Elm, **White Oak Bark**

Antidiuretic

An agent that suppresses the formation of urine.

Bedwetting, HY-C, Licorice Root

Antidote

A remedy for counteracting a poison.

Charcoal (Activated), **Lobelia**

Antifebrile

Antipyretic. See also *Febrifuge*.

Antifungal

Kills or inhibits the growth of fungus and yeast. Used for candida and other yeast infections both topically and internally. Synonymous with Fungicide.

Artemisia Combination, Black Walnut, Candida, **Candida Clear**, **Caprylic Acid Combination**, Caprylimune, Cordyceps, Eucalyptus, Garlic, Guardian, Juniper Berries, Lavender, Olive Leaf, **Pau D'Arco**, Pau D'Arco Power Pack, Paw Paw, Paw Paw Lice Remover Shampoo, Probiotics, Seasonal Defense, **Silver Shield**, Skin Detox, **Tea Tree Oil**, Thyme, Yarrow, **Yeast/Fungal Detox**

Antigalactagogue

Inhibits lactation (flow of breast milk). Opposite of galactagogue. Used for nursing mothers to help dry up breast milk when it is time to stop nursing.

Clary Sage, **Parsley**, **Sage**

Antihemorrhagic

An agent that prevents or stops hemorrhage. See *Styptic, Hemostatic.*

Antiherpetic

An agent that inhibits or counteracts the herpes virus.

L-Lysine, Uña de Gato, Uña de Gato Combination, **VS-C**

Antihistamine

Helps dry up sinus congestion by counteracting histamine. Similar to anti-allergenic, but more specific. Histamine produces swelling in mucus membranes leading to itchy, watery eyes, runny nose, swollen eustachian tubes, intestinal inflammation and other allergic reactions.

Formula Chi with Ephedra, Four, Ginger, **HistaBlock**, **Seasonal Defense**, Sinus, Sinus Support, SnorEase, **Vitamin C**

Antihypertensive

An agent that reduces high blood pressure.

Black Cohosh, **Blood Pressurex**, **Capsicum & Garlic with Parsley**, Garlic, Hawthorn Berries, **Lobelia**, **Magnesium**,

Antilipemic

An agent that counteracts high levels of lipids in the blood.

Adrenal Support, Alfalfa, Barley Grass, Cellular Energy, Fat Grabbers, Food Enzymes, Garcinia Combination, Hawthorn Berries, Licorice Root, Protease, Spirulina, Super Algae

Antilithic

An agent that prevents the formation of stone or calculus; preventing the formation of calculi in the urinary organs.

Hydrangea, **Magnesium**

Antimalarial

An agent that is therapeutically effective against malaria. see also Antiperiodic.

Antimicrobial

An agent that kills microorganisms or suppresses their multiplication or growth.

Cinnamon, Deep Relief Oil, Echinacea/Golden Seal, Eucalyptus, Frankincense, Garlic, Golden Seal, Guardian, Myrrh, Paw Paw, Pine Needle, Rosemary, **Silver Shield**, **Tea Tree Oil**, Thyme

Antimutagenic

Helps prevent the formation of cancer cells.

Burdock, **Chlorophyll**, **Thai-Go**

Antinauseous

See *Anti-emetic.*

Detoxification, **Ginger**, **Peppermint**

Antioxidant

Helps prevent free radical damage by scavenging oxygen radicals, which may help prevent aging and decay.

Alpha Lipoic Acid, Antioxidant Arsenal, Astragalus, Berry Healthy, Bilberry Fruit, Blood Pressurex, Brain-Protex, Cardio Assurance, Carotenoid Blend, Cellu-Smooth, Cellular Energy, CLA, Clove Bud, **Co-Q10**, Defense Maintenance, Germanium Combination, Ginger, Ginkgo Biloba, Ginkgo/Gotu Kola, Ginseng (Korean), Ginseng (Wild American), Grapine, **Green Tea Extract**, Hawthorn Berries, Heavy Metal Detox, HistaBlock, IF Relief, Immune Stimulator, Ionic Minerals, Lecithin, Love and Peas, Lutein, Lymphomax, Mandarin (Red), Melatonin Extra, Milk Thistle, N-acetyl Cysteine, Nature's Gold, Olive Leaf, Protector Pak, S-O-D With Gliadin, Sage, SugarReg, SUMA Combination, Sunshine Heroes Multiple Vitamin & Mineral, Sunshine Heroes Whole Foods Antioxidant, Sunshine Heros Whole Foods Papayazyme, **Super Antioxidant**, **Super ORAC**, Super Trio, **Thai-Go**, Thyme, Ultimate Build, Uña de Gato, Vitamin A & D, Vitamin C, Vitamin E,

Antiparasitic

Remedies that destroy parasites. Parasites are difficult to diagnose accurately. However, it is often advisable to take antiparasitic herbs when traveling abroad, or when one has pets or animals, or when one has digestive problems that don't respond to other approaches.

Anamu, **Artemisia Combination**, Bergamot, Black Walnut, Caprylic Acid Combination, Cinnamon, Clove Bud, Garlic, Gastro Health, Golden Seal, **Herbal Pumpkin**, **Para-Cleanse**, Parasites, **Paw Paw**, Pine Needle, Red Clover, Silver Shield

Antiperspirant

An agent that inhibits or prevents perspiration; an astringent preparation for reducing perspiration, often containing aluminum or zirconium.

Antiphlogistic

An agent that counteracts inflammation and fever.

Chamomile, IF Relief, **IF-C**, **Yarrow**

Antipruritic

An agent, usually applied topically, that relieves or prevents itching.

Burdock, **Chickweed**, Golden Seal, Oregon Grape, **Yellow Dock**

Antirheumatic

An agent that relieves or prevents rheumatism; any substance applied topically or taken internally that reduces pain and inflammation of the joints or other connective tissue.

APS II, Black Cohosh, Blue Cohosh, Eleuthero, Flax Seed Oil, Guggul, Joint Health, **Joint Support**, **Nerve Eight**, Sarsaparilla, Super GLA, Wild Yam, Yucca

Antiscorbutic

Effective in the prevention or relief of scurvy.

Rose Hips, **Vitamin C**

Antiscrofulous

An agent used to prevent or counteract infection of lymph nodes, especially cervical, and subcutaneous swellings that form into cold abscesses, multiple ulcers and draining sinus tracts. This results from contact exposure to tuberculosis (syn. Scrofuloderma).

Lymph Gland Cleanse, Lymph Gland Cleanse-HY, Lymphatic Drainage, Lymphomax

Antiseptic

Destroys or inhibits micro-organisms (germs). These terms are usually used in reference to herbs that are used on wounds or injuries rather than for internal infections.

Acne Treatment Gel, Bergamot, Black Ointment, Black Walnut, Breathe Free, Capsicum & Garlic with Parsley, Chamomile (Roman), Cinnamon, Clove Bud, Echinacea Purpurea, Eucalyptus, Fenugreek & Thyme, Frankincense, **Garlic**, Golden Salve, Golden Seal, Golden Seal/Parthenium, Helichrysum, HS II, JP-X, Juniper Berries, Kava Kava, Lavender, Mandarin (Red), Marjoram (Sweet), **Myrrh**, Neroli, Parthenium, Pine Needle, Rose Bulgaria, **Rosemary**, Sage, Sandalwood, **Silver Shield**, Skin Detox, Sunshine Brite Toothpaste, **Tea Tree Oil**, **Thyme**, Ultimate Echinacea, Uva Ursi, Vitamin C, White Oak Bark, Yarrow, Ylang Ylang, Zinc Lozenges

Antispasmodic

Relaxes or prevents muscle cramping or spasms. Used for muscle spasms in a variety of applications including: cramps and charley-horses, intestinal cramps (spastic bowel), bronchial spasms in asthma, menstrual cramps, pain during childbirth, pain due to muscle tension, tension headaches.

5-W, Anamu, APS II, Asthma, **Black Cohosh**, **Blue Cohosh**, Breathe Free, Bronchial Formula, Catnip, CBG Extract, Chamomile, Chamomile (Roman), Clary Sage, Clove Bud, CLT-X, **Cramp Relief**, Dong Quai, Geranium, Helichrysum, Herbal Sleep, Hops, **Kava Kava**, **Lobelia**, **Magnesium**, Mandarin (Red), Marjoram (Sweet), Neroli, Nerve Control, Nerve Eight, Passion Flower, Rose Bulgaria, Sandalwood, Stress-J, Thyme, Valerian Root, **Wild Yam**, Wild Yam & Chaste Tree, Wood Betony, Ylang Ylang

Antisudorific

Decreases perspiration and elimination through the skin. Used to stop night sweats and/or excessive perspiration. Closes skin pores.

Sage

Antithrombolytic

An agent that prevents or interferes with the formation of blood clots within vascular walls.

Butcher's Broom, **Vari-Gone**, **Vitamin E**

Antitoxic

Effective against a poison.

Black Cohosh, **Charcoal (Activated)**, Echinacea Purpurea, **Ultimate Echinacea**

Antitussive

Suppresses the cough reflex, reducing coughing. Used to arrest coughing when excessive coughing causes chest pain or loss of sleep.

Asthma, **Breathe EZ**, Breathe Free, Chickweed, Cough Syrup (Childrens), Cough Syrup-NT, Fenugreek & Thyme, Flax Seed Oil, **Licorice Root**, **Lobelia**, Red Clover

Antivenomous

Counteracts venom from bites and stings by absorbing it or neutralizing it. Used primarily topically for spider bites, bee stings, snake bites, etc. For topical use, moisten powders and apply as a poultice directly to the affected area. It may be necessary to change the poultice several times to obtain the full effect. (Caution: this is a first aid measure and should not replace appropriate medical assistance, especially in the case of allergies to bee stings, or the bite of poisonous snakes and spiders.)

Black Cohosh, **Charcoal (Activated)**, Echinacea Purpurea, **Lobelia**, **Nature's Fresh**, Tea Tree Oil, **Ultimate Echinacea**, White Oak Bark

Antiviral

Destroys, inhibits or helps the body to destroy viruses.

Astragalus, Black Walnut, CC-A, Chamomile, Clove Bud, Cold, Cordyceps, **Echinacea Purpurea**, Echinacea/Golden Seal, **Elderberry Defense**, **Elderberry Plus**, Eucalyptus, Garlic, Geranium, Golden Seal, Helichrysum, **Immune Stimulator**, Influenza Remedy, Licorice Root, Marjoram (Sweet), Myrrh, Olive Leaf, Oregano (Wild), Pau D'Arco, Paw Paw, Rose Bulgaria, Sandalwood, Silver Shield, Skin Detox, St. John's Wort, Super Algae, Tea Tree Oil, Thyme, **Ultimate Echinacea**, Uña de Gato, **Uña de Gato Combination**, Viral Recovery, **VS-C**, Yarrow, Zinc Lozenges

Anxiolytic

An agent that reduces anxiety.

AdaptaMax, Adrenal Pack, Adrenal Support, **Distress Remedy**, Geranium, **Kava Kava**, **Lobelia**, Neroli, **Nervous Fatigue Formula**, Nervousness, Nutri-Calm, Passion Flower, SUMA Combination

Aperient

A very mild laxative. Used to create a gentle laxative action, especially in children or elderly persons.

Aloe Vera, Burdock, Capsicum, Damiana, Ginger, Licorice Root, **Magnesium**, Sunshine Heros Whole Foods Papayazyme, **Yellow Dock**

Aperitive

Stimulating the appetite. This term is frequently applied to a pre-meal cocktail or drink.

Alfalfa, **Digestive Bitters Tonic**, **Golden Seal**

Aphrodisiac

Increases sex drive. Used for low sex drive, impotency or frigidity. Not recommended for teenagers.

Black Cohosh, C-X, Clary Sage, Clove Bud, **Damiana**, **DHEA-F**, **DHEA-M**, Eleuthero, Energ-V, Ginseng (Korean), Ginseng (Wild American), Gotu Kola, Jasmine Absolute, Kava Kava, Licorice Root, **Maca**, Patchouli, Sandalwood, SUMA Combination, **X-A**, **X-Action (Men's)**, **X-Action (Women's)**, X-Action Gel, Ylang Ylang

Appetite Stimulant

Stimulates hunger and the desire to eat. Opposite action of appetite suppressant. Used for people who are anorexic or bulimic and for children with poor appetites. They may also be helpful in wasting conditions and poor digestive function as they tend to stimulate digestive secretions. Bitters and aromatics taken in a liquid form or chewed (so they can be tasted and smelled) will generally help stimulate both appetite and digestive secretions.

Alfalfa, Anti-Gas (Chinese), Anti-Gas Formula, Bee Pollen, Bergamot, Catnip & Fennel, Chamomile, Dandelion, **Digestive Bitters Tonic**, **Golden Seal**, Licorice Root, Safflowers, **Spleen Activator**, Zinc

Appetite Suppressant

Suppresses hunger and the desire to eat. Opposite action of appetite stimulant. Remedies may supply nutrients that give satisfaction and allay appetite. They may also be bulking agents that swell in the stomach and provide a feeling of fullness. A few may affect neurotransmitters that control appetite.

5-HTP, **Appetite Control**, AS with Gymnema, Barley Grass, Fat Grabbers, Garcinia Combination, Grapefruit (Pink), Nature's Chi, Nature's Hoodia, Spirulina

Aromatic

A substance with a strong aroma or smell; odoriferous, stimulant, spicy.

Anti-Gas (Chinese), Anti-Gas Formula, Bergamot, Catnip, Chamomile, Chamomile (Roman), Cinnamon, Clary Sage, Clove Bud, Deep Relief Oil, Eucalyptus, Frankincense, Geranium, Ginger, Grapefruit (Pink), Guardian, Helichrysum, Herbal Sleep, Hops, Inner Peace Oil, Jasmine Absolute, Juniper Berries, Lavender, Lemon Oil, Mandarin (Red), Marjoram (Sweet), Myrrh, Neroli, Oregano (Wild), Patchouli, Peppermint, Pine Needle, Rose Bulgaria, Rosemary, Safflowers, Sage, Tea Tree Oil, Tei Fu Oil, Thyme, Valerian Root, Yarrow

Astringent

Contracts and tones muscle fiber and other tissue. Usually contain tannins. Used to stop bleeding (see styptic), arrest discharges (diarrhea, excess mucus, pus, etc.), tone up soft or spongy tissue (varicose veins, hemorrhoids), reduce swelling and/or counteract venom (see anti-venomous). Contraindicated in tension and dryness. Arrests digestive secretions when taken with meals, so contraindicated with meals in cases of acid or enzyme deficiency.

Acne Treatment Gel, **Bayberry**, Black Walnut, Bone/Skin Poultice, EW, Frankincense, Golden Salve, Golden Seal, Golden Seal/Parthenium, Green Tea Extract, HCP-X, Horsetail, Intestinal Soothe & Build, Juniper Berries, Menstrual-Reg, Olive Leaf, Patchouli, Pau D'Arco, Red Raspberry, Rose Hips, Sage, Thyme, **Uva Ursi**, **White Oak Bark**, **Yarrow**, Yellow Dock

Bactericidal

Destructive to bacteria. See *Antibacterial.*

Balm

A healing or soothing medication; anything that heals, soothes or mitigates pain.

Aloe Vera, **Golden Salve**, **Herbal Trim Skin Treatment**, **Nature's Fresh**

Balsamic

Healing or soothing to inflamed parts.

Aloe Vera, Marshmallow, Slippery Elm

Bitter

A bitter-tasting infusion or tonic that affects digestion or appetite by stimulating the increasing output of saliva and gastric juices. Gentian and hops are among plants used for this purpose; herbs with a bitter taste.

Alfalfa, Black Walnut, BP-X, Bronchial Formula, Burdock, Cascara Sagrada, Chamomile, Dandelion, Devil's Claw, Digestive Bitters Tonic, Enviro-Detox, Feverfew, Frankincense, FV, Golden Seal, Herbal Pumpkin, Herbal Trace Minerals, Hops, IF-C, IN-X, IN-X, LB Extract, LB-X, LBS II, LIV-J, Liver Cleanse Formula, Lung Support, Lymph Gland Cleanse-HY, Milk Thistle, Milk Thistle Combination, Mineral-Chi Tonic, Myrrh, Oregon Grape, Parsley, PBS, Red Clover Blend, Senna Combination, Skin Detox, Valerian Root, Yarrow, Yellow Dock, Yucca

Blood Building

A concept from Chinese and traditional medicine. Refers to herbs that help overcome anemia and build up the blood. Traditionally used for weak, thin pulse, pale complexion, general weakness.

Blood Build, Chlorophyll, Dong Quai, Ginseng (Korean), Ho Shou Wu, **I-X**, Iron, Nature's Chi, **Trigger Immune**

Blood Purifier

See *Alterative*

Blood Thinner

See *Anticoagulant*

Bronchial Dilator

Remedies that relax and dilate the bronchials. Used for asthma that is induced by muscular constriction of the air passages due to nervous reactions. These remedies are also good for deep cough, especially whooping cough, where there is constriction of the airways.

Asthma, **Black Cohosh**, Cordyceps, Formula Chi with Ephedra, HistaBlock, Licorice Root, **Lobelia**, Lung Support, Passion Flower, Sandalwood, Yerba Santa/Senega Combination

Calmative

A mild sedative.

Calming, **Chamomile**, Chamomile (Roman), Frankincense, Jasmine Absolute, Nervousness, Passion Flower, Sandalwood, Sunshine Heroes Calcium Plus D3

Cardiac

A remedy that tones and strengthens the heart muscle. Used to prevent heart disease or to strengthen the heart when heart disease is already present.

Astragalus, Blood Pressurex, Butcher's Broom, Capsicum, **Cardio Assurance**, **Co-Q10**, Damiana, Garlic, Ginkgo & Hawthorn, Ginseng (Korean), Ginseng (Wild American), **Hawthorn Berries**, **HS II**, Mega-Chel, Ylang Ylang

Carminative

Expels gas from the bowel or relieves bloating. Used for people who suffer from severe gas and poor digestive function. These remedies act by stimulating blood flow to the digestive organs and increasing digestive secretions. Most of the gas produced in our colon is absorbed into the blood stream and excreted via the lungs. These remedies increase absorption of gas in the intestines to relieve bloating. They may also help to increase motility (movement) along the digestive tract to push material forward and help to expel trapped gas or food material.

Anti-Gas (Chinese), **Anti-Gas Formula**, Bergamot, Capsicum, Capsicum & Garlic with Parsley, Catnip, Catnip & Fennel, Chamomile, Chlorophyll, Cinnamon, Clove Bud, Fenugreek & Thyme, Feverfew, Food Enzymes, Gall Bladder Formula, Garlic, Ginger, Hops, Juniper Berries, Lavender, Mandarin (Red), Marjoram (Sweet), Neroli, **Peppermint**, Rosemary, Safflowers, Sage, Thyme

Catalyst (Synergist)

A catalyst is a substance that causes or speeds a chemical reaction without being affected itself. A synergist is a plant medicine that increases blood flow, digestive activity and absorption of other herbs or nutrients. These are often used in small amounts in herbal formulas to enhance their effectiveness.

Capsicum, Ginger, **Licorice Root**, Lobelia

Cathartic

An agent that causes the emptying of the bowels, such as by increasing bulk or stimulating peristaltic action; also called evacuant and purgative. a substance that stimulates the movement of the bowels, which is more powerful that a laxative.

Senna Combination

Cell Proliferant

An agent that speeds tissue growth to enhance the repair of wounds and injuries. Used to speed healing in injuries.

Bone/Skin Poultice, **IGF-1**, Noni (Morinda), PLS II

Cephalalgic

An agent that can cause a headache but may also be used as a remedy to treat headache.

Black Cohosh, **Ginkgo Biloba**

Cephalic

Relating to diseases of the head; an agent to treat disorders that affect the head.

Blessed Thistle, Gotu Kola, Marjoram (Sweet), Rosemary, Sage

Cerebral Tonic

Improves memory and brain function. Used for loss of memory or concentration due to aging or senility. Also useful for increasing mental alertness and concentration in students or others who need better memory or focus. May be helpful in some cases of ADD.

Brain-Protex, Ginkgo & Hawthorn, **Ginkgo/Gotu Kola**, Gotu Kola, SUMA Combination, **Sunshine Heroes Omega 3 with DHA**

Chelating

An agent that bonds to metals to facilitate their absorption or removal from the body.

Algin, **Heavy Metal Detox**, **Mega-Chel**, Mineral-Chi Tonic

Cholagogue

Increases the flow of bile, helps prevent or dissolve gallstones. Indicated for clay colored stools, some cases of constipation and in gall bladder problems where the gallbladder is congested or has stones. Contraindicated in duodenal ulcers and intestinal inflammation. May also be contraindicated with inflammatory liver diseases.

Blessed Thistle, BP-X, Burdock, Cascara Sagrada, Cholester-Reg II, **Dandelion**, Gall Bladder Formula, Garlic, Golden Seal, Hops, Lavender, **Liver Balance**,

Liver Cleanse Formula, Mandarin (Red), **Milk Thistle**, Milk Thistle Combination, Wild Yam, **Yellow Dock**

Choleretic

A compound that promotes the flow of bile. See *Cholagogue.*

Cholinergic

Pertaining to the parasympathetic portion of the autonomic nervous system and the release of acetyl-choline as a transmitter substance. See *Parasympathomimetic.*

Cicatrisant

A cicatrisant is a remedy that helps tissues heal without scarring.

Chamomile (Roman), **Helichrysum**, **Lavender**, Mandarin (Red), Tea Tree Oil, **Vitamin E**

CNS Depressant

A remedy that can suppress the activity of the central nervous system; generally used to promote sleep or reduce pain.

Deep Relief Oil, **Hops**, Lavender, Valerian Root

Coagulant

An agent that promotes or accelerates the coagulation of blood. See also *Styptic.*

Bayberry, **Capsicum**, **Yarrow**

Condiment

Improving the flavor of food, as salt, pepper, salad, etc.

Capsicum, Garlic, Ginger, Kelp, Lemon Oil, Parsley, Sage, Stevia, Vegetable Seasoning Broth

Contraceptive

Helps prevent conception. Should be avoided when trying to get pregnant. Caution: herbs are not a very reliable method of birth control.

Uña de Gato, Wild Yam, Wild Yam & Chaste Tree

Corrects Polarity

An agent that acts upon the thymus gland to correct energy flow when the polarity is reversed so that a person can be accurately muscle tested. Also helps a person who is attracted to negative influences to be attracted to more positive influences.

Bee Pollen, Damiana, Mood Elevator, Spirulina, Super Algae, THIM-J, **Trigger Immune**

Cosmetic

Used for improving the complexion; a beautifying substance or preparation.

Acne Treatment Gel, Jojoba Oil, Nature's Fresh

Counterirritant

A remedy which causes irritation (redness) and thus draws blood away from one area of the body to another. Used topically for temporary relief of pain associated with arthritis, gout, rheumatism, strained muscles, etc. Helps to relieve pain by depleting or blocking substance P neurotransmitters. Pain relief is usually temporary, although increased blood flow may help to stimulate healing. Contraindicated with redness and swelling on the surface of the skin or when the surface of the skin is broken. Not recommended for application after a hot shower or bath. Avoid contact with sensitive areas of the body (eyes, genitals).

Capsicum, Deep Relief Oil, Tei Fu Oil, Thyme

Cytotoxic

Kills cancer cells.

Anamu, Paw Paw

Decongestant

Relieves (breaks up or loosens) respiratory congestion. Used primarily where mucus has become thick and "stuck" causing difficult breathing, sinus pressure or thick drainage. These remedies tend to thin mucus making it easier to expel.

ALJ, Breathe EZ, Breathe Free, Bronchial Formula, CC-A, Cold, Cough Syrup (Childrens), Cough Syrup-DH, Cough Syrup-LP, Cough Syrup-NT, Elderberry Defense, Elderberry Plus, Eucalyptus, EW, **Fenugreek & Thyme**, Four, Garlic, Geranium, HCP-X, HistaBlock, Lemon Oil, LH, Lobelia, Lung Support, Lymphatic Drainage, MetaboMax, Pine Needle, Rosemary, Seasonal Defense, Sinus, Sinus Support, Sore Throat/Laryngitis, Sunshine Heros Whole Foods Papayazyme, Yerba Santa/Senega Combination

Demulcent (Mucilant)

Soothes and softens tissue. Contains complex polysaccahrides which are indigestible, but hold water and absorb irritants. Mucilant is a term we coined in NSP's Manager School to describe this action. It is not found in traditional texts. Applied topically herbs with this property act as a drawing poultice to absorb irritants, reduce inflammation and swelling, and keep tissues moist and pliable. Taken internally they absorb irritants in the digestive tract, help to reduce cholesterol, act as bulk laxatives and moisten dry tissues. They also tend to act as mild nourishing foods to counteract weight loss (wasting). They are indicated in hard, dry, irritated tissue states, including inflammation of the digestive tract, dry, irritating coughs, burning or painful urination, redness and swelling topically. Also indicated in conditions where a person needs mild nourishing foods when recovering from debilitating illnesses.

They are contraindicated with bowel obstruction and should not be applied topically to deep wounds.

Aloe Vera, Chickweed, Cornsilk, Herbal CA, **Herbal Trim Skin Treatment**, Intestinal Soothe & Build, Kelp, LOCLO, **Marshmallow**, Marshmallow & Fenugreek, Mullein, Nature's Three, PLS II, Psyllium, Psyllium Hulls Combination, **Slippery Elm**

Dentifrice

A preparation, usually a paste, gel or powder, used with a toothbrush for cleaning the accessible surfaces of the teeth.

Black Walnut, **White Oak Bark**

Deobstruent

Removes obstructions from the body.

Butcher's Broom, Dong Quai, **Lobelia**, **Lymphatic Drainage**, Mega-Chel, Red Clover, Red Clover Blend, Sarsaparilla

Deodorant

Preparations used to eliminate, prevent or mask unpleasant odor.

Bowel Detox, **Chlorophyll**, **Nature's Fresh**, Neroli, Papaya Mint, Patchouli, Peppermint

Depressant

A drug that lowers nervous or functional activity; sedative.

Dermatic

An agent used to cleanse, purify, exfoliate and tone the skin.

Geranium, Rose Bulgaria

Detergent

A remedy that is cleansing to wounds, boils or ulcers. This is diffenent from commercial detergents used for laundry and household cleaning.

Golden Seal, Oregon Grape, Red Clover, Red Clover Blend, Sunshine Concentrate Cleaner

Detoxifying

Cleanses the body of toxins by promoting elimination. Related to deobsruent/lymphatic and alterative/blood purifier, but involves mild diuretic and/or laxative actions.

Alpha Lipoic Acid, Blue Vervain, Burdock, Caffeine Detox, Cellu-Tone, Cellular Energy, **CleanStart**, Detoxification, Dieter's Cleanse, Echinacea Purpurea, EverFlex, Grapefruit (Pink), Heavy Metal Detox, Indole 3 Carbinol, Kudzu/St.John's Wort, Liver Balance, Milk Thistle Combination, N-acetyl Cysteine, Nature's Gold, Probiotics, Sea Salt, **Tiao He Cleanse**, Tobacco Detox, Ultimate Echinacea, Uña de Gato, Uña de Gato Combination, Vaccine Detox,

Diaphoretic

Increases perspiration and elimination through the skin. Used to bring down fevers. In Chinese medicine these are called "surface-relieving" herbs and are used to relieve "superficial" conditions such as colds, flu, coughs, asthma, edema. Warming sudorifics are used for "wind chill" (mild fever with chills, lack of sweating, white phlegm) and cooling sudorifics are used for "wind-heat" (high fever with chills and sweating, thirst, yellow phlegm). Synonymous with sudorific.

Blue Cohosh, Blue Vervain, **Capsicum**, Catnip, Catnip & Fennel, Chamomile, Fenugreek & Thyme, **Garlic**, **Ginger**, HCP-X, Hops, Juniper Berries, Lobelia, Marjoram (Sweet), Thyme, Wild Yam, **Yarrow**

Digestant

Aids digestion of food, generally applied to remedies that contain enzymes or other elements that directly aid in the breakdown of food. Used to supplement digestive secretions and aid in the digestion of food where people are having difficulty manufacturing sufficient hydrochloric acid, enzymes and/or bile salts to breakdown their food. These remedies are especially helpful for older people with poor digestion or people in a wasting condition (where they are losing weight). Food products and plant enzymes that do this have no contraindication, except they are not necessary when digestion is good. However, Food Digestive Enzymes and Protein Digestive Aid directly substitute for the body's own secretions. Over time, this tends to lessen (rather than stimulate) the body's production of these secretions. Hence, it is best to use these products for a limited time while seeking means to rebuild the digestive organs, especially in younger clients. Elderly people, however, may need these on a more regular basis because the digestive system tends to weaken with age.

Food Enzymes, Hi Lipase, Lactase Plus, Papaya Mint, PDA, **Proactazyme Plus**, Protease, Sunshine Heros Whole Foods Papayazyme

Digestive Tonic

Strengthens the entire digestive function, stimulates secretions and tones the digestive organs. Opposite of antacid. Used for people who have dull burning pains about one hour or so after eating coupled with gas, bloating and other symptoms of poor digestion. These remedies help to stimulate the body's own production of digestive secretions, especially when chewed or taken in a liquid form

Alfalfa, Barley Grass, Blessed Thistle, Catnip & Fennel, Chamomile, Devil's Claw, **Digestive Bitters Tonic**, Liquid Cleanse, Liver Balance, Mineral-Chi Tonic, Pro-Pancreas, Red Raspberry Liquid, **Spleen Activator**, **Trigger Immune**

Discutient

dispelling or resolving tumors; scattering or causing a disappearance; a remedy that so acts.

Disinfectant

An agent that disinfects; applied particularly to agents used on inanimate objects; a substance that destroys noxious properties of decaying organic matter. See *Antiseptic*.

Clove Bud, Eucalyptus, Frankincense, Myrrh, **Silver Shield**, **Tea Tree Oil**, Thyme

Diuretic

Increases flow of urine to expel excess fluids from the body. Diuretics are used for water retention and various types of kidney problems. They vary widely in their mode of action.

Anamu, Anti-Gas (Chinese), Astragalus, Bilberry Fruit, Blue Cohosh, BP-X, Breathe EZ, Cornsilk, **Cranberry & Buchu**, Damiana, Dandelion, Dong Quai, False Unicorn, Feverfew, Golden Seal/Parthenium, Grapefruit (Pink), Herbal Trace Minerals, Hops, Horsetail, Hydrangea, Joint Support, **JP-X**, Juniper Berries, Kava Kava, KB-C, **Kidney Activator**, **Kidney Activator (Chinese)**, **Kidney Drainage**, Lymphatic Drainage, Mandarin (Red), Marjoram (Sweet), Marshmallow, Men's Formula, Olive Leaf, P-X, Parsley, Parthenium, Pro-Pancreas, Red Clover Blend, Sarsaparilla, SF, Skin Detox, Urinary Maintenance, Uva Ursi, Vitamin B-6, Wild Yam, Yarrow

Drastic

A very powerfully cathartic or violent purgative. Uncured cascara bark fits in this category, but NSP does not sell any drastics.

Drawing

Drawing refers to agents that pull morbid material out of the body.

Black Ointment, Slippery Elm

Emetic

Induces vomiting. Used for food or chemical poisoning. Call poison control center for advice before inducing vomiting in cases of chemical poisoning. With caustic agents like lye vomiting should not be induced. Emetics are also used to expel excess phlegm from the system to halt asthma attacks and relieve severe respiratory congestion.

Blue Vervain, **Lobelia**

Emmenagogue

Increases (stimulates) and regulates menstrual flow. These herbs have also been called female correctives because they help to regulate the menstrual cycle. They are generally contraindicated during pregnancy.

Black Cohosh, Blessed Thistle, **Blue Cohosh**, Blue Vervain, C-X, Clary Sage, Dong Quai, **FCS II**, **Female Comfort**, Feverfew, Golden Seal, Marjoram (Sweet), Milk Thistle, Monthly Maintenance, Nerve Control, NF-X, Noni (Morinda), Thyme, Yarrow

Emollient

A preparation that soothes and softens external tissue.

Aloe Vera, Chickweed, Golden Salve, Herbal Trim Skin Treatment, Intestinal Soothe & Build, **Jojoba Oil**, Kelp, Licorice Root, Marshmallow, Marshmallow & Fenugreek, Mullein, Nature's Three, Pro-G-Yam Cream, Psyllium, Rose Bulgaria, Slippery Elm, Triple Effect Age Relief

Emulsifier

An agent that helps break down fats to make them water soluble.

Chickweed, Fat Grabbers, **Hi Lipase**, **Lecithin**, **SF**

Escharatic

An agent that is applied topically to destroy morbid tissue. Harsh escharotics should be used with caution and are best used by experienced professional herbalists as they can cause redness, swelling, pain, drainage and scarring. The milder drawing agents available from NSP will not do this. See Drawing. A reasonably good escharatic can be made by mixing Paw Paw Cell Reg with Black Ointment. See *Drawing*.

Black Ointment, **Paw Paw**

Estrogenic

Mimics estrogen in the body. Used to help women during menopause or during the menstrual cycle when PMS symptoms are related to excess progesterone and deficient estrogen. May also have a protective effect against estrogen dependant cancers.

Black Cohosh, Clary Sage, Flash-Ease T/R, **Hops**, Licorice Root, **Phyto-Soy**, Sage

Euphoretic

An agent that produces an exaggerated feeling of physical and mental well being, especially when not justified by external reality.

Damiana, **Kava Kava**, Nature's Phenyltol with NEM

Evacuant

A remedy that empties any organ, such as a cathartic, emetic or diuretic. See listings under these properties.

Expectorant

Expels phlegm or mucus from the lungs and sinuses. Increases coughing, sneezing and stimulates drainage.

ALJ, Bayberry, Black Cohosh, Blue Vervain, **Breathe EZ**, Breathe Free, Bronchial Formula, CC-A, Chickweed, Clove Bud, Cough Syrup (Childrens), Cough Syrup-DH, Cough Syrup-LP, Cough Syrup-NT, Eucalyptus, Eyebright, Fenugreek & Thyme, Garlic, **HCP-X**, Jasmine Absolute, LH, Licorice Root, Lobelia, Lung Support, Marjoram (Sweet), Marshmallow, Marshmallow & Fenugreek, Mullein, Myrrh, Pine Needle, Red Clover, Sandalwood, Sinus Support, Slippery Elm, St. John's Wort, Thyme, Yerba Santa/Senega Combination

Febrifuge

Reduces fever and inflammation.

APS II, Bergamot, Chamomile, Echinacea Purpurea, Elderberry Plus, Feverfew, Garlic, Hops, HY-C, **IF Relief**, **IF-C**, Jasmine Absolute, Lemon Oil, Sage, St. John's Wort, Super ORAC, Thai-Go, Ultimate Echinacea, Yarrow

Female Tonic

A rather vague action referring to remedies that regulate the female hormonal cycle. See Emenogogue. These remedies help menstrual irregularities of various kinds such as heavy menstruation, scant menstruation, delayed menstruation or irregularities of the cycle. They act in different ways and an understanding of the specific types of menstrual irregularities and the specific actions of remedies is necessary for dependable results. These remedies are generally contraindicated in pregnancy, especially in the early stages. Traditionally called Female Correctives.

C-X, Dong Quai, False Unicorn, **FCS II**, **Female Comfort**, **NF-X**, Wild Yam & Chaste Tree

Food

These herbs have been eaten as vegetables.

Burdock, Dandelion, Flax Seed Oil, GreenZone, Hawthorn Berries, Nature's Gold, Parsley, Rose Hips, Spirulina, Super Algae, Tofu Moo

Fumigant

Burning substances that disinfect and counteract obnoxious odors; exposure of an area or object to disinfecting fumes.

Juniper Berries, Sage

Fungicide

Prevents and combats fungal infection; an agent that destroys fungi. See *Antifungal*.

Galactagogue

Enriches and/or increases the flow of breast milk. Opposite of antigalactogogue. Used for nursing mothers to enrich and promote the flow of breast milk. These remedies may also ease colic in nursing babies when taken by the mother to "sweeten" the milk.

Alfalfa, **Blessed Thistle**, Liver Balance, **Marshmallow**, **Milk Thistle**, Nature's Gold

Germicide

Destroys germs or microorganisms such as bacteria, etc. See *Antimicrobial*.

Glandular

Influences the glands to produce more of certain hormones. May also mimic hormones. This is a very vague and general term, since specific herbs act on specific glands. Hormonal or glandular actions need to specify which gland or hormone they act on.

5-W, AdaptaMax, **Adrenal Support**, Bee Pollen, Black Cohosh, Blood Sugar Formula, Breast Enhance, C-X, Cinnamon, Damiana, **DHEA-F**, **DHEA-M**, Dong Quai, Dulse, Energ-V, FCS II, Female Comfort, Feminine Tonic, GLD-F, HSN-W, HY-A, IGF-1, Jasmine Absolute, **Master Gland**, Men's Formula, Menopause, Menstrual, Menstrual-Reg, Monthly Maintenance, Natural Changes, Nervous Fatigue Formula, NF-X, PMS, Potassium, Pregnenolone, Pro-G-Yam Cream, Pro-Pancreas, PS II, Sarsaparilla, Saw Palmetto, Spirulina, SugarReg, SUMA Combination,

Target P-14, Thyroid Activator, **Thyroid Support**, Trigger Immune, TS II, X-A

Hallucinogenic

Causes hallucinations. Wormwood and nutmeg oil are hallucinogenic in higher doses.

Hemostatic

Stops internal bleeding or hemorrhage. Similar to styptic, but generally used internally. Caution: Seek medical assistance for serious bleeding.

Bayberry, **Capsicum**, Golden Seal, Horsetail, **Menstrual-Reg**, Red Raspberry, **White Oak Bark**

Hepatic

Strengthens liver function. Useful for people who have been exposed to chemicals or whose livers are inherently weak or damaged. Very helpful as preventatives for people who work around chemicals (painters, lab workers, farmers, etc.) to protect the liver against damage.

All Cell Detox, Blood Build, BP-X, Burdock, Cellular Build, Dandelion, Dieter's Cleanse, Digestive Bitters Tonic, Echinacea/Golden Seal, **Enviro-Detox**, Gall Bladder Formula, Guggul, Helichrysum, Herbal Beverage, I-X, LIV-J, **Liver Balance**, **Liver Cleanse Formula**, Milk Thistle, **Milk Thistle Combination**, N-acetyl Cysteine, Oregon Grape, Pau D'Arco, SAM-e, Skin Detox, Super Antioxidant, Tiao He Cleanse, Yellow Dock

Hepatoprotective

Protects the liver from toxic chemicals.

Blessed Thistle, **Milk Thistle**, **Milk Thistle Combination**, Nervous Fatigue Formula, Sarsaparilla, Vitamin C

Hypertensive

Increases blood pressure.

Hawthorn Berries, HS II, **Licorice Root**, Ylang Ylang

Hypnotic

Producing or inducing sleep; an agent that acts to induce sleep.

Herbal Sleep, Hops, **Kava Kava**, Passion Flower, Stress Pack, **Valerian Root**

Hypoglycemic

An agent that acts to lower the level of glucose in the blood.

Burdock, **Chromium GTF**, **Golden Seal**, **Nopal**, **Olive Leaf**, **PBS**, Pro-Pancreas, SugarReg, Target P-14

Hypolipidemic

Promoting the reduction of lipid concentrations in the serum (blood).

Adrenal Support, Alpha Lipoic Acid, Carbo-Grabbers, Fat Grabbers, Noni (Morinda), PBS, **Red Yeast Rice**, **SugarReg**

Hypotensive

Decreases blood pressure

Black Cohosh, **Blood Pressurex**, **Capsicum**, **Capsicum & Garlic with Parsley**, Co-Q10, Garlic, GC-X, Ginkgo/Gotu Kola, Gotu Kola, Hawthorn Berries, Herbal Sleep, HS II, Lemon Oil, **Lobelia**, Marjoram (Sweet), Olive Leaf, RG-Max, St. John's Wort, Valerian Root, Yarrow

Immune Amphoterics

Remedies that help to normalize the function of the immune system whether it is underactive (immune weakness) or overactive (autoimmune disorders). These remedies are particularly useful for autoimmune disorders. They appear to enhance communication in the immune system so that it works better without a general stimulating effect that would aggravate autoimmune conditions.

AdaptaMax, Black Currant Oil, **Black Walnut**, CBG Extract, Cellular Build, Chromium GTF, **Colostrum**, Evening Primrose Oil, Glyco Essentials, **Licorice Root**, **Mineral-Chi Tonic**, Nature's Chi, Proactazyme Plus, Protease, Sunshine Heroes Probiotic Power, Super GLA

Immune Stimulant

Stimulates the function of the immune system. Used for frequent colds and infection, lowered resistance. Contraindicated in autoimmune disorders

7-Keto, Anamu, Astragalus, Bronchial Formula, Candida Clear, Colostrum with Immune Factors, Defense Maintenance, E-Tea, **Echinacea Purpurea**, **Echinacea/Golden Seal**, Elderberry Plus, Fizz Active-Immune, Frankincense, Germanium Combination, **Immune Stimulator**, IN-X, Lymph Gland Cleanse-HY, Lymphomax, Milk Thistle, Prevention, SC Formula, Silver Shield, SUMA Combination, THIM-J, **Trigger Immune**, **Ultimate Echinacea**, Uña de Gato, Uña de Gato Combination, Yeast/Fungal Detox, Zinc Lozenges

Immunomodulator

An agent that affects the function of the immune system in a positive way; an agent that specifically or nonspecifically augments or diminishes immune responses, i.e., an adjuvant, immonostimulant or immunosuppressant.

Black Currant Oil, **Black Walnut**, Blood Build, **Colostrum**, Evening Primrose Oil, Licorice Root, **Mineral-Chi Tonic**, Nature's Gold, **Omega-3**, Super GLA, **Thai-Go**

Insecticide

A substance that destroys insects.

Anamu, Bayberry, **Black Cohosh**, Black Walnut, Feverfew, Gotu Kola, Patchouli, **Paw Paw**, Paw Paw Lice Remover Shampoo, Sage

Insulinomimetic

A substance that mimics or acts like insulin.

Golden Seal

Kidney Tonic

Strengthens kidney function and tone. In Chinese medicine the kidneys are thought to be connected to the health of the structural system, so these remedies are helpful for structural weakness, too. They stabilize the structure of the body by acting to balance mineral electrolytes through the kidneys.

Cordyceps, False Unicorn, Ho Shou Wu, Horsetail, **KB-C**, Nature's Chi, Noni (Morinda), Potassium

Laxative (bulk)

Stimulates elimination through the lower bowel by adding fiber to the intestinal tract. Safer laxatives for long term use. Indicated in high cholesterol, sluggish elimination, elimination problems during pregnancy and general detoxification.

Aloe Vera, Chickweed, CleanStart, CLT-X, **Every Body's Fiber**, Fat Grabbers, **LOCLO**, **Nature's Three**, Psyllium, **Psyllium Hulls Combination**, Slippery Elm

Laxative (general)

A medicine that acts gently on the bowels, without griping; stimulates bowel movement; an agent that acts to promote the evacuation of the bowel.

All Cell Detox, Black Walnut, Bowel Detox, Cellular Energy, Every Body's Fiber, Gentle Move, Herbal Pumpkin, **LB Extract**, **LB-X**, **LBS II**, Liquid Cleanse, LOCLO, Nature's Three, Probiotics, Psyllium, Psyllium Hulls Combination, SF, Tiao He Cleanse, Yellow Dock

Laxative (stimulant)

Stimulates elimination through the lower bowel. Used during cleanses to increase elimination. Very helpful for short term elimination problems. Useful for detoxifying the system in high fevers or other acute inflammatory diseases. Not recommended for long term elimination problems. Long term use irritates and enervates the colon.

Cascara Sagrada, CleanStart, Dieter's Cleanse, **LB Extract**, **LB-X**, **LBS II**, **Senna Combination**, Tiao He Cleanse

Lipotropic

Promoting the flow of lipids to and from the liver; acting on fat metabolism by hastening the removal of or decreasing the deposit of fat in the liver.

Chickweed, Dandelion, Juniper Berries, MSM, SAM-e, **SF**

Litholytic

An agent that dissolves calculi. See *Lithotriptic.*

Lithotriptic

Helps dissolve kidney stones. Used to help people pass kidney stones and/or to prevent their formulation. Will probably help with calcifications elsewhere in the body such as bone spurs.

Hydrangea, Joint Support, **Magnesium**, Urinary Maintenance

Low Glycemic

A low glycemic carbohydrate is one that does not spike blood sugar levels and therefore does not trigger a strong insulin response in the blood. Low glycemic carbohydrates can reduce sugar cravings, stabilize blood sugar in hypoglycemia and diabetes, aid in weight loss and help reduce chronic inflammation.

GreenZone, Nopal, Nutri-Burn, Thai-Go, Xylitol

Lung Tonic

Strengthens lung tissue and function. Used where the lungs are weak and prone to frequent infection, also for conditions where the lungs are dry and leathery and have sustained a loss of elasticity.

ALJ, **Cordyceps**, Garlic, LH, **Lung Support**

Lymphatic

Remedies that act on the lymphatic system. They cleanse, tone or improve the function of the lymph glands and vessels. Indicated with lymphatic swellings, sore throats, mumps, tonsillitis, some cases of breast swelling or tenderness and other problems where there is lymphatic congestion or stagnation.

Echinacea Purpurea, Garlic, **IN-X**, Kidney Activator (Chinese), Lobelia, **Lymph Gland Cleanse**, **Lymph Gland Cleanse-HY**, **Lymphatic Drainage**, Lymphomax, Lymphostim, Mullein, Oregon Grape, Red Clover, Red Clover Blend, Ultimate Echinacea, Yarrow

Mast Cell Stabilizer

Stabilize mast cells to reduce allergic reactions. Used in hayfever, allergenic asthma and other respiratory allergies to reduce allergic reactions.

Burdock, **HistaBlock**, **Vitamin C**

Mineralizer

Mineralizer is a term we coined at Tree of Light Publishing to describe nutritive herbs which supply trace minerals to aid in tissue healing. Used to build up structural tissues in the body by supplying nutrients to aid tissue regeneration and repair. These herbs are generally rich in calcium and silica.

Adrenal Support, Alfalfa, Barley Grass, Cellular Energy, Chickweed, **Colloidal Minerals**, Dulse, Fibralgia, **Herbal CA**, Herbal Trace Minerals, Horsetail, HSN Complex, **HSN-W**, I-X, Ionic Minerals, Kelp, Marshmallow, **Mineral-Chi Tonic**, Mullein, Nature's Gold, Parsley, Potassium, Red Clover, Skeletal Strength, Sunshine Heroes Calcium Plus D3, Target P-14, Thyroid Activator, Thyroid Support, Trace Mineral Maintenance, TS II, Ultimate Build, Yellow Dock

Moistening

Helps body tissues retain moisture. Used when tissues are dry and don't rehydrate by just drinking water.

Aloe Vera, Cellular Energy, Cough Syrup-DH, Flax Seed Oil, **HY-C**, Jojoba Oil, **Licorice Root**, Marshmallow, Rose Bulgaria, Sandalwood

Mucilaginous

See *Demulcent.*

Narcotic

A painkiller that numbs or depresses the central nerves to reduce pain sensations. Do not confuse with hallucinogenic. Narcotics are used for more serious pain than analgesics or anodynes. The herbs below are very mild narcotics.

Deep Relief Oil, **Hops**, **Kava Kava**, Lavender

Nauseant

Causing an inclination to vomit.

Lobelia, **Paw Paw**

Nervine

A remedy that strengthens the nervous system. Generally used to refer to remedies that inhibit sympathetic nerves and stimulate parasympathetic nerves to have a calming or relaxing effect. Technically speaking, an agent that does the opposite (stimulates sympathetic nerves and inhibits parasympathetic nerves) could also be called a nervine, but the term is not generally used in this manner. These remedies are used to relax muscles, reduce anxiety and tension, ease stress and aid sleep and relaxation

Adrenal Pack, APS II, Black Cohosh, Blessed Thistle, Blue Vervain, Caffeine Detox, Calming, Catnip, Catnip & Fennel, CBG Extract, CC-A, **Chamomile**, Damiana, Fenugreek & Thyme, Feverfew, Flash-Ease T/R, Focus Attention, Four, GABA Plus, Geranium, Ginger, Ginkgo & Hawthorn, Ginkgo/Gotu Kola, Ginseng (Korean), Ginseng (Wild American), Golden Seal, Gotu Kola, Grapefruit (Pink), Herbal Sleep, **Hops**, Joint Support, Lavender, Lecithin, LH, Liver Balance, **Lobelia**, Marjoram (Sweet), Monthly Maintenance, Mood Elevator, Nerve Control, Nerve Eight, Nervous Fatigue Formula, Passion Flower, Rose Bulgaria, Sage, Sandalwood, St. John's Wort, St. John's Wort with Passion Flower, Stress Pack, **Stress Relief**, **Stress-J**, Trigger Immune, TS II, **Valerian Root**, Vitamin B-12, Vitamin B-6, Vitamin B-Complex, Wood Betony, X-Action (Women's), Yarrow, Ylang Ylang

Nutritive

An herb, food or supplement that supplies essential nutrition.

Adrenal Pack, Barley Grass, Berry Healthy, Bilberry Fruit, Black Currant Oil, Carotenoid Blend, CLA, Collatrim, Colloidal Minerals, DHA, Evening Primrose Oil, EveryBody's Formula, Flax Seed Oil, Folic Acid Plus, Free Amino Acids, Glyco Essentials, **GreenZone**, Healthy Blast, Herbal Beverage, Herbal Punch, Ionic Minerals, Iron, L-Carnitine, L-Glutamine, L-Lysine, Licorice Root, Love and Peas, Nature's Gold, Nature's Prenatal, Niacin, Nutri-Burn, Pantothenic Acid, Parsley, Phyto-Soy, Recovery, RG-Max, Rose Hips, Sea Salt, Slippery Elm, Spirulina, Stress Pack, Sunshine Heroes Calcium Plus D3, Sunshine Heroes Multiple Vitamin & Mineral, Sunshine Heroes Omega 3 with DHA, Sunshine Heroes Probiotic Power, Sunshine Heroes Whole Foods Antioxidant, Sunshine Heros Whole

Foods Papayazyme, Super Algae, Super Antioxidant, Super GLA, **Super Supplemental**, Super Trio, SynerProTein, TNT, Tofu Moo, Vita Lemon, Vitamin A & D, Vitamin A & D, Vitamin B-12, Vitamin B-6, Vitamin B-Complex, Vitamin C, Vitamin D3, Vitamin E, Vitamins & Minerals, Multiple, Vitamins, Children's Multi, VitaWave, Xylitol, Zinc

Opthalmicum

A remedy for diseases of the eye.

Bilberry Fruit, **EW**, **Eyebright**, Mega-Chel, **Perfect Eyes**, Thai-Go

Oxytocic

The term oxytocic refers specifically to remedies that have an oxytocin mimicking effect. Oxytocin is the hormone responsible for uterine contractions during labor.

Blue Cohosh, Golden Seal

Palliative

A remedy to alleviate symptoms without curing.

Panacea

A remedy that has been recommended for so many different problems it appears to cure "everything."

Capsicum, Ginseng (Korean), Ginseng (Wild American), Lobelia, Nature's Fresh

Pancreatic Tonics

Strengthens and supports pancreatic function, including the production of pancreatic enzymes.

Digestive Bitters Tonic, Food Enzymes, PBS, Pro-Pancreas

Parasiticide

Destroys parasites.

All Cell Detox, **Artemisia Combination**, Black Walnut, **Garlic**, **Herbal Pumpkin**, Horsetail, SF, Yellow Dock

Parasympatholytic (Anticholinergic)

An agent that blocks or inhibits the parasympathetic nerves. These herbs are indicated in conditions where there is excess parasympathetic nervous system activity. Small pupils are a good indication of excess parasympathetic activity.

Chamomile (Roman), Horsetail, Lavender, **Licorice Root**, Thyme, Valerian Root, **Yarrow**

Parasympathomimetic

Stimulates the parasympathetic nervous system. Relaxes the nervous system. Fairly synonymous with nervine.

Fenugreek & Thyme, Marjoram (Sweet), Oregano (Wild), Rosemary, Wood Betony

Parturient

Stimulates uterine contractions to start or assist labor and delivery. If an herb has this property it should also be considered aborfacient. These remedies are used during the last five weeks of pregnancy or during labor to assist uterine contractions and the delivery of the baby. See also *Oxytocic.*

5-W, Black Cohosh, **Blue Cohosh**, Cinnamon, Clary Sage, Jasmine Absolute, Red Raspberry, Red Raspberry Liquid

Pectoral

Medicines considered proper for relieving afflictions of the chest.

ALJ, Breathe EZ, Breathe Free, Bronchial Formula, Garlic, Lung Support, Pine Needle, Rosemary, Thyme, Yerba Santa/Senega Combination

Perfume

Used for its fragrance.

Jasmine Absolute, **Lavender**, **Myrrh**, **Neroli**, **Patchouli**, **Rose Bulgaria**, **Sandalwood**, **Ylang Ylang**

Phytoestrogen

A plant substance that mimics estrogens; a plant substance that binds to estrogen receptor sites, enhancing or reducing hormonal activity.

Black Cohosh, **Breast Assured**, **Breast Enhance**, Flash-Ease T/R, **Hops**, Licorice Root, **Phyto-Soy**

Preservative

A remedy that helps prevent a food or organic substance from decaying. Many herbs were traditionally added to food to retard spoiling.

Cinnamon, Eucalyptus, Frankincense, Garlic, Juniper Berries, Myrrh, Oregano (Wild), Rosemary, Thyme

Pulmonary

An agent specific to the treatment of the lungs.

ALJ, Breathe EZ, Bronchial Formula, Garlic, Lung Support

Purgative (Cathartic)

Purgatives and cathartics are strong laxatives. Useful for occasional purging in high fevers, acute infections or occasional sluggish elimination. Not recommended for long term use (more than 30 days).

Cascara Sagrada, **LB Extract**, **LB-X**, **LBS II**, Liquid Cleanse

Refrigerant

Lowers body temperature and relieves thirst. For fever and "hot" conditions where a person feels hot, thirsty, dry or flushed, but fluids don't seem to hydrate or cool the body.

CBG Extract, **HY-C**, Licorice Root, Rose Hips, Sunshine Heroes Whole Foods Antioxidant, Super ORAC, **Thai-Go**

Relaxant

An agent that reduces tension and helps muscles to relax. Used for muscle tension and stress.

Black Cohosh, Blood Pressurex, Calming, Cramp Relief, Geranium, Jasmine Absolute, **Kava Kava**, **Lavender**, Lobelia, Neroli, Patchouli, Sarsaparilla, St. John's Wort, Stress Relief, Stress-J, **Valerian Root**

Resolvent

Discutient, dispelling or resolving tumors; an agent that disperses swelling, or effects absorption of a new growth.

Restorative

An agent that supplies some deficiency in the normal constituents of the body, either directly or by chemical reation.

Rubefacient

Produces redness of the skin; generates a localized increase in blood flow when applied to the skin, helping healing, cleansing and nourishment. These are often used to ease pain and swelling of arthritic joints.

Black Ointment, Capsicum, **Deep Relief Oil**, Tei Fu Oil

Saccharine

Containing sugar; sweetish.

Sedative

Sedates the nervous system. Has a calming effect on the body.

Calming, Catnip, Chamomile, **Herbal Sleep**, **Hops**, Jasmine Absolute, **Kava Kava**, Lobelia, Marjoram (Sweet), Melatonin Extra, Mood Elevator, Nerve Eight, Passion Flower, St. John's Wort, St. John's Wort with Passion Flower, Stress Relief, **Valerian Root**, Wood Betony, Ylang Ylang

Sialogogue

An agent that stimulates the secretion of saliva.

Capsicum, Ginger, Hydrangea, Licorice Root, Peppermint

Soothing

A remedy that reduces tissue irritation. Used when tissues are inflammed and irritated.

Bone/Skin Poultice, Cornsilk, Gastro Health, Herbal CA, Intestinal Soothe & Build, LOCLO, Marshmallow, Marshmallow & Fenugreek, Nature's Three, PLS II, Psyllium, Psyllium Hulls Combination, Rose Bulgaria, Slippery Elm

Soporific

Induces sleep. Same as Hypnotic.

Spleen Chi Tonic

Chinese term for aiding the entire digestive process of turning food into flesh. Refers to herbs that improve appetite, digestive function and promote weight gain. Stomachic is a Western term for herbs that improve digestive function in the stomach and is probably closely related, however, these remedies do more than promote digestion. They promote anabolic function. Indicated when a person is pale and thin and unable to gain muscle mass. They may lose weight during or following a debilitating illness like cancer or AIDS. This may also happen to elderly people with poor digestive function.

Ginseng (Wild American), Saw Palmetto, **Spleen Activator**, **Trigger Immune**

Stimulant

An agent that stimulates metabolism, increasing heat and energy.

Adrenal Support, Bayberry, Bee Pollen, Bronchial Formula, Capsicum, Catnip, CC-A, Cellu-Tone, Cellular Energy, Cinnamon, Clove Bud, Damiana, Deep Relief Oil, Energ-V, Exercise, Fatigue/Exhaustion, Fen-Chi, Formula Chi with Ephedra, FV, Garlic, GC-X, Geranium, Ginger, Ginseng (Korean), Ginseng (Wild American), HCP-X, HS II, Juniper Berries, Kelp, Licorice Root, Lobelia, Lymphatic Drainage, Maca, Mega-Chel, MetaboMax, MetaboStart, Mood Elevator, Neroli, Peppermint, Pine Needle, Sage, Sarsaparilla, SUMA Combination, Target Endurance, Target TS-II, Thyroid Support, X-A, X-Action (Men's), X-Action Gel, Yarrow

Stimulant (Circulatory)

An agent that stimulates circulation. Used for cold hands and feet or other symptoms of poor circulation to the extremities. Also used to aid blood flow to various areas of the body to promote tissue healing.

Capsicum, **Capsicum & Garlic with Parsley**, Cardio Assurance, HS II, Lavender, NF-X, **Niacin**, Tei Fu Oil, Thyme

Stomachic

A digestive aid and tonic, which improves stomach function and appetite.

Alfalfa, ALJ, **Anti-Gas (Chinese)**, **Anti-Gas Formula**, Artemisia Combination, Blessed Thistle, Burdock, Capsicum, Capsicum & Garlic with Parsley, Catnip, Catnip & Fennel, CC-A, Chamomile, Clove Bud, Feverfew, FV, Gall Bladder Formula, Garlic, GC-X, Ginger, Golden Seal, HS II, HY-A, Juniper Berries, Lemon Oil, LIV-J, Marjoram (Sweet), Niacin, Peppermint, Pro-Pancreas, Red Beet Formula, Safflowers, Sage, Small Intestine Detox, Spleen Activator, Stress Relief, Stress-J, Yarrow

Styptic

A powerful astringent action that closes wounds and stops bleeding. Used to stop internal and external bleeding. Caution: with serious external bleeding one should use these remedies by pouring them into a wound and then applying the standard first aid practice of applying pressure directly to the injury. For serious internal bleeding (hemorrhage) these remedies should be taken internally, preferably in liquid form, while enroute to the nearest hospital or medical facility for treatment.

Bayberry, **Capsicum**, Horsetail, Menstrual-Reg, Skin Detox, St. John's Wort, White Oak Bark, **Yarrow**

Suppurant

An agent that causes suppuration. an agent that produces, converts or drains pus.

Sweetener

An agent used to sweeten foods or beverages.

Licorice Root, **Stevia**, Xylitol

Sympatholytic

Inhibits the sympathetic nervous system. Has a calming effect in cases of stress and tension.

APS II, Chamomile (Roman), Eleuthero, Hawthorn Berries, **Herbal Sleep**, **Hops**, **Lavender**, Oregano (Wild), Passion Flower, **Valerian Root**, Ylang Ylang

Sympathomimetic

Stimulates activity of sympathetic nervous system. Could also be called a sympathetic nervine. Used to promote alertness, relieve fatigue and aid in mental concentration. Generally contraindicated with heart palpitations, anxiety, high blood pressure and high levels of stress.

Energ-V, Lemon Oil, **Licorice Root**, Pine Needle, **Rosemary**, **Sage**, Thyme

Synergist

See *Catalyst.*

Thyrotropic

Having an influence on the thyroid gland.

Dulse, Kelp, Thyroid Activator, Thyroid Support, TS II

Tonic

Refers to remedies with a general anabolic effect, they build up and strengthen organs and tissues, often causing greater structural density or greater functional strength. Used for weakened conditions of the body or specific organs.

5-W, AdaptaMax, Aloe Vera, Astragalus, Bilberry Fruit, Blessed Thistle, Blood Build, Bone/Skin Poultice, Cardio Assurance, CC-A, Co-Q10, Cordyceps, Digestive Bitters Tonic, E-Tea, Echinacea Purpurea, Eleuthero, Enviro-Detox, EW, Gentle Move, Ginkgo Biloba, Ginseng (Korean), Ginseng (Wild American), Glucosamine, Ho Shou Wu, Hops, HY-C, IGF-1, Lavender, Lung Support, Maca, Mandarin (Red), Milk Thistle, **Mineral-Chi Tonic**, Nature's Chi, Nature's Gold, Pau D'Arco, Red Beet Formula, Red Clover, Red Clover Blend, Red Raspberry, Rose Hips, Sarsaparilla, St. John's Wort, SugarReg, Thyme, **Trigger Immune**, Triple Effect Age Relief, Ultimate Echinacea, Uña de Gato Combination, Vari-Gone, Yarrow

Tranquilizer

Agents that have calming, mildly sedating and/or a muscle-relaxing effect.

Herbal Sleep, Hops, Valerian Root

Uterine

Agents that have an affinity for uterine tissue or maladies thereof.

Dong Quai, Red Raspberry, V-X, Yarrow

Uterine Tonic

Strengthens and tones the uterine muscle in preparation for childbirth.

5-W, Blue Cohosh, False Unicorn, Frankincense, Geranium, Red Raspberry, X-Action (Women's)

Vascular Tonics

Tone up varicose veins. Used for varicose veins, hemorrhoids, pain and swelling in the legs, and other conditions where venous circulation is impaired.

Butcher's Broom, Ginkgo Biloba, Rose Hips, Vari-Gone, Vitamin C

Vasoconstrictor

Constricts blood vessels to reduce circulation. Opposite of vasodilator. Used to raise low blood pressure or help reduce bleeding. Some of these remedies may also aid vasodilative headaches. Contraindicated in high blood pressure.

Menstrual-Reg

Vasodilator

Opens and relaxes blood vessels to increase circulation. Opposite of vasoconstrictor. Used for high blood pressure. Contrindicated with low blood pressure and vasodilative headaches.

Astragalus, **Blood Pressurex**, Garlic, Ginkgo Biloba, Ginkgo/Gotu Kola, Hawthorn Berries, **Magnesium**, Marjoram (Sweet), Olive Leaf, RG-Max

Vermifuge

Destroys intestinal worms. See also antiparasitics.

All Cell Detox, Aloe Vera, **Artemisia Combination**, Bayberry, Black Walnut, Chamomile, Clove Bud, Feverfew, Garlic, **Herbal Pumpkin**, Hops, Horsetail, Juniper Berries, LB Extract, LB-X, LBS II, Mullein, **Para-Cleanse**, Paw Paw, Sage, St. John's Wort, White Oak Bark

Virostatic

An agent that inhibits the replication of viruses.

Paw Paw

Virucidal

An agent that neutralizes or destroys viruses. See Antiviral.

Vulneraries (for intestinal system)

Remedies that promote healing of tissue in intestinal tract. These are indicated in ulcerations or other mechanical damage to membranes lining the digestive tract. They reduce inflammation and gut "leakiness" as well as promoting tissue healing.

Chamomile, **Every Body's Fiber**, Gastro Health, **Intestinal Soothe & Build**, **Kudzu/St.John's Wort**, **L-Glutamine**, Licorice Root, Nature's Gold, Slippery Elm, Uña de Gato, Uña de Gato Combination

Vulnerary

Helps injured tissues to heal, usually without scarring

Acne, **Aloe Vera**, Bayberry, **Bone/Skin Poultice**, Capsicum, Collatrim, Echinacea Purpurea, Eczema/Psoriasis, Eucalyptus, FV, Glucosamine, **Golden Salve**, Gotu Kola, **Healing AC Cream**, **Helichrysum**, **Herbal CA**, **Herbal Trim Skin Treatment**, Horsetail, HSN Complex, HSN-W, Joint Health, Marjoram (Sweet), Marshmallow, MSM, Mullein, **Nature's Fresh**, Nature's Gold, Nature's Phenyltol with NEM, Neroli, Noni (Morinda), PLS II, Sage, Ultimate Echinacea, Vitamin C, Yarrow

Additional Resources

Additional Resources

To Learn More About Nature's Sunshine Products Checkout the Following Resources

Tree of Light Publishing

Tree of Light Publishing offers a wide variety of materials to help you learn how to improve your health, help others improve their health and build a successful business with Nature's Sunshine Products. Check out our Herbal Hour DVDs, Gold and Silver Associate program, courses, books, charts and other materials on the treelite.com website or call us at 800-416-2887 for a free product catalog.

Sign up for our free *Nature's Field E-zine* at www.treelite.com and receive a pdf of our *Nature's Choices* brochure as a free bonus.

Heal It Yourself Article Database

Steven Horne, president and founder of Tree of Light Publishing, has a personal website with a database of "heal it yourself" articles on various health topics. Check it out at www.steven-horne.com or www.healityourself.com. He also has a free newsletter to keep you informed on new articles, webinars and other information posted to his site.

Forums

Tree of Light sponsors a discussion forum on the internet where you can post questions and receive answers and assistance from others. The forum is located at http://groups.yahoo.com/group/NSPAdvisor. This forum is exclusively NSP oriented and discussion of products from other companies is prohibited. For an NSP friendly forum that allows discussion of other products go to http://groups.yahoo.com/group/LetsTalkHealthCafe

There are also two NSP business-oriented forums for discussing business-building ideas at: http://groups.yahoo.com/group/NSPbiz and http://groups.yahoo.com/group/NSPbb.

Third-Party Educational Products

In addition to the materials available through Tree of Light, the following vendors also offer NSP friendly educational materials.

www.cobblestonehealth.com

Cobblestone Health, Ltd. is run by Judith Cobb, an NSP Manager in Canada who was a regular contributer to Tree of Light's *Nature's Field* newsletter. Many of her articles are featured on our archive CD. She offers webinars, videos and books on NSP products.

www.herballure.com

Herb Allure offers the *Hart Manual,* one of the best NSP product references. It is available both printed and on CD. They also carry other books and NSP-friendly educational tools. We offer the *Hart Manual* at www.treelite.com as well.

www.natureshealthypeople.com

NSP Senior National Manager Joan Vandergriff publishes *Nature's Treasure Chest,* a guide to using NSP's products for various ailments.

www.natutrestools.com

Sound Concepts offers booklets, audios and brochures you can use to promote NSP products and build your business. These include some publications written by Steven Horne. This is a great resource for NSP Managers and Distributors.

www.nspontape.com

NSP on Tape have recorded NSP conventions and other events for over 20 years and have hundreds of audio tapes and CDs on NSP products, including many talks given by Steven Horne. They have other useful publications, too.

www.pawpaw.tv

Richard Lund created this website to educae people about NSP's Paw Paw Cell Reg product. It features Dr. McLaughlin's lecture on Paw Paw on DVD, video or audio CD.

www.sunshinesupport.com

Sunshine Support Services offers *Recipes for Success,* a guide to using NSP products for various health problems on recipe cards or CD produced by NSP Senior National Manager Kay Lubecke. They also offer software to help you run your business.

www.wwhitman.com

Wendall Whitman Publishing offers Mark Pederson's *Nutritional Herbology,* a guide to the nutritional benefits of NSP single herbs and herbal formulas. They also publish Steven Horne's *ABC Herbal,* along with other health-related books.

Our Certified Herbal Consult Courses are a *Unique Training Program*

That Gives You a Fast-Start to a *Successful Career* Helping People Improve Their Health with Herbs and Nutritional Supplements

Tree of Light Publishing, has been producing educational materials to help people learn about NSP's quality products for 23 years. We have been a premier source for third-party product information and educational business-building tools for many of NSP's top Managers. Whatever your educational needs, we're here to assist you.

Our Certified Herbal Consultant (CHC) training program will give you the knowledge and skills you need to help people regain and maintain their health using herbs and nutritional supplements. The complete program includes the six courses listed on these pages, plus two additional courses, *Activating the Healing Response* and *Secrets of Chinese Herbs*. You can purchase coures individually or purchase the entire program for $1497, which includes our Business and Success Coaching program at no additional charge. For more information, call 800-416-2887 or visit us online at www.treelite.com.

1. Dr. Mom - Dr. Dad

Primary Healthcare Takes Place in the Home

Dr. Mom-Dr. Dad lays the foundation for the entire CHC program. It introduces the ABCs of natural healing and shows how to apply these principles to rapidly reverse injuries and quickly recovery from most common acute ailments. It also teaches the basic procedures for reversing chronic disease.

Available as a correspondence course, certified instructors throughout the country.

2 Videos
137 pages

$247.00

Item #: Cor-23

2. ABC+D Approach to Natural Healing

The Complete School of Natural Health

This course expands upon the ABC principles in Dr. Mom-Dr. Dad and shows how to apply them to nutritional consulting. It introduces the +D (or direct aid) step to the ABC system and shows how to use the latest edition of our ABC+D Charts to design an appropriate supplement program for customers and clients.

Available as a correspondence course, through certified instructors and via webinar.

2 Videos
137 pages

$247.00

Item #: Cor-24

"We recently taught Dr. Mom-Dr. Dad to our business builders as a trial run. Everyone who took the class is now a manager. What we thought complicated became simple, and they all caught on fire. When you feel confident with your products, and can offer simple solutions when people are overwhelmed by health issues and emergencies, people start thinking to call you.

"We plan to offer Dr. Mom-Dr. Dad regularly - at least quarterly in our area. We are VERY excited about the new ABC Plus D course. It is the expanded version of the School of Natural Health. Our group will be offered that one regularly as well. The other courses in Steven's certification series will be rotated every two years to offer certification through Tree of Light to anyone in our area who wants the same confidence and simple, workable consultation skills.

"Steven puts a lot of crafting into his materials. They flow toward confidence. He has put years of experiencing, thinking and refining into this series. It is evidenced by the "Ah-Ha's" you will hear yourself saying as you work through the courses. Your business will grow with your confidence. His courses have been some of the best tax write-offs I've invested in, and they have shaped my thinking toward looking for the simple truth that works."

—Susan Gingerich, NSP Area Manager and Certified Instructor

3. Practical Iridology

Iridology for Herbal and Natural Health Consulting

Practical Iridology explains how to read the iris of the eye to determine a person's constitutional strengths and weaknesses. Complete with color iris photos and four videos, the course helps students learn how to use iris information to uncover root causes of a person's health problems and guide them to appropriate solutions.

Available as a correspondence course and through certified instructors.

4 Videos
100 pages

$247.00

Item #: Cor-25

4. Practical Tools for Health Assessment

This course covers a variety of methods of assessing the body to discover which body systems need nutritional support and how to correct the underlying imbalances in the body's biological terrain. The course covers muscle testing, basic tongue and pulse analysis, glandular body typing, physical observation, pH testing and case history taking.

Available as a correspondence course and through certified instructors.

4 Videos
& manual

$247.00

Item #: Cor-26

Taking the instructional courses for Certified Herbal Consultant from the Tree of Light Publishing has been one of the smartest, most beneficial things I have done. These courses have provided a foundation of invaluable information through which I have been able to help my family and friends with their health challenges. Not only have we been able to treat day to day health needs, we have also been able to treat more serious ailments such as emphysema, a spot on the liver, and gout.

The materials provided with these classes are excellent and I continue to use them as reference guides in our day to day health challenges. Kimberly Balas and Steven Horne are extraordinary teachers who have a deep understanding of what constitutes balance in health and both are able to present the information, both written and oral, in an easy to understand manner.

Would I recommend these classes to others? Absolutely. These classes provide the foundation to understanding balance in health. Anyone who is interested in achieving and maintaining health will benefit from these classes. Tree of Light is an appropriate name for the company as Light is truly what is being offered.

—Patricia Davis, NSP Distributor and Certified Herbal Consultant

5. Nature's Pharmacy

An Energetic Approach to Healthcare with Herbs

Nature provides a rich materia medica of remedies for every human problem. Nature's Pharmacy covers basic plant chemistry, herbal energetics, and the properties, constituents, historical uses, dosages, and warnings for over 300 of the most widely used herbs in American herbalism. This course will take your understanding of herbs to a whole new level.

Will be available via correspondence, live instruction and webinar.

2 Videos
& Manual

$247.00

Item #: Cor-27

6. Herbal Preparations and Applications

There are many ways that herbs can be prepared and administered in addition to taking them in capsules. This course demonstrates the numerous ways to prepare and administer herbs both internally and externally. Preparations include infusions, decoctions, tinctures, glycerites, syrups, boluses, flower essences, etc. Application methods include enemas, baths, poultices, compresses, fomentations, soaks, etc.

Available as a correspondence course and live instruction through certified instructors.

4 Videos
& Manual

$247.00

Item #: Cor-28

Call us for a free catalog at 800-416-2887 or visit our website at www.treelite.com